International Association of Geodesy Symposia

Jeffrey T. Freymueller, Series Editor
Laura Sánchez, Series Assistant Editor

Series Editor
Jeffrey T. Freymueller
Endowed Chair for Geology of the Solid Earth
Department of Earth and Environmental Sciences
Michigan State University
East Lansing, MI, USA

Assistant Editor
Laura Sánchez
Deutsches Geodätisches Forschungsinstitut
Technische Universität München
Munich, Germany

International Association of Geodesy Symposia

Jeffrey T. Freymueller, Series Editor
Laura Sánchez, Series Assistant Editor

Symposium 114: Geodetic Theory Today
Symposium 115: GPS Trends in Precise Terrestrial, Airborne, and Spaceborne Applications
Symposium 116: Global Gravity Field and Its Temporal Variations
Symposium 117: Gravity, Geoid and Marine Geodesy
Symposium 118: Advances in Positioning and Reference Frames
Symposium 119: Geodesy on the Move
Symposium 120: Towards an Integrated Global Geodetic Observation System (IGGOS)
Symposium 121: Geodesy Beyond 2000: The Challenges of the First Decade
Symposium 122: IV Hotine-Marussi Symposium on Mathematical Geodesy
Symposium 123: Gravity, Geoid and Geodynamics 2000
Symposium 124: Vertical Reference Systems
Symposium 125: Vistas for Geodesy in the New Millennium
Symposium 126: Satellite Altimetry for Geodesy, Geophysics and Oceanography
Symposium 127: V Hotine-Marussi Symposium on Mathematical Geodesy
Symposium 128: A Window on the Future of Geodesy
Symposium 129: Gravity, Geoid and Space Missions
Symposium 130: Dynamic Planet - Monitoring and Understanding ...
Symposium 131: Geodetic Deformation Monitoring: From Geophysical to Engineering Roles
Symposium 132: VI Hotine-Marussi Symposium on Theoretical and Computational Geodesy
Symposium 133: Observing our Changing Earth
Symposium 134: Geodetic Reference Frames
Symposium 135: Gravity, Geoid and Earth Observation
Symposium 136: Geodesy for Planet Earth
Symposium 137: VII Hotine-Marussi Symposium on Mathematical Geodesy
Symposium 138: Reference Frames for Applications in Geosciences
Symposium 139: Earth on the Edge: Science for a Sustainable Planet
Symposium 140: The 1st International Workshop on the Quality of Geodetic Observation and Monitoring Systems (QuGOMS'11)
Symposium 141: Gravity, Geoid and Height Systems (GGHS2012)
Symposium 142: VIII Hotine-Marussi Symposium on Mathematical Geodesy
Symposium 143: Scientific Assembly of the International Association of Geodesy, 150 Years
Symposium 144: 3rd International Gravity Field Service (IGFS)
Symposium 145: International Symposium on Geodesy for Earthquake and Natural Hazards (GENAH)
Symposium 146: Reference Frames for Applications in Geosciences (REFAG2014)
Symposium 147: Earth and Environmental Sciences for Future Generations
Symposium 148: Gravity, Geoid and Height Systems 2016 (GGHS2016)
Symposium 149: Advancing Geodesy in a Changing World
Symposium 150: Fiducial Reference Measurements for Altimetry
Symposium 151: IX Hotine-Marussi Symposium on Mathematical Geodesy
Symposium 152: Beyond 100: The Next Century in Geodesy
Symposium 153: Terrestrial Gravimetry: Static and Mobile Measurements (TG-SMM 2019)
Symposium 154: Geodesy for a Sustainable Earth
Symposium 155: X Hotine-Marussi Symposium on Mathematical Geodesy
Symposium 156: Gravity, Positioning and Reference Frames
Symposium 157: Together Again for Geodesy

Together Again for Geodesy

Proceedings of the General Assembly of the International Association of Geodesy, Berlin, Germany, 11–20 July, 2023

Edited by

Jeffrey T. Freymueller, Laura Sánchez

Series Editor
Jeffrey T. Freymueller
Endowed Chair for Geology of the Solid Earth
Department of Earth and Environmental Sciences
Michigan State University
East Lansing, MI, USA

Assistant Editor
Laura Sánchez
Deutsches Geodätisches Forschungsinstitut
Technische Universität München
Munich, Germany

Associate Editors
Janusz Bogusz
Faculty of Civil Engineering and Geodesy
Military University of Technology in Warsaw
Warsaw, Poland

Xavier Collilieux
Institut de physique du globe de Paris
Université Paris Cité
Paris, France

E. Sinem Ince
GFZ Helmholtz Centre for Geosciences
Potsdam, Germany

Adrian Jäggi
Astronomical Institute
University of Bern
Bern, Switzerland

Robert Steven Nerem
Colorado Center for Astrodynamics Research
University of Colorado Boulder
Boulder, CO, USA

Pavel Novák
Department of Geomatics
University of West Bohemia
Pilsen, Czech Republic

Séverine Rosat
Institut Terre Environnement Strasbourg
Université de Strasbourg
Strasbourg, France

Rebekka Steffen
Lantmäteriet
Gävle, Sweden

Katarzyna Stepniak
Faculty of Geodesy, Geospatial and Civil Engineering
University of Warmia and Mazury in Olsztyn
Olsztyn, Poland

ISSN 0939-9585   ISSN 2197-9359   (electronic)
International Association of Geodesy Symposia
ISBN 978-3-031-91166-8   ISBN 978-3-031-91167-5   (eBook)
https://doi.org/10.1007/978-3-031-91167-5

© The Editor(s) (if applicable) and The Author(s) 2025. This book is an open access publication.

Open Access This book is licensed under the terms of the Creative Commons Attribution 4.0 International License (http://creativecommons.org/licenses/by/4.0/), which permits use, sharing, adaptation, distribution and reproduction in any medium or format, as long as you give appropriate credit to the original author(s) and the source, provide a link to the Creative Commons license and indicate if changes were made.

The images or other third party material in this book are included in the book's Creative Commons license, unless indicated otherwise in a credit line to the material. If material is not included in the book's Creative Commons license and your intended use is not permitted by statutory regulation or exceeds the permitted use, you will need to obtain permission directly from the copyright holder.

The use of general descriptive names, registered names, trademarks, service marks, etc. in this publication does not imply, even in the absence of a specific statement, that such names are exempt from the relevant protective laws and regulations and therefore free for general use.

The publisher, the authors and the editors are safe to assume that the advice and information in this book are believed to be true and accurate at the date of publication. Neither the publisher nor the authors or the editors give a warranty, expressed or implied, with respect to the material contained herein or for any errors or omissions that may have been made. The publisher remains neutral with regard to jurisdictional claims in published maps and institutional affiliations.

This Springer imprint is published by the registered company Springer Nature Switzerland AG
The registered company address is: Gewerbestrasse 11, 6330 Cham, Switzerland

If disposing of this product, please recycle the paper.

Preface

The International Union of Geodesy and Geophysics (IUGG) held its 28th General Assembly from 11 to 20 July in Berlin, Germany. The theme of the General Assembly was "Together again in the geosciences", as this conference gave us the opportunity to meet again in person after long years of restrictions due to the pandemic. During the General Assembly, symposia were organised by all IUGG Associations, as well as joint and union symposia, offering a wide range of oral and poster presentations. The participating Associations are the International Association of Cryospheric Sciences (IACS), International Association of Geodesy (IAG), International Association of Geomagnetism and Aeronomy (IAGA), International Association of Hydrological Sciences (IAHS), International Association of Meteorology and Atmospheric Sciences (IAMAS), International Association for the Physical Sciences of the Oceans (IAPSO), International Association for Seismology and Physics of the Earth's Interior (IASPEI), and International Association for Volcanology and Chemistry of the Earth's Interior (IAVCEI). A total of 4,884 participants from 100 countries attended the Assembly. 607 participants registered with priority for the IAG. About 3,200 oral presentations and 1,300 poster presentations were distributed in 9 Union Lectures, 170 symposia, including 50 joint inter-association symposia, and a new meeting format called *Big Themes*. The latter was introduced for the first time at this Assembly to address overarching themes related to Open Science, the International Year of Basic Sciences for Sustainable Development, the contribution of the geosciences to international policy frameworks, early career scientists, equality, diversity and inclusion, and North-South dialogue in the geosciences.

Participants of the General Assembly of the International Association of Geodesy (IAG) framed by the 28th General Assembly of the International Union of Geodesy and Geophysics (IUGG), Berlin, Germany, 11–20 July, 2023

Six geodesy-only symposia and seven joint, geodesy-lead symposia were organised with a total of 661 abstracts submitted:

- G01 Reference Systems and Frames
  - *Conveners: Christopher Kotsakis (Greece), Geoff Blewitt (USA), Johannes Böhm (Austria), Xavier Collilieux (France), Susanne Glaser (Germany)*
- G02 Static Gravity Field and Height Systems
  - *Conveners: Laura Sánchez (Germany), Hussein Abd-Elmotaal (Egypt), Roland Pail (Germany), Elmas Sinem Ince (Germany)*
- G03 Time-variable Gravity Field
  - *Conveners: Adrian Jäggi (Switzerland), Srinivas Bettadpur (USA), Frank Flechtner (Germany), Shuanggen Jin (China)*
- G04 Earth Rotation and Geodynamics
  - *Conveners: Janusz Bogusz (Poland), Chengli Huang (China), Severine Rosat (France), Michael Schindelegger (Germany)*
- G05 Multi-signal positioning, Remote Sensing and Applications
  - *Conveners: Allison Kealy (Australia), Zaminpardaz Safoora (Australia), Wielgosz Pawel (Poland), Beata Milanowska (Poland)*
- G06 Monitoring and Understanding the Dynamic Earth with Geodetic Observations
  - *Conveners: Basara Miyahara (Japan), Detlef Angermann (Germany), Allison Craddock (USA), Hansjörg Kutterer (Germany)*
- JG01 Interactions of the Solid Earth with Ice Sheets and Sea Level (IAG, IACS, IASPEI)
  - *Conveners: Rebekka Steffen (Sweden, IAG), Bert Wouters (Netherlands, IACS), Natalya Gomez (Canada, IAG/IACS), Lambert Caron (US, IAG), Doug Wiens (US, IASPEI)*
- JG02 Theory and Methods of Potential Fields (IAG, IAGA)
  - *Conveners: Dimitrios Tsoulis (Greece, IAG), Sten Claessens (Australia, IAG), Maurizio Fedi (Italy, IAGA)*
- JG03 Remote Sensing and Modelling of the Atmosphere (IAG, IAGA, IAMAS, IAVCEI)
  - *Conveners: Michael Schmidt (Germany, IAG), Ehsan Forootan (Denmark, IAG), Loren Chang (Taiwan, China, IAGA), Claudia Stubenrauch (France, IAMAS), Fabio Dioguardi (Italy, IAVCEI)*
- JG04 Satellite Gravimetry for Groundwater Monitoring (IAG, IAHS)
  - *Conveners: Adrian Jäggi (Switzerland, IAG), Andreas Güntner (Germany, IAG), Felipe de Barros (Brazil/USA, IAHS), Michelle Newcomer (USA, IAHS)*
- JG05 Geodesy for Climate Research (IAG, IAMAS, IACS, IAPSO, IAHS)
  - *Conveners: Annette Eicker (Germany, IAG), Bert Wouters (Netherlands, IACS), John T Reager (USA, IAHS), Adam Scaife (UK, IAMAS), Benoit Meyssignac (France, IAPSO)*
- JG06 Monitoring Sea Level Changes by Satellite and In-Situ Measurements (IAG, IAPSO)
  - *Conveners: Xiaoli Deng (Australia, IAG), Steve Nerem (USA, IAG), Fabio Raicich (Italy, IAPSO)*
- JG07 Modern Gravimetric Techniques for Geosciences (IAG, IAVCEI, IAPSO, IASPEI)
  - *Conveners: Jürgen Müller (Germany, IAG), Chris Hughes (UK, IAPSO), Rudolf Widmer-Schnidrig (Germany, IASPEI), Emily Montgomery-Brown (USA, IAVCEI)*

This volume contains 46 selected papers from all IAG-related symposia. All papers have been peer-reviewed and we would like to acknowledge the contribution of the Associate Editors and Reviewers. Their support is greatly appreciated.

East Lansing, MI, USA Jeffrey T. Freymueller
Munich, Germany Laura Sánchez

Contents

Part I   Reference Systems and Frames

Opportunities with VLBI Transmitters on Satellites.............................. 3
Johannes Böhm and Helene Wolf

Formation of a GNSS Network in Space Based on Simulated LEO Constellations... 9
Lukas Müller, Markus Rothacher, and Benedikt Soja

Practical Considerations of VLBI Observations to the GENESIS Mission.......... 17
David Schunck, Lucia McCallum, and Guifré Molera Calvés

Terrestrial Datum Definition Methods in VLBI Global Solutions.................. 25
Lisa Kern, Hana Krásná, Axel Nothnagel, Johannes Böhm, and Matthias Madzak

On the Potential of Accelerometers for GNSS on Satellite Positioning and Ensuing Reference Frame Determination... 33
Patrick Schreiner, Susanne Glaser, Rolf König, Karl Hans Neumayer, Shrishail Raut, and Harald Schuh

On DORIS Precise Orbit and Reference Frame Determination Based on the ITRF2020 Using Multiple Altimetry Satellite Missions........................ 45
Anton Reinhold, Patrick Schreiner, and Karl Hans Neumayer

Realisation of the Non-Rotating Terrestrial Reference Frame by an Actual Plate Kinematic and Crustal Deformation Model (APKIM2020)...................... 53
Hermann Drewes, Manuela Seitz, and Laura Sánchez

A Functional Model for Quantifying Deformation in Reference Frame Transformations.. 63
Richard Stanaway, Chris Crook, Kevin M. Kelly, and Roger Lott

Combined Global GNSS Velocity Field.. 69
A. Santamaría-Gómez, R. Rietbroek, P. Rebischung, T. Frederikse, and J. Legrand

Geophysical Loading Correction Comparison and Assessment in VLBI Analysis.... 77
Shivangi Singh, Johannes Böhm, Hana Krásná, Nagarajan Balasubramanian, and Onkar Dikshit

Exploring Non-tidal Atmospheric Loading Deformation Correction in GNSS Time Series Analysis Using GAMIT/GLOBK Software................................ 87
Fatemeh Khorrami, Yohannes Getachew Ejigu, Jyri Näränen, Arttu Raja-Halli, and Maaria Nordman

Relevance of PSInSAR Analyses at ITRF Co-location Sites..................... 103
Xavier Collilieux, Zuheir Altamimi, Jingyi Chen, Clément Courde, Zheyuan Du, Thomas Furhmann, Christoph Gisinger, Thomas Gruber, Ryan Hippenstiel, Davod Poreh, Paul Rebischung, and Yudai Sato

The DIA-Estimator for Positional Integrity: Design and Computational Challenges........ 111
P. J. G. Teunissen, S. Ciuban, C. Yin, B. G. van Noort, S. Zaminpardaz, and C. C. J. M. Tiberius

EPOS-OC, a Universal Software Tool for Satellite Geodesy at GFZ........ 119
Karl Hans Neumayer, Patrick Schreiner, Rolf König, Christoph Dahle, Susanne Glaser, Nijat Mammadaliyev, and Frank Flechtner

Part II Earth Rotation

Impact of Free Core Nutation Modeling on the Estimation of Earth Rotation Parameters from Different VLBI Session Types........ 131
Arnab Laha, Johannes Böhm, Sigrid Böhm, Hana Krásná, Nagarajan Balasubramanian, and Onkar Dikshit

Consistently Combined Earth Orientation Parameters at BKG—Extended by New VLBI Intensives Data........ 139
Lisa Klemm, Daniela Thaller, Claudia Flohrer, Anastasiia Walenta, Dieter Ullrich, and Hendrik Hellmers

Operational Forecasting of Effective Angular Momentum Functions Fourteen Days Ahead........ 147
Mostafa Kiani Shahvandi, Matthias Schartner, Junyang Gou, and Benedikt Soja

Hourly Earth Rotation Parameter Series from GPS and Galileo Observations, and Estimations of Tidal Effects........ 157
Yuting Cheng, Christian Bizouard, Sébastien Lambert, and Jean-Yves Richard

EOP Prediction Based on Multi and Single Technique Space Geodetic Solution........ 165
Sadegh Modiri, Daniela Thaller, Santiago Belda, Dzana Halilovic, Lisa Klemm, Daniel König, Hendrik Hellmers, Sabine Bachmann, Claudia Flohrer, and Anastasiia Walenta

Part III Gravity Field Modelling and Height Systems

On the Treatment of Static Gravity Field Signal for Time-Variable Gravity Field Recovery........ 179
Martin Lasser, Ulrich Meyer, Daniel Arnold, and Adrian Jäggi

Analysis of Novel Sensors and Satellite Formation Flights for Future Gravimetry Missions........ 187
Alexey Kupriyanov, Arthur Reis, Annike Knabe, Nina Fletling, Alireza HosseiniArani, Mohsen Romeshkani, Manuel Schilling, Vitali Müller, and Jürgen Müller

Automated Anomaly and Outlier Detection in GRACE and GRACE Follow-On Post-Fit Residuals Using Machine Learning........ 199
Martin Lasser, Jonas Zbinden, Ulrich Meyer, Brandon Panos, Daniel Arnold, and Adrian Jäggi

Impact of a Priori Gravity Field Models on SLR Data Processing........ 207
Linda Geisser, Ulrich Meyer, Daniel Arnold, and Adrian Jäggi

Dynamical Evaluation of Gravity Spherical Harmonic Coefficients due to Generally Shaped Polyhedra........ 213
Georgia Gavriilidou and Dimitrios Tsoulis

Optimizing Airborne Flight Line Spacing for Geoid Determination with Full Gravity Vectors... 219
Ismael Foroughi, Mehdi Goli, Stephen Ferguson, and Spiros Pagiatakis

Update of the Atmospheric Attraction Computation Service (Atmacs) for High-Precision Terrestrial Gravity Observations............................... 227
Ezequiel D. Antokoletz, Hartmut Wziontek, Thomas Klügel, Kyriakos Balidakis, and Henryk Dobslaw

Geoid Computation for the Future Circular Collider at CERN..................... 235
Julia Azumi Koch, Urs Marti, Iván Darío Herrera Pinzón, Daniel Willi, Benedikt Soja, and Markus Rothacher

Meteorite Impact Origin of Yangju Circular Structure in the Middle Part of the Korean Peninsula Estimated by Gravity Field Interpretation............... 245
Sungchan Choi, Sung-Wook Kim, Younghong Shin, and Eun-Kyeong Choi

Achievements of the GGOS Focus Area Unified Height System..................... 253
Laura Sanchez and Riccardo Barzaghi

Operational Infrastructure to Ensure the Long-Term Sustainability of the International Height Reference System and Frame (IHRS/IHRF)............ 263
Laura Sánchez, Riccardo Barzaghi, and George Vergos

Estimation of the Argentinean Vertical Datum Parameter with Respect to the International Height Reference Frame (IHRF)................................. 273
Agustín R. Gómez, Claudia N. Tocho, Ezequiel D. Antokoletz, Hernán J. Guagni, and Diego A. Piñón

Densification of the IHRF in Denmark, The Faroe Islands, and Greenland......... 281
Hergeir Teitsson, Laura Sánchez, and René Forsberg

Part IV Monitoring Sea Level Changes by Satellite and In-Situ Measurements

The Impact of Different Geophysical Corrections on Altimetry-Derived Sea Level Rise Estimates—Wet Troposphere.. 297
Denise Dettmering, Christian Schwatke, and Felix L. Müller

Bathymetry Estimation from ICESat-2 in a Region Swamped by Mud: A Case Story from Moreton Bay.. 303
Elisabet Anne Marie Hallström, Ole B. Andersen, Xiaoli Deng, and Richard Coleman

Performance Analyses of Sentinel-3A and Sentinel-3B Over Lake Issyk Kul (Kyrgyzstan)... 311
T. Schöne, J. Illigner, A. Zubovich, C. Zech, N. Stolarczuk, A. Sharshebaev, and M. Borisov

Vision of a Clock-Based Network for Absolute Sea Level Monitoring.............. 323
Asha Vincent and Jürgen Müller

Part V Monitoring and Understanding the Dynamic Earth with Geodetic Observations

Towards Clock Ties for a Global Geodetic Observing System..................... 333
Jan Kodet, Thomas Klügel, Christian Plötz, Willi Probst, Alexander Neidhardt, and Karl Ulrich Schreiber

Assessment of the Tropospheric Delay Coefficients at Co-located Sites with VGOS and GNSS.. 341
Anastasiia Walenta, Claudia Flohrer, Daniela Thaller, Rolf Dach, Stefan Schaer, Gerald Engelhardt, and Dieter Ullrich

Real-Time GNSS Integrated Water Vapor Sensing Based on Time Series Correction Deep Learning Models................................... 349
Duo Wang, Peng Yuan, and Hansjörg Kutterer

Analyzing the 3D Deformation Induced by Non-tidal Loading in GNSS Time Series in Finland.. 361
Yohannes Getachew Ejigu, Jean-Paul Boy, Arttu Raja-Halli, Fatemeh Khorrami, Jyri Naranen, and Maaria Nordman

A Geodetic Analysis of the Volume Transport in the ACC Region Based on Satellite Data.. 369
Juan A. Vargas-Alemañy, M. Isabel Vigo, David García-García, and Ferdous Zid

A Pipeline to Explore Transient Signals in GNSS Data: A Preliminary Approach Applied to the Cascadia Subduction Margin................... 381
Cristian Garcia, Benjamin Männel, Susanne Glaser, and Jonathan Bedford

Emphasizing the Value of Geodesy to Science and Society Through IAG-GGOS..... 391
Martin Sehnal, Laura Sánchez, Detlef Angermann, Allison Craddock, Basara Miyahara, and Lena Steiner

EPOS-GNSS Data Quality Monitoring Web Portal........................... 401
Fikri Bamahry, Juliette Legrand, Carine Bruyninx, and Andras Fabian

The GGXF Standard File Format for Gridded Geodetic Data.................. 411
Chris Crook, Kevin M. Kelly, Roger Lott, and Richard Stanaway

Signal Decomposition with InSAR Displacement Time Series Above a Storage Cavern Field: Example Epe (NRW, Germany)................................. 417
Alison Seidel, Malte Westerhaus, Markus Even, and Hansjörg Kutterer

Reviewers.. 427

Author Index.. 429

Subject Index... 431

Part I
Reference Systems and Frames

Opportunities with VLBI Transmitters on Satellites

Johannes Böhm and Helene Wolf

Abstract

Very Long Baseline Interferometry (VLBI) transmitters on satellites are considered for future satellites of Global Navigation Satellite Systems (GNSS) as well as for the upcoming Genesis mission of the European Space Agency (ESA). In both concepts, VLBI observations to VLBI transmitters on the satellites can contribute to orbit determination, with the geometry of observations to Genesis advantageous compared to GNSS. With Galileo as one of the GNSS, we emphasise the importance of VLBI observations for the estimation of the right ascension of the ascending node Ω, which cannot be determined from GNSS alone. We find that Ω can be estimated with accuracies better than 50 µas from 24 hour sessions with alternate observations to quasars and the Galileo satellite of interest. Station coordinates, on the other hand, can be determined from observations to three Galileo satellites with an accuracy at the centimetre-level or better from 24 hour sessions, again with alternate observations to the satellites and quasars for a better estimation of tropospheric parameters. Station coordinate results with VLBI observations to VLBI transmitters on satellites form the basis for frame tie and local tie determination. The transfer of space ties to local ties on the ground is of particular relevance for Genesis.

Keywords

Genesis · GNSS · Reference frames · VLBI · VLBI transmitter

1 Introduction and Motivation

The International Terrestrial Reference Frame (ITRF) is based on observations from Very Long Baseline Interferometry (VLBI), Global Navigation Satellite Systems (GNSS), Satellite Laser Ranging (SLR), and Doppler Orbitography and Radiopositioning Integrated by Satellite (DORIS), as well as terrestrial local tie measurements at co-location sites. In the most recent realisation ITRF2020, still a large fraction of the discrepancies between space geodetic estimates and local ties exceeds 5 mm (Altamimi et al. 2023). In order to overcome or at least mitigate these limitations, the concept of space ties (Rothacher et al. 2009) has been considered as useful tool to improve the local tie information with mission proposals submitted to space agencies, such as GRASP (Bar-Sever et al. 2009) and E-GRASP/Eratosthenes (Biancale et al. 2017). Finally in autumn 2022, the Genesis mission was approved for launch in 2027/28 at the Ministerial Conference of the European Space Agency (ESA). In addition to GNSS receiver, DORIS receiver, and SLR retro-reflector, this satellite will be equipped with a dedicated VLBI transmitter (Delva et al. 2023). Genesis will fly in a polar orbit at about 6000 km altitude. In parallel, investigations are dealing with VLBI transmitters on Galileo satellites. Galileo is the Global Navigation Satellite System of the European Union with satellites at an altitude of 23222 km in three orbital planes with an inclination of 56 degrees.

J. Böhm (✉) · H. Wolf
Department of Geodesy and Geoinformation, TU Wien, Vienna, Austria
e-mail: johannes.boehm@tuwien.ac.at; helene.wolf@tuwien.ac.at

VLBI observations to GNSS satellites at L-band have been carried out by, e.g., Haas et al. (2017) and Plank et al. (2017). Hellerschmied et al. (2018) analysed VLBI observations to the nano-satellite APOD, which was sending tones at X-band. With all real experiments so far, it can be concluded that there are technical challenges concerning scheduling, observation, correlation, fringe-fitting, and analysis. While these challenges can be overcome, there are other limitations which require conceptual improvements with dedicated VLBI transmitters on the satellites: sufficient bandwidth is needed to reach certain accuracies of the VLBI delay observable, and a second frequency is essential for ionosphere calibration. These aspects have to be considered with VLBI transmitters on future missions.

Further, several simulation studies have been carried out to investigate the potential of VLBI transmitters on satellites. For example, Plank et al. (2014) estimated station coordinates from VLBI observations to satellites at different altitudes, Anderson et al. (2018) focused on the frame tie assessment with a Genesis-like satellite, Männel (2016) investigated space ties with VLBI in general, and Klopotek et al. (2020) dealt with orbit determination including VLBI observations.

With this overview article, we address and summarise the motivation for and opportunities with upcoming satellites carrying VLBI transmitters. In particular, we provide recent estimates for the accuracies of geodetic parameters, which can be expected from observations to transmitters on Galileo satellites (Sect. 2) and Genesis (Sect. 3) to underline the opportunities with this new observation concept. Finally, we do provide conclusions and recommendations in Sect. 4.

2 VLBI Transmitters on Galileo Satellites

Geodetic VLBI is the only space geodetic technique for the estimation of UT1-UTC, which reflects the difference between the Earth Rotation Angle (ERA) and atomic time (TAI), corrected for leap seconds. Other space geodetic techniques rely on UT1-UTC values from VLBI, e.g., GNSS, SLR, and DORIS, because satellite techniques are not able to distinguish between a change in ERA and a change in right ascension of the ascending node Ω of the satellite orbit (see Fig. 1). Satellite techniques, in particular GNSS, are able to determine the negative time derivative of UT1-UTC, which is referred to as Length Of Day (LOD). Having an initial value of UT1-UTC, GNSS are able to integrate LOD values over days and weeks to have access to UT1-UTC, however with degrading accuracy over time. It is important to note here that the availability of UT1-UTC values with utmost precision is not required for positioning applications with GNSS, as errors in UT1-UTC can be compensated by changes in Ω (Dach 2022). Nevertheless, orbits of GNSS satellites have to be modelled in inertial space and UT1-UTC values from VLBI are introduced in the analysis of GNSS observations.

2.1 Estimation of Right Ascension of Ascending Node

VLBI observations to VLBI transmitters can be used to position the satellites in the quasi-inertial frame directly. In the following we focus on the estimation of Ω, as this parameter is of main interest here for the reasons mentioned above. We

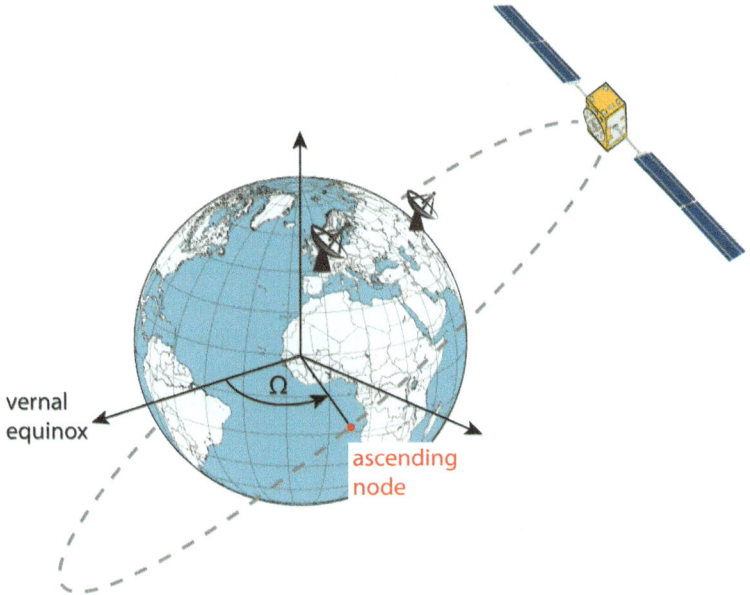

Fig. 1 Illustration of VLBI radio telescopes observing a satellite in its orbit. The absolute orientation of the satellite orbit is given by the right ascension of the ascending node Ω

Fig. 2 Network of the nine VGOS stations used in this study and ground track of the satellite GSAT0101 (E11) during 24 hours starting on January 1, 2021 00:00:00 UTC. The dots represent the position of the satellite over Earth in a fifteen minute interval

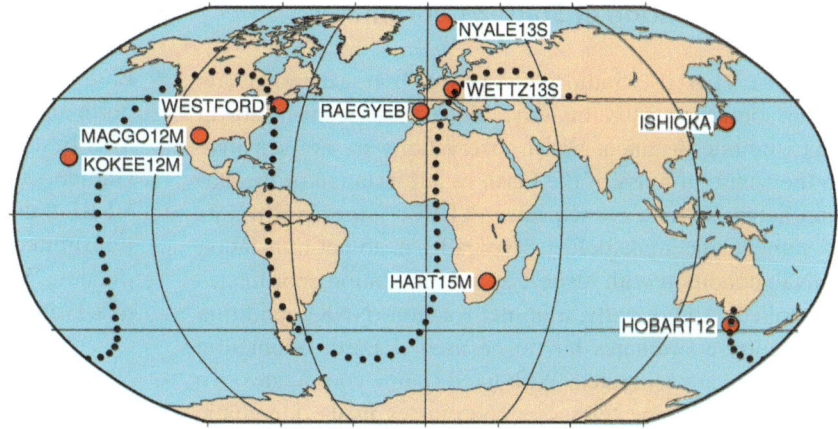

consider a network of nine VLBI Global Observing System (VGOS; Petrachenko et al. 2009) stations (see Fig. 2) and use VieSched++ (Schartner and Böhm 2019) for scheduling a 24 hour session on January 1, 2021, with observations to quasars and the Galileo satellite GSAT0101 (E11). In this study, we schedule the observations every 30 seconds for both, quasar and satellite observations, but we set different weights for satellite and quasar observations to reach certain ratios of the number of satellite observations with respect to quasar observations to test the sensitivity to this parameter.

For the Monte-Carlo simulation, we set the tropospheric refractive index structure constant C_n to $1.8 \cdot 10^{-7}$ m$^{-1/3}$ with a scale height of 2000 m for all stations. The wind speed is set to 8 m/s in eastern direction. The clock variation is described as the sum of a random walk and an integrated random walk process assuming an Allan Standard Deviation of $1 \cdot 10^{-14}$ at 50 minutes. The measurement error for both observation types, quasar and satellite observations, is simulated as white noise with 10 ps standard deviation. We simulate all schedules 1000 times and determine the repeatability (standard deviation) of the estimated parameters as measure for the accuracy which can be expected for the observation scenarios.

In this study, we fix station and radio source coordinates to their a priori values and estimate all five Earth orientation parameters (celestial pole offsets, polar motion, UT1-UTC) as constant parameters per session. Additionally, we only estimate the right ascension of the ascending node Ω of the GSAT0101 orbit once per session, while fixing all other orbital elements to their a priori values. Auxiliary tropospheric (zenith delays and gradients) and clock parameters are estimated following common practice in the least-squares adjustment of VLBI observations. All calculations are carried out with the Vienna VLBI and Satellite Software (VieVS; Böhm et al. 2018). Table 1 depicts the repeatabilities of Ω and UT1-UTC for different ratios of number of satellite observations to total number of observations.

Table 1 Repeatability of Ω and UT1-UTC for different ratios of satellite observations to total number of observations

Ratio of satellite observations in %	Repeatability of Ω in µas	Repeatability of UT1-UTC in µs
3.4	48.9	1.5
6.9	41.4	1.4
9.0	39.7	1.4
11.5	39.2	1.3
11.8	42.2	1.4
13.9	43.0	1.4
20.7	44.9	1.4

For ratios between 3.4 and 20.7 %, we achieve repeatabilities of Ω between 39 and 49 µas with the best values for ratios at about 10 % observations to satellites. We do not find improvement with a higher ratio of satellite observations, which is due to the importance of quasar observations for the estimation of tropospheric parameters. The corresponding repeatabilities of UT1-UTC are hardly affected by spending observation time for satellites and are between 1.3 and 1.5 µs. It should be noted here that 1.5 µs correspond to 22.5 µas, which is about half of the repeatability of Ω. This factor 2 can be explained by the larger number of observations to quasars and the better geometry of quasar observations.

A change of 50 µas in Ω corresponds to less than one centimetre at the altitude of the Galileo satellite and to about 3 µs in UT1-UTC. Thus, the estimation of Ω from VLBI observations provides an alternate route of providing UT1-UTC to GNSS. In a more sophisticated scenario, VLBI analysis will provide normal equations including orbital elements to combination software stacking the normal equations from all techniques. Of course and most rigorously, software packages capable of handling GNSS and VLBI observations can combine both techniques at the observation level.

2.2 Estimation of Station Coordinates

Geodetic VLBI uses radio telescopes to observe extragalactic radio sources, mostly quasars. Since these sources are at quasi-infinite distance, VLBI observations are not sensitive to the centre of mass of the Earth or a translation of the station network. As a consequence, VLBI requires constraints to remove the rank defect. Typically, a no-net-translation (NNT) condition with respect to a priori station coordinates is applied. Additionally, a no-net-rotation (NNR) condition for station coordinates has to be used if Earth orientation parameters are estimated. With fixed source coordinates, we can then determine telescope coordinates in the kinematic quasar frame.

On the other hand, VLBI observations to VLBI transmitters on GNSS satellites are sensitive to the centre of mass of the Earth, because the satellite orbits are subject to the gravity field. For simplicity, we restrict our considerations in the following to the translation of the stations with respect to the satellites, because we do not have a dynamic orbit modelling tool available in VieVS. With this approach and fixed satellite orbits and Earth orientation parameters, we estimate telescope coordinates in the satellite frame. By comparing the networks of telescope coordinates in the quasar frame and the satellite frame, we can assess the tie between those two frames. Assuming that the quasar frame has been precisely realised from decades of VLBI observations to quasars, it is our primary focus to assess the accuracy of station coordinates from VLBI observations to VLBI transmitters on GNSS satellites to get an estimate for the accuracy of frame tie realisation.

With the classical analysis of GNSS observations, we determine the position of GNSS receiving antennas in the satellite frame. With VLBI observations to well-calibrated VLBI transmitters on GNSS satellites, we are able to determine telescope positions in the dynamic satellite frame, as detailed above. In other words, we are able to transfer the space tie between the satellite antenna and the VLBI transmitter to the Earth surface yielding the local tie between the GNSS antenna and the VLBI reference point. Again, it is important to assess the accuracy of station coordinates from VLBI observations to VLBI transmitters on GNSS satellites in order to judge the potential of transferring the space tie to local tie information.

In the following, we are referring to another study by Wolf and Böhm (2023) on the optimal distribution of VLBI transmitters on Galileo satellites for frame ties. Their work is based on Monte-Carlo simulations with a network of twelve VGOS stations and the Galileo constellation. They fix source coordinates, orbits, and Earth orientation parameters. On the other hand, they estimate station coordinates with minimum constraints (no NNR/NNT conditions) as well as auxiliary tropospheric and clock parameters. The main findings are:

- Station coordinate repeatabilities from 24 hour sessions are at the centimetre-level or better.
- East coordinate components are better determined than north and up components.
- The best repeatabilities are achieved if at least three Galileo satellites, all in one plane, are equipped with a VLBI transmitter. Further Galileo satellites with VLBI transmitters do not improve the accuracy of station coordinates. With three transmitters, the best ratio of satellite observations to total number of observations is 30 to 40 %.

In summary, it is possible to estimate station coordinates from VLBI observations to VLBI transmitters on GNSS satellites with centimetre-accuracy from 24 hour sessions. With sufficiently long observing plans (months and longer), the realisation of precise frame ties and local ties will be possible. However, so far, systematic error sources have been neglected in the simulations, e.g., with the space tie.

3 VLBI Transmitters on Genesis

The main purpose of the Genesis mission (see Fig. 3) is the realisation of local ties from space ties, thereby improving the terrestrial reference frame significantly (Delva et al. 2023). Applied to VLBI, this goal can be broken down to

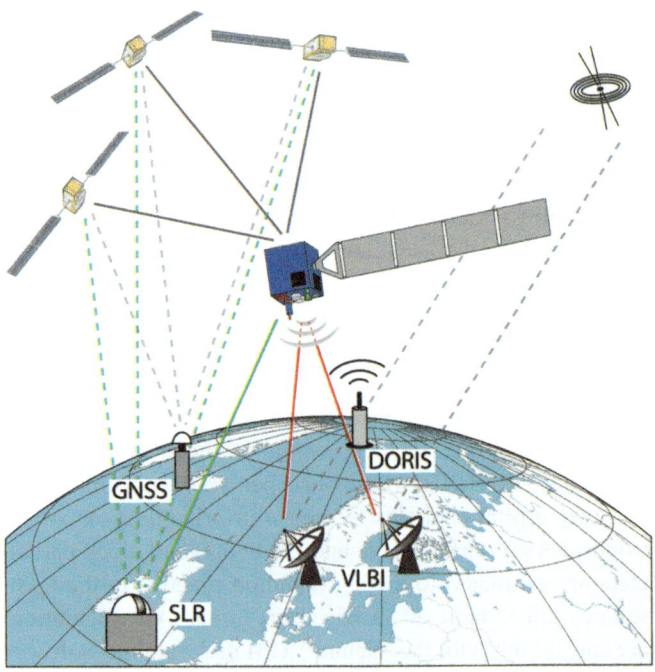

Fig. 3 Illustration of the Genesis satellite. The co-location satellite will be equipped with a VLBI transmitter, SLR retro-reflector, GNSS receiver and DORIS receiver and can therefore be observed with these four space-geodetic techniques

the estimation of station coordinates from VLBI observations to the VLBI transmitters on Genesis. Compared to VLBI transmitters on GNSS satellites, the task is more challenging, because the common visibility of Genesis is worse for baselines between VLBI stations due to the quasi-polar orbit at about 6000 km (Delva et al. 2023). And there will be only one satellite with a VLBI transmitter. On the other hand, there is a better geometry with a faster moving satellite on the sky, thereby improving the estimation of station coordinates. In any case, more simulations are required to develop the best observation scenario for Genesis possible in order to reach millimetre-accuracy of the local ties. These studies can be based on the concept by Anderson et al. (2018) and the application of realistic error descriptions and schedules.

With Genesis, there is also a better geometry for the estimation of satellite positions, compared to GNSS satellites. Wolf et al. (2022) introduced Dilution of Precision (DOP) values for kinematic positions of satellite orbits. For illustration, let us assume a radial DOP value of 2. This means that the radial component of the orbit can be determined with a precision of 2 cm if the uncertainty of the VLBI observations is 1 cm. The calculation of the DOP values is based on error propagation only. Table 2 depicts the best DOP values for a nine station VGOS network (Fig. 4, different from the one displayed above).

As expected, the radial component of the Galileo satellites cannot be well determined from VLBI observations. On the other hand, the radial component can be derived from the orbital period with Kepler's third law. In general, the DOP values for Genesis are significantly better, suggesting that VLBI can well contribute to its orbit determination. However, the DOP values do not reflect the lower number of VLBI observations, which are possible to Genesis compared to GNSS.

4 Conclusions and Recommendations

VLBI observations to VLBI transmitters on satellites add unique information for the estimation of geodetic parameters and products. As confirmed by simulations, they can be used for orbit determination or the improvement of local ties and terrestrial reference frames. Nevertheless, sophisticated simulation studies are still necessary, including realistic error modelling (e.g., with errors in the space tie or the satellite orbit) and realistic observation plans based on projected network situations. The error estimates presented here are lower bounds because of the error sources not considered.

The studies listed above are mostly simulations from a VLBI point of view. Of course, the most rigorous approach is the use of software capable of analysing all space geodetic techniques and combining the observations at the observation level. In any case, it is important to develop solutions with interfaces between VLBI software and software packages handling satellite techniques. In particular, strategies have to be developed to combine normal equations from all techniques including orbit information.

Last but certainly not least, it is about time to define the specification of the VLBI transmitters. Furthermore, the best observing concepts have to be identified. Both tasks have to be addressed in co-operation with the International VLBI Service for Geodesy and Astrometry (IVS; Nothnagel et al. 2017).

Table 2 Best possible Dilution of Precision (DOP) values for radial, along-track, and cross-track components for a nine station VGOS network

DOP	Galileo	Genesis
Radial	15.6	2.6
Along-track	1.0	0.4
Cross-track	1.3	0.4

Fig. 4 Network of the nine VGOS stations used for the determination of the DOP values

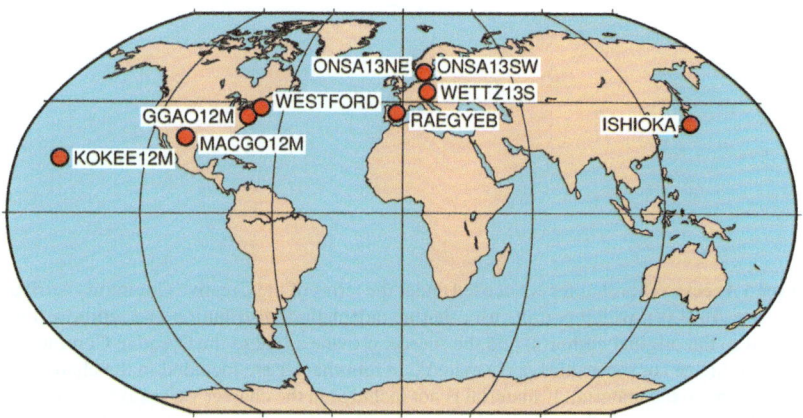

Acknowledgements We would like to thank the Austrian Science Fund (FWF) for supporting this work with project P 33925 (VLBI2Galileo).

References

Altamimi Z, Rebischung P, Collilieux X, Metivier L, Chanard K (2023) Itrf2020: an augmented reference frame refining the modeling of nonlinear station motions. J Geodesy 97(47). https://doi.org/10.1007/s00190-023-01738-w

Anderson J, Beyerle G, Glaser S, Liu L, Männel B, Nilsson T, Heinkelmann R, Schuh H (2018) Simulations of VLBI observations of a geodetic satellite providing co-location in space. J Geodesy 92:1023–1046. https://doi.org/10.1007/s00190-018-1115-5

Bar-Sever Y, Haines B, Wu S, Lemoine F, Willis P (2009) Geodetic Reference Antenna in Space (GRASP): a mission to enhance the terrestrial reference frame. Paper presented at the COSPAR Colloquium – Scientific and Fundamental Aspects of the Galileo Programme, University of Padova, Padua, Italy, 14–16 October 2009. https://doi.org/https://gssc.esa.int/education/galileo-science-colloquium/

Biancale R, et al. (2017) E-GRASP/Eratosthenes. Proposal for Earth Explorer Opportunity Mission EE-9 in response to the Call for Proposals for Earth Explorer Opportunity Mission EE-9, unpublished

Böhm J, Böhm S, Boisits J, Girdiuk A, Gruber J, Hellerschmied A, Krasna H, Landskron D, Madzak M, Mayer D, McCallum J, McCallum L, Schartner M, Teke K (2018) Vienna VLBI and Satellite Software (VieVS) for geodesy and astrometry. Publ Astronom Soc Pacific 130(986). https://doi.org/10.1088/1538-3873/aaa22b

Dach R (2022) Dependency of satellite geodesy on UT1-UTC from VLBI. In: Kyla L, Armstrong KDB, Dirk B (eds) International VLBI Service for Geodesy and Astrometry 2022 General Meeting Proceedings, pp 3–9. NASA/CP-20220018789, Helsinki, Finland

Delva P, Altamimi Z, Blazquez A, et al. (2023) GENESIS: co-location of geodetic techniques in space. Earth Planets Space 75(5). https://doi.org/10.1186/s40623-022-01752-w

Haas R, Hobiger T, Klopotek G, Kareinen N, Yang J, Combrinck L, De Witt A, Nickola M (2017) VLBI with GNSS–signals on an intercontinental baseline – A progress report. In: Haas R, Elgered G (eds) Proceedings of the 23rd European VLBI Group for Geodesy and Astrometry Working Meeting May, 2017, Gothenburg, Sweden, pp 117–121

Hellerschmied A, McCallum L, McCallum J, Sun J, Böhm J, Cao J (2018) Observing apod with the auscope vlbi array. Sensors 18(5). https://doi.org/10.3390/s18051587

Klopotek G, Hobiger T, Haas R, Otsubo T (2020) Geodetic VLBI for precise orbit determination of Earth satellites: a simulation study. J Geodesy 94:56. https://doi.org/10.1007/s00190-020-01381-9

Männel B (2016) Co-location of geodetic observation techniques in space. PhD thesis, ETH Zurich, Zürich. https://doi.org/10.3929/ethz-a-010811791

Nothnagel A, Artz T, Behrend D, Malkin Z (2017) International VLBI service for geodesy and astrometry: delivering high-quality products and embarking on observations of the next generation. J Geodesy 91(7):711–721. https://doi.org/10.1007/s00190-016-0950-5

Petrachenko B, Niell A, Behrend D, Corey B, Böhm J, Charlot P, Collioud A, Gipson J, Haas R, Hobiger T, Koyama Y, MacMillan D, Malkin Z, Nilsson T, Pany A, Tuccari G, Whitney A, Wresnik J (2009) Design aspects of the VLBI2010 system. Technical report, Progress Report of the IVS VLBI2010 Committee. In: NASA/TM-2009-214180

Plank L, Böhm J, Schuh H (2014) Precise station positions from VLBI observations to satellites: a simulation study. J Geodesy 88(7):659–673. https://doi.org/10.1007/s00190-014-0712-1

Plank L, Hellerschmied A, McCallum J, Böhm J, Lovell J (2017) VLBI observations of GNSS-satellites: from scheduling to analysis. J Geodesy 91:867–880. https://doi.org/10.1007/s00190-016-0992-8

Rothacher M, Beutler G, Behrend, D, Donnellan, A, Hinderer, J, Ma C, Noll C, Oberst J, Pearlman M, Plag H-P, Richter B, Schöne T, Tavernier G, Woodworth PL (2009). In: Plag H-P, Pearlman M (eds) The future global geodetic observing system, pp 237–272. Springer, Berlin, Heidelberg. https://doi.org/10.1007/978-3-642-02687-49

Schartner M, Böhm J (2019) VieSched++: A new VLBI scheduling software for geodesy and astrometry. Publ Astronom Soc Pacific 131:084501. https://doi.org/10.1088/1538-3873/ab1820

Wolf H, Böhm J (2023) Optimal distribution of VLBI transmitters on Galileo satellites for frame ties. Earth Planets Space 75(173). https://doi.org/10.1186/s40623-023-01926-0

Wolf H, Böhm J, Schartner M, Hugentobler U, Soja B, Nothnagel A (2022) Dilution of Precision (DOP) factors for evaluating observations to Galileo satellites with VLBI. In: Freymueller JT, Sánchez L (eds) Geodesy for a sustainable earth, pp 305–312. Springer, Cham

Open Access This chapter is licensed under the terms of the Creative Commons Attribution 4.0 International License (http://creativecommons.org/licenses/by/4.0/), which permits use, sharing, adaptation, distribution and reproduction in any medium or format, as long as you give appropriate credit to the original author(s) and the source, provide a link to the Creative Commons license and indicate if changes were made.

The images or other third party material in this chapter are included in the chapter's Creative Commons license, unless indicated otherwise in a credit line to the material. If material is not included in the chapter's Creative Commons license and your intended use is not permitted by statutory regulation or exceeds the permitted use, you will need to obtain permission directly from the copyright holder.

Formation of a GNSS Network in Space Based on Simulated LEO Constellations

Lukas Müller, Markus Rothacher, and Benedikt Soja

Abstract

Large constellations of low Earth orbit (LEO) satellites equipped with Global Navigation Satellite System (GNSS) receivers open the possibility of forming a dense and homogeneous GNSS network in space, covering the entire Earth. Based on simulations, this study investigates how geodetic Earth observation could benefit from this development. In the first part, we compare the effects of different processing strategies, parameterizations and simulated errors. The results show that for a large number and a uniform distribution of satellites in a constellation, GNSS network (double-difference) processing can improve LEO orbit determination compared to a single-satellite (zero-difference) processing, provided that the integer ambiguities have been correctly resolved. In the second part of this study, we demonstrate that in a LEO constellation with 36 uniformly distributed satellites, an accuracy of about 1 cm (3D RMS) for the a-priori LEO orbits and about 3 cm for the GNSS orbits is required to achieve the sufficient ambiguity-fixing rate necessary to take full advantage of the double-difference processing.

Keywords

GNSS · LEO · Satellite constellation

1 Introduction

The number of low Earth orbit (LEO) satellites equipped with Global Navigation Satellite System (GNSS) receivers is rapidly increasing. GNSS observations in space are no longer limited to a small number of Earth observation satellites, but large nanosatellite constellations enable the formation of a dense network of GNSS receivers around the Earth. Such a space-based GNSS network holds great potential for geodetic Earth observation: First, it allows GNSS double-difference processing in space, where GNSS signals are not affected by tropospheric refraction and where no ground observations are needed. Second, a dense and homogeneous GNSS network in space may improve the sensitivity to certain geodetic parameters by providing a better and faster changing observation geometry than a ground-based network or just a few LEO satellites.

For ground-based GNSS networks, forming double differences of simultaneous observations of two GNSS satellites from two different GNSS stations is a common practice as it eliminates the receiver and satellite clock errors and mitigates the effects of GNSS satellite orbit errors and atmospheric refraction (Langley et al. 2017). GNSS baselines in space have previously been studied based on formation-flying satellite missions such as GRACE A and B (Jäggi et al. 2007; Švehla and Rothacher 2005), TerraSAR- and TanDEM-X (Allende-Alba and Montenbruck 2016; Moon et al. 2008) and SWARM A and C (Allende-Alba et al. 2017), with baseline lengths of 220 km, 500 m and 160 km, respectively. These studies showed that the baselines between these satellites could be reconstructed at the mm or even sub-mm

L. Müller (✉) · M. Rothacher · B. Soja
Institute of Geodesy and Photogrammetry, ETH Zürich, Zurich, Switzerland
e-mail: lukamueller@ethz.ch; markus.rothacher@ethz.ch; soja@ethz.ch

level if the carrier-phase ambiguities were fixed to integer values.

In this study, we use simulated GNSS observations in LEO constellations to investigate the feasibility of forming a space-based GNSS network. We therefore process the simulated GNSS observations in both a zero-difference (undifferenced, ZD) and a double-difference (DD) mode to estimate LEO orbit parameters and examine their sensitivity to small errors in the force model. Furthermore, we analyse how well carrier-phase ambiguities can be resolved and how this depends on the quality of the a-priori orbits and the baseline geometry. From this, we draw conclusions about the required accuracy of the force model and GNSS observations to form a GNSS network in space for geodetic Earth observation.

2 Methods

2.1 Constellation Scenarios

In this study, we consider three constellation scenarios (A–C): In scenario A, we form a constellation of 16 unevenly distributed satellites by combining the orbits of existing geodetic Earth observation missions, including GRACE-FO A and B, TanDEM- and TerraSAR-X, Jason-1, -2, -3, Sentinel-1A to -3B and SWARM A, B and C. Constellation B consists of 16 satellites, which are evenly distributed over 4 orbital planes with an inclination of 60° and a phasing parameter of 0.9, according to the Walker Delta of 60°: 16/4/0.9. In constellation C, 36 satellites are evenly distributed on 6 orbital planes with an inclination of 60° and a phasing parameter of 1.0, according to the Walker Delta of 60°: 36/6/1.0. Based on these three constellations, we can study, first, the differences between a uniform and a non-uniform distribution of satellites and, second, the effect of a varying number of satellites (Table 1).

Table 1 Overview of the constellation scenarios A, B and C. The missions GRACE-FO, TanDEM- and TerraSAR-X, Jason, Sentinel and SWARM are abbreviated as G, T, J, S and SW respectively. The altitude refers to the orbit height at the equator

Constellation	Mission	#satellites	Inclination [°]	Altitude [km]
A	G	2	89	490
	T	2	97	515
	J	3	66	1322–1336
	S	6	98–99	700–810
	SW	3	87–88	435–507
B	–	16	60	530
C	–	36	60	530

2.2 Simulation and Processing Framework

Figure 1 shows a flow chart of the simulation and processing setup used in this study. The simulation of GNSS observations on board a LEO satellite requires both LEO and GNSS orbits. LEO orbits are computed based on the three constellation scenarios presented in Sect. 2.1. GNSS ephemerides are available from the Center for Orbit Determination in Europe (CODE). Using the satellite positions given in the standard product format (SP3), we determine dynamic orbits for the LEO and the GNSS satellites. This is done by estimating the six Keplerian elements of the initial state of the satellites and additional empirical parameters by fitting the dynamic trajectories to the SP3 positions. On the one hand, we estimate ground-truth orbits by using a "correct" force model for the orbit fitting, which includes the gravity potential of the Earth up to degree and order 120 and tidal potential up to degree and order 50. On the other hand, we determine a-priori orbits based on a manipulated force model. Manipulation here means that a smaller maximum degree and order is used for the spherical harmonic development of the gravity or the tidal potential, which introduces systematic errors in the a-priori orbits. The purpose of manipulating the gravitational

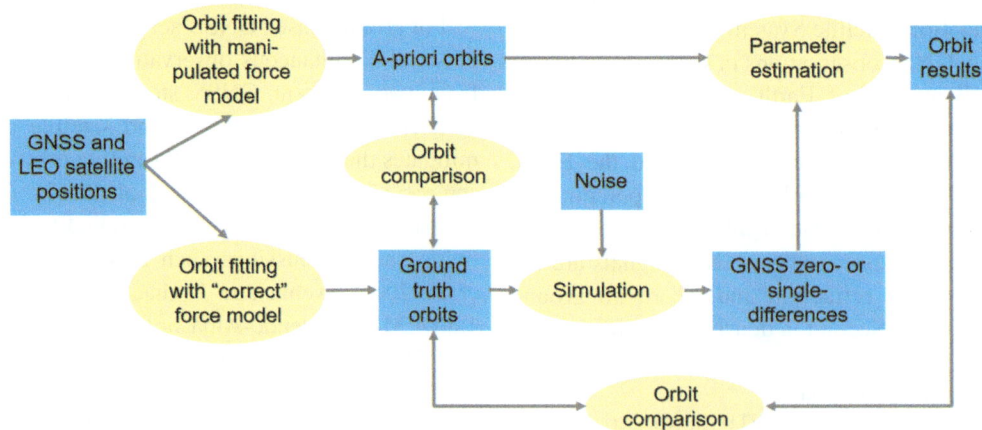

Fig. 1 Scheme for the simulation and processing of GNSS networks in space based on LEO constellations. Data sets and results are visualized by blue boxes, methods and processing steps by yellow ellipses

force model is to generate errors of a certain magnitude. These errors are intended to represent the sum of different systematic biases that may arise in a real scenario due to both gravitational and non-gravitational effects. The force model is manipulated for either the LEO or the GNSS orbit fitting. Based on the ground-truth orbits, GNSS observations are simulated as seen from a LEO satellite. White noise of 2 mm is added to the carrier-phase observations, which is about the noise level expected in real measurements. The final orbits are estimated using the a-priori orbits and the simulated observations, either in a ZD or a DD mode. In the case of the DD processing, single-difference observations (baselines) are formed prior to the parameter estimation process. To determine the magnitude of the error caused by the manipulation of the force model, we compare the a-priori orbits to the ground-truth orbits. The orbit results are compared to the ground-truth orbits to assess the impact of the simulated errors on the final orbits.

For both the simulation and the data processing we use the Bernese GNSS Software version 5.3. In the following, the length of the time interval for the simulation and data processing is 4 h 48 min. This time interval was chosen as it corresponds to approximately three orbital periods for the satellites of constellations B and C with altitudes of 530 km. The sampling rate is 1 min. Only GPS phase observations on the L1 and L2 frequencies are simulated. Multi-GNSS is not considered in this study as all relevant LEO missions track GPS only, including for example the geodetic missions of constellation A and the precise orbit determination (POD) antenna of the Spire satellites. Final orbit results are generated only for the LEO satellites, but not for the GPS satellites. We use all possible combinations of satellites to form double-differences, maximizing the number of baselines.

3 LEO Orbit Results

In this section, we first study the differences between the three constellation scenarios and between the ZD and DD approaches. For this purpose, we manipulate the force model used for the determination of the a-priori LEO orbits by neglecting the ocean tides (development up to degree 0). The estimated parameters of the LEO orbits are the six Keplerian elements plus 3D stochastic pulses in the form of instantaneous velocity changes every 15 min. In the ZD case, the carrier-phase ambiguities are estimated as float values. Also for the DD processing, no ambiguity resolution is performed at this stage, but the integer ambiguities that are known from the simulation are introduced. The resulting root mean square (RMS) errors in the a-priori orbits and the final ZD and DD solutions for the different constellation scenarios are shown in Fig. 2.

The magnitude of the simulated errors is similar for all three constellation scenarios, with a 3D RMS at the level of 65 mm to 72 mm. The RMS values vary quite strongly between the individual satellites of a constellation and a clear systematic can be identified. The more similar the orbits or the closer the satellites are to each other, the more similar the RMS. In constellation A, this is most obvious for satellites of the same mission, such as GRACE-FO A and B or TanDEM-X and TerraSAR-X. Additionally, the GRACE-FO and the SWARM satellites at altitudes of about 400 to 500 km are significantly more affected by the errors of the force model than the Jason satellites, which are at a much higher altitude of about 1300 km. In the constellations B and C, satellites of the same or adjacent orbital planes have rather similar RMS.

In the ZD solution, the RMS values are at the level of 16 to 19 mm, which is about 75 % smaller than the simulated errors. This reduction is due to the estimation of stochastic pulses, which absorb a large part of the error caused by the manipulation of the force model. Compared to the magnitude of the simulated errors in the a-priori orbits, the RMS values of the ZD solution are generally less dependent on the satellite altitude. Overall, the ZD results are similar for all three constellations, since the data are processed separately for each satellite and thus the LEO orbit results do not depend on the constellation design, but rather on the individual orbit.

The RMS values of the DD solution of about 6 to 14 mm are significantly smaller than in the ZD case. This is due to the generally larger number of observations resulting from the baseline formation and the introduction of the correct integer ambiguities in the DD approach. In constellation A, the total RMS of the DD solution is slightly higher and the errors vary more between the individual satellites than in constellation B. This might be due to the more heterogeneous distribution and the different altitudes of the satellites in constellation A. The much larger number of baseline observations in constellation C (about 430,000) than in the constellations A and B (about 76,000 to 78,000) results in a significantly smaller overall RMS. Unlike the ZD solution, the DD solution shows a clear dependency on the constellation design, where a larger number of satellites and a homogeneous distribution can improve the orbit results.

Table 2 compares the DD processing results for different simulated errors and parameterizations based on constellation C. As seen in Fig. 2, neglecting the ocean tides in the force model for the a-priori LEO orbit computation leads to an error of about 72 mm, which can be reduced to 6.7 mm if stochastic pulses are estimated in addition to the six Keplerian elements. In this case, where the remaining systematic effect is quite small, the additional noise of 2 mm has no significant further impact on the LEO orbit results.

An error of 30 mm in the GPS orbits results in an error of about 3.8 mm in the LEO orbits. Unlike the manipulated a-priori LEO orbits, an error in the GPS orbits cannot be

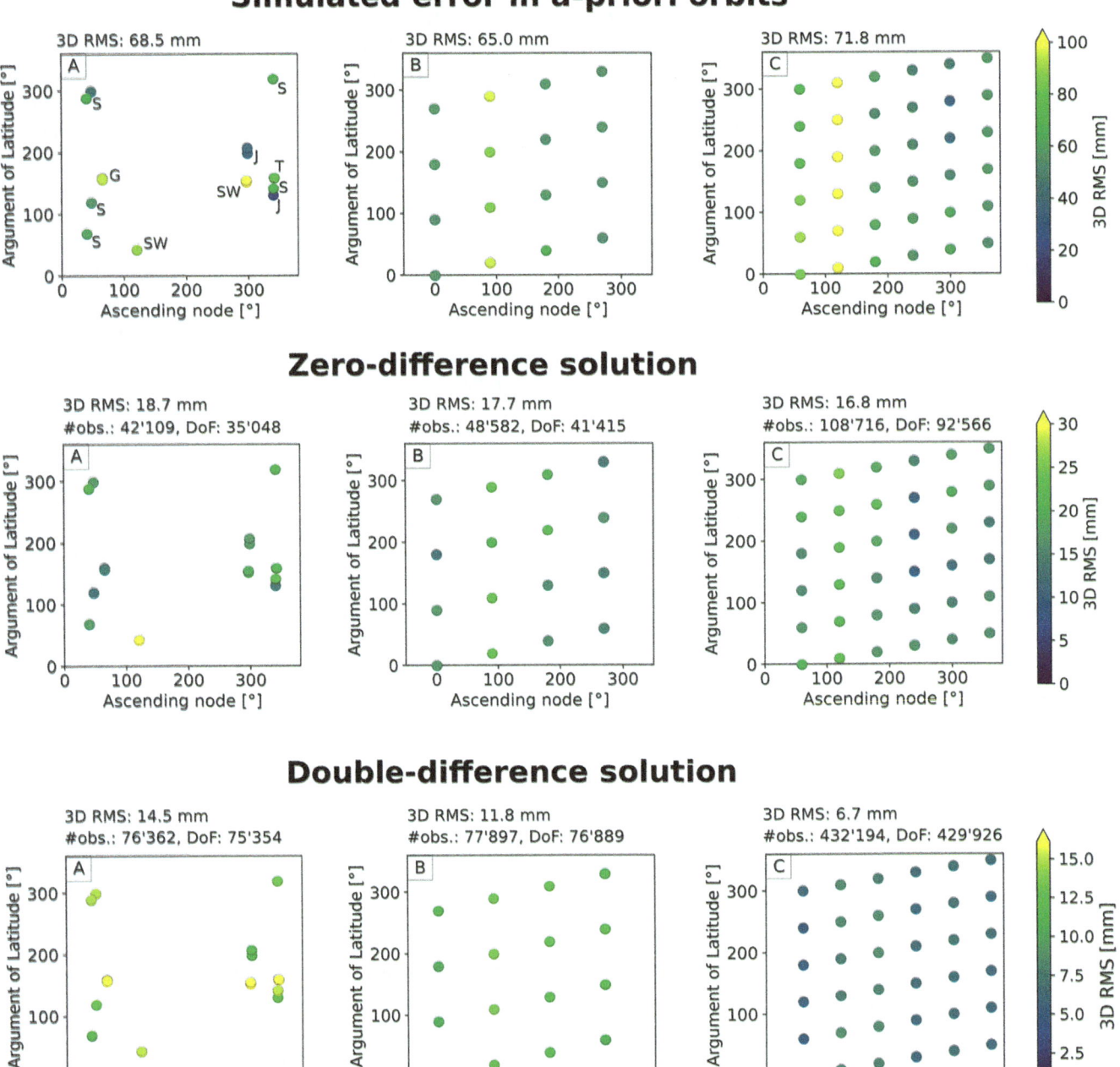

Fig. 2 3D RMS for different constellations and processing strategies, both overall and for each LEO satellite individually. In addition, the number of observations and the degree of freedom (DoF) are provided for the ZD and the DD solutions. The position of each satellite in the constellation is given by its right ascension of the ascending node and its argument of latitude at the start time of the simulation. The constellation (A,B,C) is indicated in the upper left corner of each plot. The individual missions of constellation A are labelled with the abbreviations defined in Table 1. In the case of satellite pairs where the dots overlap due to the small distance between the satellites, such as the GRACE-FO A and B or the TerraSAR- and TanDEM-X missions, both satellites involved have practically the same RMS values

Table 2 Overview of the LEO orbit results for different simulated errors and parameterizations based on constellation C. LEO-T0 means that no ocean tides (up to degree 0) were used for the a-priori LEO orbit computation. GPS-P3 means that the gravity potential used for the a-priori GPS orbit computation was only developed up to degree and order 3. K, S and A stand for the six Keplerian elements, stochastic pulses and ambiguity parameters, respectively. In the parentheses we indicate whether the RMS values refer to the LEO or the GPS orbits. The unit is mm

Introduced errors + noise	Estimated parameters	a-priori vs. ground truth (3D RMS)	Estimated vs. ground truth (3D RMS)
LEO-T0 + 0 mm	K + S	72 (LEO)	6.7 (LEO)
LEO-T0 + 2 mm	K + S	72 (LEO)	6.7 (LEO)
GPS-P3 + 2 mm	K + S	30 (GPS)	3.8 (LEO)
LEO-T0 + 2 mm	K	72 (LEO)	72.2 (LEO)
LEO-T0 + 2 mm	K + S + A	72 (LEO)	16.2 (LEO)

absorbed by stochastic pulses. This is because only the parameters of the LEO orbits are estimated, but no parameters of the GPS orbits. The fact that the error in the final LEO orbit is significantly smaller than the introduced error in the GPS orbits is due to the following reasons: First, the GPS observations coming from different directions are averaged, reducing the resulting error in the LEO orbit. Second, an error in the GPS orbits leads to a comparably smaller error in the baseline observations, as the distances between the GPS and LEO satellites are much larger than the baseline lengths (Bauersima 1983). Furthermore, the use of a correct a-priori LEO orbit does not allow a complete adjustment of the orbit to erroneous observations but forces the trajectory to be close to the ground-truth orbit.

If only the six Keplerian elements are set up as parameters, the error in the LEO orbits caused by the manipulated force model remains almost at its full magnitude. This is because the satellite trajectory cannot be adjusted to the observations due to the incorrect force model. The RMS of the LEO orbit solution is even slightly larger than the introduced systematic error due to the additional effect of the simulated noise.

If the carrier-phase ambiguities are estimated as real numbers instead of using the correct values from the simulation, the LEO orbit results are worse by a factor of two to three. The RMS of about 16.2 mm is comparable to the RMS of 16.8 mm for the ZD approach (see Fig. 2). This shows that it is crucial to fix the ambiguities to their integer values to take full advantage of the DD approach. Some recent studies, such as Arnold et al. (2018), Montenbruck et al. (2018), have demonstrated the feasibility of ambiguity resolution (AR) even at the zero-difference level by using high-quality clock and time-variable phase bias products, improving orbit determination by about 30 to 50 % compared to a float ambiguity solution. However, ZD ambiguity resolution was not realized in the frame of this paper, and therefore no ambiguity-fixed ZD solution was simulated.

4 Integer Ambiguity Resolution

In this section, carrier-phase integer ambiguity resolution (AR) is performed based on double-difference observations, which is considered the standard approach for AR, as it eliminates receiver clock and instrument errors so that the ambiguities are actual integers and no specific bias products, that may not include the full temporal variation of the biases, are necessary. We use the *sigma* strategy to resolve the narrowlane ambiguities (Dach and Walser 2015). For this purpose, the widelane ambiguities known from the simulation are introduced, assuming that they have been successfully resolved in a previous step, e.g., by processing the Melbourne-Wuebbena linear combination. In the *sigma* strategy, the narrowlane ambiguities are resolved iteratively. In each iteration step, the ambiguities are estimated as real numbers. The subset of the ambiguities with the smallest formal errors (sigmas) is then assigned to the nearest integer value if they fulfill certain criteria, i.e., if they are close to an integer and their sigma is below a given threshold. These ambiguities are then held fixed during the subsequent iterations. Thus, the probability that the remaining ambiguities can be resolved increases from iteration to iteration, maximizing the chances that a high number of ambiguities can be fixed overall. The AR is carried out without the simultaneous estimation of LEO orbit parameters.

The AR aims at fixing as many ambiguities as possible to integer values. At the same time, we would like to ensure that all of the ambiguities are resolved correctly. Since the correct integer ambiguities are known for the simulated GPS observations, we can choose the options for the *sigma* strategy such that the percentage of fixed ambiguities is maximized, while all resolved ambiguities are correct. Figure 3 shows the resulting percentages of fixed ambiguities for constellation C. We distinguish between baselines formed with satellites on the same orbit plane, the adjacent planes and other baselines. While the length of baselines within the same orbital plane is quite stable, baselines between satellites with a larger difference in the ascending node can vary considerably during one revolution.

As can be seen in Fig. 3, the AR is quite sensitive to small errors in the a-priori LEO orbits. An error of about 0.8 cm leads to quite good AR results, where for most of the baselines, in particular those with satellites on the same or adjacent orbital planes, more than 80 % of the ambiguities could be resolved. A slightly larger error of 1.4 cm causes a generally much lower percentage of ambiguity resolution. Baselines of the same orbital plane, however, still have a fairly high percentage of resolved ambiguities. With a systematic error of 3.4 cm the percentage of resolved ambiguities is for most of the baselines less than 50 %. This shows that with such a large error, it does not make sense to perform

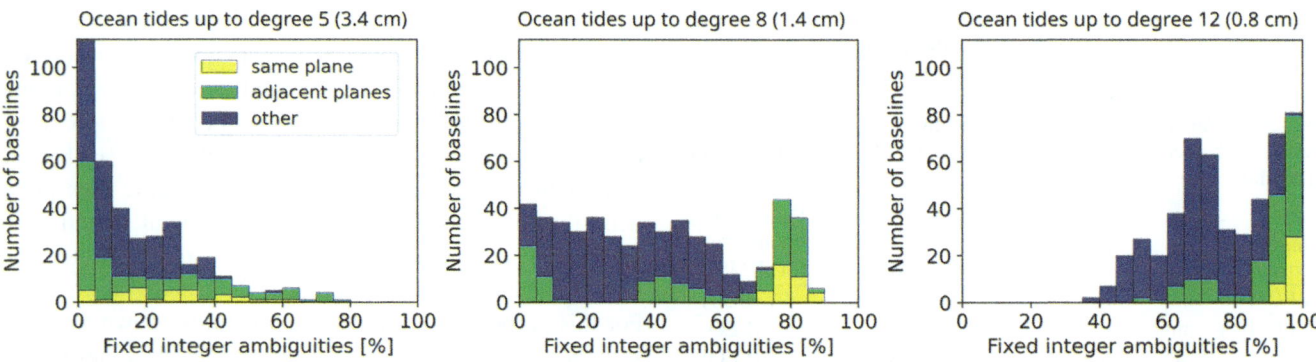

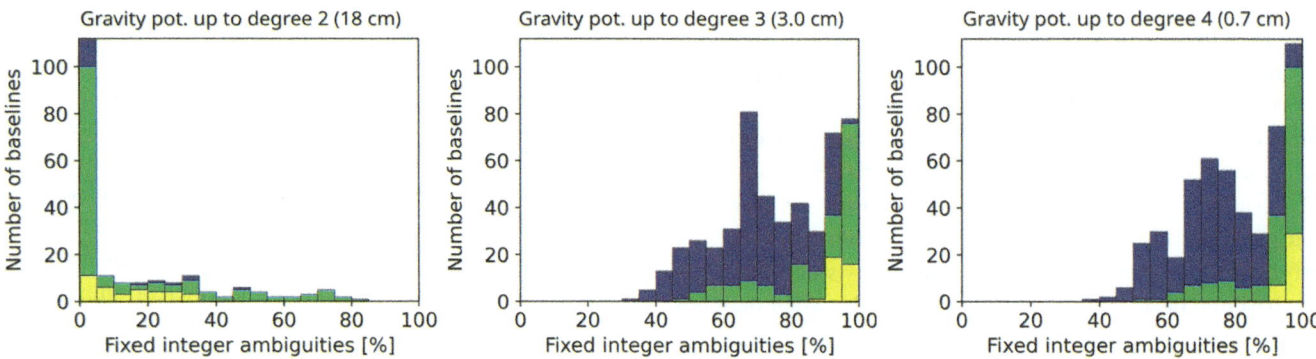

Fig. 3 Histograms of the AR results for constellation C with different simulated errors

AR on real GPS observations, where no information about the correctness of the fixed ambiguities is available.

The accuracy requirements to successfully resolve the integer ambiguities are generally lower for the GPS than for the LEO orbits. With simulated errors of 0.7 and 3.0 cm, we obtain quite good results for the AR, where all of the baselines within the same orbital plane and most of the baselines in adjacent planes have an ambiguity-fixing rate of more than 85 %. In the case of a rather large error of 18 cm in the GPS orbits, AR would not be reasonable, as most baselines, including those of the same orbital plane, have an ambiguity-fixing rate of less than 50 %.

By comparison, in studies based on real observations, the percentage of fixed integer ambiguities was usually at a level of 90 %, e.g., for the relatively short baseline between GRACE A and B (Jäggi et al. 2007). Overall, we can conclude from our results that the error due to force-model deficiencies should be approximately 1 cm or smaller for the a-priori LEO orbits and about 3 cm for the GPS orbits. These values are about what can be expected for the accuracy of reduced-dynamic LEO orbits using geodetic-type GPS receivers and final GPS orbit products (Dach et al. 2009; Mao et al. 2021; Bock et al. 2011). However, compared to the simulations presented here, which represent the ideal case, real observations may have several additional factors, such as frequent data gaps and the missing knowledge about correctly or wrongly fixed ambiguities to optimize the AR options, which further complicate the AR (Allende-Alba and Montenbruck 2016).

5 Conclusions

In this study, we have developed a framework for simulating and processing a GNSS network in space, which allows us to study the effects of different systematic biases and processing strategies for various constellation scenarios. Based on simulations, we first investigated the impact of a systematic error of 72 mm in the a-priori LEO orbits and an error of 30 mm in the GPS orbits on the estimation of LEO orbit parameters. Our results show that, especially for constellations with a large number and a uniform distribution of satellites, a GPS network solution based on double-differences can offer clear advantages compared to a single-satellite solution. However, the advantages of the double-differences only come into play if the carrier-phase integer ambiguities can be resolved. Since the correct integer ambiguities are known from the simulation, we were able to determine the maximum acceptable force-model errors for a successful ambiguity resolution, which are about 1 cm for the a-priori LEO orbits and 3 cm for the GPS orbits. While these simulated results were obtained under the assumption of ideal conditions and cannot be directly transferred to real GPS observations, they still provide a useful reference of what is feasible under the best possible circumstances. Especially regarding the strategy and the settings for the integer ambiguity resolution, the presented simulation framework can be a useful tool. In a follow-up study, the concepts and methods will be applied to GNSS observations from the Spire CubeSat constellation.

References

Allende-Alba G, Montenbruck O (2016) Robust and precise baseline determination of distributed spacecraft in LEO. Adv Space Res 57(1):46–63

Allende-Alba G, Montenbruck O, Jäggi A, Arnold D, Zangerl F (2017) Reduced-dynamic and kinematic baseline determination for the Swarm mission. GPS Sol 21:1275–1284

Arnold D, Schaer S, Villiger A, Dach R, Jäggi A (2018) Undifference ambiguity resolution for GPS-based precise orbit determination of low Earth orbiters using the new code clock and phase bias products

Bauersima I (1983) NAVSTAR/Global Positioning System (GPS)(II). Radiointerferometrische Satellitenbeobachtungen. Mitt. Satell.-Beobachtungsstn. Zimmerwald 10

Bock H, Jäggi A, Meyer U, Visser P, van den IJssel J, van Helleputte T, Heinze M, Hugentobler U (2011) GPS-derived orbits for the GOCE satellite. J Geodesy 85:807–818

Dach R, Brockmann E, Schaer S, Beutler G, Meindl M, Prange L, Bock H, Jäggi A, Ostini L (2009) GNSS processing at CODE: status report. J Geodesy 83:353–365

Dach R, Walser P (2015) Bernese GNSS Software Version 5.2. Citeseer

Jäggi A, Hugentobler U, Bock H, Beutler G (2007) Precise orbit determination for GRACE using undifferenced or doubly differenced GPS data. Adv Space Res 39(10):1612–1619

Langley RB, Teunissen PJG, Montenbruck O (2017) Introduction to GNSS. In: Peter JG Teunissen, O Montenbruck, (eds), Springer Handbook of Global Navigation Satellite Systems, chapter 1, pp 3–22. Springer

Mao X, Arnold D, Girardin V, Villiger A, Jäggi A (2021) Dynamic GPS-based LEO orbit determination with 1 cm precision using the Bernese GNSS Software. Adv Space Res 67(2):788–805

Montenbruck O, Hackel S, Jäggi A (2018) Precise orbit determination of the Sentinel-3a altimetry satellite using ambiguity-fixed GPS carrier phase observations. J Geodesy 92(7):711–726

Moon Y, König R, Michalak G, Rothacher M (2008) Precise orbit and baseline determination for TerraSAR-X and TanDEM-X. In: IGARSS 2008-2008 IEEE International Geoscience and Remote Sensing Symposium, vol 2, pp II–121. IEEE

Švehla D, Rothacher M (2005) Kinematic precise orbit determination for gravity field determination. In: A Window on the Future of Geodesy: Proceedings of the International Association of Geodesy IAG General Assembly Sapporo, Japan, June 30–July 11, 2003, pp 181–188. Springer

Open Access This chapter is licensed under the terms of the Creative Commons Attribution 4.0 International License (http://creativecommons.org/licenses/by/4.0/), which permits use, sharing, adaptation, distribution and reproduction in any medium or format, as long as you give appropriate credit to the original author(s) and the source, provide a link to the Creative Commons license and indicate if changes were made.

The images or other third party material in this chapter are included in the chapter's Creative Commons license, unless indicated otherwise in a credit line to the material. If material is not included in the chapter's Creative Commons license and your intended use is not permitted by statutory regulation or exceeds the permitted use, you will need to obtain permission directly from the copyright holder.

Practical Considerations of VLBI Observations to the GENESIS Mission

David Schunck, Lucia McCallum, and Guifré Molera Calvés

Abstract

With the GENESIS proposal accepted, this study reevaluates the implementability of incorporating VLBI observations of satellites into geodetic VLBI experiments. Observations of NavIC system satellites were carried out using the 12-m AuScope radio telescopes in Hobart and Katherine. The primary focus is on scrutinizing the necessary efforts within the VLBI community aimed at effectively supporting the GENESIS satellite mission. Our investigation identifies limitations in the existing processing pipelines, particularly in the generation of station-specific procedure and local control files, as well as in satellite tracking support within the antenna control units, resulting in step-wise tracking rather than continuous tracking. Additionally, we have conducted an analysis to ascertain the effective visibility of the GENESIS satellite within both current and future VLBI networks. Our findings align with the envisioned visibility criteria of GENESIS when more VGOS-type stations are integrated into the current network. In this case, the satellite becomes visible from at least two stations with long baselines for approximately 75.6% of the time during experiments and 21.5% of the time in a year.

Keywords

GENESIS mission · Networks · Visibility · VLBI

1 Introduction

The realization of the International Terrestrial Reference Frame (ITRF) is based on the observations of the four space geodetic techniques Very Long Baseline Interferometry (VLBI), Satellite Laser Ranging (SLR), Global Navigation Satellite Systems (GNSS) and Doppler Orbitography and Radiopositioning Integrated by Satellites (DORIS) (Altamimi et al. 2023). The individual station networks are tied together at co-location sites where two or more geodetic instruments of different techniques are operated. In local surveys, so-called local ties, the differential coordinates between the instrument reference points are determined. However, when the tie vectors are computed from global-scale space geodesy estimates, significant discrepancies to terrestrial local ties can be found. Most of these discrepancies are believed to be caused by systematic errors in the techniques (Delva et al. 2023).

The GENESIS mission, a component of the FutureNAV program of the European Space Agency (ESA), will be a dynamic space geodetic observatory carrying geodetic instruments of all four space geodetic techniques that are referenced to one another through carefully calibrated space ties (Delva et al. 2023). The mission has a target launch date in 2027. The purpose of the GENESIS satellite is the in-orbit combination and co-location of the techniques in space to solve the biases between the different geodetic techniques to improve the ITRF accuracy and stability.

Given that GNSS, DORIS, and SLR are space geodetic techniques reliant on satellite observations, the facilitation

D. Schunck (✉) · L. McCallum · G. M. Calvés
School of Natural Sciences, University of Tasmania, Hobart, TAS, Australia
e-mail: david.schunck@utas.edu.au; lucia.mccallum@utas.edu.au; guifre.moleracalves@utas.edu.au

of observations to the GENESIS satellite can be readily implemented. In contrast to the satellite-based techniques, VLBI observations to satellites are not standard and are not performed routinely yet. Several test experiments have been conducted. Tornatore et al. (2014) assessed VLBI's capability to observe GNSS signals using Medicina and Onsala radio telescopes, highlighting the need for improved scheduling and tracking automation. Haas et al. (2014) successfully conducted a GNSS-VLBI test experiment, using two software correlators and applying three delay models. Hellerschmied et al. (2014) developed the Satellite Scheduling Module for the Vienna VLBI and Satellite Software (VieVS), allowing a standardized VLBI schedule generation which has been tested on the Wettzell-Onsala baseline. Plank et al. (2017) integrated these features to streamline observations, data correlation, and time delay observable generation. Hellerschmied et al. (2018) operationalized this pipeline for observations to the Chinese APOD-A nano satellite (Tang et al. 2020) using the AuScope VLBI array in Australia.

With the GENESIS proposal being accepted, we see the urge of revisiting the processing pipeline, established in Plank et al. (2017) and Hellerschmied et al. (2018), after almost 6 years to reassess the implementability of satellite experiments into geodetic VLBI experiments. A significant advancement was achieved in the scheduling of satellite observations with the development of the scheduling software VieSched++ (Schartner and Böhm 2019). Satellite observations can now be effortlessly planned and optimized, and if necessary, seamlessly integrated into mixed schedules that contain quasar sources. Furthermore, VieSched++ supports the revised VEX format (VEX 2.0[1]) which provides the possibility to include TLE[2] orbit data directly to the control files. This is important as it complements the satellite tracking capabilities of the NASA Field System,[3] enabling continuous satellite tracking.

In this work, we present a VLBI experiment to satellites using the 12-m AuScope radio telescopes in Hobart and Katherine. In the absence of operational satellites equipped with dedicated VLBI transmitters, our observations focus on the S-band signals emitted by satellites of the Navigation with Indian Constellation (NavIC) system. To our knowledge, this experiment is the first VLBI observation to satellites that were scheduled using the VieSched++ scheduling software. Similar to Plank et al. (2017), we utilize as many available procedures and established programs as possible. Our investigations are focused on scrutinizing the necessary efforts within the VLBI community aimed at effectively supporting the GENESIS satellite mission. An overview of the experiment is given in Sect. 2.1. The experiment is described in Sect. 2.2. The results and future work are discussed in Sect. 2.3. We consider this investigation important towards the final aim of realizing actual frame ties.

Besides inquiring into the difficulty of the implementability of satellite observations, it is also essential to examine how often VLBI observations to the GENESIS satellite could be carried out. Unlike the continuous observations of GNSS, VLBI is a schedule-driven technique, with only a few sessions each week, which constrains the available observation time for the GENESIS satellite. Concerning this matter, the GENESIS proposal (Delva et al. 2023) states:

> "The adoption of the circular orbit [...] enables [...] long baseline VLBI observations (with more than 6500 km) over 75% of the time" (in the following, this statement will be referred to as the "GENESIS Baseline Statement").

As this statement lacks details concerning aspects such as the utilized VLBI network, the cadence of experiments, or the type of experiment, we examine it in terms of various aspects. With the International VLBI Service for Geodesy and Astrometry (IVS) being in transition from the legacy S/X system to the VLBI Global Observing System (VGOS), we conduct practical considerations with regard to the visibility of the GENESIS satellite using VLBI networks of both systems. Furthermore, we investigate the global distribution of VLBI observations to GENESIS. The explanation of the visibility study is provided in Sect. 3.1. The results are presented in Sect. 3.2. Conclusion and outlook are presented in Sect. 3.3.

2 VLBI Observations of NavIC Satellites

2.1 Overview

In view of the upcoming GENESIS mission, we assess the processing pipeline of VLBI observations to satellites currently being in place at the AuScope radio telescopes. To achieve this, we executed a test experiment in June 2023, using the 12-meter antennas located in Hobart and Katherine. In the test experiment, we observed the S-band signal of the NavIC satellites IRNSS-1D and IRNSS-1G together with three bright quasar sources. Within this work we performed the steps of scheduling, tracking, recording, correlation, and fringe fitting. It is crucial to note that no in-depth analysis has been executed at this point. The focus is on the challenges we were confronted with in the generation of the data.

---

[1] https://safe.nrao.edu/wiki/bin/view/VLBA/Vex2doc.
[2] https://celestrak.org/NORAD/documentation/tle-fmt.php.
[3] https://nvi-inc.github.io/fs/.

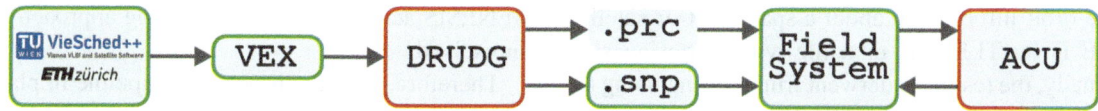

Fig. 1 Applied pipeline for preparing and performing the VLBI experiment. A red frame indicates the missing satellite support. A green frame shows that a software or file format already supports satellite observations

2.2 Experiment

Figure 1 visualizes the descriptions of the pipeline for preparing and performing the VLBI experiment. While a red frame indicates the missing support for satellite observations, a green frame shows that a software or file format already supports satellite observations.

In this study, the scheduling of the experiment has been performed with VieSched++, a commonly used software to schedule VLBI experiments to quasars and satellites. We generated an experiment of the duration of one hour including three bright quasar sources as well as two NavIC satellites. 12 scans to each of the two satellites were scheduled for the single-baseline experiment between Hobart and Katherine, using a fixed scan length of 60 seconds. Following the scheduling, a VEX file is generated in the revised VEX 2.0 file format. This enhanced format incorporates satellite scans specified with TLE orbit data. This augmentation paves the way to directly include non-siderial source types in the antenna control files.

In standard VLBI experiments, the DRUDG program is universally employed across participating stations to derive a procedure file (.prc) and a local control file (.snp) from the VEX schedule file. The .prc file contains all essential information for configuring the recording setup with the antenna back-end. The .snp file is noted in the Standard Notification for Astronomical Procedures (SNAP) format, containing commands for the NASA Field System to govern antenna movement, recording management, and logging of antenna information. However, the current version of DRUDG does not yet encompass the capability to generate local control files from VEX 2.0 files. While the .prc file can be created manually, the .snp file can be derived from the VEX file via workarounds.

One alternative approach is to manipulate the VEX file by treating the satellite as a static source. This entails splitting the satellite scan into shorter scans, wherein the satellite is no longer represented by TLE data but rather by its right ascension and declination. This enables DRUDG processing, albeit necessitating distinct VEX files for each station due to differing right ascension and declination values for near-field targets across stations. Furthermore, DRUDG treats each scan as an individual recording, resulting in several short data streams instead of a single continuous one.

We followed another workaround, entailing the introduction of a script to replace the DRUDG program. Our script is capable to translate quasar and satellite scans from VEX 2.0 files into commands in the .snp file. We can differentiate between step-wise and continuous tracking of satellites. In case of the step-wise satellite tracking, a single scan is recorded in which the Field System's source command is sequentially updated with right ascension and declination values for each antenna. Additionally, Field System versions 9.11.2 and higher are able to process TLE data and offer three different tracking modes using the satellite SNAP command. The *radc* mode works on all antennas and treats satellites as a normal source drifting off if the satellite is not stationary on the celestial sphere. For antennas that implement the fixed azimuth-elevation pointing, the *azel* mode is useful for geostationary and slow-moving satellites. Both tracking modes allow a periodic re-commanding, realizing a step-wise tracking of the satellite. The *track* mode allows to continuously follow an arbitrary satellite. All tracking commands can be created with our script. However, continuous tracking is only supported for stations that implement satellite tracking in their Antenna Control Unit (ACU). At the Geodetic Observatory Wettzell, the 20-m and both 13.2-m antennas are equipped with an ACU from Vertex Antennentechnik GmbH, enabling operation in a *Two Line Track Mode* (Hellerschmied et al. 2014). In contrast, the AuScope antennas lack satellite support within their ACU. In this experiment, step-wise tracking was conducted with a re-positioning interval of 10 seconds. The satellite position commanded at the beginning of each interval referred to the center satellite position of this 10 second orbit arc.

NavIC satellites emit signals at L5-band at 1176.450 MHz and at S-band at 2492.028 MHz. With the equipped wide-band receiver and S-band signal chain in operation (McCallum et al 2022), the 12-m radio telescopes in Katherine and Hobart are capable of recording the S-band signal. The data was recorded with a bandwidth of 32 MHz in 2-bit mode with a center frequency at the NavIC signal in S-band with X- and Y-polarization.

The correlation was performed with the DiFX software correlator (Deller et al. 2011). To handle observations of near-field targets, the correlator uses the SPICE[4] toolkit from NASA to formulate a delay model. Two of many ways to

---
[4] https://naif.jpl.nasa.gov/naif/.

incorporate orbit information about a spacecraft or satellite into SPICE is by TLE data or a file with Cartesian state vectors. Finally, the results underwent fringe-fitting using the *fourfit* program, a component of the Haystack Observatory Processing Software (HOPS[5]) package.

2.3 Discussion and Future Work

We conducted a VLBI experiment with quasar and satellite sources to assess the processing pipeline for VLBI observations to satellites currently in place at the AuScope radio telescopes. By successfully performing the first VLBI experiment containing satellite sources that was scheduled with VieSched++, we validated the software's benefit for future endeavors in this field. However, because the current version of DRUDG does not support dedicated satellite commands, a way has to be found to include the satellite SNAP commands in the control files. For the AuScope telescopes, this was resolved via a Python script. Another weakness in the pipeline is the missing satellite support in the ACUs of the antennas in Katherine and Hobart. As a result, a continuous tracking of the satellites is not supported via the Field System. Consequently, in the presented experiment, we perform a step-wise tracking of the satellites with 10 second re-commanding of the position. Notably, while the NavIC satellites maintain an orbital height of approximately 36000 km, the forthcoming GENESIS satellite will operate within a Low Earth Orbit (LEO), at an altitude of roughly 6000 km. The enhanced velocity of the GENESIS satellite as it moves across the sky necessitates shorter update intervals for its positions. As with a narrower beam maintaining a satellite accurately within the field of view is increasingly challenging, high-frequency signals and large antennas require shorter update intervals. Based on the proposed GENESIS orbit, we calculate angular rates for different elevations at the Mount Pleasant radio observatory in Hobart. In combination with the half power beam width (HPBW, e.g., Hellerschmied 2018), we derive minimum update rates. For elevations up to 85 degrees, the required update rates are 1.4 seconds for the 26m antenna (HOBART26, Ho) and 3.1 seconds for the 12m antenna (HOBART12, Hb). Performing VLBI observations to the LEO satellite APOD-A with the AuScope VLBI array, Hellerschmied et al. (2018) state that intervals as short as 1 second led to the Field System becoming overcharged by the high rate of positioning commands for both the step-wise tracking and the TLE tracking modes implemented in the Field System. While the calculated slew rates for modern VGOS and even legacy antennas are not critical for following GENESIS across the sky, the tracking approach itself is still unresolved.

Therefore, with the processing pipeline in place, we plan to investigate more on the nuanced impacts of different tracking methods on the results. In view of the upcoming GENESIS mission and considering that VLBI observations to satellites can be performed with short integration times, we are primed to scrutinize and compare various tracking approaches - encompassing continuous, step-wise, and even static observations. Static observations are possible provided the signals of the VLBI transmitter onboard GENESIS are bright enough to allow integration times of about 1 second. Static tracking could be realized as presented in Parker et al. (2019). While the non-continuous tracking of satellites would simplify the endeavors of global VLBI observations to satellites, it could also lead to problems. According to Tornatore et al. (2011), not keeping the satellite exactly centered within the antenna beam all the time during an observation may lead to glitches in the resulting data or the residuals with a period equal to the duration of the stepping interval.

3 Visibility of the GENESIS Satellite

3.1 Concept

The objective of this study is to analyze the effective visibility of the GENESIS satellite with both present and future VLBI networks. The inputs to the study are the GENESIS orbit, IVS station coordinates, and a list of experiments. Based on these inputs, visibilities are calculated with an elevation cutoff of 5 degrees.

In addition to our comprehensive analysis of overall visibility, we also touch upon the GENESIS Baseline Statement highlighted in Sect. 1. Contrasting with short baseline VLBI observations spanning just a few hundred kilometers, long baselines have the advantage of more effectively interlinking the station network while enhancing the observation geometry between the stations and the satellite. Both of these attributes are important within the framework of space ties. Therefore, we additionally analyze the visibilities, taking into account the requirement of baselines measuring 6500 km or greater. *Case All* is in the following referred to as the case in which all baselines are considered. *Case Long* will be referred to as the case in which only baselines with a length of 6500 km or greater are considered.

Within the scope of this study, we examine three distinct scenarios. In Scenario A, we analyze the cumulative visibility of the GENESIS satellite across all VLBI experiments listed in the IVS Master Schedule 2023. It contains 223 individual experiments. In Scenario B, we calculate the effective visibility corresponding to presently built VGOS-

---

[5] https://www.haystack.mit.edu/haystack-observatory-postprocessing-system-hops/.

type stations with fully functional signal chains. Scenario C further incorporates VGOS-type stations in the developmental phase, poised to attain operational status by the expected time of launch of the GENESIS satellite in 2027. The networks for Scenario B and C are based on the work by Behrend et al. (2023). These scenarios only contain a single experiment.

3.2 Results

Figure 2 visualizes the results for all scenarios by means of two variables. *Observability* colors the ground track of the GENESIS satellite based on the amount of stations for which the satellite is above the horizon. The observability reflects the quantity of observations that can be generated at certain orbit positions of the GENESIS satellite. *Uptime* colors the stations based on the percentage of time the satellite is above the horizon while the specific station is actively participating in an experiment. Stations that are part of a dense network and are closer to the poles should show larger uptimes and, therefore, are able to observe GENESIS more often.

Table 1 contains values for the cumulative GENESIS satellite visibility for the Scenarios A, B and C for *Case All* and *Case Long*. The satellite is considered to be visible when at least two stations see the satellite at the same time. The visibility is represented as three quantities. *V1* expresses the absolute, cumulative visibility in days. *V2* describes the

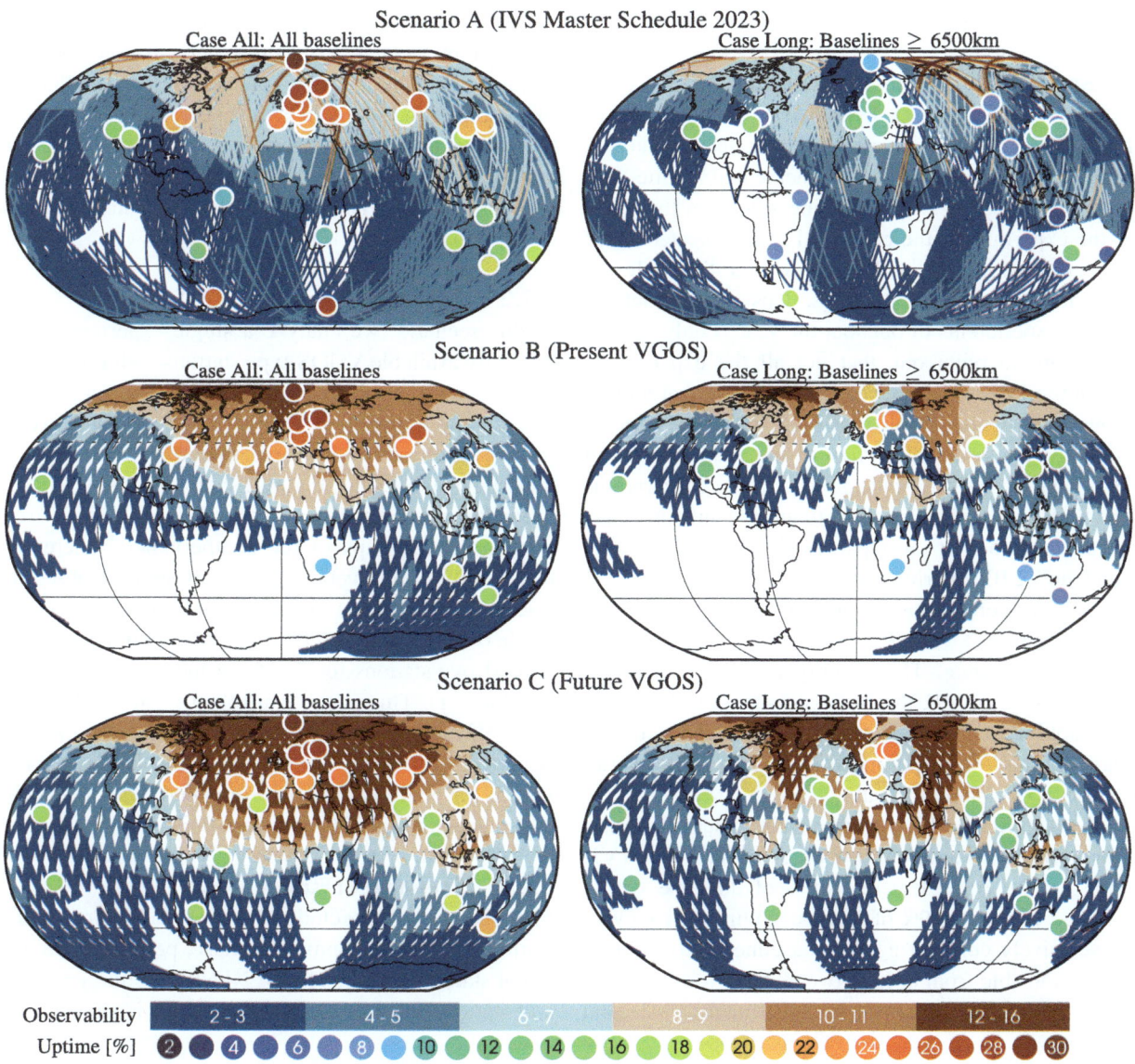

Fig. 2 GENESIS satellite ground tracks and stations for Scenarios A, B and C. *Observability*, displayed as colored ground tracks, reflects the number of stations seeing the satellite at this time and orbit position. *Uptime*, displayed as colored station positions, reflects the temporal visibility as percentage while the respective station is actively participating in experiments and disregarding idle times

Table 1 Cumulative GENESIS satellite visibility for the Scenarios A, B and C in days, percentage within experiments, and percentage within a year. The visibility is valid when at least two stations see the satellite. For bold values, the satellite visibility in days and percentage in year is based on the assumption of performing an experiment with the respective stations with a cadence of two times a week

Scenario	Case All			Case Long		
	V1 [days]	V2 [% of exp]	V3 [% of year]	V1 [days]	V2 [% of exp]	V3 [% of year]
A	98.2	54.9	26.9	49.7	27.8	13.6
B	**74.8**	71.9	**20.5**	57.4	55.2	**15.7**
C	**93.0**	89.4	**25.5**	**78.6**	75.6	**21.5**

cumulative visibility relative to the experiment duration in percentage. $V3$ is the cumulative visibility relative to the time of the year, including the idle times, in percentage. While Scenario A contains a schedule with experiments for a whole year, no schedule but just a single experiment is given for Scenario B and C. We assume that the experiment is conducted with a cadence of two times a week.

3.2.1 Scenario A: IVS Master Schedule 2023

We proceed by assuming a hypothetical scenario where the GENESIS mission has already become operational and observations to the satellite are implemented in the IVS Master Schedule 2023. When focusing solely on the duration of the experiments, the combined time allocated to all the outlined experiments in the IVS Master Schedule for 2023 amounts to merely 178.7 days. This means, that the coverage of experiments throughout the year is only 49.0%. It is important to point out, that not all the experiments outlined in the IVS Master Schedule are of a global nature. Numerous experiments involve only a handful of stations. Consequently, a substantial portion of the year, amounting to 168.3 days or 51.0%, undergo no active observations from any IVS station. During these times, the antennas may undergo maintenance or are repurposed for activities other than geodetic VLBI experiments.

In Fig. 2 the percentage uptime for all stations in Scenario A range from 8% to 30% in *Case All* and from 3% to 18% in *Case Long*. This is only a relative measure. For example, in *Case All*, the Antarctic station OHIGGINS shows an uptime of 25.5%. However, because the station is only participating in 17 experiments over the year, its overall contribution is comparably small.

In *Case All*, GENESIS is visible from at least two stations for a cumulative amount of time of 98.2 days (V1, see Table 1), the equivalent of 54.9% of the time within experiments (V2) and 26.9% of the time within the year (V3). When exclusively considering long baselines (*Case Long*), GENESIS is visible from at least two stations for 49.7 days (V1), the equivalent of 27.8% of experiments (V2) and 13.6% of the year (V3). It is also apparent that large gaps with no observations to GENESIS are present, especially in the southern hemisphere.

Figure 2 additionally displays the GENESIS observability. The colors of the ground track reflect the number of stations seeing the satellite at a specific time and orbit position. The figure shows that the common visibility of stations is larger in the northern hemisphere than in the southern hemisphere. This can be explained by the smaller number of stations in the southern hemisphere. It significantly affects the number of baseline observations. For example, a common visibility of 3 stations leads to 3 baseline observations, while a common visibility of 7 stations leads to 21 baseline observations.

3.2.2 Scenario B: Present VGOS

In this scenario, we analyze a single experiment using all presently available VGOS-type stations with fully-functional signal chains. As the GENESIS orbit is not a repeat orbit, we run the experiment for 10 days and then average to receive results for a 24 h experiment to reduce the bias of the orbit on the results. Figure 2 shows the GENESIS satellite ground tracks and stations. For both *Case All* and *Case Long*, there are large gaps in the southern hemisphere and the observability, the overall number of stations observing the satellite at the same time, is considerably larger in the northern hemisphere. In *Case All*, the satellite is visible from at least two stations for 71.9% of the experiment time (V2, see Table 1). This is the equivalent of about 20.5% of the year (V3) when assuming an experiment with this network is performed twice a week over a year. In *Case Long*, it is visible for 55.2% of the time in experiments (V2) which is the equivalent of 15.7% in a year (V3).

3.2.3 Scenario C: Future VGOS

In this scenario, we include all present and future VGOS-type stations in the developmental phase, poised to attain operational status by 2027. The GENESIS satellite ground tracks and stations for this 10 day experiment is shown in Fig. 2.

In comparison to Scenario B, the additional stations lead to significantly better coverage of the southern hemisphere for both *Case All* and *Case Long*. Furthermore, the station uptimes are considerably improved, especially with the additional stations in the southern hemisphere. In *Case All*, the satellite is visible from at least two stations for 89.4% of the experiment time (V2, see Table 1). This is the equivalent of about 25.5% of the year (V3) when assuming an experiment with this network is performed twice a week over a year. In *Case Long*, it is visible for 75.6% of the time in experiments (V2) which is the equivalent of 21.5% in a year (V3).

3.3 Conclusion and Outlook

The GENESIS Baseline Statement says, that the GENESIS satellite will be visible over 75% of the time on long baselines over 6500 km. When only considering long baselines (*Case Long*) with a future VGOS network (Scenario C) the satellite is visible 75.6% of the time during experiments indeed. However, the IVS Master Schedule 2023 (Scenario A) has shown that VLBI experiments are performed only 49% of the year. In 2023, large network VGOS experiments are performed twice a month with intentions to increase the cadence to weekly. When assuming an experiment cadence of two per week, a proposed global VGOS network (Scenario C) would be able to observe the GENESIS satellite 25.5% of the year (93.0 days). In the case of only long baselines (*Case Long*), the yearly visibility reduces to 21.5% (78.6 days). The GENESIS satellite has a minimum operational lifetime of 3 years. Whether the overall observational volume will be enough, to tie the VLBI frame to the satellite frame with the required accuracy is a matter we want to answer in future work by the means of a dedicated GENESIS simulation study. With the legacy S/X system still carrying out the majority of experiments in the IVS Master Schedule 2023, it could be useful that the VLBI transmitter onboard the GENESIS mission is also going to cater the frequencies of the legacy S/X system. Furthermore, the analysis underlines the importance of future VGOS-type stations in the southern hemisphere. These would close large gaps and established stations increase the amount of observation time.

Acknowledgements This work has been funded by the Australian Research Council (ARC) with the project DE180100245. I, David Schunck, am grateful for the support from the University of Tasmania in my PhD candidature.

References

Altamimi Z, Rebischung P, Collilieux X, et al (2023) ITRF2020: an augmented reference frame refining the modeling of nonlinear station motions. J Geod 97:47. https://doi.org/10.1007/s00190-023-01738-w

Behrend D, Ruszczyk C, Elosegui P, et al (2023) The VGOS high road: From inception and prototyping to operations to maturation and beyond. In: 26th European VLBI Group for Geodesy and Astronomy Working Meeting (poster)

Deller AT, Brisken WF, Phillips CJ, et al (2011) Difx-2: A more flexible, efficient, robust, and powerful software correlator. PASP 123:275–287

Delva P, Altamimi Z, Blazquez A, et al (2023) GENESIS: co-location of geodetic techniques in space. Earth Planets Space 75(5). https://doi.org/10.1186/s40623-022-01752-w

Haas R, Neidhardt A, Kodet J, et al (2014) The Wettzell-Onsala G130128 experiment - VLBI-observations of a GLONASS satellite. In: Behrend D, Baver KD, Armstrong KL (eds) IVS 2014 General Meeting Proceedings - VGOS: The New VLBI Network. Science Press, Beijing, pp 451–455

Hellerschmied A (2018) Satellite observations with VLBI. PhD thesis, Technische Universität Wien. https://doi.org/10.34726/hss.2018.28517

Hellerschmied A, Plank L, Neidhardt A, et al (2014) Observing satellites with VLBI radio telescopes - practical realization at Wettzell. In: Behrend D, Baver KD, Armstrong KL (eds) IVS 2014 General Meeting Proceedings - VGOS: The New VLBI Network. Science Press, Beijing, pp 441–445

Hellerschmied A, McCallum L, McCallum J, et al (2018) Observing APOD with the AuScope VLBI Array. Sensors 18(5):1587. https://doi.org/10.3390/s18051587

McCallum L, Chin Chuan L, Krsn H, et al (2022) The Australian mixed-mode observing program. J Geod 96:Article Number 67. https://doi.org/10.1007/s00190-022-01657-2

Parker AL, McCallum L, Featherstone WE, et al (2019) The potential for unifying global-scale satellite measurements of ground displacements using radio telescopes. Geophys Res Lett 46(21):11841–11849. https://doi.org/10.1029/2019GL084915

Plank L, Hellerschmied A, McCallum J, et al (2017) VLBI observations of GNSS-satellites: from scheduling to analysis. J Geod 91:867–880. https://doi.org/10.1007/s00190-016-0992-8

Schartner M, Böhm J (2019) VieSched++: A new VLBI scheduling software for geodesy and astrometry. PASP 131:084501

Tang G, Li X, Cao J, et al (2020) APOD mission status and preliminary results. Sci China Earth Sci 63:257–266. https://doi.org/10.1007/s11430-018-9362-6

Tornatore V, Haas R, Duev D, et al (2011) Single baseline GLONASS observations with VLBI: data processing and first results. In: Proceedings of the 20th Meeting of the Euoprean VLBI Group for Geodesy and Astrometry, pp 162–165

Tornatore V, Haas R, Casey S, et al (2014) Direct VLBI observations of global navigation satellite system signals. In: Rizos C, Willis P (eds) Earth on the Edge: Science for a Sustainable Planet, Proceedings of the IAG General Assembly. Springer, Berlin/Heidelberg, pp 247–252

Open Access This chapter is licensed under the terms of the Creative Commons Attribution 4.0 International License (http://creativecommons.org/licenses/by/4.0/), which permits use, sharing, adaptation, distribution and reproduction in any medium or format, as long as you give appropriate credit to the original author(s) and the source, provide a link to the Creative Commons license and indicate if changes were made.

The images or other third party material in this chapter are included in the chapter's Creative Commons license, unless indicated otherwise in a credit line to the material. If material is not included in the chapter's Creative Commons license and your intended use is not permitted by statutory regulation or exceeds the permitted use, you will need to obtain permission directly from the copyright holder.

Terrestrial Datum Definition Methods in VLBI Global Solutions

Lisa Kern, Hana Krásná, Axel Nothnagel, Johannes Böhm, and Matthias Madzak

Abstract

A *geodetic datum* describes the origin, orientation and scale of a station network, typically with respect to a reference frame. In the analysis process of Very Long Baseline Interferometry (VLBI) observations, the introduction of a *geodetic datum* is inevitable for the determination of precise reference frames and Earth orientation parameters (EOP). In general, several methods of datum definition exist within the VLBI community, including Helmert rendering and the no-net-translation/no-net-rotation (NNT/NNR) approach. While the first introduces conditions with quasi-infinite weight, the NNT/NNR method can be controlled by the selection of formal errors. Evaluations of the CONT17 legacy-1 campaign and a longer time series of IVS 24-hour sessions show that the variance information (formal errors) of the estimated terrestrial reference frames based on the different methods can differ in the mm to almost cm range. Neglecting this issue could lead to potential issues when combining or comparing solutions from different analysis centers.

Keywords

Geodetic datum · Global solution · Reference systems · Very Long Baseline Interferometry (VLBI)

1 Introduction

Very Long Baseline Interferometry (VLBI) observations, which are the difference in arrival time of signals from extragalactic radio sources at radio telescopes on Earth, enable the determination of precise terrestrial (TRF) and celestial reference frames (CRF) as well as of Earth orientation parameters (EOP). Besides inter-technique combination, which combines the advantages of all space-geodetic techniques, e.g., in the realization of the International Terrestrial Reference Frame (ITRF; Altamimi et al. (2022)), VLBI-only global solutions, both single analysis center and combined multi analysis center combinations, are a necessary step for internal quality control and data interpretation. Global solutions are constructed from as many observing sessions as possible. In the context of datum definition, it does not matter whether the global solution is derived from data of a single analysis center alone or by a combination of data from several analysis centers, e.g., the combined solutions by the International VLBI Service for Geodesy and Astrometry (IVS). As an example for a single analysis center product, the most recent VIE2023sx global solution provided by the Vienna IVS analysis center (VIE) consists of over 7300 sessions and over 20 million observations (Krásná et al. 2023).

For the construction of a global solution, datum-free normal equation systems (NEQs) are stacked. These are derived from single-session analyses and are stored in solution inde-

---

All the authors contributed equally to this work.

L. Kern (✉) · H. Krásná · A. Nothnagel · J. Böhm · M. Madzak
Department of Geodesy and Geoinformation, TU Wien, Vienna, Austria
e-mail: lisa.kern@tuwien.ac.at; hana.krasna@tuwien.ac.at; axel.nothnagel@tuwien.ac.at; johannes.boehm@tuwien.ac.at; matthias.madzak@tuwien.ac.at

pendent exchange (SINEX) format. Typically parameters that are time-dependent (e.g., clocks, atmospheric parameters, EOP) are reduced session-wise, whereas global parameters, that are constant over several sessions, are kept in the NEQ system. Subsequently, the NEQ of the sessions are stacked, i.e., common parameters are added, forming one global NEQ system. By the inversion of the global NEQ system, the global parameters (e.g., station coordinates and source positions) and their corresponding variance information can be estimated.

However, since VLBI observations are relative and only describe the network geometry (configuration) of a three-dimensional VLBI station network with no origin or orientation, the global NEQ is singular and hence not yet invertible. The *geodetic datum* contains all the definitions needed (three translations and three rotations) to locate this stiff station network at an origin and with a specific orientation by applying a Helmert transformation. Since VLBI observations rely on the propagation of microwave signals and consequently on the speed of light, no external information on the seventh parameter of this three-dimensional similarity transformation, the scale, is necessary. Note that *transformation* does not, in this case, mean that the form of the network is distorted since we only consider rigid transformations (Nothnagel 2023). Therefore, six datum constraints are introduced in the adjustment process to compensate for the degree of freedom of the VLBI NEQ and to make the NEQ system regular and solvable (*minimum constraints*; Sillard and Boucher (2001)). Hence, this process is often referred to as the *regularization* of the NEQ. In the case of determining a kinematic solution, the transformation model is extended by the rates of the datum parameters, leading to twelve necessary constraints.

Within the VLBI community, different methods of terrestrial datum realization exist, which will be presented in Sect. 2. Differences between the methods lead to results that are in general not identical and could lead to potential issues when comparing results from different VLBI analysis or combination centers. On the basis of two datasets (see Sect. 3), TRFs are computed using different methods and are compared in Sect. 4.

2 Terrestrial Datum Realization

It must be noted that the *geodetic datum* can be introduced in multiple ways and referred to any reference frame. However, in VLBI analysis, the datum is mostly applied using a conventional reference frame (e.g., ITRF). Since the reference frame is assumed to be accurate and only small changes due to potential new and better observations and models are expected, the vectorial residuals $\Delta \mathbf{x}$ of the transformation from the VLBI network frame $\mathbf{x}$ to the reference frame $\tilde{\mathbf{x}}$ should generally be close to zero for most of the stations. However, in the event of earthquakes occurring after an ITRF release, these differences can become significantly large.

Furthermore, usually, a subset of stations is used to define the datum (*partial inner constraints*; Blaha (1971); Dermanis (1994)) which will be referred to as *datum stations*. These stations should have a long observation history and are chosen to provide relatively consistent global coverage. Note that for the sake of simplicity, any usage of $\mathbf{x}$ in the following Sects. 2.1 and 2.2 only refers to station positions and that the NEQ solely carries information for determining these global parameters. As already mentioned, when determining a kinematic reference frame, the transformation model is extended by the rates of the datum parameters. The corresponding equations can be found in Nothnagel (2023).

In general, there are two expressions used in the following, *conditions* strictly force the model onto the VLBI configuration (they have quasi-infinite weight; see Sect. 2.1), whereas *constraints* can be controlled by formal errors, which are used to populate a covariance matrix for the generation of a regular NEQ system (see Sect. 2.2, Eqs. 8, 11).

2.1 Helmert Rendering

The Helmert similarity transformation is a widely used approach to relate two frames by shifting along and rotating around the coordinates axis. The coordinates of the N selected datum stations in the VLBI network frame $\mathbf{x}$ and the coordinates in the reference frame $\tilde{\mathbf{x}}$ are related as follows

$$\tilde{\mathbf{x}} = \begin{pmatrix} 1 & -r_z & r_y \\ r_z & 1 & -r_x \\ -r_y & r_x & 1 \end{pmatrix} \cdot \mathbf{x} + \begin{pmatrix} t_x \\ t_y \\ t_z \end{pmatrix} \quad (1)$$

with t_x, t_y, t_z, r_x, r_y and r_z being the translation t and rotation r parameters. By re-shaping and ordering, the observation equations $\Delta \mathbf{x} = \mathbf{x} - \tilde{\mathbf{x}}$ read

$$\begin{pmatrix} \Delta x_i \\ \Delta y_i \\ \Delta z_i \end{pmatrix} = \begin{pmatrix} 1 & 0 & 0 & 0 & -\tilde{z}_i & \tilde{y}_i \\ 0 & 1 & 0 & \tilde{z}_i & 0 & -\tilde{x}_i \\ 0 & 0 & 1 & -\tilde{y}_i & \tilde{x}_i & 0 \end{pmatrix} \cdot \begin{pmatrix} t_x \\ t_y \\ t_z \\ r_x \\ r_y \\ r_z \end{pmatrix} \rightarrow \Delta \mathbf{x} = \mathbf{B} \cdot \xi. \quad (2)$$

Setting up these equations per datum station i and vertically stacking them, $\mathbf{B}$ is the Jacobi matrix of the Helmert parameters with dimensions $3N \times 6$. Finally, $\mathbf{B}$ is used to render the datum-free normal equation matrix $\mathbf{N}_{free}$, by the expansion of the NEQ system and forcing the Helmert parameters to be zero ($\xi = 0$).

$$\begin{pmatrix} \mathbf{N}_{free} & \mathbf{B} \\ \mathbf{B}^T & 0 \end{pmatrix} \begin{pmatrix} \Delta \mathbf{x} \\ \xi = 0 \end{pmatrix} = \begin{pmatrix} \mathbf{b}_{free} \\ 0 \end{pmatrix} \quad (3)$$

The corresponding covariance matrix of the estimated parameters is

$$\mathbf{C}_{\mathbf{x},HR} = \sigma^2 \mathbf{N}_{free}^{-1} \quad (4)$$

with σ^2 being the a posteriori variance of unit weight. The subscript *HR* denotes the method, Helmert rendering. This is the classical way of adding conditions to the adjustment, also known as *Gauss-Markov model with restrictions/conditions*.

2.2 NNT/NNR

A currently widely used approach includes the introduction of no-net-translation (NNT) and no-net-rotation (NNR) constraints (see Eqs. 5 and 6 respectively) to N datum stations, which are used to map a set of telescopes to a conventional reference set. **r** represents the position vector in Cartesian three-dimensional coordinates.

$$\sum_{i=1}^{N} \Delta \mathbf{x} = \sum_{i=1}^{N} \begin{pmatrix} \Delta x_i \\ \Delta y_i \\ \Delta z_i \end{pmatrix} = \begin{pmatrix} 0 \\ 0 \\ 0 \end{pmatrix} \quad (5)$$

$$\sum_{i=1}^{N} (\mathbf{r} \times \Delta \mathbf{x}) = \sum_{i=1}^{N} \left(\begin{pmatrix} \tilde{x}_i \\ \tilde{y}_i \\ \tilde{z}_i \end{pmatrix} \times \begin{pmatrix} \Delta x_i \\ \Delta y_i \\ \Delta z_i \end{pmatrix} \right) = \begin{pmatrix} 0 \\ 0 \\ 0 \end{pmatrix} \quad (6)$$

By forming the partial derivatives of the translation and rotation constraints, they can be combined into one composite constraint

$$\sum_{i=1}^{N} \begin{pmatrix} 1 & 0 & 0 \\ 0 & 1 & 0 \\ 0 & 0 & 1 \\ 0 & -\tilde{z}_i & \tilde{y}_i \\ \tilde{z}_i & 0 & -\tilde{x}_i \\ -\tilde{y}_i & \tilde{x}_i & 0 \end{pmatrix} \cdot \begin{pmatrix} \Delta x_i \\ \Delta y_i \\ \Delta z_i \end{pmatrix} = \mathbf{0} \rightarrow \sum_{i=1}^{N} \mathbf{B}^T \cdot \Delta \mathbf{x} = \mathbf{0} \quad (7)$$

leading to the constraint matrix **B** with dimensions $3N \times 6$ which is a vertically concatenated matrix of the individual B_i values and is used to resolve the rank defect of the datum-free normal equation matrix $\mathbf{N}_{free}$. When comparing the condition/constraint matrices from Eq. 2 (Helmert rendering) and Eq. 7 (NNT/NNR), it can be noted that they are identical and should lead to the same results. However, as stated in the beginning, in comparison to the Helmert rendering, it is possible to incorporate formal errors for the constraints. Hence, a covariance matrix $\boldsymbol{\Sigma}$ with dimensions 6×6, which specifies the impact of datum constraints on the solution, is applied to generate a regular normal equation matrix forming the datum matrix $\mathbf{N}_B$, whereas the right-hand side vector of the datum $\mathbf{b}_B$ is zero:

$$\mathbf{N}_B = \mathbf{B} \boldsymbol{\Sigma}^{-1} \mathbf{B}^T. \quad (8)$$

The datum matrix can also be determined in another way by forming the solution equation for the transformation parameters ξ

$$\xi = (\mathbf{B}^T \mathbf{B})^{-1} \mathbf{B}^T \Delta \mathbf{x} \quad (9)$$

and imposing the constraints by

$$\mathbf{0} = \mathbf{H} \Delta \mathbf{x} \quad (10)$$

which leads to a new constraint matrix $\mathbf{H} = (\mathbf{B}^T \mathbf{B})^{-1} \mathbf{B}^T$ with the dimensions $6 \times 3N$ which can be again used to generate a weighted regular datum matrix using the covariance matrix $\boldsymbol{\Sigma}$:

$$\mathbf{N}_H = \mathbf{H}^T \boldsymbol{\Sigma}^{-1} \mathbf{H}. \quad (11)$$

In both cases, so-called pseudo-observations are introduced and new regular NEQ systems are compiled (Eq. 12). Numerically, these two approaches should lead to the same results (Kotsakis 2012), however, the implementation of Eq. 11 is preferred due to better numerical stability. The NEQ systems of the pseudo-observations are then added to the NEQ of the real VLBI observations:

$$\mathbf{N} = \mathbf{N}_{free} + \mathbf{N}_{B/H}. \quad (12)$$

As further shown in Kotsakis (2012), the estimated parameters of the minimal constrained NEQ system are independent of the introduced covariance matrix from a mathematical point of view. However, the covariance matrix of the final NEQ system shows a dependency (see Eq. 13).

$$\mathbf{C}_{\mathbf{x},B/H} = \sigma^2 (\mathbf{N}_{free} + \mathbf{N}_{B/H})^{-1} \mathbf{N}_{free} (\mathbf{N}_{free} + \mathbf{N}_{B/H})^{-1} \quad (13)$$

For more details, see Kotsakis (2012, 2013). Nevertheless, in the case of an over-constrained NEQ system, the choice of $\boldsymbol{\Sigma}$ has an impact on the estimated parameters (Kotsakis and Chatzinikos 2017).

To summarize, there are three possible approaches to introduce a terrestrial *geodetic datum*. For obvious reasons, we exclude the *3-2-1 method* where six coordinate components of three stations are simply fixed to their a priori value (Nothnagel 2023). First, by enlarging the datum-free normal equation matrix system with conditions (Helmert rendering, see Eq. 3) or second, by adding the squared and weighted constraint matrix **B** (NNT/NNR, see Eq. 8) to the singular NEQ system or third, by using the discussed **H** (NNT/NNR, see Eq. 11).

3 Data

In the following, VLBI-only TRFs using the three different terrestrial datum realization methods are estimated based on two datasets:

- *dataset #1*: 24-hour sessions (1108 sessions, 66 stations, time frame: January 01 2015–December 31 2020)
- *dataset #2*: CONT17 legacy-1 sessions (15 sessions, 14 stations, time frame: November 28–December 12 2017) (Behrend et al. 2020)

Hence, the differences caused by the different datum methods imposed on a VLBI network over a short time frame can be compared to a combination of many global sessions over a longer time period. Furthermore, due to the longer time frame of *dataset #1*, station velocities are estimated using the transformation model extended by the rates of the datum parameters (Nothnagel 2023). It has to be stated here again, that typically, global solutions combine the data of thousands of VLBI sessions, however, the combination process is computationally expensive and the focus of the study is on highlighting the differences of the datum methods on the TRF determination rather than calculating highly precise reference frames. Thus, only the differences between the estimated station positions **dx** and their formal errors **mx** as well as between the estimated station velocities **dvx** and their formal errors **mvx** (in the case of *dataset #1*) are displayed in Sect. 4.

In this study, the list of datum stations is taken from the most recent VIE solution (Kraśná et al. 2023) in the case of both datasets. In Fig. 1, the station network of both experiments is displayed. The most recent ITRS realization, ITRF2020 (Altamimi et al. 2022), is selected as the a priori reference frame in the process of regularization and potential station position discontinuities have been adopted. Source positions are fixed to their ICRF3 position (Charlot et al. 2020). Furthermore, EOP, clocks and atmospheric parameters as well as baseline-dependent clock offsets (Krásná et al. 2021) are reduced session-wise. In general, by using the same input data for all solutions, the analysis directly shows the impact of the different datum realization methods on the TRF determination.

4 Results

Since Helmert rendering is the strictest approach through application of conditions and since we routinely use this method in our VLBI single-session and multi-session analysis at VIE with the Vienna VLBI and Satellite Software (VieVS) (Böhm et al. 2018), these solutions are chosen as a reference and are compared to the two NNT/NNR approaches in the following.

In Figs. 2 and 3 the reference solution is compared to the NNT/NNR approach using the datum matrix from Eq. 8 and in Figs. 4 and 5 it is compared to the NNT/NNR method using the **H** from Eq. 11. As already mentioned, due to the differences in the magnitude of the elements of **B** and **H**, different formal errors must be introduced to achieve a comparable strength of datum realization. Therefore, a formal error of 10 mm has been incorporated in the NNT/NNR method using **B** and 1 mm when using **H**. When considering the rates of datum parameters, formal errors of 10 mm/yr and 1 mm/yr are introduced respectively. In both cases, the resulting differences of *dataset #1* are displayed on the left and those of *dataset #2* on the right of the following figures. No station velocities are estimated for *dataset #2*. Stations marked with a star are used to define the datum in the global adjustment.

As expected, no differences are visible in the estimated corrections to the a priori values shown in Fig. 2. Differences only show up in the formal errors of the coordinate components. Figure 3 shows comparable results with regard to the impact of the datum realization methods on the estimated station velocities and their formal errors. In general, the southern VLBI stations of *dataset #1* show significantly larger differences in formal errors **mx** and **mvx**, which may indicate that the selection of stations contributing to the datum realization is not optimal for the stations in the southern hemisphere, which is a well-known issue within the VLBI community.

When comparing the second NNT/NNR approach (see Figs. 4 and 5) with the reference solution, again no differences in the estimated corrections to the a priori values are visible. However, in this case, the formal errors **mx** and **mvx** are more uniform than in Figs. 2 and 3. This method seems to provide greater stability across all stations, which appears reasonable, as this method exhibits greater numerical stability. In general, by introducing tighter con-

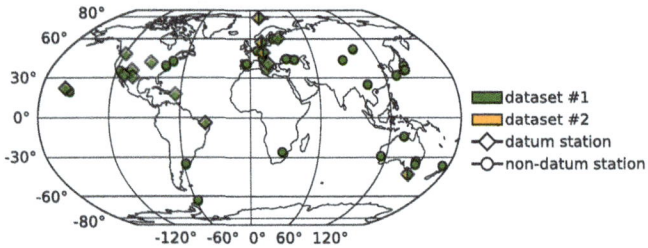

Fig. 1 Station networks of *dataset #1* (green) and *dataset #2* (orange). Datum stations are represented as diamonds, reduced stations as pentagons and remaining stations as circles. If a station occurs in both networks, the marker is split in two

Terrestrial Datum Definition Methods in VLBI Global Solutions

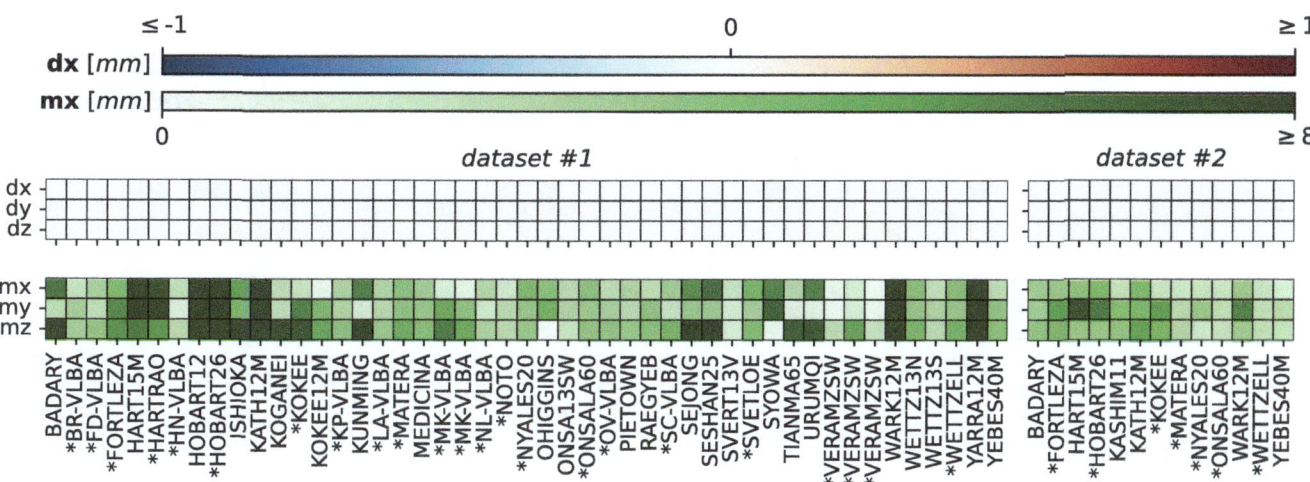

Fig. 2 Heatmap of differences in station coordinates **dx** and formal errors **mx** comparing the NNT/NNR approach using the **B** matrix and a formal error of 10 mm with the reference solution (Helmert rendering). The left column shows the results of *dataset #1* and the right column of *dataset #2*. Stations marked with a star are datum stations. The repeated listing of stations is due to discontinuities where each occurrence relates to a different interval

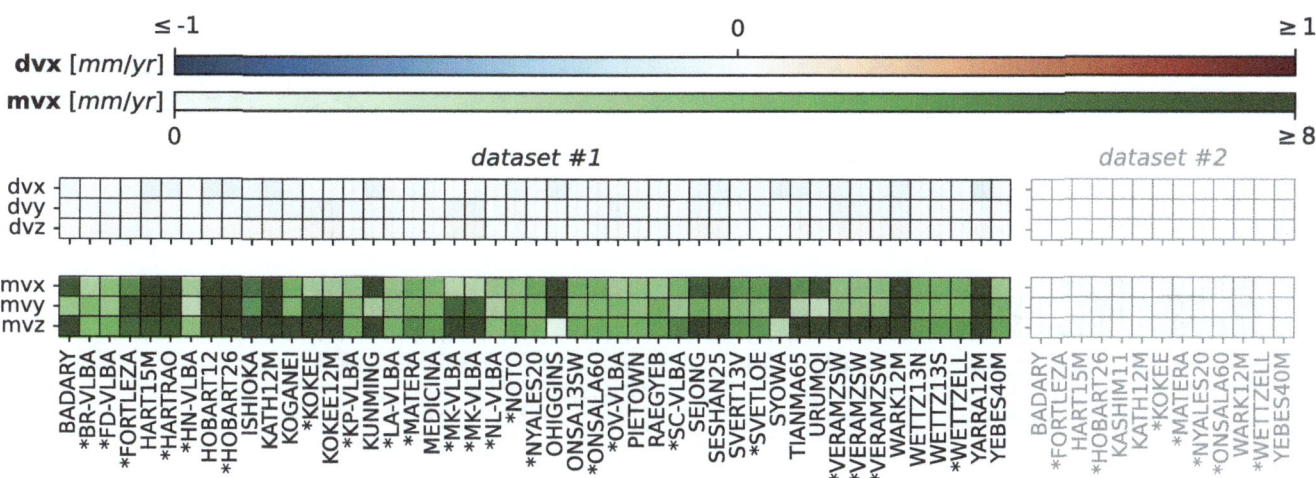

Fig. 3 Heatmap of differences in station velocities **dvx** and formal errors **mvx** comparing the NNT/NNR approach using the **B** matrix and a formal error of 10 mm/yr with the reference solution (Helmert rendering). The left column shows the results of *dataset #1*. No station velocities are estimated for *dataset #2*. Stations marked with a star are datum stations. The repeated listing of stations is due to discontinuities where each occurrence relates to a different interval

straints and therefore respectively increasing the weight of the constraints, the formal errors of the NNT/NNR method can be further decreased.

To sum up, the influence of the datum realization methods on the TRF solution depends on the selection of datum stations and the formal error chosen in the NNT/NNR approach. However, the covariance matrix of the constraints only influences the covariance matrix of the estimated parameters, and therefore their formal errors, and not the values themselves. Tighter constraints push the results towards those of the strict Helmert rendering conditions. The dependency of the formal errors of the estimated parameters in the case of minimal constraints on the chosen method and introduced formal errors makes the comparison of solutions a potential problem and demonstrates the importance of transparency in the VLBI community. To show the influence of the chosen formal error, the maximum absolute differences of the formal errors **mx** and **mvx** using the **H** matrix are presented in Table 1 which emphasizes that reducing the formal errors of the constraints pushes the results towards those of the Helmert rendering conditions.

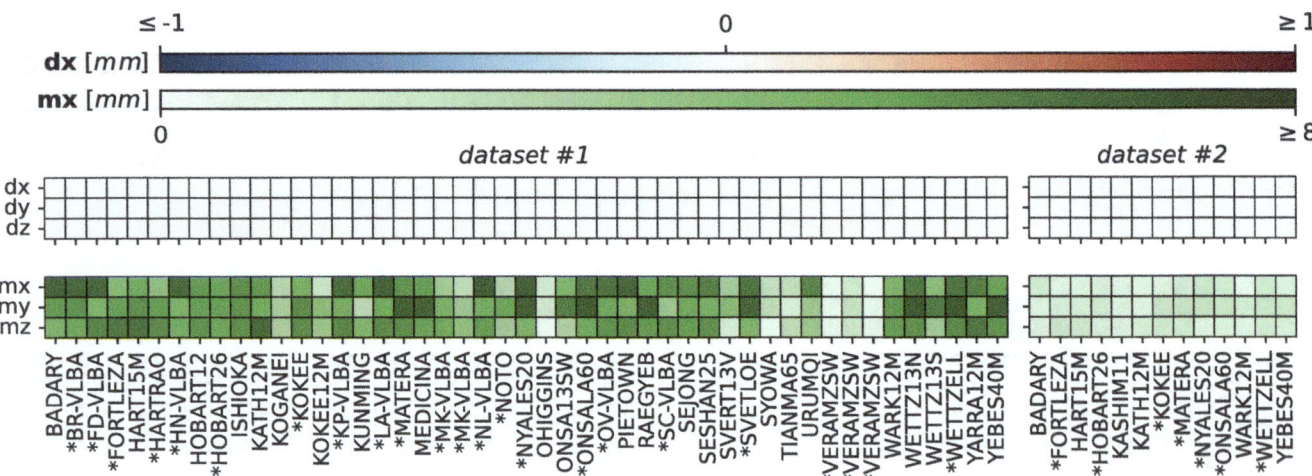

Fig. 4 Heatmap of differences in station coordinates **dx** and formal errors **mx** comparing the NNT/NNR approach using the **H** matrix and a formal error of 1 mm with the reference solution (Helmert rendering). The left column shows the results of *dataset #1* and the right column of *dataset #2*. Stations marked with a star are datum stations. The repeated listing of stations is due to discontinuities where each occurrence relates to a different interval

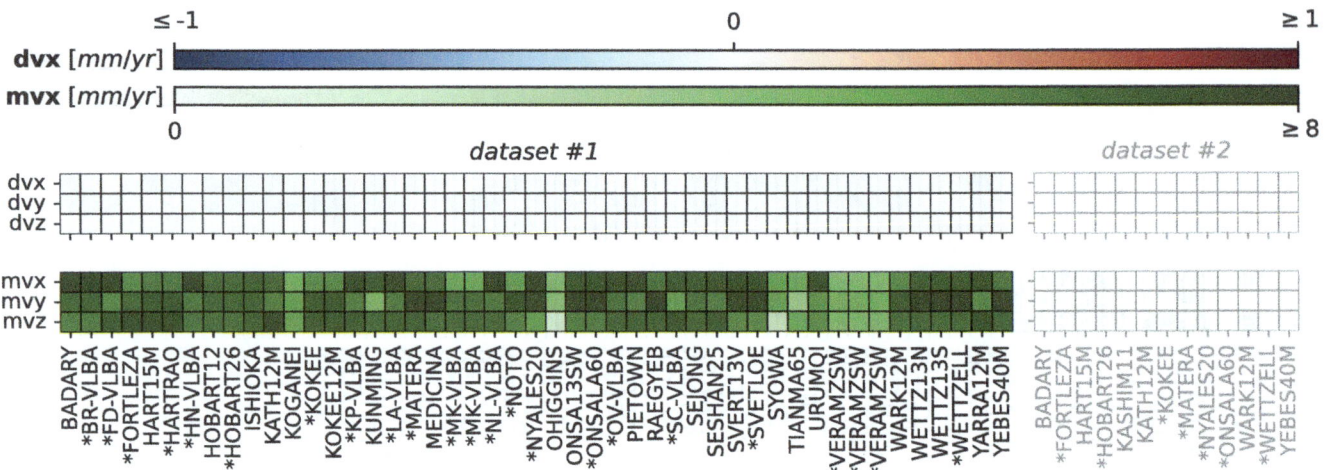

Fig. 5 Heatmap of differences in station velocities **dvx** and formal errors **mvx** comparing the NNT/NNR approach using the **H** matrix and a formal error of 1 mm/yr with the reference solution (Helmert rendering). The left column shows the results of *dataset #1*. No station velocities are estimated for *dataset #2*. Stations marked with a star are datum stations. The repeated listing of stations is due to discontinuities where each occurrence relates to a different interval

Table 1 Maximum absolute differences in the formal errors **mx** and **mvx** comparing the NNT/NNR approach using the **H** matrix and different formal errors for the datum constraints (values and rates) with the reference solution (Helmert rendering) given in mm. No station velocities are estimated for *dataset #2*

Formal error		Dataset #1		Dataset #2	
		mx	mvx	mx	mvx
1 mm	1 mm/yr	6.9	7.6	1.8	–
0.1 mm	0.1 mm/yr	0.3	0.5	0.1	–

5 Conclusion

The introduction of a *geodetic datum* is inevitable when analyzing VLBI sessions. Different methods of datum realization do not lead to significant differences in the computed TRFs when introducing minimal constraints. These findings are in agreement with the theoretical investigation by Kotsakis (2012). For both datasets, large differences in the formal errors of the estimated parameters of almost up to one cm between the methods can be seen when using a formal error of 10 mm (10 mm/yr) in the case of **B** and respectively 1 mm (1 mm/yr) in the case of **H**. This indicates that tighter constraints need to be imposed in the NNT/NNR approach to achieve the same effect as when the Helmert rendering is performed. In general, the NNT/NNR method utilizing **H** is always preferable to using **B**. It not only demonstrated superior stability across all stations in the current investigation but is also considered more stable numerically. However, until now it is not reported by VLBI analysis centers which method and whether formal errors are applied to the datum constraints. This study aims to demonstrate the importance of transparency in the VLBI community, as comparing solutions, especially their variance information, can be problematic due to the use of different methods.

Acknowledgements We are grateful to all parties that contributed to the success of the CONT17 campaign. A detailed listing of all institutions is provided under https://ivscc.gsfc.nasa.gov/program/cont17/. We would like to thank the editor as well as Miltiadis Chatzinikos and an anonymous reviewer for the constructive comments and valuable suggestions.

Conflict of Interest The authors declare that they have no conflict of interest.

Authors' Contributions All authors designed the study. LK performed the evaluations and analysis. All authors worked on the theoretical considerations, discussed the results and contributed on the final manuscript.

References

Altamimi Z, Rebischung P, Collilieux X, et al (2022) ITRF2020: main results and key performance indicators. EGU General Assembly 2022 3958. https://doi.org/10.5194/egusphere-egu22-3958

Behrend D, Thomas C, Gipson J, et al (2020) On the organization of CONT17. J Geodesy 94(100). https://doi.org/10.1007/s00190-020-01436-x

Blaha G (1971) Inner adjustment constraints with emphasis on range observations. Department of Geodetic Science, The Ohio State University, OSU Report No. 148, Columbus

Böhm J, Böhm S, Boisits J, et al (2018) Vienna VLBI and Satellite Software (VieVS) for Geodesy and Astrometry. Publ Astron Soc Pac 130(986):044503. https://doi.org/10.1088/1538-3873/aaa22b

Charlot P, Jacobs CS, Gordon D, et al (2020) The third realization of the International Celestial Reference Frame by very long baseline interferometry. Astron Astrophys 644:A159. https://doi.org/10.1051/0004-6361/202038368

Dermanis A (1994) The photogrammetric inner constraints. ISPRS J Photogramm Remote Sens 49:25–39. https://doi.org/10.1016/0924-2716(94)90053-1

Kotsakis C (2012) Reference frame stability and nonlinear distortion in minimum-constrained network adjustment. J Geodesy 86:755–774. https://doi.org/10.1007/s00190-012-0555-6

Kotsakis C (2013) Generalized inner constraints for geodetic network densification problems. J Geodesy 87:661–673. https://doi.org/10.1007/s00190-013-0637-0

Kotsakis C, Chatzinikos M (2017) Rank defect analysis and the realization of proper singularity in normal equations of geodetic networks. J Geodesy 91:627–652. https://doi.org/10.1007/s00190-016-0989-3

Krásná H, Jaron F, Gruber J, et al (2021) Baseline-dependent clock offsets 287 in VLBI data analysis. J Geodesy 95(126). https://doi.org/10.1007/s00190-021-01579-5

Krásná H, Baldreich L, Böhm J, et al (2023) VLBI celestial and terrestrial reference frames VIE2022b. Astron Astrophys 679:A53. https://doi.org/10.1051/0004-6361/202245434

Nothnagel A (2023) Elements of geodetic and astrometric very long baseline interferometry. Tech. rep., Technische Universität Wien, Austria. https://www.vlbi.at/data/publications/Nothnagel_Elements_of_VLBI.pdf

Sillard P, Boucher C (2001) A review of algebraic constraints in terrestrial reference frame datum definition. J Geodesy 75:63–73. https://doi.org/10.1007/s001900100166

Open Access This chapter is licensed under the terms of the Creative Commons Attribution 4.0 International License (http://creativecommons.org/licenses/by/4.0/), which permits use, sharing, adaptation, distribution and reproduction in any medium or format, as long as you give appropriate credit to the original author(s) and the source, provide a link to the Creative Commons license and indicate if changes were made.

The images or other third party material in this chapter are included in the chapter's Creative Commons license, unless indicated otherwise in a credit line to the material. If material is not included in the chapter's Creative Commons license and your intended use is not permitted by statutory regulation or exceeds the permitted use, you will need to obtain permission directly from the copyright holder.

On the Potential of Accelerometers for GNSS on Satellite Positioning and Ensuing Reference Frame Determination

Patrick Schreiner, Susanne Glaser, Rolf König, Karl Hans Neumayer, Shrishail Raut, and Harald Schuh

Abstract

Solar Radiation Pressure (SRP) is the largest non-conservative force acting on Global Navigation Satellite Systems (GNSS) satellites. Modeling this force is still one of the challenging tasks in precise orbit determination (POD) of GNSS satellites and therefore also for subsequent applications as geodetic reference frame determination. Commonly used methods for SRP modeling are empirical or analytical ones, as well as combinations of the two. These points give rise to the motivation whether and how alternative observation techniques can improve future GNSS and support them in aspects of POD, reference frame determination and other subsequent applications. For this purpose, we analyze the potential of accelerometers onboard of each Galileo satellite by using simulations for different accelerometer specifications and evaluate the effect on position and clock estimates of the satellite vehicle, as well as the effect on derived Terrestrial Reference Frames (TRF). We thereby see, by assuming accelerometer sensitivities which are already available, the possibility to decorrelate the clock estimates from radial orbit position determinations. The advantages for GNSS based positioning are limited, since radial orbit errors and clock errors almost perfectly compensate. Promising potential for improvements for derived TRF and geocenter determination can be seen, which would bring us one step closer to achieving the accuracy requirements of a global TRF, defined by the Global Geodetic Observing System (GGOS).

Keywords

Accelerometer · Clock · Geocenter · GNSS · POD · Scale · TRF

1 Introduction

Global Navigation Satellite Systems (GNSS) are nowadays an indispensable space geodetic observation technique, which has become a standard in Earth observation, as well as for precise positioning. More importantly, the reference to which measurements are made and certain measurements are derived must meet high accuracy standards. For this purpose, the Global Geodetic Observing System (GGOS (Gross et al. 2009)) has defined goals for the accuracy requirement of global Terrestrial Reference Frames (TRF), i.e. 1 mm accuracy and 1 mm per decade long-term stability for the reference defining parameters: origin, scale, orientation. One fundamental basis for GNSS derived reference frame determination is to keep especially radial orbital errors at a minimum. An obstacle hereby is the strong correlation between the estimated satellite clocks and the radial position of the satellites. Thereby, Solar Radiation Pressure (SRP) is the largest non-conservative force acting on GNSS satellites (Rebischung et al. 2013). Modeling this force is still one of the challenging tasks in precise orbit determination (POD) of GNSS satellites and, therefore, also for subsequent applications as geodetic reference frame determination. Currently commonly used methods for SRP modeling are empirical models, such as ECOM (Springer et al. 1999) in the extended version (Arnold et al. 2015), which was improved with the perspective of reducing oscillations in the geocenter estimates. Another method is the analytical computation of the disturbance force by the use of macro models of the satellite. Box-wing models are often used for this purpose, which have some residual inaccuracy regarding the shape of the satellite, which, according to current studies, can be reduced to percentages in the single digits (Bury et al. 2020). Ray-tracing models try to counteract this by using highly accurate 3D models (Darugna et al. 2018), but often certain dimensions, such as final shapes or surface textures are not publicly available. Another possibility are hybrid approaches, where e.g., the analytically calculated force based on a box-wing model provides the basis and in addition, empirical parameterizations like ECOM are estimated. Over the recent years, correlations between SRP parameters and geocenter coordinates have been studied, which significantly complicate the estimation of geocenter coordinates (Meindl et al. 2013; Rodriguez-Solano et al. 2014; Arnold et al. 2015; Bury et al. 2019; Glaser et al. 2020; Zajdel et al. 2020). An alternative option would be to use an accelerometer to directly measure accelerations. Ash (2002) suggested this device for use onboard GNSS satellites. This concept has also been discussed by the Galileo Scientific Advisory Committee (Vespe and Rothacher 2014) with numerous advantages, such as: improved derivation of models for non-conservative forces, improved orbit predictions, less frequent need to update the satellites ephemerides, better separation of the orbits and clocks and the possibility for autonomous orbit determination of the satellite. Further investigations were also carried out by Kalarus et al. (2016) and Lucchesi et al. (2016), who present some requirements and preliminary results. In this study, the potential of accelerometers is investigated by simulations. GNSS observations and accelerometer observations are simulated for a Galileo-like constellation over a period of three years. The basis for the simulations of Galileo observations is provided by a real data evaluation. For the simulation of accelerometer observations, we adopt the specifications of the GRACE-FO accelerometer, as well as those of a future accelerometer type, which will be available in the next few years. We additionally improve this performance by reducing the noise level by one order of magnitude and finally show where the absolute potential of accelerometers can lie. Further effects, such as possible drift behavior or scaling of the accelerometer were not considered for the time being, but these are planned for future studies. The subsequent evaluation first analyzes the positional accuracy of the satellites obtained by the POD, as well as that of the estimated clocks. Finally, station coordinates, geocenter and Earth Rotation Parameters (ERPs) are estimated and a TRF is generated for the simulation period. The advantages for GNSS satellite POD and derived TRF and geocenter determination are shown by Helmert parameters for station coordinates and ERPs and improvement of formal errors of estimated parameters. We will investigate how accelerometers on GNSS satellites can bring us one step closer to achieving the accuracy requirements to a TRF, defined by the GGOS.

2 Current and Future Accelerometers in Satellite Geodesy

Accelerometers are instruments that can measure one-dimensional acceleration for a certain axis. In satellite geodesy, they have been used in the last decades on prominent satellite missions such as CHAMP, GRACE, GOCE, and GRACE-FO. In dynamic orbit determination, the big difference to the processing with models is that the modeling of surface forces like SRP, albedo, and others can be replaced by accelerometer measurements. These are usually used as true values of the non-gravitational acceleration in the equation of motion. For this purpose various accelerometer models are available, which were differently constructed and show different accuracy characteristics. The quality of an accelerometer can be categorized e.g., on the basis of an amplitude spectrum. The measurement error is represented as acceleration over the frequency range. For the accelerometer, which was installed on the GRACE-FO satellites (Kornfeld et al. 2019), this is shown in Fig. 1 in blue. This accelerometer has two more sensitive axes and one less sensitive axis. The noise characteristics of the two spectra are comparable, but are about one to two orders of magnitude different. Recent developments in accelerometers are also aimed at reducing the cost factor while keeping or even improving sensitivity, that is fundamentally limited by the thermal fluctuation of the test mass. With mechanical fused-silica oscillator based

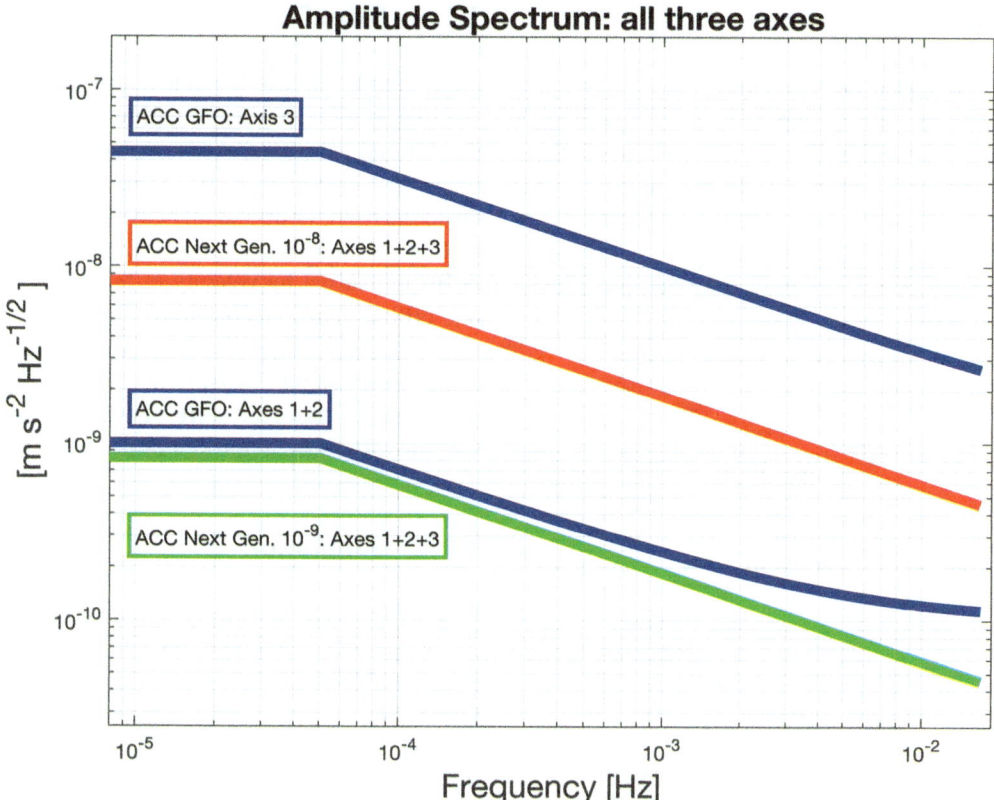

Fig. 1 Amplitude spectra of the simulated accelerometer scenarios. Stochastic model for GRACE-FO-like case in blue, next-generation accelerometer case with 10^{-8} in red and with 10^{-9} in green

accelerometers, however, it could be shown that a sensitivity close to this limit can be achieved (Guzmán et al. 2014, 2018; Hines et al. 2020, 2021, 2022). The current sensitivity can be described as (personal communication with Felipe Guzmán):

$$\log 10(a) = -0.4999 \cdot \log 10(f) - 10.23 \qquad (1)$$

with a the acceleration noise in $\mathrm{ms}^{-2}/\sqrt{\mathrm{Hz}}$ and f the Fourier frequency. This is therefore about one order of magnitude worse than that of the sensitive axes of the GRACE-FO accelerometer, as shown in red in Fig. 1. For the following tests, we simulate this accelerometer scenario with three equally sensitive axes. As it can be expected (Hines et al. 2023) that this sensitivity can even be significantly improved in the next few years, we add another scenario with three accelerometers in each axis featuring a one order of magnitude higher sensitivity representing the next generation of accelerometer for space geodesy, shown in green in Fig. 1. Even if it will be possible to achieve this sensitivity in the next few years, we also performed a noise-free simulation of the accelerometer to show what sensitivity is really needed for GNSS-based applications.

In order to make the Power Spectral Density (PSD) characteristics consistent within each scenario, the same edge frequency of the functional model of 5×10^{-5} Hz is assumed in each case. The frequency behavior is relevant up to one orbital cycle of the satellite, which corresponds to about 2×10^{-5} Hz in the case of Galileo and is therefore below the edge frequency. This range is assumed to be constant in the simulation. Below certain edge frequencies, the performance exhibits different properties, such as an approximately linear increase or stagnation or also decrease. Although, since the accelerometer built for GOCE has shown that sensitivities two orders of magnitude better than the best axis of the GRACE-FO accelerometer are possible (ESTEC 2000), we have decided to assume the range of $< 5 \times 10^{-5}$ Hz as constant. It is important to note that the assumptions about accelerometers on satellites are based on previous satellite missions, such as GRACE. Unlike GNSS satellites, these satellites do not have continuously moving elements that could affect the position of the satellite's center of mass. These effects may influence the measurement properties and must be considered in data processing and calibration. However, these influences were not taken into account in our study.

3 Simulation Procedure

The simulations are based on a previous setup for the proposed "Kepler" constellation (Giorgi et al. 2019; Glaser et al. 2020). We assume a Galileo-like constellation with three orbital planes, each separated by 120 degrees in the ascending node, and 8 satellites per plane. The dynamic models are listed in Table 1. Satellite vehicle specific information, such as the optical properties for a box-wing model, were taken from EUSPA (2023).

Table 1 Setup of the simulation

General	
Software	EPOS-OC (Neumayer et al. 2024)
Arc-length	24h
Simulation time span	3 years
Dynamic model	
Gravity field	GOCO06s 120×120 time variable 120×120 (Kvas et al. 2021)
Atmospheric tide	BB2003 (Biancale and Bode 2006)
Earth tide	Wahr model (Wahr 1981) IERS Conventions 2010 (Petit and Luzum 2010)
Ocean tide model	FES2014 30×30 (Lyard et al. 2021)
Ocean pole tide	Desai 30×30 (Desai 2002)
Albedo with IR	Internal implementation
Ephemeris	DE430 (Folkner et al. 2014)
Moon	FERRARI77 (Ferrari 1977)
Geometric model	
Earth Orientation Parameters	EOP20C04
Mean pole	Linear mean pole (Ries 2017)
Station coordinates	IGS20 without seasonal harmonics
Satellite configuration	
Attitude	Attitude law
Satellite mass	According to IGS metadata
Observation correction models	
Antenna phase pattern	igs20 ANTEX
Troposphere model	Global Mapping Function (Boehm et al. 2006)
Carrier phase wind-up	Applied
Simulation	
Satellite model	Macro model used for SRP and albedo with IR
Recovery	
– with empirical parameterization	
Satellite model	Macro model removed for SRP
– with accelerometer readings	
Satellite model	Macro model removed for SRP and albedo with IR

3.1 Orbit Design

According to the Galileo constellation design, each satellite shall be in its nominal slot within a 2 degree window to its neighboring satellite, which corresponds to about 1000 km (Navarro-Reyes et al. 2009). As in the simulation, we want to avoid the need for orbital maneuvers, it is necessary to ensure that the inter-satellite distances are appropriate. Intersatellite drifts are caused by velocity differences of satellites within one orbital plane. Accordingly, appropriate initial state elements can be found by an optimization process. By doing so a maximum angular separation of 0.15 degrees after three years can be ensured, corresponding to a maximum inter-satellite difference to the slot of less than 100 km.

3.2 Simulation

GFZ's EPOS-OC software (Neumayer et al. 2024) is used for all simulations. First, Galileo stations were selected for the station network according to the clusters of the International GNSS Service (IGS) core network. Then, stations offering a co-location with DORIS or SLR were added to enable the co-location of the techniques on the ground for upcoming work. Finally, based on a global optimization, further stations were added so that they are as evenly globally distributed as possible (see Fig. 2). Thus, a total of 146 stations are available for which Galileo observations are simulated. Code and phase observations are simulated at both frequencies E1 and E5a with a step size of 30 seconds to all visible satellites above an elevation of 20 degrees. Random white noise equivalent to 50 cm and 5 mm in the ionosphere-free linear combination of code and phase is applied to the observations. Ambiguities are simulated so that the total number of ambiguities per 24 hours arc is about 5000. Simulated accelerometer observations are first written directly without noise based on the calculated disturbance forces of the integration with a stepsize to the integration of 30 seconds. Based on the different stochastic models described in Sect. 2, colored noise tables are generated over the full simulation time span of three years. After that the noise-free accelerometer data is endowed with colored noise. The three scenarios based on the GRACE-FO accelerometer model are called GFO in the following. The less sensitive axis is aligned to X, Y, or Z in each of the three scenarios in the satellite-fixed reference frame. The scenarios for the mechanical fused silica accelerometer are referred to as Next Generation (Next Gen.), in the original characteristics as Next Gen. 10^{-8} (as this corresponds approximately to the level of the upper edge of this stochastic model), improved by one order of magnitude as Next Gen. 10^{-9}.

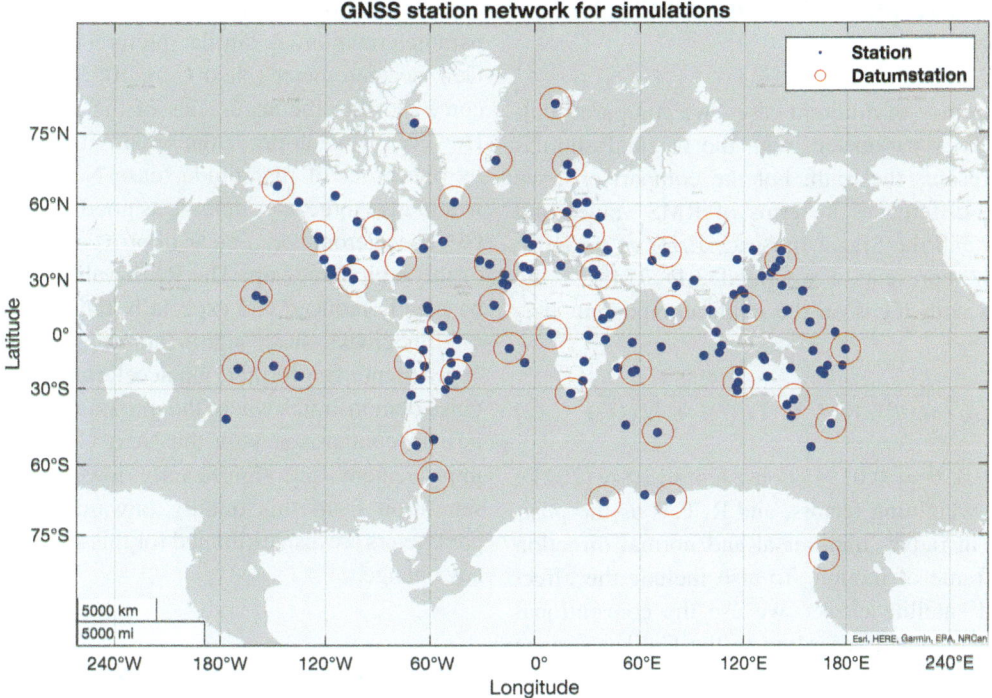

Fig. 2 Ground station network for simulation (blue) and stations used for datum definition (red)

3.3 Orbit Determination and Parameter Estimation

In the simulation, the accelerations due to SRP and albedo (including infrared (IR)) of the satellites are calculated based on the box-wing model. In the recovery, we compare different methods – empirical SRP modeling, accelerometers – only and a combination of the two. In all cases of the recovery, the calculation of the SRP based on the box-wing model is omitted. In the cases where simulated accelerometer observations are used, the calculation of influences due to albedo is also omitted. This is necessary because the simulated accelerometer observations contain the accelerations for both SRP and albedo, and the focus of empirically modeled methods should exclusively be on the influence of direct SRP. For the empirical modeling, we use the well-known nine parameter extended ECOM (hereafter named ECOM-2(9)) (Arnold et al. 2015) modeling (compare Table 2).

Table 2 Parameterization used for the different scenarios. D,Y,B indicate the direction, the number indicates the number per revolution for the sine and cosine terms

Scenario	Orbit parameters
ECOM-2(9)	D2D4B1
All acc. GFO	Three biases
All acc. Next Gen.	Three biases
Acc. w/o noise	Three biases
Acc. w/o noise+ECOM-2(9)	Three biases + D2D4B1

For each of these scenarios, a POD run is performed. Initial state elements of the satellites are estimated with the parameterization from Table 2, station coordinates (with 1/10 mm No-Net-Rotation (NNR),-Translation (NNT), and -Scale (NNS) condition) and for each epoch clocks for all satellites and all ground stations, except a fixed ground reference clock, one tropospheric scaling factor per station per passage, as there is no tropospheric mismodeling. Ambiguities are estimated and afterward fixed, using a double difference approach. The results of this run provide the basis for the orbit comparisons in Sec. 4.

Starting from this converged state, a special normal equation mode can be used with EPOS-OC. This makes it possible to perform one further iteration from the converged state with all parameters, in which additional parameters can be estimated and set up in the normal equation. This makes it possible to create datum-free normal equation, without the need for a priori information for e.g. station velocities. These daily normal equations are then reduced by parameters that are out of scope of this research (e.g. troposphere) and accumulated over the entire simulation period. This setup results in a normal equation containing a set of station positions and velocities per station, and with daily resolution ERPs, geocenter coordinates, and parameters of the orbit (e.g., initial state). To solve this normal equation, a NNR and NNT condition with a strength of 1/10 mm is applied to the datum stations shown in Fig. 2.

4 On the Overall Orbital Fit and Clock Estimate

To assess the accuracy of different cases, we compare satellite orbits and clock parameters with the original simulation, which represents the truth. For the comparisons, we use the position differences in terms of RMS values and mean deviations and the Signal In Space Error of the orbit ($SISE_{orb}$), which represents a weighted 3-D RMS of the satellite position with focus on the radial component. It is expressed as:

$$SISE_{orb} = \sqrt{(\omega \cdot R(t))^2 + \theta \cdot (T(t)^2 + N(t)^2)} \quad (2)$$

with $\omega = 0.96910$, $\theta = 0.01545$ consistent to ESA (2021), which represent weighting factors, and R, T, N as the position differences in radial, transversal and normal direction in the satellite frame of motion. To also include the effect of the estimated satellite clocks, we use the conventional SISE, which is calculated according to the Galileo service definition (ESA 2021). In this way, the contribution of orbit and clock errors on the modeled range can be assessed for precise point positioning applications. The calculated statistical properties of orbit comparison for each scenario are shown in Table 3.

4.1 Orbit Comparison

Comparing the statistics of the orbit comparison in Table 3, larger RMS values of all components of the orbit and also of the clock can be found for the three scenarios assuming a GFO accelerometer with different orientations in comparison to the empirical ECOM-2(9) modeling. The best orientation of a GFO accelerometer would be to position the less sensitive axis in the radial direction (corresponding to the Z direction of the satellite fixed reference frame according to the IGS convention), which led to the best values. Systematic offsets are lower for all GFO cases, but the general accuracy is not as good as with the empirical ECOM-2(9) parameterization. A similar picture is shown for the Next Gen. accelerometer (Next Gen. 10^{-8}), which also does not come close to the accuracies of the ECOM-2(9) modeling. However, if we improve the sensitivity of this accelerometer by one order of magnitude (case Next Gen. 10^{-9}), which would give three axes that are as good as the good axes of the GFO accelerometer, a clear improvement in all components of the orbit shows up. The RMS values of the position can be greatly reduced and especially the RMS of the estimated satellite clocks shows significant improvements. To take this experiment even further, the accelerometer scenario without noise demonstrates where the maximum achievable potential lies. In comparison with the Next Gen. 10^{-9} scenario, the improvements are significantly lower, showing a kind of saturation from this quality onwards. If the ECOM-2(9) parameters are also estimated for this scenario, there is hardly any change.

4.2 Satellite Clock Comparison

The orbit comparisons revealed that by using an accelerometer that is sensitive enough, it is possible to improve the clock estimate of the satellite. Montenbruck et al. (2014) showed that there is a need and possibility for improvement of the estimated clocks depending on the beta angle and that this can be achieved by, e.g., suitable a priori box-wing models in a hybrid setup. These effects are particularly evident when the satellite is in an eclipse period (beta angle of about ±10 degrees). Figure 3 shows as an example the radial position differences and clock differences over a 24 hours arc, at a beta angle of +8.2 degrees. The signal of the radial orbit difference, which follows the Sun's elongation angle, is clearly recognizable. The estimated clock difference is strongly anticorrelated, which is also the reason for the lower SISE of this scenario (0.43 cm) compared to the $SISE_{orb}$ (1.25 cm). The GFO accelerometer scenarios and Next Gen. 10^{-8} exhibit biases in the radial component and

Table 3 Orbit differences of the recovered orbits in comparison to simulated. All values are given in centimeters. Mean values over whole time span of three years in radial (R), transversal (T), and normal (N) direction. "ls to SV-X" means: less sensitive accelerometer axis orientated to IGS Satellite Vehicle (SV) reference frame X axis

Scenario	RMS					SV clock	Mean		
	R	T	N	SISE	$SISE_{orb}$	RMS	R	T	N
ECOM-2(9)	1.51	0.42	0.38	0.43	1.25	1.57	0.84	0.00	0.01
Acc. GFO – ls to SV-X	8.20	3.06	2.97	0.93	5.98	8.14	0.04	0.00	0.02
Acc. GFO – ls to SV-Y	6.72	2.67	2.26	0.85	5.16	6.68	0.16	−0.00	0.01
Acc. GFO – ls to SV-Z	3.78	1.51	1.13	0.58	2.93	3.77	−0.03	0.00	−0.00
Acc. Next Gen. 10^{-8}	2.04	0.76	0.68	0.47	1.60	2.08	−0.01	0.00	0.00
Acc. Next Gen. 10^{-9}	0.21	0.10	0.08	0.41	0.12	0.57	0.02	0.00	−0.00
Acc. w/o noise	0.09	0.05	0.04	0.41	0.07	0.48	0.00	0.00	−0.00
Acc. w/o noise + ECOM-2(9)	0.08	0.07	0.07	0.41	0.06	0.48	−0.00	−0.00	−0.00

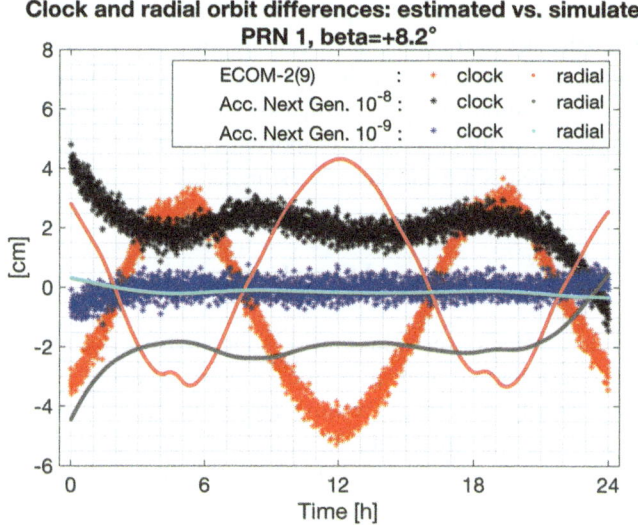

Fig. 3 Estimated clock and radial orbit component compared to simulated for one PRN with a beta angle of 8.2°

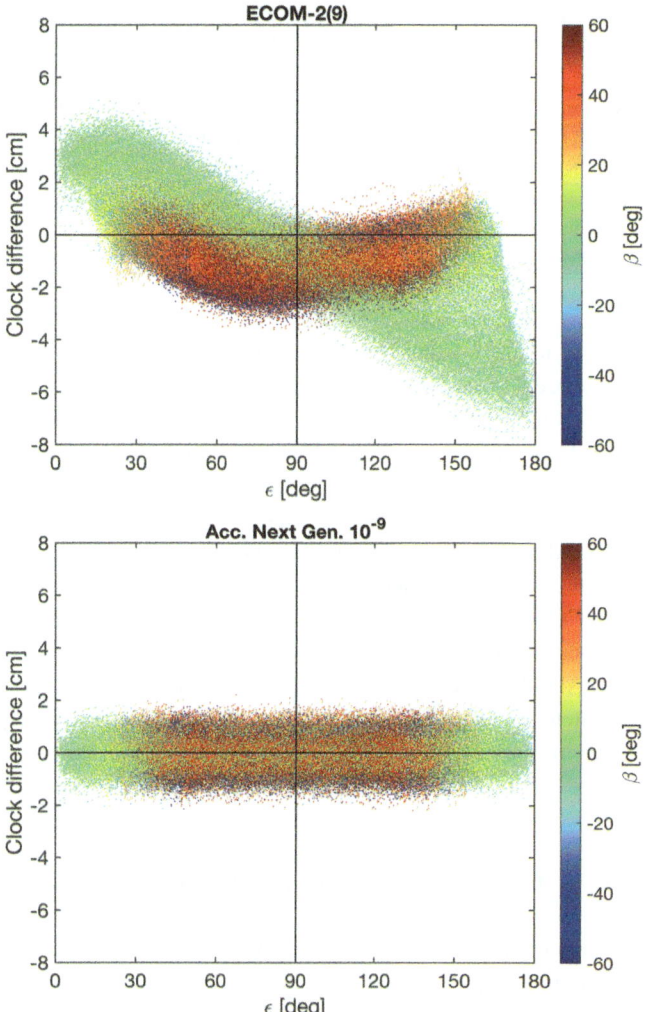

Fig. 4 Clock difference for PRN 1 in cm over the Sun elongation angle ϵ. Different β angles according to colorbar. Top figure shows the ECOM-2(9) case, bottom figure the results for the Next Gen. 10^{-9} accelerometer case

the clock. However, the Next Gen. 10^{-9} scenario shows an immense improvement. The differences of the radial orbit component and the estimated clock show no bias and no systematic signals. The noise-free scenario shows only minor further improvement. Figure 4 illustrates this effect of the accelerometers to improve the clock estimate related to the Sun elongation angle for the entire beta period of PRN 1 over three years compared to the ECOM-2(9) scenario. The ECOM-2(9) scenario shows major clock differences for low beta angles and elongation angles, which are very small or very large. For other beta angles, a negative clock difference on average is visible. In comparison, the analysis of the clock differences for the Next Gen. 10^{-9} scenario shows a very uniform noise of the clock differences over the entire range of the elongation and beta angle.

5 Reference Frame Determination

According to the procedure described in Sec. 3, TRF solutions have been generated for the cases ECOM-2(9), Next Gen. 10^{-9}, Acc. w/o noise, and Acc. w/o noise + ECOM-2(9).

5.1 TRF Error Assessment

We compare the derived station coordinates and velocities to a priori, using a 14 parameter Helmert transformation. Only the ECOM-2(9) case shows minor systematic rotations and scale differences, which are in the low submillimeter range. All other cases show no difference to a priori. Possible improvements are examined, by changes in mean formal errors of derived station coordinates, velocities, ERPs and geocenter coordinates in comparison to the ECOM-2(9) case, which represents the reference. For all accelerometer cases improvements for station coordinates in X and Y direction are at the level of up to 10%, for Z direction no changes can be noted. Changes for station velocities are significantly lower, with maximum improvements in the low single-digit percentage range. The additional ECOM-2(9) parameterization shows no influence here for station coordinates and velocities. For the pole coordinates, the improvements are 6 and 8% for X and Y pole, respectively, for both accelerometer cases. However, the additional estimation of ECOM-2(9) parameters to the acc. w/o noise case results in the loss of the entire improvement, and formal errors being similar to those of the ECOM-2(9) case. The greatest improvements can be

Table 4 Mean formal errors of geocenter coordinates over all data points within the time span of 3 years. All values given in millimeters

Scenario	Mean		
	$\sigma(GCC_x)$	$\sigma(GCC_y)$	$\sigma(GCC_z)$
ECOM-2(9)	0.27	0.27	0.82
Acc. Next Gen. 10^{-9}	0.25	0.26	0.25
Acc. noise free	0.25	0.26	0.25
Acc. noise free + ECOM-2(9)	0.28	0.28	0.61

seen in length of day (LOD), where the formal errors can be reduced to about 40%. Again, the co-estimation of the ECOM-2(9) parameters to the noise free accelerometer case leads to an increase in formal errors, which are just at the level of 70% of the ECOM-2(9) case. A similar picture can be seen for the geocenter coordinates. Slight improvements in X and Y direction show up, but the greatest improvement is in the Z direction, which can be reduced from 0.8 mm in the ECOM-2(9) case to about 0.25 mm. The values in X and Y component are for all scenarios at a similar level of 0.25 mm. Also here, the additional estimation of the ECOM-2(9) parameters causes the formal errors to increase significantly to 0.6 mm, especially in the Z direction (see Table 4).

5.2 ERP and Geocenter Estimation

The estimated ERPs for the ECOM-2(9) case show slight offsets in LOD and Y pole of about 0.4 mm each. A minor trend can also be noted in the Y pole, which is significant according to 3 sigma of the mean estimate (cf. Table 5). In both the pole coordinates, a period of 119 days is observed (see Fig. 5), which is the symmetric repetition rate of the orbital planes of the Galileo constellation. The standard deviation of the pole coordinates is about 0.1 mm, whereas for LOD it is over one centimeter. The accelerometer cases can eliminate all biases and have no significant periods in the ERPs. The scatter of the pole coordinates can be reduced to 40%, in LOD remarkably from about 12 mm to 0.1 mm and below. The inclusion of the ECOM-2(9) parameters does not result in the mentioned periods to reappear. For the geocenter coordinates, the scatter for the ECOM-2(9) case is 0.6, 0.8, and 21.2 mm in X, Y, and Z, respectively. In the X direction, there is a significant bias of almost 1.5 mm. The PSD shows a period of 89 days (see Fig. 5) in the X and Y directions; much stronger is the already mentioned period of 119 days in the Z direction. These periods and their amplitudes are also known from real data evaluations (e.g. Zajdel et al. (2020)). Here, the accelerometer cases can significantly reduce the scatter too, especially in the Z direction to less than 0.4 mm. In addition, the periods that occur can be completely removed from all geocenter components, even if the ECOM-2(9) parameters are additionally estimated.

6 Summary and Conclusion

Accelerometer observations were simulated for a Galileo-like constellation over a period of three years. Colored noise based on various existing and future accelerometer sensitivities was applied to the simulated observations. The evaluations show that an accelerometer should for all three axes preferably offer at least the sensitivity of the good axes of the accelerometer used on the GRACE-FO satellites. This allows a complete decorrelation of the estimated satellite clocks from the radial satellite position. This can also be improved by using appropriate hybrid approaches for GNSS SRP modeling. Improvements for the determination of station positions and velocities are less distinct. However, ERPs and geocenter coordinates show immense improvements, especially for LOD and the geocenter Z component. Here, the scatter can be reduced from centimeter to submillimeter level and all significant periods related to SRP mismodeling can be removed. The additional estimation of ECOM parameters is possible and does not cause signals to reappear in the ERPs or geocenter coordinates. Benefits for post-processed GNSS based positioning are less evident in this study, as the SISE can only be improved marginally. Furthermore, accelerometers would also offer the potential for measuring, e.g., thermal thrust forces, whose effect is less well understood and predictable, especially in eclipse phases. It should be mentioned, that we estimated one parameter per axis to account for accelerometer biases. However, biases are not the only error source, as accelerometers may also require scaling factors. They can be calibrated over longer periods, such as a month, but have to be somehow derived as they can be related to the small misalignment of the accelerometer axes, and secular drifts, which are typically related to aging of elements, voltage fluctuations, etc. Moreover, accelerometers are very sensitive to thermal effects which may introduce once-per-revolution variations of recordings – special care must be taken when placing such a sensitive device on a GNSS satellite. This will be a topic for further investigation. It is likely that in the first phase only a limited number of satellites will be equipped with an accelerometer.

Table 5 Statistics of the derived ERPs and geocenter coordinates. Mean and standard deviation of the time series and trend with its standard deviation over all data points within the time span of 3 years

Scenario	Pole$_x$				Pole$_y$				LOD			
	Mean	Std	Trend/a	Std	Mean	Std	Trend/a	Std	Mean	Std	Trend/a	Std
ECOM-2(9)	−0.026	0.121	−0.008	0.004	−0.420	0.118	−0.015	0.004	0.412	11.742	−0.495	0.416
Next Gen. 10^{-9}	0.046	0.044	0.002	0.002	0.007	0.042	−0.000	0.002	−0.003	0.106	−0.007	0.004
w/o noise	0.046	0.021	0.003	0.001	−0.006	0.020	0.000	0.001	0.000	0.047	−0.004	0.002
w/o noise + ECOM-2(9)	0.013	0.038	0.001	0.001	−0.006	0.043	−0.000	0.002	0.001	0.059	−0.006	0.002
	GCC$_x$				GCC$_y$				GCC$_z$			
ECOM-2(9)	−1.448	0.559	−0.008	0.020	−0.038	0.790	0.027	0.028	0.893	21.218	−0.459	0.752
Next Gen. 10^{-9}	0.010	0.287	−0.006	0.010	0.001	0.301	−0.006	0.011	0.013	0.409	0.013	0.014
w/o noise	−0.002	0.243	−0.002	0.009	0.004	0.248	−0.009	0.009	0.017	0.236	0.002	0.008
w/o noise + ECOM-2(9)	0.006	0.273	−0.001	0.010	0.011	0.265	−0.009	0.009	0.026	0.473	0.014	0.017

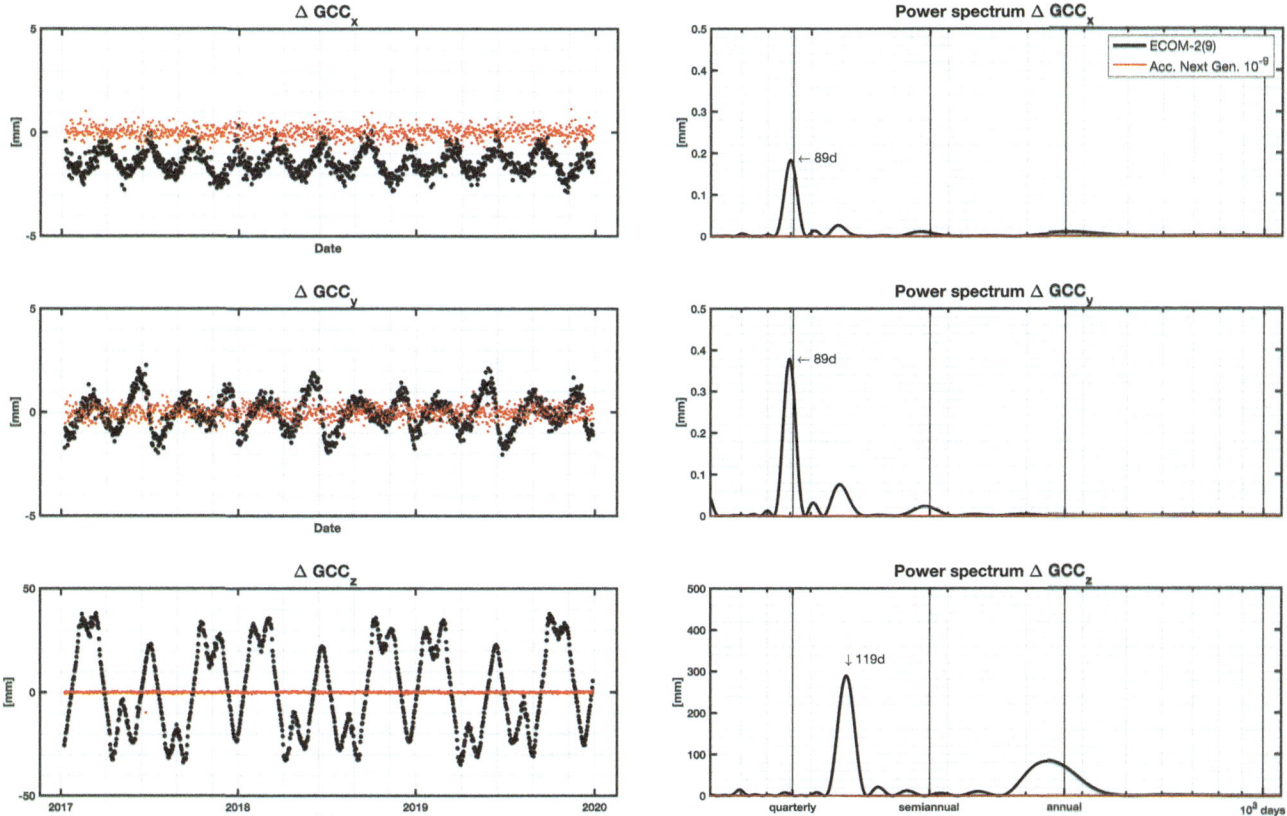

Fig. 5 Difference of the estimated geocenter coordinates for the ECOM-2(9) (black) and accelerometer Next Gen. 10^{-9} (red) case and the corresponding power spectrum

Acknowledgements This work has been supported by the German Research Foundation (DFG) under Grant number GL 1028/1-1 (NextGNSS4GGOS – Next generation GNSS constellations for GGOS-compliant geodetic solutions).

References

Arnold D, Meindl M, Beutler G, Dach R, Schaer S, Lutz S, Prange L, Sośnica K, Mervart L, Jäggi A (2015) CODE's new solar radiation pressure model for GNSS orbit determination. J Geodesy 89(8):775–791. https://doi.org/10.1007/s00190-015-0814-4

Ash ME (2002) Equipping GPS satellites with accelerometers and satellite-to-satellite observables. In: Proceedings of the 2002 National Technical Meeting of The Institute of Navigation, pp 750–761

Biancale R, Bode A (2006) Mean annual and seasonal atmospheric tide models based on 3-hourly and 6-hourly ECMWF surface pressure data. Technical Report STR06/01, Deutsches GeoForschungsZentrum GFZ

Boehm J, Niell A, Tregoning P, Schuh H (2006) Global Mapping Function (GMF): A new empirical mapping function based on numerical weather model data. Geophys Res Lett 33(7). https://doi.org/10.1029/2005gl025546

Bury G, Sośnica K, Zajdel R, Strugarek D (2020) Toward the 1-cm Galileo orbits: challenges in modeling of perturbing forces. J Geodesy 94(16). https://doi.org/10.1007/s00190-020-01342-2

Bury G, Zajdel R, Sośnica K (2019) Accounting for perturbing forces acting on Galileo using a box-wing model. GPS Solut 23(74). https://doi.org/10.1007/s10291-019-0860-0.

Darugna F, Steigenberger P, Montenbruck O, Casotto S (2018) Raytracing solar radiation pressure modeling for QZS-1. Adv Space Res 62(4):935–943. https://doi.org/10.1016/j.asr.2018.05.036

Desai SD (2002) Observing the pole tide with satellite altimetry. J Geophys Res 107(C11):7-1 – 7-13. https://doi.org/10.1029/2001jc001224

ESA (2021) Galileo - Open Service - Service Definition Document, Issue 1.2. https://www.gsc-europa.eu/sites/default/files/sites/all/files/Galileo-OS-SDDv1.2.pdf. Accessed: 2023-10-31

ESTEC (2000) GOCE Mission Requirements Documents. techreport GO-RS-ESA-SY-0001, ESA

EUSPA (2023) Galileo Satellite Metadata. https://www.gsc-europa.eu/support-to-developers/galileo-satellite-metadata. Accessed: 2023-10-31

Ferrari AJ (1977) Lunar gravity: A harmonic analysis. J Geophys Res 82(20):3065–3084. https://doi.org/10.1029/jb082i020p03065

Folkner WM, Williams JG, Boggs DH, Park RS, Kuchynka P (2014) The planetary and lunar ephemerides DE430 and DE431. In: Interplanetary Network Progress Report, pp 42–196

Giorgi G, Schmidt T, Trainotti C, Mata-Calvo R, Fuchs C, Hoque M, Berdermann J, Furthner J, Günther C, Schuldt T, Sanjuan J, Gohlke M, Oswald M, Braxmaier C, Balidakis K, Dick G, Flechtner F, Ge M, Glaser S, König R, Michalak G, Murböck M, Semmling M, Schuh H (2019) Advanced technologies for satellite navigation and geodesy. Adv Space Res 64(6):1256–1273. https://doi.org/10.1016/j.asr.2019.06.010

Glaser S, Michalak G, Männel B, König R, Neumayer KH, Schuh H (2020) Reference system origin and scale realization within the future GNSS constellation "Kepler". J Geodesy 94(117). https://doi.org/10.1007/s00190-020-01441-0

Gross R, Beutler G, Plag HP (2009) Integrated scientific and societal user requirements and functional specifications for the GGOS, pp 209–224. Springer, Berlin, Heidelberg

Guzmán F, Kumanchik L, Jon J, Pratt T, Jacob M (2014) High sensitivity optomechanical reference accelerometer over 10 kHz. Appl Phys Lett 104(22). https://doi.org/10.1063/1.4881936

Guzmán F, Kumanchik LM, Spannagel R, Braxmaier C (2018) Compact fully monolithic optomechanical accelerometer. Appl Phys. https://doi.org/10.48550/ARXIV.1811.01049

Hines A, Nelson A, Richardson L, Valdes G, Guzmán F (2021) Advancements in optomechanical resonators for novel inertial sensors. In: Doyle KB, Ellis JD, Youngworth RN, Sasián JM (Eds) Optomechanics and optical alignment. SPIE

Hines A, Nelson A, Zhang Y, Valdes G, Sanjuan J, Guzmán F (2023) Compact optomechanical accelerometers for use in gravitational wave detectors. Appl Phys Lett 122(9). https://doi.org/10.1063/5.0142108

Hines A, Nelson A, Zhang Y, Valdes G, Sanjuan J, Stoddart J, Guzmán F (2022) sep. Optomechanical accelerometers for geodesy. Remote Sens 14(17): 4389. https://doi.org/https://doi.org/10.3390/rs14174389

Hines A, Richardson L, Wisniewski H, Guzmán F (2020) Optomechanical inertial sensors. Appl Opt 59(22):G167. https://doi.org/10.1364/ao.393061

Kalarus M, Sośnica K, Wielgosz A, Liwosz T, Zielioski JB (2016) Possible advantages of equipping GNSS satellites with on-board accelerometers. http://www.igig.up.wroc.pl/IAG2016/download/KalarusPossible.pdf. Accessed: 2024-02-27

Kornfeld RP, Arnold BW, Gross MA, Dahya NT, Klipstein WM, Gath PF, Bettadpur S (2019) GRACE-FO: The gravity recovery and climate experiment follow-on mission. J Spacecraft Rockets 56(3):931–951. https://doi.org/https://doi.org/10.2514/1.A34326

Kvas A, Brockmann JM, Krauss S, Schubert T, Gruber T, Meyer U, Mayer-Gürr T, Schuh WD, Jäggi A, Pail R (2021) GOCO06s – a satellite-only global gravity field model. Earth Syst Sci Data 13(1): 99–118. https://doi.org/10.5194/essd-13-99-2021

Lucchesi DM, Santoli F, Peron R, Fiorenza E, Lefevre C, Lucente M, Magnafico C, Iafolla VA, Kalarus M, Zielinski J (2016) Non-gravitational accelerations measurements by means of an on-board accelerometer for the Second Generation Galileo Global Navigation Satellite System. In: 2016 IEEE Metrology for Aerospace (MetroAeroSpace), pp 423–433. https://doi.org/10.1109/metroaerospace.2016.7573253

Lyard FH, Allain DJ, Cancet M, Carr'ere L, Picot N (2021) FES2014 global ocean tide atlas: design and performance. Ocean Sci 17(3):615–649. https://doi.org/10.5194/os-17-615-2021

Meindl M, Beutler G, Thaller D, Dach R, Jäggi A (2013) Geocenter coordinates estimated from GNSS data as viewed by perturbation theory. Adv Space Res 51(7):1047–1064. https://doi.org/10.1016/j.asr.2012.10.026

Montenbruck O, Steigenberger P, Hugentobler U (2014) Enhanced solar radiation pressure modeling for Galileo satellites. J Geodesy 89:283–297. https://doi.org/10.1007/s00190-014-0774-0

Navarro-Reyes D, Notarantonio A, Taini G (2009) Galileo constellation: evaluation of station keeping strategies. https://api.semanticscholar.org/CorpusID:204923425. Accessed: 2024-01-26

Neumayer KH, Schreiner P, König R, Dahle C, Glaser S, Mammadaliyev N, Flechtner F (2024) EPOS-OC, a universal software tool for satellite geodesy at GFZ. In: Freymueller J, Sánchez L (eds) Proceedings of the IAG Symposia at IUGG Berlin. Springer. https://link.springer.com/chapter/10.1007/1345_2024_260

Petit G, Luzum B (2010) IERS Technical Note No. 36. Verlag des Bundesamts für Kartographie und Geodäsie, Frankfurt am Main, Germany

Rebischung P, Altamimi Z, Springer T (2013) A collinearity diagnosis of the GNSS geocenter determination. J Geodesy 88(1):65–85. https://doi.org/10.1007/s00190-013-0669-5

Ries J (2017) Conventional Model Update for Rotational Deformation. Presented at AGU Fall Meeting 2017, New Orleans, Louisiana, USA

Rodriguez-Solano CJ, Hugentobler U, Steigenberger P, Bloßfeld M, Fritsche M (2014) Reducing the draconitic errors in GNSS geodetic products. J Geodesy 88(6):559–574. https://doi.org/10.1007/s00190-014-0704-1

Springer T, Beutler G, Rothacher M (1999) A new solar radiation pressure model for GPS. Adv Space Res 23(4):673–676. https://doi.org/10.1016/s0273-1177(99)00158-1

Vespe F, Rothacher M (2014) On-board Accelerometry on GALILEO Satellites: Technical Notes. https://doi.org/10.13140/RG.2.1.2718.7048

Wahr JM (1981) The forced nutations of an elliptical, rotating, elastic and oceanless earth. Geophys J Roy Astronom Soc 64(3):705–727. https://doi.org/10.1111/j.1365-246x.1981.tb02691.x

Zajdel R, Sośnica K, Bury G (2020) Geocenter coordinates derived from multi-GNSS: a look into the role of solar radiation pressure modeling. GPS Solut 25(1). https://doi.org/10.1007/s10291-020-01037-3

Open Access This chapter is licensed under the terms of the Creative Commons Attribution 4.0 International License (http://creativecommons.org/licenses/by/4.0/), which permits use, sharing, adaptation, distribution and reproduction in any medium or format, as long as you give appropriate credit to the original author(s) and the source, provide a link to the Creative Commons license and indicate if changes were made.

The images or other third party material in this chapter are included in the chapter's Creative Commons license, unless indicated otherwise in a credit line to the material. If material is not included in the chapter's Creative Commons license and your intended use is not permitted by statutory regulation or exceeds the permitted use, you will need to obtain permission directly from the copyright holder.

On DORIS Precise Orbit and Reference Frame Determination Based on the ITRF2020 Using Multiple Altimetry Satellite Missions

Anton Reinhold, Patrick Schreiner, and Karl Hans Neumayer

Abstract

Following extensive evaluations, the latest realization of the International Terrestrial Reference System (ITRS), the International Terrestrial Reference Frame (ITRF) 2020 (ITRF2020), was published at the end of last year. For operational application, certain extensions of an ITRF are generated by the services of the different space geodetic techniques. The extension of the ITRF2020 for the Doppler Orbitography and Radiopositioning Integrated by Satellite (DORIS) technique, is the recently released DPOD2020, which is generated by the International DORIS Service (IDS). In this study we exhibit the differences that we see in the application of the DPOD2020. For this purpose, we use altimetry satellites equipped with a DORIS receiver in a setup using the latest DPOD2014 and DPOD2020. Initially we performed Precise Orbit Determination (POD) and evaluate the differences we see internally, in terms of the orbital fit, as well as changes in the derived orbit. Subsequently, weekly local terrestrial reference frames (TRFs) are computed for each single satellite as well as a combined solution to evaluate the impact on derived station coordinates and Earth Rotation Parameters (ERPs). The following generated TRF solutions are evaluated with respect to the reference frame defining parameters, i.e. origin, scale, and orientation, in comparison to the a priori TRF and as differences between ITRF2014 and ITRF2020 solutions. The processed orbits show comparable results w.r.t. orbital fits and orbit comparisons between both solutions. The local TRF's show also overall good agreement between the ITRF2020 and ITRF2014 solutions with no systematic bias.

Keywords

DORIS · ITRF2020 · Precise orbit determination · Reference frame

The authors "Patrick Schreiner" and "Karl Hans Neumayer" have contributed equally to this work.

A. Reinhold (✉)
GFZ German Research Centre for Geosciences, Potsdam, Germany

SpaceTech GmbH STI, Immenstaad am Bodensee, Germany
e-mail: anton.reinhold@gfz-potsdam.de

P. Schreiner · K. H. Neumayer
GFZ German Research Centre for Geosciences, Potsdam, Germany
e-mail: patrick.schreiner@gfz-potsdam.de; neumayer@gfz-potsdam.de

1 Introduction

DPOD2020 is the latest realization of the International Terrestrial Reference Frame 2020 based on the Doppler Orbitography and Radiopositioning Integrated by Satellite (DORIS) technique (Moreaux et al. 2023). For the purpose of the validation of DPOD2020 and to display the difference seen while switching from DPOD2014 (Moreaux et al. 2019) we performed an analysis of the latest orbit products generated with EPOS-OC (Earth Parameter and Orbit System – Orbital Computation) (Neumayer et al. 2024). EPOS-OC is an universal software package from GFZ which can be used for Precise Orbit Determination (POD) and reference frame determination. It is continuously developed and is capable of handling all four main space geodetic techniques: DORIS, Global Navigation Satellite Systems (GNSS), Satellite Laser Ranging (SLR) and Very Long Baseline Interferometry (VLBI). In the framework of this study we have used 10 most known altimetry missions: TOPEX/Poseidon (TOP), Jason-1 (JA1), Envisat (ENV), OSTM/Jason-2 (JA2), CryoSat-2 (CS2), Saral/AltiKa (SRL), Jason-3 (JA3), Sentinel-3A (S3A),Sentinel-3B (S3B) and Sentinel-6A Michael Freilich (S6A). All these satellites are equipped with a DORIS receiver and an SLR retroreflector.

2 Precise Orbit Determination

2.1 Model Definition

The software for orbit computation and reference frame determination is GFZ's EPOS-OC. For this study we have processed about 30 years of data for the time span from 1993 starting with the TOPEX/Poseidon mission to the end of 2023 including 10 altimetry missions. We have set the arc length to 7 days according to GPS weeks and we use highly dynamic parametrization with once per revolution sine and cosine empirical acceleration coefficients in transverse and normal directions every 48 hours. An overview of applied models and standards is given in the Table 1 (Schreiner et al. 2023). For SLR range bias we are not using any a priori values given in the respective Data Handling Files for SLRF2014/SLRF2020 (ILRS 2020, 2023; Pearlman et al. 2019), but estimate a bias once per station and arc. The used macromodels are partly self-derived partly follow the recommendations of the IDS. In this study we use SLR and DORIS observations and process SLR+DORIS combined orbits as well as DORIS-only orbits. SLR and DORIS observations are thereby obtained from the Crustal Dynamics Data Information System (CDDIS) (Noll 2010; Noll et al. 2018) and from the EUROLAS Data Center (EDC) (Schwatke 2012). DORIS observation are given in the DOPPLER count format for TOPEX/Poseidon, Jason-1 and Envisat, while starting from OSTM/Jason-2 the DORIS observations are given in the DORIS/RINEX format (Lemoine et al. 2016). To handle these RINEX observations in EPOS-OC a new preprocessing procedure was implemented (Schreiner et al. 2023). The SLR+DORIS combined orbits are mainly used for quality validation in terms of internal orbital fit but also in order to take care of possible DORIS time bias. The approach to correct the DORIS time bias is thereby as follow. In the first step a combined SLR+DORIS orbit determination is performed and initial data screening is undertaken. Starting from this point the SLR observations are downweighted with weight close to zero so the resulting orbit is following the DORIS time system. The orbit is then kept fixed while the

Table 1 Models and standards used for the Precise Orbit Determination

	ITRF2014	ITRF2020	Reference
Dynamic model			
A priori gravity field	GOCO06s 120x120 time variable 120x120		Kvas et al. (2021)
Ocean tide model	FES2014 100x100		Lyard et al. (2021)
AOD	AOD1B RL06 180x180		Dobslaw et al. (2017)
Atmosphere	MSIS-E-90		Hedin (1991)
Geometric model			
EOP	IERS C04 14	IERS C04 20	Bizouard et al. (2018)
Station coordinates	DPOD2014 v. 5.5	DPOD2020 v. 11	Moreaux et al. (2019, 2023)
	SLRF2014	SLRF2020	ILRS (2020), ILRS (2023)
			Pearlman et al. (2019)
Post-seismic deformation	ITRF2014	ITRF2020	Altamimi et al. (2016, 2023)
Observations correction models			
Frequency bias and drift	Once per station and pass for all DORIS stations estimated		
Range bias	Once per station and arc estimated		
Time bias	One global bias per arc estimated		
Troposphere model	DORIS: VMF-1		Boehm et al. (2006)
	SLR: MP 2004		Mendes and Pavlis (2004)
Troposphere bias	Once per station and pass estimated		
Parametrization			
Atmospheric drag	1 per 3h		
Albedo with IR	1 linear scaling per arc estimated		
Empirical acc.	1/rev cosine and sine every 48 h in T and N directions		
Solar radiation	1/day		

SLR observations are normal weighted again and only a time bias value as a mean value over all stations is estimated for the SLR observations. Eventually the value of the estimated SLR time bias is applied to the DORIS observations with a reversed sign to bring the DORIS observations as close as possible to the SLR time system (Schreiner et al. 2023). A new DORIS-only time bias corrected orbit is calculated in the last step of POD for the determination of the reference frame.

2.2 Orbital Fit

To validate the internal orbit quality a statistical evaluation of SLR and DORIS residuals has been performed. Technique-wise RMS values together with mean and number of accepted observations as a statistical key parameters are given in the Tables 2 and 3 for SLR+DORIS combined orbits. The SLR observations are downweighted with weight close to zero and used as an independent technique for quality validation only. The RMS values for SLR observations are typically around 1 cm for most of the missions while the RMS values for DORIS are around 0.44 mm/s. Due to significant changes in the station network between ITRF2014 and ITRF2020 in particular reflected in the different number of stations but also different observation times for certain stations, the total number of accepted observations differs between the ITRF2014 and ITRF2020 solutions. An overview about the number of observing stations is given in the Fig. 1. Here we can see an increase in DORIS observing stations for DPOD2020 especially between 2001 and 2006. While for SLRF2020 there is an almost constant number of around 20 stations.

2.3 Orbit Comparisons

Orbit comparisons between DPOD2014 and DPOD2020 based solutions was performed to validate the differences in the orbit quality. The RMS values together with mean for radial, normal and transverse components are shown in the Table 4. Thereby the first value indicates the mean value of the corresponding time series while the second value shows the standard deviation and therefore the scatter of the time series. For the radial component we have the mean and standard deviation of the mean range close to zero while the RMS values are in single-digit range and amount to 1–2 mm which indicate good agreement between both orbit solutions with no systematic bias. However the RMS values for the transverse component are the highest with apparently small offsets in sub millimeter range highlighted by the mean values, which is most likely caused by the time bias. For additional validation of external orbit quality an comparison with CNES-SSALTO (Coutin-Faye et al. 2000) POE-F (Precision Orbit Ephemeris version F) orbits obtained from CDDIS (Noll 2010) was exemplary performed for Envisat and Sentinel-3A with the results given in the Table 5.

Table 2 SLR post fit residuals and number of observations

Satellite	SLRF2014			SLRF2020		
	RMS [cm]	Mean [cm]	No. Obs.	RMS [cm]	Mean [cm]	No. Obs.
TOPEX	1.65	0.01	1,646,137	1.69	0.02	1,673,558
Jason-1	0.99	−0.01	1,332,323	1.03	−0.01	1,421,439
Envisat	1.19	−0.02	679,657	1.25	−0.01	722,777
Jason-2	0.99	−0.01	1,652,896	1.01	0.00	1,689,579
Cryosat-2	1.40	0.08	663,681	1.42	0.08	689,738
Saral	1.18	−0.14	503,068	1.04	−0.01	510,930
Jason-3	1.58	0.07	952,115	1.52	0.00	962,317
Sentinel-3A	0.99	0.01	370,490	1.02	−0.01	382,487
Sentinel-3B	1.07	−0.01	229,402	1.07	−0.02	230,655
Sentinel-6A	0.99	−0.01	203,292	0.97	0.00	204,468

Table 3 DORIS post fit residuals and number of observations

Satellite	DPOD2014			DPOD2020		
	RMS [mm/s]	Mean [mm/s]	No. Obs.	RMS [mm/s]	Mean [mm/s]	No. Obs.
TOPEX	0.47	0.00	19,208,581	0.47	0.00	18,291,758
Jason-1	0.33	0.00	39,495,711	0.33	0.00	32,244,317
Envisat	0.42	0.00	22,861,364	0.42	0.00	21,870,523
Jason-2	0.34	0.00	57,699,818	0.34	0.00	52,347,797
Cryosat-2	0.39	0.00	36,612,501	0.39	0.00	36,726,049
Saral	0.38	0.00	29,873,758	0.38	0.00	27,118,485
Jason-3	0.40	0.00	41,143,092	0.40	0.00	36,993,763
Sentinel-3A	0.40	0.00	22,828,991	0.41	0.00	21,125,879
Sentinel-3B	0.42	0.00	14,807,196	0.44	0.00	14,806,047
Sentinel-6A	0.39	0.00	11,608,890	0.39	0.00	11,150,218

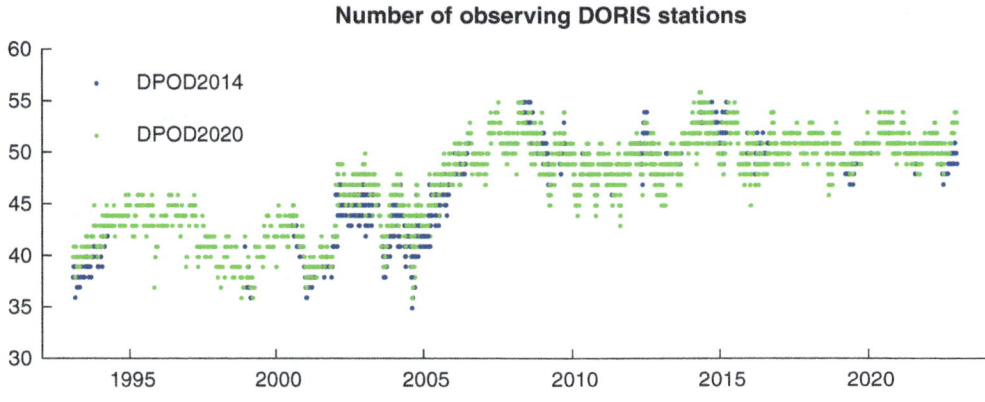

Fig. 1 Total number of DORIS and SLR observing stations

Table 4 Orbital comparison between DPOD2014 and DPOD2020 solutions

Satellite	Radial		Normal		Transverse	
	Mean [mm]	RMS [mm]	Mean [mm]	RMS [mm]	Mean [mm]	RMS [mm]
Topex	0.0±0.0	1.8±0.9	0.0±0.5	5.1±3.5	0.2±4.5	7.0±5.2
Jason-1	0.0±0.1	1.5±0.8	0.1±0.5	3.9±3.1	−1.5±3.4	6.6±3.7
Envisat	0.0±0.1	1.4±0.9	0.0±0.1	4.3±3.1	−0.9±2.9	5.4±3.5
Jason-2	0.0±0.1	1.0±0.4	0.1±0.4	3.0±1.4	0.5±2.0	4.7±1.6
Cryosat-2	0.0±0.0	0.9±0.4	0.0±0.1	3.7±2.1	0.3±2.5	5.2±2.8
Saral	0.0±0.0	1.3±0.7	0.1±0.7	4.9±8.9	0.1±1.6	5.4±1.9
Jason-3	0.0±0.1	1.3±0.5	0.0±0.3	4.3±1.9	0.1±1.6	5.0±1.5
Sentinel-3A	0.0±0.0	1.1±0.3	0.0±0.1	3.8±1.7	−0.6±2.2	4.9±1.6
Sentinel-3B	0.0±0.0	1.4±0.8	0.1±0.1	4.9±3.8	−0.9±3.0	6.6±2.7
Sentinel-6A	0.0±0.0	1.5±0.5	0.0±0.1	5.3±2.1	−0.9±2.3	5.2±1.9

While the mean values are close to zero for the most cases except the transverse component of Envisat, the RMS values for normal and transverse component are between 2–3 cm and around 1 cm for the radial component. This indicate further a good orbit quality.

3 Estimation of Station Positions and Earth Rotation Parameters

3.1 Solution Strategy

For the estimation of local TRF's we solve station coordinates as well as pole coordinates and *LOD*. Hereby we use the following strategy. In the first step a POD of DORIS-only orbits is performed until the convergence is reached. In the next step the station coordinates and ERPs are estimated and normal equation matrices (NEQ) are set up. Here we are using loose constrains for stations and ERPs with $\sigma = 1$ m. Eventually in the last step we solve the generated NEQ's for station coordinates and ERPs with minimal No-Net-Rotation constraints with $\sigma = 1$ mm. Hereby the NEQ's will be solved in an iterative process in which the stations with 3D position correction above 5 cm are excluded from the datum-defining network to ensure the most reliable and accurate solution.

We are generating weekly local TRF's for each satellite as a single satellite solution as well as a combined solution. For the combined solution we accumulate the NEQ of each

Table 5 Orbital comparison between DPOD2014 resp. DPOD2020 and CNES-SSALTO POE-F solutions for Envisat and Sentinel-3A

Satellite	Radial Mean [mm]	Radial RMS [mm]	Normal Mean [mm]	Normal RMS [mm]	Transverse Mean [mm]	Transverse RMS [mm]
DPOD2014						
Envisat	−0.4±0.3	7.3±3.0	3.0±0.5	21.1±6.8	7.2±8.6	21.7±6.7
Sentinel-3A	−0.2±0.3	12.4±3.7	1.7±0.4	22.4±5.3	2.0±5.4	30.3±6.9
DPOD2020						
Envisat	−0.4±0.3	7.2±3.0	2.9±0.5	20.8±6.6	8.0±8.2	21.8±6.7
Sentinel-3A	0.0±0.4	8.2±1.6	1.6±0.6	15.8±4.4	−0.3±4.3	20.9±3.0

Table 6 Comparison of derivered ERPs, translation and scale as difference between ITRF2014 and ITRF2020 single solutions. Mean and standard deviation

Satellite	ΔLOD (mas)	$\Delta XPole$ (mas)	$\Delta YPole$ (mas)	ΔTx (mm)	ΔTy (mm)	ΔTz (mm)	$\Delta Scale$ (mm)
Topex	0.01±0.29	0.02±0.11	−0.06±0.14	1.32±0.78	−0.36±0.61	0.53±1.60	1.00±1.41
Jason-1	0.00±0.15	−0.12±0.08	0.04±0.08	0.75±0.57	−0.14±0.47	0.60±1.88	−0.09±1.51
Envisat	0.02±0.21	−0.11±0.10	0.03±0.10	0.92±0.40	−0.24±0.44	1.52±2.92	0.35±1.42
Jason-2	−0.01±0.14	−0.18±0.07	0.14±0.07	1.03±0.39	−0.15±0.47	1.21±2.19	−0.21±0.90
Cryosat-2	0.00±0.25	−0.24±0.13	0.15±0.08	0.85±0.36	−0.43±0.54	2.08±9.90	−0.70±0.66
Saral	0.02±0.27	−0.29±0.13	0.13±0.10	1.02±0.37	−0.64±0.64	3.70±7.96	−0.66±0.69
Jason-3	0.00±0.18	−0.25±0.09	0.12±0.10	1.08±0.54	−0.85±0.55	1.07±3.83	−0.04±1.49
Sentinel-3A	0.03±0.14	−0.27±0.08	0.13±0.10	1.20±0.41	−0.85±0.47	0.52±2.94	−0.83±0.59
Sentinel-3B	0.02±0.16	−0.29±0.07	0.09±0.10	0.79±0.52	−0.77±0.53	−0.03±3.84	−0.94±0.81
Sentinel-6A	0.00±0.20	−0.27±0.06	0.03±0.05	1.39±0.55	−1.49±0.42	1.14±4.04	−0.18±0.90

Table 7 Comparison of derived ERPs as differences between ITRF2014 resp. ITRF2020 and corresponding a priori solutions, as well as differences between ITRF2014 and ITRF2020 combined solutions. Translation and scale from transformation to the a priori station network for ITRF2014 and ITRF2020 solutions. ΔTx, ΔTy, ΔTz, $\Delta Scale$ as differences between corresponding parameters from ITRF2014 and ITRF2020 solutions. Mean and standard deviation

Solution	ΔLOD (mas)	$\Delta XPole$ (mas)	$\Delta YPole$ (mas)	Tx (mm)	Ty (mm)	Tz (mm)	$Scale$ (mm)
2002–2008							
ITRF2014	−0.02±0.30	−0.05±0.22	0.08±0.17	1.00±1.83	−0.14±1.72	−0.49±3.97	−1.34±3.19
ITRF2020	−0.02±0.32	−0.09±0.22	0.04±0.18	0.09±1.82	0.07±1.84	−0.70±3.94	−1.41±3.33
2008–2023							
ITRF2014	0.00±0.14	0.05±0.15	−0.06±0.15	1.87±1.64	−0.08±1.42	−0.10±2.07	−1.16±1.28
ITRF2020	0.00±0.12	0.06±0.15	−0.13±0.15	0.98±1.60	0.33±1.68	−0.28±2.34	−0.77±1.32
				ΔTx (mm)	ΔTy (mm)	ΔTz (mm)	$\Delta Scale$ (mm)
2002–2008							
ITRF2014-ITRF2020	0.00±0.20	0.04±0.06	0.04±0.06	0.92±0.56	−0.24±0.63	0.30±1.49	0.13±1.04
2008–2023							
ITRF2014-ITRF2020	0.00±0.15	−0.01±0.08	0.07±0.06	0.89±0.30	−0.43±0.57	0.17±1.43	−0.39±0.54

satellite to a corresponding global NEQ and solve it accordingly. The validation of the generated TRF's is performed through statistical analysis of ERPs and the reference frame defining parameters (Scale, Translation and Rotation) from the Helmert transformation to the a priori station network.

3.2 Single Satellite Solution

The statistical evaluation of single satellite solutions is given as differences between ITRF2014 and ITRF2020 solutions in Table 6. Thereby in case of ERPs the differences between estimated corrections to the a priori values of the corresponding solution IERS C04 has been formed to make the a priori independent comparison between both solutions. The first value in the statistics indicates the mean while the second one refers to the standard deviation. It should be also noted that the scale was converted to a length unit in mm by multiplying the former scale values with the mean earth radius to illustrate the magnitude of the scale bias. While the ΔLOD was converted to arc seconds for better comparison with the pole coordinates. The results show good agreement between both solutions with no systematic bias.

3.3 Combined Solution

The results from statistical evaluation of combined solutions are given in Table 7. The evaluation of ERPs is based on differences between the estimated parameters and a priori values as well as differences between ITRF2014 and ITRF2020 solutions as comparisons between estimated corrections to the a priori values of the corresponding IERS C04 solution analogous to the single solution. For the translation and scale the results are given in form of statistical evaluations of the corresponding parameters for each of the solutions, as well as direct comparison between ITRF2014 and ITRF2020 solutions. Thereby the first value refers to mean and the second one indicates the scatter of the time series in form of standard deviation. Due to the fact that before the launch of Envisat only TOPEX/Poseidon solution is available the combine solution begins with 2002. Further the evaluation was split in 2 different time periods: 2002–2008 and 2008–2023. These time periods represents natural break points in DORIS system performance caused by launch of OSTM/Jason-2. The mean values for ERPs are close to zero so no systematic bias can be seen. The results for translation and scale show no systematic bias as well although there is an increase in standard deviation for the ITRF2020 solution which can be most likely connected to an increase of observing stations in the DPOD2020 network. Overall there is an clear improvement for the latter time period 2008–2023. The direct comparison between ITRF2014 and ITRF2020 solutions shows a good agreement between both solutions with no systematic bias.

4 Conclusion and Outlook

We have investigated the differences we can see in the application of ITRF2020 in comparison with ITRF2014 by viewing POD and generation of local terrestrial reference frames. The processed orbits show comparable results by orbital fit and through orbit comparison between ITRF2014 and ITRF2020 solutions. While the generated local TRFs show good agreement between both solutions with no systematic bias. However the combined ITRF2020 solution shows an increase in standard deviation of scale most likely due an extended station network in DPOD2020. Overall the ITRF2020 solution shows good agreement to ITRF2014 solution and no systematic bias in comparisons can be detected.

References

Altamimi Z, Rebischung P, Collilieux X, Métivier L, Chanard K (2023) ITRF2020: an augmented reference frame refining the modeling of nonlinear station motions. J Geodesy 97(5):Article number 47. https://doi.org/10.1007/s00190-023-01738-w

Altamimi Z, Rebischung P, Métivier L, Collilieux X (2016) ITRF2014: A new release of the international terrestrial reference frame modeling nonlinear station motions. J Geophys Res Solid Earth 121(8):6109–6131. https://doi.org/https://doi.org/10.1002/2016jb013098

Bizouard C, Lambert S, Gattano C, Becker O, Richard JY (2018) The IERS EOP 14c04 solution for earth orientation parameters consistent with ITRF 2014. J Geodesy 93(5):621–633. https://doi.org/10.1007/s00190-018-1186-3

Boehm J, Werl B, Schuh H (2006) Troposphere mapping functions for GPS and very long baseline interferometry from european centre for medium-range weather forecasts operational analysis data. J Geophys Res Solid Earth 111(B2). https://doi.org/10.1029/2005jb003629

Coutin-Faye S, Noubel J, Boutonnet G (2000) SSALTO: a new ground segment for a new generation of altimetry satellites. AVISO Newslett 7

Dobslaw H, Bergmann-Wolf I, Dill R, Poropat L, Thomas M, Dahle C, Esselborn S, König R, Flechtner F (2017) A new high-resolution model of non-tidal atmosphere and ocean mass variability for de-aliasing of satellite gravity observations: AOD1b RL06. Geophys J Int 211(1):263–269. https://doi.org/10.1093/gji/ggx302

Hedin AE (1991) Extension of the MSIS thermosphere model into the middle and lower atmosphere. J Geophys Res Space Phys 96(A2):1159–1172. https://doi.org/10.1029/90ja02125

ILRS (2020) SLRF 2014 version: 2030 from 200428

ILRS (2023) SLRF 2020 version: 230322 from 230328

Kvas A, Brockmann JM, Krauss S, Schubert T, Gruber T, Meyer U, Mayer-Gürr T, Schuh WD, Jäggi A, Pail R (2021) GOCO06s – a satellite-only global gravity field model. Earth Syst Sci Data 13(1):99–118. https://doi.org/10.5194/essd-13-99-2021

Lemoine JM, Capdeville H, Soudarin L (2016) Precise orbit determination and station position estimation using doris rinex data. Adv Space Res 58(12):2677–2690. https://doi.org/10.1016/j.asr.2016.06.024

Lyard FH, Allain DJ, Cancet M, Carrère L, Picot N (2021) FES2014 global ocean tide atlas: design and performance. Ocean Sci 17(3):615–649. https://doi.org/10.5194/os-17-615-2021

Mendes VB, Pavlis EC (2004) High-accuracy zenith delay prediction at optical wavelengths. Geophys Re Lett 31(14). https://doi.org/10.1029/2004gl020308

Moreaux G, Lemoine FG, Capdeville H, Otten M, Štěpánek P, Saunier J, Ferrage P (2023) The international DORIS service contribution to ITRF2020. Adv Space Res 72(1):65–91. https://doi.org/10.1016/j.asr.2022.07.012

Moreaux G, Willis P, Lemoine FG, Zelensky NP, Couhert A, Lakbir HA, Ferrage P (2019) DPOD2014: A new DORIS extension of ITRF2014 for precise orbit determination. Adv Space Res 63(1):118–138. https://doi.org/10.1016/j.asr.2018.08.043

Neumayer KH, Schreiner P, König R, Dahle C, Flechtner F (2024) EPOS-OC a universal software tool for satellite geodesy at GFZ. International Association of Geodesy Symposia. Submitted to: Under revision

Noll CE (2010) The crustal dynamics data information system: A resource to support scientific analysis using space geodesy. Adv Space Res 45(12):1421–1440. https://doi.org/10.1016/j.asr.2010.01.018

Noll CE, Ricklefs R, Horvath J, Mueller H, Schwatke C, Torrence M (2018) Information resources supporting scientific research for the international laser ranging service. J Geodesy 93(11):2211–2225. https://doi.org/10.1007/s00190-018-1207-2

Pearlman MR, Noll CE, Pavlis EC, Lemoine FG, Combrink L, Degnan JJ, Kirchner G, Schreiber U (2019) The ilrs: approaching 20 years and planning for the future. J Geodesy 93(11):2161–2180. https://doi.org/10.1007/s00190-019-01241-1

Schreiner P, König R, Neumayer KH, Reinhold A (2023) On precise orbit determination based on DORIS, GPS and SLR using sentinel-3a/b and -6a and subsequent reference frame determination based on DORISonly. Adv Space Res 72(1):47–64. https://doi.org/10.1016/j.asr.2023.04.002

Schwatke C (2012) Eurolas data center (edc) - a new website for tracking the slr data flow. Presented at EGU General Assembly 2012, Vienna, Austria

Open Access This chapter is licensed under the terms of the Creative Commons Attribution 4.0 International License (http://creativecommons.org/licenses/by/4.0/), which permits use, sharing, adaptation, distribution and reproduction in any medium or format, as long as you give appropriate credit to the original author(s) and the source, provide a link to the Creative Commons license and indicate if changes were made.

The images or other third party material in this chapter are included in the chapter's Creative Commons license, unless indicated otherwise in a credit line to the material. If material is not included in the chapter's Creative Commons license and your intended use is not permitted by statutory regulation or exceeds the permitted use, you will need to obtain permission directly from the copyright holder.

Realisation of the Non-Rotating Terrestrial Reference Frame by an Actual Plate Kinematic and Crustal Deformation Model (APKIM2020)

Hermann Drewes, Manuela Seitz, and Laura Sánchez

Abstract

Since 1991, the International Terrestrial Reference Frame (ITRF) includes the global time evolution of station positions (velocities) in addition to the three-dimensional Cartesian station positions at a fixed reference epoch. The orientation of the velocities refers to a kinematic model of rigid tectonic plates derived from geophysical observations over millions of years. For consistency with other geodetic parameters (e.g., Earth orientation), the models must be aligned to actual no-net-rotation of the whole Earth surface. Because of deviations of present-day velocities and neglect of non-rigid surface deformations, e.g., in seismic zones, the geophysical models are not valid for today. This paper describes a further developed method of estimating a non-rotating terrestrial reference frame from space geodetic observations. Different to previous estimations of geodetic no-net-rotation models, regional inter-plate and intra-plate crustal deformations are included, and instead of using the irregularly distributed observed station velocities, an evenly distributed grid throughout the Earth is interpolated by least squares collocation. Due to significant changes of the station velocities from one ITRF to another, NNR models must be computed for each ITRF. Here it is done for the ITRF2020.

Keywords

Earth surface deformation · Lithosphere plate model · No net rotation (NNR) · Plate rotation · Terrestrial Reference Frame

1 Introduction

The International Terrestrial Reference System (ITRS) is a spatial reference system co-rotating with the Earth in its diurnal motion in space (Petit and Luzum 2010). It is realised by the International Terrestrial Reference Frame (ITRF), which includes three-dimensional Cartesian coordinates of station positions and velocities estimated by a combination of Global Navigation Satellite Systems (GNSS), Satellite Laser Ranging (SLR), Very Long Baseline Interferometry (VLBI), and Doppler Orbitography and Radiopositioning Integrated by Satellite (DORIS) observations (e.g., Altamimi et al. 2023a). The station velocities, which are of most interest in this paper, are at present referring totally or partially to a global plate velocity model derived from geophysical observations covering a period of million years. Starting with ITRF91 (Boucher et al. 1992), the velocity field was aligned to the NNR-NUVEL-1 model (Argus and Gordon 1991). For the ITRF92 (Boucher et al. 1993) and ITRF94 (Boucher et al. 1996) it was changed to NNR-NUVEL-1A (DeMets et al. 1994), using the 7 rates of the transformation parameters (3 translations, 3 rotations, 1 scale, Petit and Luzum 2010; Boucher et al. 1996). The following ITRF solutions, in orientation only starting from ITRF2000 (Altamimi et al. 2002), were then aligned to the respective previous ITRF, so that all later ITRF are referring (indirectly) to the NNR-NUVEL-1A; see e.g., Altamimi et al. (2007, 2012, 2023a).

H. Drewes (✉) · M. Seitz · L. Sánchez
Deutsches Geodätisches Forschungsinstitut, Technische Universität München (DGFI-TUM), Munich, Germany
e-mail: h.drewes@tum.de

NNR stands for "no net rotation" of the entire Earth surface, which was computed especially for geodetic use from the original geophysical models NUVEL-1 and NUVEL-1A, respectively, which refer to the fixed Pacific plate.

The kinematic orientation of the station velocities is important, because Earth Orientation Parameters (EOP), i.e., Polar Motion (PM) and Difference to Universal Time (DUT) or Length of Day (LOD) depend on the station velocities. One may shift any part of the plate motions to the EOP and vice versa. For instance, using the original NUVEL-1A referring to the Pacific plate instead of NNR-NUVEL-1A would change the EOP by nearly 10 cm/a, which is the velocity of the Pacific plate in the NNR-NUVEL-1A. Because the ITRF velocities are changing significantly (more than three times of their formal errors) from one ITRF to the following, it is a fundamental task to compute a model with global no net rotation for each new ITRF solution.

2 Geophysical Plate Motion Models

According to a Theorem of Euler, the plate motions are estimated as a two-dimensional rotation on a sphere around an oblique axis given by its pole with latitude Ω, longitude Λ and rotational velocity ω (Drewes 1982). The NNR-NUVEL-1A is based on data of the latest three million years, namely velocities of the sea floor spreading along the mid ocean ridges, azimuths of the transform faults caused by the sea floor spreading and azimuths of earthquake slip vectors. Because of short term variations (e.g., caused by seismic activities), these long-period geophysical data are not valid for ITRF periods of a few decades. Furthermore, NNR-NUVEL-1A includes only 11 large plates and does not consider any non-rigid crustal deformations, e.g., the Circum-Pacific Belt or the Mediterranean Belt (from the Azores to East Asia), which are consequently disregarded in the NUVEL-1A NNR condition. Later geophysical models include more plates (e.g., DeMets et al. 2010) and are considering crustal deformation zones, e.g., PB2002 (Bird 2003) with 52 plates (the NUVEL-1A plates being accepted with the original parameters) and 13 orogens (Fig. 1). However, they also refer to the average motion over millions of years and are not suitable for present-day surface velocity models.

Subtracting the NNR-NUVEL-1A plate motions from the ITRF2020 (Altamimi et al. 2023a) velocities reveals large systematic features caused by the geophysical plate motion model and the effect of the deformation zones, mainly along the Circum-Pacific Belt and the Mediterranean Belt (Fig. 2).

3 Plate Kinematic Models from Geodetic Observations

Similarly to the geophysical models over geologic times, present-day plate kinematic models are computed using the velocities estimated from the observations of the geodetic space techniques by various authors (e.g., Altamimi et al. 2003, 2012, 2023b; Drewes 1990, 2009; Kreemer and Holt

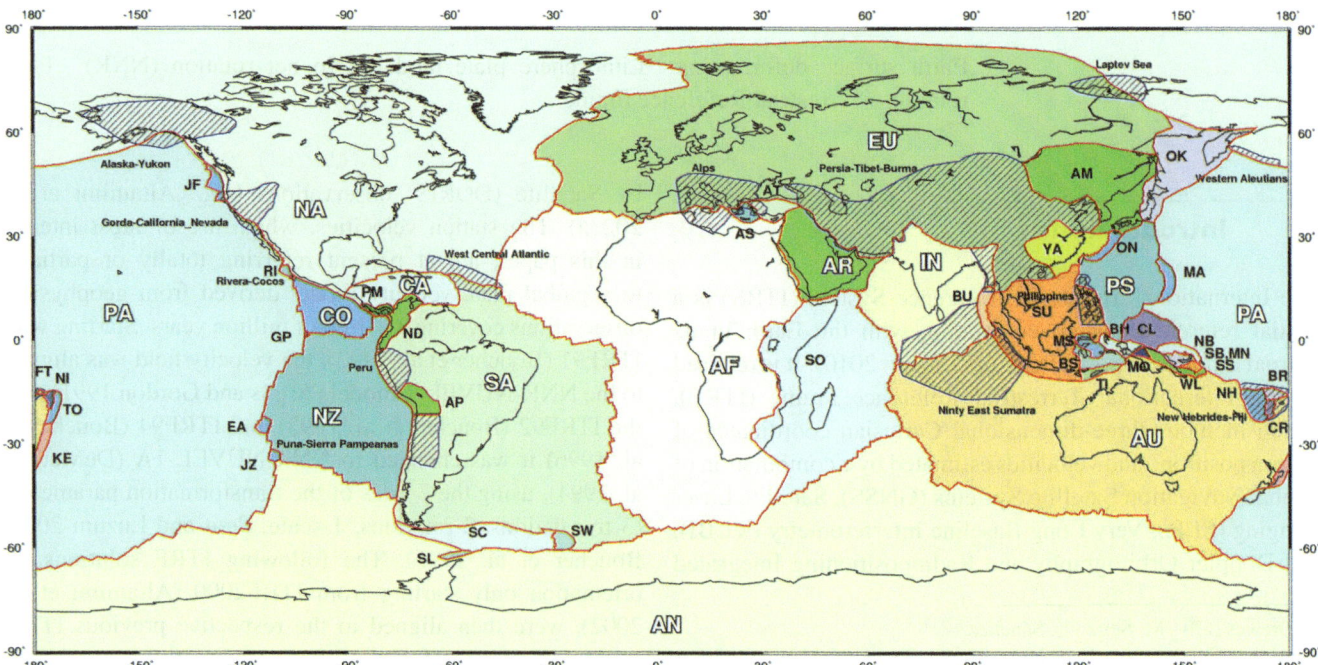

Fig. 1 Geophysical rigid plate and orogenic deformation model PB2002 (after Bird 2003)

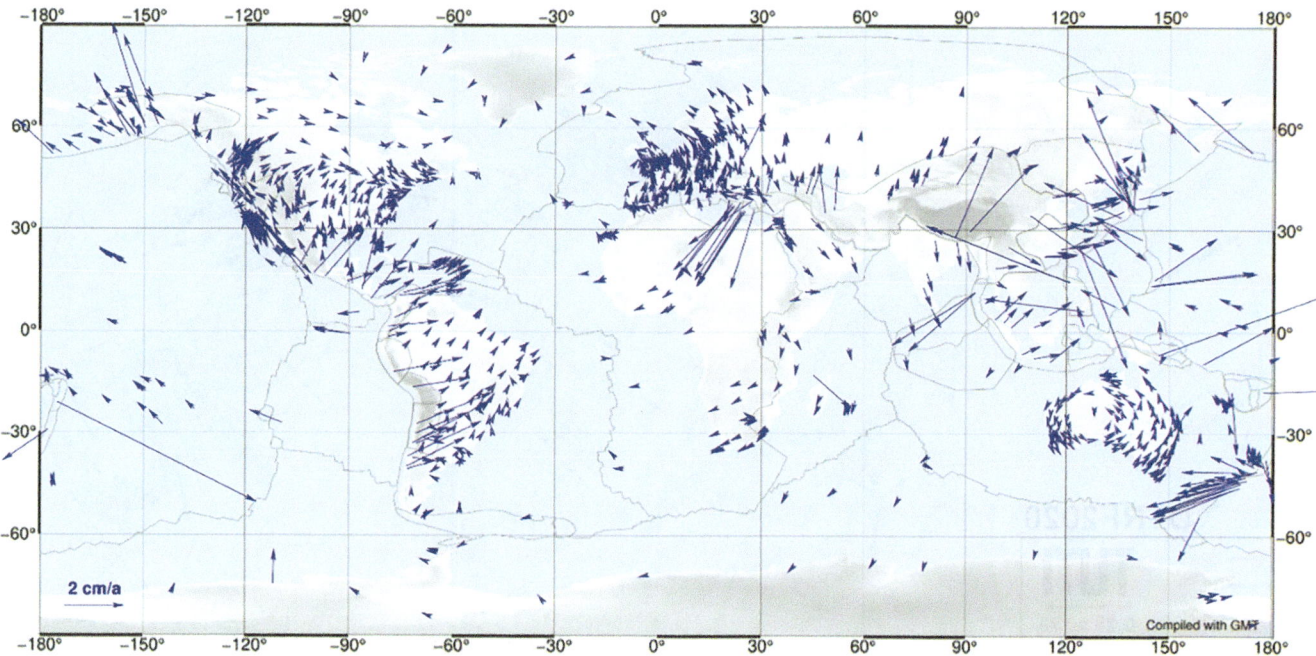

Fig. 2 ITRF2020 velocities with the NNR-NUVEL-1A plate motions removed

2001; Kreemer et al. 2006, 2014). According to the Theorem of Euler, the input data must be the horizontal station velocities in (dφ/dt and dλ/dt) in terms of latitude (φ) and longitude (λ) and the unknowns are the rotation poles (Φ, Λ, ω) on a sphere, see Eq. (1a and 1b). A two-dimensional horizontal adjustment is appropriate to avoid the effect of the less precise vertical velocities estimated from space-geodetic observations, which cannot be separated in the Cartesian representation of dX/dt, dY/dt, dZ/dt.

$$(d\varphi/dt)_k = \omega_i \cdot \cos\,\Phi_i \cdot \sin(\lambda_k - \Lambda_i) \quad (1a)$$

$$(d\lambda/dt)_k = \omega_i \cdot (\sin\,\Phi_i - \cos(\lambda_k - \Lambda_i) \cdot \tan\varphi_k \cdot \cos\,\Phi_i) \quad (1b)$$

Subindexes k and i represent station and plates, respectively. Actual Plate Kinematic Models (APKIM) from geodetic data are computed since 1988 (Drewes 1990). The present Actual Plate Kinematic Model (APKIM2020) is based on the geometry of the tectonic structures of PB2002 (Fig. 1), but it is modified according to patterns evidenced by the geodetic data. Individual sites or regional clusters obviously not moving with a rigid plate are not included in the plate rotation parameters estimation but considered as regional deformations and included in the deformation models (see below). For comparison purpose, there are two computations of the rigid plate model based on ITRF2020 (Altamimi et al. 2023a) and DTRF2020 (Seitz et al. 2023), respectively. As the velocities are given for various periods at the same site, only the latest period is taken if observing at least for one year. If there are different velocities from the observation techniques (GNSS, SLR, VLBI, DORIS), only the most precise is used. In total, we have 1,198 velocities from ITRF2020 and 1,254 velocities from DTRF2020 (Fig. 3). Iterative adjustments were done for 23 plates occupied by at least 3 stations of the Terrestrial Reference Frame. "Non-fitting" station velocities are eliminated according to the 3-sigma-criterion because they are mainly due to local deformations. These adjustments are iteratively repeated. The results of the least squares adjustment are given in Table 1 in comparison with the NNR-NUVEL-1A.

The novel approach applied in the present paper is the homogeneous modelling of rigid plate motions, inter-plate and intra-plate crustal deformations. While previous models were directly using the observed station velocities, which are mostly located in the northern hemisphere and therefore represent in general the North American and European plate motions and deformations and depend directly on the selection of sites to be included, we apply here a uniformly distributed velocity field over the total Earth surface computed in a global grid by least square collocation. Only those station velocities are included in the plate parameter estimation, which do not get significant residuals (greater than three sigma) in the least squares adjustment, the other stations are considered as deformations (see below).

We see an agreement with less than 3 sigma difference between nearly all APKIM parameters of DTRF and ITRF, but a disagreement of the geodetic solutions with respect to nearly all NNR-NUVEL-1A parameters (c.f. e.g. Kreemer

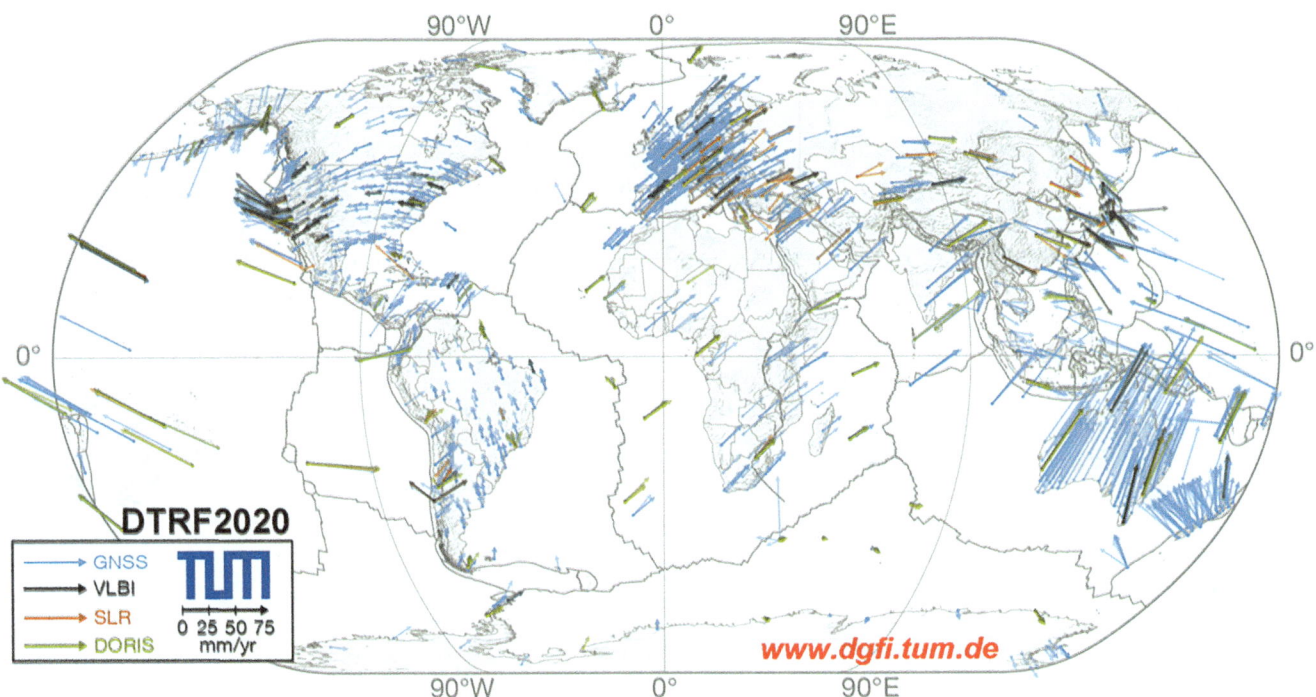

Fig. 3 Station velocities of the DTRF2020 (Seitz et al. 2023)

Table 1 Comparison of the DTRF and ITRF plate rotation parameters with the NNR-NUVEL-1A (*italic* numbers indicate differences greater 3 sigma w.r.t. DTRF2020; 1°/Ma = 3.6 mas/a)

Plate	APKIM2020 (DTRF2020)			APKIM2020 (ITRF2020)			NNR-NUVEL-1A		
	Φ [°]	Λ [°]	ω [°/Ma]	Φ [°]	Λ [°]	ω [°/Ma]	Φ [°]	Λ [°]	ω [°/Ma]
Africa	49.587±0.080	278.376±0.228	0.2645±0.0004	49.612±0.100	279.191±0.315	0.2652±0.0005	*50.57*	*286.04*	*0.291*
Antarctica	58.153±0.177	231.432±0.225	0.2074±0.0012	59.026±0.277	232.302±0.436	*0.2184±0.0022*	*62.99*	*244.24*	*0.238*
Arabia	49.240±0.293	8.333±0.760	0.6038±0.0116	49.044±0.335	8.487±0.968	0.6152±0.0124	*45.23*	*355.54*	*0.546*
Australia	32.270±0.033	38.115±0.072	0.6291±0.0003	*32.418±0.032*	37.890±0.074	*0.6311±0.0003*	*33.85*	*33.17*	*0.646*
Caribbean	31.934±0.777	266.419±2.042	0.3044±0.0172	31.850±0.778	267.531±1.865	0.3135±0.0170	*25.00*	266.99	*0.214*
Eurasia	55.844±0.137	261.457±0.232	0.2586±0.0006	55.580±0.149	261.019±0.248	0.2585±0.0006	*50.62*	*247.73*	*0.234*
India	51.304±0.246	5.748±2.399	0.5265±0.0068	51.570±0.199	6.815±2.038	0.5315±0.0059	*45.51*	0.34	*0.545*
N. America	−6.333±0.228	271.907±0.096	0.1874±0.0007	−7.172±0.281	272.23±0.114	0.1861±0.0008	*−2.43*	*274.10*	*0.207*
Nazca	43.199±0.960	257.764±0.419	0.6362±0.0091	45.347±1.618	257.229±0.708	0.6326±0.0171	*47.80*	*259.87*	*0.743*
Pacific	−62.687±0.049	110.271±0.199	0.6811±0.0005	−62.623±0.054	*111.362±0.237*	0.6796±0.0007	−63.04	*107.33*	*0.641*
S. America	−18.751±0.281	226.313±0.681	0.1196±0.0005	−19.117±0.302	228.012±0.689	0.1197±0.0005	*−25.35*	*235.58*	*0.116*

and Holt 2001; Altamimi et al. 2003, 2023b; Drewes 2009), This is a clear indication that we need a geodetic NNR realisation.

4 Continuous Intra-Plate and Inter-Plate Deformations

In addition to the rigid plate motions, we need to model the deformations within and between the plates, caused by environmental (e.g., climate, weather), geophysical (e.g., seismicity, tectonics) or man-made (e.g., withdrawal of fluids) effects. For this purpose, we subtract the estimated plate motion from the observed station velocities to compute a global deformation model. This is done by a least squares collocation approach (e.g., Drewes 2009; Steffen et al. 2022). In the following, we only present the results obtained for the DTRF, as the results for the ITRF are very similar.

The correlation between the observed velocities v_i, v_k at the (adjacent) geodetic stations i, k is determined under the stationarity condition over a domain defined by

$$C_{obs}(d_{ik}) = E\{v_i \cdot v_k\}, \qquad (2)$$

E is the statistical expectation and d_{ik} is the distance between i and k. A correlation matrix of the observed velocities (C_{obs}) is then set up from the empirical covariance functions in North (C_{NN}), East (C_{EE}) and between North and East

(C_{NE}) depending on the distance between the stations. The same is done for the correlations between observed and new velocities to be predicted (C_{new}). The covariance functions must fulfil the conditions of homogeneity (independence of the geographic position), isotropy (independence of the direction) and stationarity (independent of time); see e.g. Drewes (2009) and Steffen et al. (2022). This is done by removing the plate motions from the observed velocities. The C_{obs} values are classified in Δd_j class intervals and the respective cross-covariance $C_{obs}(\Delta d_j)$ and auto-covariance $C_{obs}(d=0) = C_0$ are determined using:

$$C_{obs}\left(\Delta d_j\right) = \frac{1}{n_j}\sum_{i<k}^{j} v_i \cdot v_k \ ;$$
$$C_{obs}(d=0) = C_0 = \frac{1}{n}\sum_{i=1}^{n} v_i^2, \qquad (3)$$

n stands for the number of stations available at the defined domain, while n_j represents the number of stations available at each class interval Δd_j. After estimating the discrete empirical covariance values with Eq. (3), they are approximated by a continuous function $C(d_{ik})$; which in this case is the exponential function:

$$C(d_{ik}) = a\, e^{-b\cdot d_{ik}}. \qquad (4)$$

The function parameters a and b are estimated by a least-squares adjustment. C_{obs} is symmetrical and its main diagonal ($i=k$) contains the values C_0. Fulfilling the stationarity condition, the elements of C_{new} are computed using the same Eq. (4) as a function of the distance between the grid node to be interpolated and the geodetic stations. The formula for the prediction is then:

$$\mathbf{v}_{pred} = \mathbf{C}_{new}^{T}\, \mathbf{C}_{obs}^{-1}\, \mathbf{v}_{obs} \qquad (5)$$

$\mathbf{v}_{pred}$ represents the velocities to be predicted at the continuous grid. $\mathbf{v}_{obs}$ are the observed velocities with the plate motions removed. The latter are shown for DTRF2020 in Fig. 4. The deformations are evident, but they are smaller than with respect to NNR-NUVEL-1A, see Fig. 2.

To get a homogeneous global distribution of the deformations for computing the common rotation and fulfilling the no-net-rotation condition, we interpolate a grid of points equally distributed over the entire Earth surface. The latitudinal distance between the grid points is $\Delta\varphi = 5°$ and the longitudinal distance is $\Delta\lambda = 5°/\cos\varphi$ rounded to integer parts of 360°. The result is shown in Fig. 5. W.r.t. Figure 4 we see a clear smoothening because very local deformations are disregarded. The precision of the interpolation is less than ±5 mm/a for 93% of the grid points, only in regions with strong deformation, the precision decreases to ±2 cm/a in North and ±1 cm/a in East direction. We then add the rigid plate motions computed from the estimated plate parameters (Table 1) and get the global velocity field over the complete Earth surface (Fig. 6).

From this global velocity field, we estimate one rotation vector representing the common rotation of the entire reference frame. The rotation of the DTRF2020 results in 0.125 ± 0.031 mas/a counterclockwise around a pole at latitude $\varphi = -23.689°$ and longitude $\lambda = 95.195°$. This

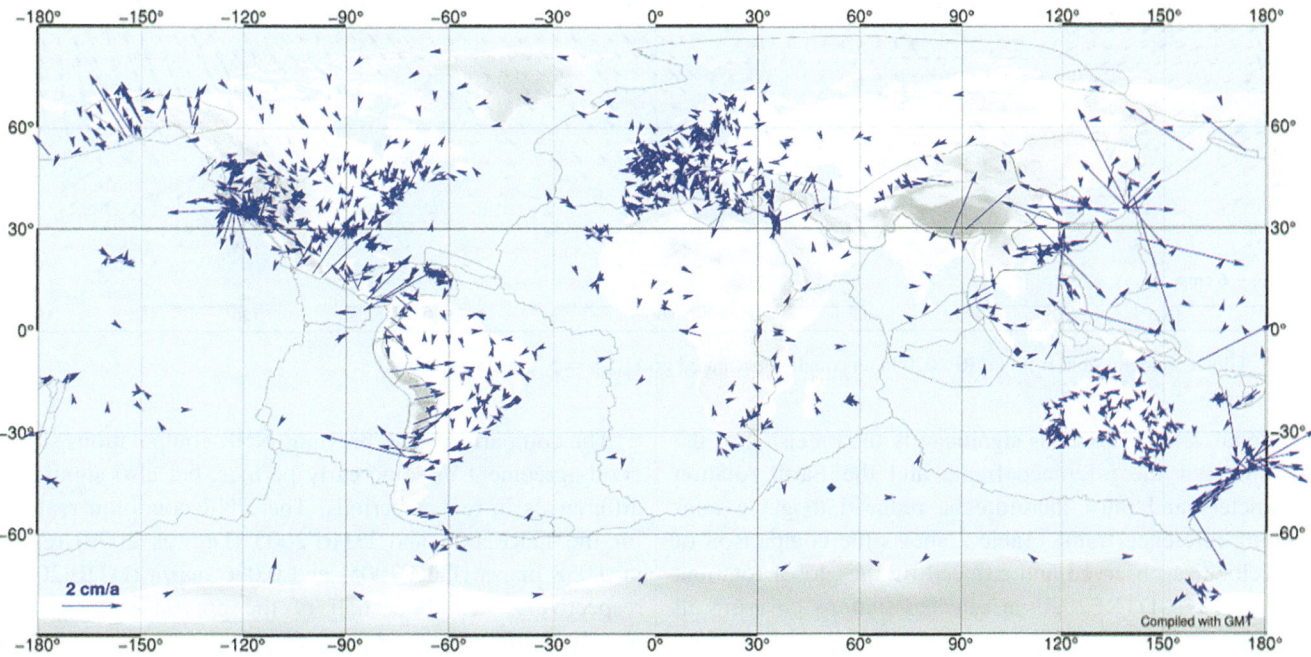

Fig. 4 DTRF2020 velocities with the plate motions removed (parameters given in Table 1)

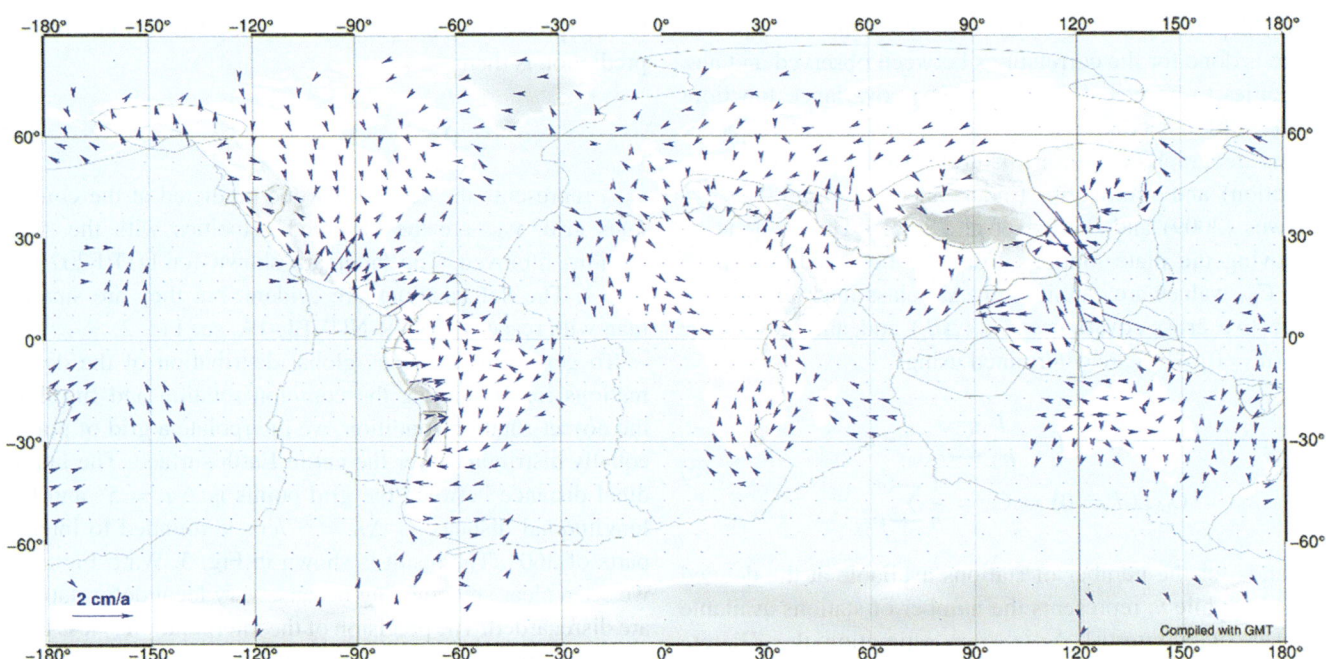

Fig. 5 DGFI2020 deformations in an equally distributed global grid from least squares collocation

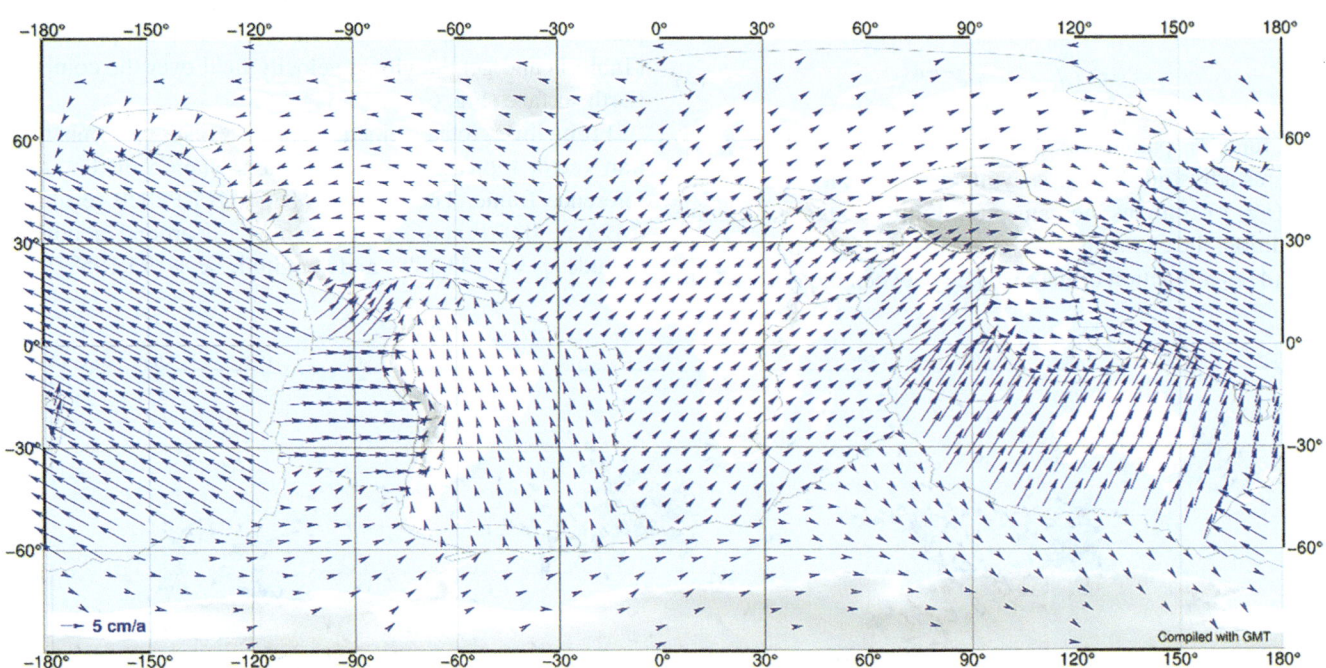

Fig. 6 Global velocity field from DTRF2020 in an equally distributed grid ($\Delta\varphi = 5°$, $\Delta\lambda = 5°/\cos\varphi$)

rotational velocity exceeds significantly the precision of the velocities of the reference frame and the Earth rotation parameters and must therefore be reduced to get a non-rotating reference frame. Table 2 shows the comparison of the velocities observed and reduced by the global rotation, i.e., the NNR-DTRF2020, at selected stations covering all the Earth surface. Figure 7 shows the NNR corrections, i.e., the differences between observed and reduced velocities.

The comparison with previous NNR computations shows good agreement in some early periods, but also significant differences in recent periods. The NNR condition realized for the ITRF2005 and DTRF2005 (Drewes 2009) results in 0.058 mas/a (ITRF2005) and 0.036 mas/a (DTRF2005), respectively, i.e. about half of the present value, which may be attributed to the later displacements caused by the magnitude 9 earthquakes in 2010 and 2011. Kreemer et al.

Table 2 Velocities observed and reduced for global rotation in some selected stations (NNR effect) (Techniques: D = DORIS, L = Laser ranging, P = GNSS, V = Very long baseline interferometry)

Domes no., Techn, Station name	Φ [deg.]	Λ [deg.]	DTRF observed v(N) [m/a]	v(E)	NNR-DTRF2020 v(N) [m/a]	v(E)	Diff. obs. − NNR v(N) [m/a]	v(E)
43001 D Thule	76.54	−68.82	0.0046	−0.0229	0.0056	−0.0259	−0.0010	0.0030
12360 P Tixi Seism.	71.63	128.87	−0.0116	0.0166	−0.0136	0.0199	0.0020	−0.0033
49204 P Missoula	46.93	−114.11	−0.0095	−0.0139	−0.0112	−0.0151	0.0017	0.0012
40169 P Shediac	46.22	−64.55	0.0082	−0.0155	0.0095	−0.0168	−0.0013	0.0013
19821 P Royan	45.64	−1.02	0.0159	0.0184	0.0194	0.0192	−0.0035	−0.0008
25603 L Baikonur	45.70	63.34	0.0046	0.0279	0.0065	0.0311	−0.0019	−0.0033
21611 P Changchun	43.79	125.44	−0.0128	0.0274	−0.0145	0.0306	0.0017	−0.0032
49979 P Hilo airport	19.73	−155.05	0.0347	−0.0629	0.0313	−0.0618	0.0034	−0.0011
40503 D Socorro I.	18.74	−110.95	0.0217	−0.0558	0.0202	−0.0554	0.0015	−0.0004
43201 R Ste. Croix	17.77	−64.58	0.0007	−0.0136	0.0019	−0.0131	−0.0012	−0.0005
39601 P Palmeira	16.73	−22.94	0.0156	0.0190	0.0187	0.0200	−0.0031	−0.0011
33812 P CGGN	10.12	9.12	0.0195	0.0246	0.0231	0.0262	−0.0036	−0.0016
25001 P Yibal	22.19	56.11	0.0312	0.0340	0.0335	0.0364	−0.0023	−0.0024
21904 P Bangkok	13.74	100.53	−0.0098	0.0243	−0.0101	0.0266	0.0003	−0.0023
50506 P Kwajalein	8.72	167.73	0.0297	−0.0694	0.0264	−0.0677	0.0033	−0.0017
92902 P Futuna	−14.31	−178.12	0.0342	−0.0599	0.0306	−0.0585	0.0036	−0.0014
92301 P Rikitea	−23.13	−134.96	0.0316	−0.0676	0.0289	−0.0653	0.0027	−0.0023
42005 P Santa Cruz	−0.74	−90.30	0.0107	0.0492	0.0104	0.0508	0.0003	−0.0016
41606 P Brasilia	−15.95	−47.88	0.0128	−0.0045	0.0149	−0.0022	−0.0021	−0.0023
30606 P St.-Helena	−15.94	−5.67	0.0183	0.0227	0.0217	0.0243	−0.0034	−0.0016
33402 P Zomba	−15.38	35.32	0.0165	0.0202	0.0196	0.0212	−0.0031	−0.0010
50127 P Coco Islnd	−12.19	96.83	0.0533	0.0433	0.0532	0.0441	0.0001	−0.0008
59984 P Larrimah	−15.57	133.21	0.0588	0.0346	0.0566	0.0353	0.0022	−0.0007
50253 P Owenga	−44.02	−176.37	0.0343	−0.0395	0.0307	−0.0385	0.0036	−0.0010
41715 P Coyhaique	−45.51	−71.89	0.0094	0.0015	0.0102	0.0050	−0.0008	−0.0035
42701 P King Edw.	−54.30	−36.51	0.0139	−0.0027	0.0166	0.0002	−0.0027	−0.0029
30313 D Marion Isl	−46.88	37.86	0.0010	0.0049	0.0039	0.0045	−0.0030	0.0004
91201 P Pt. Stanvac	−49.35	70.26	−0.0040	0.0048	−0.0025	0.0034	−0.0015	0.0014
50144 P BURNIE	−41.05	145.92	0.0561	0.0165	0.0533	0.0162	0.0028	0.0003
66018 D Belgrano	−77.87	−34.63	0.0126	0.0058	0.0153	0.0084	−0.0027	−0.0026
66061 P FLEMING	−77.53	160.27	−0.0121	0.0084	−0.0153	0.0072	0.0032	0.0012

(2006) modelled NNR velocities for the ITRF2000 from over 5,700 station velocities distributed over all continents by three-dimensional Cartesian rotation vectors (ω). The NNR rotations result in ω $\omega_x = 0.029$ mas/a, $\omega_y = -0.044$ mas/a and $\omega_z = 0.067$ mas/a, i.e. they are in the same order of magnitude. The updated version derived from IGS08 (Kreemer et al. 2014) comes up to 0.073 mas/a around a pole located at 41.6_N and 47.7 W. As only few stations are available in the oceans, one must consider that the continental plate motions and deformations are dominant. Altamimi et al. (2023b) compared the plate motion models derived from ITRF2014 and ITRF2020 and found a very small three-dimensional Cartesian rotation around the three axes X, Y, Z resulting in +3, −4 and +4 micro-arc-seconds. This indicates that the NNR vectors are mainly based on the inter-plate and intra-plate deformations (cf. Fig. 5).

Plate models inferred from station velocities depend on, among others, the extension of the geodetic observation time series, standards applied in the analysis of the geodetic data, the quality and quantity of geodetic stations and on regional or global effects affecting the linear motion of the stations. The actual plate kinematic and deformation model (APKIM) is therefore only valid for the time interval represented by the observations of the reference frames. They should be updated after any major update of the TRF solutions. This is of particular importance in regions with strong seismic activity, where the station velocities may considerably change after a strong earthquake. An example is the Circum-Pacific-Belt, where intensive activities are going on in East Asia and North and South America. An example of the changing velocities is shown in Fig. 8. We are computing velocity models for South America since 2003, and we are detecting the strong

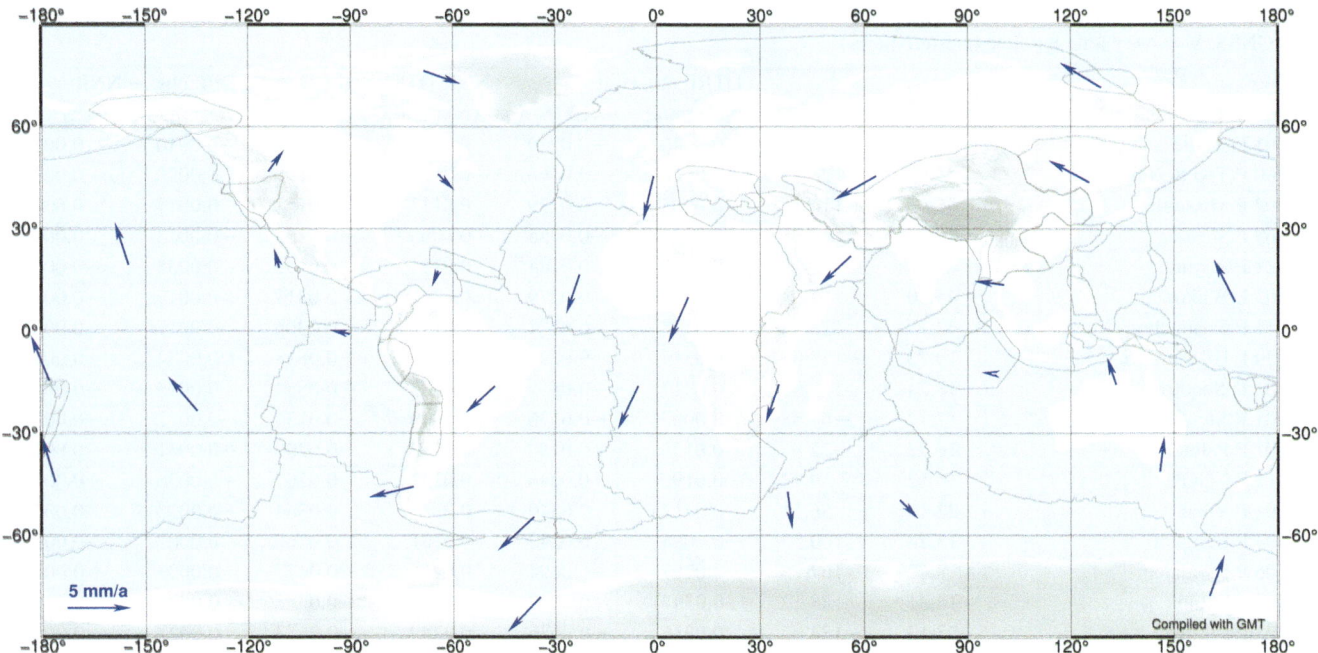

Fig. 7 DTRF NNR effect: Differences between velocities observed and reduced by global rotation

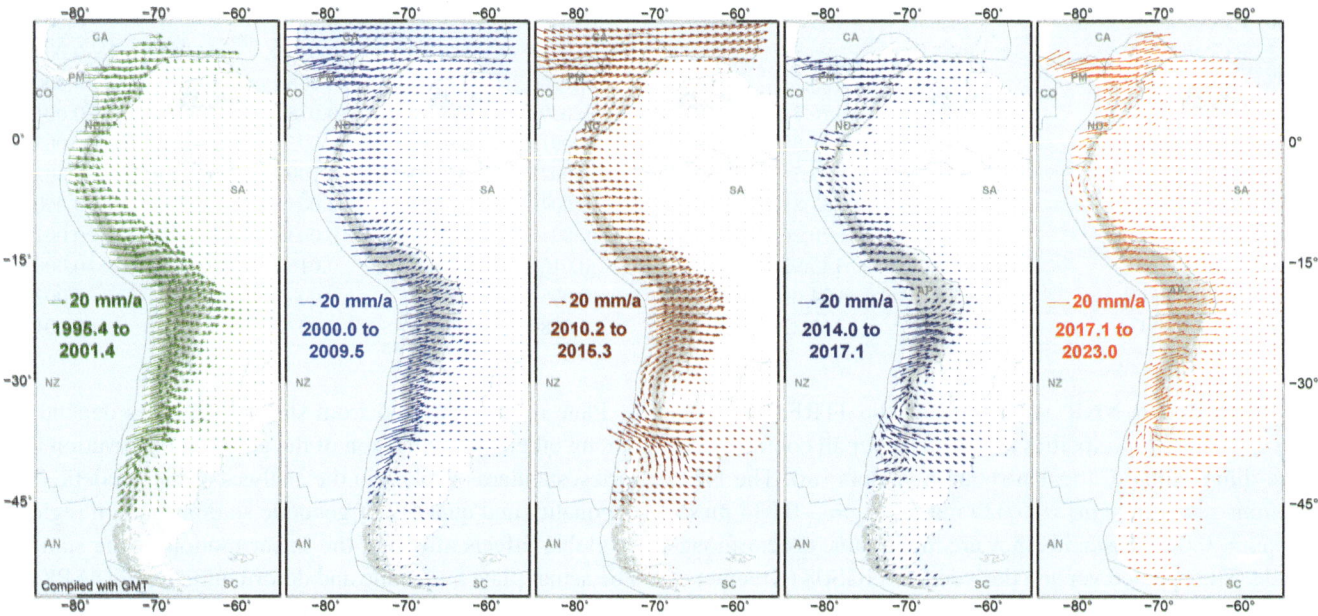

Fig. 8 Regional Velocity Model for South America (VEMOS) 1993.0–2002.0, 2000–2009.6, 2010.2–2015.2, 2014.0–2017.1, and 2017.1–2022.0; see e.g., Sánchez and Drewes (2020) and Sánchez et al. (2022)

effect of the seismicity in the Andean deformation zone. The frequent occurrence of strong earthquakes poses a problem in modelling linear station motions, as post-seismic effects are superimposed, and the parameters of the post-seismic functions show large differences.

5 Conclusions and Perspectives

Using the International Terrestrial Reference Frame for scientific purposes or precise applications by ITRF users requires a global no-net-rotation (NNR) model of surface velocities. Otherwise, the interpretation, e.g., of the EOP, is affected by the variation of the reference frame caused by irregular station motions. After the large earthquakes in Chile 2010 and Japan 2011 with surface displacements of several meters, there were publications on the changes in the EOP. In this case one must first compute a NNR reference frame over the complete Earth surface and then interpret the EOP. The accuracy of the current EOP estimates is at present below the NNR effect, but a trend in the long-term plate motions would be transferred to polar motion and UT1.

The ITRF may be used in geodesy as a basis for global studies and models, but not for representing the detailed regional crustal deformation, which requires frequent solutions of continental reference frames (e.g. AFREF, APREF, EUREF, NAREF, SIRGAS). Due to frequent changes of station velocities caused by geophysical and environmental events (e.g., seismicity, withdrawing of fluids) a frequent realization of the ITRF is required. The present repetition rate (at present every 6 years) is insufficient. One could think in a monthly ITRF, where the non-constant velocities $d\varphi/dt$, $d\lambda/dt$ may be replaced by monthly coordinate differences $\Delta\varphi/\Delta t$, $\Delta\lambda/\Delta t$ with respect to a defined epoch t (e.g., 2020.0).

References

Altamimi Z, Boucher C, Sillard P (2002) New trends for the realization of the International Terrestrial Reference System. Adv Space Res 30:175–184. https://doi.org/10.1016/S0273-1177(02)00282-X

Altamimi Z, Sillard P, Boucher C (2003) The impact of a no-net-rotation condition on ITRF2000. Geophys Res Lett 30(1–4):GL016279

Altamimi Z, Collilieux X, Legrand J, Garayt B, Boucher C (2007) ITRF2005: a new release of the International Terrestrial Reference Frame based on time series of station positions and earth orientation parameters. J Geophys Res 112(B09401):19 pp. https://doi.org/10.1029/2007JB004949

Altamimi Z, Métivier L, Collilieux X (2012) ITRF2008 plate motion model. J Geophys Res 117:B07402. https://doi.org/10.1029/2011JB008930

Altamimi Z, Rebischung P, Collilieux X, Métivier L, Chanard K (2023a) ITRF2020: an augmented reference frame refining the modeling of nonlinear station motions. J Geod 85(97):47. https://doi.org/10.1007/s00190-023-01738-w

Altamimi Z, Métivier L, Rebischung P, Collilieux X, Chanard K, Barnéoud J (2023b) ITRF2020 Plate Motion Model. Geophys Res Lett 50:e2023GL106373. https://doi.org/10.1029/2023GL106373

Argus DF, Gordon RG (1991) No-net-rotation model of current plate velocities incorporating plate motion model NUVEL-1. Geophys Res Lett 18:2039–2042

Bird P (2003) An updated digital model of plate boundaries, G^3. Geochem Geophys Geosyst 4(3):1027, 52 pp. https://doi.org/10.1029/2001GC000252

Boucher C, Altamimi Z, Duhem L (1992) ITRF 91 and its associated velocity field (IERS technical note; 12). Central Bureau of IERS - Observatoire de Paris, Paris, 143 p

Boucher C, Altamimi Z, Duhem L (1993) ITRF 92 and its associated velocity field (IERS technical note; 15). Central Bureau of IERS - Observatoire de Paris, Paris, iv, 164 p

Boucher C, Altamimi Z, Feissel M, Sillard P (1996) Results and analysis of the ITRF94. IERS technical note no. 20. https://www.iers.org/TN31

DeMets C, Gordon R, Argus DF, Stein S (1994) Effect of recent revisions to the geomagnetic reversal time scale on estimates of current plate motions. Geophys Res Lett 21:2191–2194. https://doi.org/10.1029/94GL02118

DeMets C, Gordon RG, Argus DF (2010) Geologically current plate motions. Geophys J Int 181:1–80. https://doi.org/10.1111/j.1365-246X.2009.04491.x

Drewes H (1982) A geodetic approach for the recovery of global kinematic plate parameters. Bull Géod (56):70–79

Drewes H (1990) Global plate motion parameters derived from actual space geodetic observations. IAG Symposia, No. 101, pp 30–37, Springer

Drewes H (2009) The actual plate kinematic and crustal deformation model (APKIM2005) as basis for a non-rotating ITRF. IAG Symposia, Springer, vol 134, pp 95–99. https://doi.org/10.1007/978-3-642-00860-3_15

Kreemer C, Holt WE (2001) A no-net-rotation model of present-day surface motions. Geophys Res Lett 28:4407–4410

Kreemer C, Lavallée DA, Blewitt G, Holt WE (2006) On the stability of a geodetic no-net-rotation frame and its implication for the International Terrestrial Reference Frame. Geophys Res Lett 33(17):5 pp. https://doi.org/10.1029/2006GL027058

Kreemer C, Blewitt G, Klein EC (2014) A geodetic plate motion and global strain rate model, G3. Geochem Geophys Geosyst 15:3849–3889. https://doi.org/10.1002/2014GC005407

Petit G, Luzum B (2010) IERS Conventions (2010), IERS technical note no. 36, Frankfurt am Main, 179 pp.

Sánchez L, Drewes H (2020) Geodetic monitoring of the variable surface deformation in Latin America. IAG Symposia. https://link.springer.com/chapter/10.1007/1345_2020_91. https://doi.org/10.1007/1345_2020_91

Sánchez L, Drewes H, Kehm A, Seitz M (2022) SIRGAS reference frame analysis at DGFI–TUM. J Geodetic Sci 12:92–119. https://doi.org/10.1515/jogs-2022-0138

Seitz M, Bloßfeld M, Angermann D, Glomsda M, Rudenko S, Zeitlhöfler J, Seitz F (2023) DTRF2020: ITRS 2020 realization of DGFI-TUM (data). Zenodo. https://doi.org/10.5281/zenodo.8220524

Steffen R, Legrand J, Agren J, Steffen H, Lidberg M (2022) HV-LSC-ex2: velocity field interpolation using extended least-squares collocation. J Geod 96:15. https://doi.org/10.1007/s00190-022-01601-4

Open Access This chapter is licensed under the terms of the Creative Commons Attribution 4.0 International License (http://creativecommons.org/licenses/by/4.0/), which permits use, sharing, adaptation, distribution and reproduction in any medium or format, as long as you give appropriate credit to the original author(s) and the source, provide a link to the Creative Commons license and indicate if changes were made.

The images or other third party material in this chapter are included in the chapter's Creative Commons license, unless indicated otherwise in a credit line to the material. If material is not included in the chapter's Creative Commons license and your intended use is not permitted by statutory regulation or exceeds the permitted use, you will need to obtain permission directly from the copyright holder.

A Functional Model for Quantifying Deformation in Reference Frame Transformations

Richard Stanaway, Chris Crook, Kevin M. Kelly, and Roger Lott

Abstract

IAG Commission 1 Working Group 1.3.1 in association with the Open Geospatial Consortium (OGC) have developed a functional model for crustal deformation (FMCD) and an associated Geodetic Grid Exchange Format (GGXF) for quantifying and disseminating deformation information for use in time-dependent reference frame transformations.

The FMCD provides a framework within which producers and users of deformation models can describe crustal displacement and velocity data using robust grid formats such as GGXF. Using the FMCD and GGXF combined, positional displacements can be readily applied in point motion coordinate operations and non-conformal time-dependent transformations. This approach is essential in deforming zones where conformal time-dependent transformation approaches do not adequately handle crustal deformation.

This paper describes application of the FMCD in typical cases including: (1) transformation of GNSS PPP positions (e.g. in an IGS20 frame) to a national geodetic datum in a deforming zone and (2) transformation between reference frames across earthquake events that resulted in significant coseismic and postseismic crustal displacement. The FMCD and associated GGXF provide a framework for developers of geodetic software such as those used in GIS, GNSS processing and positioning to better handle complex deformation.

Keywords

Deformation model · GGXF · Reference frame · Time-dependent transformation

1 Introduction

IAG Commission 1 Working Group 1.3.1 *"Time-dependent transformations between reference frames in deforming zones"* has been working in close collaboration with the Open Geospatial Consortium (OGC) Coordinate Reference System (CRS) Domain and Standards working groups (DWG and SWG) to develop specifications and standards for a functional model of crustal deformation (FMCD) and associated geodetic grid exchange format (GGXF) (Crook et al. 2024) for dissemination of deformation models used in geodetic applications such as GNSS positioning and GIS. This work has been strongly motivated by the need to align precise GNSS positioning in global geodetic reference frames such as ITRF with existing spatial data sets defined in an earlier epoch of ITRF or other defined coordinate reference system (CRS). Currently, GNSS positioning in a global frame across many plate boundary deforming zones is localised to a working CRS using *ad-hoc* site transformations or other vendor-specific and proprietary positioning

R. Stanaway (✉)
Department of Surveying and Land Studies, Papua New Guinea University of Technology, Lae, Papua New Guinea
e-mail: richard.stanaway@pnguot.ac.pg

C. Crook
Toitū Te Whenua Land Information New Zealand, Wellington, New Zealand

K. M. Kelly
Environmental Systems Research Institute, Redlands, CA, USA

R. Lott
International Association of Oil and Gas Producers, London, UK

calibration techniques. In these settings, misalignment of GNSS positioning and existing spatial data is undesirable but also often inevitable if a site transformation is not applied or not robustly estimated and validated. Furthermore, vendor-specific site transformations are not generally supported or easily configured by other vendors, GNSS analysis or GIS software. There has been a rapid uptake in GNSS precise point positioning (PPP) and the improved precision and latency has necessitated a better approach for coordinate transformations in deforming zones. At the same time, in more seismically active regions a more rigorous strategy is also required for integration of spatial data acquired over multiple seismic events. A typical example is the need to integrate post-earthquake recovery surveys with pre-earthquake digital cadastral data, engineering designs, sub-surface utility surveys and Building Information Models (BIM). Within stable plate settings, ITRF positions can be transformed to different epochs using plate motion models (PMM) and time-dependent conformal transformations such as a 14-parameter model with high precision. However, when rigidity of the underlying plate is assumed in plate boundary zones and regions affected by glacial isostatic adjustment (GIA), interseismic strain degrades the performance of a conformal transformation. In these settings, a grid-based time-dependent transformation model can be used to estimate displacement of the crust between different epochs. Crustal velocities can be estimated by interpolation of a velocity model grid and episodic displacements resulting from seismic activity can be represented as both coseismic displacement and postseismic parametric grids for each event. These grids enable transformation of positions across seismic epochs to support alignment of positioning and spatial data at any valid epoch within deforming zones. The use of coseismic grids is also advantageous within stable plate regions in the event of intraplate earthquakes or other deformation events.

In deforming zones, deformation models including velocity grids have already been implemented by many geodetic agencies to support positioning applications and geodetic datum maintenance. Examples include:

- The use of Horizontal Time-Dependent Positioning (HTDP) for reference frame transformations in plate boundary regions of the United States of America since 1992 (Snay 1999; Pearson and Snay 2013; Snay et al. 2016)
- Incorporation of a deformation model comprised of velocity, coseismic and postseismic displacement grids in New Zealand's national geodetic reference frame (NZGD2000) since its inception (Grant and Blick 1998; Beavan and Haines 1998; Crook et al. 2016).
- Use of a velocity grid within Canada's NRCan-PPP service since 2011 (Craymer et al. 2011; Robin et al. 2020)
- Implementation of a composite interseismic and postseismic velocity grid (POS2JGD) in Japan. This is updated every 3 months to account for the decay of the ongoing significant postseismic deformation from the 2011 Tohoku earthquake and other large events (Yamashita et al. 2022).
- Implementation of the Nordic Geodetic Commission (NKG) time-dependent transformation model (currently NKG2020) for transformations from ITRF to ETRF within the Nordic countries also accounting for glacial isostatic adjustment (Häkli et al. 2023)

The proliferation of different deformation model formats and transformation workflows by different geodetic agencies and geodetic surveyors has been an impediment to the standardisation of models and development of software in geodetic positioning and GIS applications to support these different formats and approaches.

This paper summarises the recent work of IAG WG 1.3.1 in close collaboration with OGC. The WG has aimed to provide a scientifically robust yet workable strategy for effective implementation of time-dependent transformations in deforming zones. The different types of reference frames used in practice are also summarised with typical scenarios of application.

2 Classification of Reference Frames in Practice

The application of a deformation model for RF coordinate transformations varies according to how the RF are classified. To broadly summarise, three main types of geodetic reference frames are used in practice: kinematic, semi-kinematic and static. These are characterised by the way the reference frames are defined, which in turn defines how coordinates of objects fixed to the earth's deforming crust are represented.

- A kinematic reference frame such as ITRF is characterised by continuously changing coordinates with respect to a time-invariant terrestrial reference system. Such a frame reflects geodynamic phenomena such as plate motion, GIA, subsidence and surface creep.
- Kinematic frames can be further classified as having either an absolute no-net-rotation (NNR) condition (including ITRF and similarly aligned GNSS reference frames) or, be referenced to a stable plate with a rotation rate implicitly defined as zero (e.g. the ETRS/ETRF in Europe).

The latter approach is known as a plate-fixed reference frame (PFRF) with interseismic velocities close to zero within the stable portion of the plate, but becoming (often significantly) non-zero in the plate boundary zone, or areas affected by

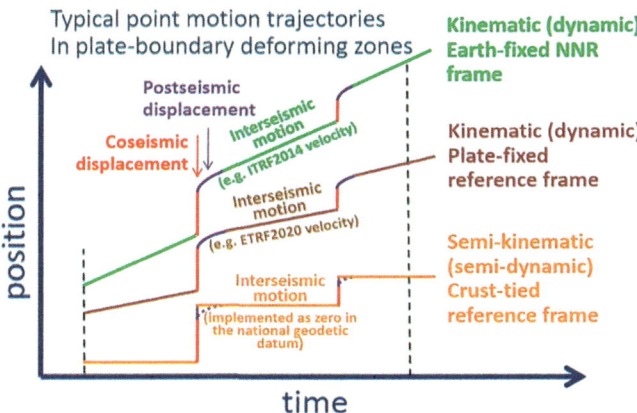

Fig. 1 Typical point motion trajectories in deforming zones for different classifications of reference frames

intraplate deformation such as GIA. In contrast, a semi-kinematic reference frame maintains continuous coordinate stability at a specified reference epoch (with the exception of episodic deformation events and geodetic adjustments) and incorporates a deformation model to estimate displacement (strain) within the geodetic network over time. The characterisation of these different types of reference frames in terms of point motion trajectories is illustrated in Fig. 1.

The coordinates in a semi-kinematic reference frame are not necessarily fixed at the frame reference epoch if there is subsequent displacement due to seismic activity which is mapped back to the (earlier) reference epoch. The coordinate system is fixed to the interseismic crust and can also be referred to as a crust-tied reference frame (CTRF) to distinguish it from a PFRF where coordinates do change in the plate boundary zone as a result of interseismic strain. In this regard, the CTRF approach has some merit as it has no requirement for a stable plate to be referenced and is therefore useful in countries or regions that lie within broad or complex deformation zones including microplates and semi-rigid crustal blocks located between adjacent larger plates; examples are: New Zealand, Japan, Greece, Indonesia and Papua New Guinea. The key benefit for users of the semi-kinematic or CTRF approach is that coordinate stability for "ground-fixed" features such as property boundary corners in a digital cadastre are maintained during the interseismic period. The longevity of CTRF is limited by a trade-off between dimensional tolerances (derived from CTRF coordinates fixed at the reference epoch) and interseismic strain accumulation after the reference epoch. The requirement for updates is driven by the highest tolerances expected of any end user application (e.g., civil engineering and construction projects) (Stanaway et al. 2012) or legal requirements. In areas with high strain rates, the reference epoch and associated coordinates of a semi-kinematic reference frame may require regular updates to account for interseismic strain

where certain derived dimensional tolerances are exceeded by holding coordinates fixed at a reference epoch. In any case, end users requiring high precision relative positioning will still need to account for deformation of the coordinate system.

A static reference frame does not generally allow changes in the datum definition, although redefinition of the datum is often warranted after significant deformation events (e.g., earthquakes, mining subsidence).

3 Transformation Approaches Between Different Classifications of Reference Frames

Transformation of coordinates between static reference frames has been typically achieved by conformal parametric transformation methods. Many earlier reference frames have known distortions (the differences between tabulated and high precision observations for higher order points in the frame). A grid transformation such as the widely used NTv2 (NRCan 1995) has been used for higher precision transformations either where the source or target (or both) reference frames have known and modelled distortions or where there is unmodelled deformation of the earlier datum (e.g. the NZGD49 to NZGD2000 transformation).

Development of kinematic reference frames has meant the need for the static conformal transformations to be augmented with rate-of-change parameters for each of the static parameters (e.g., a static 7 parameter transformation becomes a time-dependent 14 parameter transformation). The reference epoch of the static parameters in a time-dependent transformation is sometimes considered to be an additional (15th) parameter. Time-dependent conformal transformations assume uniform changes in scale, translation and rotation between the reference frames involved. In stable plate settings, the rotation of a plate-fixed reference frame within a global NNR frame is adequately modelled by this approach. In the case where there is no translation or scale change between the two frames, the plate motion model parameters in a geodetic Cartesian system can be directly mapped into a 14 parameter transformation as rotation rate parameters with zeros for all of the other parameters (Stanaway et al. 2017). This approach has been adopted for transformations between ITRF2014 and Australia's GDA2020 geodetic datum (ICSM 2021).

In regions characterised by significant crustal deformation, such conformal transformation methodologies are not applicable for higher precision transformations and instead grid-based velocity and coseismic displacement models have been widely used as mentioned previously.

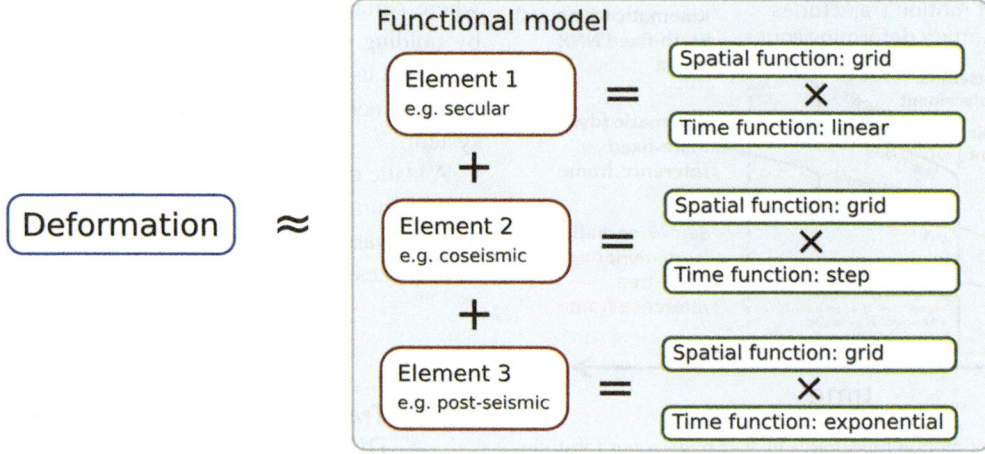

Fig. 2 High level structure of the deformation model function model (FMCD)

4 Functional Model for Crustal Deformation

The functional model for crustal deformation (FMCD) developed by the OGC and IAG WG provides a logical framework for different elements (sub-models) of a deformation model to be selected and combined for time-dependent coordinate transformations to account for different geodynamic phenomena. The total deformation is represented as the sum of the model elements each of which is defined by a spatial representation of displacement that is scaled by a time function (Fig. 2). This estimates the displacement within the temporal and spatial limits of the model, which supports the transformation of points between epochs within a kinematic RF.

How the different deformation elements are combined for coordinate transformations between different classifications of RF is illustrated in Fig. 3. In this diagram the trajectory of a point in a kinematic RF (green line) is related to that of the same point's trajectory in a crust-tied RF (indicated by the solid orange line), showing the effect of a coseismic step in the trajectory.

Many different deformation elements are supported in the standard functional model for crustal deformation including:

- secular displacement (grid model with a linear time function to describe the velocity)
- coseismic displacement (grid model with a step time function)
- postseismic deformation (grid model with an exponential or logarithmic time function).

Other time functions including ramp, acceleration, hyperbolic tangent and cyclic time functions to model other episodic geodynamic phenomena (e.g., major earthquakes,

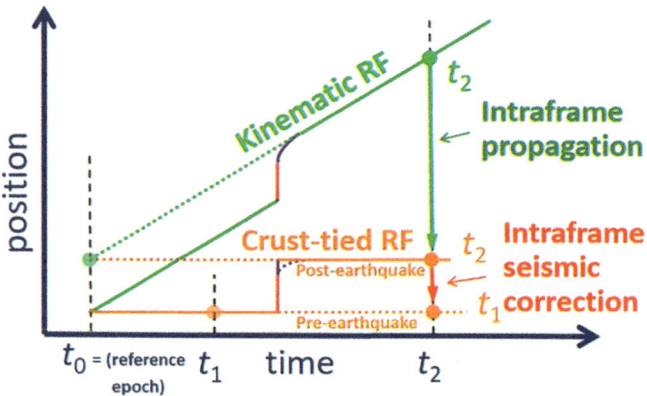

Fig. 3 Application of a deformation model for intraframe propagation and transformation between different classifications of RF showing typical point trajectories for a common point in both types of RF

velocity changes, slow-slip events and periodic deformation) are also supported.

The spatial function defines the displacement at the nodes of a grid. The displacement at other points within the grid is interpolated between these, typically using a bilinear interpolation method. The model supports nesting of grids with different data resolution to improve spatial accuracy where the deformation field is more complex (such as close to active faulting) or where there is a need for great accuracy (for example in built environments and urban areas). This aspect is particularly important in the vicinity of displaced faults where interpolation of a coarse resolution grid can result in unreliable estimation of displacement.

Other representations of deformation, such as irregular triangulated structures, were considered as alternative ways to model the spatially varying complexity of the deformation model. These representations were not included in the current version of the specification as they are more complex and not currently used by geodetic agencies.

Table 1 Time-dependent transformation scenarios between different classifications of reference frames. Velocity grids may be defined in either plate-fixed or absolute-NNR reference frames

Target RF→ Source RF↓	Kinematic NNR RF	Plate-fixed RF	Crust-tied ("static") RF post-earthquake(s)	Crust-tied ("static") RF pre-earthquake(s)
Kinematic NNR RF	Null[a]	PMM/14 par.[a]	FMCD[b] velocity grid	FMCD[b] velocity grid FMCD coseismic grid(s)
Plate-fixed RF	PMM/14 par.[a]	Null	FMCD intraplate velocity grid	FMCD intraplate velocity grid FMCD coseismic grid(s)
Crust-tied ("static") RF post-earthquake(s)	FMCD velocity grid	FMCD intraplate velocity grid	Null	FMCD coseismic grid(s)
Crust-tied ("static") RF Pre-earthquake(s)	FMCD[b] velocity grid FMCD coseismic grid(s)	FMCD[b] intraplate velocity grid FMCD coseismic grid(s)	FMCD coseismic grid(s)	Null

[a]At same transformation epoch—velocity grid required for transformation between different epochs
[b]Where the crust-tied RF is realised by a plate-fixed RF at a defined reference epoch (e.g. ETRF), the recommended approach is a concatenated transformation (14-parameter or PMM transformation followed by an intraplate velocity transformation)

The deformation model also can include gridded values for uncertainty of displacements that may be used to estimate the uncertainties of coordinates derived using the deformation model.

The FMCD describes how deformation is modelled. The associated GGXF standard describes how this is encoded for publication. It supports both a text based and binary format for encoding.

5 FMCD Transformation Scenarios

The way the FMCD time function is applied for a coordinate transformation depends on the classification of the source and target reference frames (Table 1).

For example, transformation of spatial data across an earthquake event within a CTRF would only require application of coseismic displacement and postseismic grids as the secular interseismic movement of the crust is not applicable. By contrast, transforming an ITRF aligned position from PPP (e.g., IGS20) to a local pre-earthquake reference frame requires both the velocity model time function and the coseismic displacement step (in reverse). If the post-earthquake position was required, only the velocity model would be applied. Transformation between an absolute NNR reference frame and plate-fixed reference frame to a different epoch is by a 14 parameter or PMM transformation and a supplementary velocity model transformation if the PFRF velocity is non-zero. Such a transformation may be achieved in a single step if the velocity model definition is common to both frames, This single-step approach is not rigorous, particularly in the case of a coarse velocity grid generated from a high rotation rate PMM which can result in interpolation errors. The 14 parameter and PMM transformation methods are described in more detail in existing literature.

6 Conclusions and Future Development

The FMCD provides a generic way of representing surface deformation for use in coordinate transformations. It supports both secular velocity and episodic deformation in a simple framework, and together with GGXF allows geodetic agencies to publish deformation models in a common format. Both the FMCD and the GGXF specifications have recently been adopted as OGC standards. The standards documents can be accessed on Github (OGC 2024a, b). It is envisaged that development of equivalent ISO standards may proceed in the future.

References

Beavan J, Haines A (1998) Revised horizontal velocity model for the New Zealand geodetic datum. Client Report 43865B. Institute of Geological and Nuclear Sciences, Lower Hutt

Craymer M, Henton J, Piraszewksi M, Lapelle E (2011) An updated GPS velocity field for Canada. AGU fall meeting. American Geophysical Union, San Francisco

Crook C, Donnelly N, Beavan J, Pearson C (2016) From geophysics to geodetic datum: updating the NZGD2000 deformation model. N Z J Geol Geophys 59(1):22–32. https://doi.org/10.1080/00288306.2015.1100641

Crook C, Kelly K, Lott R, Stanaway R (2024) The GGXF standard file format for gridded geodetic data. IAG Symposium, Berlin 2023. https://doi.org/10.1007/1345_2024_254

Grant D, Blick G (1998) A new geocentric datum for New Zealand (NZGD2000). N Z Surv 288:40–42

Häkli P, Evers K, Jivall L, Nilsson T, Himle S, Kollo K, Liepiņš I, Paršeliūnas E, Vestøl O, Lidberg M (2023) NKG2020 transformation: an updated transformation between dynamic and static reference frames in the Nordic and Baltic countries. J Geodetic Sci 13(1):20220155. https://doi.org/10.1515/jogs-2022-0155

Intergovernmental Committee on Surveying and Mapping (ICSM) (2021) Geocentric Datum of Australia 2020 technical manual, version 1.7

NRCan (1995) NTv2 National Transformation Version 2, geomatics Canada (now NrCan), September 1995

OGC (2024a) Abstract specification topic 24 - Functional Model for Crustal Deformation v1.0. Open Geospatial Consortium, 29 April 2024. https://docs.ogc.org/as/22-010r4/22-010r4.html. https://doi.org/10.62973/22-010r4. Accessed 6 June 2024

OGC (2024b) The GGXF geodetic data grid exhange format v1.0. Open Geospatial Consortium, 29 April 2024. https://docs.ogc.org/is/22-051r7/22-051r7.pdf. https://doi.org/10.62973/22-051r7. Accessed 6 June 2024

Pearson C, Snay R (2013) Introducing HTDP 3.1 to transform coordinates across time and spatial reference frames, GPS solutions. https://doi.org/10.1007/s10291-012-0255-y

Robin CMI, Craymer M, Ferland R, James TS, Lapelle E, Piraszewski M, Zhao Y (2020) NAD83v70VG: a new national crustal velocity model for Canada; geomatics Canada, open file 0062, zip file. https://doi.org/10.4095/327592

Snay R (1999) Using the HTDP software to transform spatial coordinates across time and between reference frames by Richard A. Snay. Surv Land Inf Syst 59(1):15–25

Snay R, Freymueller J, Craymer M, Pearson C, Saleh J (2016) Modeling 3-D crustal velocities in the United States and Canada. J Geophys Res Solid Earth 121:5365–5388

Stanaway R, Roberts C, Blick G, Crook C (2012) Four dimensional deformation modelling, the link between international, regional and local reference frames. In Proceedings of FIG working week 2012, Rome, 6–10 May

Stanaway R, Roberts CA, Rizos C, Donnelly N, Crook C, Haasdyk J (2017) Defining a local reference frame using a plate motion model and deformation model. In: Proceedings of IAG Commission 1 Symposium 2014: Reference frames for applications in geosciences (REFAG2014), 13–17 October 2014, Kirchberg, Luxembourg. Springer, Berlin

Yamashita T, Marvit K, Tanaka M, Kagawa A (2022) Crustal deformation model to support high-level utilization of PPP - POS2JGD and its future perspectives. In: Proceedings from Japan Geoscience Union Meeting 2022

Open Access This chapter is licensed under the terms of the Creative Commons Attribution 4.0 International License (http://creativecommons.org/licenses/by/4.0/), which permits use, sharing, adaptation, distribution and reproduction in any medium or format, as long as you give appropriate credit to the original author(s) and the source, provide a link to the Creative Commons license and indicate if changes were made.

The images or other third party material in this chapter are included in the chapter's Creative Commons license, unless indicated otherwise in a credit line to the material. If material is not included in the chapter's Creative Commons license and your intended use is not permitted by statutory regulation or exceeds the permitted use, you will need to obtain permission directly from the copyright holder.

Combined Global GNSS Velocity Field

A. Santamaría-Gómez, R. Rietbroek, P. Rebischung, T. Frederikse, and J. Legrand

Abstract

A global combined GNSS velocity field with almost 13,400 sites has been derived by the International Association of Geodesy's Joint Working Group 3.2. The combined field is aligned to the ITRF2020 and gathers global and regional velocity fields computed by nineteen groups using different approaches. In addition to the combined velocities and their uncertainties, the combination also provides the alignment of each velocity field to the ITRF2020, the scaling of their velocity uncertainty and the estimated repeatability of the velocity estimates across the different groups at almost 3,000 sites. The median repeatability is at the level of 0.17 and 0.27 mm/yr for the horizontal and vertical velocities. Up to 11 % of the sites show poor velocity repeatability exceeding 3 times the median values.

Keywords

Crustal motion · Geodynamics · GNSS velocities · Reference frame

Supplementary Information The online version contains supplementary material available at https://doi.org/10.1007/1345_2024_263.

A. Santamaría-Gómez (✉)
Géosciences Environnement Toulouse, Université Paul Sabatier, CNES, CNRS, IRD, UPS, Toulouse, France
e-mail: alvaro.santamaria@get.omp.eu

R. Rietbroek
ITC Faculty of Geo-information Science and Earth Observation, Department of Water Resources, Enschede, The Netherlands

P. Rebischung
Université Paris Cité, Institut de physique du globe de Paris, CNRS, IGN, Paris, France

Univ Gustave Eiffel, ENSG, IGN, Paris, France

T. Frederikse
Jet Propulsion Laboratory, California Institute of Technology, Pasadena, CA, USA

Planet Labs PBC, San Francisco, CA, USA

J. Legrand
Royal Observatory of Belgium, Brussels, Belgium

1 Introduction

GNSS velocities estimated by different groups differ due to several factors including the different strategies to compute the station positions from the raw GNSS observations, the completeness of the time series, their level of noise, the removed position discontinuities, and the alignment to a terrestrial reference frame. Among these factors, the detection and handling of position discontinuities that populate the GNSS time series has probably the biggest impact on the velocity estimates (Williams 2003; Griffiths and Ray 2016). Even when using the same GNSS position time series, it is common for different analysts to provide different velocity estimates and uncertainties, mainly due to the different choices of handling position discontinuities (Gazeaux et al. 2013).

Position discontinuities can be classified in four groups. The first group includes apparent position discontinuities generated when the GNSS antennas are replaced. They are an indication that the phase center corrections obtained from antenna calibrations do not reflect in general the actual antenna radiation pattern on the field, creating an apparent position change when antennas are replaced (Wanninger 2009). The second group includes a wide range of situations, often poorly understood, related to changes of the antenna environment, the signal tracking or the settings of the receiver's frontend. The third group includes those situations where the position discontinuity is generated when estimating the station positions from the raw GNSS observations, for instance due to an incorrect use of the station's metadata: outdated antenna calibration, wrong antenna model, wrong antenna orientation or wrong antenna eccentricity. Discontinuities from the previous three groups can introduce changes in the estimated station position that do not reflect any position change of the station's benchmark. The last fourth group of discontinuities are generated by earthquakes, which can offset the position of the station's benchmark if estimated in a conventional coordinate frame that is not affected by the earthquake itself. Not all the antenna changes, nor all the earthquakes, will introduce a significant position discontinuity. This fact, and the potential occurrences of discontinuities from the second and third groups, implies that different analysts will remove different sets of position discontinuities, even if they use exactly the same series (Gazeaux et al. 2013).

The International Association of Geodesy's Joint Working Group 3.2 "Global combined GNSS velocity field" (2019–2023) was created to support the scientific community using GNSS velocities in fields such as tectonics, sea-level change, land subsidence and global isostatic adjustment (GIA) modelling. The objective of this JWG is to combine and compare the available global and regional GNSS velocity fields obtained by different groups from both network and precise point positioning (PPP) solutions.

This contribution presents the combined global GNSS velocity field, the methodology followed to combine the input velocity fields and the results of their comparison.

2 Input GNSS Velocity Fields

Nineteen GNSS velocity fields have been gathered and used in the combination (Table 1). The geographical distribution of each velocity field is shown in Figure 1 of the supplemental material.

The input velocity fields considered for the combination include:

- The International Terrestrial Reference Frame 2020 velocity field (Altamimi et al. 2023), which is the velocity field used as datum to constrain the origin, orientation and scale

Table 1 List of the input velocity fields with the number of sites retained in the combination, the geographical extension and the file format

Id	#sites	Coverage	Format
APREF	720	Global	Table
CWU	2102	Central & North America	Table
EOST	986	Europe	Table
EPND	2528	Europe	Table
EUREF	227	Europe	SINEX
INGV	594	Europe/Africa	Table
ITRF2020	881	Global	SINEX
JPL	2434	Global	Table
LTK	581	Europe/Africa	SINEX
NCL	697	Global	Table
NGL	11546	Global	Table
NGS	1748	Global	Table
NMT	633	Central & North America	Table
NRCAN	593	North America	SINEX
PBO	2233	Central & North America	Table
SIRGAS	152	Central & South America	SINEX
SOPAC	978	Global	Table
UGA	547	Europe	Table
ULR	503	Global	Table

  of the combined velocity field, i.e., no global drift exists between the ITRF2020 and the combined velocity field. This velocity field is available at https://itrf.ign.fr/ftp/pub/itrf/itrf2020/ITRF2020-IGS-TRF.SNX.gz (Dec 6 2021)
- Solutions from several IAG regional commissions, like EUREF (Legrand 2022), available at https://epncb.oma.be/ftp/station/coord/EPN/EPN_IGb14_C2235.SNX.Z (Oct 13 2023); APREF (John Dawson, personal communication, Jan 31 2022) and SIRGAS (Sánchez and Drewes 2020), available at https://hs.pangaea.de/model/VEMOS2017/SIR17P01.zip (Oct 13 2023). The velocity field of the NAREF commission was not finalized in time, but its main components were included in the combination, namely the velocity fields from NGS (MYCS2, Phillip McFarland, personal communication, May 9 2022) and NRCAN (CBN, Michael Craymer, personal communication, May 5 2022).
- Several global solutions such as the JPL (repro2018a, Heflin et al. 2020), available at https://sideshow.jpl.nasa.gov/post/tables/table2.html (Jan 4 2022); NCL (NCL20, Vardić et al. 2022), available at https://doi.pangaea.de/10.1594/PANGAEA.935079 (May 5 2022); NGL (MIDAS5, Blewitt et al. 2018), available at http://geodesy.unr.edu/velocities/midas.IGS14.txt (Oct 14 2023); SOPAC/ESESES (Bock et al. 2021), available at https://cddis.nasa.gov/archive/GPS_Explorer/archive/time_series/V1/2022/GLB_Clean_TrendNeuTimeSeries_sopac_20220117.tar.gz (Jan 17 2022); and ULR (ULR7a, Gravelle et al. 2023), available at https://www.sonel.org/IMG/txt/ulr7_vertical_velocities.txt (Oct 20 2023).

- Several regional solutions, mostly in Europe and North America, such as the EOST (SPOTGINS, Michel et al. 2021; personal communication, Jan 7 2022), EPND (D2200, Kenyeres et al. 2019), available at https://epnd.sgo-penc.hu/downloads/D2200/EPND_D2200_IGS14NEU.VEL (Oct 14 2023); the EPOS solutions from INGV (v2.0, INGV RING 2016), LTK, and UGA (Socquet and Janex 2019), available at https://gnssproducts.epos.ubi.pt/filemanager (Jan 20 2022); the EarthScope solutions (Herring et al. 2016) from CWU, NMT and PBO velocity fields, available at https://gage-data.earthscope.org/archive/gnss/products/velocity (Feb 17 2024).

The inclusion of large global velocity fields in the combination, like NGL and JPL (see Table 1), helps making more robust the alignment of sparse regional velocity fields, like SIRGAS, by maximizing the number of sites used in the alignment to the ITRF2020. The number of common sites between each input velocity field is given in Table 1 of the supplemental material. Velocity fields with a geographical extension smaller than the continental scale were not considered in the combination due to limitations in their accurate alignment to a global reference frame (see Table 3 of the supplemental material).

The numbers of sites given in Table 1 correspond to the sites that were considered from each input velocity field. The sites retained for the combination have at least 5 years of data and a constant velocity, i.e., sites with velocity discontinuities were excluded. The threshold of 5 years of data was considered as a tradeoff between including the maximum number of sites and minimizing the secular velocity errors due to interannual deformation (Santamaría-Gómez and Mémin 2015). Other criteria to exclude sites were null velocity uncertainty or velocity uncertainty larger than 1 mm/yr, which should reject sites with strong non-linearity, excessive amount of discontinuities or substantial missing data.

The sites retained in each input velocity field were verified for duplicates, i.e., sites with the same 4-character IDs located at different places. Comparison of the coordinates of all the sites resulted in 223 duplicated sites being renamed (see Table 2 of the supplemental material).

After the verification of duplicated sites, the coordinates of each site were still slightly different across the input velocity fields, mainly due to the chosen reference epoch, which was unknown in most of the input velocity fields. The coordinates of the sites in each velocity field were replaced by a set of common coordinates. This way, the weighting of each velocity field in the combination process will only include the contribution of the velocity residuals.

A final preprocessing of the input velocity fields involved their conversion into SINEX format when necessary. Most of the input velocity fields are provided as ASCII tables in topocentric coordinates (see Table 1). These velocity fields were transformed into Cartesian coordinates while keeping the velocity covariances between the coordinates. Still in the case of velocity fields provided as ASCII tables, the velocity covariances between sites are unknown and were set to zero. For the velocity fields already provided in SINEX format, the covariances between sites were also set to zero to avoid down weighting these velocity fields.

3 Combination Method

The velocity field combination was carried out using the CATREF software (Altamimi et al. 2016) and involved four steps.

In the first step we attributed an a priori weight to each input velocity field, i.e., before any comparison of the velocity estimates. The approach followed to compute the velocity uncertainties varies among the input velocity fields, and they do not necessarily reflect realistic velocity errors in an absolute sense. They can nevertheless be assumed to reflect the relative proportions of the velocity errors of different sites within a given velocity field. Therefore, variance factors were computed to harmonize the uncertainties of the input velocity fields. The a priori variance factors were estimated with the starting assumption that all input velocity fields have the same average velocity error. Assuming an average velocity error of 0.1 mm/yr and 0.3 mm/yr for horizontal and vertical velocities respectively, the average 3D velocity error was set to $\sqrt{0.1^2 + 0.1^2 + 0.3^2} = \sqrt{0.11}$ mm/yr. The a priori variance factor of each input velocity field i was thus computed as:

$$VF_i = \frac{0.11}{\sigma_{3D_i}^2} \quad (1)$$

where $\sigma_{3D_i}^2$ represents the average squared 3D velocity uncertainty of the velocity field i. Due to the different numbers and geographic distributions of sites among the input velocity fields, the average 3D uncertainty was obtained from a subset of common sites between the input velocity fields.

In the second step, the input velocity fields were iteratively combined using the variance factors from the first step. Drifts in the origin, orientation and scale of each input velocity field were estimated with respect to the ITRF2020 velocity field. After each iteration, sites with velocity residuals larger than 0.7 or 2 mm/yr for the horizontal and vertical velocities, respectively, were considered as outliers and removed from the next iteration. These thresholds were chosen as being 7 times the a priori average velocity error in the first step, so that only extreme velocity errors were removed. Only one site per velocity field and only one velocity estimate per site present in several velocity fields was removed at each iteration.

Once there were no more outliers to eliminate, in a third step, a posteriori variance factors were estimated based on the variance of the residuals of each velocity field. The outliers iteratively removed in the second step ensure the gaussianity of the velocity residuals, from which the a posteriori variance factors were obtained.

In the fourth and last step, a final iterative combination was carried out using the a posteriori variance factors from the third step to weight the input velocity fields. Due to the change of the weights of the input velocity fields, all the previously removed outliers were included and assessed again. Velocity outliers were removed using similar criteria as in the second step. For each iteration, we removed one velocity outlier per site, but only for those sites for which the outlier velocity could be identified unambiguously, i.e., sites having more than two velocity estimates available and a velocity residual that is significantly larger than the rest. This way, the combined velocity field includes sites for which the estimated velocity among the input velocity fields is ambiguous, which is reflected by a larger repeatability and uncertainty of the combined velocity. In addition, since changes of the weight of each input velocity field barely affects their estimated alignment, the alignment parameters (origin, orientation and scale drifts) of the velocity fields were held fixed to the ones obtained in the second step. This way, multiple velocity outliers could be removed from each input velocity field in the same iteration, which drastically reduced the number of iterations needed with no loss of rigor in the results. In total, 223 velocity outliers were removed, ranging between 0 and 55 outliers among the input velocity fields (see Table 7 of the supplemental material). The final combined velocity field was obtained from the last iteration once all the velocity outliers were removed.

4 Results of the Combination

Table 2 shows the square-roots of the a posteriori variance factors used to weight each input velocity field. Applying these (square-root) variance factors to each input velocity field allows comparing them in terms of their average 3D velocity uncertainty (also in Table 2), which is inversely equivalent to their relative weight in the combination. Most of the input velocity fields have comparable a posteriori 3D median velocity uncertainties, mostly in the range 0.2–0.4 mm/yr. The two rightmost columns of Table 2 shows the WRMS of the 3D velocity differences between each input velocity field and the combined field, with and without correcting their alignment bias (see Table 3 of the supplemental material). Only stations present in several velocity

Table 2 Square-roots of the a posteriori variance factors of each input velocity field, their median weighted 3D velocity uncertainty, in mm/yr, the 3D velocity WRMS with respect to the combined field, in mm/yr, and also including the alignment bias (raw WRMS)

Solution	$\sqrt{VF}$	σ_{3D}	WRMS	Raw WRMS
APREF	3.0	0.33	0.36	0.38
CWU	0.7	0.33	0.18	0.20
EOST	1.0	0.24	0.15	0.27
EPND	1.4	0.33	0.05	0.13
EUREF	7.2	0.22	0.09	0.14
INGV	0.5	0.24	0.14	0.18
ITRF2020	1.0	0.24	0.16	0.16
JPL	1.0	0.33	0.17	0.20
LTK	4.2	0.35	0.22	0.32
NCL	0.5	0.42	0.19	0.22
NGL	0.3	0.24	0.14	0.18
NGS	0.6	0.35	0.23	0.25
NMT	0.5	0.33	0.16	0.21
NRCAN	2.1	0.33	0.19	0.25
PBO	0.6	0.33	0.18	0.24
SIRGAS	49.6	0.52	0.79	0.79
SOPAC	1.0	0.37	0.24	0.28
UGA	0.5	0.24	0.12	0.18
ULR	0.8	0.24	0.14	0.24

fields were used to compute the WRMS. This WRMS can be interpreted as the distance between the combined and each input velocity field, also taking into account their misalignment.

The alignment parameters of the input velocity fields with respect to the ITRF2020 are well determined, with uncertainties in general smaller than 0.05 mm/yr, rising up to the level of 0.1 mm/yr for some of the regional velocity fields with a small number of sites (see Table 3 of the supplemental material). We also noticed a larger translation and scale drift in some of the regional velocity fields, most likely due, in part, to the correlation between the seven transformation parameters for velocity fields of small geographical extent.

Figures 1 and 2 show the combined horizontal and vertical velocity fields, respectively. The velocity estimates and their uncertainties can be found in Table 4 of the supplemental material. The combined field includes velocity estimates at 13,395 sites, from which 2,877 were present in more than four input velocity fields. These sites were used to assess the quality of the combined velocity field via the repeatability of site velocity across the input fields. The velocity repeatability can be considered as an alternative assessment of the velocity precision compared to the formal velocity uncertainty obtained from variance propagation of the models fitted to the GNSS position time series. For instance, the velocity repeatability includes the effect of the different time series used by the different groups and the different models fitted to them. However, the velocity repeatability is only available at those sites for which several velocity estimates are available in the input velocity fields. The velocity estimates for 7,207

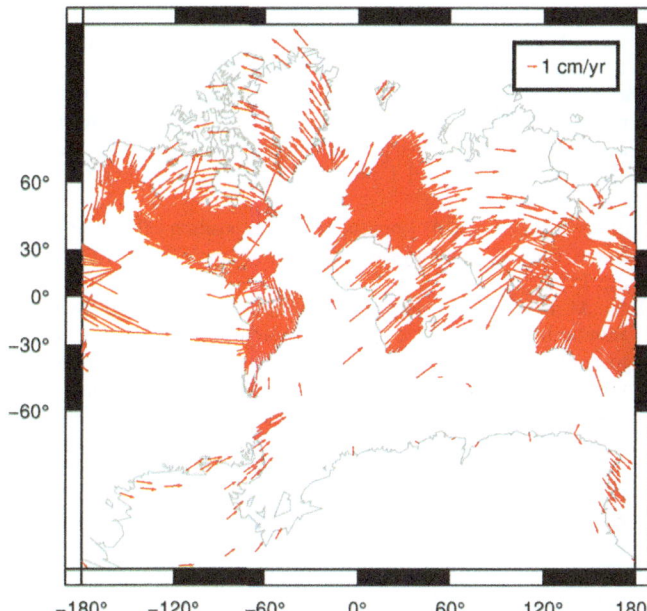

Fig. 1 Horizontal velocities of the global combined field

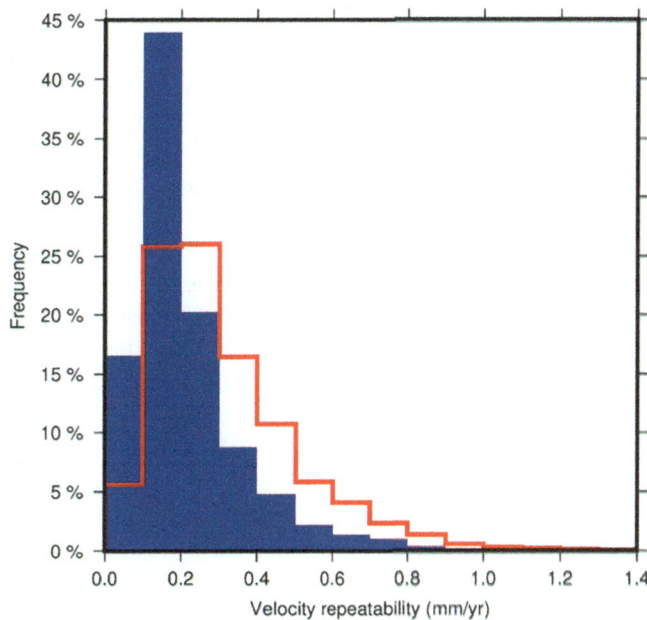

Fig. 3 Histograms of repeatability in mm/yr of the horizontal (in blue) and vertical (in red) velocities

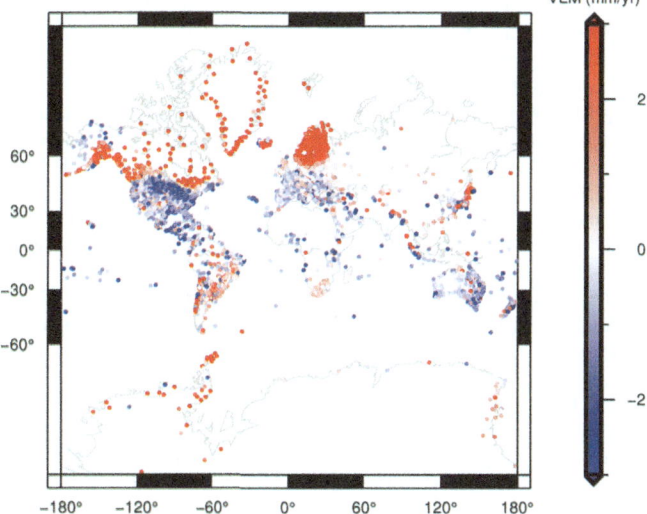

Fig. 2 Vertical velocities of the global combined field

sites are provided by a single velocity field, most of them by the NGL (5,791) and the EPND (921) velocity fields.

Figure 3 shows the histograms of the horizontal and vertical velocity repeatabilities (WRMS of the velocity residuals) per site. The median repeatability of the velocity estimates is 0.17 and 0.27 mm/yr for horizontal and vertical velocities, respectively; in close agreement with the values used as a priori average velocity errors. The estimated velocity repeatability per site can be found in Table 5 of the supplemental material. The residuals of the velocity combination per site and per input velocity field can be found in Table 6 of the supplemental material.

The combination also helped to identify 327 sites out of 2,877 (11 %) for which the velocity repeatability is either larger than 0.45 mm/yr for horizontal velocities or larger than 0.75 mm/yr for vertical velocities (3 times the median values). The velocity repeatability per site was obtained after removing velocity outliers with deviations of several mm/yr, exceeding 1 cm/yr in a few cases. The rate of removed velocity outliers is typically less than 1% of the estimated velocities, but rises up to 10% for some velocity fields (see Table 7 of the supplemental material). There are several reasons that may explain a poor velocity repeatability, like noisy time series, the removal of different discontinuities and/or non-linear deformation due to post-seismic, hydrological loading or glacier discharge, among other processes (Riva et al. 2017; Gobron et al. 2021; Young et al. 2023). These sites should be used with caution, especially if their estimated linear velocities are extrapolated for long periods of time, as is common practice in sea-level (Wöppelmann et al. 2014) and GIA studies (King et al. 2022). By simple extrapolation, around 11% of the sites of the combined velocity field, but also of each input velocity field, might have poorly estimated velocities that may or may not be reflected by their formal velocity uncertainty.

5 Conclusions and Outlook

The International Association of Geodesy's Joint Working Group 3.2 derived a global combined GNSS velocity field aligned to the ITRF2020 that gathers nineteen global and regional GNSS velocity fields computed by different groups. The combined field provides velocity estimates for 13,395 sites with at least 5 years of observations, completing by 1,849 sites the largest individual velocity field available.

More importantly, the combination allowed comparing the velocity estimates from the different groups and assessing their repeatability at 2,877 sites for which more than four velocity estimates were available. A median velocity repeatability at the level of 0.15 and 0.25 mm/yr was obtained for the horizontal and vertical velocities, respectively. However, up to 11% of the sites show poor velocity repeatability exceeding 3 times the median values. The estimated velocities at these sites should be used with caution.

For almost half the sites in the combined velocity field, the quality of the velocity estimates is given exclusively by the formal velocity uncertainty from a single velocity field. In this regard, the combination also provides estimated variance factors, based on the velocity repeatability across the input fields, which can be used to adjust the formal uncertainties of each input velocity field.

Acknowledgements The authors are thankful to the individuals and groups that contributed with the velocity fields used in the combination. This material is based on services provided by the GAGE Facility, operated by EarthScope Consortium, with support from the National Science Foundation, the National Aeronautics and Space Administration, and the U.S. Geological Survey under NSF Cooperative Agreement EAR-1724794.

References

Altamimi Z, Rebischung P, Métivier L, Collilieux X (2016) ITRF2014: a new release of the international terrestrial reference frame modeling nonlinear station motions. J Geophys Res Solid Earth 121(8):6109–6131. https://doi.org/10.1002/2016JB013098

Altamimi Z, Rebischung P, Collilieux X, Métivier L, Chanard K (2023) ITRF2020: an augmented reference frame refining the modeling of nonlinear station motions. J Geod 97(47). https://doi.org/10.1007/s00190-023-01738-w

Blewitt G, Hammond WC, Kreemer C (2018) Harnessing the GPS data explosion for interdisciplinary science. Eos 99. https://doi.org/10.1029/2018EO104623

Bock Y, Moore AW, Argus D, Fang P, Golriz D, Guns K, Jiang S, Kedar S, Knox SA, Liu Z, Sullivan A (2021) Extended solid earth science ESDR system (ES3): algorithm theoretical basis document, NASA MEaSUREs project #NNH17ZDA001N., http://garner.ucsd.edu/pub/measuresESESES_products/ATBD/ESESES-ATBD.pdf.

Gazeaux J, Williams S, King M, Bos M, Dach R, Deo M, Moore AW, Ostini L, Petrie E, Roggero M, Teferle FN, Olivares G, Webb FH (2013) Detecting offsets in GPS time series: first results from the detection of Offsets in GPS Experiment. J Geophys Res Solid Earth 118:2397–2407. https://doi.org/10.1002/jgrb.50152

Gobron K, Rebischung P, Van Camp M, Demoulin A, de Viron O (2021) Influence of aperiodic non-tidal atmospheric and oceanic loading deformations on the stochastic properties of global GNSS vertical land motion time series. J Geophys Res Solid Earth 126:e2021JB022370. https://doi.org/10.1029/2021JB022370

Gravelle M, Wöppelmann G, Gobron K, Altamimi Z, Guichard M, Herring T, Rebischung P (2023) The ULR-repro3 GPS data reanalysis and its estimates of vertical land motion at tide gauges for sea level science. Earth Syst Sci Data 15:497–509. https://doi.org/10.5194/essd-15-497-2023

Griffiths J, Ray J (2016) Impacts of GNSS position offsets on global frame stability. Geophys J Int 204:1. https://doi.org/10.1093/gji/ggv455

Heflin M, Donnellan A, Parker J, Lyzenga G, Moore A, Ludwig LG, Rundle J, Wang J, Pierce M (2020) Automated estimation and tools to extract positions, velocities, breaks, and seasonal terms from daily GNSS measurements: illuminating nonlinear Salton Trough deformation. Earth Space Sci 7. https://doi.org/10.1029/2019EA000644

Herring TA, Melbourne TI, Murray MH, Floyd MA, Szeliga WM, King RW, Phillips DA, Puskas CM, Santillan M, Wang L (2016) Plate boundary observatory and related networks: GPS data analysis methods and geodetic products. Rev Geophys 54. https://doi.org/10.1002/2016RG000529

INGV Ring Working Group (2016) RETE Integrata Nazionale GNSS, https://doi.org/10.13127/RING

Kenyeres A, Bellet JG, Bruyninx C, Caporali A, De Doncker F, Droscak B, Duret A, Franke P, Georgiev I, Bingley R, Huisman L, Jivall L, Khoda O, Kollo K, Kurt AI, Lahtinen S, Legrand J, Magyar B, Mesmaker D, Morozova K, Nagl J, Ozdemir S, Papanikolaouo X, Parseulinas E, Stangl G, Tangen OB, Valdes M, Ryczywolski M, Zurutuza J, Weber M (2019) Regional integration of long-term national dense GNSS network solutions. GPS Solut 23:122. https://doi.org/10.1007/s10291-019-0902-7

King MA, Watson CS, White D (2022) GPS rates of vertical bedrock motion suggest late Holocene ice-sheet readvance in a critical sector of East Antarctica. Geophys Res Lett 49. https://doi.org/10.1029/2021GL097232

Legrand J (2022) EPN multi-year position and velocity solution C2235. Available from Royal Observatory of Belgium, https://doi.org/10.24414/ROB-EUREF-C2235

Michel A, Santamaría-Gómez A, Boy J-P, Perosanz F, Loyer S (2021) Analysis of GNSS displacements in Europe and their comparison with hydrological loading models. Remote Sens. https://doi.org/10.3390/rs13224523

Riva REM, Frederikse T, King MA, Marzeion B, van den Broeke MR (2017) Brief communication: the global signature of post-1900 land ice wastage on vertical land motion. The Cryosphere 11(1327–1332):2017. https://doi.org/10.5194/tc-11-1327-2017

Sánchez L, Drewes H (2020) SIRGAS reference frame realization SIR17P01. https://doi.org/10.1594/PANGAEA.912349

Santamaría-Gómez A, Mémin A (2015) Geodetic secular velocity errors due to interannual surface loading deformation. Geophys J Int 202:2. https://doi.org/10.1093/gji/ggv190

Socquet A, Janex G (2019) GNSS position and velocity solutions calculated in the framework of the EPOS initiative with IGS final products. CNRS, OSUG, ISTERRE. https://doi.org/10.17178/GNSS.products.EPOS.2019

Vardić K, Clarke PJ, Whitehouse PL (2022) A GNSS velocity field for crustal deformation studies: The influence of glacial isostatic adjustment on plate motion models. Geophys J Int 231(1):426–458. https://doi.org/10.1093/gji/ggac047

Wanninger L (2009) Correction of apparent position shifts caused by GNSS antenna changes. GPS Solut 13:133–139. https://doi.org/10.1007/s10291-008-0106-z

Williams SDP (2003) Offsets in global positioning system time series. J Geophys Res 108(B6):2310. https://doi.org/10.1029/2002JB002156

Wöppelmann G, Marcos M, Santamaría-Gómez A, Martín-Míguez B, Bouin M-N, Gravelle M (2014) Evidence for a differential sea level rise between hemispheres over the twentieth century. Geophys Res Lett 41:1639–1643. https://doi.org/10.1002/2013GL059039

Young ZM, Kreemer C, Hammond WC, Blewitt G (2023) Interseismic strain accumulation between the Colorado Plateau and the Eastern California Shear Zone: implications for the Seismic Hazard near Las Vegas, Nevada. Bull Seismol Soc Am 113(2). https://doi.org/10.1785/0120220136

Open Access This chapter is licensed under the terms of the Creative Commons Attribution 4.0 International License (http://creativecommons.org/licenses/by/4.0/), which permits use, sharing, adaptation, distribution and reproduction in any medium or format, as long as you give appropriate credit to the original author(s) and the source, provide a link to the Creative Commons license and indicate if changes were made.

The images or other third party material in this chapter are included in the chapter's Creative Commons license, unless indicated otherwise in a credit line to the material. If material is not included in the chapter's Creative Commons license and your intended use is not permitted by statutory regulation or exceeds the permitted use, you will need to obtain permission directly from the copyright holder.

Geophysical Loading Correction Comparison and Assessment in VLBI Analysis

Shivangi Singh , Johannes Böhm , Hana Krásná ,
Nagarajan Balasubramanian, and Onkar Dikshit

Abstract

The Earth's crust experiences deformation caused by a range of geophysical phenomena, including the motion of tectonic plates and the redistribution of surface fluids like the atmosphere, oceans, and continental water. These natural processes result in substantial changes in the Earth's crust load, leading to the displacement of geodetic sites and alterations in station coordinates over time scales that can vary from yearly to sub-diurnal periods. Geophysical models are employed in Very Long Baseline Interferometry (VLBI) analysis to consider loading effects resulting from the global movement of the geophysical fluids to accurately estimate parameters of interest. Given VLBI's significance as a key technique for terrestrial reference frame determination, the accuracy of geophysical models becomes paramount. This study focuses on comparing elastic surface loading products, specifically on the corresponding changes in station coordinates. Non-tidal surface loading (NTSL) data is obtained from different loading services, such as VieAPL, EOST, IMLS, and ESMGFZ. Notably, VieAPL exclusively provides non-tidal atmospheric loading (NTAL), while EOST, IMLS, and ESMGFZ provide all three NTSL components—NTAL, non-tidal oceanic loading, and hydrological loading. The analysis of 20 years data of NTSL (from 2001 to 2020), extracted from these services demonstrates consistency among them, except for the hydrological loading component of ESMGFZ. The implementation of NTSL models in VLBI analysis has revealed that baseline length repeatability shows improvements or remains stable in 90.25% of the baselines for IMLS, 89.02% for EOST, and 86.18% for ESMGFZ. Additionally, the application of NTSL models leads to an improvement in the standard deviation of station height by 65% in both EOST and IMLS, and by 61.25% in the case of ESMGFZ. We also investigate the variance reduction coefficients, demonstrating the distinctions in loading corrections offered by various services.

Keywords

EOST · ESMGFZ · Geophysical models · IMLS · NTSL · VieAPL · VieVS · VLBI

1 Introduction

Very Long Baseline Interferometry (VLBI) is a sophisticated geospatial measurement technique that can be used to examine the Earth's geophysical properties. Geophysical models are crucial in the VLBI analysis, helping to account for and adjust various factors that influence VLBI observations. These models are indispensable for extracting

S. Singh (✉) · N. Balasubramanian · O. Dikshit
Geoinformatics, Department of Civil Engineering, IIT Kanpur, Kanpur, Uttar Pradesh, India
e-mail: shivangi@iitk.ac.in

J. Böhm · H. Krásná
Department of Geodesy and Geoinformation, TU Wien, Vienna, Austria

© The Author(s) 2024
J. T. Freymueller, L. Sánchez (eds.), *Together Again for Geodesy*,
International Association of Geodesy Symposia 157, https://doi.org/10.1007/1345_2024_257

crucial geophysical insights. Despite concerted efforts to address various influencing factors, unaccounted non-linear effects persist in the analysis of station position time series, significantly impairing the accuracy of VLBI products. Non-tidal surface loading (NTSL) is one of the geophysical effects which have not been traditionally corrected in VLBI and other geodetic measurements. NTSL refers to the non-tidal redistribution of mass within Earth's fluid components (atmosphere, oceans, and land water storage) leading to elastic deformations in the Earth's crust. These deformations are non-linear and lead to instantaneous displacements of reference points, occasionally reaching magnitudes of several centimeters (Wijaya et al. 2013).

Non-tidal atmospheric loading (NTAL) is considered in the operational solutions of International VLBI Service for Geodesy and Astrometry (IVS) analysis centers. However, since other services such as the International GNSS Service (IGS), International DORIS Service (IDS), and International Laser Ranging Service (ILRS) do not currently incorporate NTAL into their operational solutions, it has been removed again from the official IVS contribution to the current International Terrestrial Reference Frame (ITRF2020) as indicated in the International Earth Rotation and Reference Systems Service (IERS) annual report (https://www.iers.org/IERS/EN/Publications/AnnualReports/AnnualReports.html). Consequently, ITRF2020 was generated without correcting station deformations caused by non-tidal loading but it accounts for annual and semi-annual seasonal signals present in the station position time series (Altamimi et al. 2023). In the International Terrestrial Reference System (ITRS) realization by Jet Propulsion Laboratory, NASA in the USA (e.g., JTRF2020), NTSL data is used to estimate the station-dependent variance of the position process noise (Abbondanza et al. 2017; Gross et al. 2022). However, in the latest ITRS realization by Deutsches Geodätisches Forschungsinstitut der Technischen Universität München in Germany (i.e., DTRF2020), all three NTSL components derived by the Global Geophysical Fluid Center (GGFC, http://loading.u-strasbg.fr/GGFC/) are consistently incorporated at the normal equation level (Seitz et al. 2023).

Several research studies have extensively investigated the implications of correcting NTSL effects in the realm of space geodesy. Noteworthy among these, Tregoning and van Dam (2005) demonstrated a significant 77% reduction in the values of the weighted root mean square (WRMS) of the height component through corrections for the NTAL effect. Emphasizing the importance of precise correction of atmospheric loading at the observational level, Böhm et al. (2009) highlighted their vital role in VLBI studies. These corrections play a key role in VLBI analysis, as effect of neglected corrections due to non-tidal loading at specific stations can be transferred to other stations in the network. Williams and Penna (2011) studied the effects of NTAL and non-tidal oceanic loading (NTOL) on GPS height time series around the southern North Sea. The investigation unveiled that the displacements resulting from NTOL were comparable to those caused by NTAL in this area. Incorporating corrections for both these factors resulted in a significant reduction in the GPS heights variance, approximately ranging from 12 to 22 square millimeters. This reduction corresponded to 20–30% decrease in GPS height root mean square (RMS) values.

The utilization of non-tidal loading models has been demonstrated to effectively mitigate systematic effects, including, e.g., yearly patterns in station height data. Additionally, it decreases the repeatability of station coordinates and reduces residual annual amplitudes of the station coordinate time series, as evidenced in studies conducted by Männel et al. (2019) or Glomsda et al. (2020). Moreover, the incorporation of non-tidal loading models into VLBI analysis has been found to decrease the standard deviation of daily estimated vertical coordinates and has notable implications on the Earth orientation parameters, as highlighted in the research by Roggenbuck et al. (2015). Over the recent years, the majority of studies have indicated a reduction in baseline length repeatability (BLR) following the application of NTSL. However, the estimates of computed baseline length are influenced by noise in VLBI measurements, and the computed loading estimates are also affected by noise, arising from errors in surface pressure values or limitations in the Green's functions utilized in the loading computation (van Dam and Herring 1994). Thus, to address the noise, it is necessary to address the statistical properties of the computed BLR and NTSL products.

The GGFC within IERS provides essential data and models of geodetic effects (rotation, gravity, and deformation of Earth) which are influenced by fluid movements over time. GGFC's insights are crucial for geodetic observations dependent on the Earth's internal and surface fluid dynamics. The GGFC offers NTSL data through various services listed on their website, including:

1. GGOS atmosphere at TU Wien (https://vmf.geo.tuwien.ac.at/)
2. EOST loading services (http://loading.u-strasbg.fr/)
3. GFZ products (http://rz-vm115.gfz-potsdam.de:8080/repository)
4. University of Luxembourg products (https://geophy.uni.lu/)

In our research, we delve into NTSL data and its constituents. We source NTSL data from multiple services listed on the GGFC website, excluding the University of Luxembourg data, which is available only until 2015. Furthermore, we incorporate data from the International Mass Loading Service (IMLS, http://massloading.net/), which is not given on the GGFC website. Section 2 compares the NTSL models employed by these services for extracting displacement data, encompassing all loading components. Section 3 details the results obtained by comparing the NTSL data and by integrating NTSL into VLBI analysis as adjustments to station coordinates at the observation level. Additionally, we attempted to address the statistical properties of the computed BLR before and after applying the loading by calculating the noise associated with it. Section 4 concludes by exploring additional pertinent aspects within this context, such as the variance reduction coefficients.

2 Non-tidal Surface Loading

NTSL pertains to the correction of systematic effects resulting from time-dependent mass fluctuations in geophysical fluids. NTSL encompasses NTAL, NTOL and hydrological loading (HYDL). The surface displacements due to atmospheric pressure occur with a period of approximately 2 weeks. Variations in HYDL become important with an annual period, reflecting a yearly cycle. The NTOL, on the other hand, varies approximately with a period of 1 month (Schuh et al. 2003).

Among the four services examined in our study—VieAPL (Vienna Atmospheric Pressure Loading), ESMGFZ (Earth System Modelling group at GFZ), IMLS, and EOST (École & observatoire des sciences de la Terre)—NTAL data is provided by all. Conversely, NTOL and HYDL data is exclusively supplied by ESMGFZ, IMLS, and EOST. Each service derives loading products from various models. The specific NTSL components calculated from their corresponding models by each service are outlined below.

2.1 NTSL Models

NTAL captures the effects of temporal mass changes in the Earth's atmosphere, impacting the solid crust. These mass changes stem from dynamic alterations in atmospheric pressure caused by meteorological events and other contributing factors. As a consequence, the variations in pressure exert force on the Earth's crust, establishing a negative correlation between atmospheric pressure and crustal deformation (van Dam and Herring 1994; Petrov and Boy 2004).

HYDL pertains to the Earth's crust deformation caused by fluctuations in continental water storage. This phenomenon results in vertical component variations of 3–15 mm and horizontal component variations of 1–2 mm. The hydrology signal exhibits yearly and half-yearly patterns, coupled with interannual fluctuations (Eriksson and MacMillan 2014). Incorporating hydrology loading corrections in VLBI analysis has been proven to significantly decrease BLR and site vertical coordinate repeatability in a considerable number of cases (Schuh et al. 2003).

NTOL involves the deformation of the Earth's crust induced by mass redistribution within oceans, excluding tidal forces. These non-tidal factors encompass atmospheric pressure fluctuations, oceanic currents, changes in water temperature and salinity, glacial isostatic adjustment, and human activities. Table 1 presents information about the

Table 1 Attributes of models employed by different services tailored for individual loading components

Loading	Service	Models	Spatial resolution	Time steps	Data availability
NTAL	EOST	ECMWF Operational_IB	$0.5° \times 0.5°$	3 h	2000–present
		ECMWF Operational_TUGO-m	$0.25° \times 0.25°$	3 h	2002–2017
		ERA interim	$0.5° \times 0.5°$	6 h	1979–present
	IMLS	GEOSFPIT	$2' \times 2'$	3 h	2000–present
		MERRA-2	$2' \times 2'$	6 h	1980–present
	ESMGFZ	ECMWF	$0.5° \times 0.5°$	3 h	1976–present
	VieAPL	ECMWF	$1° \times 1°$	6 h	1994–present
HYDL	EOST	ERA interim	—	6 h	1979–2016
		GLDAS	$0.25° \times 0.25°$	3 h	2000–2016
		GLDAS2	$0.25° \times 0.25°$	3 h	2000–present
		MERRA-2	$0.5° \times 0.625°$	1 h	1980–present
		MERRA-land	$0.5° \times 0.67°$	1 h	1980–2016
		NASA GSFC iterated global mascon	$1° \times 1°$	Monthly	2002–2021
	IMLS	GEOS-FPIT	$2' \times 2'$	3 h	2000–present
		MERRA-2	$2' \times 2'$	3 h	1980–present
	ESMGFZ	LSDM	$0.5° \times 0.5°$	24 h	1976–present
NTOL	EOST	ECCO1	$1° \times 1°$	12 h	1993–2021
		ECCO2	$0.25° \times 0.25°$	24 h	1992–present
		GLORYS2v3	$0.25° \times 0.25°$	24 h	1992–2013
	IMLS	MPIOM06	$2' \times 2'$	3 h	1980–present
	ESMGFZ	MPIOM	$1° \times 1°$	3 h	1976–present

characteristics of models employed by various services specific to each loading component, namely NTAL, NTOL and HYDL.

2.2 Model Comparison

In our study, the choice of models for distinct loading categories and services relies on considerations such as data availability period, temporal intervals, update frequency, and spatial resolution. VieAPL's NTAL data relies on surface pressure derived from European Centre for Medium-Range Weather Forecasts (ECMWF) operational datasets (https://www.ecmwf.int/).

They provide non-tidal radial and horizontal displacements, while using the inverted barometer (IB) assumption to account for atmospheric pressure effects above the ocean. ESMGFZ offers Homogenized Atmospheric Surface Pressure Time-Series, a methodology introduced by Dobslaw (2016). This approach allows for the utilization of the up-to-date data from the deterministic model of operational ECMWF while concurrently ensuring the re-analysis datasets stability in the long-term. The NTAL data provided by the IMLS is based on the MERRA-2 (Modern-Era Retrospective Analysis for Research and Applications, Version 2) model and relies on the spherical harmonic transformation (Petrov 2015). It incorporates the integration of aerosol observations, and improvements in stratospheric modelling, including enhanced representation of cryospheric processes and ozone (Gelaro et al. 2017). The NTAL data provided by EOST is generated from the surface pressure data derived from the ECMWF operational model. The fundamental assumption underlying this computation is that the ocean reacts to pressure forcing in a manner that is congruent with the inverted barometer effect.

Oceanic mass loads by ESMGFZ are calculated based on the data of non-tidal ocean bottom pressure collected at three-hour intervals and obtained from the ocean model MPIOM (Jungclaus et al. 2013). To focus solely on the non-tidal aspect, the impact of oceanic tides is removed using harmonic analysis involving 12 primary tidal constituents. It's essential to highlight that MPIOM consistently utilizes atmospheric data from ECMWF, which includes information on wind and atmospheric pressure (Dill et al. 2022). The NTOL data from IMLS is generated by applying spherical harmonics transform to the ocean bottom pressure obtained using the MPIOM06 model (Jungclaus et al. 2013). EOST considered ECCO1 (Estimating the Circulation and Climate of the Ocean, Phase I, http://www.ecco-group.org) for estimation of NTOL.

The hydrological loading data by ESMGFZ is calculated from the Land Surface Discharge Model (LSDM). The LSDM incorporates the simulation of multiple elements, such as snow accumulation, shallow groundwater, soil moisture, and the presence of surface water in lakes and rivers (Dill 2008). The HYDL data provided by the IMLS is based on MERRA-2 and is computed using the spherical harmonic transform approach. It improves precision through the rectification of precipitation by employing observed data, thereby enhancing estimations of land surface hydrology. The enhancement of accuracy is further achieved through modifications made to the land model, specifically in terms of the representation of rainfall (Reichle et al. 2017). The HYDL data for EOST is calculated from the GLDAS2 model. It brings together data from both satellites and ground-based sources, considering factors like soil moisture, snow, and canopy water in its calculations. Notably, it excludes permanent ice-covered regions such as Greenland, Alaska, and mountain glaciers from its analysis (Rodell et al. 2004).

3 Methodology

We identify a total of 163 common VLBI stations (also ITRF sites) for which data is provided by all the four services considered in our study. NTSL data spanning 20 years (2001-2020) are extracted in the center-of-mass (CM; Blewitt 2003) frame for the selected models. To facilitate comparison, data from each service is standardized into a format identical to that of the VieAPL service, as the comparison is intended to be conducted within the geodetic analysis software VieVS (Vienna VLBI and Satellite Software, Böhm et al. 2018). As a result, each set of NTSL data is compiled into yearly files for all 163 stations with a 6-h time step, structured in the following order: station name, modified Julian date (MJD), displacement in up direction (m), displacement in east direction (m), and displacement in north direction (m).

3.1 Data Comparison

To compare the NTSL data extracted from each service, we initially plot a station displacement time series graph for NTAL and then for the combined NTSL components. It is crucial to focus on the NTAL data for comparison since VieAPL exclusively provides this information. The displacement time series graph generated for NTAL data exhibits consistent agreement across all the services. Similar results were also presented in the research by Roggenbuck et al. (2015) and Glomsda et al. (2020). This consistency might be attributed to the fact that, except for the IMLS MERRA-2 model, ECMWF model was selected for the remaining three services, which satisfied the model selection criteria explained above.

Out of the 163 stations being analysed for the site displacement time series of the combined NTSL, ESMGFZ demonstrates a continuous downward trend in the up direc-

Geophysical Loading Correction Comparison and Assessment in VLBI Analysis

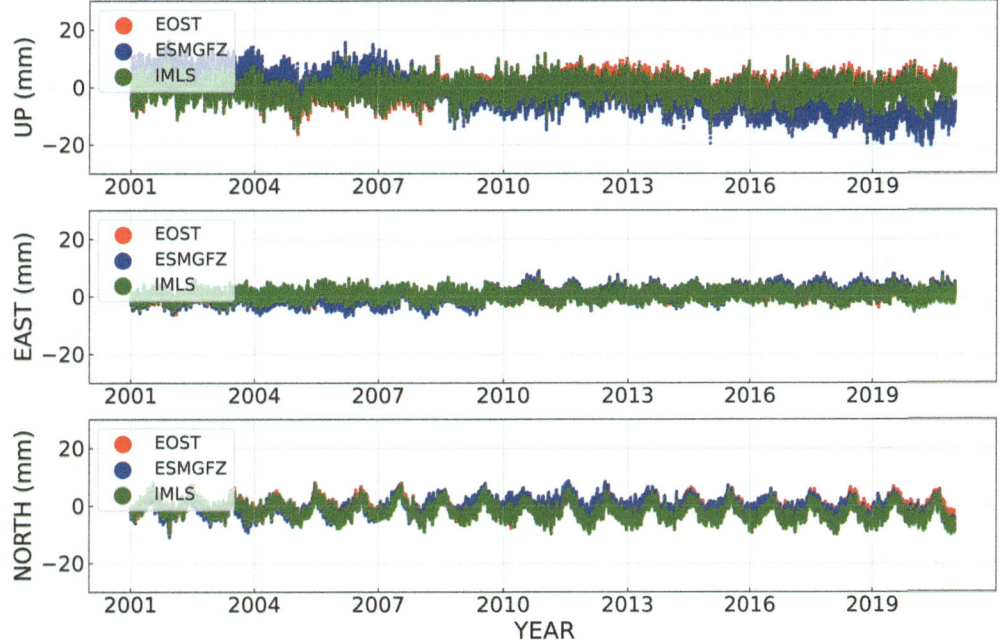

Fig. 1 Time series of site displacement due to all NTSL in CM-frame at the station FD-VLBA

tion for 37 stations throughout the entire 20-year period. Similar trends in ESMGFZ data were also highlighted by Glomsda et al. (2022). Figure 1 presents a 20-year site displacement time series depicting the combined NTSL components in the CM-frame, showcasing movements in the up, east, and north directions at the FD-VLBA station.

However, when examining individual loading components, the time series graphs for NTAL and NTOL across all services demonstrate agreement, but this trend is only evident in the HYDL time series graph of ESMGFZ. To address the consistent downward trend observed in the up direction for 37 stations in the HYDL component of ESMFZ, we employ a Butterworth high-pass filter (Balidakis 2019). Figures 2 and 3 display the HYDL component of ESMGFZ at the FD-VLBA station before and after applying the filter respectively.

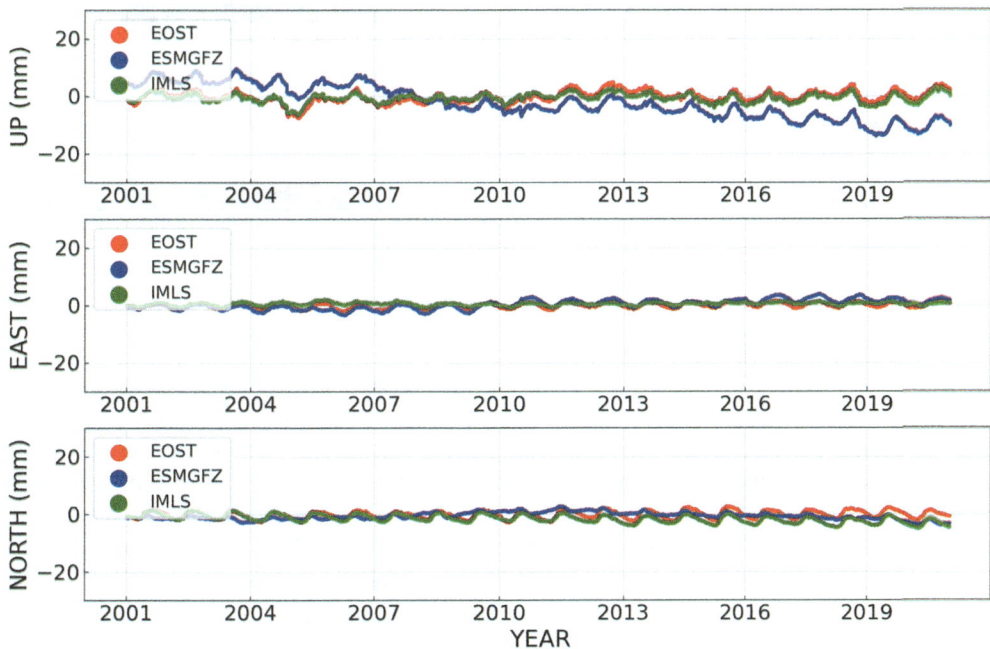

Fig. 2 Time series of site displacement resulting from HYDL at FD-VLBA station before detrending the ESMGFZ data

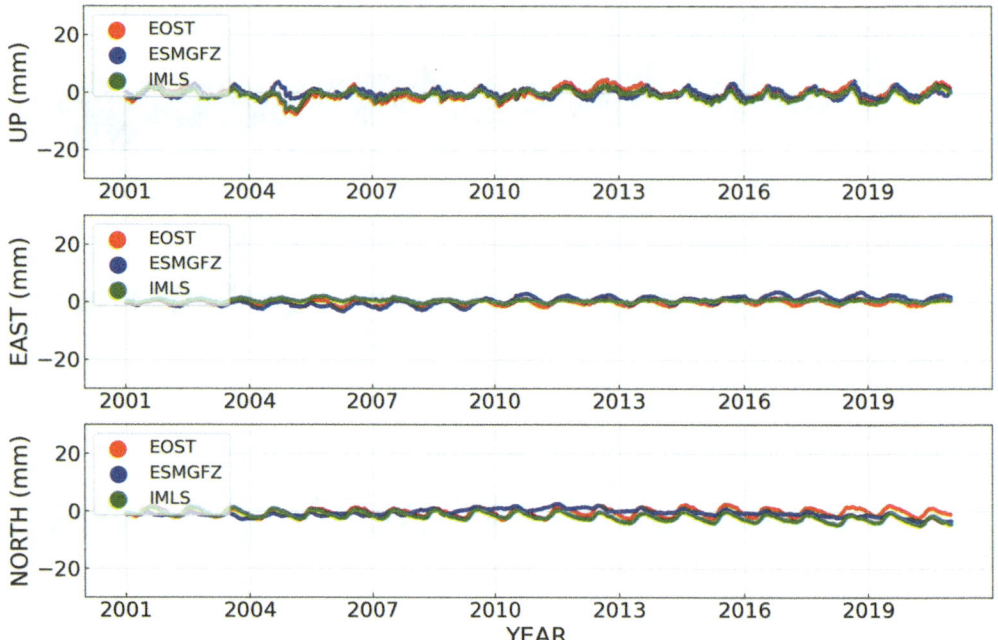

Fig. 3 Time series of site displacement resulting from HYDL at FD-VLBA station after detrending the ESMGFZ data

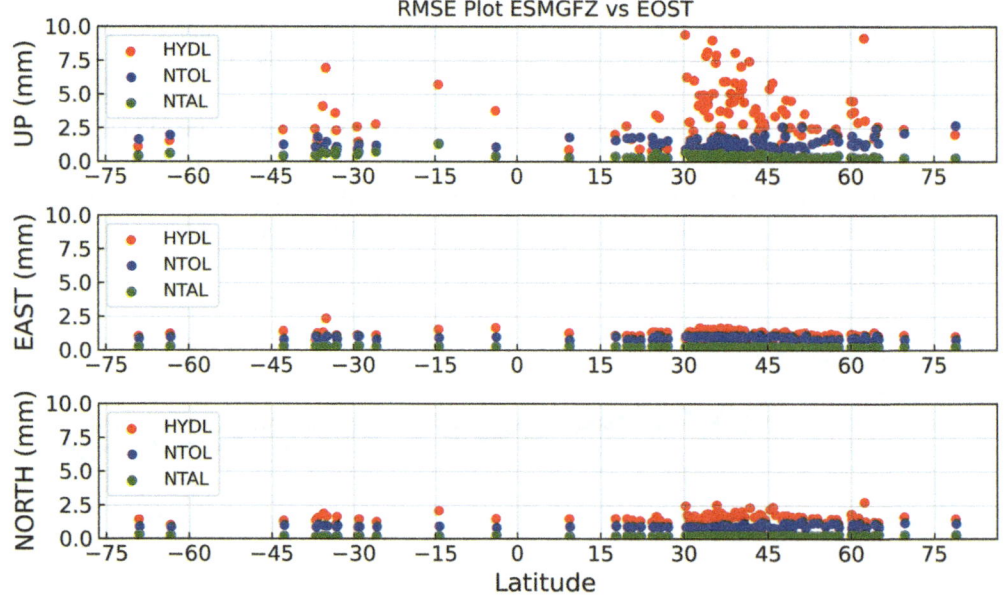

Fig. 4 RMSE values of differences of site displacements due to each NTSL component between ESMGFZ (before detrending) and EOST

To evaluate the 20-year data from all services based on loading categories (both before and after detrending the ESMGFZ HYDL component), we generate plots illustrating Root Mean Square Error (RMSE) values, indicating the disparities in site displacement for 163 VLBI stations between two services (see Figs. 4 and 5). These RMSE values are organized based on the latitude of each corresponding VLBI station.

Notably, large RMSE values exceeding 5 mm are observed, primarily within the latitude range of 30°N to 65°N, particularly in the up direction. Among the different loading components, the NTAL component exhibits the least variation between the two services, while the HYDL component displays the most substantial differences.

However, after detrending the data, for the HYDL component of ESMGFZ vs. EOST, the average RMSE value for the up direction reduces from 3.59 mm to 2.66 mm, and the maximum RMSE value reduces from 9.45 mm to 7.48 mm. In the case of ESMGFZ vs. IMLS, the average RMSE value for the up direction reduces from 3.43 mm to 2.66 mm,

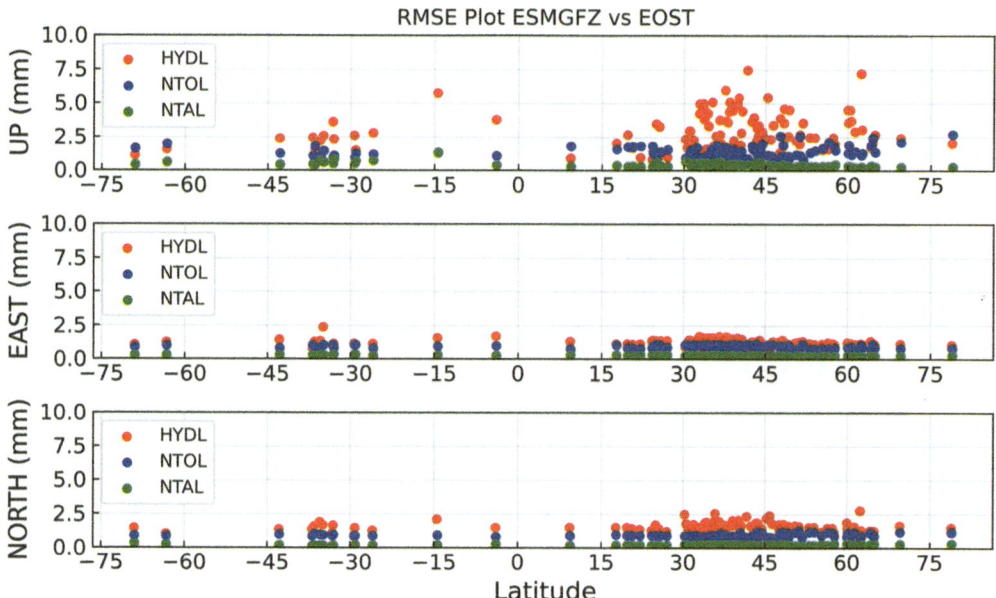

Fig. 5 RMSE values of differences of site displacements due to each NTSL component between ESMGFZ (after detrending) and EOST

Table 2 Average and maximum RMSE values in up direction

Loading	Services	Average RMSE (mm)	Max RMSE (mm)
NTAL	ESMGFZ vs. EOST	0.45	1.31
	ESMGFZ vs. IMLS	0.30	0.73
	IMLS vs. EOST	0.50	1.28
NTOL	ESMGFZ vs. EOST	1.35	2.70
	ESMGFZ vs. IMLS	0.24	1.18
	IMLS vs. EOST	1.30	2.77
HYDL	ESMGFZ vs. EOST	3.59	9.45
	Detrended ESMGFZ vs. EOST	2.66	7.48
	ESMGFZ vs. IMLS	3.43	10.10
	Detrended ESMGFZ vs. IMLS	2.66	9.91
	IMLS vs. EOST	1.90	10.40

and the maximum RMSE value reduces from 10.10 mm to 9.91 mm (refer to Table 2). These disparities can be attributed to the use of distinct models with varying resolutions by different services. Additionally, the separate treatment of Sea Level Loading (SLEL) for achieving global mass conservation, as implemented by ESMGFZ, may contribute to this observed variation. In contrast, other services incorporate partial mass conservation in both NTOL and HYDL, which could influence the level of agreement in these components.

3.2 NTSL Impact on VLBI Analysis

We examined the influence of NTSL effect-induced displacement on VLBI data analysis by integrating these displacements as adjustments to station coordinates. In our study, the NTSL is applied at the observation level. The entire process is carried out using VieVS, utilizing a 20-year process list of R1/R4 sessions along with the OPT files.[1] We followed the parameterization described in Krásná et al. (2023). We excluded VieAPL from our analysis of the NTSL's impact on VLBI analysis because it only offers NTAL. For ESMGFZ, detrended data is used. We calculated the station height standard deviation before and after applying NTSL for each service, considering 80 stations. The results reveal that 65% of station height standard deviation decreased after the application of NTSL in the cases of EOST and IMLS. However, for ESMGFZ, the improvement is slightly smaller (61.25%).

We then assessed the BLR before and after incorporating NTSL, aiming to determine any enhancements in BLR due to the utilization of NTSL models (refer to Fig. 6), focusing on 246 baselines. Baseline lengths remain constant regardless of rotation and translation, making them unaffected by net-translation and net-rotation constraints (Petrov and Boy 2004). Since the computation of BLR scatter involves inherent statistical noise, it is essential to recognize that the total loading contribution typically accounts for only 10-20% of the baseline length scatter. Meaningful comparisons require considering baseline measurements repeated at least 100 times (van Dam and Herring 1994). Consequently, we consider baselines that are observed in a minimum of 100 sessions. The outcomes revealed that 90.25% of baselines exhibited improvement or remained unchanged (i.e., difference in BLR before and after the application of loading correction is 0) when using IMLS data, while 89.02%

---

[1] https://github.com/TUW-VieVS/VLBI_OPT.

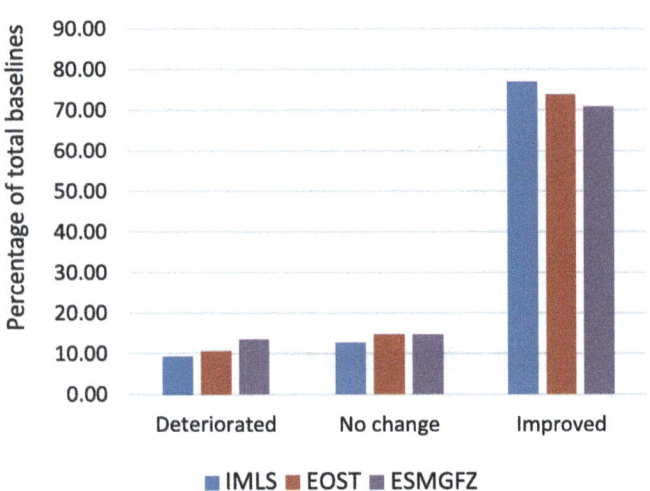

Fig. 6 Percentage of total 246 baselines deteriorated, not changed, or improved after applying all NTSL

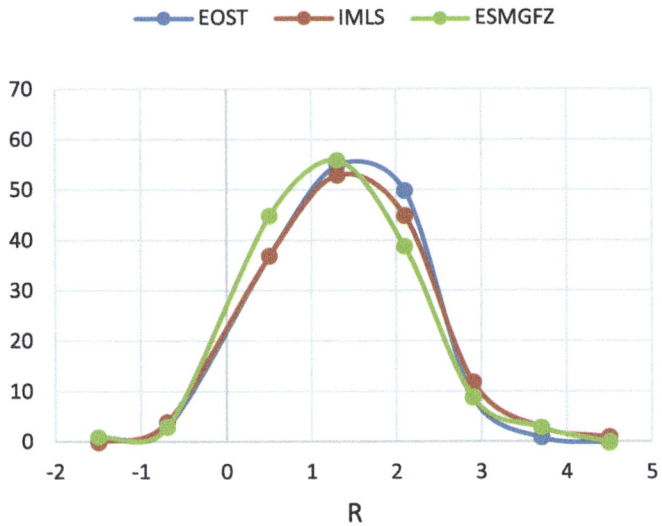

Fig. 7 Histogram of the distribution of R

of baselines showed improvement or remained unchanged with EOST data. Additionally, 86.18% of baselines demonstrate improvement or remained unchanged when utilizing ESMGFZ data.

The BLR calculation offers estimates of the standard deviation of the observed quantities. Therefore, it is crucial to delve into the statistical aspects associated with the calculated BLR (van Dam and Herring 1994). To achieve this, we compute the variance reduction coefficient, denoted as R, as introduced by Petrov and Boy (2004). This coefficient is calculated using the following formula:

$$R = \frac{\Delta\sigma^2 + \sigma_m^2}{2\sigma_m^2}$$

In this context, $\Delta\sigma^2 = \sigma_1^2 - \sigma_2^2$, where σ_1^2 is the variance of baseline length residuals before applying the model and σ_2^2 is the variance of baseline length residuals after applying the model. σ_m^2, on the other hand, represents the variance of the loading model signal. If the value of R is 1, it signifies a perfect model where the considered signal is not correlated with any other unmodeled effects. When the baseline length repeatability lacks any signal that aligns with the model, applying the loading model will increase the variance by an amount equivalent to the signal's variance within that model. This results in the value of R being equal to 0.

We identified 156 baselines (observed in more than 100 sessions) for which BLR is simulated only accounting for the loading displacement series (yielding σ_m^2). Also for these baselines, BLR is calculated with VieVS before and after applying cumulative NTSL for all three services. R values are then calculated for these 156 baselines. The weighted mean values of R were determined to be 1.14 for IMLS, 1.08 for EOST and 0.98 for ESMGFZ (see Fig. 7).

The variance reduction coefficient, as calculated by van Dam and Herring (1994) for 22 baselines using data from 1979-1992, was 0.62. Likewise, Petrov and Boy (2004) analysed the R value for 69 baselines using data spanning from 1980 to 2002, reporting it as 0.97. However, both studies only considered atmospheric loading correction, in contrast to our approach where we incorporated cumulative NTSL.

After applying NTSL, we observe that out of 156 baselines, 93 for EOST, 91 for IMLS, and 76 for ESMGFZ baselines exhibit a more significant decrease in variance of the baseline length measurements than expected (i.e., $R > 1$) given the variance of the corresponding loading model signal. Furthermore, 45 baselines for EOST, 49 baselines for IMLS, and 55 baselines for ESMGFZ display a lesser decrease in variance of the baseline length measurements than predicted based on the variance of the corresponding loading model signal (i.e., $0.5 \leq R \leq 1$). Moreover, 18 baselines for EOST, 16 baselines for IMLS, and 25 baselines for ESMGFZ show an increase in the variance of the baseline length measurements after application of NTSL (i.e., $R < 0.5$).

To delve deeper into our analysis, we examined statistical properties associated with the estimated BLR before and after applying the NTSL. The baseline length residual before applying NTSL (δb_i) is the sum of the true non-tidal loading contribution (l_i) and noise in the VLBI measurement (v_i). In addition, non-tidal loading contribution (p_i) computed from the analysis of the various NTSL models is sum of the true non-tidal loading contribution (l_i) and the noise (w_i) associated with it. The baseline length residual after applying NTSL (δc_i) can be calculated by subtracting the non-tidal loading contribution from the baseline length residual before application of the NTSL. For calculating the baseline noise variance (σ_v^2), pressure signal variance (σ_f^2), and pressure noise variance (σ_w^2), we employed the equations derived in the appendix of van Dam and Herring's (1994) paper. Our analysis followed a similar approach, but we extended it for 156 baselines. We observe that baselines exhibiting greater improvement than expected ($R > 1$) have a negative value for σ_w^2. Although a negative value for σ_w^2 is physically impossible, it can occur due to the random noise effects of the baseline length repeatability scatters (since the baselines are observed only in a finite number of sessions) or if the noise in the load signal is correlated with the error in the baseline length estimates (van Dam and Herring 1994).

4 Summary and Conclusions

Among the 163 stations analysed in our NTSL data comparison, at 37 stations a consistent downward trend in the up direction is observed over the entire 20-year period with ESMGFZ. Consequently, we utilize detrended ESMGFZ data in our study to eliminate the long-term trend in the up direction.

The RMSE plot among services for each loading component indicates the most significant variations within the HYDL. However, after detrending the ESMGFZ HYDL, the average and maximum RMSE values between ESMGFZ vs EOST and ESMGFZ vs IMLS are reduced.

The application of NTSL led to improvements in both BLR and the standard deviation of station height, with comparable outcomes obtained using EOST and IMLS loading products. Nevertheless, the R values indicate that the application of ESMGFZ NTSL resulted in more baselines showing expected reduction in variance (i.e., $0.5 \leq R \leq 1$).

Acknowledgements We express our sincere gratitude to Robert Dill and Leonid Petrov for their expert assistance in extracting loading data from the ESMGFZ and IMLS service. Special thanks to Kyriakos Balidakis for his valuable suggestions regarding ESMGFZ data. This research was made possible through the joint support of the Austrian Agency for International Cooperation in Education and Research (OeAD-GmbH) and the National Centre for Geodesy (NCG), India.

References

Abbondanza C, Chin TM, Gross RS, Heflin MB, Parker JW, Soja BS, van Dam T, Wu X (2017) JTRF2014, the JPL Kalman filter and smoother realization of the international terrestrial reference system. J Geophys Res-Sol Ea 122:8474–8510. https://doi.org/10.1002/2017JB014360

Altamimi Z, Rebischung P, Collilieux X et al (2023) ITRF2020: an augmented reference frame refining the modeling of nonlinear station motions. J Geod 97:47. https://doi.org/10.1007/s00190-023-01738-w

Balidakis K (2019) On the development and impact of propagation delay and geophysical loading on space geodetic technique data analysis, PhD thesis, (scientific technical report; 19/11). GFZ German Research Centre for Geosciences., 292 p, Potsdam. https://doi.org/10.2312/GFZ.b103-19114

Blewitt G (2003) Self-consistency in reference frames, geocenter definition, and surface loading of the solid earth. J Geophys Res 108(B2):2103. https://doi.org/10.1029/2002JB002082

Böhm J, Heinkelmann R, Mendes Cerveira PJ, Pany A, Schuh H (2009) Atmospheric loading corrections at the observation level in VLBI analysis. J Geod 83(11):1107. https://doi.org/10.1007/s00190-009-0329-y

Böhm J, Böhm S, Boisits J, Girdiuk A, Gruber J, Hellerschmied A, Krásná H, Landskron D, Madzak M, Mayer D, McCallum J, McCallum L, Schartner M, Teke K (2018) Vienna VLBI and satellite software (VieVS) for geodesy and astrometry. Publ Astron Soc Pac 130(986):044503. https://doi.org/10.1088/1538-3873/aaa22b

Dill R (2008) Hydrological model LSDM for operational earth rotation and gravity field variations, (scientific technical report STR; 08/09). Deutsches GeoForschungsZentrum GFZ., 35 p, Potsdam. https://doi.org/10.2312/GFZ.b103-08095

Dill R, Dobslaw H, Thomas M (2022) ESMGFZ products for earth rotation prediction. Artificial Satellites. https://doi.org/10.2478/arsa-2022-0022

Dobslaw H (2016) Homogenizing surface pressure time-series from operational numerical weather prediction models for geodetic applications. J Geodetic Sci 6(1). https://doi.org/10.1515/jogs-2016-0004

Eriksson D, MacMillan DS (2014) Continental hydrology loading observed by VLBI measurements. J Geod 88:675–690. https://doi.org/10.1007/s00190-014-0713-0

Gelaro R, McCarty W, Suarez MJ, Todling R, Molod A, Takacs LL, Randles CA, Darmenov A, Bosilovich MG, Reichle R, Wargan K, Coy L, Cullather R, Draper C, Akella S, Buchard V, Conaty A, da Silva AM, Gu W, Kim GK, Koster R, Lucchesi R, Merkova D, Nielsen JE, Partyka G, Pawson S, Putman W, Rienecker M, Schubert SD, Sienkiewicz M, Zhao B (2017) The modern-era retrospective analysis for research and applications, version 2 (MERRA-2). J Clim 30(14):5419–5454. https://doi.org/10.1175/JCLI-D-16-0758.1

Glomsda M, Bloßfeld M, Seitz M et al (2020) Benefits of non-tidal loading applied at distinct levels in VLBI analysis. J Geod 94:90. https://doi.org/10.1007/s00190-020-01418-z

Glomsda M, Bloßfeld M, Seitz M, Angermann D, Seitz F (2022) Comparison of non-tidal loading data for application in a secular terrestrial reference frame. Earth Planets Space 74(1):1–22. https://doi.org/10.1186/s40623-022-01634-1

Gross R, Abbondanza C, Chin TM, Heflin M, Parker J (2022) A sequentially estimated terrestrial reference frame: JTRF2020, EGU General Assembly 2022, Vienna, Austria, 23–27 May 2022, EGU22-3221, https://doi.org/10.5194/egusphere-egu22-3221

Jungclaus JH, Fischer N, Haak H, Lohmann K, Marotzke J, Matei D, Mikolajewicz U, Notz D, von Storch JS (2013) Characteristics of the ocean simulations in the max Planck Institute Ocean model (MPIOM) the ocean component of the MPI-earth system model. J Adv Modeling Earth Sys 5(2):422–446. https://doi.org/10.1002/jame.20023

Krásná H, Baldreich L, Boehm J, Böhm S, Gruber J, Hellerschmied A, Jaron F, Kern L, Mayer D, Nothnagel A, Panzenboeck O, Wolf H (2023) VLBI celestial and terrestrial reference frames VIE2022b. Astronomy Astrophys 679. https://doi.org/10.1051/0004-6361/202245434

Männel B, Dobslaw H, Dill R et al (2019) Correcting surface loading at the observation level: impact on global GNSS and VLBI station networks. J Geod 93:2003–2017. https://doi.org/10.1007/s00190-019-01298-y

Petrov L (2015) The international mass loading service. In: REFAG 2014: proceedings of the IAG commission 1 symposium Kirchberg,

Luxembourg, 13–17 October, 2014. Springer International Publishing, Cham, pp 79–83. https://doi.org/10.1007/1345_2015_218

Petrov L, Boy JP (2004) Study of the atmospheric pressure loading signal in VLBI observations. J Geophys Res 109(B03405):1–14. https://doi.org/10.1029/2003jb002500

Reichle RH, Draper CS, Liu Q, Girotto M, Mahanama SPP, Koster RD, De Lannoy GJM (2017) Assessment of MERRA-2 land surface hydrology estimates. J Clim 30:2937–2960. https://doi.org/10.1175/JCLI-D-16-0720.1

Rodell M, Houser PR, Jambor U, Gottschalck J, Mitchell K, Meng C, Arsenault K, Cosgrove B, Radakovich J, Bosilovich M, Entin JK, Walker JP, Lohmann D, Toll D (2004) The global land data assimilation system. Bull Am Meteorol Soc 85(3):381–394. https://doi.org/10.1175/BAMS-85-3-381

Roggenbuck O, Thaller D, Engelhardt G, Franke S, Dach R, Steigenberger P, Steigenberger P (2015) Loading-induced deformation due to atmosphere, ocean, and hydrology: model comparisons and the impact on global SLR, VLBI, and GNSS solutions. https://doi.org/10.1007/1345_2015_214

Schuh H, Estermann G, Crétaux JF, Bergé-Nguyen M, van Dam T (2003) Investigation of hydrological and atmospheric loading by space geodetic techniques. In: Hwang C, Shum CK, Li J (eds) Satellite altimetry for geodesy, geophysics and oceanography. International Association of Geodesy Symposia, vol 126. Springer, Berlin. https://doi.org/10.1007/978-3-642-18861-9_15

Seitz M, Bloßfeld M, Angermann D, Glomsda M, Rudenko S, Zeitlhöfler J, Seitz F (2023) DTRF2020: ITRS 2020 realization of DGFI-TUM. Data Set. https://doi.org/10.5281/zenodo.8220524

Tregoning P, van Dam T (2005) Effects of atmospheric pressure loading and seven-parameter transformations on estimates of geocenter motion and station heights from space geodetic observations. J Geophys Res. https://doi.org/10.1029/2004JB003334

van Dam TM, Herring TA (1994) Detection of atmospheric pressure loading using very long baseline interferometry measurements. J Geophys Res 99:4505–4517. https://doi.org/10.1029/93JB02758

Wijaya D, Böhm J, Karbon M, Krasna H, Schuh H (2013) Atmospheric pressure loading. Springer, Berlin, pp 137–157. https://doi.org/10.1007/978-3-642-36932-2

Williams SDP, Penna NT (2011) Non-Tidal Ocean loading effects on geodetic GPS heights. Geophys Res Lett 38(9). https://doi.org/10.1029/2011GL046940

Open Access This chapter is licensed under the terms of the Creative Commons Attribution 4.0 International License (http://creativecommons.org/licenses/by/4.0/), which permits use, sharing, adaptation, distribution and reproduction in any medium or format, as long as you give appropriate credit to the original author(s) and the source, provide a link to the Creative Commons license and indicate if changes were made.

The images or other third party material in this chapter are included in the chapter's Creative Commons license, unless indicated otherwise in a credit line to the material. If material is not included in the chapter's Creative Commons license and your intended use is not permitted by statutory regulation or exceeds the permitted use, you will need to obtain permission directly from the copyright holder.

Exploring Non-tidal Atmospheric Loading Deformation Correction in GNSS Time Series Analysis Using GAMIT/GLOBK Software

Fatemeh Khorrami, Yohannes Getachew Ejigu, Jyri Näränen, Arttu Raja-Halli, and Maaria Nordman

Abstract

This study investigates the effects of non-tidal atmospheric loading on GNSS time series for a network covering the Nordic countries, with a specific focus on Finland. We processed a 5-month dataset from the year 2015 using GAMIT/GLOBK software, implementing two distinct non-tidal atmospheric loading grid models, namely 'atmfilt' and 'atmdisp'. Our results reveal that both grid models yield similar improvements in the variability of GNSS coordinate time series, albeit with a slightly better performance for 'atmdisp' grid. Our results show that implementing these built-in models in the time series analysis yields up to a 14% improvement (reduction in scatter) in the vertical component for 75% of the selected stations. However, the enhancement diminishes for the horizontal components (increase in scatter), exacerbating the eastern component of time series. The corrections lead to a 10% improvement of the North component. We also examined the effectiveness of the loading corrections by comparing our processing-level corrected time series to the daily averaged time series improved by the loading model provided by EOST loading service as a post-processing approach. Given the relatively short 5-month duration of the time series, drawing definitive conclusions when comparing models is challenging. However, it is evident that the GNSS time series exhibits distinct variations related to atmospheric loading in their vertical positions across the various models that were examined.

Keywords

Finland · GAMIT/GLOBK · GNSS · Non-tidal atmospheric loading

1 Introduction

The surface of the Earth is undergoing constant deformation at different spatial and temporal scales because of a range of factors, such as tectonic and volcanic activities, melting ice sheets, and redistribution of the Earth's masses. Some of these mass movements are periodic, for example, tides in the oceans and in the atmosphere, but there are also non-periodic displacements generated by mass redistribution, for example, in the atmosphere and oceans, and due to hydrological processes (e.g., Darwin 1882; van Dam and Wahr 1987; Schuh et al. 2003). Tidal corrections to geodetic GNSS time series are performed routinely, following recommendations and guidelines of the IERS Conventions (Petit and Luzum 2010), while there is currently no conventional model recommended for correcting non-tidal loading (NTL) deformations. NTL is usually divided into three components: non-tidal atmospheric loading (NTAL), non-tidal ocean loading (NTOL), and loading due to changes in hydrology (HL). At high latitudes, our understanding of local and regional deformations is

F. Khorrami (✉) · Y. G. Ejigu
School of Engineering, Aalto University, Espoo, Finland
e-mail: fatemeh.khorrami@aalto.fi

J. Näränen · A. Raja-Halli
Finnish Geospatial Research Institute FGI, Espoo, Finland

M. Nordman
School of Engineering, Aalto University, Espoo, Finland
Finnish Geospatial Research Institute FGI, Espoo, Finland

limited because of insufficient data. Some global studies (e.g., Mémin et al. 2020) revealed that high latitude stations are subject to distinct loading effects when compared to low-latitude stations. Therefore, examining the GNSS position time series is necessary to uncover unmodeled (nonlinear) effects, which hinders the achievement of highly accurate reference frames and a deeper understanding of geophysical processes.

Different approaches can be used to implement NTL corrections in GNSS time series. Most NTL corrections are done by employing daily averaged corrections on daily coordinate time series at post-processing level (e.g., Mémin et al. 2020). The alternative approach is to correct NTL deformations at the processing level (e.g., Gégout et al. 2010). In recent years, various studies have examined the impact of NTL deformation on GNSS coordinates at processing level using different packages for GNSS data processing such as GIPSY, Bernese, and EPOS (GFZ package) (e.g., van Dam et al. 1994; Haritonova 2019, 2021; Männel et al. 2019).

This study focuses on validating the effectiveness of built-in corrections of Non-Tidal Atmospheric Loading (NTAL) on GNSS (in this research GPS-only) time series using GAMIT/GLOBK software, in Finland and the surrounding high-latitude regions. More precisely, we concentrated on 60 GNSS sites situated between 55°N and 70°N latitude. At the time of the study, there were two built-in grid models available for removing non-tidal atmospheric effects in GAMIT/GLOBK software. In the first stage of our research, we utilized these NTAL grid models for removing NTAL. As a result, we obtained two specific corrected solutions, allowing us to evaluate each model comprehensively.

During the past decade, model-based NTL corrections have been developed widely by different Earth System Modeling groups, including the School and Observatory of Earth Sciences (EOST, Univ. Strasbourg) loading service (http://loading.u-strasbg.fr). Thus, in the second step, we corrected daily time series with the NTAL corrections from EOST service and compared them to our time series of step 1. In this first case study, these two approaches, within GNSS-processing and post-processing, allow us to compare the usability and effectiveness of the NTAL corrections, a topic that has not been studied in detail before.

In this study, we aim to understand the potential of loading models on GNSS time series within processing level. Also, this research examines whether the incorporation of loading corrections demonstrate similar pattern of positioning improvement at both processing and post-processing level. Due to the short dataset, clear results may be limited. The study is organised as follows: the second section introduces the data and methods used for our study, the third section shows the results and discussion, and the last section is left for summary and conclusions.

2 Data and Methods

In the following, we explain briefly the methodology employed for generating GAMIT/GLOBK position time series improved by applying appropriate NTAL grid corrections ('atmfilt' and 'atmdisp'). At the time of our research, these NTAL corrections were the only grid models available within the software. We also explain the NTAL models provided by EOST that can be used on position time series generated by GAMIT/GLOBK.

2.1 GNSS Data Processing

For the first part of our study, we processed a dataset that included 60 continuous GNSS stations belonging to FinnRef (Finnish Permanent GNSS Network) and/or EPN (EUREF Permanent GNSS Network). Figure 1 shows the location of the stations, specifically all permanent GNSS stations located in Finland, used for processing. We chose a 5-month period starting from January 1, 2015, because in our preliminary visual inspection, the loading correction showed a large variability during that time. The data were processed by double-differenced ionospheric-free phase observations using GAMIT/GLOBK (Ver. 10.71) software (Herring et al. 2018).

Processing was carried out in two steps: (1) generating phase carrier reduction in a loosely constrained solution and (2) GAMIT solutions combination using Kalman filter (GLOBK program) to extract daily coordinate time series in IGB14 [The IGS realization of ITRF2014 (Altamimi et al. 2016)] reference frame. In the first step, the atmospheric zenith delay was estimated every 2-h using a piecewise linear model. We imposed a 10-degree cut-off elevation and employed the IGS14.atx antenna phase center model. Additionally, the Vienna Mapping Function (VMF1) (Boehm et al. 2006) was used for tropospheric mapping, and the FES2004 ocean loading model (Scherneck 1991; Lyard et al. 2006) was utilized. The processing scheme in this study was mainly based on three distinct types of solutions: (1) the first solution referred to as *ref_sol*, where no NTAL correction was applied, (2) the *disp_sol* denoted the second solution, wherein the 'atmdisp' grid model in the Center of Figure (CF) reference frame was utilized to eliminate the NTAL effects and (3) the third solution, termed *filt_sol*, involved the application of an alternative NTAL correction known as 'atmfilt' in the same reference frame. The grid models namely 'atmfilt' and 'atmdisp' are available on ftp://everest.mit.edu/pub/GRIDS and provided in three different geodetic reference frames: CM (center of mass of the whole Earth), CF (center of figure of the outer surface of the solid Earth) and CE (center of mass of the solid Earth without mass load). We used models in CF as our time series were

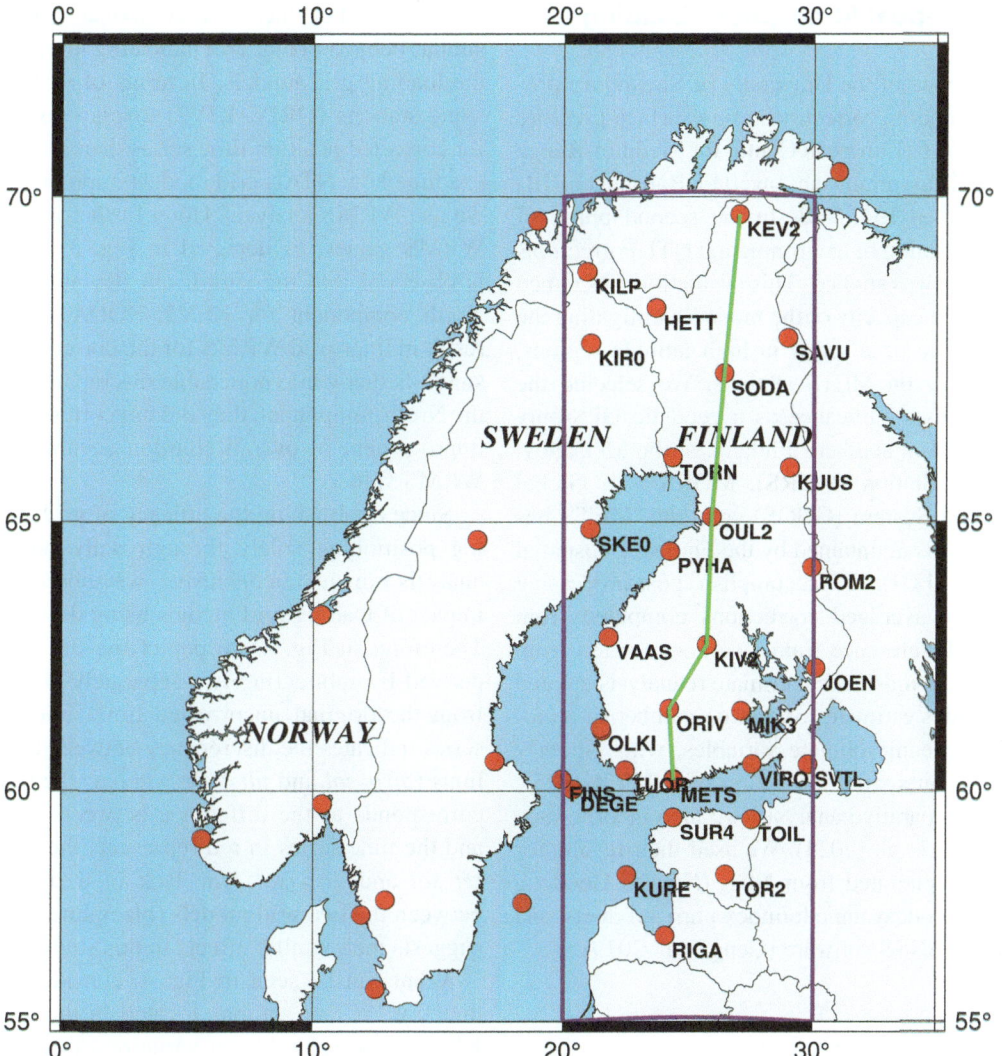

Fig. 1 Distribution of permanent GNSS sites used in this research. Sites belong to FinnRef and EPN. The purple box indicates the selected area for WRMS reduction analysis and the stations along the green line are shown later in Fig. 2 and Fig. 4

transferred into ITRF2014 (Blewitt 2003). These NTAL grids were produced by Tregoning and van Dam (2005) based on the NCEP (National Centers for Environmental Prediction) global pressure data, with 6-hourly global temporal sampling on a 2.5° × 2.5° regular grid using inverted barometer assumption. The 'atmfilt' grid is a model applying a filter to separate out tidal atmospheric loading from non-tidal atmospheric loading. This filter removes the tidal energy from the non-tidal part. The 'atmdisp' model includes both tidal and non-tidal components. To avoid double removal of tidal atmospheric corrections, the tidal atmospheric corrections are not used elsewhere in the processing. We refer the reader to Tregoning and van Dam (2005) for more details regarding these grid models. Then for generating daily position time series, we used 11 IGS core stations to define reference frame for stabilization using GLOBK software.

As a result, we obtained three discrete position time series for each station. We cleaned the time series by identifying the outliers by visual inspection and removing them by re-processing GAMIT solution. In addition to the visual comparison of time series, we also evaluated the effectiveness of NTAL models by comparing the Weighted Root Mean Square (WRMS) reduction for different solutions (i.e. *disp_sol* and *filt_sol*). WRMS reduction was calculated by Eq. (1) in which $WRMS_{ref}$ and $WRMS_{corrected}$ indicate WRMS for time series with and without corrections respectively.

$$WRMS_{reduction} (\%) = \frac{WRMS_{ref} - WRMS_{corrected}}{WRMS_{ref}} \times 100 \quad (1)$$

2.2 EOST Non-tidal Atmospheric Loading

EOST Loading Service of the University of Strasbourg provides atmospheric loading corrections using surface-pressure data from the ECMWF (European Centre for Medium-Range Weather Forecasts), assuming an Inverted Barometer (IB) ocean response, called ERA5-IB. In the second phase of our study, the contribution of environmental NTL corrections introduced by EOST was studied. This evaluation was aimed at comprehending the capacity of the model in mitigating the variability of geodetic time series in high latitude regions, with a specific focus on METS stations. We selected the METS site due to its significance as a geodetic GPS-only IGS core site located in southern Finland, at the Metsähovi Geodetic Research Station (MGRS). MGRS is a Global Geodetic Observing System (GGOS) core site. METS has been established and is maintained by the Finnish Geospatial Research Institute (FGI). To accomplish post-processing approach, the daily averaged corrections computed from ERA5-IB in the CF reference frame were subtracted from *ref_sol*. ERA5-IB, the most recent climate reanalysis created by ECMWF, provides estimates of a large number of atmospheric, land and oceanic climate variables, with 1-hourly temporal sampling on spatial resolution of $0.25° \times 0.25°$. To enhance our comparative analysis, we also incorporated the findings of Ejigu et al. (2024). We used their results for METS time series generated from NGL (Nevada Geodetic Laboratory, http://geodesy.unr.edu/index.php) products and PRIDE-AR ver 2.2 GNSS software (Geng et al. 2019).

3 Results and Discussion

To explore the impact of correction grid models on positioning variations, we illustrate our findings using GPS position time series for some stations in Fig. 2. To gain an insight into how models are influenced by geography (moving toward the North Pole), we choose specific stations along a south-to-north path in Finland. METS, ORIV, KIV2, OUL2, SODA, and KEV2 are selected with roughly the same longitude (green line in Fig. 1). The *ref_sol*, *disp_sol* and *filt_sol* solutions obtained by GAMIT/GLOBK software are shown in black, orange, and blue lines respectively in Fig. 2a–c. The initial finding derived from the daily coordinate time series in Fig. 2 indicates that both NTAL grid models exhibit nearly identical improvements in displacements as evidenced by the close alignment of the orange and blue time series in Fig. 2. For the North component, discernible deviations are apparent in the corrected time series compared to the reference solution, for stations situated at higher latitudes such as KEV2, SODA, and OUL2. However, similar deviations in the vertical and East components for northern stations are not readily discernible. For the East component, all solutions show similar behavior (Fig. 2b), indicating minimal influence from the loading grid models. In terms of vertical displacement, some stations (ORIV, KIV2) display less overall variation for corrected position time series than reference solution. To examine how NTAL grid models improve position accuracy, we use WRMS analysis. Upon further investigation into the WRMS values, as depicted in Fig. 3 for selected sites, it is observed that the significant deviations observed in the North component for KEV2, SODA, and OUL2 did not result in improved WRMS for the corrected time series. This suggests that while noticeable discrepancies were evident in the North component, they did not correspond to a significant improvement in overall solution accuracy as measured by WRMS values.

Since establishing the efficacy of grid models in improving positioning solely through daily position time series analysis remains inconclusive, we subsequently explore the impact of loading grid models using differential time series. Therefore, in Fig. 4 we depict the differential time series derived by subtracting two separately corrected time series from the original uncorrected time series. The black time series indicates the discrepancy between two corrected solutions (*disp_sol* and *filt_sol*), whereas the time series in blue corresponds to the difference between *ref_sol* and *filt_sol*, and the time series in red represents the difference between *ref_sol* and *disp_sol*. The lack of a significant difference between the two grid models (black line) (less than 0.5 mm) suggests their similar effects in this study.

What can be seen in Fig. 4. considering south-to-north direction, is that stations located in the northern area, like KEV2, experience bigger variation as compared to southern sites, like METS in their horizontal time series. However, this discernible pattern of further changes is not observed for the vertical component when we move from south to north. Although an increasing pattern of south-to-north orientation is absent across vertical movement, all stations' differential time series exhibit an evident influence on positioning during the snow season, specifically for height deformation, This result is in line with the study of van Dam and Wahr (1987). They demonstrated that displacements are generally larger at higher latitudes and during the winter months due to larger atmospheric pressure variations. Some of the variability may be attributed to the accumulation of snow atop GNSS antennas, as mentioned in Lahtinen et al. (2022). As expected, the vertical component experiences a larger improvement in comparison to the horizontal components shown previously by e.g., Mémin et al. (2020). Moreover, the differential time series (red and blue lines in Fig. 4) follow an identical pattern across all stations, particularly in the vertical components. This conformity can be attributed to the low resolution of the grid models used in these results.

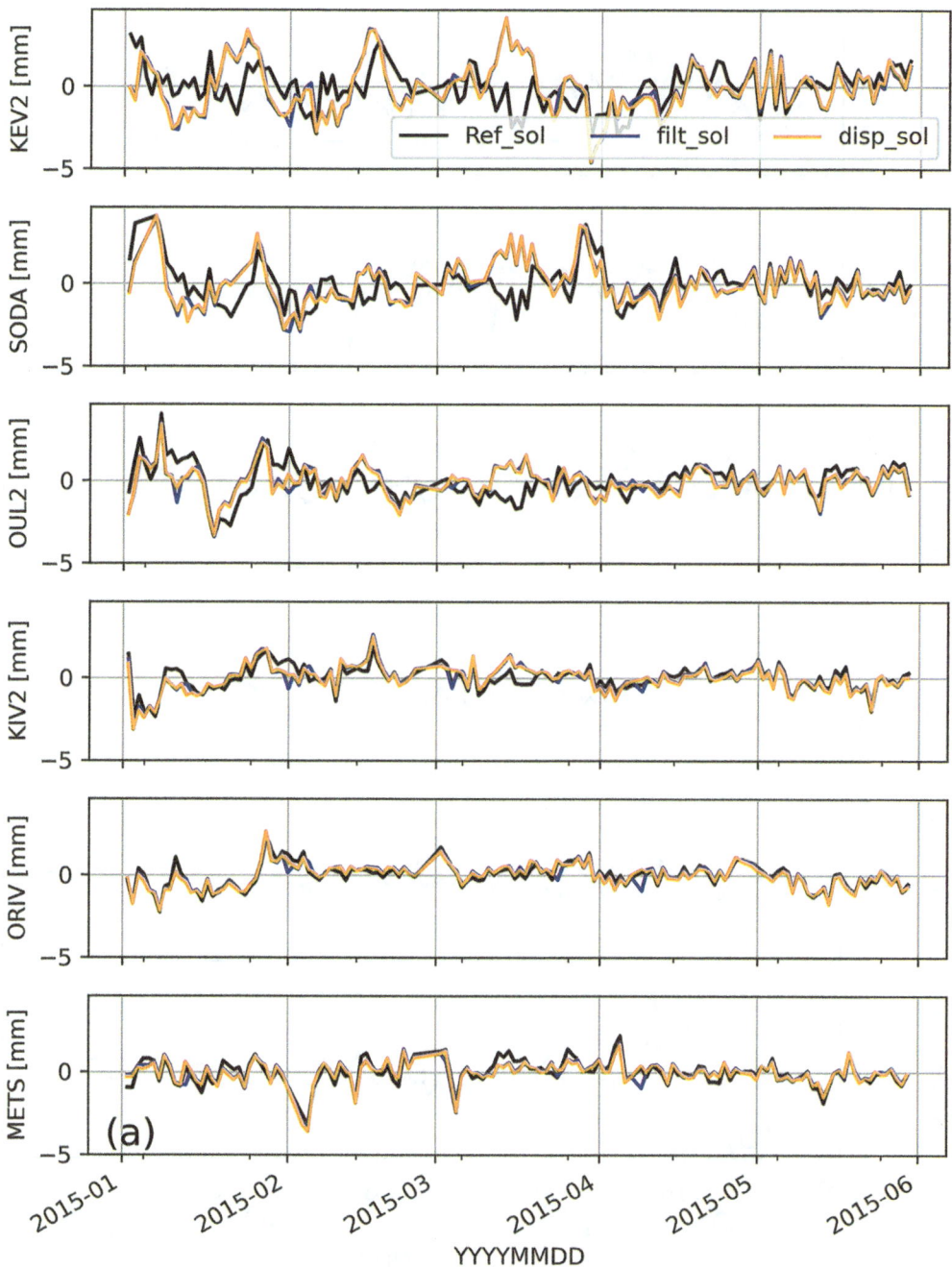

Fig. 2 Position time series obtained from GAMIT/GLOBK analysis. Black, orange, and blue lines indicate Ref_sol, disp_sol and filt_sol solutions respectively. From bottom to top, stations sorted from south-to-north direction (green line in Fig. 1). (**a**), (**b**) and (**c**) represent north, east, and vertical components respectively for selected sites. Note the different scales for North, East and Up components

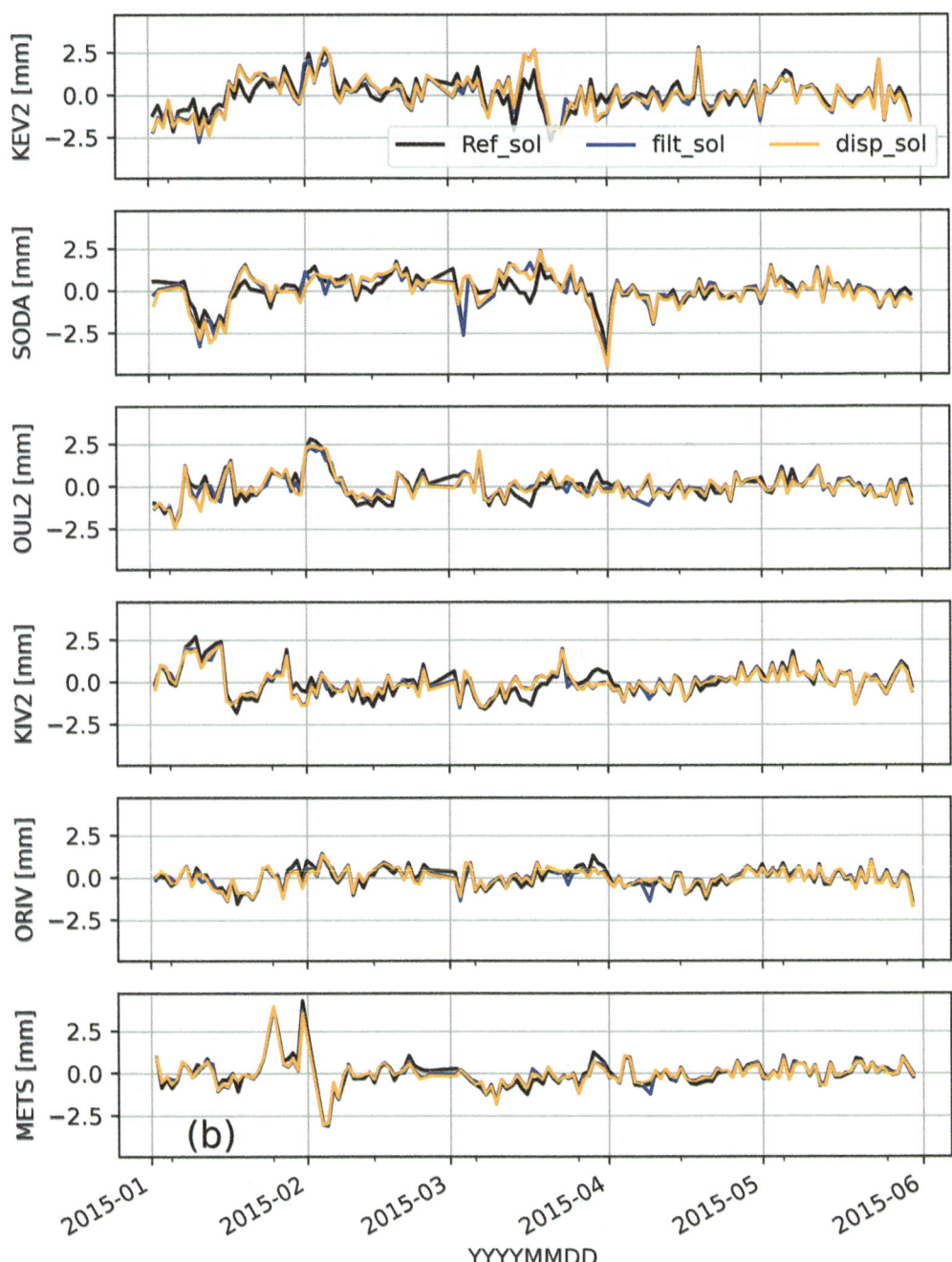

Fig. 2 continued

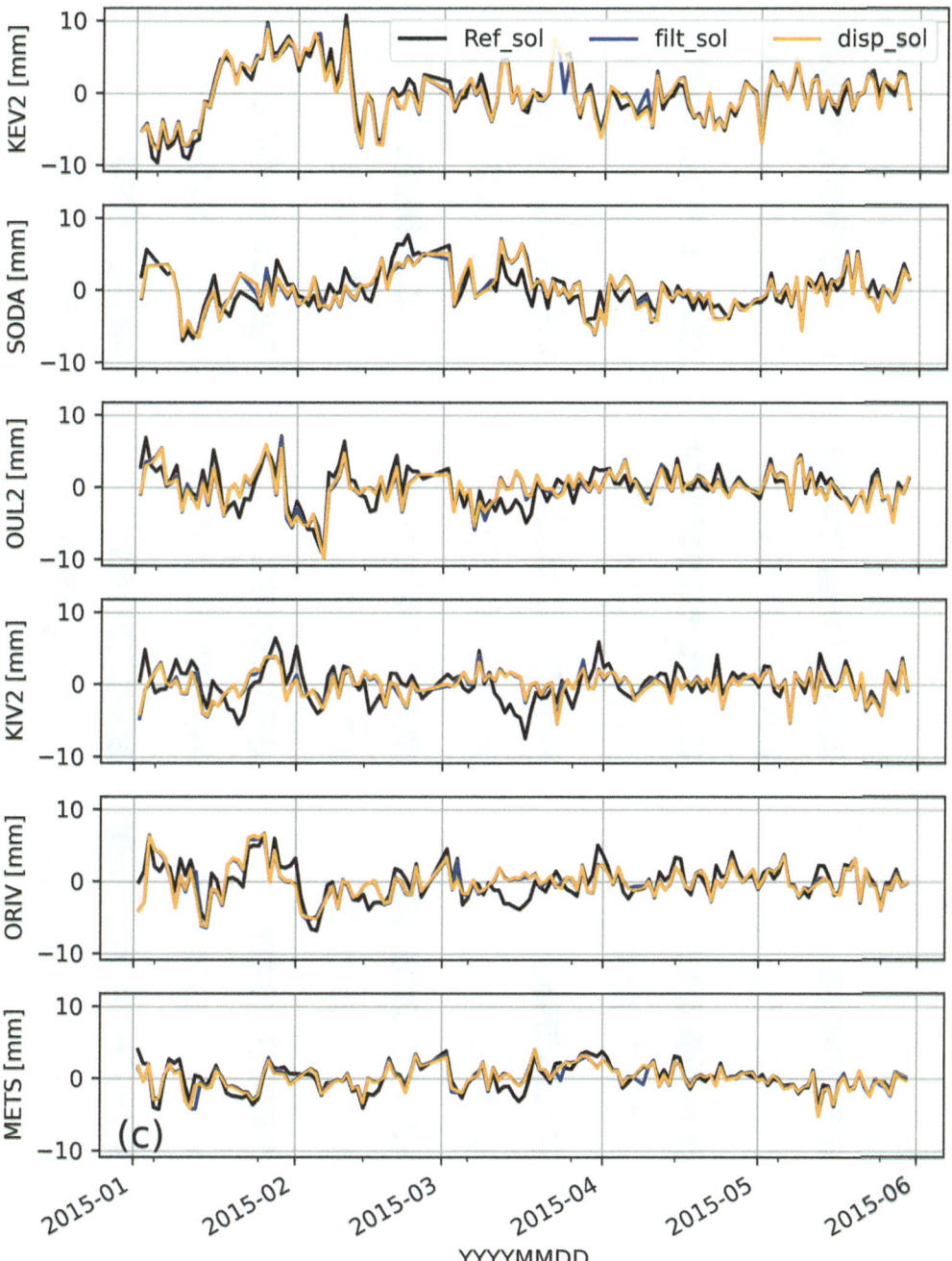

Fig. 2 continued

We also evaluated the effectiveness of the NTAL factor by plotting the Weighted Root Mean Square (WRMS) reduction (Eq. 1, Fig. 5) for the 30 stations in the purple box in Fig. 1. The higher RMS reduction value corresponds to better performance in the loading models. Figure 5a shows 'atmfilt' and 'atmdisp' grids have identical potential for improvement of positioning, albeit with a slightly better performance for the 'atmdisp'' grid. As shown in Fig. 5a, stations are sorted from south to north.

Examining Fig. 5b, 50% of stations' north displacements are improved, i.e., the RMS is reduced, up to 10% while roughly 75% of stations experience deterioration for their East component. This implies that implementing 'atmdisp' and 'atmfilt' grid models fails to improve position time series in the East direction. Accurately understanding the cause of the differing behavior of the two horizontal components is challenging. This discrepancy may be attributed to the low spatial resolution of 'atmdisp' and 'atmfilt', which might be insufficient to precisely capture the complex interactions

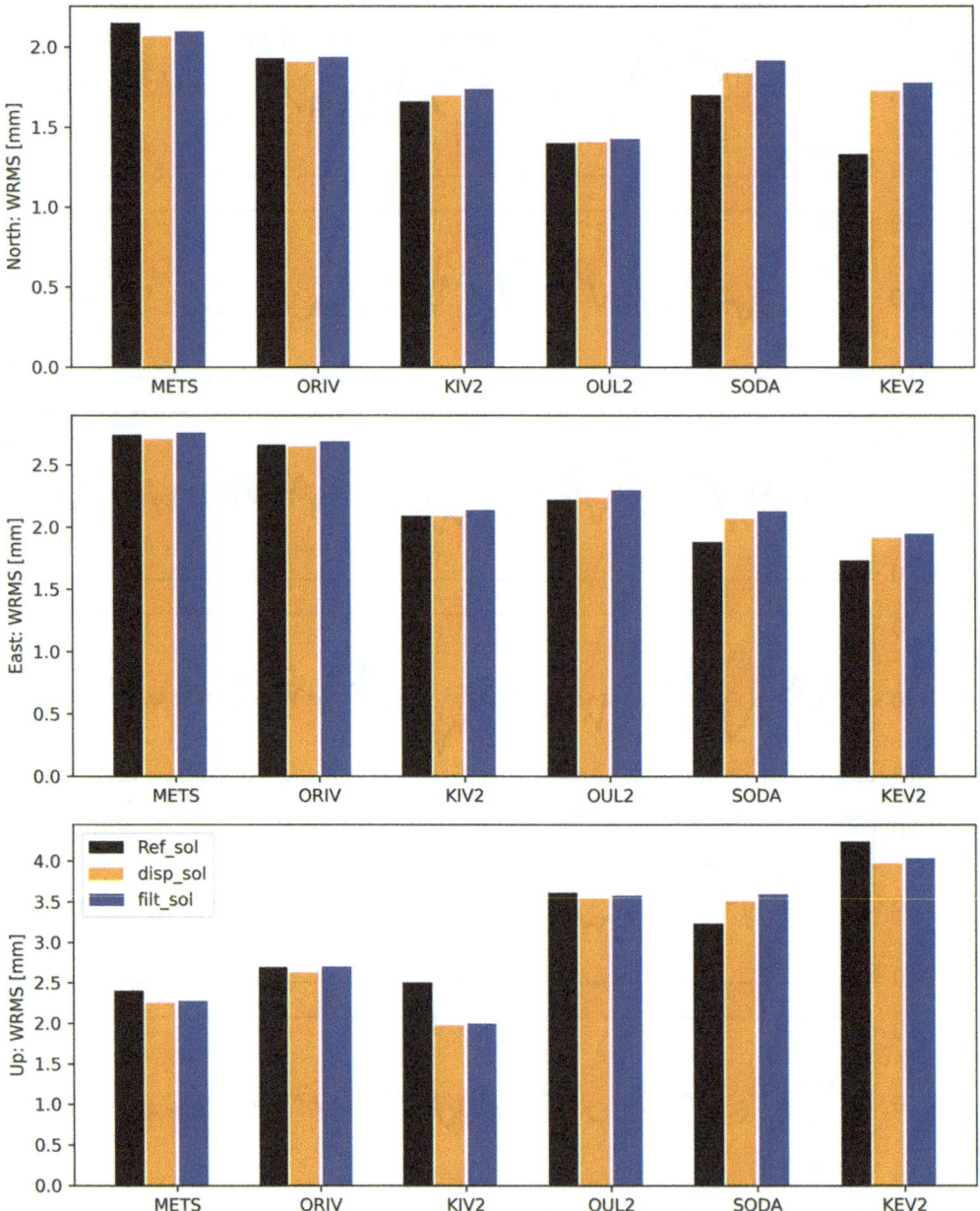

Fig. 3 WRMS values for six selected stations. Black, orange, blue bar lines indicate ref_sol, disp_sol, filt_sol solutions respectively

between the Baltic Sea and atmospheric pressure. Furthermore, it is visible that vertical component experience larger improvement compared to North and East components (Fig. 5a). Upon closer examination of Fig. 5b, most sites (75% in case of disp_sol) exhibit more enhancement (up to 15%) for vertical positions. Li et al. (2020) demonstrated that NTAL models perform better at stations in higher latitudes because changes in atmospheric pressure have a more significant impact in high latitude regions compared to lower latitude. In their research, NTAL models reduced measurement variability by about 30% in northern stations (latitude >34°N) but only about 5% in southern stations (latitude <30°N).

They believed that this observed latitude-dependent disparity in performance can arise from the substantial atmospheric pressure changes caused by monsoons in subtropical regions. However, these findings may not be directly applicable to this study, as we lack a clear conclusion regarding the correlation between our models' performance and latitude-dependent RMS reduction for our specific monitoring stations.

In the second phase of our study, we undertook a comparative analysis to assess the influence of implementing NTAL models at both the data processing and post-processing levels. This comparison helps us reveal the difference between approaches where non-tidal correction is applied within the

Fig. 4 Differences between time series obtained from GAMIT/GLOBK analysis. Black line indicates difference between disp_sol and filt_sol. Blue and red lines show difference between reference solution and improved one. From bottom to top, stations sorted from south-to-north direction (green line in Fig. 1). (**a**), (**b**) and (**c**) represent north, east, and vertical components respectively for selected sites. Note the different scale for North, East and Up components

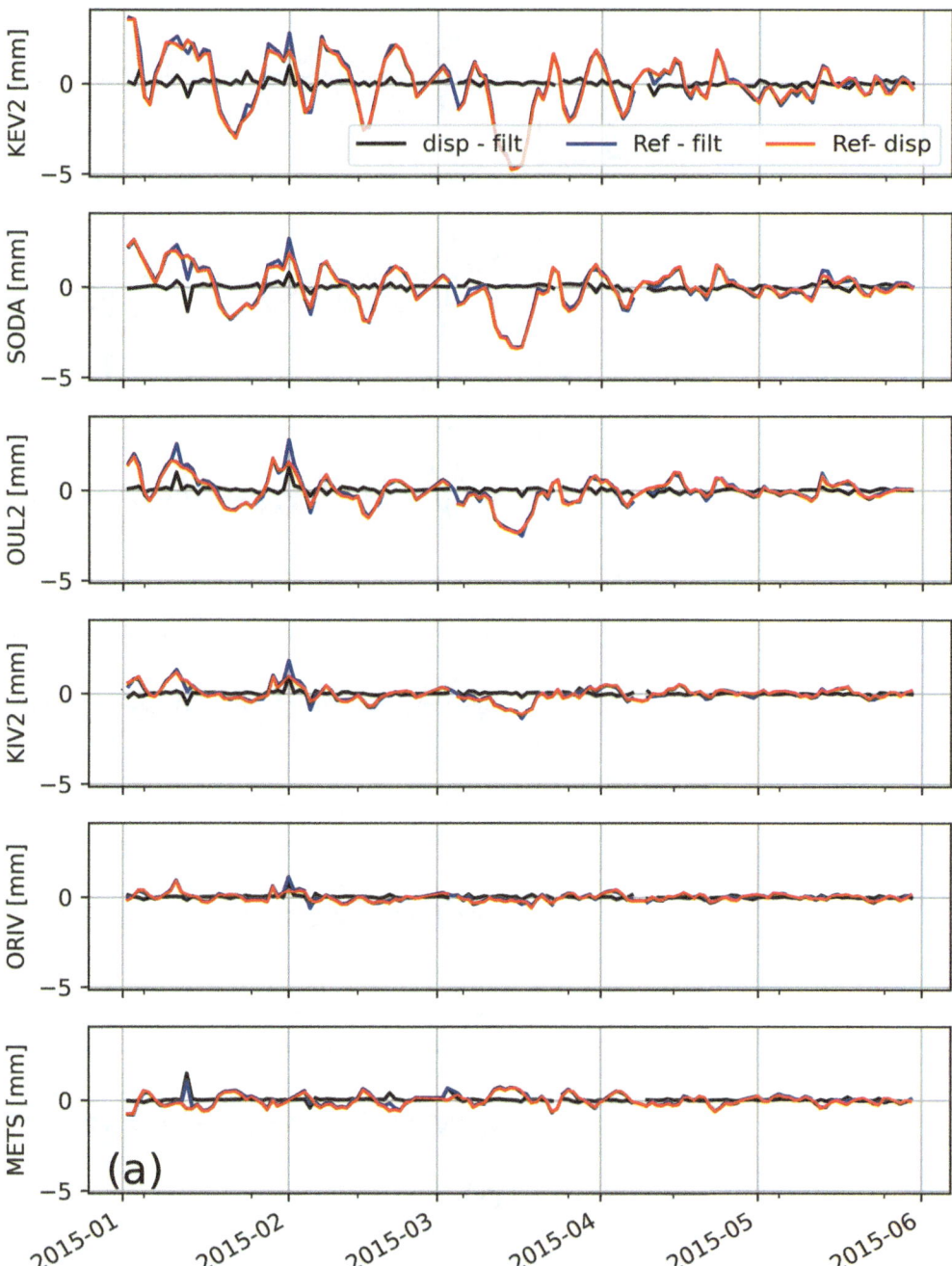

data processing rather than as a daily averaged coordinate. Herein, we present our results exclusively for the METS site.

Initially, we applied the ERA5-IB model, the NTAL product developed by the EOST loading service, to METS time series. The ERA5-IB was subtracted from the daily-averaged *ref_sol* time series obtained using GAMIT/GLOBK processing and led to the corrected METS time series at the post-processing level (dotted green line in Fig. 6). In the context of position time series correction at the data processing level, we selected *disp_sol*, corrected by 'atmdisp' grid. We chose *disp_sol* because it showed a slightly better improvement compared to the 'atmfilt' grid, as seen in Fig. 5.

The time series, following corrections by the 'atmdisp' grid model and the ERA5-IB model (orange line and dotted green line, respectively, in Fig. 6), exhibit close compatibility for horizontal components. However, a pronounced difference becomes apparent in the vertical variations during winter, within the initial first 3 months of the year 2015, exceeding 10 mm for peak-to-peak of two corrected time series (orange and green dotted lines in Fig. 6). Notably, from March onward, this disparity gradually diminishes, leading to a convergence of the two correction methodologies. It is also worth noting that the spatial resolution of the two models (NCEP and ERA5) is different; therefore, some of the

Fig. 4 (continued)

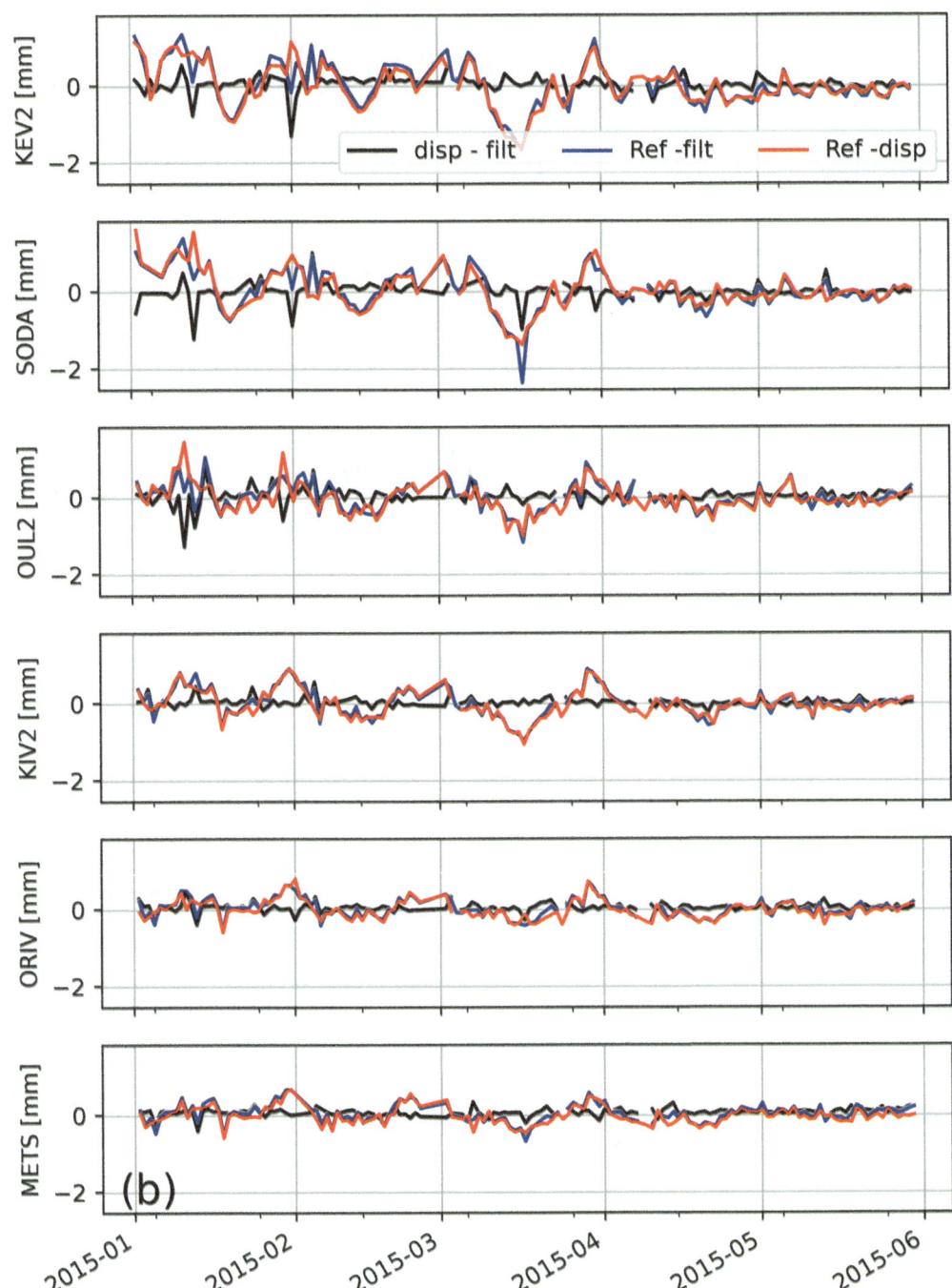

inconsistencies may stem from that. Additionally, limitations in dataset processing prevent us from definitively concluding that the greater variability observed in the corrected vertical displacement by ERA5 during the winter season may contribute to diminished enhancements in positioning accuracy.

In a later stage, we looked at how integrating the ERA5-IB model affected across various time series solutions during the post-processing stage. To achieve this, we utilized daily averaged displacements derived from the ERA5-IB model on METS time series solutions generated by the PRIDE software (https://github.com/PrideLab/PRIDE-PPPAR), as well as the NGL (http://geodesy.unr.edu/index.php) (full results in Ejigu et al. 2024) and GAMIT/GLOBK. NGL and PRIDE time series have similar processing strategies based on precise point positioning (PPP), while GAMIT processes data in double difference (DD) mode. Although direct comparison of results was not feasible; our objective was to determine whether these methodologies produce similar improvement patterns for GNSS time series. To facilitate clearer visualization, we plotted corrected and uncorrected time series separately, employing an offset to corrected ones. NTAL corrected METS time series for three solutions (GAMIT,

Fig. 4 (continued)

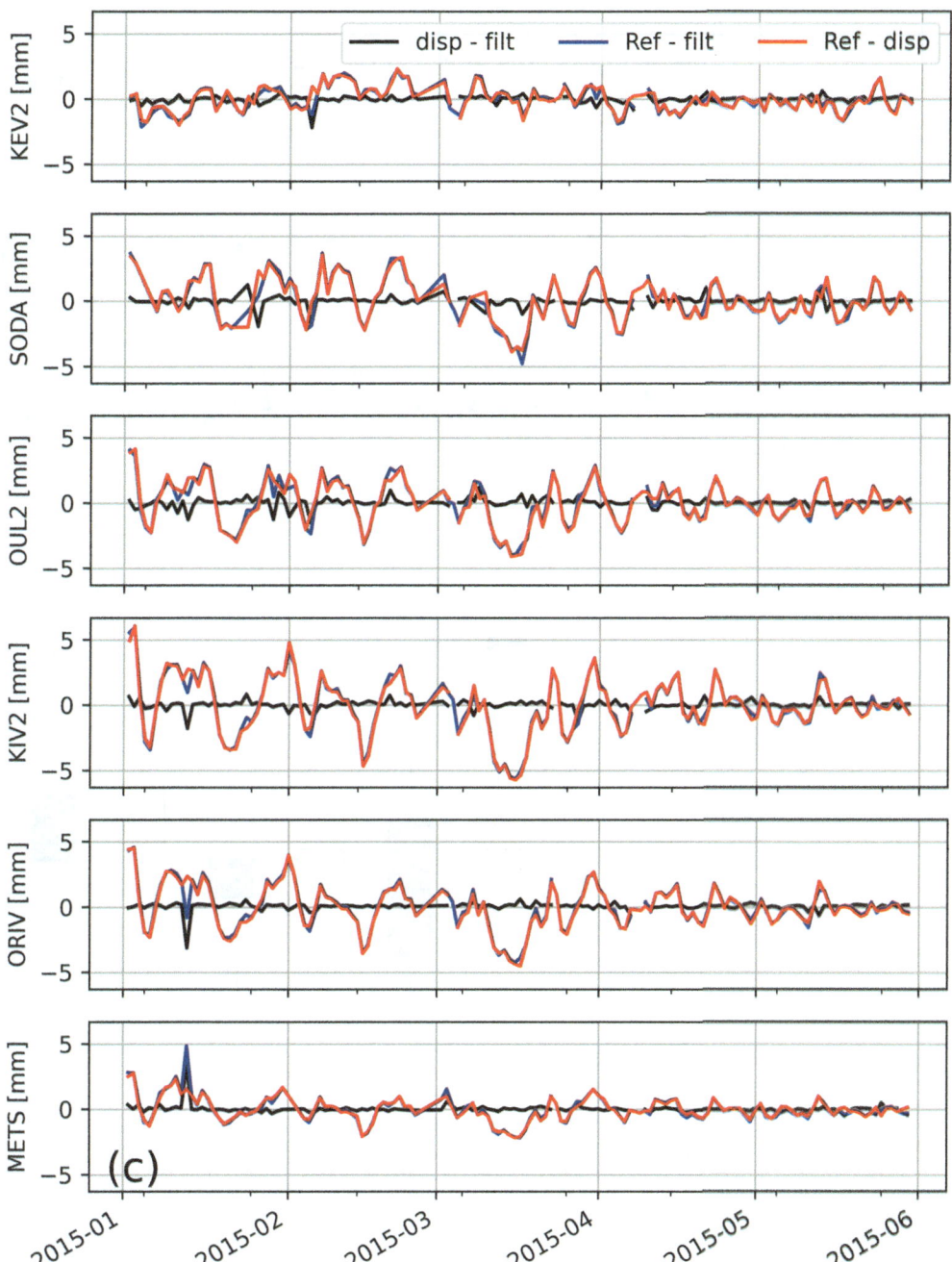

NGL and PRIDE) can be seen in Fig. 7. The black, green, and red lines depict uncorrected solutions, while the purple, olive and violet dashed lines represent the corrected time series. It is evident that the ERA5-IB model significantly influences vertical deformation across all solutions during the winter months. The application of ERA5-IB correction leads to a noticeable reduction in the height variability of the NGL and PRIDE time series, while the GAMIT corrected data exhibit a larger variance for height displacement. This implies that there is an inconsistent behavior of corrected solutions with respect to vertical displacement over snow season. This inconsistency can be attributed to the newer version of the ERA5 model compared to models derived from NCEP. Consequently, the ERA5 model is probably more capable of capturing the variability of height displacement during winter time. The updated model uses better algorithms and higher resolution data, improving its ability to represent the complex interactions and atmospheric dynamics of the winter season. However, from March onwards, all solutions exhibit a roughly similar pattern of enhancement for both horizontal and vertical components. Thus, we can conclude that regardless of the processing strategy, we see similar patterns leading us to believe that the cause for the variability is the same and can be corrected for. One further detail is that

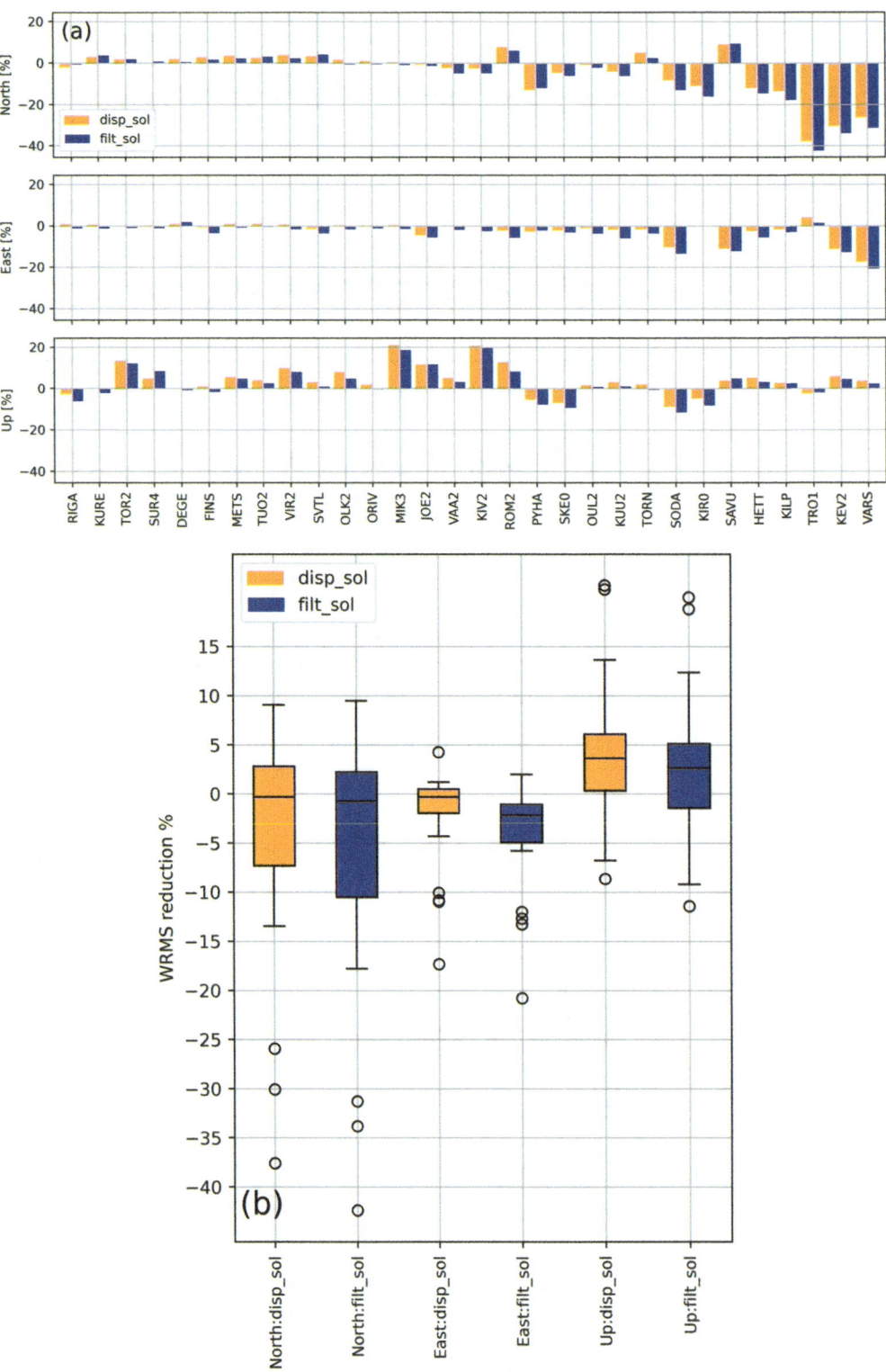

Fig. 5 Comparing effectiveness of *disp_sol* and *filt_sol* resulted from GAMIT/GLOBK analysis. (**a**) WRMS reduction for 30 stations located in 20°- 30°E (purple box in Fig. 1) and sorted from south to north direction. Orange bars indicate *disp_sol* and blue bars the *filt_sol*. (**b**) Box plots of the distribution of WRMS values for the North, East, and Up components for 30 stations, comparing the performance of *disp_sol* and *filt_sol*. The boxes in the plot (**b**) represents the 50% of results whereas the whiskers show the whole data range. The individual points illustrate the outliers

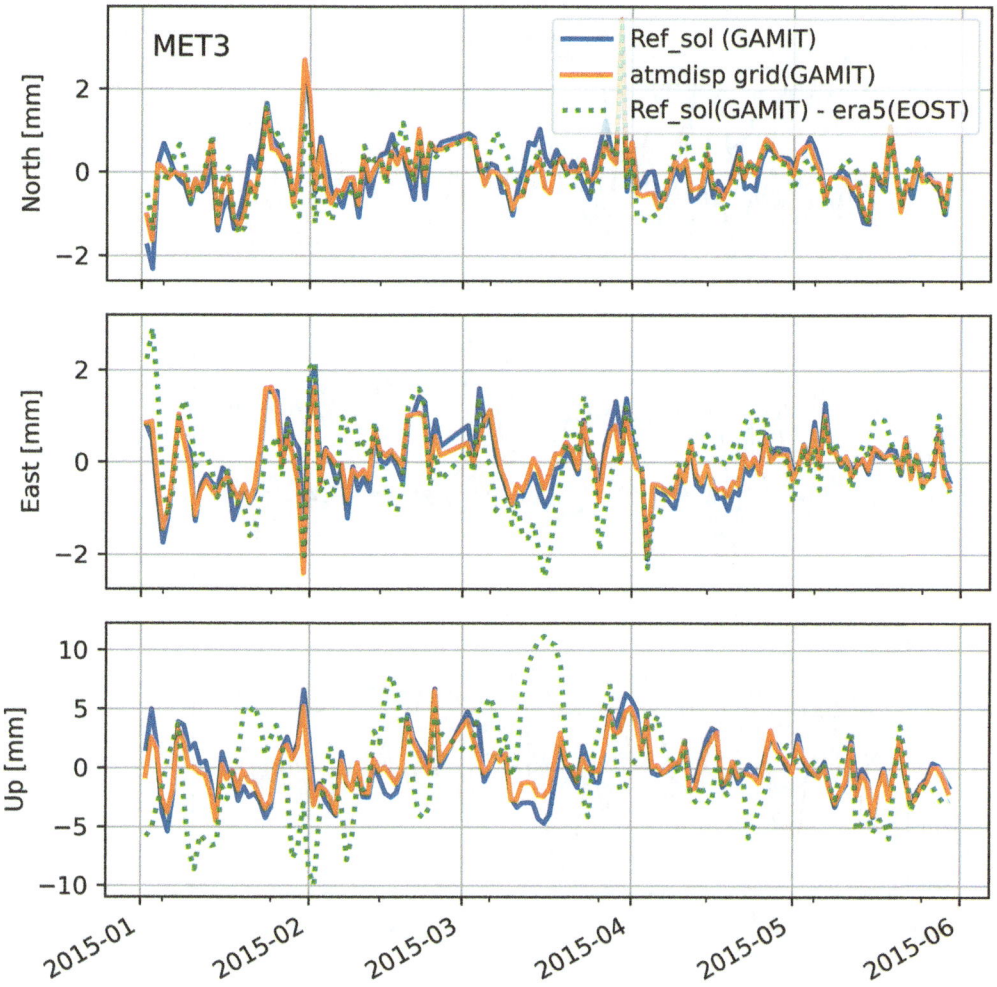

Fig. 6 Comparison between application of NTAL corrections at processing level (GAMIT/GLOBK processing) and on position time series (EOST model). Blue line: *ref_sol*. The orange line: *disp_sol*. Dotted green line: *ref_sol* corrected with EOST NTAL loading correction

both NTAL corrections use IB response, which might not be optimal for shallow seas, such as the Baltic Sea. Understanding the Baltic Sea response in spatial and temporal domain requires further research and is out-of-scope for this short study.

4 Conclusions and Outlook

This research investigated the potential improvement in position GNSS time series by integrating NTAL during the GAMIT/GLOBK processing stage, focusing on high-latitude regions like Finland. Our research reveals the following key findings: (1) 'atmdisp' and 'atmfilt', two NTAL grid models utilized by GAMIT/GLOBK, have similar impacts on GNSS time series. (2) Due to larger pressure variations, the greatest impact of NTAL models can be seen on height displacements during the winter months at higher latitudes. (3) Incorporating NTAL grids results in greater enhancements in vertical movements rather than horizontal displacements. (4) Applying the daily averaged NTAL correction in the post-processing stage yields similar effects on GNSS time series positioning as using grid models in GAMIT processing. The similarity is evident in horizontal displacements but differs for time series heights in winter. It is difficult to conclusively determine which approach provides a more reliable interpretation due to the dissimilar spatial resolution between the models and our short-period time series processing. Moreover, the assumption regarding the behavior of oceans as inverted barometers in NTAL corrections present further

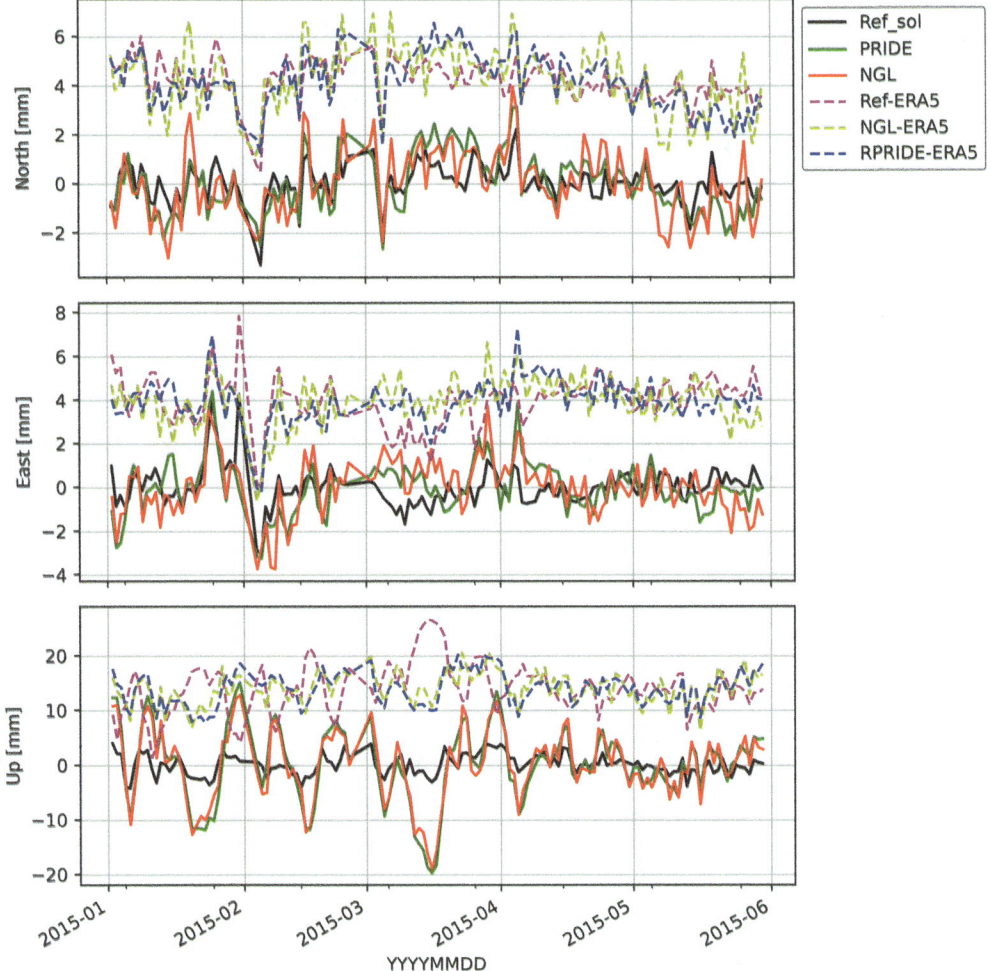

Fig. 7 Comparative analysis of applying NTAL model, ERA5-IB, on METS time series at post-processing stage. The red, green, and black lines: solutions generated by NGL, PRIDE and GAMIT respectively without NTAL correction. The violet, purple, and olive dash lines: corrected METS time series by ERA5-IB. Graphical representation of uncorrected (solid lines) and corrected (dashed lines) solutions, each plotted with an offset to enhance clarity

constraints on the conclusions drawn. For future research, we will prolong the duration of GAMIT/GLOBK processing for minimum of a 2-year period to observe the yearly variability and apply a new grid model correction implementing in GAMIT program. The new grid, sourced from the Vienna Mapping Function (VMF), is based on the ECMWF numerical weather model and features a higher spatial resolution of $1° \times 1°$, compared to the previous NCEP-derived grids used in this paper, which had a resolution of $2.5° \times 2.5°$.

Acknowledgements We would like to express our gratitude to Sonja Lahtinen of FGI for generously providing the FinnRef data for the year 2015 and for their invaluable support throughout the course of this research. We also acknowledge the EOST service for providing loading model productions.

References

Altamimi Z, Rebischung P, Métivier L, Collilieux X (2016) ITRF2014: a new release of the International Terrestrial Reference Frame modeling nonlinear station motions. J Geophys Res 121(8):6109–6131

Blewitt G (2003) Self-consistency in reference frames, geocenter definition, and surface loading of the solid Earth. J Geophys Res Solid Earth 108(B2)

Boehm J, Werl B, Schuh H (2006) Troposphere mapping functions for GPS and very long baseline interferometry from European Centre for Medium-Range Weather Forecasts operational analysis data. J Geophys Res 111(B)

Darwin GH (1882) On variations in the vertical due to elasticity of the Earth's surface. Science 14(90):409–427

Ejigu YG, Boy, JP, Raja-Halli A, Khorrami F, Naranen J, Nordman M (2024) Analyzing the 3D deformation induced by nontidal loading in GNSS time series in Finland. In: International Association of Geodesy Symposia. Springer, Berlin, Heidelberg. https://doi.org/10.1007/1345_2024_259

Gégout P, Boy JP, Hinderer J, Ferhat G (2010) Modeling and observation of loading contribution to time-variable GPS sites positions. Int Assoc Geod Symp 135(8):651–959

Geng J, Chen X, Pan Y, Mao S, Li C, Zhou J, Zhang K (2019) PRIDE PPP-AR: an open-source software for GNSS PPP ambiguity resolution. GPS Solut 23:1–10

Haritonova D (2019) The impact of the Baltic Sea non-tidal loading on GNSS station coordinate time series: the case of Latvia. Baltic J Mod Comput 7(4):541–549. https://doi.org/10.22364/bjmc.2019.7.4.07

Haritonova D (2021) Non-tidal loading of the Baltic Sea in Latvian GNSS time series. J Appl Geod 15(4):293–304. https://doi.org/10.1515/jag-2021-0024

Herring TA, King RW, Floyd MA, McClusky SC (2018) Introduction to GAMIT/GLOBK, Release 10.7. http://geoweb.mit.edu/gg/Intro_GG.pdf

Lahtinen S, Jivall L, Häkli P, Nordman M (2022) Updated GNSS velocity solution in the Nordic and Baltic countries with a semi-automatic offset detection method. GPS Solut 26(1):9. https://doi.org/10.1007/s10291-021-01194-z

Li C, Huang S, Chen Q, van Dam TM, Fok H, Zhao Q, Wang X (2020) Quantitative evaluation of environmental loading induced displacement products for correcting GNSS time series in CMONOC. Remote Sens 12(4):594

Lyard F, Lefevre F, Letellier T (2006) Modelling the global ocean tides: modern insights from FES2004. Ocean Dyn 56. https://doi.org/10.1007/s10236-006-0086-x

Männel B, Dobslaw H, Dill R, Glaser S, Balidakis K, Thomas M, Schuh H (2019) Correcting surface loading at the observation level: impact on global GNSS and VLBI station networks. J Geod 93(10):2003–2017. https://doi.org/10.1007/s00190-019-01298-y

Mémin A, Boy JP, Santamaria-Gomez A (2020) Correcting GPS measurements for non-tidal loading. GPS Solut 24:1–13

Petit G, Luzum B (2010) IERS conventions 2010. IERS technical note 36. Verlag des Bundesamts für Kartographie und Geodäsie

Scherneck HG (1991) A parametrized solid earth tide model and ocean tide loading effects for global geodetic baseline measurements. Geophys J Int 106:677–694

Schuh H, Estermann G, Crétaux JF, Bergé-Nguyen M, van Dam TM (2003) Investigation of hydrological and atmospheric loading by space geodetic techniques. In: Satellite altimetry for geodesy, geophysics and oceanography: proceedings of the international workshop on satellite altimetry, a joint workshop of IAG Section III Special Study Group SSG3. 186 and IAG Section II, September 8–13, 2002, Wuhan, China. Springer, Heidelberg, pp 123–132

Tregoning P, van Dam TM (2005) Atmospheric pressure loading corrections applied to GPS data at the observation level. Geophys Res Lett 32:L22310. https://doi.org/10.1029/2005GL024104

van Dam TM, Wahr JM (1987) Displacements of the Earth's surface due to atmospheric loading: effects on gravity and baseline measurements. J Geophys Res 92(B2):1281–1286

van Dam TM, Blewitt G, Heflin MB (1994) Atmospheric pressure loading effects on global positioning system coordinate determinations. J Geophys Res Solid Earth 99:23939–23950

Open Access This chapter is licensed under the terms of the Creative Commons Attribution 4.0 International License (http://creativecommons.org/licenses/by/4.0/), which permits use, sharing, adaptation, distribution and reproduction in any medium or format, as long as you give appropriate credit to the original author(s) and the source, provide a link to the Creative Commons license and indicate if changes were made.

The images or other third party material in this chapter are included in the chapter's Creative Commons license, unless indicated otherwise in a credit line to the material. If material is not included in the chapter's Creative Commons license and your intended use is not permitted by statutory regulation or exceeds the permitted use, you will need to obtain permission directly from the copyright holder.

Relevance of PSInSAR Analyses at ITRF Co-location Sites

Xavier Collilieux, Zuheir Altamimi, Jingyi Chen, Clément Courde, Zheyuan Du, Thomas Furhmann, Christoph Gisinger, Thomas Gruber, Ryan Hippenstiel, Davod Poreh, Paul Rebischung, and Yudai Sato

Abstract

The PSInSAR (Persistent Scatterer Interferometric Synthetic Aperture Radar) technique allows determining deformation maps over large areas. In this paper, we investigate the applicability of PSInSAR analyses for ITRF co-location sites characterized by spatial extents varying between 20 m and 3 km. Although PSInSAR shows some limitations such as spatial resolution and sparse Persistent Scatterer distribution, this technology can be used to determine relative motion between geodetic instrumentation at sufficient spatial detail, specifically for large sites. The spatial resolution varies from 3×22 m [rg $\times$ az] from typical Sentinel 1A/1B products (IW mode) to 0.6×0.25 m [rg $\times$ az] for staring spotlight mode of TerraSAR-X/Tandem-X. As an illustration, C-band PSInSAR results derived by the European Ground Motion Service (EGMS) from Sentinel 1A/1B images have been investigated for the five largest ITRF co-location sites in Europe. Maximum relative velocity differences have been found to be smaller than 2.0 mm/yr. Moreover, as high-resolution X-band SAR images show great potential for mapping deformations at high resolution, an inventory of already available TerraSAR-X/Tandem-X images at ITRF co-location sites has been established. Based on this, five candidate sites are proposed for further PSInSAR analyses using X-band data.

Keywords

Co-location · InSAR · Local tie · Space geodetic techniques · Terrestrial Reference Frame

---

X. Collilieux (✉) · Z. Altamimi · P. Rebischung
Université Paris Cité, Institut de physique du globe de Paris, CNRS, IGN, Paris, France

ENSG-Géomatique, IGN, Marne-la-Vallée, France
e-mail: xavier.collilieux@ensg.eu

J. Chen
University of Texas at Austin, Department of Aerospace Engineering & Engineering Mechanics, Austin, TX, USA

C. Courde
Université Côte d'Azur, CNRS, Observatoire de la Côte d'Azur, IRD, Géoazur, Observatoire de Calern, Caussols, France

Z. Du
Geoscience Australia, School of Civil and Environmental Engineering, University of New South Wales, Sydney, NSW, Australia

T. Furhmann
Airbus Defence & Space, Munich, Germany

C. Gisinger
German Aerospace Center, Remote Sensing Technology Institute, Wessling, Germany

T. Gruber
Technical University of Munich, Department of Aerospace and Geodesy, Munich, Germany

R. Hippenstiel
National Geodetic Survey, Field Operations, Chesapeake, VA, USA

D. Poreh
Universita degli Studi di Napoli Federico II, Dipartimento di Ingegneria Elettrica e delle Tecnologie dell'Informazione, Naples, Italy

Y. Sato
Geospatial Information Authority of Japan, Ibaraki, Japan

© The Author(s) 2024
J. T. Freymueller, L. Sánchez (eds.), *Together Again for Geodesy*,
International Association of Geodesy Symposia 157, https://doi.org/10.1007/1345_2024_269

1 Introduction

The scientific community has recognized the need for a highly accurate terrestrial reference frame (TRF) for Earth Science applications. Current determinations of the International Terrestrial Reference System are made by combining data from various space geodetic techniques, namely Satellite Laser Ranging (SLR), Very Long Baseline Interferometry (VLBI), Doppler Orbitography and Radiopositioning Integrated by satellite (DORIS), Global Navigation Satellite Systems (GNSS), and terrestrial measurements from local tie surveys at co-location sites. For most of the sites, such local tie surveys are not performed on a regular basis. Thus, the assumption of equal velocity at co-location sites or the detection of some discontinuities in position time series cannot be generally verified independently from space geodesy data.

Zerbini et al. (2007) (see their Fig. 5) have shown the first Persistent Scatterer Interferometric Synthetic Aperture Radar (PSInSAR) result, also named Persistent Scatterer Interferometry (PSI), at the Medicina co-location site (Italy). As the site is located in a rural area, too few high-quality InSAR measurement points were obtained to derive the relative velocity between the GNSS and VLBI stations. However, due to the availability of new InSAR data at many sites with extended spatial coverage and shorter satellite revisit cycle, it raises the question of whether this data could serve as a supplementary resource for local tie measurements at sites lacking frequent terrestrial surveys. This analysis motivated the creation of the study group "SG 1.2.1: Relevance of PSInSAR analyses at ITRF co-location sites" of the International Association of Geodesy (IAG) in 2020.

This paper aims at reporting the conclusions and main findings of the study group. First, the strengths and weaknesses of the PSInSAR technique for relative velocity determination at co-location sites are listed. Then, some PSInSAR results at co-location sites are reviewed based on Sentinel 1A/1B radar images. Finally, perspectives are given to process and interpret X-band SAR images in the future.

2 Strengths and Limitations

Interferometric Synthetic Aperture Radar is a technique that allows determining displacements that occurred between two SAR acquisitions along the radar Line-Of-Sight (LOS) direction. In the PSInSAR approach, a large set of radar images—acquired with similar acquisition geometries—are jointly processed to estimate surface displacements at selected high-quality pixels, the so-called Persistent Scatterers (PS) (Crosetto et al. 2016). PS pixels typically have high temporal coherence, which indicates the stability of the reflection characteristics associated with the measurement points (Kotzerke et al. 2022), but PS spatial distribution is not homogenous and often correlated with landcover types. 1D-displacement time series at image acquisition times can be derived for every PS in the LOS direction of the satellite (slant measurements). Recent satellite missions often have regular short revisit cycles (Sentinel-1 satellites revisit the same area every 6 or 12 days), and thus it is possible to monitor abrupt displacement or velocity changes with PSInSAR. Several acquisition geometries may be available, for example from descending or ascending orbits, sometimes acquired from several incidence angles. However, as SAR satellites have near-polar orbits, the sensitivity to South-North displacements is very low, and only 2D displacements (i.e. in the East-West and vertical directions) can be reconstructed from PSInSAR with sufficiently high accuracy (Fuhrmann and Garthwaite 2019). The typical precision of estimated velocities is up to 1 mm/yr (Crosetto et al. 2016; Ferretti et al. 2007). PS processing requires satellite orbit (constrained by GPS), digital elevation model (DEM) and atmospheric corrections. There are many factors that influence the quality of the results, including but not limited to quality of PS candidates, temporal/spatial baselines, phase unwrapping strategy, residual atmospheric errors, and overall data analysis strategy (software). It is worth noting that PSInSAR algorithms predominantly rely on data-driven approaches to estimate not only pixel displacements but also additional corrections such as atmospheric delays or satellite orbit errors, without requiring additional auxiliary data from other geodetic techniques, with the exception of a priori orbits. Furthermore, the technique can be used for relative measurement of deformations w.r.t. a presumably stable reference location within the image extent without the need for integration into an accurate spatial reference frame.

The main drawback of PS techniques for our application is that PSInSAR determines ground, building roof or monument motion, and not necessarily the geodetic instrument reference point displacements themselves. For example, SLR instruments are located inside a building under an open dome and may be anchored more deeply than the building to which the dome is fixed. Additionally, individual instruments, such as concrete-based VLBI monuments or GNSS pillars, may be anchored at varying depths. For this reason, it is crucial to acquire PS directly on these monuments themselves. However, the availability of PS close to geodetic stations is not guaranteed due to the pixel selection process. Fortunately, PS reflection points can be constructed by installing corner reflectors (CR) or transponders. Gruber et al. (2020, 2022) studied transponders (active devices) and reported that those are easy to install and much smaller than conventional CR. They can also observe both ascending and descending arcs but some limitations have been pointed out such as phase center correction, radio license constraint, software adaptation and possible interference with existing geodetic infrastructure (GNSS, DORIS, SLR, VLBI). As a well-known alternative, passive CR can be installed. At least five ITRF co-location sites currently host CR: Grasse (Collilieux et al. 2022), Metsähovi, O'Higgins, Wettzell, Yarragadee (Carman 2018; Balss et al. 2018).

The spatial resolution of SAR images ranges from several tens of meters to several decimeters. As the typical size of co-location sites lies between 10 m to several kilometers, small sites require high-resolution images whereas lowest resolution could still inform on relative displacements at larger sites.

As a conclusion, PSInSAR currently does not provide a measure as reliable as regular local tie surveys for our application but the availability of images and products makes it worth investigating.

3 PSInSAR Results

3.1 Sentinel 1A/1B: Ground Motion Services

A few publicly available services provide PSInSAR results at national or continental level. Most of the products rely on the Sentinel 1A/1B mission. The spatial resolution of Sentinel 1A/1B images is about 3 × 22 m [rg × az] in Interferometric Wide swath (IW) mode and 3 × 5 m [rg × az] in StripMap mode (SM). Only four of the ITRF sites are covered by the national services known to the authors: Onsala (Swedish National Space Agency 2023); Effelsberg, Potsdam, Wettzell, (BodenBewegungsdienst Deutschland; BGR 2023). Displacement results for eight co-location sites are available over Japan from Small Baseline Subset (SBAS) InSAR analysis of ALOS-2 images (L-band) provided by Geospatial Information Authority of Japan (2023) but the pixel size is about 100 m and the resolution is not sufficient for our application.

We studied European Ground Motion Service (EGMS; Copernicus 2023) results for the 22 covered ITRF co-location sites (three are located in oversea territories). Displacement time series and velocities spanning February 2015 to December 2021 are provided in the line of sight (LOS) of the satellite for each ascending and descending orbit. East/West and vertical motions also reconstructed from these two orbit results are not investigated here given that they require spatial interpolation of individual LOS results.

A visual inspection of Level 2B (L2b) products for those areas has been carried out using the EGMS online visualization tools. Level 2B products are georeferenced to ETRF2000 using GNSS permanent station coordinates (Kotzerke et al. 2022) but are still provided in the InSAR satellite LOS. No displacement has been clearly evidenced on the velocity maps available on the EGMS portal at any of the co-location sites. We further analyzed the velocity differences within the five largest co-location sites: Metsähovi, Potsdam, Reykjavik, Toulouse, Wettzell. Non-calibrated L2a products were analyzed. Those are constrained by InSAR measurements only, and referenced to a local reference point (Kotzerke et al. 2022). As a result, each orbit arc has to be assessed independently. As an example, Fig. 1a shows the raw L2a product relative velocities for the Wettzell site from a descending orbit. No significant motion is evidenced close to instruments in this figure.

For the five considered sites, 88% of stations have a measurement point available at a distance less than 20 m (considering all available orbits), see Table 1. Note that the absolute 3D position accuracy of the measurement points is less than 10 m (Kotzerke et al. 2022). Table 1 also reports the maximum relative LOS velocities w.r.t. one GNSS station of the site (closest PS). Although the standard deviation of PSInSAR velocities is between 0.1 mm/yr and 0.3 mm/yr, maximum velocity differences can be as large as 3.5 mm/yr. We note that some of the selected PS show a rather low temporal coherence. If a coherence threshold of 0.8 is chosen to filter the PS, the average number of available PS in the vicinity of the stations drops to 48%. But the velocity consistency increases, as indicated in Table 1, with maximum velocity differences of 1.4 mm/yr in Toulouse and 1.7 mm/yr in Wettzell. (The Reykjavik site is discussed in Sect. 3.2). However, the velocities of these specific PS are found to be inconsistent with the velocities of the closest high coherence PS. This example shows that the selection of relevant PS around a geodetic instrument is an important step when relating the PS-derived displacements to a potential movement of the geodetic instruments.

3.2 Sentinel 1A/1B: Examples

We investigated further EGMS displacement time series at Metsähovi since the site is composed of two sub-sites separated by 2.8 km. The GNSS stations of each sub-site show inconsistent seasonal displacements in the ITRF2020 input data (Altamimi et al. 2023), see Fig. 2a. The relative velocity of the GNSS stations of the two sub-sites are compared to the GNSS time series computed by the International GNSS Service (IGS) and projected in the LOS of the SAR satellite (ascending orbit). The individual seasonal signals predicted by the ITRF2020 analyses (Collilieux et al. 2023) are also shown. Unfortunately, as SAR images acquired during snow cover periods are excluded from EGMS products (Kotzerke et al. 2022), it is not possible to confirm GNSS seasonal displacements. However, the observed trend is consistent with space geodesy results.

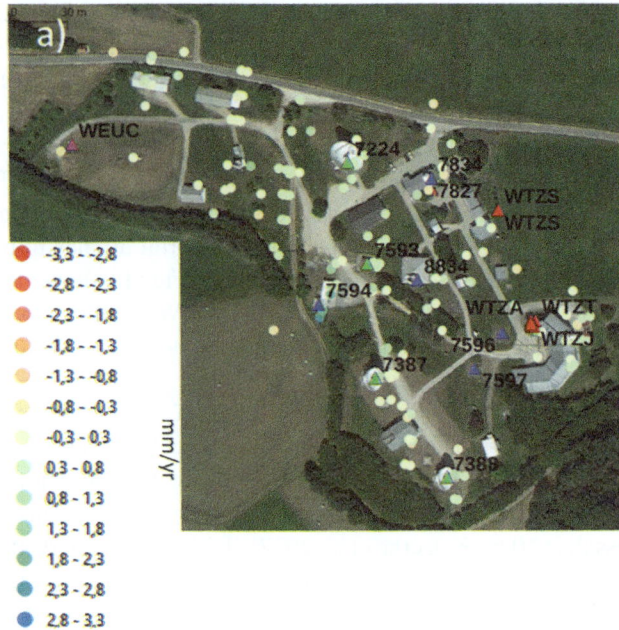
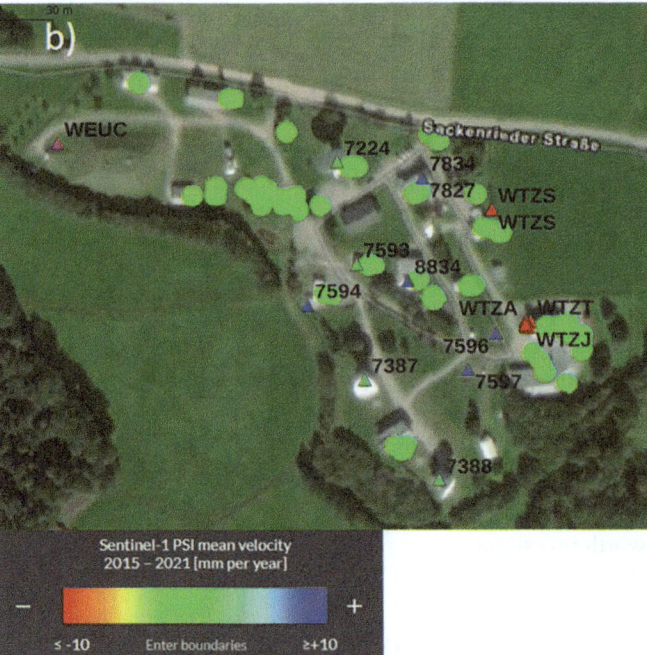

Fig. 1 PSInSAR LOS velocities at the Wettzell Observatory from two different processings of Sentinel 1A/1B data, descending orbit 095. (**a**) EGMS level 2A products, 2015/02 to 2021/12, (**b**) German ground motion service, 2015/04–2021/12. Note the different color scales in each figure. The space geodetic stations have been added as colored triangles

Table 1 Statistics on available PS at the five largest co-location sites covered by EGMS products

Site	% of stations with available PS (distance <20 m) Between brackets: only PS with temporal coherence >0.8	Max velocity differences between closest PS and the closest PS to GNSS (all orbits) (distance <20 m) Between brackets: only PS with temporal coherence >0.8, all distances
Metsähovi	81% (10%)	1.8 (0.5) mm/yr
Potsdam	94% (81%)	0.7 (0.7) mm/yr
Reykjavik	100% (67%)	1.4 (2.0) mm/yr
Toulouse	100% (100%)	0.7 (1.4) mm/yr
Wettzell	85% (44%)	3.5 (1.7) mm/yr
All 5 sites	88% (48%)	3.5 (2.0) mm/yr

As shown in Table 1, there is a significant LOS velocity difference between DORIS and GNSS at Reykjavik. Figure 2b shows the LOS relative EGMS displacements between the two DORIS sites and the Reykjavik GNSS station separated by 2.4 km. A negative trend of about −2 mm/yr is observed for the DORIS PS in the ascending orbit and is explained by displacement changes during the very last time-segment in the EGMS products. This displacement shows a clear spatial pattern. Excluding this most recent period, velocity differences from PSInSAR are small. They confirm the absence of significant motion between DORIS and GNSS stations during the ITRF2020 data period (ended in 2021.0). As an indication, the projected GNSS displacement in the satellite LOS (ascending orbit) has been computed in an absolute frame (IGS20) and reported in Fig. 2c. A velocity change is visible following the M5.6 Earthquake in February 2021 (6 km SE of Vogar, Iceland), although nothing is visible in the descending orbit (not shown). Unfortunately, the DORIS station REZB was decommissioned in September 2020, before the observed motion, so it is not possible to derive the GNSS-DORIS relative motion that could be compared to PSInSAR during this interesting period. This latter result illustrates the potential of PSInSAR results to provide a meaningful constraint for large co-location sites.

As a final remark, we compare two PSInSAR processings for the Wettzell co-location site for the same orbit arc on Fig. 1a, b. Figure 1b shows a screenshot of the German ground motion service PSInSAR result. The periods of considered data are not exactly equal but overlap significantly (see Fig. 1 caption). This figure shows that the PS distribution and LOS velocities depend on the PSInSAR algorithm. In rural locations such as the DORIS station vicinity on the left side of each figure, PS are not always detected. Specific algorithms allow getting a measurement point in this specific environment as discussed by Wang and Chen (2022), which may improve the density of detected PS in future work.

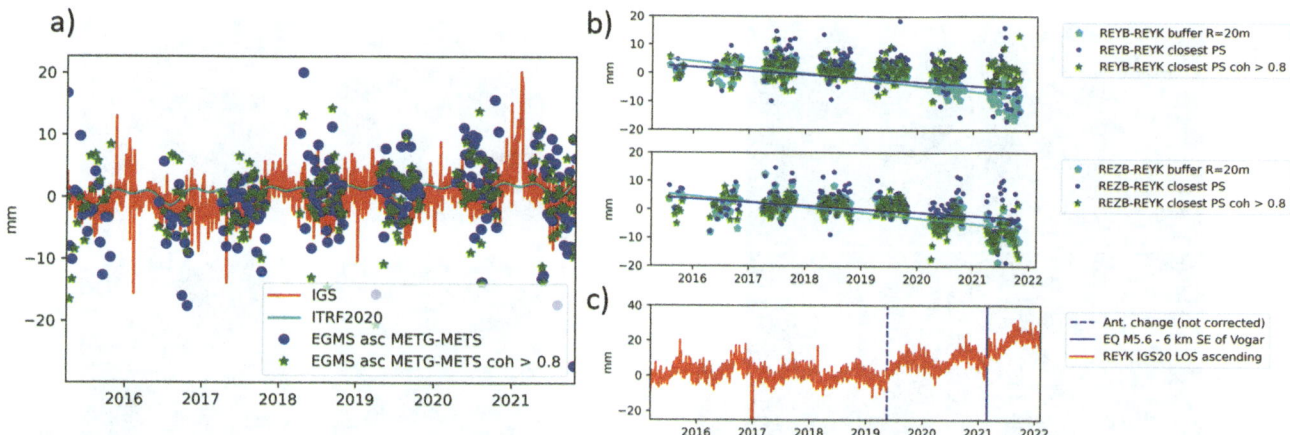

Fig. 2 (**a**) LOS displacement differences (EGMS L2a product) between the closest PS to the METG and METS GNSS stations (blue circles), same but using the closest PS with temporal coherence >0.8 (green stars), difference between GNSS station position time series from IGS projected to line of sight (red), predicted displacements from ITRF2020 input data analyses (light blue). (**b**) LOS displacement differences (EGMS L2a product) between the closest PS to the REYK (GNSS) and REYB (DORIS) stations (top) and to the REYK (GNSS) and REZB (DORIS) stations (bottom) for ascending orbit (EGMS L2a product; blue), same but using the closest PS with temporal coherence >0.8 (green), and by averaging PS displacements in 20 m radius circles (light blue). (**c**) REYK IGS station position time series (in IGS20 reference frame) projected in LOS, same orbit as **b**. The series in **c** has been detrended over the period 2013.33–2019.36 (antenna change epochs)

3.3 X-band

It is possible to derive PSInSAR results at a higher spatial resolution using SAR images from other missions. Poreh and Pirasteh (2020) studied ground deformation at the Medicina co-location site from the end of 2009 to the end of 2011 using CosmoSkyMed X-band images in StripMap/HIMAGE mode, resolution 2.5 × 2.5 m [rg × az]. At this site, a VLBI telescope and a GNSS station are installed. Unfortunately, as found by Zerbini et al. (2007) with ERS C-band SAR images, the density of obtained PS is not sufficient to study relative motion between the instruments. No PS has been found on the VLBI telescope likely due to continuous VLBI telescope motions.

Figure 3 shows X-band PSInSAR results at the Mount Stromlo site (Australia) which hosts GNSS, DORIS and SLR stations. Four years of TerraSAR-X images acquired in StripMap (SM) mode (descending orbit) spanning 26-09-2011 to 25-12-2015 have been processed using the Gamma (Werner et al. 2000) and STAMPS software packages (Hooper et al. 2012). The pixel size in this mode is about 3.5 m. All PS velocities in the displayed area show an agreement within ±1 mm/yr. It is worth noting that a PS was detected at the exact location of the STR2 pillar, but unfortunately no PS was selected at the other GNSS station STR1.

The TerraSAR-X (TSX) and TanDEM-X (TDX) missions are able to provide even higher resolutions: about 0.6 × 1.0 m [rg × az] for High Resolution SpotLight (HS) mode and about 0.6 × 0.25 m [rg × az] for Staring Spotlight (SS) mode. Figure 4a shows an amplitude image of the Yarragadee site (Australia) from SS mode which includes the VLBI (bottom

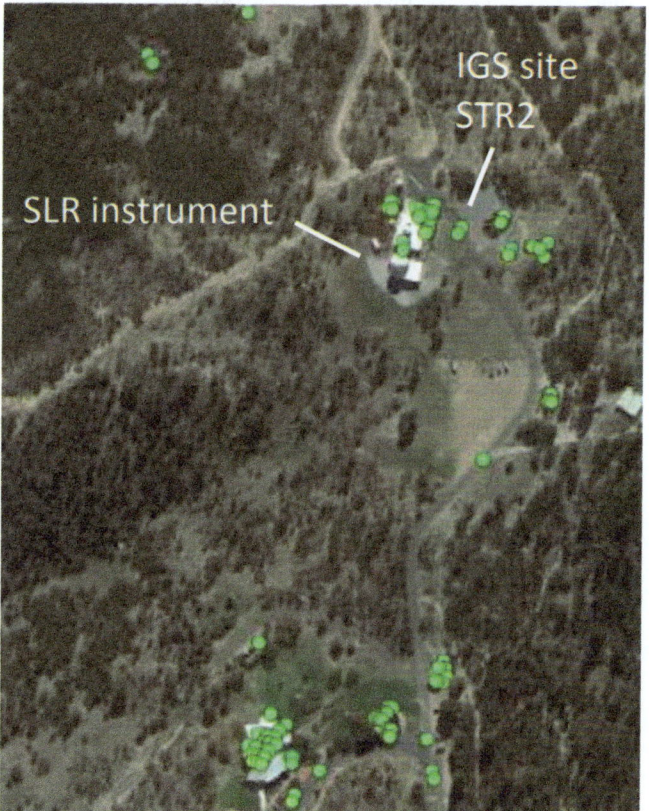

Fig. 3 PS distribution over the Mount Stromlo site (Australia) as a result of the PSInSAR analysis of TerraSAR-X images spanning 26-09-2011 to 25-12-2015 in StripMap (SM) mode

of the image) and SLR stations (top of the image). The details of the man-made infrastructure are clearly evidenced, which shows interesting perspectives to obtain a high density of

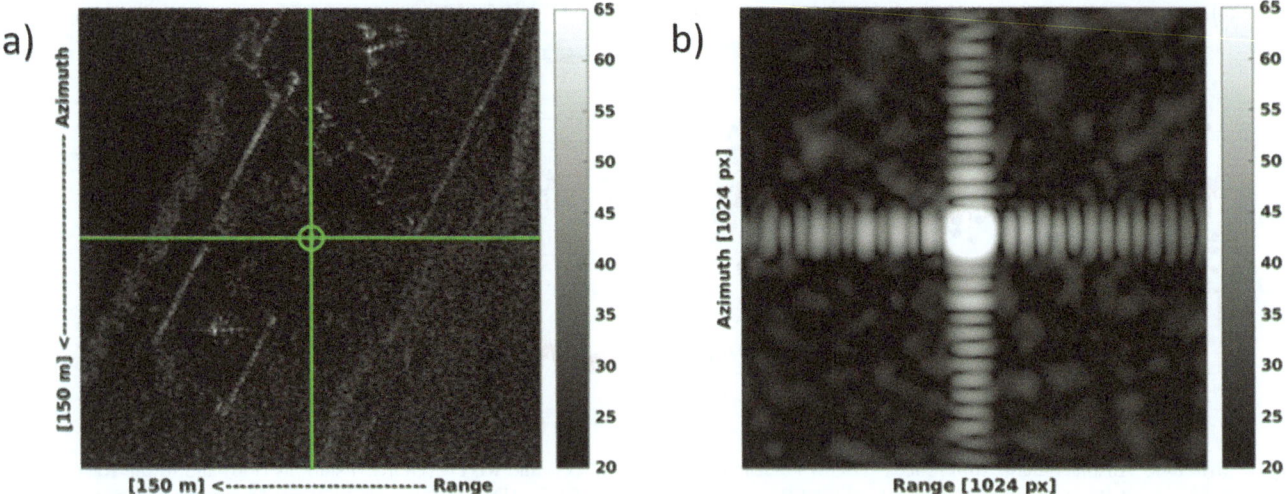

Fig. 4 (a) Crop of amplitude image (sigma-nought) from TerraSAR-X at Yarragadee co-location site in Staring SpotLight mode (0.6 m × 0.25 m [rg × az]). (b) Oversampled close up of the corner reflector (CR)

Table 2 Inventory of TDX/TSX images at ITRF co-location sites for Staring SpotLight (SS) (0.6 × 0.25 m [rg × az]) and High Resolution SpotLight (HS) (0.6 × 1.0 m [rg × az]) modes over the period 2008–2022

Site	Mode	Orbit (#)	Period
Yarragadee	SS	Ascending (147)Descending (144)	31/03/19->30/10/2208/03/19->17/12/22
	HS	Ascending (60)Descending (172)	11/03/18->21/10/1910/03/18->18/12/22
Wettzell	SS	Descending (17)	07/02/15->21/03/16
	HS	Ascending (425)Descending (205)	30/10/12->30/12/2203/04/11->01/01/23
Wuhan	SS	Descending (23)	28/06/14->15/08/17
Simeiz	SS	Ascending (13)Descending (13)	29/03/18->19/08/1801/04/18->22/08/18
O'Higgins	HS	Ascending (520)Descending (322)	05/03/08->08/12/2208/10/08->01/01/23
Metsähovi	HS	Descending (469)	12/05/13->15/12/22

PS on this site. Figure 4b shows the amplitude image of the CR that has been installed nearby and will provide a highly reliable PS with very low background scatter.

An inventory of all available TSX/TDX images at ITRF co-location sites has been developed. A web map has been developed to easily review each co-location site at https://www.arcgis.com/apps/dashboards/f9e56ed0713141c48986240faeefe684. Five co-location sites have more than 15 TSX/TDX high resolution images (SS or HS mode) available in ascending or descending arcs: Metsähovi, O'Higgins, Yarragadee, Wettzell, Wuhan. Statistics on these images are reported in Table 2. The number of candidate sites is much larger using TSX/TDX stripmap mode, resolution 2 × 3 m [rg × az].

4 Conclusion and Perspectives

This paper discussed the advantages of using PSInSAR techniques at ITRF co-location sites to supplement local tie measurements. Results obtained with Sentinel 1A/1B images were discussed and illustrated that PSInSAR is capable of providing information on relative deformation for large co-location sites. As EGMS products are updated annually, this analysis will be worth repeating regularly. Finally, X-band radar images showed a great potential for this application. A significant set of high-resolution TSX/TDX SAR images at co-location sites is already available and would gain to be processed to provide a final conclusion to this study.

Acknowledgements This study contributes to the IdEx Université de Paris ANR-18-IDEX-0001. This work is partly funded by the Centre National d'Etudes Spatiales (CNES), under TOSCA grant. Copernicus Land Monitoring Service products used in this study were produced with funding by the European Union. We are grateful to Lukas Ruesch and Amy Parker who were members of the IAG study group SG 1.2.1 (2020–2023) for fruitful discussions. We also thank the two anonymous reviewers for their relevant suggestions.

References

Altamimi Z, Rebischung P, Collilieux X, Métivier L, Chanard K (2023) ITRF2020: an augmented reference frame refining the modeling of nonlinear station motions. J Geodesy 97:47. https://doi.org/10.1007/s00190-023-01738-w

Balss U, Gisinger C, Eineder M (2018) Measurements on the absolute 2-D and 3-D localization accuracy of TerraSAR-X. Remote Sens 10:656. https://doi.org/10.3390/rs10040656

BGR (2023) BGR - BodenBewegungsdienst Deutschland, Bundesanstalt für Geowissenschaften und Rohstoffe, available at https://bodenbewegungsdienst.bgr.de, visited in June 2023

Carman R (2018) Status and recent upgrades at Yarragadee / MOBLAS-5, poster presented at 21st International Workshop on Laser Ranging Canberra

Collilieux X, Courde C, Fruneau B, Aimar M, Schmidt G, Delprat I, Defresne M-A, Pesce D, Bergerault F, Wöppelmann G (2022) Validation of a Corner Reflector installation at Côte d'Azur multi-technique geodetic Observatory. Adv Space Res 70(2):360–370. https://doi.org/10.1016/j.asr.2022.04.050

Collilieux X, Altamimi Z, Rebischung P, de la Serve M, Métivier L, Chanard K, Boy J-P (2023) A review of space geodetic technique seasonal displacements based on ITRF2020 results, REFAG 2022 proceedings, International Association of Geodesy Symposia

Copernicus (2023) Copernicus Land Monitoring Service, available at https://land.copernicus.eu/, visited in June 2023

Crosetto M, Monserrat O, Cuevas-González M, Devanthéry N, Crippa B (2016) Persistent scatterer interferometry: a review. ISPRS J Photogramm Remote Sens 115:78–89

Ferretti A et al (2007) Submillimeter accuracy of InSAR time series: experimental validation. IEEE Trans Geosci Remote Sens 45(5):1142–1153. https://doi.org/10.1109/TGRS.2007.894440

Fuhrmann T, Garthwaite M (2019) Resolving three-dimensional surface motion with InSAR: constraints from multi-geometry data fusion. Remote Sens 11(3):241. https://doi.org/10.3390/rs11030241

Geospatial Information Authority of Japan (2023) InSAR map server, available at https://maps.gsi.go.jp/#5/37.387617/138.735352/&base=english&ls=english%7Curgent_sbas_japan_merge_qu_u16&blend=0&disp=11&lcd=urgent_sbas_japan_merge_qu_u16&vs=c0g1j0h0k0l0u0t0z0r0s0m0f1&d=m, visited in June 2023

Gruber T et al (2020) Geodetic SAR for height system unification and sea level research—observation concept and preliminary results in the Baltic Sea. Remote Sens 12:3747. https://doi.org/10.3390/rs12223747

Gruber T et al (2022) Geodetic SAR for height system unification and sea level research—results in the Baltic Sea Test Network. Remote Sens 14:3250. https://doi.org/10.3390/rs14143250

Hooper A, Bekaert D, Spaans K, Arikan M (2012) Recent advances in SAR interferometry time series analysis for measuring crustal deformation. Tectonophysics 514–517:1–13. https://doi.org/10.1016/j.tecto.2011.10.013

Kotzerke P, Siegmund R, Langenwalter J (2022) End-to-end implementation and operation of the European Ground Motion Service (EGMS), Product User Manual, v1.6, EGMS-D4-PUM-SC1-2.0-007

Poreh D, Pirasteh S (2020) InSAR observations and analysis of the Medicina Geodetic Observatory and CosmoSkyMed images. Nat Hazards 103(3):3145–3161

Swedish National Space Agency (2023) InSAR Sweden, available at http://insar.rymdstyrelsen.se/, visited in June 2023

Wang K, Chen J (2022) Accurate persistent scatterer identification based on phase similarity of radar pixels. IEEE Trans Geosci Remote Sens 60(1–13):5118513. https://doi.org/10.1109/TGRS.2022.3210868

Werner C, Wegmüller U, Strozzi T, Wiesmann A (2000) GAMMA SAR and interferometric processing software. In European Space Agency, (Special Publication) ESA SP. 461, 211–219

Zerbini S, Richter B, Rocca F, van Dam T, Matonti F (2007) A combination of space and terrestrial geodetic techniques to monitor land subsidence: case study, the southeastern Po Plain, Italy. J Geophys Res 112:B05401. https://doi.org/10.1029/2006JB004338

Open Access This chapter is licensed under the terms of the Creative Commons Attribution 4.0 International License (http://creativecommons.org/licenses/by/4.0/), which permits use, sharing, adaptation, distribution and reproduction in any medium or format, as long as you give appropriate credit to the original author(s) and the source, provide a link to the Creative Commons license and indicate if changes were made.

The images or other third party material in this chapter are included in the chapter's Creative Commons license, unless indicated otherwise in a credit line to the material. If material is not included in the chapter's Creative Commons license and your intended use is not permitted by statutory regulation or exceeds the permitted use, you will need to obtain permission directly from the copyright holder.

The DIA-Estimator for Positional Integrity: Design and Computational Challenges

P. J. G. Teunissen, S. Ciuban, C. Yin, B. G. van Noort, S. Zaminpardaz, and C. C. J. M. Tiberius

Abstract

The geodetic method of positional data processing is usually not one of position estimation only, nor one of model testing only, but usually one in which estimation and testing are combined. The Detection, Identification and Adaptation (DIA)-estimator captures the statistical intricacies of this combination, providing a unifying framework for rigorous analyses of positional integrity and quality control procedures. However, to be able to establish fit-for-purpose quality control, not only solutions for the forward problem (quality of control) need to be available, but also for the inverse problem (control of quality). With the DIA-estimator and its multi-modal probability density function (PDF), we have solutions available for the forward problem, but not yet for the inverse problem. That is, no objective methods and strategies are currently available that allow one to design DIA-estimators specifically for given fit-for-purpose quality criteria. In this invited contribution we present and illustrate some of the underlying design and computational challenges that are brought forward by the complexities of the inverse problem. This relates, amongst others, to the DIA-variables, such as the chosen partitioning of the misclosure space, and to the 'winner-takes-all' structure of the DIA-class of estimators currently employed. To appreciate the fundamental differences with the traditional estimation-only approaches, we also show how the position probability distribution, and therefore the quality of positioning, is affected and driven by the combination of estimation and testing. For an underpinning of the design and computational challenges various numerical and graphical examples are presented.

Keywords

Control of quality · Detection, Identification and Adaptation (DIA) · DIA-estimator · Multi-modal probability distribution · Positional integrity · Quality of control

1 Introduction

In Teunissen (2022) the following two statements on measurement data processing in our geodetic discipline were put forward:

Statement 1 (**The mean-variance addiction**): In practice, we have the habit of considering only the first two moments of the (position) estimator, mean and variance, rather than considering its entire Probability Density Function (PDF).

P. J. G. Teunissen (✉)
Delft University of Technology, Delft, The Netherlands

Curtin University, Perth, WA, Australia
e-mail: p.j.g.teunissen@tudelft.nl

S. Ciuban · C. Yin · B. G. van Noort · C. C. J. M. Tiberius
Delft University of Technology, Delft, The Netherlands

S. Zaminpardaz
RMIT University, Melbourne, VIC, Australia

Statement 2 (**The estimation-testing marriage**): Our data processing always consists of both estimation and testing. One can think of outlier detection (data screening), GNSS cycle slip detection, and testing for the presence of a postulated deformation behaviour in deformation analysis.

In this contribution we will further substantiate these statements, discuss their consequences and highlight the challenges that they bring to our general data processing and analysis tasks.

Restricting the analysis and interpretation of data processing results to considering the mean and variance, as put forward in the first statement, would be fine provided that the parameter estimator distribution is normal (Gaussian), which is true in case the estimator is linear in the data (with normally distributed data), or, if the central limit theorem applies. The first statement is however not justified in practice, since the parameter estimator is obtained as described by statement 2, and estimation is effectively *conditioned* on testing.

Teunissen (2018) introduced a unifying framework to capture the combination of parameter estimation and statistical hypothesis testing. Through a canonical model formulation and a partitioning of the misclosure space, the entire testing and estimation procedure is integrated into a single DIA-estimator. In practice the distribution of the estimator under a certain identified hypothesis is used, *without* accounting for the conditioning process (testing) that led to the decision to use the identified hypothesis. Data processing procedures effectively used in practice are typically conditional ones, but the corresponding conditional distribution for the estimator is *not* used to describe and evaluate its quality.

Zaminpardaz and Teunissen (2022) consequently point out that confidence regions, commonly used in practice, do generally not truly reflect the confidence one can have in the resulting estimator. Instead, the combined uncertainty of hypothesis testing and parameter estimation has to be taken into account. Also, the traditional confidence ellipses do not provide adequate and sufficient spatial information about the actual distribution of the estimator. This intricacy is long-standing in statistics and has already been touched upon in the past by statisticians, e.g., Meeks and D'agostino (1983), Chatfield (1995), and Kabaila (2009).

In safety-critical applications, integrity risk is a key-metric of the estimator's quality. It is the probability that the estimator lies outside a safety region around the unknown parameter (vector), typically being one or more position coordinates. In Zaminpardaz et al. (2019) a tendency is observed that the integrity risk computed without regard to the preceding testing procedure is smaller than the risk computed using conditional distributions instead. A too optimistic description of the integrity risk may not sufficiently safeguard against possibly hazardous situations. Underestimating the integrity risk cannot be acceptable in safety-critical applications and the use of the rigorous approach to evaluate the integrity risk is recommended.

2 Estimation and Testing: DIA-Estimator

To illustrate the integrated estimation and testing principle we put forward the problem of misspecifications in linear(ized) functional models. We have the following model under the null hypothesis $\mathcal{H}_0$ which is opposed against k alternative hypotheses $\mathcal{H}_i$

$$\begin{aligned}\mathcal{H}_0 &: \mathsf{E}(\underline{y}) = Ax, \ \mathsf{D}(\underline{y}) = Q_{yy} \\ \mathcal{H}_i &: \mathsf{E}(\underline{y}) = Ax + C_i b_i, \ \mathsf{D}(\underline{y}) = Q_{yy}\end{aligned} \quad (1)$$

for $i = 1, ..., k$ with $\mathsf{E}(.)$ the expectation operator, $\underline{y} \in \mathbb{R}^m$ the normally distributed random vector of observables, an underscore denotes a random variable/vector, $A \in \mathbb{R}^{m \times n}$ of rank n, $x \in \mathbb{R}^n$ the to-be-estimated unknown parameter vector, $\mathsf{D}(.)$ the dispersion operator and $Q_{yy} \in \mathbb{R}^{m \times m}$ the given positive-definite variance matrix of $\underline{y}$. Under $\mathcal{H}_i$, matrix $C_i \in \mathbb{R}^{m \times q_i}$ models the misspecification signature (e.g., one or multi-dimensional faults or blunders), and $b_i \in \mathbb{R}^{q_i}$ is the vector of (observation) biases. The augmented matrix $[A \ C_i]$ is assumed to be of full rank, equals to $n + q_i$. The redundancy under $\mathcal{H}_0$ is $r = m - \mathrm{rank}(A)$.

In Teunissen (2018) the Tienstra-transformation is applied to $\underline{y} \in \mathbb{R}^m$ to obtain the Best Linear Unbiased Estimator (BLUE) $\underline{\hat{x}}_0 \in \mathbb{R}^n$ (under $\mathcal{H}_0$) and the vector of misclosures $\underline{t} \in \mathbb{R}^r$. The vector of misclosures is at the basis of statistical hypothesis testing: test statistics can be formulated as functions of the misclosure vector. It is defined as $\underline{t} = B^T \underline{y}$ with the variance-covariance matrix $Q_{tt} = B^T Q_{yy} B$, such that $B^T A = 0$, where $\underline{t} \in \mathbb{R}^r$ and $B \in \mathbb{R}^{m \times r}$ has $\mathrm{rank}(B) = r$.

In Fig. 1 we present an example of a partitioned $\mathbb{R}^r$. Each outcome for the vector of misclosures $\underline{t}$ is linked to one of the hypotheses through the corresponding partition. In practice the Overall Model Test (OMT) is commonly used for detection, and next, w-tests are used for identification. Assuming the case of datasnooping (C_i's are the canonical unit vectors), the partitioning of $\mathbb{R}^r$ can be done as follows,

$$\begin{aligned}\mathcal{P}_0 &= \left\{ \underline{t} \in \mathbb{R}^r \mid ||\underline{t}||^2_{Q_{tt}} \leq \tau^2 \right\}, \\ \mathcal{P}_{i \neq 0} &= \left\{ \underline{t} \in \mathbb{R}^r, \underline{t} \notin \mathcal{P}_0 \mid |\underline{w}_i| = \max_{j = \{1,...,k\}} |\underline{w}_j| \right\},\end{aligned} \quad (2)$$

where $||\underline{t}||^2_{Q_{tt}} = \underline{t}^T Q_{tt}^{-1} \underline{t}$, τ^2 is the critical value for a chosen level of significance α for the OMT, and $\underline{w}_i$ is the w-test statistic, which was introduced by Baarda (1968). If detection occurs ($\mathcal{H}_0$ is rejected) then the hypothesis $\mathcal{H}_i$ is identified as most likely if $\underline{t} \in \mathcal{P}_i$ and then the BLUE $\underline{\hat{x}}_i$ (under $\mathcal{H}_i$) is provided as an output.

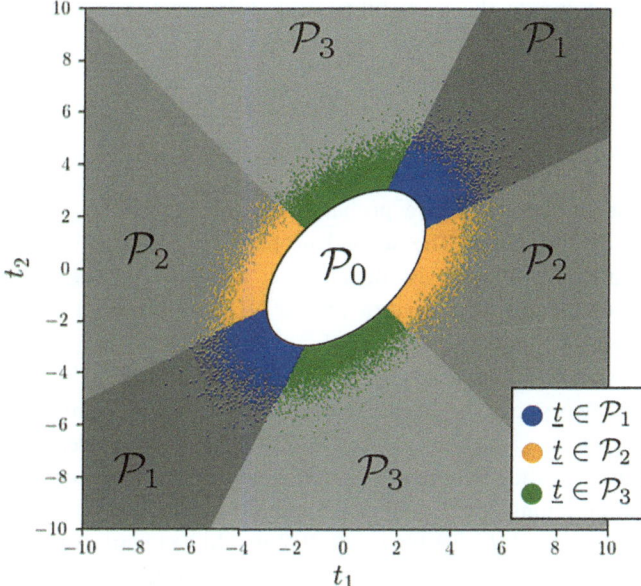

Fig. 1 Misclosure space partitioning for datasnooping when $\alpha = 10^{-1}$, $k = 3$, $n = 1$, $m = 3$, $r = 2$, $q_i = 1$ (for $i = 1, 2, 3$), $A = e_3 = [1\ 1\ 1]^T$, $Q_{yy} = I_3$ and $C_i = c_i$ are the canonical unit vectors. The number of pseudo-random samples drawn from $f_{\underline{t}}(t|\mathcal{H}_0)$ is 10^6 and the ones landing in $\mathcal{P}_0$ are not displayed

In Fig. 1 the pseudo-random samples drawn from the PDF of $\underline{t}$ are distributed across the partitions. Obviously, in advance we do not know in which partition the vector of misclosures will land, and hence all possibilities need to be taken into account when evaluating the quality of the resulting conditional estimator.

The BLUEs $\hat{\underline{x}}_0$ and $\hat{\underline{x}}_i$ are provided based on in which partition $\underline{t}$ will land. The DIA-estimator was introduced in Teunissen (2018) as

$$\overline{\underline{x}} = \begin{cases} \hat{\underline{x}}_0 & \text{if } \underline{t} \in \mathcal{P}_0 \\ \hat{\underline{x}}_i & \text{if } \underline{t} \in \mathcal{P}_i \end{cases} \quad (3)$$

which can be expressed in a more compact-form

$$\overline{\underline{x}} = \sum_{i=0}^{k} \hat{\underline{x}}_i\, p_i(\underline{t}), \quad (4)$$

with $p_i(\underline{t}) = 1$ if $\underline{t} \in \mathcal{P}_i$ and zero otherwise. In this expression, $\hat{\underline{x}}_i$ carries the uncertainty of estimation, and $p_i(\underline{t})$ carries the uncertainty of testing. Theorem 1 in section 3 of Teunissen (2018) states the PDF of $\overline{\underline{x}}$ under a hypothesis $\mathcal{H}$ as

$$f_{\overline{\underline{x}}}(\theta|\mathcal{H}) = \sum_{i=0}^{k} \int_{\mathcal{P}_i} f_{\hat{\underline{x}}_0}(\theta + L_i \tau|\mathcal{H})\, f_{\underline{t}}(\tau|\mathcal{H})\, d\tau, \quad (5)$$

where $L_i = 0$ for $i = 0$ and $L_i = A^+ C_i C_{t_i}^+$ for $i > 0$. The BLUE-inverse of A is $A^+ = (A^T Q_{yy}^{-1} A)^{-1} A^T Q_{yy}^{-1}$ and that of $C_{t_i} = B^T C_i$ is $C_{t_i}^+ = (C_{t_i}^T Q_{tt}^{-1} C_{t_i})^{-1} C_{t_i}^T Q_{tt}^{-1}$. The L_i term links the BLUEs under $\mathcal{H}_{i \neq 0}$ and $\mathcal{H}_0$ as $\hat{\underline{x}}_i = \hat{\underline{x}}_0 - L_i \underline{t}$.

Under $\mathcal{H}_0$, the PDF of $\overline{\underline{x}}$ becomes

$$f_{\overline{\underline{x}}}(\theta|\mathcal{H}_0) = f_{\hat{\underline{x}}_0}(\theta|\mathcal{H}_0)\mathrm{P}_{\mathrm{CA}} + \sum_{i=1}^{k} f_{\overline{\underline{x}}|\mathrm{FA}_i}(\theta|\mathrm{FA}_i)\mathrm{P}_{\mathrm{FA}_i} \quad (6)$$

with the probability of Correct Acceptance (CA) $\mathrm{P}_{\mathrm{CA}} = \int_{\mathcal{P}_0} f_{\underline{t}}(\tau|\mathcal{H}_0)\,d\tau$, and the probabilities of False Alarm (FA) $\mathrm{P}_{\mathrm{FA}_i} = \int_{\mathcal{P}_i} f_{\underline{t}}(\tau|\mathcal{H}_0)\,d\tau$, and $\mathrm{P}_{\mathrm{CA}} + \sum_{i=1}^{k} \mathrm{P}_{\mathrm{FA}_i} = 1$. Similarly, under an alternative hypothesis $\mathcal{H}_i$ we have

$$\begin{aligned}f_{\overline{\underline{x}}}(\theta|\mathcal{H}_i, b_i) &= f_{\hat{\underline{x}}_0}(\theta|\mathcal{H}_i, b_i)\mathrm{P}_{\mathrm{MD}_i} \\ &+ f_{\overline{\underline{x}}|\mathrm{CI}_i}(\theta|\mathrm{CI}_i)\mathrm{P}_{\mathrm{CI}_i} \\ &+ \sum_{j \neq 0, i}^{k} f_{\overline{\underline{x}}|\mathrm{WI}_{i,j}}(\theta|\mathrm{WI}_{i,j})\mathrm{P}_{\mathrm{WI}_{i,j}} \end{aligned} \quad (7)$$

where one faces the probability of Missed Detection (MD) $\mathrm{P}_{\mathrm{MD}_i} = \int_{\mathcal{P}_0} f_{\underline{t}}(\tau|\mathcal{H}_i, b_i)\,d\tau$, Correct Identification (CI) $\mathrm{P}_{\mathrm{CI}_i} = \int_{\mathcal{P}_i} f_{\underline{t}}(\tau|\mathcal{H}_i, b_i)\,d\tau$ and Wrong Identification (WI) $\mathrm{P}_{\mathrm{WI}_i} = \sum_{j \neq 0, i}^{k} \mathrm{P}_{\mathrm{WI}_{i,j}} = \sum_{j \neq 0, i}^{k} \int_{\mathcal{P}_j} f_{\underline{t}}(\tau|\mathcal{H}_i, b_i)\,d\tau$; under a certain $\mathcal{H}_i$ one has $\mathrm{P}_{\mathrm{MD}_i} + \mathrm{P}_{\mathrm{CI}_i} + \mathrm{P}_{\mathrm{WI}_i} = 1$. Note that $f_{\overline{\underline{x}}}(\theta|\mathcal{H}_i, b_i)$ is also a function of the unknown bias b_i. As outlined in Teunissen (2018), $f_{\overline{\underline{x}}}(\theta|\mathcal{H})$ is a non-Gaussian multi-modal PDF, whereas usually $f_{\hat{\underline{x}}_0}(\theta|\mathcal{H}_0)$ or $f_{\hat{\underline{x}}_i}(\theta|\mathcal{H}_i, b_i)$, typically Gaussian, are considered in practice.

With the above exposition we arrive at two conclusions. First, applying linear estimation to Gaussian distributed data, and applying testing as well, leads to a **non-Gaussian PDF** for the resulting estimator. Second, unbiased estimation, together with testing actually leads to **biased** estimation under an alternative hypothesis, though the bias without performing any testing is always larger than the bias with testing (i.e. larger than the bias in the DIA-estimator).

Hence, when interpreting and analysing data processing results, we should no longer focus on just mean and variance, but instead consider the entire distribution of the estimator. With the development so far, we formulate three challenges ahead.

1. How to adequately evaluate the quality of the DIA-estimator? What measures do we need to consider? And, we have to acknowledge the fact that the estimator's PDF, under $\mathcal{H}_1, \ldots, \mathcal{H}_k$, depends on unknown biases. This, by the way, is not different from the traditional situation in which for instance the detection power of a test-statistic depends on the bias size, or conversely, the analysis is carried out for a pre-set power as with the Minimal Detectable Bias (MDB) Baarda (1968).

2. How to design an appropriate DIA-estimator for a specific application given user demands or requirements? On account of parameter estimation we are used to work with optimal estimation, BLUE and Maximum Likelihood (ML). On account of statistical hypothesis testing we work with the Generalized Likelihood Ratio (GLR) in linear(ized) models, which is Uniformly Most Powerful Invariant (UMPI) in binary testing, that is testing $\mathcal{H}_0$ against a single alternative hypothesis.
3. How to efficiently compute measures of quality from a numerical perspective? This quickly comes to evaluation of high dimensional integrals in terms of misclosures and working with non-Gaussian, multi-modal distributions in terms of the parameter estimators. As an example of the dimension of the misclosure space, we mention that a single point positioning model, with a single receiver and with for instance tri-constellation GNSS the number of satellites can be 24, and hence, the redundancy will be 20 (assuming inter-system biases have been sufficiently calibrated).

In the next two sections examples are shown of the DIA-estimator PDF bias dependency and consequent challenges in evaluating the quality of the estimator are pointed out.

3   Example 1

Let us assume that we have 3 observables of an unknown quantity x, which are independent and of the same precision σ. With $\underline{y} = [\underline{y}_1, \underline{y}_2, \underline{y}_3]^T$ the vector of three observables, the null and alternative hypotheses are formulated as

$$\begin{aligned}\mathcal{H}_0 &: \mathsf{E}(\underline{y}) = e_3 x, \quad \mathsf{D}(\underline{y}) = \sigma^2 I_3 \\ \mathcal{H}_i &: \mathsf{E}(\underline{y}) = e_3 x + c_i b_i, \quad \mathsf{D}(\underline{y}) = \sigma^2 I_3\end{aligned} \quad (8)$$

for $i = 1, 2, 3$, with $e_3 \in \mathbb{R}^3$ the vector of ones, $I_3 \in \mathbb{R}^{3\times 3}$ the identity matrix, and $c_i \in \mathbb{R}^3$ the canonical unit vector having one as its ith element and zeros otherwise. The alternative hypothesis $\mathcal{H}_i$ (for $i = 1, 2, 3$), hence $k = 3$, describes a bias in the ith observation.

Under $\mathcal{H}_0$, there are $r = 3 - 1 = 2$ redundancies, resulting in a two-dimensional misclosure space, $\underline{t} \in \mathbb{R}^2$. To test the four hypotheses in (8), we first use the OMT to check the validity of $\mathcal{H}_0$, which, upon the rejection of $\mathcal{H}_0$, is followed by identification of the potential bias using w-tests. The misclosure space partitioning induced by this testing procedure is formulated in (2), an example of which is illustrated in Fig. 1. To evaluate the quality of the resulting DIA-estimator of x, i.e. $\underline{\bar{x}}$, considering its entire PDF, one can compute the probability of $\underline{\bar{x}}$ lying in the x-centered interval $\mathcal{B}_{x,a} = [x - a\sigma, x + a\sigma]$ as

$$\begin{aligned}\text{Under } \mathcal{H}_0 &\quad : \mathsf{P}\,(\underline{\bar{x}} \in \mathcal{B}_{x,a} | \mathcal{H}_0) \\ \text{Under } \mathcal{H}_{i\neq 0} &\quad : \mathsf{P}\,(\underline{\bar{x}} \in \mathcal{B}_{x,a} | \mathcal{H}_i, b_i)\end{aligned} \quad (9)$$

The higher the above probabilities, the higher the quality of the estimator. The second probability, defined under an alternative hypothesis, depends on the bias b_i. Therefore, when judging the quality of the DIA-estimator, one may take a conservative route by considering a bias value b_i which *minimizes* the probability $\mathsf{P}\,(\underline{\bar{x}} \in \mathcal{B}_{x,a} | \mathcal{H}_i, b_i)$, i.e., worst-case scenario. Note, if b_i is set to zero, then $\mathsf{P}\,(\underline{\bar{x}} \in \mathcal{B}_{x,a} | \mathcal{H}_i, b_i = 0) = \mathsf{P}\,(\underline{\bar{x}} \in \mathcal{B}_{x,a} | \mathcal{H}_0)$. Furthermore, due to the symmetry of $\mathcal{H}_0$-model and making all observations with the same precision, the probabilities $\mathsf{P}\,(\underline{\bar{x}} \in \mathcal{B}_{x,a} | \mathcal{H}_i, b_i)$ for $i = 1, 2, 3$ would be identical for the same biases $b_1 = b_2 = b_3$. Therefore, the results obtained in this section hold true for any of the alternative hypotheses.

Figure 2 shows, for $\alpha = \mathsf{P}_{\text{FA}} = 0.01$ and $a = 1$, the probability $\mathsf{P}\,(\underline{\bar{x}} \in \mathcal{B}_{x,a} | \mathcal{H}_i, b_i)$ as a function of bias-to-noise ratio $\frac{b_i}{\sigma}$ in *blue*. When $\frac{b_i}{\sigma}$ increases, the probability $\mathsf{P}\,(\underline{\bar{x}} \in \mathcal{B}_{x,a} | \mathcal{H}_i, b_i)$ first shows a decreasing behavior, followed by an increase, and finally its behavior stabilizes. This can be understood by analysing the individual components forming the blue curve. Using the total probability rule, the probability $\mathsf{P}\,(\underline{\bar{x}} \in \mathcal{B}_{x,a} | \mathcal{H}_i, b_i)$ can be decomposed as

$$\begin{aligned}\mathsf{P}(\underline{\bar{x}} \in \mathcal{B}_{x,a} | \mathcal{H}_i, b_i) = \\ \mathsf{P}\,(\underline{\hat{x}}_0 \in \mathcal{B}_{x,a} | \mathcal{H}_i, b_i) \mathsf{P}\,(\underline{t} \in \mathcal{P}_0 | \mathcal{H}_i, b_i) \\ + \mathsf{P}\,(\underline{\hat{x}}_i \in \mathcal{B}_{x,a}, \underline{t} \in \mathcal{P}_i | \mathcal{H}_i, b_i) \\ + \sum_{j \notin \{0,i\}} \mathsf{P}\left(\underline{\hat{x}}_j \in \mathcal{B}_{x,a}, \underline{t} \in \mathcal{P}_j \Big| \mathcal{H}_i, b_i\right)\end{aligned} \quad (10)$$

The three terms on the right-hand side of the above equation, as a function of $\frac{b_i}{\sigma}$, behave as follows.

- The first term consists of two probabilities both of which are decreasing functions of $\frac{b_i}{\sigma}$, so is their product which is shown in *red* in Fig. 2a.
- The second term is the volume under the joint PDF $f_{\underline{\hat{x}}_i, \underline{t}}(\theta, t | \mathcal{H}_i, b_i)$ over the intersection of $\mathcal{B}_{x,1}$ and $\mathcal{P}_i$. This intersection defines two disconnected sub-regions in $\mathbb{R}^3$ symmetric around the origin. Increasing $\frac{b_i}{\sigma}$, the PDF $f_{\underline{\hat{x}}_i, \underline{t}}(\theta, t | \mathcal{H}_i, b_i)$ moves away from the origin along $[0, c_{t_i}^T]^T$, thus its probability mass increases inside the intersection of $\mathcal{B}_{x,1}$ and $\mathcal{P}_i$. Therefore, $\mathsf{P}\,(\underline{\hat{x}}_i \in \mathcal{B}_{x,1}, \underline{t} \in \mathcal{P}_i | \mathcal{H}_i, b_i)$, shown in *yellow* in Fig. 2a, is an increasing function of $\frac{b_i}{\sigma}$.

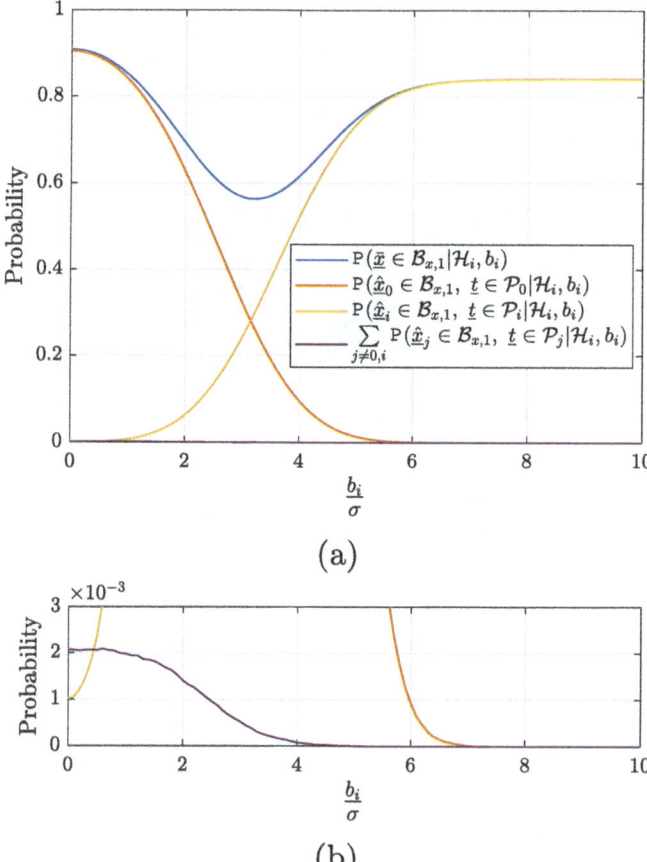

(a)

(b)

Fig. 2 (a) Illustration of $\mathsf{P}\left(\underline{\bar{x}} \in \mathcal{B}_{x,a} | \mathcal{H}_i, b_i\right)$ and its components (cf. (10)) as a function of bias-to-noise ratio $\frac{b_i}{\sigma}$ for $\alpha = 0.01$ and $a = 1$. (b) Zoom in on the *purple* curve in (a)

- The last term is the sum of two probabilities each of which is the volume under the joint PDF $f_{\underline{\hat{x}}_j \neq 0,i \cdot \underline{t}}(\theta, t | \mathcal{H}_i, b_i)$ over the intersection of $\mathcal{B}_{x,1}$ and $\mathcal{P}_j$. This intersection again defines two disconnected sub-regions in $\mathbb{R}^3$ symmetric around the origin. Increasing $\frac{b_i}{\sigma}$, the PDF $f_{\underline{\hat{x}}_j, \underline{t}}(\theta, t | \mathcal{H}_i, b_i)$ moves away from the origin along $[A^+ c_i - L_j c_{t_i}, c_{t_i}^T]^T$. Its probability mass might initially increase inside the intersection of $\mathcal{B}_{x,1}$ and $\mathcal{P}_j$, but in general it has a decreasing behavior if $\frac{b_i}{\sigma}$ increases as the *purple* curve in Fig. 2b shows.

When $\frac{b_i}{\sigma}$ is close to zero, then due to the small value of $\alpha = 0.01$, the majority of the probability mass of $f_{\underline{t}}(t | \mathcal{H}_i, b_i)$ will lie in $\mathcal{P}_0$ which explains the proximity of the blue curve to the red curve at the beginning. On the other hand, when $\frac{b_i}{\sigma} \to \infty$, the probability mass of $f_{\underline{t}}(t | \mathcal{H}_i, b_i)$ gets closer to 1 inside $\mathcal{P}_i$, i.e. $\mathsf{P}(\underline{t} \in \mathcal{P}_i | \mathcal{H}_i, b_i \to \infty) \to 1$ and $\mathsf{P}(\underline{t} \in \mathcal{P}_{j \neq i} | \mathcal{H}_i, b_i \to \infty) \to 0$, which explains the proximity of the blue curve to the yellow curve at the end. The probability $\mathsf{P}\left(\underline{\bar{x}} \in \mathcal{B}_{x,a} | \mathcal{H}_i, b_i\right)$, in addition to $\frac{b_i}{\sigma}$, also depends on a. Figure 3 shows the colormap of the

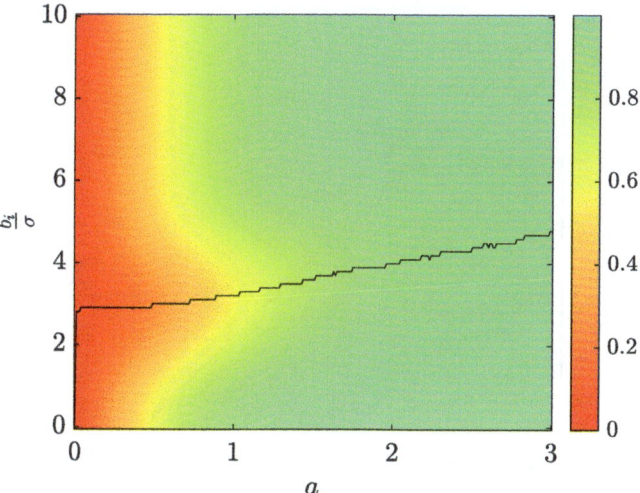

Fig. 3 Illustration of $\mathsf{P}\left(\underline{\bar{x}} \in \mathcal{B}_{x,a} | \mathcal{H}_i, b_i\right)$ as a function of bias-to-noise ratio $\frac{b_i}{\sigma}$ and a, for $\alpha = 0.01$. The black curve shows $\frac{1}{\sigma} \underset{b_i}{\operatorname{argmin}} \mathsf{P}\left(\underline{\bar{x}} \in \mathcal{B}_{x,a} | \mathcal{H}_i, b_i\right)$ as a function of a

probability $\mathsf{P}\left(\underline{\bar{x}} \in \mathcal{B}_{x,a} | \mathcal{H}_i, b_i\right)$ as a function of $\frac{b_i}{\sigma}$ vertically, and a horizontally. In addition, the black curve shows $\frac{1}{\sigma} \underset{b_i}{\operatorname{argmin}} \mathsf{P}\left(\underline{\bar{x}} \in \mathcal{B}_{x,a} | \mathcal{H}_i, b_i\right)$ as a function of a, showing that the worst-case bias varies as a changes.

4 Example 2

Let us consider the scenario of distance-based positioning in a 2D space, as shown in Fig. 4. The aim is to estimate the position of the receiver with distance observables $\underline{y}_i$ and analyze the distributional properties of the obtained position estimator after hypothesis testing. There are four reference stations (triangles in the figure) with coordinates $(-100, 99)$, $(100, 100)$, $(100, 0)$, and $(100, -100)$ in meters for stations 1 through 4, respectively. The receiver is located exactly at the origin.

Through linearization of the observation equations, taking the origin as initial state of the receiver, the unknown position vector $x \in \mathbb{R}^2$ can be related to the observables as (1), with the design matrix $A \in \mathbb{R}^{4 \times 2}$ and potential faults $C_i b_i$. The full column-rank A-matrix is:

$$A = \begin{bmatrix} 0.711 & -0.704 \\ -\frac{1}{2}\sqrt{2} & -\frac{1}{2}\sqrt{2} \\ -1 & 0 \\ -\frac{1}{2}\sqrt{2} & \frac{1}{2}\sqrt{2} \end{bmatrix}. \qquad (11)$$

Notice that the entries in A corresponding to the linearization for $\underline{y}_1$ slightly deviate from the $\frac{1}{2}\sqrt{2}$ values since the coordinates of station 1 are $(-100, 99)$m instead of $(-100, 100)$m.

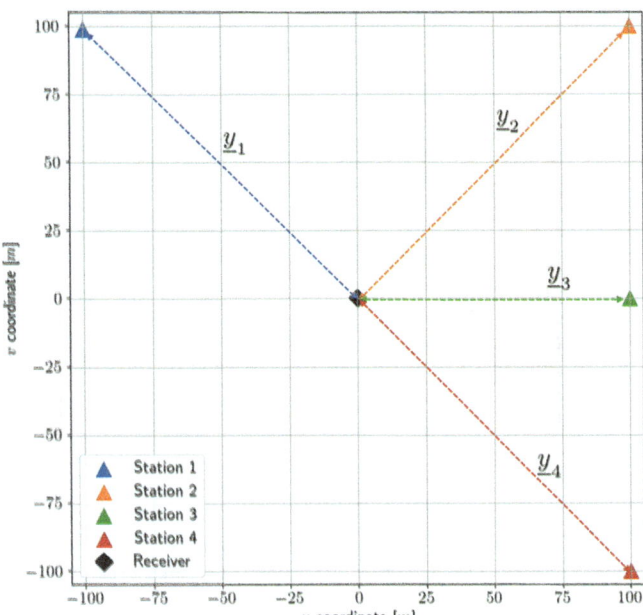

Fig. 4 Measurement setup, four reference stations (coordinates in text) and one receiver at the origin. The reference stations and the receiver are assumed to be synchronized

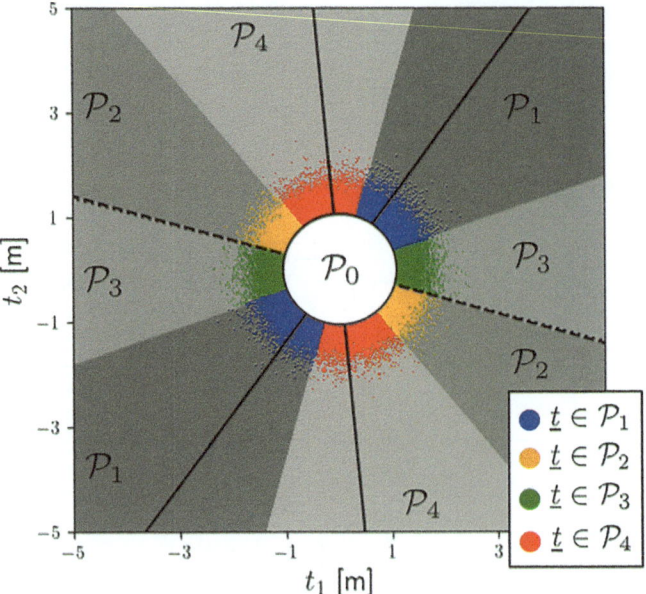

Fig. 5 Partitioning of the misclosure space corresponding to the setup from Fig. 4, the hypotheses defined in (11), and $\alpha = 0.1$. The continuous and dashed lines are vector spans of the $c_{t_i} = B^T c_i$ vectors, or fault vectors, for the $k = 4$ different alternative hypotheses. The number of pseudo-random samples drawn from $f_{\underline{t}}(t|\mathcal{H}_0)$ is 10^6 and the ones landing in $\mathcal{P}_0$ are not displayed

These coordinates were chosen intentionally, which will become clear when we explain the testing procedure and identifiability of possible faults. We assume observables are uncorrelated and normally distributed with $Q_{yy} = \sigma^2 I_4$, $\sigma = 0.5$ m.

Four different model misspecification scenarios are considered: a bias in any of the four observations. Therefore, we have five hypotheses, see (1). The model under $\mathcal{H}_0$ consists of four observations ($m = 4$) and two state parameters ($n = 2$), which leaves a redundancy of $r = m - n = 2$. As a result, we can carry out hypothesis testing to detect and identify the most likely model errors. In the model under $\mathcal{H}_i$, for $i = 1, \ldots, 4$, c_i is a canonical unit vector having 1 as its ith entry and zeros elsewhere, $q_i = 1$.

As before, hypothesis testing is done based on misclosure vector $\underline{t}$. We find a full-rank matrix B, such that $B^T A = 0$ and $\underline{t} = B^T \underline{y}$. The testing procedure is completely analogous to (2) with $k = 4$. The level of significance for the OMT is set to $\alpha = 0.1$. Consequently, the misclosure space partitioning can be found in Fig. 5.

It looks like in the figure there are only three fault lines $c_{t_i} = B^T c_i$. There are in fact four: the fault lines for hypothesis 2 and 3 are almost identical and thus lie very close together. In Table 1 we show the angles (in degrees) between the four different fault lines.

The angle between the two fault lines c_{t_2} and c_{t_3} is almost 180 deg. As a result, these two hypotheses are very poorly identifiable. In Fig. 6, we have plotted the probabilities of missed detection (MD), correct identification (CI$_2$) and

Table 1 Angles between the four fault lines from Fig. 5 in degrees. The closer the angle is to $0°$ or $180°$, the poorer the identifiability of the two corresponding hypotheses. The opposite is true for angles close to $90°$

$\angle(c_{t_i}, c_{t_j})$	c_{t_2}	c_{t_3}	c_{t_4}
c_{t_1}	110.4	69.2	41.4
c_{t_2}	0	179.6	69.0
c_{t_3}	179.6	0	110.6

wrong identification (WI$_{2,i}$, $i = 1, 3, 4$) as a function of the size of bias b_i given that $\mathcal{H}_2$ is correct.

Poor identifiability for $\mathcal{H}_2$ and $\mathcal{H}_3$ is caused by the geometry; in particular due to the coordinates of station 1. They are slightly different from $(-100, 100)$m, namely $(-100, 99)$m as we want to show what drastic effects a set of poorly identifiable hypotheses can have on the position estimator.

Important to note here is that if station 1 would have been given the coordinates $(-100\,\text{m}, 100\,\text{m})$, the two fault lines for hypothesis 2 and 3 would have been exactly the same. Then, we would not be able to identify a fault in hypothesis 2 or 3 at all: for any vector $\underline{t}$ we would then have $|\underline{w}_2| = |\underline{w}_3|$. In that case, when we decide to reject $\mathcal{H}_0$ and $\underline{t}$ lies in the regions $\mathcal{P}_2$ or $\mathcal{P}_3$ from Fig. 5, we cannot tell which is the most likely hypothesis that has occurred.

A way to understand is as follows. When a line connecting stations 1 and 4 goes through the origin, we can rotate the measurement setup accordingly such that this line will be

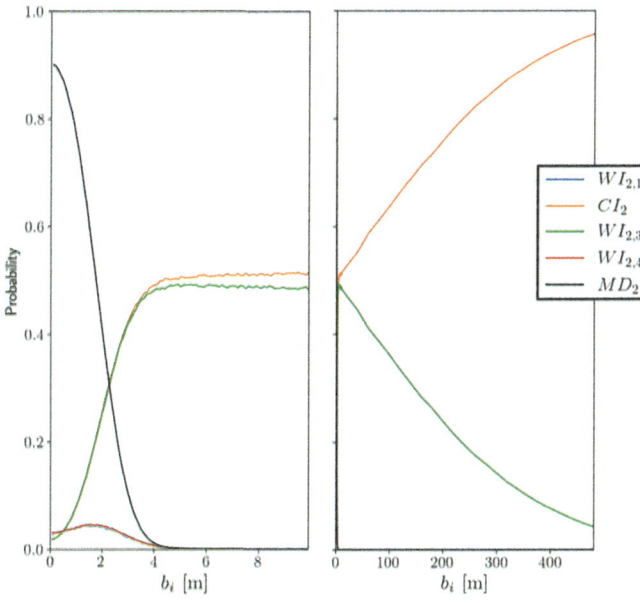

Fig. 6 Probability of missed detection, correct and wrong identification under $\mathcal{H}_2$ as a function of the size of bias b_i, for $i = 2$. The graph at right is essentially the same graph, but with a limit for b_i of 500 m

the new u' axis. Since stations 1 and 4 lie directly on the u' coordinate axis, they provide direct observations of the u' coordinate of the receiver. Station 2 lies on the other coordinate axis v', perpendicular to u', while station 3 does not lie on either of the coordinate axis. As a result, the observation from station 2 will provide a direct measure of the v' coordinate of the receiver, while the observation from station 3 contains information on both coordinates. Hence, there are now three stations (1, 3 and 4) which provide a measure for the u' axis and only two for the v' axis. Hence, there is not enough redundancy to conclude which of stations 2 or 3 will be faulty in case a fault has occurred in either one.

The PDF of $\underline{\bar{x}}$ under a certain hypothesis $\mathcal{H}_i$ depends on b_i. Figure 7 presents illustrations of the DIA distributions under $\mathcal{H}_4$ with six different b_4 values, $[0, 0.8, 2.4, 3.2, 4.0, 7.2]$m. The left sub-figure for each b_4 value exhibits the samples of the conditional DIA distribution $f_{\underline{\bar{x}}}(\theta|\mathcal{H}_4, b_4)$; the samples are obtained by Monte Carlo simulation. Firstly, 10^5 samples of observable $\underline{y}$ are generated. Then, samples of $\underline{\hat{x}}_i$ and $\underline{t}$ are computed from the samples of $\underline{y}$. Finally, the samples of the conditional DIA estimator, which are made up of samples of estimators $\underline{\hat{x}}_i$, are obtained by (4). The colours of the samples in the figure

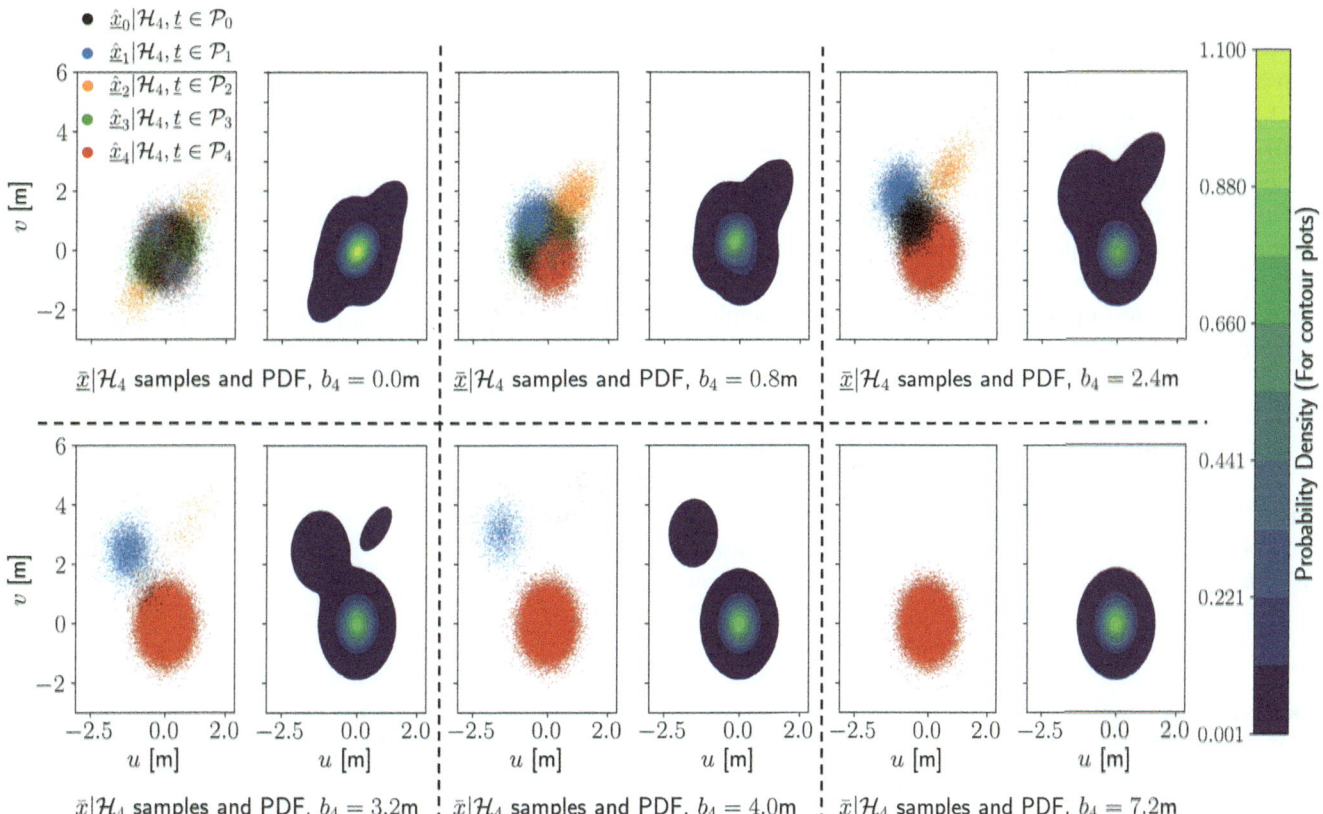

Fig. 7 Samples and contour plots of $f_{\underline{\bar{x}}}(\theta|\mathcal{H}_4, b_4)$ for the setup of **Example 2** and six b_4 values. 10^5 samples of DIA estimator are generated for each b_4 value. The colors of the samples indicate which estimator the samples are taken from. The colorbar shows the probability density corresponding to the contour plots, where only the probability density larger than 10^{-3} is taken into account

indicate which estimator the samples are taken from. The right sub-figure is the contour plot of the PDF computed with (5), and only the probability density larger than 10^{-3} is taken into account.

In Fig. 7, the irregular shapes of the DIA distribution are observed in the contour plots since the DIA distribution is a combination of several distributions with different shapes, as shown in the corresponding sample plots. When $b = 0$ m, around $(1 - \alpha) = 90\%$ samples are taken from $\hat{\underline{x}}_0$, around 10% samples are taken from other estimators. Despite its irregular distribution, the DIA estimator $\overline{\underline{x}}|\mathcal{H}_4$ with $b_4 = 0$ m is unbiased. With the increase of b_4, more samples are taken from $\hat{\underline{x}}_4$ since the fault is easier to identify when its magnitude is larger, and the DIA estimator $\overline{\underline{x}}|\mathcal{H}_4$ turns to be biased since all the estimators other than $\hat{\underline{x}}_4$ are biased when $b_4 > 0$ m. In the plots of $b_4 = 3.2$ m and 4.0 m, samples from the estimators other than $\hat{\underline{x}}_1$ (in blue) and $\hat{\underline{x}}_4$ (in red) are hard to be found. This can be explained by Fig. 5. With the increase of b_4, the cloud of points of $\underline{t}$ samples moves along the fault line c_{t_4}. It can happen that most of $\underline{t}$ samples lie in $\mathcal{P}_4$ and $\mathcal{P}_1$ and few, or even no samples lie in $\mathcal{P}_0$, $\mathcal{P}_2$ and $\mathcal{P}_3$, since the fault vector of $\mathcal{H}_4$ is closest to that of $\mathcal{H}_1$. As a result, most of the conditional DIA estimator samples are taken from $\hat{\underline{x}}_4$ and $\hat{\underline{x}}_1$. When b_4 is large enough, as shown in the plots for $b_4 = 7.2$ m, the probability of the correct identification of the fault is close to 1; thus, all the samples are taken from $\hat{\underline{x}}_4$, and the conditional DIA estimator $\overline{\underline{x}}|\mathcal{H}_4$ with $b_4 = 7.2$ m is nearly unbiased.

The distribution of the DIA estimator may have irregular shapes and is very different from the distributions of the individual estimators $\hat{\underline{x}}_i$. Thus, the distributions of individual $\hat{\underline{x}}_i$ do not properly reflect the statistical property of the testing and estimation procedure. Hence, the DIA estimator $\overline{\underline{x}}$, instead of the normally distributed $\hat{\underline{x}}_i$, should be used when statistical testing and estimation are jointly applied, especially in safety-critical applications.

5 Conclusions and Recommendations

Once the DIA-estimator was put forward, we gave two conclusions. First, applying linear estimation to Gaussian distributed data, and applying testing as well, leads to a non-Gaussian PDF for the resulting estimator. Second, unbiased estimation, together with testing actually leads to biased estimation, under an alternative hypothesis.

Our recommendation reads to free ourselves from the 'mean-variance addiction', and to consider instead the entire distribution of the estimator.

We identified three challenges ahead. First is how to evaluate the quality of the DIA-estimator. Thereby proper justice to the multi-modal distribution is required. Second is how to design an appropriate DIA-estimator, which is a quest for new optimal estimators. And third, how to numerically efficiently compute evaluations of the DIA-estimator quality, where we foresee a big role to be played by advanced simulation methods.

Acknowledgements Sebastian Ciuban and Chengyu Yin work in the 'I-GNSS positioning for assisted and automated driving' project (18305) funded by the Dutch Research Council (NWO), and Bob van Noort works in the 'Integrity of Galileo satellite navigation for aviation' project funded by the Netherlands Space Office (NSO).

References

Baarda W (1968) A testing procedure for use in geodetic networks. Netherlands Geodetic Commision, Publ. on geodesy, New Series 2(5)

Chatfield C (1995) Model uncertainty, data mining and statistical inference. J Roy Stat Soc 158:3

Kabaila P (2009) The coverage properties of confidence regions after model selection. Int Stat Rev 77:3

Meeks SL, D'agostino RB (1983) A note on the use of confidence limits following rejection of a null hypothesis. Am Stat 37(2):134–136

Teunissen PJG (2018) Distributional theory for the DIA method. J Geodesy 92(1):59–80

Teunissen PJG (2022) Geodetic inference: a selection of some challenging topics. In: Vening-Meinesz Medal Lecture of EGU General Assembly Conference Abstracts (pp. EGU22-13344)

Zaminpardaz S, Teunissen PJG (2022) On the computation of confidence regions and error ellipses: a critical appraisal. J Geodesy 96(2):10

Zaminpardaz S, Teunissen PJG, Tiberius CCJM (2019) Risking to underestimate the integrity risk. GPS Solut 23:1–16

Open Access This chapter is licensed under the terms of the Creative Commons Attribution 4.0 International License (http://creativecommons.org/licenses/by/4.0/), which permits use, sharing, adaptation, distribution and reproduction in any medium or format, as long as you give appropriate credit to the original author(s) and the source, provide a link to the Creative Commons license and indicate if changes were made.

The images or other third party material in this chapter are included in the chapter's Creative Commons license, unless indicated otherwise in a credit line to the material. If material is not included in the chapter's Creative Commons license and your intended use is not permitted by statutory regulation or exceeds the permitted use, you will need to obtain permission directly from the copyright holder.

EPOS-OC, a Universal Software Tool for Satellite Geodesy at GFZ

Karl Hans Neumayer, Patrick Schreiner, Rolf König, Christoph Dahle, Susanne Glaser, Nijat Mammadaliyev, and Frank Flechtner

Abstract

The adjustment of parameters from different observations describing the state and change of system Earth has been conducted at the Helmholtz Centre Potsdam—GFZ German Research Centre for Geosciences via satellite observations for many decades. Satellite Laser Ranging (SLR) is used to establish ground station coordinates and their drifts as well as Earth Rotation Parameters (ERPs). Doppler Orbitography and Radiopositioning Integrated by Satellite (DORIS), Global Navigation Satellite System (GNSS), SLR and Very Long Baseline Interferometry (VLBI) observations are combined to contribute to the development of an International Terrestrial Reference Frame (ITRF) with the highest precision possible. The Earth's gravity field and its temporal variations are adjusted analyzing orbit perturbations of Low Earth Orbiting (LEO) satellites, where the corresponding trajectories are obtained from Global Positioning System (GPS), on-board accelerometers (ACC) or gradiometers, star tracker (STR) and inter-satellite ranging observations. Apart from real data analysis, numerous simulation studies are conducted, e.g. to investigate the performance of Next Generation Gravity Missions or possible improvements of terrestrial reference frames by space-tie satellites. Also, we contribute to testing the theory of general relativity by analysing observations of the Laser Geodynamic Satellites (LAGEOS). All that would not be possible without a universal software tool that is central to all these activities. In this paper we give a short overview of our program package Earth Parameter and Orbit System (EPOS) with its core module for precise orbit computation (OC) EPOS-OC. We briefly describe its main features and give examples on Precise Orbit Determination (POD) of Earth satellites, describe how the program is used for determination of ERPs, station coordinates, reference frames and the adjustment of Earth's gravity field using real-world data and within simulation studies. We finally show that EPOS-OC is also a useful tool to test some predictions of the theory of General Relativity.

Keywords

EPOS-OC software · Parameter estimation · Precise orbit determination · Satellite geodesy

K. H. Neumayer (✉) · P. Schreiner · C. Dahle · N. Mammadaliyev
Helmholtz Centre Potsdam GFZ German Research Centre for Geosciences, Potsdam, Germany
e-mail: neumayer@gfz-potsdam.de

R. König
Technische Universität Berlin, Chair of Physical Geodesy, Berlin, Germany

S. Glaser
Helmholtz Centre Potsdam GFZ German Research Centre for Geosciences, Potsdam, Germany

Institute of Geodesy and Geoinformation, University of Bonn, Bonn, Germany

1 Introduction

The Helmholtz Centre Potsdam—GFZ German Research Centre for Geosciences is the national centre for geoscience research. GFZ's key mission is to secure a profound understanding of the systems and processes of solid Earth, to develop strategies and options for action in addressing global change and its impacts on a regional level, to understand natural hazards and to minimize associated risks, or to evaluate the influence of human activity on system Earth. In pursuit of our mission, we have developed a comprehensive spectrum of expertise in geodesy, geophysics, physics and mathematics.

The task of geodesy is to measure in detail the shape and rotation of the Earth, its orientation in space, its surface, and its gravitational field. For that purpose, GFZ's Sect. 1.2 "Global Geomonitoring and Gravity Field" analyses and interprets data from various national and international satellite missions for remote sensing of the Earth. As an example, dedicated gravity satellite missions such as GRACE, GOCE or GRACE-FO, to a large extent developed and operated by Sect. 1.2, enable the study of mass distribution and mass transport in system Earth on various spatial and temporal scales. In this context we also make use of various Low Earth Orbiting (LEO) and other geodetic satellites to estimate dynamic parameters such as the long-wavelength part of the gravity field, Earth orientation parameters, and global ground station coordinates based on a huge suite of space-geodetic tracking techniques such as Doppler Orbitography and Radiopositioning Integrated by Satellite (DORIS), Global Navigation Satellite Systems (GNSS) and Satellite Laser Ranging (SLR).

Consistent analysis and modelling of all existing space-borne gravity sensors and space geodetic techniques as well as simulation studies for future gravity mission concepts or the International Terrestrial Reference Frame (ITRF) in view of the Global Geodetic Observing System (GGOS) are integratively performed with our in-house software package Earth Parameter and Orbit System (EPOS) which follows the standards of the most actual IERS conventions (Petit and Luzum 2010). This allows us to contribute to the international services of the IAG (International Association of Geodesy). The GFZ branch Oberpfaffenhofen is an Analysis Center (AC) for the ILRS (International Laser Ranging Service) and an Associate Analyis Center (AAC) for the IDS (International DORIS Service).

F. Flechtner
Helmholtz Centre Potsdam GFZ German Research Centre for Geosciences, Potsdam, Germany

Technische Universität Berlin, Chair of Physical Geodesy, Berlin, Germany

EPOS-OC is the Orbit Computation module of EPOS and has common ancestors with the GINS software of GRGS/CNES (Marty et al. 2011) and the DOGS software of DGFI-TUM (Kwak et al. 2017). The accuracy of the entire orbit and gravity field determination functionality of EPOS-OC is continuously evaluated against other international state-of-the-art precise orbit determination (POD) software packages (e.g. Lasser et al. 2020). EPOS-OC is coded in Fortran 90. Its source consists of some 350 individual files and approx. 100,000 lines of code, which are under rigorous source code control. Coded footprints which are automatically produced when running EPOS-OC allow to re-construct the original source code, and past releases can be re-established at a time granularity of 1 day for debugging purposes. EPOS-OC can run in adjustment, simulation and pure integration mode. It can process linear side constraints and constraints of the no-net translation/scaling/rotation type. On demand, it produces design and normal equations (NEQs) in various formats for subsequent processing (see Fig. 1).

The following chapters provide a detailed overview of the available observation techniques and applications like precise orbit determination, gravity field determination, Earth rotation and reference frame determination, and an extraordinary application of EPOS-OC in the field of the theory of general relativity.

2 Available Observation Types and Adjustable Parameters

EPOS-OC is among the very few software packages that can process all four space-geodetic techniques on the observation level: DORIS, GNSS, SLR, Very Long Baseline Interferometry (VLBI, with far-distant radio sources as well as radio sources on a satellite). GNSS that can be processed comprehend the Global Positioning System (GPS), GLObalnaja NAwigazionnaja Sputnikowaja Sistema (GLONASS), Galileo, Satellite Based Augmentation Systems (SBAS), BeiDou and Quasi-Zenith Satellite System (QZSS). Further applicable observation types are: microwave and laser inter-satellite links (i.e., range, range-rate and range-acceleration measurements), orbit positions and velocities, altimeter ranges and crossover differences, measured satellite surface accelerations, historical Doppler measurements, PRARE range and range-rate measurements (Falck et al. 2004), and other observations like right ascension and declination.

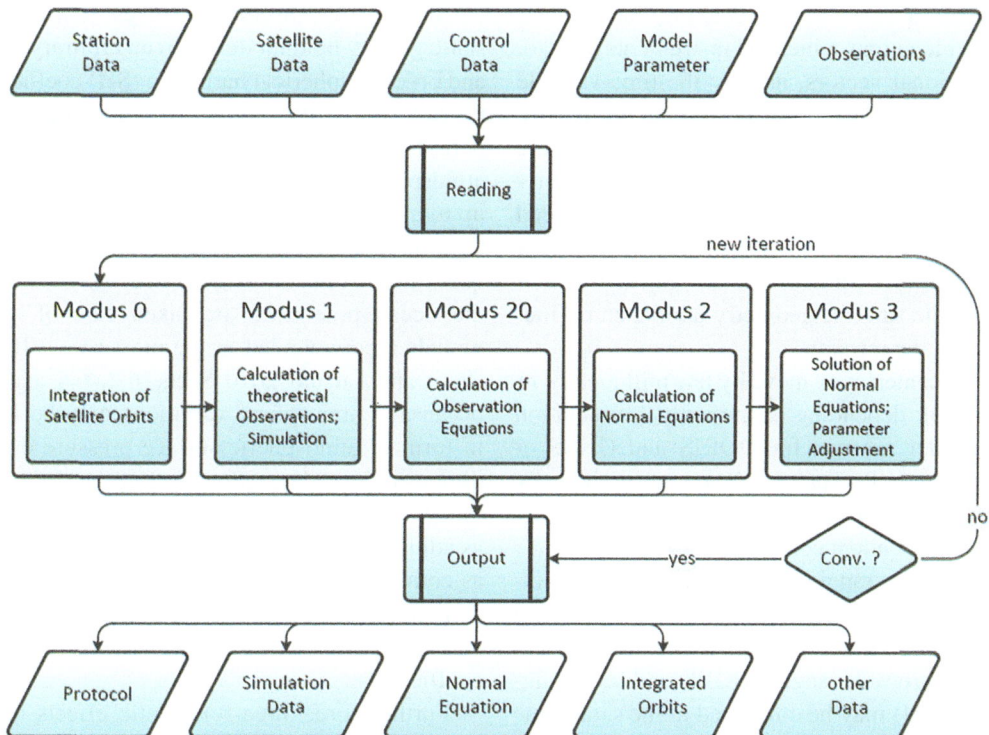

Fig. 1 EPOS-OC high level flow chart (Conv. = Convergence)

For the main space-geodetic techniques, a multitude of observation-specific corrections and adjustable parameters are available:

For corresponding ground stations, coordinates and velocities can be estimated. In order to obtain the movement of the geometrical center defined by a ground station network with respect to the center-of-mass of the Earth, geocenter corrections can be introduced and also solved for. These are constants as well as annual and semi-annual trigonometric terms in all three spatial directions. Additionally, vertical and horizontal loading effects on the station coordinates due to ocean tides, ocean pole tides, solid Earth tides and non-tidal atmosphere and ocean mass variations (AOD) can be considered, as well as the influence of post-seismic deformation corrections (Altamimi et al. 2023).

Regarding the signal paths of a given measurement technique, relativistic corrections can be applied.

For the tropospheric signal path correction of SLR, there are classical models (Marini and Murray 1973; Mendes et al. 2002; Mendes and Pavlis 2004) as well as newer developments like the Potsdam Mapping Function (Balidakis 2019) available. For DORIS, GNSS, and VLBI, there are seven different tropospheric signal path range corrections (Hopfield 1969; Yionoulis 1970; Saastamoinen 1973; Davis et al. 1985; Niell 1996; Böhm et al. 2006a, 2006b) implemented. All tropospheric corrections are scaled by adjustable factors that are station-specific constants—global or passage-wise—in the case of DORIS and SLR or station-specific continuous and piece-wise linear functions in time for GNSS and VLBI.

For GNSS, if deemed necessary, second-order ionospheric corrections may be activated (Fritsche et al. 2005).

SLR reflector-specific—in the sense as depending on the satellite—corrections are implemented for satellites like TOPEX/Poseidon, GEOS-1/2, PEOLE, DIADEME-1/2, BEACON B/C, GEOS-C, METEOR-3. For spherical satellites like LAGEOS time and station dependent range corrections (Otsubo and Appleby 2003) as available from the ILRS (ILRS 2023) can be handled. For all satellites, in particular for altimetry and gravity missions, azimuth-and elevation-dependent correction masks can be applied.

Furthermore, a great variety of signal biases and corrections can be adjusted for the various observation types.

In case of SLR and DORIS we can solve for range and time biases. For DORIS, being a measurement of the Doppler type, frequency offsets and drifts can be adjusted. For VLBI, we have implemented adjustable offsets of station receiver clocks that are piecewise-linear and continuous in time. In the case of GNSS, the clock offsets of both senders and receivers are implemented as adjustable parameters. The exact location of instrument antenna phase centers for both satellites—for senders and receivers—and ground stations are quantities that affect the computation of the signal path and the signal travel time. By convention, a geometric origin point in the satellite body is chosen, which on the one hand is the actual center of mass of the satellite, and on the other hand the exact locations which the instruments refer to. Those three-dimensional vectors, namely the time-variable center of mass, and the measurement instrument references, are adjustable vectors in EPOS-OC. Such adjustments are crucial for scenarios of instrument co-locations: The center of mass can change if on-board fuel is used up or provided nominal values for instrument phase center positions may be inaccurate or simply wrong. Thus, the adjustability of dedicated points in the satellite geometry admits an on-the-fly calibration during the mission.

Instrument phase centers may move by few millimeters, or even fractions thereof, depending of the mutual orientation of receiver and sender antenna for DORIS and GNSS or reflector geometry for SLR. That is taken care of either by azimuth/elevation masks for SLR, DORIS and GNSS, or by azimuth−/elevation-dependent corrections.

Inter-satellite links are ranging instruments between single satellite pairs like GRACE, GRACE-FO, and planned successor missions of similar type. Different measurement types like K-band microwave ranging (KBR) and laser ranging interferometry (LRI) may be employed at the same time between the GRACE-FO satellite pair. Adjustable empirical parameters for both KBR and LRI, like piecewise linear functions and trigonometric terms (Kim 2000, Sect. 3.3.2), have been implemented. For LRI, in the context of in-flight calibration, we can also adjust a scaling factor which is a piecewise-linear and continuous function in time. Also, a sophisticated satellite interconnectivity has been added to EPOS-OC allowing to simultaneously use multiple links between an arbitrary number of satellite pairs of any given imaginable topology. One example of such a scenario is the proposed KEPLER constellation of the Advanced Technologies for Navigation and Geodesy (ADVANTAGE) project (Glaser et al. 2020). Here, it is assumed to have up to three orbital planes with evenly distributed satellites in each plane that are interconnected by highly accurate range and time inter-satellite optical links.

3 Precise Orbit Determination

EPOS-OC enables POD of satellites with highest accuracy using any desired combination of implemented observation techniques. This is possible both dynamically and, for GNSS, also geometrically. For this purpose, a large number of conservative and non-conservative force models and correction models are implemented. In addition, it is also possible to use accelerometer (ACC, see Fig. 2) observations instead of non-conservative force models.

In the following, more details regarding the satellite dynamics are provided.

Among the conservative forces, first there is the Earth's static gravity field model up to an arbitrary maximum degree and order of spherical harmonic (SH) coefficients which may be endowed with drifts as well as annual and semi-annual periodic terms. Second, there are models for oceanic and atmospheric tides with an arbitrary set of constituents, also up to any maximum degree and order. Just as for the gravity field, their expansion coefficients are potentially adjustable quantities. Solid Earth tides, the solid Earth pole tides and the ocean pole tides are taken care of according to the models recommended in the most recent IERS conventions (Petit and Luzum 2010 Sects. 6.2, 6.4 and 6.5). Non-tidal ocean and atmosphere variations may be introduced either in form of atmospheric surface pressure data or in form of dimensionless SH coefficients as provided by so-called AOD models (e.g., Shihora et al. 2022). The latter can be made as adjustable parameters and forwarded to external programs as components of NEQs. Sun, Moon and planets are treated as point masses, with the option to apply for the Moon an extended gravity field, like for the Earth, in form of SH coefficients.

Furthermore, three relativistic effects are taken care of in the orbit dynamics. The first implements the relativistic modification of a Kepler orbit if the Earth was a point mass. The second is the Lense-Thirring effect that models the deviations of a satellite trajectory due to an extended rotating central body. The third, the deSitter effect, models trajectory deviations due to the circular motion of the Earth around the Sun. Those three relativistic effects may be switched on and off by dedicated flags for experimental purposes.

Concerning non-conservative forces, satellites can be modelled in EPOS-OC either by spheres or by a satellite body that is defined by a convex polyhedron, together with movable solar panels. In the somewhat simplistic default mode, those solar panels are assumed to be always oriented towards the Sun. For some dedicated satellites, e.g. Jason-1/2/3, the angular orientation of the solar panels with respect to the satellite body may be given by an empirical time series (Rudenko et al. 2017).

The atmospheric drag is handled by adjustable scaling factors that are global constants or polygons in time. The atmospheric density, which enters the air drag as a factor, may be chosen from currently nine different models.

Solar radiation pressure is taken care of either analytically by a box-wing model, or for GNSS satellites also by the class of semi-empirical models along the lines of Fliegel and Gallini (1992, 1996). Also, here, these models are scaled by factors that are either global constants or polygons in time.

In addition to that, also parameters for completely empirical models, as in the case of e.g. ECOM (Arnold et al. 2015), can be estimated. These parameters are polygons in time, together with trigonometric polynomials of arbitrary degree

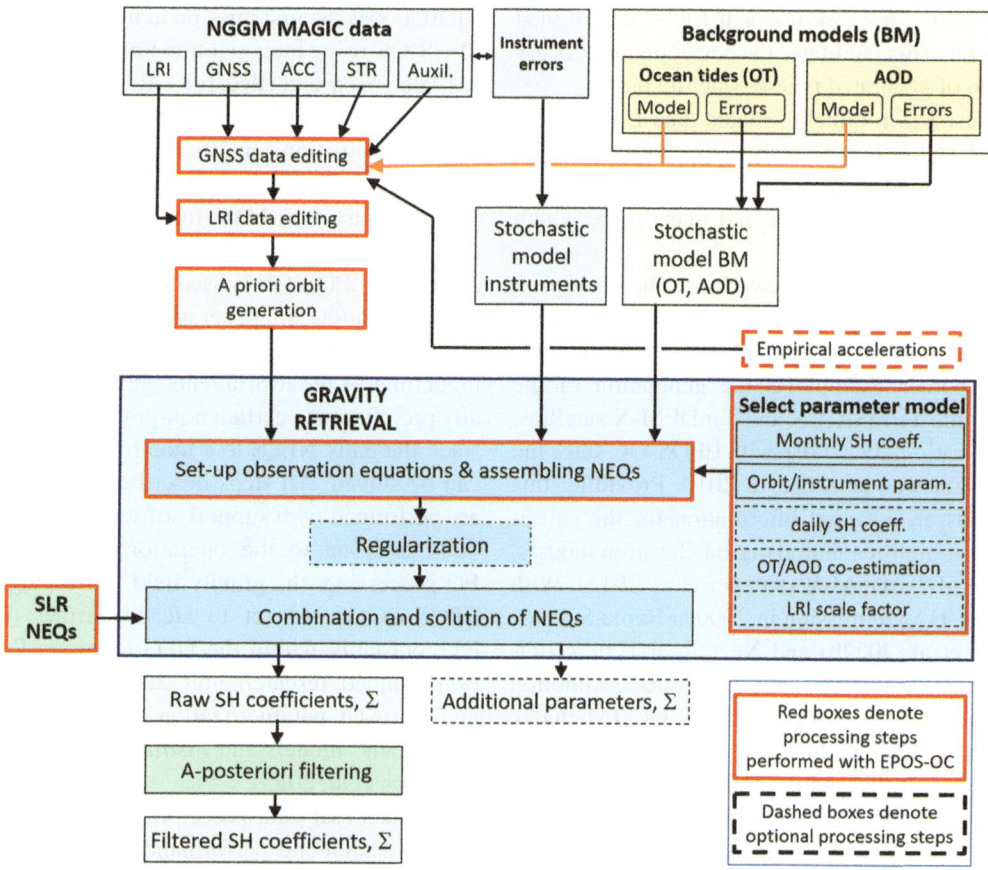

Fig. 2 GFZ processing scheme for gravity field retrieval; processing steps performed with EPOS-OC are marked by red boxes

in the argument of latitude. In addition to that, also various hybrid combinations of all these are possible.

For both, air drag and solar radiation pressure, the application of the ANGARA model (Doornbos 2001) is possible. In this approach, angle-dependent lift and drag factors are pre-computed externally by ray-tracing of air particles or radiation photons in a kind of numerically simulated wind-tunnel. Those angle-dependent tables are then read by EPOS-OC and used to infer the air drag or the radiation pressure from the orientation of the satellite body against the external particle stream. Furthermore, models for the Earth albedo and for the infrared radiation of the Earth (Borderies 1990; Borderies and Longaretti 1990; Knocke 1989) can be used. Concurrently to modelled surface forces, it is possible to use the measurements of a satellite's on-board accelerometer. Those are calibrated by adjustable scaling factors and bias vectors. Instead of scaling factors in all three spatial directions of the satellite body, an entire scaling factor matrix may be adjusted, thereby eliminating possible sensor crosstalk and axis misalignment. All those calibration parameters are applied either as polygons in time or spline functions. As the knowledge of the satellite body orientation in space is crucial to model the effect of non-conservative forces, a multitude of satellite-specific orientation algorithms has been implemented for, e.g., the satellite missions CHAMP, GRACE, GRACE-FO, Jason-1/2/3, TOPEX/Poseidon, COSMIC, Swarm, TerraSAR-X/ TanDEM-X, Cryosat-2, as well as for ESA's Sentinel satellites, and GNSS satellites. Alternatively, the satellite body orientation can be provided via star tracker quaternions that are fed into EPOS-OC in form of numerical time series.

It is also possible to set up empirical acceleration parameters in all three spatial directions. Available types of these parameters are constant and piecewise-linear accelerations, as well as trigonometric polygons at frequencies of multiples per satellite revolution.

The trajectory is obtained from the orbit dynamics with a high-order implicit Adams-Cowell predictor-corrector method in the Kulikov variant. The same device is used to integrate the variational equations, being the differential equations that are satisfied by the partial derivatives of satellite positions and velocities with respect to the dynamical parameters.

Together with the space vehicle trajectory, all above-mentioned accelerations can be written to output files in the Earth-fixed and the inertial reference frame, but also in the radial/cross-track/along-track orthogonal tripod determined by the satellite orbit. This facilitates the development process and the validation of generated precise satellite orbits.

EPOS-OC is used to routinely process data from GNSS satellites, ground stations and satellite missions with different application backgrounds, such as geodetic satellites, gravity field and altimetry missions, and many more which contribute to global geomonitoring.

As one example, in the framework of the European Union's Earth observation program Copernicus, GFZ generates orbit solutions for a large number of Sentinel satellites (GMV 2023). Another example is the generation of the baseline between the TerraSAR-X and TanDEM-X satellites which is done operationally at GFZ with EPOS-OC since the TanDEM-X mission was launched in 2010. Providing this baseline product is an essential information for the subsequent generation of high-resolution digital elevation models, the main purpose of TanDEM-X (Krieger et al. 2013). With the launch of the CHAMP mission in 2000, a Rapid Science (RSO, Schreiner et al. 2022b) and Near Real-Time (NRT, Schreiner et al. 2022a) orbit chain for radio occultation measurements is operated based on EPOS-OC (Michalak and König 2010a, 2010b). At first, NRT GPS constellations are generated operationally and orbits for the LEO satellite are calculated within the shortest time after each dump contact of the satellite at GFZ's Ny-Ålesund station (time from dump to final LEO orbit less than 90 min). These are subsequently used to derive calculations of temperature or humidity profiles which contribute to international weather forecasts (Wickert et al. 2004). A big advantage is that the whole orbit processing for the GPS constellation and the LEO can be done consistently with one software.

4 Determination of the Earth's Gravity Field

As also SH coefficients of the Earth's gravity field are solve-for parameters, EPOS-OC has been and is extensively used for global gravity field determination. During the past decades, the focus was first on the generation of static Earth gravity field models, based on satellite missions like GFZ-1 (König et al. 1999), CHAMP (Reigber et al. 2003) and GOCE (Bruinsma et al. 2014). Since 2002, the main focus switched to the computation of time-variable gravity field models using data from the twin-satellite missions GRACE and its successor GRACE-FO. For both these missions, GFZ is part of the US-German Science Data System (SDS), and EPOS-OC became the key software to derive GRACE/GRACE-FO monthly gravity field solutions, the so-called SDS Level-2 products (Dahle et al. 2019). With the KBR measurements being the main observation type, GRACE was the first mission to make use of the capability of EPOS-OC using inter-satellite link observations. This feature has later been extended for GRACE-FO to also enable the processing of the novel LRI measurements. Monthly gravity field determination consists of several processing steps: First, EPOS-OC is used in adjustment mode to iteratively generate daily precise orbits. In the final processing step, when gravity field coefficients are eventually set up as parameters to be estimated, EPOS-OC is used in a dedicated mode where only design equations are generated. This allows to further manipulate these design equations, e.g., to apply proper stochastic modelling of the instruments, generate daily NEQs, optionally pre-eliminate certain non-gravity parameters, and finally stack the daily NEQs to a monthly NEQ system which then can be solved. All steps described in the previous sentence are performed with support software other than EPOS-OC.

In addition to the operational SDS GRACE/GRACE-FO processing, the gravity field processing chain including EPOS-OC is subject to steady further development, as, e.g., currently within the DFG (German Research Foundation) funded research unit 2,736 NEROGRAV aiming at an improved parametrization and stochastic modelling of background models and instrument data (Hauk et al. 2023; Murböck et al. 2023).

Besides real data processing, various simulation studies for the design and performance of future gravity field missions have been based on EPOS-OC, such as

- ESA-funded NGGM/MAGIC Science Support Study (2021–2023, Final Report not public),
- ESA-funded Third Party Mission Program Science Study (2021–2023, Final Report not public),
- ESA Earth Explorer 10 Proposal "MOBILE" (Pail et al. 2019)
- Pre-launch analysis of the GRACE-FO LRI performance for scientific applications (Flechtner et al. 2016), or
- BMBF funded Geotechnologien project "Future Gravity Field Satellite Mission" (Flechtner et al. 2014).

An overview of GFZ's processing scheme for time-variable gravity field determination, as developed within NEROGRAV for the upcoming GRACE/GRACE-FO Level-2 product release and already adapted to the planned NGGM/MAGIC constellation, is provided in Fig. 2.

5 Earth Rotation and Reference Frames

The so-called three pillars of geodesy, i.e. geokinematics, the Earth's gravity field and Earth rotation are all connected by the ITRF. By determination of satellite orbits and station coordinates as well as gravity field parameters with EPOS-OC, a contribution of GFZ to the monitoring of the shape, respectively the gravity potential of the Earth is constantly made. Additionally, GFZ also continuously contributes to various IAG services for the determination of the Earth's rotation. Currently, EPOS-OC is used to generate products for two IAG services:

Since 2003, GFZ operates an Analysis Center (AC) of the ILRS and delivers NEQs in SINEX format for Earth Rotation Parameters (ERPs) and station coordinate time series on a regular basis. This service can be supported because EPOS-OC can adjust the coordinates of the Earth rotation axis as well as the coordinates of the celestial intermediate pole, together with either UT1 or LOD (length-of-day).

For the IDS, an associate AC is operated at GFZ, which is currently a candidate to become a regular AC. For this purpose, various satellite missions equipped with a DORIS receiver are evaluated and station coordinates and ERPs are estimated. Solutions for validation are regularly provided to the IDS Combination Centre. The processed orbits are made freely available at GFZ's Information System and Data Service (ISDC) for the scientific community (Schreiner et al. 2022a).

By the continuous contributions to the IAG services the quality of EPOS-OC derived Earth system parameters is maintained and it is ensured that the latest models and standards according to the currently valid IERS conventions are implemented.

Since very recently, EPOS-OC is also able to process VLBI data, either from radio sources that are far away in space, e.g. quasars, but also from radio sources that are located on a satellite close to Earth. VLBI, in a combination with the other space geodetic techniques, is an important contribution to the ITRF. For the functionality with the far away radio sources, EPOS-OC has been developed along the logic of the VieVS software (Böhm et al. 2018) which served as a reference for consistency tests. The functionality to process radio sources on satellites via POD is specific to EPOS-OC. Two different methods have been implemented: The first one is a series expansion (Klioner Sergei 1991), the second one is a GFZ in-house development without any simplifying assumptions. Both have shown to reach an accuracy of about 1 mm for the baseline length based on simulated data. The VLBI-specific parameters are ground station coordinates and their velocities, declinations and right ascensions of the radio sources, clock offset parameters, and tropospheric scaling factors. Combining VLBI to far away radio sources with VLBI to a satellite allows to solve for geocenter motion. Thus, with VLBI included, EPOS-OC could in principle produce ERPs completely on its own, without having to rely on external data services.

For many years now, various projects at GFZ have been using EPOS-OC to investigate, on the basis of simulations, how it is possible to improve the accuracy of the reference frame. For this purpose, e.g., extensive simulations of space-tie satellites have been performed. Space-tie satellites are a concept where the four main space geodetic techniques are combined on one satellite. For this purpose, the proposed GENESIS mission (Delva et al. 2023) was recently approved by the EU. With EPOS-OC, GFZ is able to consistently process all these four techniques within one software package.

6 Test of the Theory of General Relativity

The precession of the orbital node of a particle orbiting a rotating mass is known as Lense-Thirring effect (LTE, Lense and Thirring 1918) and is a manifestation of the general relativistic phenomenon of dragging of inertial frames, also known as frame-dragging. The measurement of the LTE by using the node drifts of the Laser GEOdynamic Satellite LAGEOS, launched in 1976 (NASA 2023) as the first spacecraft dedicated to high precision SLR, and a yet to be launched similar satellite in a complementary orbit was firstly proposed by Ciufolini (1984). With the help of LAGEOS-2 (ASI 2023) launched in 1992, and the advent of the early GRACE-based Earth gravity field models (here EIGEN-GRACE02S, Reigber et al. 2005) an accuracy of about 10% in measuring the LTE could be achieved by Ciufolini and Pavlis (2004). For a rigorous validation, the authors repeated their approach with different software packages, and thus they also used EPOS-OC. In that context, a fertile cooperation ensued. EPOS-OC and its entire processing environment has been installed at the University of Rome La Sapienza and has been used for a series of related projects and investigations (e.g. Ciufolini et al. 2011, 2013, 2016, 2019a).

With the Laser RELativity Satellite LARES launched in 2012 (LARES-Mission 2023), an improvement of the accuracy of the LTE test down to a few percent was possible (Ciufolini et al. 2019b). We also investigated the potential of the GALILEO satellite constellation (ESA-GALILEO 2023) to contribute in testing the LTE (Moreno et al. 2014). It turned out that the accuracy in radiation pressure modelling for GALILEO makes the recovery of the Lense-Thirring effect presently impossible. The chance for a future success lies in better knowledge of the satellite surface properties, general improvements of GALILEO precise orbit determination and an averaging process over the full constellation. That is why we will keep up our interest into this during the GALILEO operational phase.

7 Summary

The GFZ in-house developed software package EPOS-OC is a universal tool to perform numerous applications in the fields of POD, gravity field retrieval, and estimation of

ERPs, station coordinates and geocenter and to contribute to the further development of the ITRF. EPOS-OC is able to combine heterogenous data at the observation level, such as Doppler tracking (DORIS), GNSS, SLR, VLBI, inter-satellite ranging between an arbitrary number of satellites, or onboard accelerometer and gradiometer data. Various dynamic and geometric parameter groups can be adjusted in parallel. With EPOS-OC, a large amount of real-world data from various satellite missions is routinely processed at GFZ and derived products are provided to the international community. In parallel, the software is also an indispensable tool to simulate the performance of upcoming satellite concepts such as Next Generation Gravity Missions or space-tie satellites.

References

Altamimi Z, Rebischung P, Collileux X et al (2023) ITRF 2020: an augmented reference frame refining the modeling of nonlinear station motions. J Geod 97:43. https://doi.org/10.1007/s00190-023-01738-w

Arnold D, Meindl M, Beutler G, Dach R, Schaer S, Lutz S, Prange L, Sosnica K, Mervart L, Jäggi A (2015) CODE's new solar radiation pressure model for GNSS orbit determination. J Geod 89:775–791. https://doi.org/10.1007/s00190-015-0814-4

ASI (2023) LAGEOS-2, https://www.asi.it/en/earth-science/lageos-2, Retrieved on 2023-10-30

Balidakis K (2019) On the development and impact of propagation delay and geophysical loading on space geodetic technique data, analysis, PhD dissertation, https://doi.org/10.14279/depositonce-9125

Böhm J, Niell AE, Tregoning P, Schuh H (2006a) Global mapping function (GMF): a new empirical mapping function based on numerical weather model data. Geophys Res Lett 33. https://doi.org/10.1029/2005GL025546

Böhm J, Werl B, Schuh H (2006b) Troposphere mapping functions for GPS and very long baseline interferometry from European Centre for Medium-Range Weather Forecasts operational analysis data. J Geophys Res Solid Earth 111(B2). https://doi.org/10.1029/2005JB003629

Böhm J, Böhm S, Boisits J, Girdiuk A, Gruber J, Hellerschmied A, Krasna H, Landskron D, Madzak M, Mayer D, McCallum J, McCallum L, Schartner M, Teke K (2018) Vienna VLBI and satellite software (VieVS) for geodesy and astrometry. Publ Astron Soc Pac 130(986). https://doi.org/10.1088/1538-3873/aaa22b

Borderies N (1990) A general model of the planetary radiation pressure on a satellite with a complex shape. Celestial Mech Dyn Astr 49:99–110. https://doi.org/10.1007/BF00048583

Borderies N, Longaretti P (1990) A new treatment of the albedo radiation pressure in the case of a uniform albedo and of a spherical satellite. Celestial Mech Dyn Astr 49:69–98. https://doi.org/10.1007/BF00048582

Bruinsma S, Förste C, Abrikosov O, Lemoine JM, Marty JC, Mulet S, Rio MH, Bonvalot S (2014) ESA's satellite-only gravity field model via the direct approach based on all GOCE data. Geophys Res Lett 41(21):7508–7514. https://doi.org/10.1002/2014GL062045

Ciufolini I (1984) Theory and experiments in general relativity and other metric theories. Ph.D. dissertation, Univ. of Texas, Austin, Ann Arbor, Michigan

Ciufolini I, Pavlis E (2004) A confirmation of the general relativistic prediction of the Lense-Thirring effect. Nature 431:958–960. https://doi.org/10.1038/nature03007

Ciufolini I, Paolozzi A, Pavlis EC, Ries J, König R, Matzner R, Sindoni G, Neumayer KH (2011) Testing gravitational physics with satellite laser ranging. Eur Phys J Plus 1026:72. https://doi.org/10.1140/epjp/i2011-11072-2

Ciufolini I, Moreno MB, Paolozzi A, König R, Sindoni G, Michalak G, Pavlis E (2013) Monte Carlo simulations of the LARES space experiment to test general relativity and fundamental physics. Classical Quantum Gravity 30:235009. https://doi.org/10.1088/0264-9381/30/23/235009

Ciufolini I, Paolozzi A, Pavlis E, König R, Ries J, Gurzadyan V, Matzner R, Penrose R, Sindoni G, Paris C, Khachatryan H, Mirzoyan S (2016) A test of general relativity using the LARES and LAGEOS satellites and a GRACE Earth's gravity model. Eur Phys J C 76:120

Ciufolini I, Matzner R, Paolozzi A, Pavlis EC, Sindoni G, Ries J, Gurzadyan V, König R (2019a) Satellite laser-ranging as a probe of fundamental physics. Nat Sci Rep 9:15881. https://doi.org/10.1038/s41598-019-52183-9

Ciufolini I, Paolozzi A, Pavlis EC, Sindoni G, Ries J, Matzner R, König R, Paris C, Gurzadyan V, Penrose R (2019b) An improved test of the general relativistic effect of frame-dragging using the LARES and LAGEOS satellites. Eur Phys J C 79:872. https://doi.org/10.1140/epjc/s10052-019-7386-z

Dahle C, Murböck M, Michalak G, Neumayer K, Abrykosov O, Reinhold A, König R, Sulzbach R, Förste C (2019) The GFZ GRACE RL06 monthly gravity field time series, processing details and quality assessment. Remote Sensing 11(18):2116. https://doi.org/10.3390/rs11182116

Davis JL, Herring A, Shapiro II, Rogers AEE, Elgered G (1985) Geodesy by radio interferometry: effects of modeling errors on estimates of baseline length. Radio Sci 20(6):1593–1697. https://doi.org/10.1029/RS020i006p01593

Delva P, Altamimi Z, Blazquez A et al (2023) GENESIS: co-location of geodetic techniques in space. Earth Planets Space 75:5. https://doi.org/10.1186/s40623-022-01752-w

Doornbos E (2001) Modelling of non-gravitational forces for ERS-2 and ENVISAT, Master Thesis, Delft Institute for Earth-Oriented Space Research

ESA-Galileo (2023) Galileo. https://www.esa.int/Applications/Navigation/Galileo, last retrieved 2023-10-29

Falck C, Flechtner F, Massmann FH, Raimondo JC, Reigber Ch, Scherbatschenko A (2004) Betrieb des PRARE Bodensegments für ERS-2, Abschlussbericht 2003, Scientific Technical Report STR04/20, Helmholtz-Zentrum Potsdam, Deutsches GeoForschungsZentrum, https://gfzpublic.gfz-potsdam.de/rest/items/item_217144_3/component/file_217145/content, last visited 2023-10-29

Flechtner F, Sneeuw N, Schuh WD (eds) (2014) Observation of the System Earth from Space—CHAMP, GRACE, GOCE and future missions, (GEOTECHNOLOGIEN Science Report ; 20) (Advanced Technologies in Earth Sciences). Springer., XV, 230 p, Berlin. https://doi.org/10.1007/978-3-642-32135-1

Flechtner F, Neumayer KH, Dahle C, Dobslaw H, Fagiolini E, Raimondo JC, Güntner A (2016) What can we expect from the GRACE-FO laser ranging interferometer for earth science applications? Surv Geophys 37:453–470. https://doi.org/10.1007/s10712-015-9338-y

Fliegel H, Gallini TE (1992) Global positioning system radiation force model for geodetic applications. J Geophys Res 97(B1):559–568. https://doi.org/10.1029/91JB02564

Fliegel HF, Gallini TE (1996) Solar force modeling of block IIR global positioning system satellites. J Spacecr Rockets 33(6):863–866. https://doi.org/10.2514/3.26851

Fritsche M, Dietrich R, Knöfel C, Rülke A, Vey S, Rothacher M, Steigenberger P (2005) Impact of higher order ionospheric terms on GPS estimates. Geophys Res Lett 32(23). https://doi.org/10.1029/2005GL024342

Glaser S, Michalak G, Männel B, König R, Neumayer KH, Schuh H (2020) Reference system origin and scale realization with the future GNSS constellation "Kepler". J Geod 94:117. https://doi.org/10.1007/s00190-020-01441-0

GMV (2023) Copernicus POD Regular Service Review May.—Aug. 2023. Technical Report GMV. https://sentinels.copernicus.eu/-web/sentinel/technical-guides/sentinel-3-altimetry/pod/documentation, Last retrieval 2023-10-30

Hauk M, Wilms J, Sulzbach R, Panafidina N, Hart-Davis M, Dahle C, Müller V, Murböck M, Flechtner F (2023) Satellite gravity field recovery using variance-covariance information from ocean tide models, earth and space. Science 10:e2023EA003098. https://doi.org/10.1029/2023EA003098

Hopfield HS (1969) Two-quartic tropospheric refractivity profile for correcting satellite data. J Geophys Res Ocean Atmos 74(18):4487–4499. https://doi.org/10.1029/JC074i018p04487

ILRS (2023). https://ilrs.gsfc.nasa.gov/, last retrieved 2023-10-29

Kim J (2000) Simulation Study of a Low-Low Satellite-to-Satellite Tracking Mission, PhD dissertation, The University of Texas at Austin

Klioner Sergei A (1991) General relativistic model of VLBI observables. In: Carter WE (ed) Proceedings of AGU chapman conference on geodetic VLBI: monitoring global change, NOAA technical report NOS 137 NGS 49. American Geophysical Union, Washington DC, pp 188–202. https://www.researchgate.net/publication/253171626

Knocke P (1989) Earth radiation pressure effects on satellites, CSR-89-1, May 1989, PhD dissertation

König R, Chen Z, Reigber C, Schwintzer P (1999) Improvement in global gravity field recovery using GFZ-1 satellite laser tracking data. J Geod 73:398–406. https://doi.org/10.1007/s001900050259

Krieger G, Zink M, Bachmann M, Braeutigam B, Breit H, Fiedler H, Fritz T, Hajnsek I, Hueso GJ, Kahle R, Koenig R, Schaettler B, Schulze D, Wermuth M, Wessel B, Moreira A (2013) TanDEM-X. In: D'Errico M (ed) Distributed space Mission for earth system monitoring. Springer, Cham, pp 387–436. https://doi.org/10.1007/978-1-4614-4541-8_12

Kwak Y, Gerstl M, Seitz F (2017) DOGS-RI: new VLBI analysis software at DGFI-TUM, https://www.semanticscholar.org/paper/DOGS-RI%3A-new-VLBI-analysis-software-at-DGFI-TUM-Kwak-Gerstl/de08d3b313041c074b5abcad819dde81bd78bc68, last visit on 27.10.2023

LARES-Mission (2023) LARES e LARES-2 the laser relativity satellites, https://www.lares-mission.com, retrieved 2023-10-09

Lasser M, Meyer U, Jäggi A, Mayer-Gürr T, Kvas A, Neumayer KH, Dahle C, Flechtner F, Lemoine JM, Koch I, Weigelt M, Flury J (2020) Benchmark data for verifying background model implementations in orbit and gravity field determination software. Adv Geosci 55:1–11. https://doi.org/10.5194/adgeo-55-1-2020

Lense J, Thirring H (1918) Über den Einfluss der Eigenrotation der Zentralkörper auf die Bewegung der Planeten und Monde nach der Einsteinschen Gravitationstheorie. Phys Z 19:156–163

Marini JK, Murray CW (1973) Correction of laser range tracking data for atmospheric refraction at elevations above 10 degrees, NASA technical memorandum NASA-TM-X-70555. Goddard Space Flight Center, Greenbelt. https://ntrs.nasa.gov/citations/19740007037

Marty JC, Loyer S, Perosanz F, Mercier F, Bracher G, Legresy B, Portier L, Capdeville H, Fund F, Lemoine JM, Biancale R (2011). https://www.semanticscholar.org/paper/GINS%3A-THE-CNES-GRGS-GNSS-SCIENTIFIC-SOFTWARE-Marty-Loyer/72c0dd872c6fcc4e9d704d75a6dbd3a6df0d5e76. Last visited 27.10.2023

Mendes VB, Pavlis EC (2004) High-accuracy zenith delay prediction at optical wavelengths. Geophys Res Lett Solid Earth 31(14). https://doi.org/10.1029/2004GL020308

Mendes VB, Prates G, Pavlis EC, Pavlis DE, Langley RB (2002) Improved mapping functions for atmospheric refraction correction in SLR. Geophys Res Lett 29(10). https://doi.org/10.1029/2001GL014394

Michalak G, König R (2010a) Rapid science orbits for CHAMP and GRACE Radio occultation data analysis. In: Flechtner F, Gruber T, Guentner A, Mandea M, Rothacher M, Schoene T, Wickert J (eds) System earth via geodetic-geophysical space techniques. Advanced Technologies in Earth Sciences. Springer, Berlin, pp 67–77. https://doi.org/10.1007/978-3-642-10228-8

Michalak G, König R (2010b) Near-real time satellite orbit determination for GPS radio occultation with CHAMP and GRACE. In: Flechtner F, Gruber T, Guentner A, Mandea M, Rothacher M, Schoene T, Wickert J (eds) System earth via geodetic-geophysical space techniques. Springer, Berlin, pp 443–454. https://doi.org/10.1007/978-3-642-10228-8

Moreno MB, König R, Michalak G (2014) Preliminary study for the measurement of the Lense-Thirring effect with the Galileo satellites. Acta Futura 7:87–96. https://doi.org/10.2420/AF07.2013.87

Murböck M, Abrykosov P, Dahle C, Hauk M, Pail R, Flechtner F (2023) In-orbit performance of the GRACE accelerometers and microwave ranging instrument. Remote Sens 15(3):563. https://doi.org/10.3390/rs15030563

NASA (2023) LAGEOS: laser GEOdynamic satellite, https://lageos.gsfc.nasa.gov/index.html, last visit on 27.10.2023

Niell AE (1996) Global mapping functions for the atmosphere delay at radio wavelengths. J Geophys Res Solid Earth 101:3227–3246. https://doi.org/10.1029/95JB03048

Otsubo T, Appleby GM (2003) System-dependent center-of-mass correction for spherical geodetic satellites. J Geophys Res 108(B4). https://doi.org/10.1029/2002JB002209

Pail R, Bamber J, Biancale R, Bingham R, Braitenberg C, Flechtner F, Gruber T, Güntner A, Savenije H, Authors M (2019) Mass variation observing system by high low inter-satellite links (MOBILE) - a new concept for sustained observation of mass transport from space. J Geodetic Sci 9(1):48–58. https://doi.org/10.1515/jogs-2019-0006

Petit G, Luzum B (2010) IERS conventions 2010, IERS technical note No. 36, https://www.iers.org/IERS/EN/Publications/TechnicalNotes/tn36.html

Reigber C, Schwintzer P, Neumayer KH, Barthelmes F, König R, Förste C, Balmino G, Biancale R, Lemoine JM, Loyer S, Bruinsma S, Perosanz F, Fayard T (2003) The CHAMP-only earth gravity field model EIGEN-2. Adv Space Res 31(8):1883–1888

Reigber C, Schmidt R, Flechtner F, König R, Meyer U, Neumayer KH, Schwintzer P, Zhu SY (2005) An earth gravity field model complete to degree and order 150 from GRACE: EIGEN-GRACE02S. J Geodyn 39(1):1–10., ISSN 0264-3707. https://doi.org/10.1016/j.jog.2004.07.001

Rudenko S, Neumayer KH, Dettmering D, Esselborn S, Schöne T, Raimondo JC (2017) Improvements in precise orbits of Altimetriy satellites and their impact on mean sea level monitoring. IEEE Trans Geosci Remote Sens 55(6):3382–3395., June 2017. https://doi.org/10.1109/TGRS.2017.2670061

Saastamoinen J. (1973), Contribution to the theory of atmospheric refraction, part II: refraction corrections in satellite geodesy, Bull Géod 107, 13–14, https://doi.org/https://doi.org/10.1007/BF02522083

Schreiner P, Reinhold A, König R, Neumayer KH, Flechtner F (2022a) GFZ Precise Science Orbit Products for satellites equipped with DORIS receiver (version 2). GFZ Data Services. https://doi.org/10.5880/GFZ_ORBIT/PSO/GFZ_IDS_v02

Schreiner P, Neumayer KH, König R (2022b) GFZ rapid science orbits. GFZ data services, https://doi.org/10.5880/GFZ_ORBIT/RSO

Shihora L, Balidakis K, Dill R, Dahle C, Ghobadi-Far K, Bonin J, Dobslaw H (2022) Non-tidal background modeling for satellite gravimetry based on operational ECWMF and ERA5 reanalysis data: AOD1B RL07. J Geophys Res Solid Earth 127:e2022JB024360. https://doi.org/10.1029/2022JB024360

Wickert J, Schmidt T, Beyerle G, König R, Reigber C, Jakowski N (2004) The radio occultation experiment aboard CHAMP: operational data analysis and validation of vertical atmospheric profiles. J Meteorol Soc Jpn 82(1B):381–395

Yionoulis SM (1970) Algorithm to compute tropospheric refraction effects on range measurements. J Geophys Res Ocean Atmos 75(36):7636–7637. https://doi.org/10.1029/JC075i036P07636

Open Access This chapter is licensed under the terms of the Creative Commons Attribution 4.0 International License (http://creativecommons.org/licenses/by/4.0/), which permits use, sharing, adaptation, distribution and reproduction in any medium or format, as long as you give appropriate credit to the original author(s) and the source, provide a link to the Creative Commons license and indicate if changes were made.

The images or other third party material in this chapter are included in the chapter's Creative Commons license, unless indicated otherwise in a credit line to the material. If material is not included in the chapter's Creative Commons license and your intended use is not permitted by statutory regulation or exceeds the permitted use, you will need to obtain permission directly from the copyright holder.

Part II

Earth Rotation

Impact of Free Core Nutation Modeling on the Estimation of Earth Rotation Parameters from Different VLBI Session Types

Arnab Laha, Johannes Böhm, Sigrid Böhm, Hana Krásná, Nagarajan Balasubramanian, and Onkar Dikshit

Abstract

Free Core Nutation (FCN) arises from complex geophysical processes causing misalignment between the mantle and the liquid core, and exhibits a retrograde motion with a period of about 431 days as observed by Very Long Baseline Interferometry (VLBI) as part of the celestial pole offsets (CPO). This study assesses the influence of using an empirical model of FCN on estimating Earth Rotation Parameters (ERP) from different types of VLBI sessions, i.e., 24-hour S/X sessions (2001–2022), 24-hour VGOS sessions (2019–2022), and Intensive sessions (2001–2022). To evaluate the impact, a priori values of CPO from the IERS Bulletin A series and the FCN model by Belda et al. (2016) are used, and the estimated polar motion and UT1-UTC values are compared against the IERS 20 C04 EOP solution. The results indicate that the sole application of the empirical FCN model does not degrade the WRMS values but introduces time-dependent systematic differences in ERP. The comparison of S/X and VGOS sessions indicates that ERP estimated using the Belda model in VGOS sessions demonstrate slightly lower WRMS values.

Keywords

ERP · FCN · VLBI

1 Introduction

Earth's response to torque applied by celestial bodies deviates slightly from that of a solid body due to complex geophysical processes involving atmospheric and oceanic mass circulation, a deformable mantle, and the presence of a liquid outer core. This leads to a misalignment between the Earth's mantle and the rotational axis of the ellipsoidal liquid core within the visco-elastic Earth, resulting in free core nutation (FCN) in the celestial reference frame (CRF) or nearly diurnal free wobble (NDFW) in the terrestrial reference frame. In the celestial reference frame, FCN is a retrograde long periodic motion of the Earth's figure axis with a period of approximately 431 days and an average amplitude of about 100 μas (Krásná et al. 2013). The presence of FCN is found in the motion of the celestial intermediate pole (CIP), observed by Very Long Baseline Interferometry (VLBI). Celestial pole offsets (CPO), denoted by dX and dY, signify the differences between the observed and the calculated position using the conventional International Astronomical Union (IAU) models of the CIP.

Several studies have focused on determining the FCN period through various methods, such as gravimetric (Vondrák and Ron 2017) and VLBI (Krásná et al. 2013) observations. Through gravimetric observations, an FCN period of 431±0.9 sidereal days (Defraigne et al. 1994) and 428 days (Florsch and Hinderer 2000) were estimated. Additionally, joint inversion of VLBI and gravimetric

A. Laha (✉) · N. Balasubramanian · O. Dikshit
Geoinformatics, Department of Civil Engineering, Indian Institute of Technology Kanpur, Kanpur, India
e-mail: alaha@iitk.ac.in; nagaraj@iitk.ac.in; onkar@iitk.ac.in

J. Böhm · S. Böhm · H. Krásná
Higher Geodesy, Department of Geodesy and Geoinformation, TU Wien, Vienna, Austria
e-mail: johannes.boehm@geo.tuwien.ac.at; sigrid.boehm@geo.tuwien.ac.at; hana.krasna@geo.tuwien.ac.at

data estimates the FCN period of 430±0.2 days (Ziegler et al. 2020). IAU standards, including nutation IAU 2000 (Mathews et al. 2002) and precession IAU 2006 (Capitaine and Wallace 2006), were established with a period of 430.21±0.28 days (Vondrák and Ron 2017). Empirical models were used to directly fit the FCN free mode period and amplitude from VLBI observed nutation residuals, resulting in various FCN periods, including 430.55±0.11 days (Vondrák et al. 2005), 429 sidereal days (Amoruso et al. 2012), 431.18±0.10 sidereal days (Krásná et al. 2013), 441±4.5 sidereal days (Chao and Hsieh 2015), and 430.28±0.04 mean solar days (Vondrák and Ron 2017). The effect of these different FCN periods extends to astrometric results (Malkin 2017) and UT1 intensive estimates (Nothnagel and Schnell 2008) (Malkin 2011). Nothnagel and Schnell (2008) highlighted that an error in the polar motion (PM) and nutation of 1 mas can introduce an error of 30 µas in the estimation of UT1 from intensives. Similarly, Belda et al. (2016) noted the importance of complementing FCN models with the IAU 2006/2000A precession/nutation theory, as the absence of such modeling could lead to significant deviations (400 µas) in the CIP coordinates. Further, Belda et al. (2017) state that the selection of different a priori Earth Orientation Parameters (EOP) and terrestrial reference frame (TRF) configurations significantly impact the estimates of CPO. Similarly, alterations in the initial CPO may bear consequential effects on the Earth Rotation Parameters (ERP) (PM and UT1-UTC) estimations.

In this comparative study, we conducted a preliminary analysis of the influence of the FCN model on ERP, estimated from VLBI 24-hour sessions. While 24-hour VLBI sessions offer accurate ERP estimations, they are typically available only after 8–10 days of observations. To get timely UT1 estimates, the International VLBI Service for Geodesy and Astrometry (IVS) introduced special sessions known as intensive sessions (Nothnagel et al. 2017). Moreover, IVS has introduced the next-generation geodetic VLBI system, VGOS, with higher observing frequencies and an improved hardware system (Niell et al. 2018). This study also analysed the impact of FCN model on intensive sessions and a comparison was made between S/X and VGOS 24-hour sessions.

2 Data and Approach

2.1 Free Core Nutation Model

In this study, we have used an empirical FCN model (Eq. 1) with higher temporal resolution calculated by Belda et al. (2016). The contribution of the FCN to the CPO are X_{FCN} and Y_{FCN}.

$$\begin{aligned} X_{FCN} &= A_C \cos(\sigma_{FCN} t) - A_S \sin(\sigma_{FCN} t) \\ dX_{belda} &= X_{FCN} + X_0 \\ Y_{FCN} &= A_S \cos(\sigma_{FCN} t) + A_C \sin(\sigma_{FCN} t) \\ dY_{belda} &= Y_{FCN} + Y_0 \end{aligned} \quad (1)$$

where, $\sigma_{FCN} = 2\pi/P$ is the frequency of FCN in the CRF, A is the amplitude, t is the time relative to J2000, P is the period, and X_0 and Y_0 are constant offsets (low-frequency part).

A sliding window size of 400 days with a minimal displacement (one-day step) is used to improve the model's accuracy. According to Belda et al. (2016), the model shows slight improvement of 1 µas and 5 µas with respect to Malkin (2013) and Krásná et al. (2013) models, respectively, in terms of standard deviations of the residuals CPO-FCN after each respective FCN model is removed from the CPO. Additionally, Belda et al. (2017) demonstrate that the precession rate and the scatter of the residuals show minimum RMS values when IAU 2006/200A model plus Belda et al. (2016) FCN model is used as a priori CPO.

The comparison of the CPO from IERS Bulletin A and one computed using the Belda FCN model (Belda) is shown in Fig. 1. It is clearly visible that IERS Bulletin A seems to have high-frequency terms, and the low-frequency part still remains in the FCN model with approximately the same magnitude and opposite signs for dX and dY. Since the CPO series from IERS Bulletin A and Belda data are based on the same VLBI data, they are highly correlated with one another, with a correlation coefficient of 0.90 for dX and 0.92 for dY.

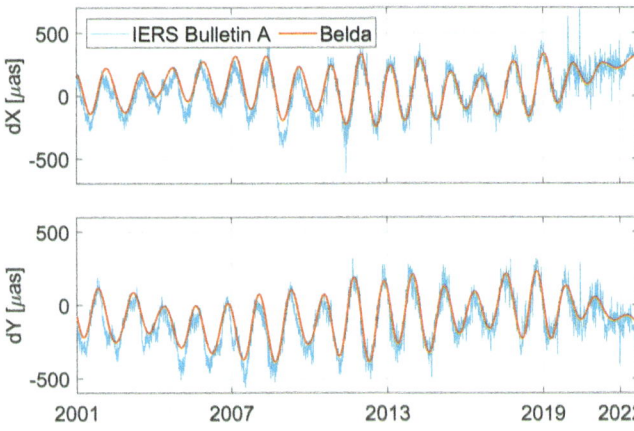

Fig. 1 Comparison of CPO from IERS Bulletin A and one computed using Belda FCN model

2.2 Parametrization and Analysis

We utilized 24-hour and intensive sessions from Jan 2001 to Sep 2022, and VGOS 24-hour sessions from Jan 2019 to Sep 2022 (Fig. 2), all observed by the IVS.

These sessions were separately analyzed using the Vienna VLBI and Satellite Software, VieVS (Böhm et al. 2018), resulting in three different ERP series (24-hour, intensives and VGOS), introducing IERS Bulletin A as a priori values for EOP. The ERP values (24-hour) were estimated at 1440 min intervals (1 day) and UT1-UTC (intensive) was estimated at 360 min (6 hr) intervals with relative constraints of 10 mas (0.7 ms) and 0.01 μs, respectively.

Additionally, station coordinates were also estimated, and the datum was defined by applying no-net-translation (NNT) and no-net-rotation (NNR) conditions for stations with continuous observations in the ITRF2020. To evaluate the impact of FCN on the estimation of ERP, dX_{belda} and dY_{belda} (Eq. 1) are used as a priori values in place of CPO of IERS Bulletin A. These different time series were assessed in terms of weighted root mean square (WRMS) value with respect to IERS 20 C04 EOP solution using Eq. 2.

$$WRMS = \sqrt{\frac{\sum_{i=1}^{N} \frac{(a_i - b_i)^2}{c_i^2 + d_i^2}}{\sum_{i=1}^{N} \frac{1}{c_i^2 + d_i^2}}} \quad (2)$$

where, a_i and b_i are the values of EOP (x_p, y_p, UT1-UTC) from the different calculated EOP series and IERS 20 C04 solution, respectively. c_i and d_i denote the respective standard error with i referring to the epoch.

The estimated time series through VieVS were discontinuous and certain epochs exhibit multiple values. To address these concerns, an epoch-wise comparison was conducted to utilize available data, and ERP values from the same epoch were chosen based on their minimum standard error. On the other hand, intensive sessions had a shorter duration, typically lasting 1 to 2 hours, and, as a result, the UT1-UTC estimates lacked a specific epoch throughout the time series. Therefore, UT1-UTC values were linearly interpolated to align with the IERS 20 C04 epoch, specifically at 00:00 UTC.

3 Results and Discussion

Our analysis focuses on three distinct scenarios, all processed uniformly with variations solely in the a priori values of CPO. The first and second scenarios explore the impact of the FCN model on 24-hour and intensive VLBI sessions, respectively, while the third scenario delves into S/X and VGOS sessions.

3.1 Comparison of 24-hour and Intensive Sessions

Figure 3a compares WRMS values of ERP, revealing no distinct impact of the FCN model on ERP estimation. However, FCN modeling introduces time-dependent systematic difference (Fig. 3b), as the high-frequency component is smoothed in the FCN model, and low-frequency terms accumulated in X_0 and Y_0 are similar to IERS Bulletin A (Fig. 1). However, the WRMS value of UT1-UTC from 24-hour sessions remains approximately similar ($\approx 20\,\mu s$) (Fig. 3a) with respect to IERS 20 C04 EOP solution and a difference of $20\,\mu s$ arises when a priori values of CPO from IERS Bulletin A and Belda FCN model are used. For PM, x_p and y_p exhibit a similar WRMS value, with a consistent systematic difference (Fig. 3b). However, y_p displays larger systematic differences with WRMS value slightly lesser when the Belda FCN model is used to estimate ERP. The difference in PM exhibits a periodic signal pattern until 2011, with one significant harmonic term with a period of about 1 year, possibly due to the better coherence of the FCN model with

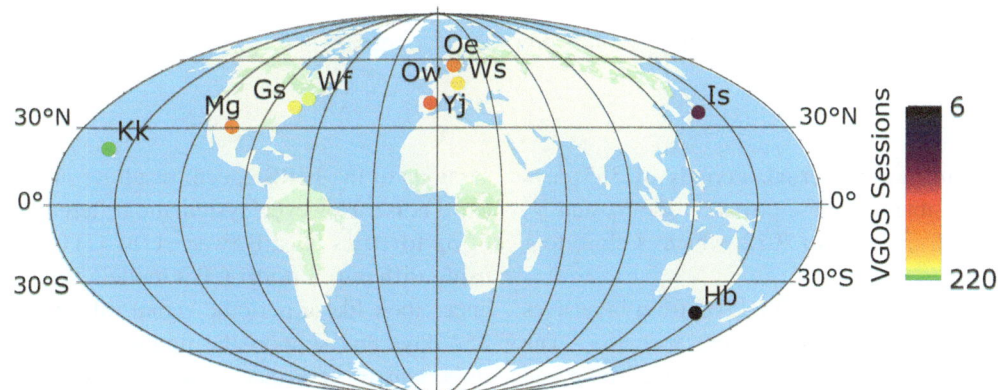

Fig. 2 Operational VGOS stations from 2019 to 2022. Color code indicates the number of sessions per station

Fig. 3 ERP estimated from 24-hour sessions using a priori values of CPO from IERS Bulletin A (IERS) and using Belda FCN model (Belda). (**a**) WRMS values of ERP in comparison to IERS 20 C04. (**b**) Difference between estimated ERP. Blue line represents the end of 2011. Spectra of these differences are shown unitless

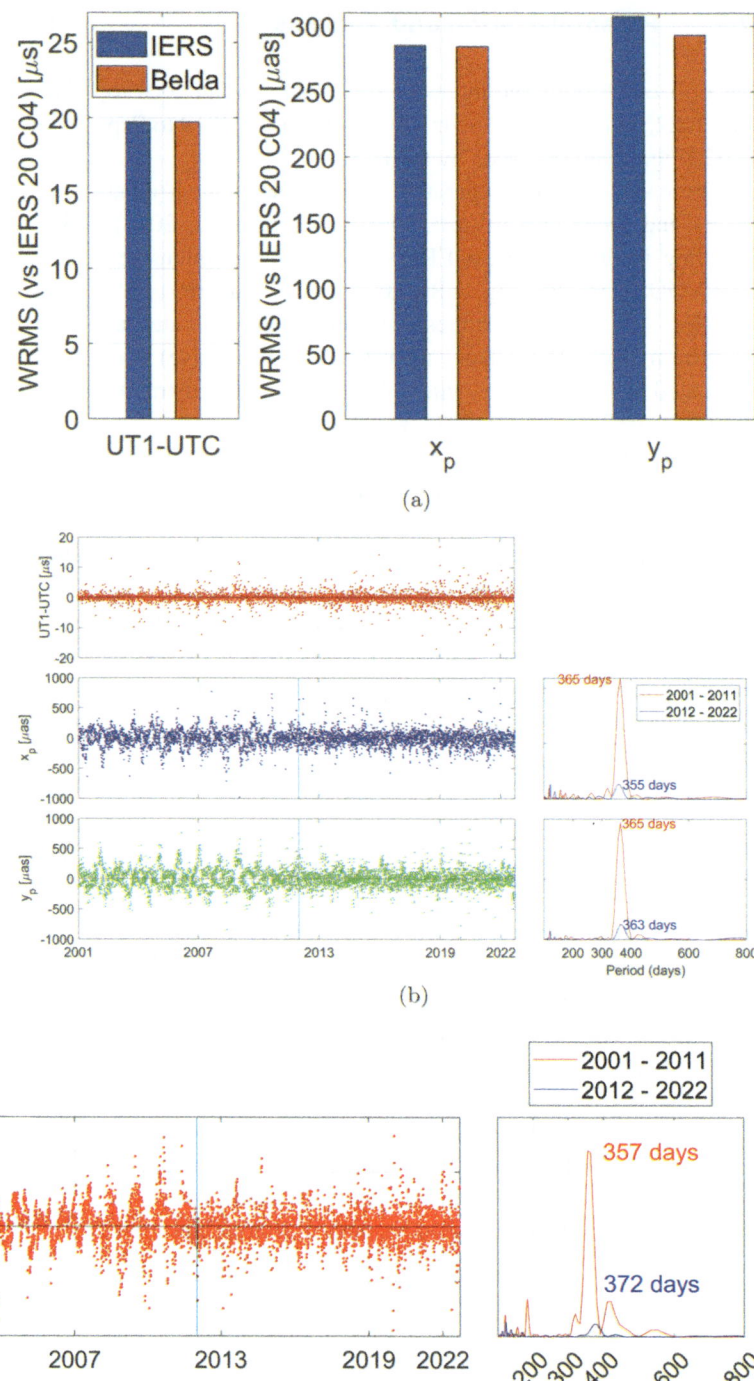

Fig. 4 Difference of the UT1-UTC estimated from intensive sessions using a priori values of CPO from IERS Bulletin A and Belda FCN model. Blue line represents the end of 2011. Spectra of these differences are shown unitless

IERS Bulletin A during that period. Periods of 355 days and 363 days are visible with less power in the x_p and y_p signal, respectively, from 2012 to 2022, which is close to 1 year.

On the other hand, estimating UT1-UTC from intensives yields a WRMS value of 34 µs for both scenarios. The difference in WRMS values for UT1-UTC from 24-hour and intensive sessions arises due to the interpolation of intensive sessions to the IERS epoch, resulting in the introduction of extra noise. However, systematic differences with amplitude of up to 10 µs exist between UT1-UTC estimates obtained using different a priori CPO models (Fig. 4). These differences look like a periodic signal till 2011 with a period of 357 days and support the findings of Malkin (2011). It is important to note that some periodicity still remains from 2012 to 2022 with less power.

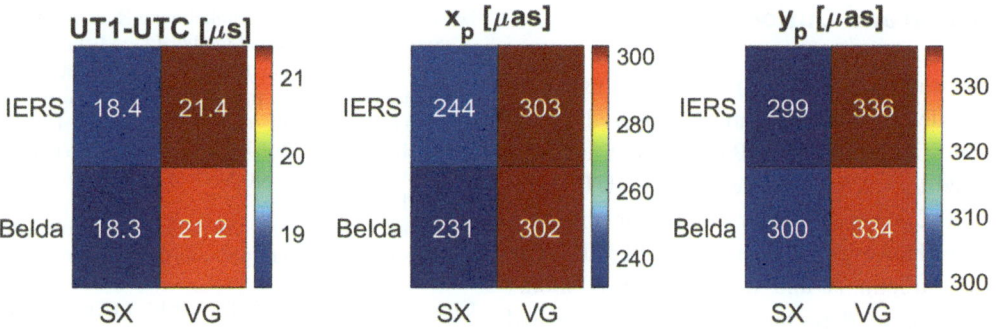

Fig. 5 WRMS values of ERP estimated from 24-hour S/X and VGOS sessions using a priori values of CPO from IERS Bulletin A (IERS) and using Belda FCN model (Belda) with respect to IERS 20 C04

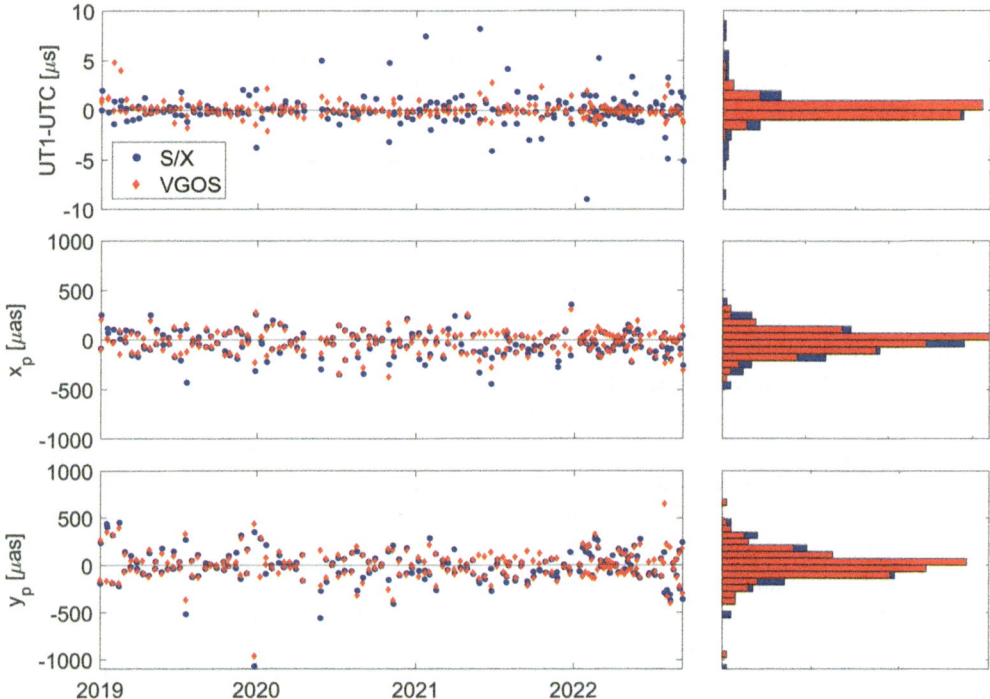

Fig. 6 Difference of the ERP estimated from S/X and VGOS sessions using a priori values of CPO from IERS Bulletin A and Belda FCN model, along with histogram

3.2 Comparison of S/X and VGOS Sessions

24-hour S/X and VGOS sessions of same epoch were estimated in a uniform way and the WRMS values are compared with respect to IERS 20 C04 EOP solution. Figure 5 shows that the WRMS values for ERP estimated from VGOS sessions are notably higher than those from S/X sessions with respect to the IERS 20 C04 EOP solution, irrespective of any CPO model. This difference can be attributed to the northern hemisphere bias of the VGOS station network and the presence of fewer stations compared to the S/X network (Glomsda et al. 2023). The WRMS values for UT1-UTC are similar for both S/X and VGOS sessions (Fig. 5), however S/X sessions show higher difference between the UT1-UTC estimated using CPO as a priori from IERS Bulletin A and Belda FCN model than VGOS sessions (Fig. 6). In case of PM, the difference in WRMS values for S/X sessions using different CPO a priori series are slightly higher from VGOS sessions, 13 μas for x_p. However, the difference for PM estimated from S/X and VGOS sessions have similar amplitude and variations (Fig. 6). VGOS sessions in this time frame show that ERP estimated using Belda FCN CPO as a priori have smaller WRMS values than using IERS Bulletin A CPO as a priori. Except for y_p, remaining ERP also show smaller WRMS values when using Belda FCN model.

4 Conclusion and Outlook

This preliminary study shows the influence of using an empirical FCN model on ERP estimation compared to discrete values of celestial pole offsets. Based on our analysis, it can be concluded that WRMS values for UT1-UTC with respect to IERS 20 C04 do not differ in 24-hour, intensive and VGOS sessions. However, differences in UT1-UTC with CPO from IERS Bulletin A and the Belda FCN model show amplitudes of up to $20\,\mu s$ and $10\,\mu s$ for 24-hour and intensive sessions, respectively. S/X and VGOS sessions show similar deviation in UT1-UTC estimation from both the a priori models. PM exhibits similar WRMS values for 24-hour sessions (2001 to 2022) and VGOS sessions (2019 to 2022) but show significant changes for S/X sessions in x_p. Finally, it is observed that Belda FCN model does not impact the estimation of ERP for 24-hour sessions and intensive sessions but causes time-dependent systematic differences in PM estimated from 24-hour and in UT1-UTC from intensive sessions. Sinusoidal behaviour with an annual period is seen till 2011 in UT1-UTC estimates from intensive sessions, confirming with the results of Kern et al. (2023) and in PM estimates from 24-hour sessions. These sinusoidal variations are present in or are not represented by the a priori CPO model, which is reflected in the estimated ERP. The study emphasizes the importance of choosing the optimal CPO model aligned with the IVS EOP series due to lower WRMS values during the recent years for S/X sessions. In the non-operational processing of the multi-year time series, crucial for the densification of ERP series, Belda FCN model can be used as a priori CPO. The explanation of the systematic differences in PM and UT1-UTC (from intensive sessions) would be worth a special study. Additionally, future studies could consider different FCN models (e.g., Krásná et al. (2013) and Malkin (2013)) to assess their impact on ERP since 2010 and extend the analysis to assess the FCN impact on the estimation of all EOP.

Acknowledgements The authors thank Santiago Belda for providing the FCN model. They also thank OeAD for providing financial support to the first author, AL, through the Ernst Mach grant, facilitating his stay in Vienna. Additionally, AL extends his heartfelt appreciation to the research team at TU Wien for their valuable research insights and for providing computational resources. He also acknowledges the National Centre for Geodesy, IITK and IUGG travel grant for their support in covering expenses related to IUGG.

References

Amoruso A, Botta V, Crescentini L (2012) Free core resonance parameters from strain data: sensitivity analysis and results from the gran sasso (italy) extensometers. Geophys J Int 189(2):923–936

Belda S, Ferrándiz JM, Heinkelmann R, et al (2016) Testing a new free core nutation empirical model. J Geodynam 94:59–67

Belda S, Heinkelmann R, Ferrándiz JM, et al (2017) An improved empirical harmonic model of the celestial intermediate pole offsets from a global VLBI solution. Astronom J 154(4):166

Böhm J, Böhm S, Boisits J, et al (2018) Vienna VLBI and satellite software (VIEVS) for geodesy and astrometry. Publ Astronom Soc Pac 130(986):044503

Capitaine N, Wallace PT (2006) High precision methods for locating the celestial intermediate pole and origin. Astron Astrophys 450(2):855–872

Chao BF, Hsieh Y (2015) The earth's free core nutation: Formulation of dynamics and estimation of eigenperiod from the very-long-baseline interferometry data. Earth Planet Sci Lett 432:483–492

Defraigne P, Dehant V, Hinderer J (1994) Stacking gravity tide measurements and nutation observations in order to determine the complex eigenfrequency of the nearly diurnal free wobble. J Geophys Res Solid Earth 99(B5):9203–9213

Florsch N, Hinderer J (2000) Bayesian estimation of the free core nutation parameters from the analysis of precise tidal gravity data. Phys Earth Planet Interiors 117(1–4):21–35

Glomsda M, Seitz M, Angermann D (2023) Comparison of simultaneous VGOS and legacy VLBI sessions. In: KL Armstrong, D Behrend, KD Baver (eds), IVS 2022 General Meeting Proceedings

Kern L, Schartner M, Böhm J, et al (2023) On the importance of accurate pole and station coordinates for VLBI intensive baselines. J Geodesy 97(10):97

Krásná H, Böhm J, Schuh H (2013) Free core nutation observed by VLBI. Astron Astrophys 555:A29

Malkin Z (2011) The impact of celestial pole offset modelling on VLBI UT1 intensive results. J Geodesy 85:617–622

Malkin Z (2013) Free core nutation and geomagnetic jerks. J Geodynam 72:53–58

Malkin Z (2017) Joint analysis of celestial pole offset and free core nutation series. J Geodesy 91(7):839–848

Mathews PM, Herring TA, Buffett BA (2002) Modeling of nutation and precession: New nutation series for nonrigid earth and insights into the earth's interior. J Geophys Res Solid Earth 107(B4):ETG-3

Niell A, Barrett J, Burns A, et al (2018) Demonstration of a broadband very long baseline interferometer system: a new instrument for high-precision space geodesy. Radio Sci 53(10):1269–1291

Nothnagel A, Schnell D (2008) The impact of errors in polar motion and nutation on UT1 determinations from VLBI intensive observations. J Geodesy 82(12):863–869

Nothnagel A, Artz T, Behrend D, et al (2017) International VLBI service for geodesy and astrometry: Delivering high-quality products and embarking on observations of the next generation. J Geodesy 91(7):711–721

Vondrák J, Ron C (2017) New method for determining free core nutation parameters, considering geophysical effects. Astron Astrophys 604:A56

Vondrák J, Weber R, Ron C (2005) Free core nutation: direct observations and resonance effects. Astron Astrophys 444(1):297–303

Ziegler Y, Lambert SB, Nurul Huda I, et al (2020) Contribution of a joint Bayesian inversion of VLBI and gravimetric data to the estimation of the free inner core nutation and free core nutation resonance parameters. Geophys J Int 222(2):845–860

Open Access This chapter is licensed under the terms of the Creative Commons Attribution 4.0 International License (http://creativecommons.org/licenses/by/4.0/), which permits use, sharing, adaptation, distribution and reproduction in any medium or format, as long as you give appropriate credit to the original author(s) and the source, provide a link to the Creative Commons license and indicate if changes were made.

The images or other third party material in this chapter are included in the chapter's Creative Commons license, unless indicated otherwise in a credit line to the material. If material is not included in the chapter's Creative Commons license and your intended use is not permitted by statutory regulation or exceeds the permitted use, you will need to obtain permission directly from the copyright holder.

Consistently Combined Earth Orientation Parameters at BKG—Extended by New VLBI Intensives Data

Lisa Klemm, Daniela Thaller, Claudia Flohrer, Anastasiia Walenta, Dieter Ullrich, and Hendrik Hellmers

Abstract

The Earth Orientation Parameters (EOPs) describe the rotation between the Terrestrial Reference Frame and the Celestial Reference Frame and represent an essential component of the Global Geodetic Reference Frame. This study presents the current activities of BKG in the area of combined processing of GNSS and VLBI data in one common adjustment with the main objective to generate a consistent combined EOP time series. In earlier studies, we have investigated different combination approaches using VLBI and GNSS data. We generate EOP series with latencies of about one to 14 days, depending on the input data we used. In this way, a significant improvement in accuracy compared to the individual technique-specific solutions was achieved, especially for the highly variable component dUT1. The combination process starts at the level of normal equations using an EOP parameterization with piece-wise linear offsets and a temporal resolution of one day. Our main objective is to generate a continuous, daily and regular EOP product with the shortest possible latency. The requirement for achieving these characteristics is the daily and rapid availability of the input data. In particular, the VLBI Intensive (INT) sessions play an important role in the precise and rapid estimation of the UT1-UTC component. Since 2020, an increasing number of VLBI Global Observing System (VGOS) INT campaigns has been conducted in addition to the legacy S/X INT sessions. The VGOS network is under continuous extension and the accuracy and latency of the VGOS INT sessions are at least at the level of the legacy S/X sessions. Therefore, an inclusion of the VGOS INT data is beneficial for rapid EOP estimation. The integration of the VGOS data into the combination process results in a constant slight decrease of the Weighted Root Mean Square (WRMS) level of the UT1-UTC residuals in comparison to the external EOP series. The growing number of available INT sessions with independent networks, up to four per day, increases the continuity and reliability of the combined EOP solution.

Keywords

Combination · EOP · GGRF · GNSS · NEQ level · UT1-UTC · VGOS · VLBI

1 Introduction

The Earth Orientation Parameters (EOPs) are a set of five time-dependent parameters that describe the rotation between the Terrestrial Reference Frame (TRF) and the Celestial Reference Frame (CRF). They are represented by the difference dUT1 between Universal Time UT1

L. Klemm (✉) · D. Thaller · C. Flohrer · A. Walenta · D. Ullrich · H. Hellmers
Federal Agency for Cartography and Geodesy (BKG), Frankfurt am Main, Germany
e-mail: Lisa.Klemm@bkg.bund.de

and Coordinated Universal Time UTC, the polar motion components x_p and y_p, and the celestial pole offsets dX and dY (Petit and Luzum 2010). Just as important as the EOPs themselves are their temporal derivatives. Among these, the temporal derivative of the most variable component dUT1, the so-called Length-of-Day (LOD), plays an important role. Together with the TRF and CRF, the EOPs represent an essential component of the Global Geodetic Reference Frame (GGRF) (Plag et al. 2009). Knowledge of accurate EOPs is not only important for reference frames on Earth and in space. They are needed for a multitude of applications, e.g., precise positioning, satellite navigation, and precise orbit determination (Gambis and Luzum 2011).

Four space-geodetic techniques contribute to the permanent monitoring of the Earth's orientation: the Very Long Baseline Interferometry (VLBI) based on the observation of extra-galactic radio sources and the satellite-based techniques Global Navigation Satellite Systems (GNSS), Satellite Laser Ranging (SLR) and Doppler Orbitography and Radiopositioning Integrated by Satellite (DORIS) (Petit and Luzum 2010).

The polar motion components as well as LOD can be observed by all four techniques. The time component dUT1 is dominated by significant and unpredictable variations and is, thus, the most variable component among the EOPs. Together with the celestial pole offsets, the dUT1 component can only be measured with VLBI. It is thus the only technique capable of monitoring all EOP components including their time derivatives (Dermanis and Mueller 1978; Artz et al. 2011; Thaller 2008). The International VLBI Service for Geodesy and Astrometry (IVS) organizes two different types of geodetic VLBI sessions: Firstly, the 24-hour sessions conducted twice a week, capable of monitoring all EOP components. Secondly, the daily so-called Intensive (INT) sessions, that are specifically designed for rapid estimation of dUT1 (Nothnagel et al. 2017; Robertson et al. 1985; Leek 2015).

The four space-geodetic techniques as well as the individual VLBI session types have different characteristics in terms of EOP sensitivity, latency and data continuity. Therefore, a multi-technique combination is essential to generate the best possible continuous and rapidly available EOP solution.

At BKG, we aim to cover the entire EOP processing chain, which is mainly composed of three elements: (1) the data analysis of the different space-geodetic techniques, (2) the EOP estimation by consitent combination of the preprocessed single-technique data, and (3) the subsequent EOP prediction based on the combined EOP series.

We operate the analysis centers (ACs) for VLBI and SLR in-house and we are also involved in the AC of the International GNSS Service (IGS) of the CODE[1] (Center for Orbit Determination in Europe) consortium. Through this direct link to the analysis, we can individually optimize the solution set-up for the different space-geodetic data analysis in order to achieve the highest consistency possible. This allows to start the combination process at the level of normal equations (NEQ) leading to highly consistent results that are comparable to those from a combination process at the observation level. As described previously, the contributions of GNSS and VLBI INT to the EOP determination are complementary, and both NEQs are available daily with short latency of two days or less. These characteristics allow the rapid estimation of dUT1, pole coordinates and their time derivatives by combining both sets of observation data. The consistent combined EOP time series has a daily equidistant temporal resolution and a latency of less than two days. The additional combination with VLBI 24-hour rapid turn-around sessions further stabilizes all EOPs twice per week and enables the estimation of high-precision EOPs including the celestial pole offsets with a latency of about two weeks.

In previous studies, we have investigated different combination approaches using legacy S/X VLBI and GNSS data. We generate EOP series with latencies of about one to 14 days, depending on the input data we used. In this way, a significant improvement in accuracy compared to the individual technique-specific solutions was achieved, especially for the highly variable component dUT1 (Lengert et al. 2022; Klemm et al. 2023). In recent years, mainly due to the development of the VLBI Global Observing System (VGOS), numerous new VLBI INT sessions have been added to the VLBI observing schedule. This has significantly increased the number of INT data available per day and improved the continuity of the INT data series which is an important criterion for the reliability of our combined EOP products. In this paper, we will discuss in more detail the new VLBI INT data, that we have now integrated into our combination process. In this context, we will compare the characteristics of the different INT session types in terms of latency and quality of the resulting dUT1 estimates. Additionally, we present the challenges of extending the EOP combination process with the data of the new VLBI observing program.

---

[1] CODE is a consortium of the Astronomical Institute of the University of Bern (AIUB, Bern, Switzerland), the Swiss Federal Office of Topography (swisstopo, Wabern, Switzerland), the Federal Agency for Cartography and Geodesy (BKG, Frankfurt a. M., Germany), and the Institut für Astronomische und Physikalische Geodäsie, Technische Universität München (IAPG/TUM, Munich, Germany) https://www.aiub.unibe.ch/research/code___analysis_center/index_eng.html.

2 The VLBI Observing Program

The VLBI processing chain, from observations to the combined analysis, is not entirely automated. Therefore, continuous observations around the clock, as with satellite-based techniques, are currently not feasible (Nothnagel et al. 2017). For this reason, VLBI observations have to be organized session-wise with a subset of radio telescopes. These sessions are typically categorized into either 24-hour or INT campaigns. In order to provide EOPs within a 15-day period from the end of recording to results, the IVS organizes twice per week operational 24-h sessions, called rapid (RAP) turnaround sessions. The RAP sessions start their observations every Monday (R1) and Thursday (R4) at 17:00 UTC and 18:30 UTC, respectively, with a globally distributed network of up to 14 antennas. These sessions are capable of monitoring all EOP components and their derivatives, but have a comparatively long latency due to the large number of observations and the computationally intense data analysis workflow (Nothnagel et al. 2017). For a daily monitoring of the dUT1 component close to real-time, the IVS organizes additional daily INT sessions of one hour duration. These sessions are characterized by small networks of only two or three radio telescopes with a large East-West extension, which are particularly sensitive to dUT1. The big advantage of these sessions is the short latency of one to two days when the dUT1 estimates become available after the end of the session. However, the INT sessions are intended for the monitoring of dUT1 only. Hence, they are not able to monitor the other components of the EOPs (Nothnagel et al. 2017).

Our main objective is to generate an EOP product that is characterized by a continuous, daily and regular resolution and the shortest possible latency, especially for the highly variable dUT1. The mandatory requirement for achieving these characteristics is the daily and rapid availability of the input data, especially of the VLBI INT sessions. However, we noticed that the series of daily SINEX (Solution INdependent EXchange) files has some gaps in the past. For example, in 2018 there are no SINEX files available for altogether 60 days (see Fig. 1). The reasons for data gaps are manifold and can be found throughout the entire VLBI processing chain, i.e., from observation to analysis. The most common reason is the inability to observe the session due to technical problems of a telescope. Another common case is that the session is observable, but difficulties during the observation do not allow proper correlation. Consequently, in these cases the AC lacks the necessary database for analysis and is therefore unable to generate a SINEX file. Data gaps in the INT series are problematic for the continuity and reliability of combined EOPs. With the development of VGOS, the next generation of VLBI, the issue of missing data has been improved significantly in recent years. The VGOS system is characterized by small, fast-slewing 12–13 meter class antennas observing in at least four bands from 2.5 GHz to 14 GHz with a high data rate of 8 Gbps and more. Overall, the VGOS system is designed to provide significantly more precise geodetic results compared to the legacy two-band system. As a result, VGOS is able to meet the scientific requirements of the Global Geodetic Observing System (GGOS) (Petrachenko et al. 2012; Niell et al. 2018; Beutler and Rummel 2012). The installation of the first VGOS antennas started in 2017, and the first VGOS observations in operational IVS sessions were conducted in 2020. Currently, up to twelve stations are participating in the VGOS operational sessions, with plans to expand the global network to up to 25 stations in the coming years. It is intended that VGOS will replace the legacy S/X system as the operational system in the next few years (Behrend et al. 2022).

Since 2020, a growing number of VGOS INT sessions has been conducted in addition to the legacy S/X INT sessions. In 2023, the number of planned VGOS INT sessions is nearly

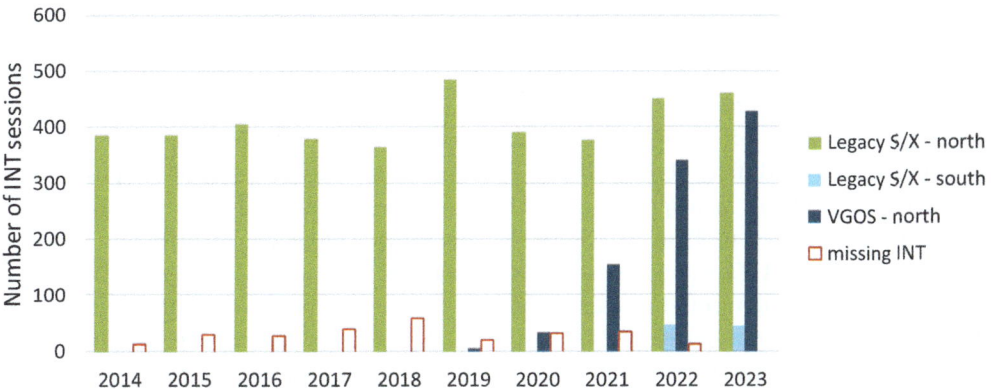

Fig. 1 Number of three different types of Intensives sessions per year: the legacy S/X sessions with networks in the northern (green) and southern (light blue) hemispheres, and the new VGOS sessions. Additionally, the number of days with missing Intensives SINEX file in the IVS data center is shown in red

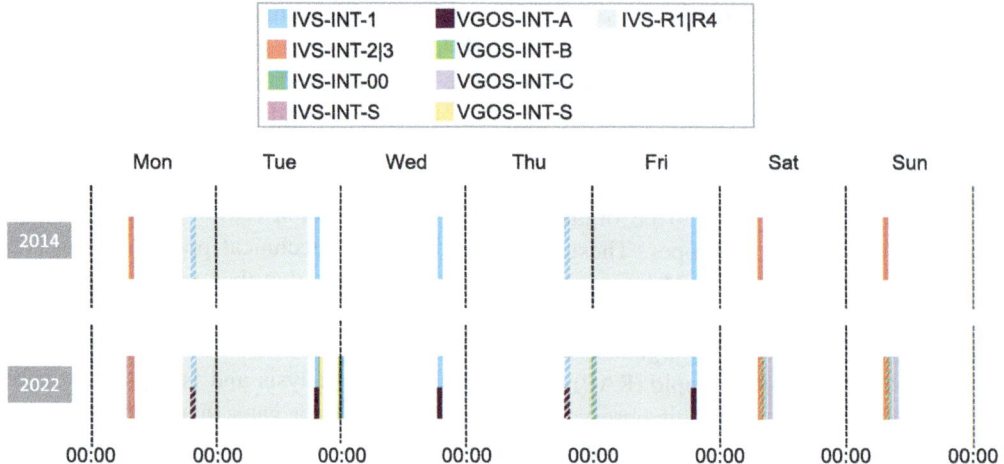

Fig. 2 Weekly session distribution of the different VLBI sessions in 2014 (**a**) and 2022 (**b**): legacy S/X INT of the northern (green) and southern (light blue) hemispheres, Rapid (R1/4) (light orange), and the new VGOS INT (dark blue)

equal to the number of legacy S/X sessions (see Fig. 1). After an initial delay mainly in the correlation process, the latency of the VGOS INT data has been reduced to one to two days since mid-2022. With this improvement, the VGOS INT sessions become suitable for the generation of rapid EOP products. Currently, there are primarily four different VGOS-INT sessions being regularly observed (see Fig. 2):

- VGOS-INT-A, observed on the baseline KOKEE12M-WETTZ13S, runs parallel to IVS-INT-1 from Monday to Friday.
- VGOS-INT-B, between the telecopes ONSA13NE-ONSA13SW-ISHIOKA, operates in parallel with IVS-INT-2.
- VGOS-INT-C, on the baseline KOKEE12M-WETTZ13S, scheduled on Saturdays and Sundays.
- VGOS-INT-S, between GGAO12M-WETTZ13S, operates on Tuesdays.

In order to investigate the performance of the different INT session types, we analyzed the session-wise dUT1 solutions at the mid-session epochs and compare the estimates with the *IERS Bulletin A* series (Luzum and Gambis 2014) (see Fig. 3). For more detailed information on this session-wise solution we refer to Klemm et al. (2023). The Weighted Root Mean Square (WRMS) values of the dUT1 residuals between mid-2014 and 2022 for the different legacy S/X INT sessions of the northern and southern hemispheres and the VGOS INT sessions are summarized in Table 1. It should be noted that the accuracy of the dUT1 estimates is highly correlated with the quality of the station coordinates. For the calculation of ITRF2020, sessions up to epoch 2021.0 have been considered. Some new VGOS stations have only participated in a few 24h-VGOS sessions up to this point. Therefore, they do not have reliable coordinates and velocities. Using an updated VLBI TRF (VTRF) considering all 24h-VGOS sessions until mid-2023, the WRMS values of dUT1 residuals for VGOS-INT-A, VGOS-INT-B, VGOS-INT-C and VGOS-INT-S are 16.5 µs, 22.0 µs, 24.3 µs, 27.3 µs, respectively. As expected, the WRMS values of the dUT1 residuals are lower on average when an updated TRF is used as a priori. Since the expected potential of the VGOS INT sessions in terms of EOP accuracy is higher, it is expected to improve significantly with the availability of more accurate coordinates through updated TRF versions in the coming years. Similar to the stations of the legacy S/X INT sessions (IVS-INT-1, IVS-INT-2, IVS-INT-3, IVS-INT-00), all VGOS telescopes participating in the INT sessions are located in the northern hemisphere. Starting in 2022, a weekly southern hemisphere INT observing program (IVS-INT-S) between Australia (HOBART12, YARRA12M) and South Africa (HART15M) was added to the observation schedule (see Figs. 1–3). The accuracy of the estimated dUT1 values for the IVS-INT-S session is the lowest among all INT sessions, with a WRMS of 28 µs. A closer examination of the coordinate time series and the associated TRF model of the participating stations HOBART12 and HART15M reveals significant variations of the coordinate estimates relative to the model. These inaccuracies in the TRF have a direct impact on the accuracy of the estimated dUT1 values. Nevertheless, establishing an INT session in the southern hemisphere may help to detect possible systematic effects due to northern hemisphere dominated network geometry and provide an independent way to validate results of established INT sessions (Böhm et al. 2022).

As a result, up to four different INT sessions per day are on the observation schedule, and the INT series is almost uninterrupted (see Figs. 1 and 2). The VLBI network is under continuous expansion and the accuracy and reliability of the new INT sessions are at least at the level of the legacy S/X sessions (see Fig. 3). Therefore, an inclusion of the

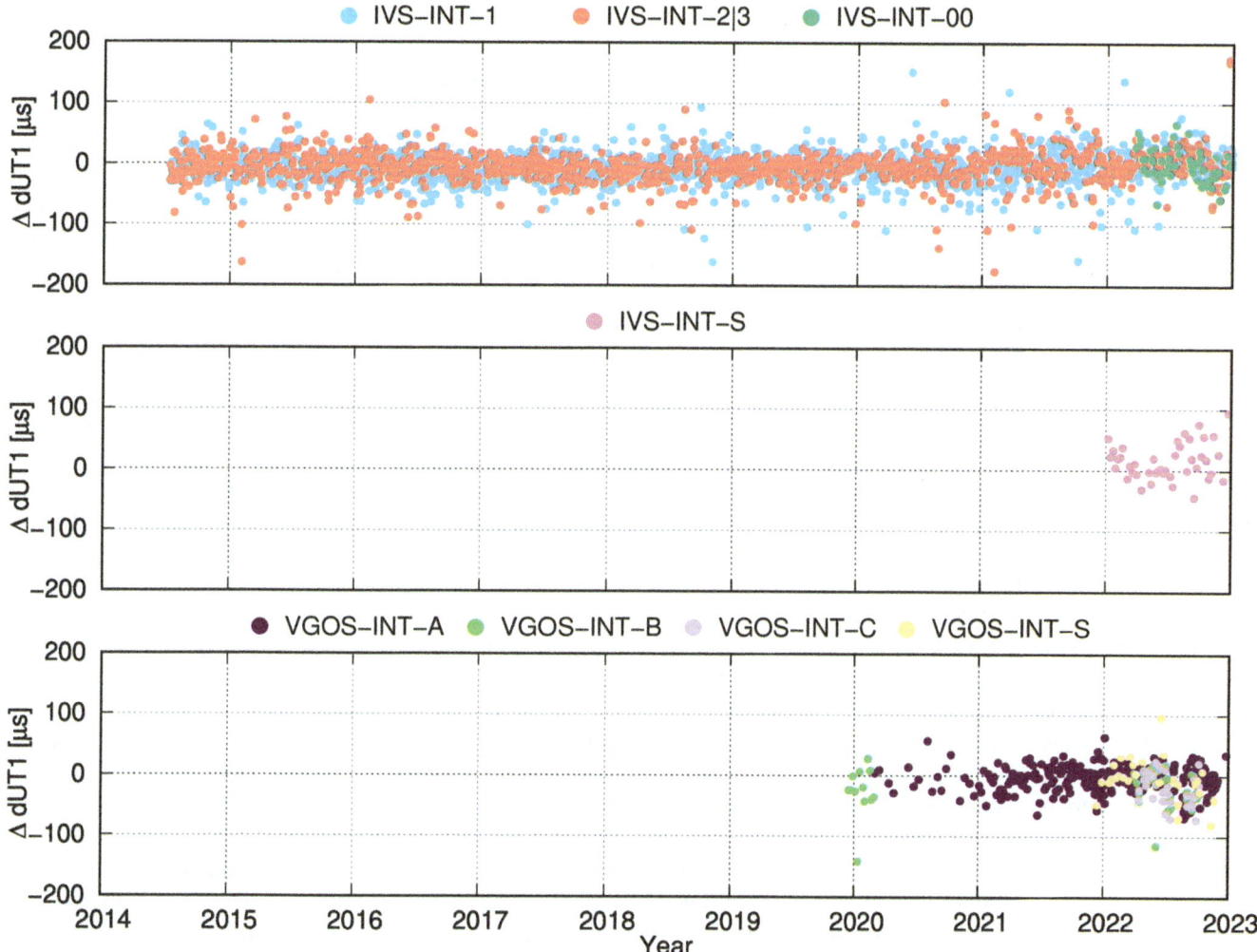

Fig. 3 dUT1 time series resulting from the different VLBI INT solution w.r.t. IERS Bulletin A, analyzed at mid-session epochs

Table 1 Comparison of dUT1 solutions resulting from the different VLBI INT sessions w.r.t. *IERS Bulletin A*. The WRMS of the differences are computed at mid-session epochs; in [μs]

Session	WRMS [μs]
IVS-INT-1	17.4
IVS-INT-2+3	20.6
IVS-INT-00	21.6
IVS-INT-S	28.0
VGOS-INT-A	16.5
VGOS-INT-B	25.0
VGOS-INT-C	28.0
VGOS-INT-S	27.6

new VGOS and the southern hemisphere INT data into the combination process is expected to be beneficial for the rapid EOP estimation. In the following, we will show the impact of the new INT data on the combined EOP series.

3 Data Input and Combination Methodology

The combination process is based on a least squares algorithm using VLBI and GNSS NEQs obtained via SINEX files from BKG IVS AC and CODE IGS AC, respectively (Engelhardt et al. 2021; Dach et al. 2020). This study covers the period between mid-2014 and 2022. In earlier publications, we already described in detail the entire combination process, including the input data, the chosen parameterization, the preparation of the NEQ systems, the a priori values used, the types of constraints for the different solutions, and the estimated parameters. In this paper, we briefly summarize the combination methodology and the different combined EOP solutions. For more detailed information we refer to Lengert et al. (2022), Klemm et al. (2023).

Each technique-specific NEQ underwent several transformation steps to obtain consistent and datum-free input for the combination process. First, the parameterization of the VLBI EOPs was adapted to that of the GNSS EOPs by converting the offsets/drift parameterization at mid-session epochs into continuous piece-wise linear (PWL) polynomials with offsets at the day boundaries. In a next step, the a priori values of the EOPs and the station coordinates were transformed to a consistent set. We chose the *IERS 20 C04* and *ITRF2020* series, respectively (Luzum and Gambis 2014; Altamimi et al. 2023). For GNSS, a LOD bias correction had to be performed. The LOD bias is caused by deficiencies in the orbit modelling of the GNSS satellites and causes a drift in the estimated GNSS dUT1 time series (Meindl et al. 2014).

For the inter-technique combination, the NEQ systems of seven consecutive days were accumulated to one multi-day inter-technique combined NEQ system. A constant technique-specific scaling factor proportional to the technique- or session-specific observation duration was used.

We compare two different combination approaches: First, a 7-day combination of GNSS with VLBI INT-only for rapid EOP estimation with a latency of about two days. This solution is labeled "7d-RAP". Second, a 7-day combination of GNSS, VLBI INT, and VLBI R1/R4 for an EOP estimation with a latency of about two weeks. This solution is labeled "7d-FIN". The NEQ systems can be solved for the parameters to be estimated, i.e. primarily the EOPs, after applying datum conditions and other technique-specific constraints summarized in Table 2. The EOP time series were generated using a 7-day sliding window shifted over the daily NEQs. The NEQs within the window were combined into a multi-day NEQ system before the window was shifted by one day and the combination was repeated. This procedure was iterated over the entire series of daily NEQs between mid-2014 and 2022.

As a result, for both combination approaches 7d-RAP and 7d-FIN, we obtained EOP time series consisting of 7-day arcs shifted by one day. For the processing we use the *Combination and Solution* package of the *DGFI Orbit and Geodetic parameter estimation Software* (DOGS-CS), developed and maintained at DGFI-TUM (Deutsches Geodätisches Forschungsinstitut, Technische Universität München) (Gerstl et al. 2004).

4 Resulting EOP Series

In order to validate the estimated multi-day EOP series, we generated sub-series by extracting the EOPs at 0 h and 24 h of the same day from each 7-day arc. As a result, we obtained seven sub-series, one for each analysis day, ranging from -6 to 0, where $d = -6$ is the first day and $d = 0$ is the last day of the 7-day solution. The estimated EOPs at day boundaries were interpolated to a mean epoch, at 12:00 UTC, to obtain one validation epoch per day. We study the Weighted Root Mean Square (WRMS) of the EOP residuals with respect to the *IERS Bulletin A* series. The reciprocal value of each EOP variance was used to determine the weighting factors for the WRMS calculation. Since this study focuses on the integration of the new VLBI INT data, we will concentrate on the resulting dUT1 series in the following. Table 3 summarizes the WRMS values of the dUT1 residuals with respect to the *IERS Bulletin A* series for the two combination approaches, 7d-RAP and 7d-FIN, respectively. Both combination approaches and the comparison of the resulting EOP series have already been discussed in Lengert et al. (2022). In the following, the discussion will focus on the impact of the new VLBI INT data. Two different solutions have been compared in each case: A combined solution containing only legacy S/X INT data of the northern hemisphere ("S/X-north only") and a solution containing all available INT data ("all INT"). The comparison of both solutions shows a constant slight reduction of all WRMS values of about 0.4 μs and 0.2 μs when using all INT data for the two combination approaches 7d-RAP and 7d-FIN. We conclude that the integration of the VGOS and southern hemisphere INT data has a slight stabilizing effect on the combined solution, although the accuracies of the session-wise analyzed VGOS and southern hemisphere INT data are lower compared to the legacy northern hemisphere INT data (see Chap. 2). In addition, the reliability of the combination increases. The risk of data gaps and the impact of individual INT sessions with low accuracy are reduced by increasing the number of available INT sessions from one up to four sessions per day with different baselines. One reason why the 7d-FIN solution benefits less from the integration of the VGOS data than the 7d-RAP solution could be the datum

Table 2 Summary of the a priori values and the respective constraints used for the combined solutions ([a]parameter group is pre-eliminated)

		A priori	7d-RAP	7d-FIN
EOP	dUT1	IERS 20 C04	None	None
	x_p, y_p	IERS 20 C04	None	None
	$\delta X, \delta Y$	IERS 20 C04	Fixed[a]	None
Station coordinates	X,Y,Z	ITRF2020	Minimum	Minimum
Source coordinates	α, δ	ICRF3	Fixed[a]	Fixed

Table 3 Comparison of dUT1 resulting from the different solution types w.r.t. *IERS Bulletin A*. The WRMS of the differences are computed at 12:00 UTC epochs; in [μs]

Analysis day d	−6	−5	−4	−3	−2	−1	0
7d-RAP—S/X-north only	16.7	15.1	14.8	14.7	14.8	15.3	16.9
7d-RAP—all INT	16.2	14.7	14.4	14.4	14.5	15.0	16.5
7d-FIN—S/X-north only	15.4	14.9	14.7	14.7	14.8	15.0	16.0
7d-FIN—all INT	15.2	14.7	14.5	14.5	14.6	14.8	15.9

definition. For the 7d-RAP solution, the VLBI network needs to be fixed to the a priori values due to the short observation duration and the small sub-networks of the INT sessions. For the 7d-FIN solution, minimum constraints are usually applied to the VLBI network. Since the radio telescopes of the legacy S/X INT sessions are usually involved in the 24-h RAP campaigns, the sub-networks can be stacked and estimated in a common adjustment. The newly added VGOS telescopes are completely decoupled from the 24-h RAP observations and need to be fixed to their a priori coordinates. This approach introduces a slight network inconsistency into the datum definition and directly affects the EOP solution.

We have additionally validated our EOP series against the *IERS 20 C04* series. The consistency of both combined EOP series (7d-RAP, 7d-FIN) with the *IERS 20 C04* series is slightly lower, but the behavior of the WRMS values is similar to that with respect to *IERS Bulletin A*.

5 Conclusion and Outlook

This paper presents the combined processing of VLBI and GNSS data in a joint adjustment with the focus on the integration of new VLBI INT data into the combination process. The combination is based on homogenized, datum-free NEQs provided via SINEX files from the BKG IVS AC and CODE IGS AC, which allows a combination on the NEQ level. We use the DOGS-CS software, developed and maintained at DGFI-TUM. The aim is to combine the strengths of both techniques to estimate EOP series characterized by daily, continuous and temporally regular resolution and short latency of two days or less. For this purpose, the fast availability of dUT1 information, which is provided by VLBI INT sessions, is essential. We discussed the different types of VLBI INT sessions and compared their characteristics in terms of latency and quality of the resulting dUT1 estimates. Beginning in 2020, an increasing number of VGOS INT sessions was conducted in addition to legacy S/X INT sessions, resulting in nearly equal numbers of VGOS and legacy INT sessions in 2023. Since the VGOS network is under continuous extension and the accuracy and latency of the VGOS INT sessions are at least at the level of the legacy S/X sessions (with the tendency to further improve) an inclusion is beneficial for rapid EOP estimation.

In summary, the integration of the new VGOS INT data into the combination results in a constant slight decrease of the WRMS level of the dUT1 residuals in comparison to external EOP series. The growing number of available INT sessions with independent networks (up to four per day) increases the reliability of the combined solution. The characteristics of the combined EOP series give us the opportunity to use them as training data set for prediction algorithms. We are currently working on the development of an EOP prediction center using our rapid EOP product as input in order to extend the EOP series with predicted values (Modiri et al. 2023).

In the near future we plan to include operational 24-h VGOS campaigns into the combination process to improve the consistency of the datum definition of the 7d-FIN solution. We also plan to extend the combination of GNSS and VLBI data by adding SLR data in order to exploit the benefit of the combination to its maximum extent.

References

Altamimi Z, Rebischung P, Collilieux X, Métivier L, Chanard K (2023) ITRF2020: An augmented reference frame refining the modeling of nonlinear station motions. J Geodesy 97(5):47

Artz T, Bernhard L, Nothnagel A, Steigenberger P, Tesmer S (2011) Methodology for the combination of sub-daily Earth rotation from GPS and VLBI observations. J Geodesy 86:221–239. https://doi.org/10.1007/S00190-011-0512-9

Behrend D, Ruszczyk C, Elosegui P, Weston S (2022) Status of the VGOS Infrastructure Rollout. In: IVS general meeting

Beutler G, Rummel R (2012) Scientific rationale and development of the global geodetic observing system. In: Geodesy for Planet Earth: Proceedings of the 2009 IAG Symposium, Buenos Aires, Argentina, 31 August 31–4 September 2009, pp 987–993. Springer

Böhm S, Böhm J, Gruber J, Kern L, McCallum J, McCallum L, McCarthy T, Quick J, Schartner M (2022) Probing a southern hemisphere VLBI intensive baseline configuration for UT1 determination. Earth Planets Space 74(1):1–16

Dach R, Schaer S, Arnold D, Kalarus M, Prange L, Stebler P, Villiger A, Jäggi A (2020) CODE rapid product series for the IGS. Published by Astronomical Institute, University of Bern. https://doi.org/10.7892/boris.75854.4

Dermanis A, Mueller II (1978) Earth rotation and network geometry optimization for very long baseline interferometers. Bull Geodesique 52:131–158. https://doi.org/10.1007/BF02521695

Engelhardt G, Girdiuk A, Goltz M, Ullrich D (2021) BKG VLBI analysis center. International VLBI service for geodesy and astrometry 2019+2020 Biennial report, edited by D. Behrend, K. L. Armstrong, and K. D. Baver, NASA/TP-2020-219041

Gambis D, Luzum B (2011) Earth rotation monitoring, UT1 determination and prediction. Metrologia 48(4):165. https://doi.org/10.1088/0026-1394/48/4/S06

Gerstl M, Kelm R, Müller H, Ehrnsperger W (2004) DOGS-CS Kombination und Lösung großer Gleichungssysteme. Deutsches Geodätisches Forschungsinstitut (DGFI)

Klemm L, Thaller D, Flohrer C, Hellmers H, Bloßfeld M, Kehm A, Dach R (2023) Single- and multi-day combination of VLBI and GNSS data for consistent estimation of low-latency earth rotation parameters. Submitted to JoG

Leek J (2015) The application of impact factors to scheduling VLBI Intensive sessions with twin telescopes. PhD thesis, Rheinische Friedrich-Wilhelms-Universität Bonn (February 2015). https://hdl.handle.net/20.500.11811/6226

Lengert L, Thaller D, Flohrer C, Hellmers H, Girdiuk A (2022) On the improvement of combined EOP series by adding 24-h VLBI sessions to VLBI intensives and GNSS Data, pp 1–8. Springer, Berlin, Heidelberg. https://doi.org/10.1007/13452022175

Luzum B, Gambis D (2014) Explanatory supplement to IERS bulleting A and bulletin B/C04. ftp://hpiers.obspm.fr/iers/bul/bulbnew/bulletinb.pdf

Meindl M, Beutler G, Thaller D, Dach R, Schaer S, Jäggi A (2014) A comment on the article "A collinearity diagnosis of the GNSS geocenter determination" by P. Rebischung, Z. Altamimi, and T. Springer. J Geodesy 89:189–194. https://doi.org/10.1007/s00190-014-0765-1

Modiri S, Thaller D, Belda S, Halilovic D, Klemm L, König D, Hellmers H, Bachmann S, Flohrer C, Walenta A (2023) Closing the gap: A redesigned prediction package for enhanced accuracy of EOP prediction using multi-technique combined and technique-specific EOP. Submitted to IAGS

Niell A, Barrett J, Burns A, Cappallo R, Corey B, Derome M, Eckert C, Elosegui P, McWhirter R, Poirier M, Rajagopalan G, Rogers A, Ruszczyk C, SooHoo J, Titus M, Whitney A, Behrend D, Bolotin S, Gipson J, Gordon D, Himwich E, Petrachenko B (2018) Demonstration of a broadband very long baseline interferometer system: a new instrument for high-precision space geodesy. Radio Sci 53(10):1269–1291. https://doi.org/10.1029/2018RS006617

Nothnagel A, Artz T, Behrend D, Malkin Z (2017) International VLBI service for geodesy and astrometry: delivering high-quality products and embarking on observations of the next generation. J Geodesy 91:711–721. https://doi.org/10.1007/s00190-016-0950-5

Petit G, Luzum B (eds.) (2010) IERS technical note, No. 36. International earth rotation and reference systems service, Central Bureau. Verlag des Bundesamts für Kartographie und Geodäsie, Frankfurt am Main. http://www.iers.org/TN36

Petrachenko W, Niell A, Corey B, Behrend D, Schuh H, Wresnik J (2012) VLBI2010: next generation VLBI system for geodesy and astrometry. In: Geodesy for Planet Earth: Proceedings of the 2009 IAG Symposium, Buenos Aires, Argentina, 31 August 31–4 September 2009, pp 999–1005. Springer

Plag H-P, Rothacher M, Pearlman M, Neilan R, Ma C (2009) The global geodetic observing system. In: Advances in geosciences. World Scientific. https://doi.org/10.1142/97898128361820008

Robertson DS, Carter WE, Campbell J, Schuh H (1985) Daily Earth rotation determinations from IRIS very long baseline interferometry. Nature 316(6027):424–427

Thaller D (2008) Inter-technique combination based on homogeneous normal equation systems including station coordinates, Earth orientation and troposphere parameters. PhD thesis, Scientific technical report STR 08/15, Deutsches Geo-ForschungsZentrum. https://doi.org/10.2312/GFZ.b103-08153

Open Access This chapter is licensed under the terms of the Creative Commons Attribution 4.0 International License (http://creativecommons.org/licenses/by/4.0/), which permits use, sharing, adaptation, distribution and reproduction in any medium or format, as long as you give appropriate credit to the original author(s) and the source, provide a link to the Creative Commons license and indicate if changes were made.

The images or other third party material in this chapter are included in the chapter's Creative Commons license, unless indicated otherwise in a credit line to the material. If material is not included in the chapter's Creative Commons license and your intended use is not permitted by statutory regulation or exceeds the permitted use, you will need to obtain permission directly from the copyright holder.

Operational Forecasting of Effective Angular Momentum Functions Fourteen Days Ahead

Mostafa Kiani Shahvandi, Matthias Schartner, Junyang Gou, and Benedikt Soja

Abstract

Forecasts of Earth's Effective Angular Momentum functions (EAM) are used for different applications, including prediction of Earth Orientation Parameters (EOPs). Since May 2021, the Chair of Space Geodesy at ETH Zurich has been operationally providing accurate EAM forecasts. These forecasts cover the domain of atmosphere, ocean, hydrology, and sea level. They are based on the EAM forecasts by GFZ Potsdam but are corrected and extended to cover a forecasting horizon of two weeks using machine learning techniques. Here, we present a summary of the methodology and the results achieved during the past two years. We demonstrate the enhanced accuracy of our improved EAM functions of up to 50%. Furthermore, we demonstrate the impact on the potential application of utilizing EAM forecasts in the form of ultra-short-term prediction of length of day, where an improved accuracy of up to 19% has been achieved. The improved EAM forecasting product is updated daily and available at https://gpc.ethz.ch/EAM/.

Keywords

Earth orientation parameters · Effective angular momentum functions · Forecasting · Machine learning

1 Introduction

Effective Angular Momentum (EAM) functions of the Earth are important quantities, directly related to the mass redistribution within the Earth system (Barnes et al. 1983). This mass redistribution occurs primarily in four different domains (Dobslaw et al. 2010), namely, (1) atmosphere: Atmospheric Angular Momentum (AAM); (2) oceans: Oceanic Angular Momentum (OAM); (3) land water: Hydrological Angular Momentum (HAM); (4) sea surface: Sea level Angular Momentum (SLAM).

AAM is based on the dynamics of the atmosphere (Gross 2015), which is directly affected by mass exchange in the troposphere and tropospheric winds. Climatic events such as El Niño Southern Oscillation and North Atlantic Oscillation further impact AAM (Yu et al. 2021). Oceanic mass distribution changes, currents and circulations affect OAM. Large-scale continental mass redistributions and changes in terrestrial water storage determine HAM. Finally, the periodic changes in the water level in the oceans and seas result in SLAM.

EAM functions are defined through integral formulas that are based on the variable mass density, relative motion, and the Earth's geophysical parameters (Dobslaw et al. 2010). Since the mass redistributions affect the inertia tensor of the Earth (which has three main components in the terrestrial frame), the EAM functions are presented in a three-dimensional form with components denoted here as (x, y, z). The angular momentum is either related to mass change or the motion with respect to the rotating reference frame (Gross 2015). Therefore, each of the EAM functions (except for SLAM) are further divided into the mass and motion

M. K. Shahvandi (✉) · M. Schartner · J. Gou · B. Soja
Institute of Geodesy and Photogrammetry, ETH Zurich, Zürich, Switzerland
e-mail: mkiani@ethz.ch

terms. This is related to the physical effects behind EAM, namely pressure (mass) and wind plus current (motion). SLAM does not include motion terms because it is almost completely dominated by pressure resulting from changes in sea level (Dobslaw and Dill 2018, 2019).

Determination of the EAM functions is a demanding task because it requires assimilation of various data sources and then numerically computing the integrals defining the EAM functions. Several institutions have undertaken this task, including SYstèmes de Référence Temps-Espace, Technische Universität Wien, Jet Propulsion Laboratory, and GFZ German Research Centre for Geosciences (Dobslaw et al. 2010). It is essential to note that there are sometimes major differences between the EAM functions provided by different institutions. The reason can be in the treatment of different EAM or data sources used for the computations. For instance, some institutions (e.g., Special Bureau for the Oceans) may not provide SLAM as they incorporate it into OAM (Dobslaw et al. 2010). Among the institutions that provide EAM, GFZ is often the preferred choice because they provide operational EAM products to the present time, which include 6-day and 90-day forecasts (Dill et al. 2019).

For the GFZ EAM data products, the temporal resolution of the observations and forecasts depends on the data sources used for the analysis and determination of EAM. AAM and OAM are provided in 3-h intervals (subdaily), while HAM and SLAM are provided in 24-h intervals (daily). The GFZ EAM forecasts can be categorized into two groups: (1) short-term (up to 6 days); (2) long-term (up to 90 days). The short-term forecasts are physics-based, namely the results of using processed 6-day forecasts of numerical weather data to generate AAM, and use the forcing models for the rest of EAM. This means, for instance, the forecasts of numerical weather models from the European Centre for Medium-Range Weather Forecast (Uppala et al. 2005), which is one of the primary sources for the determination of EAM, especially AAM, are converted to EAM functions in the same manner as the observations. On the contrary, long-term forecasts are determined using mathematical forecasting methods, including an auto-regressive approach (Dill et al. 2019). Generally, the first forecast category is of higher quality because of its physical basis. As a result, we compare our results only with the six-day-ahead forecasts.

The forecasts provided by GFZ are not entirely consistent with their corresponding observations, due to inaccuracy in the utilized weather model forecasts and deficiencies in the mathematical forecasting methods. The differences between the forecasts and observations describe some unmodelled signals and show periodic and trend terms. They are, however, challenging to model physically since they are related to some omissions at the level of forecast data. Instead, a viable approach would be to use data-driven methods for modelling these signals (Dill et al. 2021; Kiani Shahvandi et al. 2022a). Considering the success of machine learning in geodetic time series modelling and prediction (Reiterer et al. 2010; Reichstein et al. 2019; Kiani Shahvandi and Soja 2021, 2022a, b; Kiani Shahvandi et al. 2022a, b, 2023; Gou et al. 2023), we developed our methodology based on this approach.

The redistribution of mass results in changes in the Earth's tensor of inertia (Gross 2015), expressed via EAM functions. On the other hand, temporal variations of Earth Orientation parameters (EOPs) are directly related to the variations in Earth's tensor of inertia. Therefore, EAM plays a pivotal role in the analysis and prediction of EOPs. Because of the highly dynamic nature of the atmosphere, AAM is generally believed to cause variations in length of day (LOD) on the shortest time scales (subdaily to seasonal, Gross 2015). Therefore, forecasting AAM is an important task for accurate LOD prediction. Furthermore, OAM is particularly important in modelling and prediction of polar motion (Kiani Shahvandi et al. 2022b), because the ocean is one of the main causes of the Chandler wobble (Gross 2000).

The benefit of using forecast data in the prediction of EOPs is shown in several studies. For instance, as shown in Kiani Shahvandi et al. (2022b) EAM provides valuable sources of physical information for the prediction of polar motion, which can help improve the prediction performance of machine learning algorithms. Another example of the importance of EAM functions in the prediction of EOPs is the successful use of forecasts of the axial component of AAM (mass and motion terms combined) to improve 10-day prediction of LOD (Gou et al. 2023).

Therefore, improving the accuracy of EAM forecasts is of high importance for the prediction performance of EOPs. Highly accurate predictions of polar motion and LOD help in applications such as spacecraft navigation and real-time positioning (Dobslaw and Dill 2019). All these imply that working on an improved forecasting framework is well justifiable.

Note that in Gou et al. (2023) the improved EAM forecasts are briefly touched upon. Nevertheless, no comprehensive analyses are presented regarding the EAM forecasting accuracy. Therefore, our contributions in the present work compared to Gou et al. (2023) are: (1) presenting the methodology in more detail; and (2) providing users with a detailed reference, including assessments of the forecasting quality. The rest of this paper is organized as follows: Sect. 2 describes the methodology; Sect. 3 presents the EAM forecasts and their accuracy; Sect. 4 analyzes the impact of our EAM forecasts on LOD predictions; Sect. 5 gives the conclusions.

2 Methodology

The methodology used in our framework is summarized in Fig. 1. The inputs to the framework are the GFZ EAM observations and their six-day-ahead forecasts. The framework aims to correct and extend the GFZ EAM forecasts to provide higher accuracy and a longer forecasting horizon. Since occasionally there are outliers in the inputs, which can significantly affect the prediction performance, we implement a simple moving median filter: if the absolute value of EAM at a certain time epoch is at least $c = 10$ times bigger than the absolute value of the median in a sliding window of 6 days around the time epoch (that is six days before and six days after the time epoch), then the value is replaced with the value of this median. The value of $c = 10$ is chosen since it is effective in keeping as much information in the original EAM series as possible while also removing gross errors. After this data cleaning step, we use the approach of Neural Ordinary Differential Equations (Neural ODEs; Chen et al. 2018). The algorithm used here is based on a variation of the algorithm presented in Kiani Shahvandi et al. (2022b). The mathematical formulation of the algorithm presented here is similar to Gou et al. (2023). The input sequence length is $n = 2$, i.e., to predict the next EAM value (time step $t + 1$), we need the value at the current and previous time steps (t and $t - 1$, respectively). The output of this algorithm, denoted by r_{t+1} is derived from a simple linear transformation with weights and biases denoted as W_3 and b_3, respectively:

$$r_{t+1} = W_3 h_{t+1} + b_3 \tag{1}$$

where h_{t+1} is the so-called hidden state at time step $t + 1$. h_{t+1} is related to the hidden state at time step t, h_t, via first order Neural ODEs and a Recurrent Neural Network (RNN; Rumelhart et al. 1986) with weights and biases W_2, U_2 and b_2 as in Eq. (2). Note that the mathematical form of RNN used here is the so-called Elman RNN (Elman 1990).

$$\frac{dh_{t+1}}{dt} = \text{RNN}(h_t, W_2, U_2, b_2)$$

$$RNN(x) = \sigma(W_2 x + U_2 h_{t-1} + b_2), \quad \sigma(x) = \frac{e^x - e^{-x}}{e^x + e^{-x}} \tag{2}$$

The hidden state h_t itself is related to the inputs via a dense layer with tangent hyperbolic activation function (denoted by D, Goodfellow et al. 2016) with weights and biases W_1 and b_1. As mentioned, the inputs are the observations (denoted by o) and forecasts (denoted by f) at time steps t, $t - 1$, i.e., $o_t, o_{t-1}, f_t, f_{t-1}$. This is summarized in Eq. (3).

$$h_t = D(o_t, o_{t-1}, f_t, f_{t-1}, W_1, b_1)$$

$$D(x) = \sigma(W_1 x + b_1) \tag{3}$$

In the optimization phase, the difference between the output of the model r_{t+1} and the corresponding EAM observations (i.e., o_{t+1}) should be minimized. This is done using the squared loss function, given in Eq. (4).

$$(r_{t+1} - o_{t+1})^2 \to min \tag{4}$$

Note that we train an individual network for each of the (x, y, z) components of AAM, OAM, and HAM mass and motion terms, and SLAM mass term only, thus 21 models in total. This allows us to focus on one component in each model and achieve more accurate results. Note that the time resolution of EAM functions implies that in Eq. (1) each of f_t, f_{t-1} consists of either 48 values (6-day forecasts with three-hourly resolution) for AAM and OAM, or 6 values for HAM and SLAM. The output dimension is also either 112 (14-day forecast with three-hourly resolution) for AAM and OAM, or 14 for HAM and SLAM.

3 EAM Forecasting Accuracy

Here we present the forecasting accuracy of our operational EAM forecasts from May 20, 2021 to May 20, 2023. The accuracy is quantified by Mean Absolute Error (MAE) metric, defined as in Eq. (5).

$$\text{MAE}_k = \frac{1}{N} \sum_{i=1}^{N} \left| f^{ETH}_{i,k} - o_i \right| \tag{5}$$

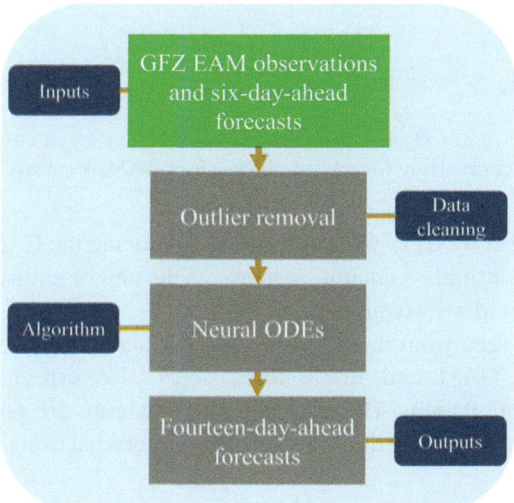

Fig. 1 Methodology for forecasting EAM functions used in our operational framework

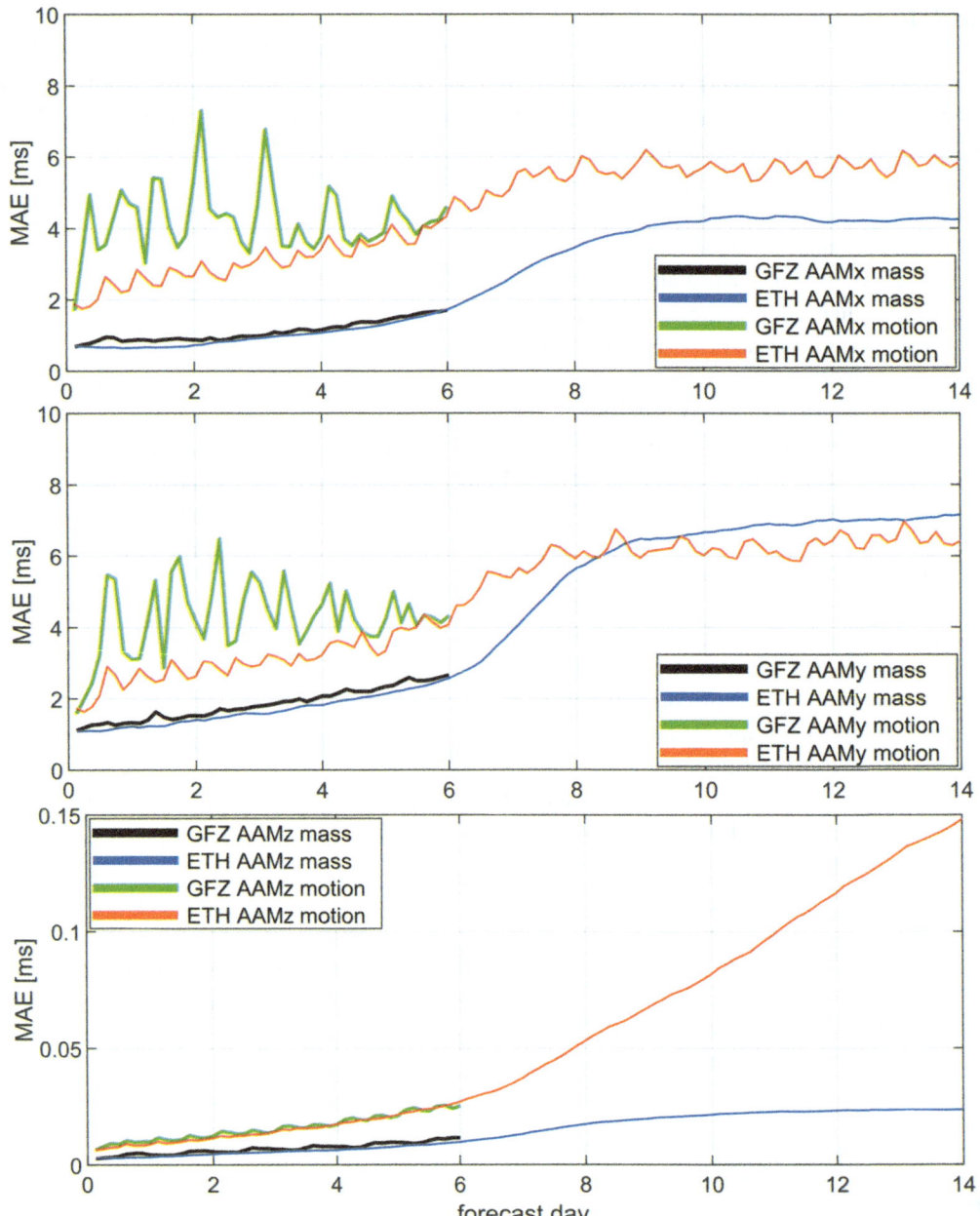

Fig. 2 Forecasting accuracy of our AAM forecasts in terms of MAE in the unit of ms. Shown also is the forecasting accuracy of GFZ AAM forecasts. Both the mass and motion terms are shown. The upper, middle, and lower panels are for x, y, z components of AAM, respectively

where $N = 731$ is the total number of prediction days during the two-year period, and $f^{ETH}_{i,k}$ is the kth step of our forecasts of the ith day ($k = 1, \ldots, 112$ for AAM and OAM, while $k = 1, \ldots, 14$ for HAM and SLAM) while o_i is the corresponding observation. In Figs. 2, 3, 4 and 5, we show the prediction performance of our improved forecasts and a comparison with that of the original forecasts provided by GFZ. Note that since we provide the EAM data in dimensionless quantities, to assign the unit of milliseconds (ms) to MAE we multiply by 6.4×10^7.

As seen from these figures, we managed to improve the AAM, OAM, and SLAM forecasts of GFZ, especially for the motion terms. This is because motion terms are generally more variable and thus more difficult to forecast using simple

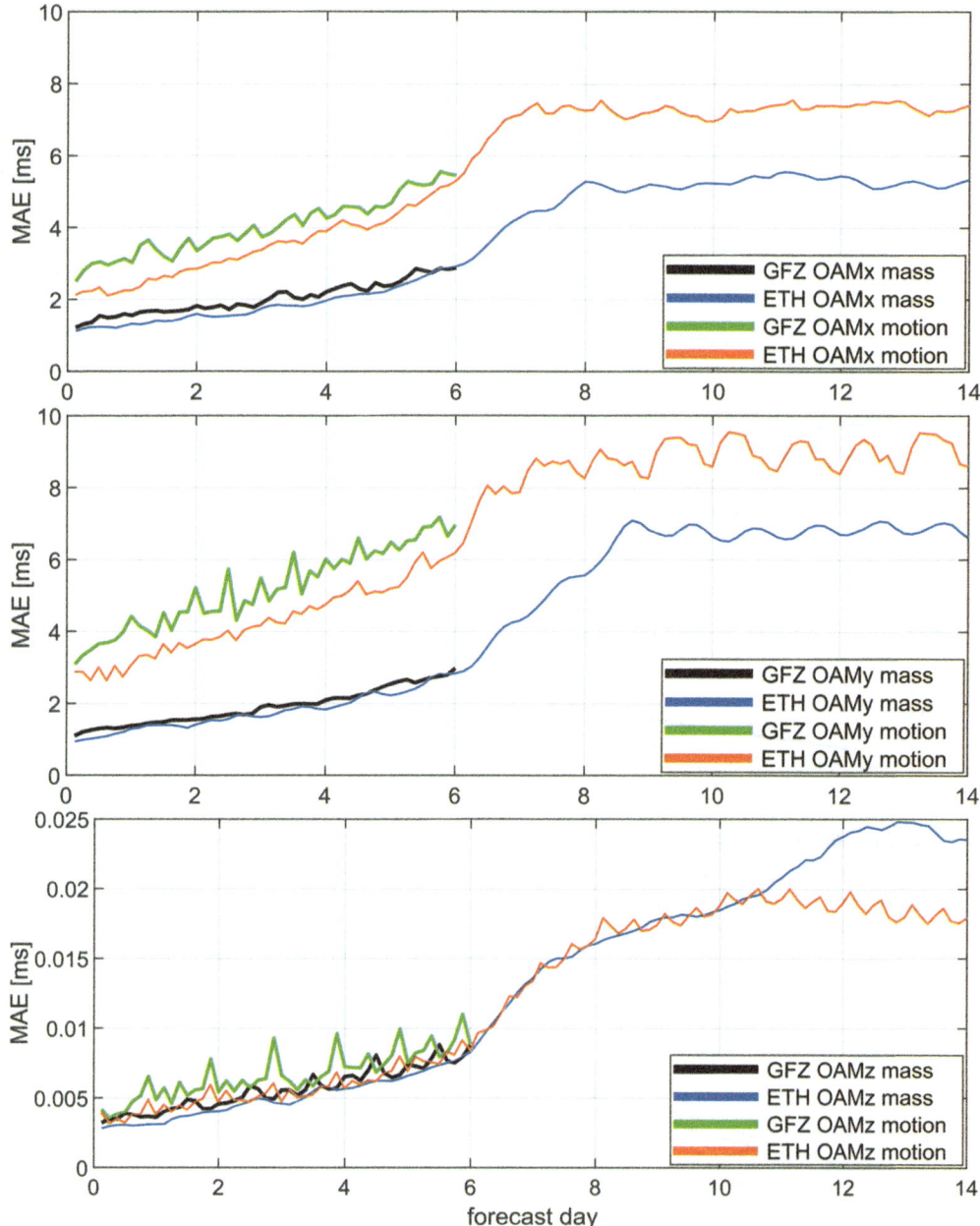

Fig. 3 Forecasting accuracy of our OAM forecasts in terms of MAE in the unit of ms. Shown also is the forecasting accuracy of GFZ OAM forecasts. Both the mass and motion terms are shown. The upper, middle, and lower panels are for x, y, z components of OAM respectively

models. Machine learning models can consider different relationships between inputs and outputs and therefore, significantly improve the prediction performance of motion terms by up to 50%. The situation is different for HAM, where we cannot improve the prediction performance, especially for the motion terms. A possible reason is that the GFZ HAM forecasts are based on models that are already accurate (called LSDM; Dill 2008). Note, however, that the amplitude of HAM is less than those of AAM and OAM. Furthermore, HAM motion terms are around 1000 times smaller than HAM mass terms. This implies that HAM motion terms are far less important compared to mass terms.

Although we cannot improve the forecasting accuracy of HAM, its impact on applications utilizing EAM functions such as EOP prediction is small. In fact, for the prediction of EOPs, AAM and OAM are the most important components of EAM (Kiani Shahvandi et al. 2022b), for which we achieve significant improvements of the GFZ forecasts. In the event that highest-accuracy HAM forecasts are required, users are advised to use the plain GFZ HAM forecasts

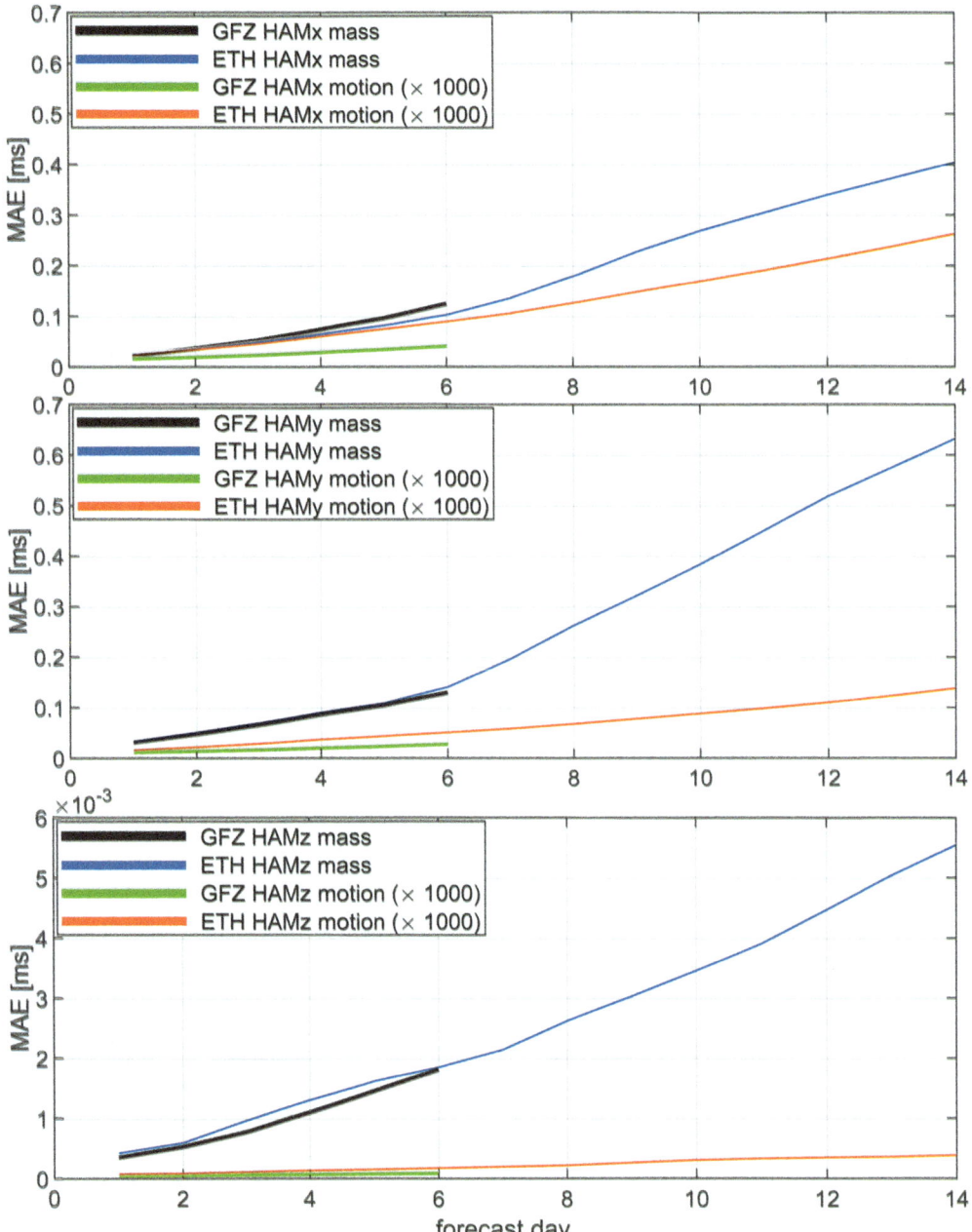

Fig. 4 Forecasting accuracy of our HAM forecasts in terms of MAE in the unit of ms. Shown also is the forecasting accuracy of GFZ HAM forecasts. Both the mass and motion terms are shown. The upper, middle, and lower panels are for x, y, z components of HAM respectively. Note that motion terms are multiplied by 1000 for better visualization (actual values are 1000 times smaller)

instead. On the other hand, if longer forecasting horizons are needed, ETH HAM forecasts can be utilized.

It is important to mention that recently, GFZ has also started to provide the corrected forecasts of AAM motion terms using machine learning (Dill et al. 2021). These products are available in the range of our study and therefore, we compare the forecasting accuracy of our forecasts with theirs. This comparison is shown in Fig. 6. The results imply that the forecasting accuracy of GFZ is slightly better than ours in (x, y) components (on average around 12% and 16%,

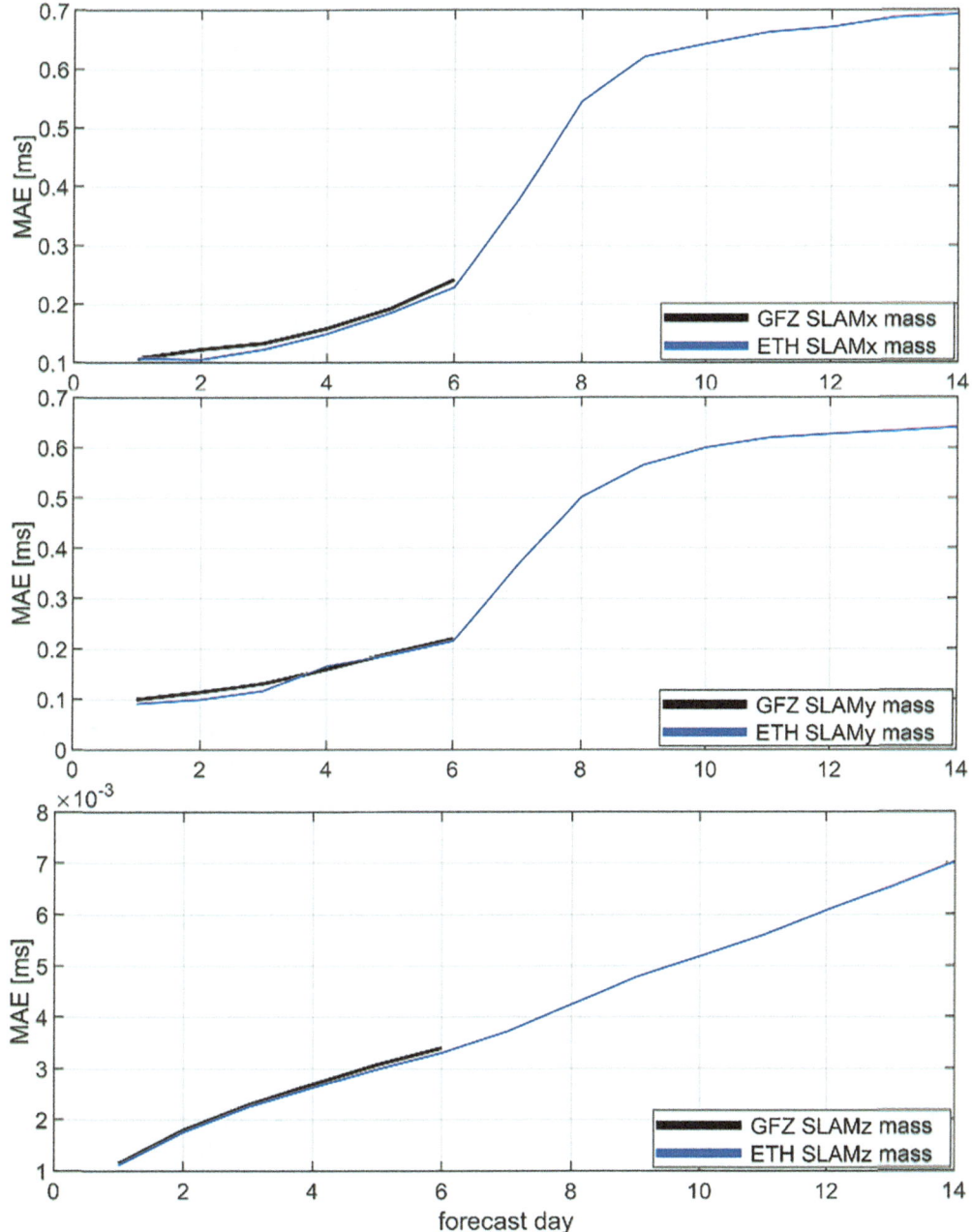

Fig. 5 Forecasting accuracy of our SLAM forecasts (only motion terms) in terms of MAE in the unit of ms. Shown also is the forecasting accuracy of GFZ SLAM forecasts. The upper, middle, and lower panels are for x, y, z components of SLAM respectively

respectively), whereas in z component our forecasts are slightly more accurate (on average around 7%). Note that the differences are small in all (x, y, z), thus we can say the forecasting accuracy of the two institutes is comparable. It should be noted that GFZ does this correction only for AAM motion terms, whereas we do this for all the EAM components and mass and motion terms.

It is important to note that the decrease in the accuracy after day 6 is because the GFZ EAM forecast data used in the algorithm are 6-day-ahead. As a result, from day 7 onward the correlation between the input and output in the algorithm is reduced and the accuracy decreases rapidly.

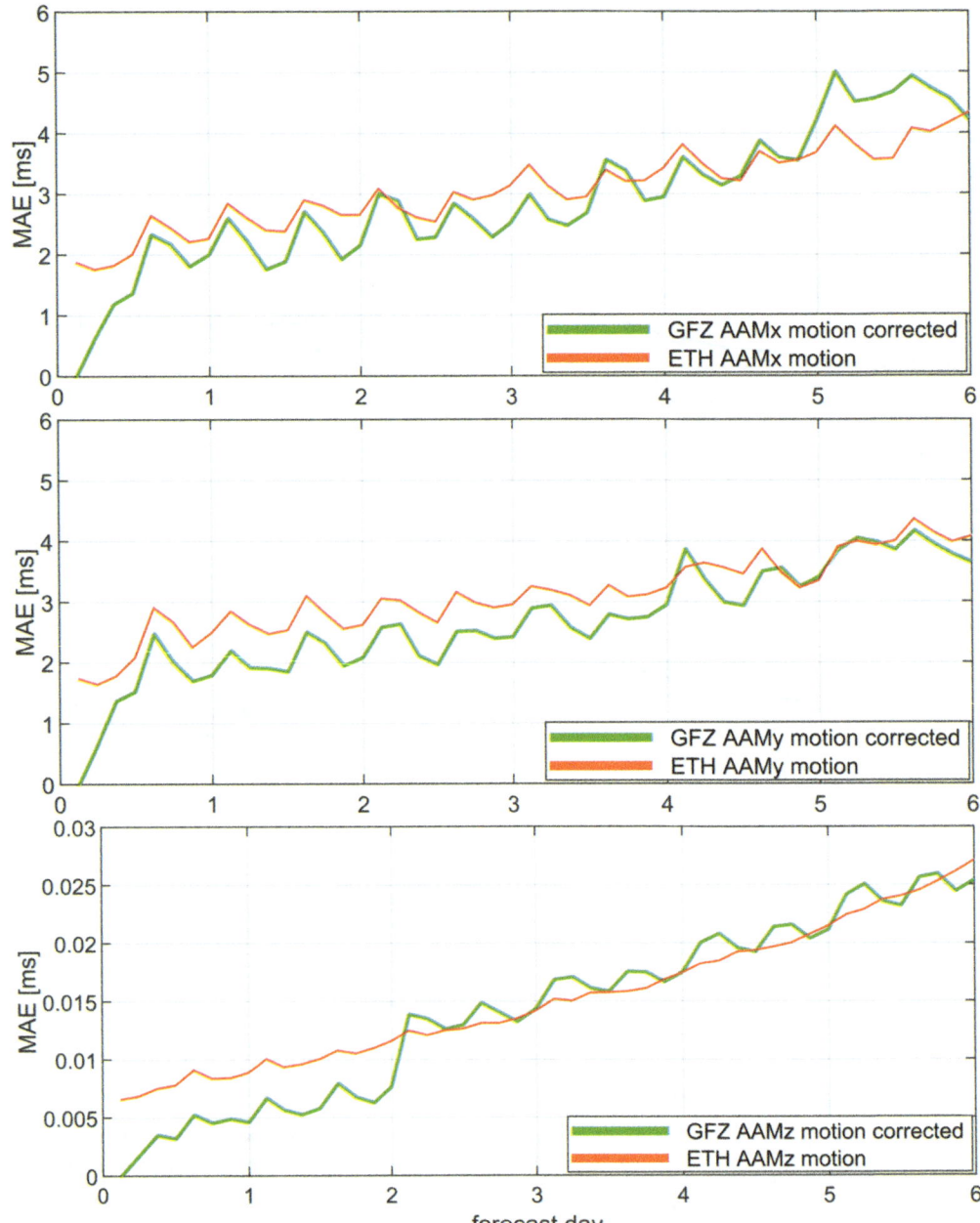

Fig. 6 Forecasting accuracy of our AAM forecasts in the first six days in terms of MAE in the unit of ms. Shown also is the forecasting accuracy of GFZ AAM forecasts that are corrected with machine learning algorithms (Dill et al. 2021). Only the motion terms are shown. The upper, middle, and lower panels are for *x*, *y*, *z* components of AAM, respectively

4 Impact on the Prediction of LOD

As an independent metric to assess the improved quality of the EAM forecasts on the prediction of EOPs, we focus on the application of ultra-short-term prediction of LOD utilizing EAM forecasts. Gou et al. (2023) have shown that the improved EAM forecasts can result in significant improvement in LOD prediction. Here we use the same model (i.e., Encoder Decoder LSTM, EDLSTM) but under operational conditions. The derived operational LOD predictions are available at https://gpc.ethz.ch/EOP/. The prediction interval is the same as that of EAM, i.e., May 20, 2021 to May 20, 2023. The results are shown in Fig. 7. Also, prediction performance and the improvements on different days are summarized in Table 1. As can be seen from this figure, up to 19% improvement can be obtained w.r.t. to the case of using the original GFZ EAM forecasts, even under fully opera-

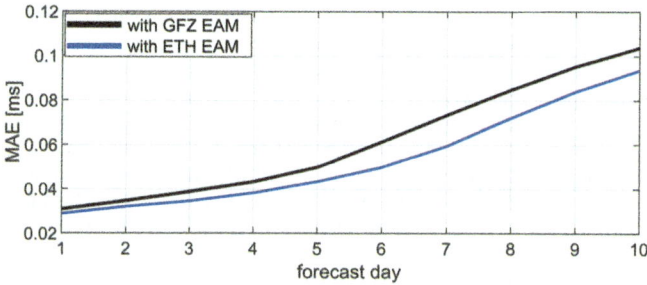

Fig. 7 Comparison between prediction accuracy of LOD using EDL-STM method in two cases: using GFZ EAM and ETH EAM forecasts. The results are in MAE in terms of ms

Table 1 The improvements of LOD prediction in days 1 to 10 when our improved EAM forecasts are used in contrast to the original GFZ EAM forecasts

Prediction day	MAE with GFZ EAM (μs)	MAE with ETH EAM (μs)	Improvement (%)
1	30.8	29.0	5.8
2	34.6	32.2	6.9
3	38.7	34.7	10.3
4	43.2	38.4	11.1
5	49.8	43.5	12.7
6	61.3	50.0	18.4
7	73.4	59.7	18.7
8	84.8	72.4	14.6
9	95.3	84.2	9.5
10	103.8	93.9	7.9

tional conditions. This further confirms that the improved forecasts have superior accuracy for the application of LOD prediction (Gou et al. 2023).

5 Conclusions and Outlook

We present a summary of the methodology and achievements of the operational EAM forecast products provided by the Chair of Space Geodesy at ETH Zurich. During the past two years, the high accuracy and benefit of the prediction of EOPs have been demonstrated. We achieved an improvement in EAM forecasting accuracy of up to 50% compared to the GFZ EAM forecasts. The benefits of these more accurate EAM forecasts and a longer forecasting horizon for LOD prediction with an improvement of up to 19% have been demonstrated (Gou et al. 2023). We provide our forecasts publicly at our Geodetic Prediction Center (Soja et al. 2022) at https://gpc.ethz.ch/EAM/ to benefit the geodetic community. We invite interested parties to use these forecasts for various applications, including more accurate prediction of EOPs.

Acknowledgements We acknowledge GFZ for providing their EAM products, without which this study would not have been possible.

References

Barnes RTH, Hide R, White AA, Wilson CA (1983) Atmospheric angular momentum fluctuations, length-of-day changes and polar motion. Proc R Soc Lond A 387:31–73. https://doi.org/10.1098/rspa.1983.0050

Chen RTQ, Rubanova Y, Bettencourt J, Duvenaud D (2018) Neural ordinary differential equations. In: 32nd Conference on Neural Information Processing Systems. https://doi.org/10.48550/arXiv.1806.07366

Dill R (2008) Hydrological model LSDM for operational Earth rotation and gravity field variations, (Scientific Technical Report STR; 08/09). Deutsches GeoForschungsZentrum GFZ, Potsdam. https://doi.org/10.2312/GFZ.b103-08095

Dill R, Dobslaw H, Thomas M (2019) Improved 90-day Earth orientation predictions from angular momentum forecasts of atmosphere, ocean, and terrestrial hydrosphere. J Geod 93:287–295. https://doi.org/10.1007/s00190-018-1158-7

Dill R, Saynisch-Wagner J, Irrgang C, Thomas M (2021) Improving atmospheric angular momentum forecasts by machine learning. Earth Space Sci 8(12). https://doi.org/10.1029/2021EA002070

Dobslaw H, Dill R (2018) Predicting Earth orientation changes from global forecasts of atmosphere-hydrosphere dynamics. Adv Space Res 61(4):1047–1054. https://doi.org/10.1016/j.asr.2017.11.044

Dobslaw H, Dill R (2019) Effective angular momentum functions from Earth system modelling at GeoForschungsZentrum in Potsdam, GFZ German Research Centre for Geosciences, Department 1: Geodesy, Section 1.3: Earth System Modelling. http://rz-vm115.gfz-potsdam.de:8080/repository/entry/show?entryid=e8e59d73-c0c2-4a9d-b53b-f2cd70f85e28

Dobslaw H, Dill R, Groetzsch A, Brzeziński A, Thomas M (2010) Seasonal polar motion excitation from numerical models of atmosphere, ocean, and continental hydrosphere. J Geophys Res: Solid Earth 115. https://doi.org/10.1029/2009JB007127

Elman JL (1990) Finding structure in time. Cogn Sci: Multidiscip J 14:179–211. https://doi.org/10.1207/s15516709cog1402_1

Goodfellow I, Bengio Y, Courville A (2016) Deep learning. MIT Press

Gou J, Kiani Shahvandi M, Hohensinn R, Soja B (2023) Ultra-short-term prediction of LOD using LSTM neural networks. J Geod 97. https://doi.org/10.1007/s00190-023-01745-x

Gross RS (2000) The excitation of the Chandler wobble. Geophys Res Lett 27(15):2329–2332. https://doi.org/10.1029/2000GL011450

Gross RS (2015) Earth rotation variations - long period. Treatise Geophys 3:215–261. https://doi.org/10.1016/B978-0-444-53802-4.00059-2

Kiani Shahvandi M, Soja B (2021) Modified deep transformers for GNSS time series prediction. In: 2021 IEEE International Geoscience and Remote Sensing Symposium. https://doi.org/10.1109/IGARSS47720.2021.9554764

Kiani Shahvandi M, Soja B (2022a) Small geodetic datasets and deep networks: attention-based residual LSTM autoencoder stacking for geodetic time series. In: International Conference on Machine Learning, Optimization, and Data Science. https://doi.org/10.1007/978-3-030-95467-3_22

Kiani Shahvandi M, Soja B (2022b) Inclusion of data uncertainty in machine learning and its application in geodetic data science, with case studies for the prediction of Earth orientation parameters and GNSS station coordinate time series. Adv Space Res 70(3):563–575. https://doi.org/10.1016/j.asr.2022.05.042

Kiani Shahvandi M, Gou J, Schartner M, Soja B (2022a) Data driven approaches for the prediction of Earth's effective angular momentum functions. In: 2022 IEEE International Geoscience and Remote Sensing Symposium. https://doi.org/10.1109/IGARSS46834.2022.9883545

Kiani Shahvandi M, Schartner M, Soja B (2022b) Neural ODE differential learning and its application in polar motion prediction. J Geophys Res: Solid Earth 127(11). https://doi.org/10.1029/2022JB024775

Kiani Shahvandi M, Dill R, Dobslaw H, Kehm A, Bloßfeld M, Schartner M, Mishra S, Soja B (2023) Geophysically informed machine learning for improving rapid estimation and short-term prediction of earth orientation parameters. J Geophys Res: Solid Earth 128(10). https://doi.org/10.1029/2023JB026720

Reichstein M, Camps-Valls G, Stevens B, Jung M, Denzler J, Carvalhais N, Prabhat (2019) Deep learning and process understanding for data-driven Earth system science. Nature 566:195–204. https://doi.org/10.1038/s41586-019-0912-1

Reiterer A, Egly U, Vicovac T, Mai E, Moafipoor S, Grejner-Brzezinska DA, Toth CK (2010) Application of artificial intelligence in Geodesy-A review of theoretical foundations and practical examples. J Appl Geod 4:201–217. https://doi.org/10.1515/jag.2010.020

Rumelhart DE, Hinton GE, Williams RJ (1986) Learning internal representations by error propagation. In: Parallel distributed processing: explorations in the microstructure of cognition, vol 1, pp 318–362. https://ieeexplore.ieee.org/document/6302929

Soja B, Kiani Shahvandi M, Schartner M, Gou J, Kłopotek G, Crocetti L, Awadaljeed M (2022) The new geodetic prediction center at ETH Zurich, EGU General Assembly 2022. https://doi.org/10.5194/egusphere-egu22-9285

Uppala SM, Kållberg PW, Simmons AJ, Andrae U, Bechtold VDC, Fiorino M, Gibson JK, Haseler J, Hernandez A, Kelly GA, Li X, Onogi K, Saarinen S, Sokka N, Allan RP, Andersson E, Arpe K, Balmaseda MA, Beljaars ACM, Van De Berg L, Bidlot J, Bormann N, Caires S, Chevallier F, Dethof A, Dragosavac M, Fisher M, Fuentes M, Hagemann S, Hólm E, Hoskins BJ, Isaksen L, Janssen PAEM, Jenne R, Mcnally AP, Mahfouf J-F, Morcrette J-J, Rayner NA, Saunders RW, Simon P, Sterl A, Trenberth KE, Untch A, Vasiljevic D, Viterbo P, Woollen J (2005) The ERA-40 re-analysis. Q J R Meteorol Soc 131(612):2961–3012. https://doi.org/10.1256/qj.04.176

Yu N, Liu H, Chen G, Chen W, Ray J, Wen H, Chao N (2021) Analysis of relationships between ENSO events and atmospheric angular momentum variations. Earth Space Sci 8(12). https://doi.org/10.1029/2021EA002030

Open Access This chapter is licensed under the terms of the Creative Commons Attribution 4.0 International License (http://creativecommons.org/licenses/by/4.0/), which permits use, sharing, adaptation, distribution and reproduction in any medium or format, as long as you give appropriate credit to the original author(s) and the source, provide a link to the Creative Commons license and indicate if changes were made.

The images or other third party material in this chapter are included in the chapter's Creative Commons license, unless indicated otherwise in a credit line to the material. If material is not included in the chapter's Creative Commons license and your intended use is not permitted by statutory regulation or exceeds the permitted use, you will need to obtain permission directly from the copyright holder.

Hourly Earth Rotation Parameter Series from GPS and Galileo Observations, and Estimations of Tidal Effects

Yuting Cheng, Christian Bizouard, Sébastien Lambert, and Jean-Yves Richard

Abstract

We use the GINS/DYNAMO software to produce hourly time series of Earth Rotation Parameters (ERP) from 2017 to 2022. Data from the American constellation GPS and the European constellation Galileo are used. Single solutions and combined solutions are produced and analyzed. The best spectral coherence between constellations lies in the retrograde semi-diurnal band. We also perform least-squares adjustments for main tidal frequencies and compare with those of previous works. A sliding window analysis reveals time variation of amplitudes of several main tides when adjusting on a selected set of frequencies.

Keywords

GNSS · High time-resolution ERP · Tidal effects · Space geodesy

1 Introduction

Since the adoption of the new precession/nutation model (Petit and Luzum 2010), the rotation of the Earth has been described by a set of five parameters: the Celestial Pole Offsets (CPO) in the celestial reference frame (dX and dY) as corrections to the precession/nutation model, UT1-UTC for non-uniform change of the Earth's rotation speed (expressed as variations of a time scale), and pole coordinates in the terrestrial reference frame (x_p and y_p) for polar motion. The Global Navigation Satellite System (GNSS) has a major contribution to the determination of the pole coordinates thanks to its numerous ground stations and frequent observations.

Attempts of estimating Earth Rotation Parameters (ERP) with sub-daily time resolution started in 1985 with Carter and Robertson (1985) for UT1, followed by similar experiments for polar motion (Herring 1993; Ma et al. 1993; Sovers et al. 1993) using VLBI data. For GNSS, it has begun shortly after the Global Positioning System (GPS) became operational in Earth rotation monitoring (Hefty et al. 2000). It is quite a natural idea considering the frequency and continuity of GPS observations which make it possible to look into sub-daily effects of ocean tides, atmospheric tides and hydrological mass redistribution on polar motion and UT1 or LOD. The time resolution is usually chosen to be between 2 hours (e.g. Hefty et al. (2000); Zajdel et al. (2021)) and 15 minutes (e.g. Sibois et al. (2017)). It is suggested that polar motion parameters be modeled as piece-wise constants for intervals shorter than 30 minutes, and piece-wise linear for longer intervals (Peng et al. 2024). The noise level of pole coordinates in general estimated to be $\approx 60\,\mu$as for a time resolution of 15 minutes (Sibois et al. 2017) or $\approx 25\,\mu$as for a time resolution of 2 hours. The spectral noise floor is estimated at the level of $\approx 5\,\mu$as (Sibois et al. 2017; Zajdel et al. 2021).

The American Global Positioning System (GPS) has been operational since 1994 and has so far a dominant role in the GNSS technique. With the development of the European Galileo system, we now have a competitive constellation to compare with the GPS constellation in terms of systematics and artifacts.

Y. Cheng (✉) · C. Bizouard · S. Lambert · J.-Y. Richard
SYRTE, Observatoire de Paris, Paris, France
e-mail: yu-ting.cheng@obspm.fr; christian.bizouard@obspm.fr; sebastien.lambert@obspm.fr; jean-yves.richard@obspm.fr

2 Data Reduction

We use data pre-processed by the French analysis center GRG in Repro 3 campaign organized by the International GNSS Service (IGS). The time span is 2017 to 2022. Integer ambiguities are resolved in this step (Katsigianni et al. 2019). Then we use the GINS/DYNAMO software (Chassaing and Roumiguier 1965) to produce hourly time series of UT1 and pole coordinates. UT1 is fixed one point per day at midnight, and the celestial pole offsets are fixed to model values. A-priori values of pole coordinates are taken from the 14C04 series (Bizouard et al. 2019) then interpolated with cubic spline to match the hourly resolution. Sub-diurnal tidal oscillations of pole coordinates and UT1 are also taken into account through the Desai-Sibois (Desai and Sibois 2016) model. No-Net-Rotation constrain is applied on about 50 stations for combined solutions and GPS only solutions. For Galileo only solution, this number goes down to around 30. These stations are selected based on a list of clusters to ensure a balanced global distribution to better constrain the ground network.

Table 1 gives general statistics of all solutions in this work. The presented parameters are residuals of pole coordinates with respect to the C04 value plus the subdaily corrections given by Desai and Sibois (2016). GPS has a larger bias compared with Galileo, while Galileo has a larger weighted root mean square (WRMS) value. GNSS solutions with fully fixed UT1 and partially fixed UT1 present similar spectral feature with an overall coherence above 0.8 as reported Fig. 1. This indicates that fixing UT1 one point per day is enough for hourly ERP solutions. Figure 2 shows the numbers of used measurements and weighted mean residuals after convergence for both single-constellation solutions and the combined solution. GPS has been stable while Galileo gaining a more important role over time.

Table 1 Statistics of all solutions with different strategies for UT1 fixing

Constellations	Pole coordinate	Bias(µas) x	y	WRMS(µas) x	y
GPS+Galileo	UT1 all fixed	12.60	−19.60	79.92	82.70
GPS+Galileo	UT1 fixed at 0h	18.92	−20.09	70.84	70.42
GPS only	UT1 fixed at 0h	21.50	−19.51	86.23	86.76
Galileo only	UT1 fixed at 0h	−0.11	−3.49	120.34	119.99

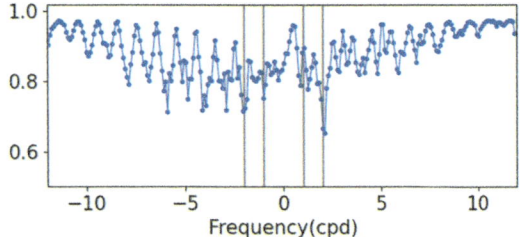

Fig. 1 Spectral coherence of two different approaches of treating UT1: (a) UT1 fixed only at midnight (b) UT1 is completely fixed to the a-priori value

3 Tidal Signals

3.1 Recovering Absent Frequencies of the FES2014 Model

We have implemented another sub-diurnal correction model based on the FES2014b ocean tidal model (Lyard et al. 2021) into the GINS software. Including only 6 diurnal and 13

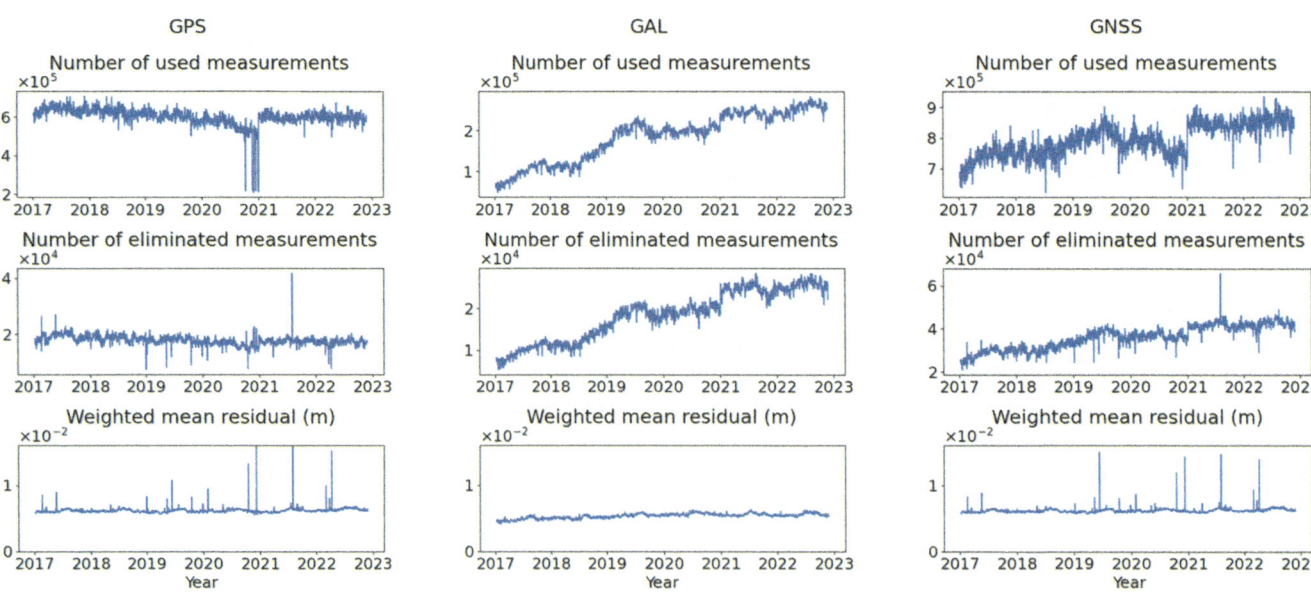

Fig. 2 Statistics of observations used in each solution

semi-diurnal tidal frequencies, that model does not include several significant oscillations with amplitudes above 1.5 µas presented in the Desai-Sibois model. We list them in Table 2. Thus, the estimated residuals of the pole coordinates with respect to FES2014 model should recover these extra harmonics, so that we have an opportunity to look into the ability of each constellation to recover each significant tidal frequencies absent in FES2014 model.

In Fig. 3, we plot the spectra of the residuals against FES 2014b (FES) and against Desai-Sibois (DS) models. Considering Desai-Sibois model reliable, Galileo may have a better ability to recover prograde diurnal signals, whereas GPS might better retrieve the prograde semidiurnal signals. Their performances are similar in the retrograde semidiurnal band.

3.2 Spectral Coherence

We calculate the spectral coherence between two single-constellation solutions. Figure 4 shows that the best coherence between two single constellation solutions, apart from the blocked retrograde diurnal band, lies in the prograde diurnal band, and yet the value of the coherence function does not exceed 0.2. When we look into details of the spectra, amplitudes of main tidal frequencies in the retrograde semidiurnal band have a better coherence between constellations. The spectral coherence is probably degraded by the artifacts around -2 cpd.

When we look at how the spectral coherence between two constellations changes between 2-year periods 2017–2019 and 2020–2022 (see Fig. 5), we notice an improvement in the last period in the high frequency band, namely frequencies above 5 cpd in both prograde and retrograde bands, also for frequencies between ± 1 cpd and between -2 and -1 cpd.

Table 2 Significant absent frequencies in FES2014 model, but present in the Desai-Sibois model given with their corresponding amplitudes

Name	Period (hour)	Pro (µas)	Retro (µas)
N2−	12.65	0.58	1.56
M2−	12.42	2.89	9.67
K2+	11.96	2.06	11.03
σ1	27.84	5.24	–
ρ1	26.72	5.59	–
Q1−	26.87	5.62	–
O1−	25.82	27.09	–
M1	24.83	10.34	–
K1−	23.94	3.34	–
K1+	23.93	23.02	–

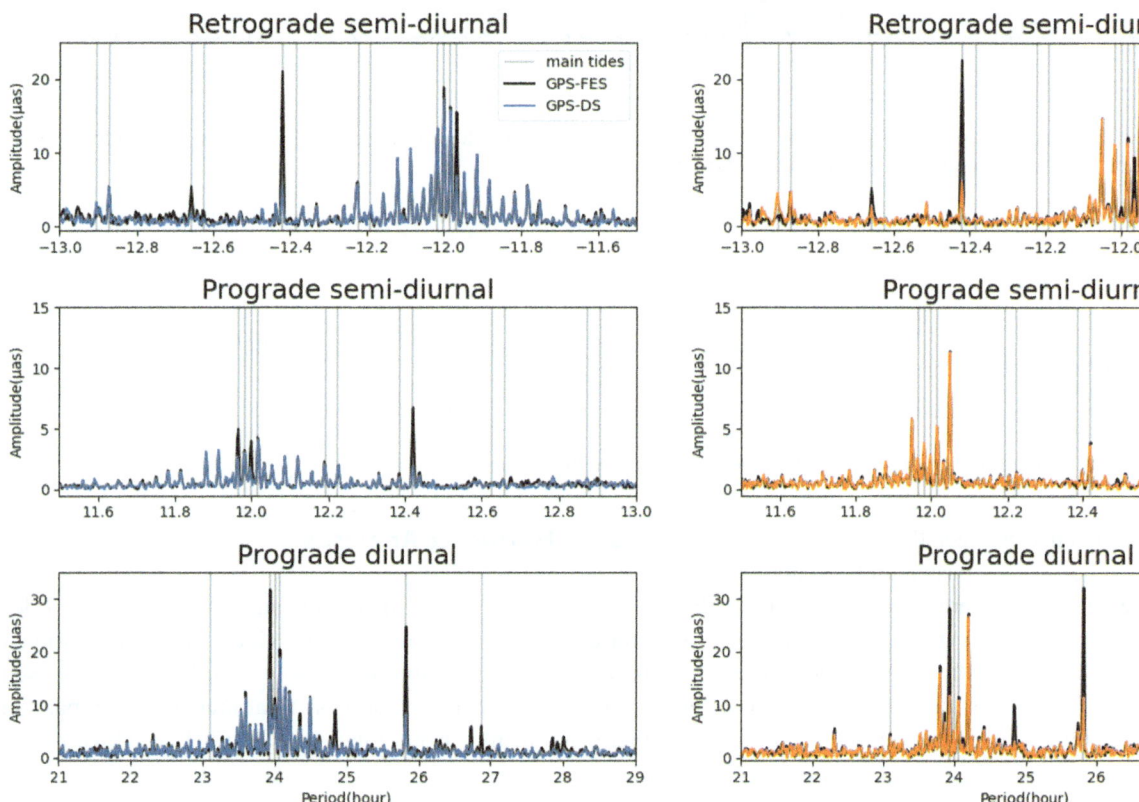

Fig. 3 Spectra of residual pole coordinates with respect to FES2014 and Desai-Sibois models: GPS only solution on the left and Galileo only solution on the right. Light greylines gives main tidal frequencies and yellow lines gives significant tidal frequencies that are absent in the FES2014 model

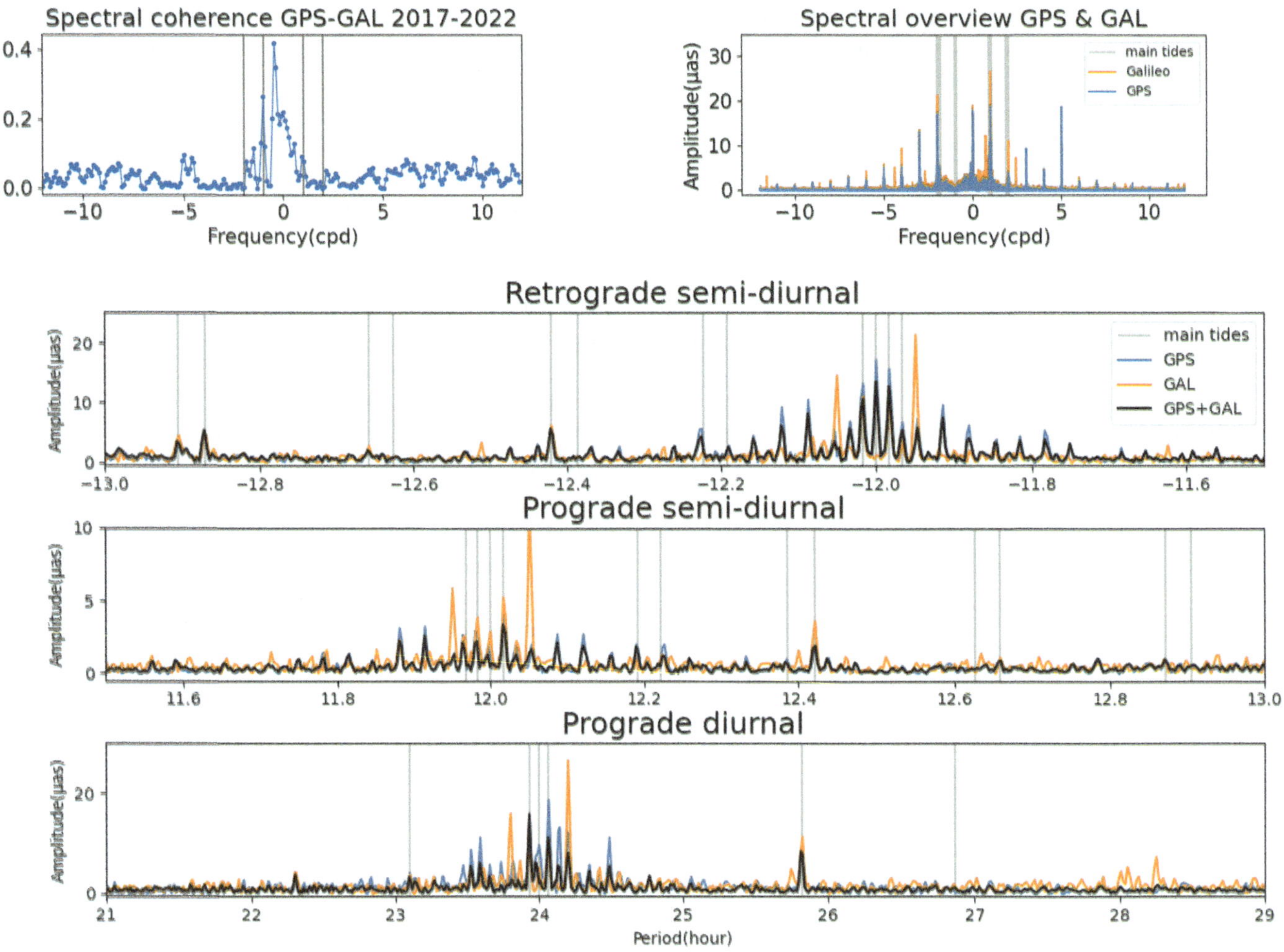

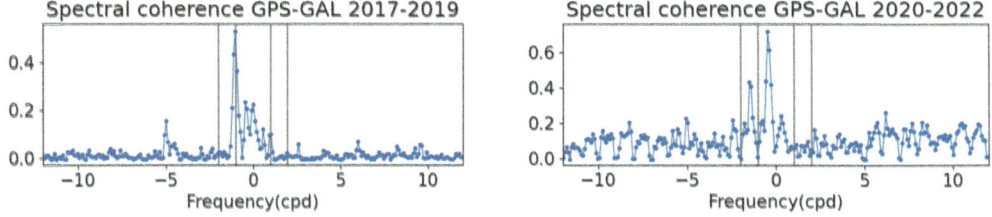

Fig. 4 Comparison of GPS and Galileo single solutions of the entire time span. Light grey lines mark main tidal frequencies. The lower panel gives the spectra of GPS and Galileo single solutions in the three tidal bands

Fig. 5 Spectral coherence between GPS and Galileo single solutions of different time spans

In the spectral coherence between the time intervals mentioned above of same single-constellation solutions (see Fig. 6), most of the visible peaks are those of system specific artifacts. We notice that GPS has much more peaks of this kind than Galileo. Some of them, such as the one at 2.4 cpd (10 h) in the Galileo series can be attributed to orbit modelling errors, but further investigation is needed.

3.3 Harmonic Analyses

We perform a least-squares adjustment on 33 main tides and compare the results with previous works. Adjustment is applied on the sub-diurnal pole coordinates, composed of the residuals and the a-priori tidal model. In comparison there are results from Sibois et al. (2017), adjusting 30

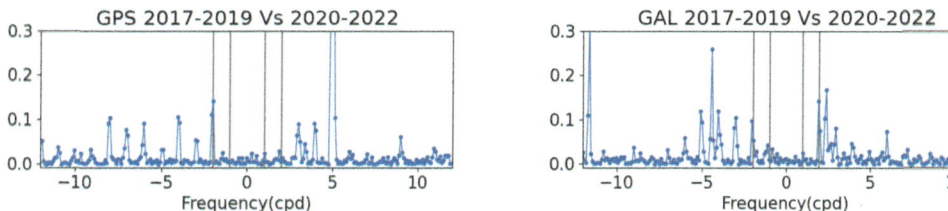

Fig. 6 Spectral coherence of each single constellation solutions between the time intervals 2017–2019 and 2020–2022: GPS (left) and Galileo (right)

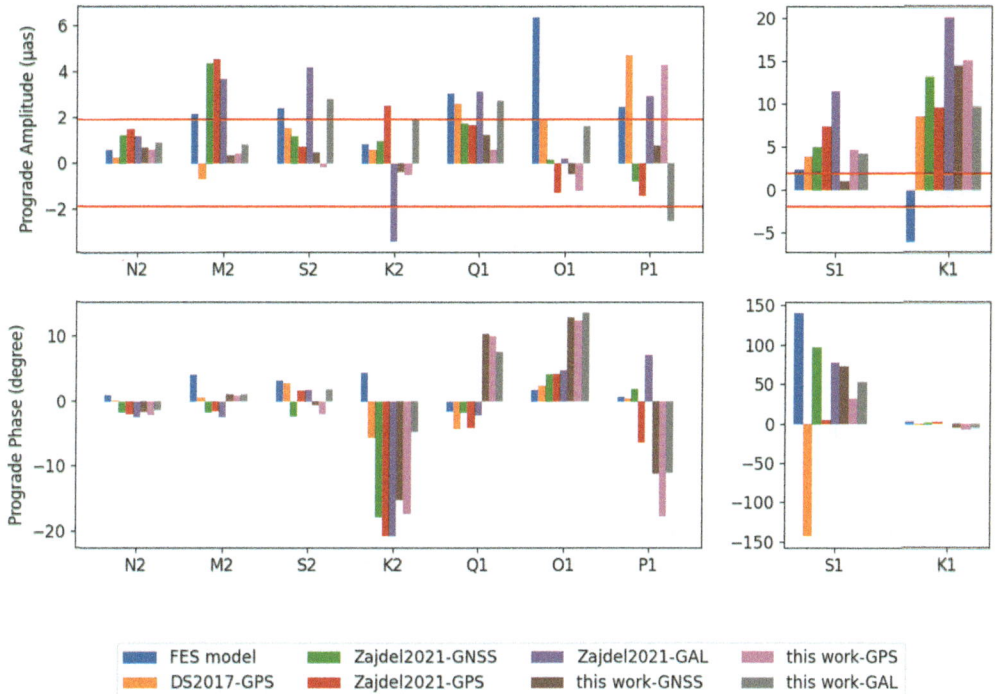

Fig. 7 Prograde amplitudes and phases of main tides of different solutions compared with Desai-Sibois model value. Red lines mark 3σ value of the estimations of this work

main tides with GPS data from 2004 to 2013, and Zajdel et al. (2021) adjusting 38 main tides with GPS, Galileo and GLONASS data from 2017 to 2019. In Figs. 7 and 8, we show the corresponding amplitudes and phases of nine most significant tides, where model values of DS2016 model have been subtracted to focus on the differences. The 3σ value of the estimations in this work (1.89 µas) is marked with red lines in the bar-plots of amplitude differences. We keep the retrograde diurnal band in this plot to show the leakage in the harmonic estimation, and it is confirmed insignificant. Largest differences are observed at frequencies close to 1 cpd (S_1 and K_1) or 2 cpd (S_2 and K_2), whereas they do not exceed 20 µas. Only the corresponding numerical values for the GNSS solution of this work are given in Table 3 for reference due to the page limit of this article. It is worth noting that in contrary to Zajdel et al. (2021) and this work, Sibois et al. (2017) did not add any constraint on the retrograde diurnal band. Although Sibois et al. (2017) did not give estimated values of the retrograde diurnal band, the amplitudes showed on their spectra do not exceed 20 µas. Differences in data time spans can also cause dispersion in estimated amplitudes of tidal frequencies. We performed a sliding window analysis to trace the possible time variations of amplitudes of main tides. The length of window is chosen to be 600 days to be able to separate all frequencies included in the fitting model, and the sliding step is set to 30 days. Figure 9 gives the results of the sliding window analysis for K1 and K2 that show a significant trend with respect to the error bars of the overall estimation. Estimated values on the entire period and their uncertainties are marked by red lines. Similar trends are revealed by both constellations independently, but there are differences in for example K2, where variations revealed by Galileo show strong annual periodicity. It can be attributed to orbital modeling issues such as for the solar radiation pressure (List et al. 2015; Zajdel, et al. 2023). The primary potential cause of the trend is the beating effect from side

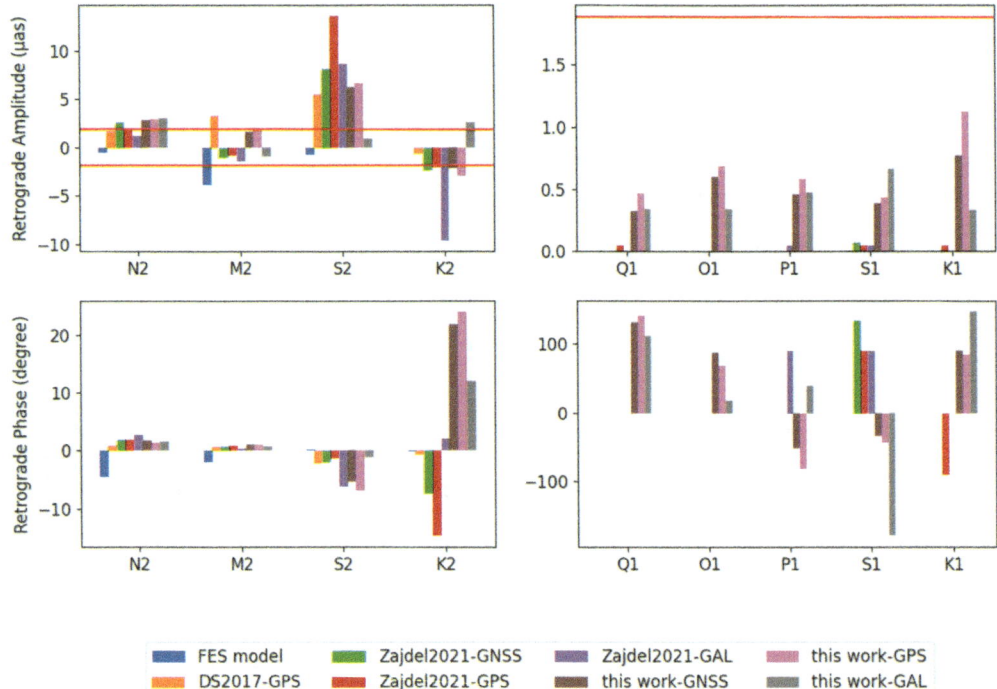

Fig. 8 Retrograde amplitudes and phases of main tides of different solutions compared with Desai-Sibois model value. Red lines mark 3σ value of the estimations of this work

Table 3 Prograde and retrograde components of fitted tidal terms of GNSS solution (retrograde diurnal blocked), formal error of the amplitudes is 0.63 μas

Name (day)	A_{pro} (μas)	ϕ_{pro} (degree)	A_{retro} (μas)	ϕ_{retro} (degree)
Q_1	31.04	−3.45	0.32	132.42
O_1	143.08	−15.51	0.60	87.14
P_1	53.05	−46.36	0.46	−52.35
S_1	2.33	−144.97	0.39	−33.83
K_1	183.65	136.89	0.77	91.28
N_2	16.31	131.32	44.58	−85.67
M_2	77.77	120.20	261.03	−87.02
S_2	25.06	90.36	136.55	−62.91
K_2	6.51	84.92	34.87	−42.89

bands that are not included in the fitting model. Including these frequencies in the fitting model can be tricky and needs further investigation. As a consequence, it is important to keep in mind when comparing similar work, that the time span of the data set can also make a difference in the estimated harmonics.

4 Summary

By estimating hourly ERP from GNSS observations, we obtain single-constellation solutions for GPS and Galileo, as well as combined solutions. Studies show that for hourly ERP computation from GNSS observations, fixing UT1 one point per day is viable. Abilities of recovering significant tidal signals of both constellations are investigated by choosing an incomplete a-priori model of subdiurnal corrections.

Spectral coherence is calculated between single-constellation solutions, and spectral details are also presented for comparison, which shows that the best agreement of the two constellations lies in the retrograde semidiurnal band. We also study the time evolution of the spectral coherence between the two constellations by calculating it with the first and the last two years of the time series. It shows an improvement of agreement in the last two years of the time series comparing with the first two years. When comparing time series of different time span of a single-constellation solution, GPS shows more persisting artifacts than Galileo.

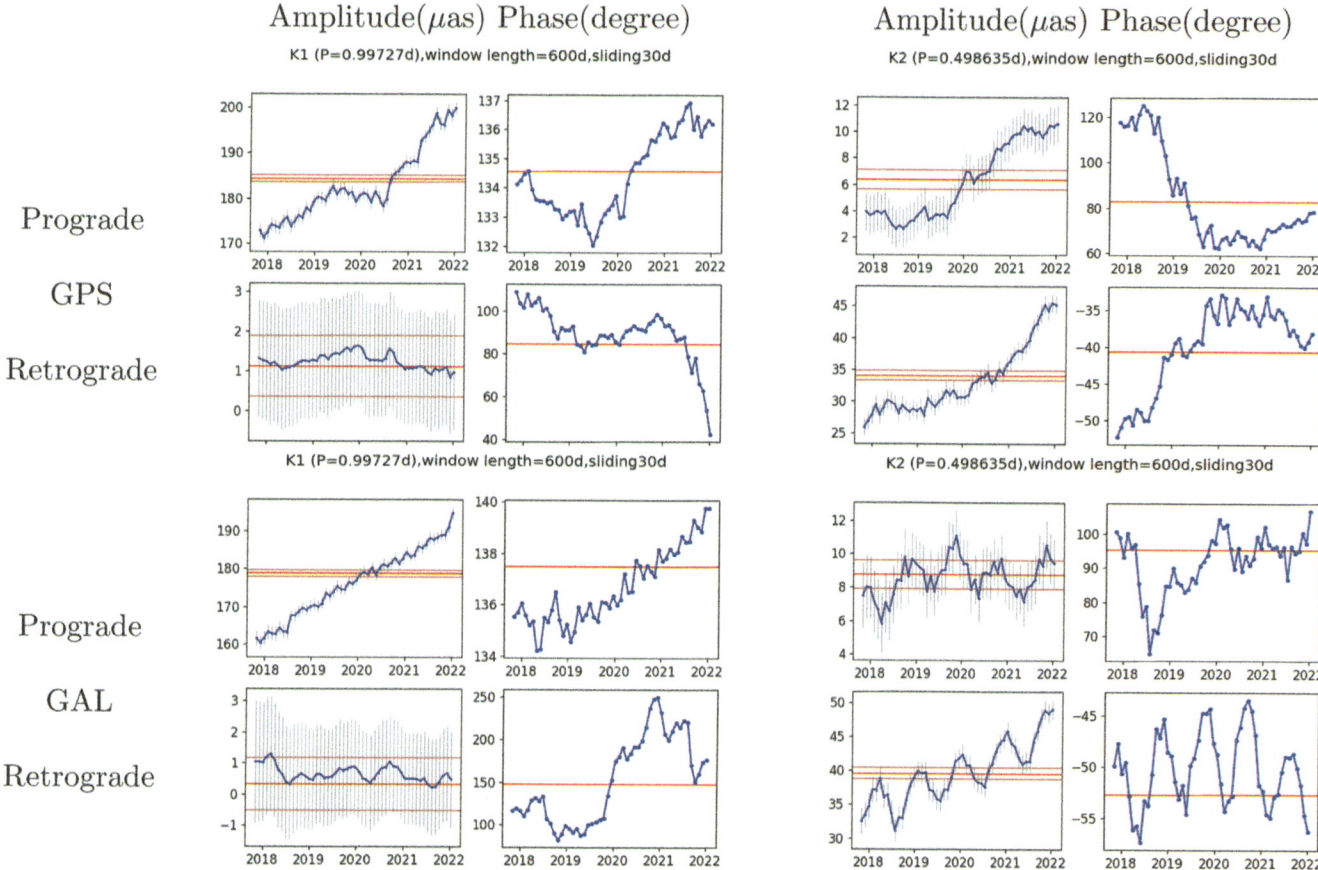

Fig. 9 Examples of sliding window analyses results: diurnal band

Harmonic analyses for significant tides are performed over the entire period and compared with prior works. A sliding window analysis reveals apparent time variations of amplitudes of certain tidal frequencies that need to be further investigated.

References

Bizouard C, Lambert S, Gattano C, Becker O, Richard J (2019) The IERS EOP 14C04 solution for Earth orientation parameters consistent with ITRF 2014. J Geodesy 93:621–633

Carter W, Robertson D (1985) High-frequency variations in the rotation of the Earth. IEEE Trans Geosci Remote Sens GE-23:369–372

Chassaing J, Roumiguier (1965) A description d'un programme de corrections différentielles utilisant l'intégration numérique; programme GIN (in French)

Desai S, Sibois A (2016) Evaluating predicted diurnal and semidiurnal tidal variations in polar motion with GPS-based observations. J Geophys Res Solid Earth 121:5237–5256

Fang Z, Xu T, Nie W, Yang Y, Li M (2023) Earth rotation parameters from GPS and BDS: Contributions from MEO and IGSO satellites. Adv Space Res 71:3091–3108

Hefty J, Rothacher M, Springer T, Weber R, Beutler G (2000) Analysis of the first year of Earth rotation parameters with a sub-daily resolution gained at the CODE processing center of the IGS. J Geodesy 74:479–487

Herring T (1993) Diurnal and semidiurnal variations in Earth rotation. Adv Space Res 13:281–290

Katsigianni G, Loyer S, Perosanz F, Mercier F, Zajdel R, Sośnica K (2019) Improving Galileo orbit determination using zero-difference ambiguity fixing in a Multi-GNSS processing. Adv Space Res 63:2952–2963

List M, Bremer S, Rievers B, Selig H, Others (2015) Modelling of solar radiation pressure effects: parameter analysis for the microscope mission. Int J Aerospace Eng 2015:Article ID 928206

Lyard F, Allain D, Cancet M, Carrère L, Picot N (2021) FES2014 global ocean tide atlas: design and performance. Ocean Sci 17:615–649

Ma C, Gipson J, Gordon D, Caprette D, Ryan J (1993) GSFC submission to the IERS 1992 annual report, IERS Tech. (Note 14, R-27, Obs. de Pads, Pads, 1993)

Panafidina N, Hugentobler U, Seitz M (2016) Interaction between subdaily Earth rotation parameters and GPS orbits. In: IAG 150 Years: Proceedings Of The IAG Scientific Assembly In Postdam, Germany, 2013, pp 159–167

Peng Y, Lou Y, Dai X, Shi C (2024) Analysis of sub-daily polar motion derived from GPS with different temporal resolutions. GPS Sol 28:33

Petit G, Luzum B (2010) IERS conventions. IERS Tech Note 36:1

Sibois A, Desai S, Bertiger W, Haines B (2017) Analysis of decade-long time series of GPS-based polar motion estimates at 15-min temporal resolution. J Geodesy 91:965–983

Sovers O, Jacobs C, Gross R (1993) Measuring rapid ocean tidal Earth orientation variations with very long baseline interferometry. J Geophys Res Solid Earth 98:19959–19971

Zajdel R, Sośnica K, Bury G, Dach R, Prange L, Kazmierski K (2021) Sub-daily polar motion from GPS, GLONASS, and Galileo. J Geodesy 95:3

Zajdel R, Masoumi S, Sośnica K, Gałdyn F, Strugarek D, Bury G (2023) Combination and SLR validation of IGS Repro3 orbits for ITRF2020. J Geodesy 97:87

Open Access This chapter is licensed under the terms of the Creative Commons Attribution 4.0 International License (http://creativecommons.org/licenses/by/4.0/), which permits use, sharing, adaptation, distribution and reproduction in any medium or format, as long as you give appropriate credit to the original author(s) and the source, provide a link to the Creative Commons license and indicate if changes were made.

The images or other third party material in this chapter are included in the chapter's Creative Commons license, unless indicated otherwise in a credit line to the material. If material is not included in the chapter's Creative Commons license and your intended use is not permitted by statutory regulation or exceeds the permitted use, you will need to obtain permission directly from the copyright holder.

EOP Prediction Based on Multi and Single Technique Space Geodetic Solution

Sadegh Modiri, Daniela Thaller, Santiago Belda, Dzana Halilovic, Lisa Klemm, Daniel König, Hendrik Hellmers, Sabine Bachmann, Claudia Flohrer, and Anastasiia Walenta

Abstract

Real-time Earth Orientation Parameters (EOP) are crucial in various space geodetic applications, from satellite navigation to weather forecasting. This study introduces a refined prediction package leveraging diverse EOP series from the Federal Agency of Cartography and Geodesy (BKG), including rapid and final series, Satellite Laser Ranging (SLR) series, and International Earth Rotation and Reference Systems Service (IERS) C04. Our approach yields substantial improvements in EOP prediction accuracy. Results highlight superior performance in critical parameters such as Polar Motion, (UT1-UTC) dUT1, and Length of Day (LOD) predictions. Notably, our predictions surpass benchmarks from the Second EOP Prediction Comparison Campaign (2nd EOP-PCC)" organized by International Association of Geodesy (IAG) and IERS, showcasing the effectiveness of our methodology. Additionally, BKG's Rapid EOP stands out with remarkable accuracy, featuring a shorter latency of 1 to 2 days. This study contributes to our understanding of Earth's rotational dynamics. It provides practical advancements in real-time EOP predictions, demonstrating the potential impact on a wide range of scientific and operational applications.

Keywords

BKG · EOP · Prediction · Real time

1 Introduction

There are numerous geophysical factors that have an influence on the rotation and orientation of the Earth in space. his implies that our planet's rotation is not constant and undergoes several variations (Seitz and Schuh 2010; Fodor et al. 2019; Ferrándiz et al. 2020; Modiri et al. 2021; Malkin et al. 2022; Raut et al. 2022).Understanding these variations is crucial for establishing precise astronomical and geodetic reference systems, and it underlies applications ranging from satellite orbit determination to space navigation (Kalarus et al. 2010). Earth Orientation Parameters (EOP), which consist of five angles (polar motion (x, y), Earth rotational angle (ERA) or Ut1-UTC (dUT1), celestial pole motion) that relate points in the terrestrial reference system to the celestial reference system, serve as key indicators of these critical changes.

Real-time EOP are essential for a variety of applications, including high-precision satellite navigation and positioning, monitoring of water vapor, interplanetary spacecraft missions, as well as radio astronomy observation.

While space geodetic techniques, such as Very Long Baseline Interferometry (VLBI), Satellite Laser Ranging (SLR), Lunar Laser Ranging (LLR), Doppler Orbitography and Radiopositioning Integrated by Satellite (DORIS), and Global Navigation Satellite Systems (GNSS), have signifi-

S. Modiri (✉) · D. Thaller · D. Halilovic · L. Klemm · D. König · H. Hellmers · S. Bachmann · C. Flohrer · A. Walenta
Department Geodesy, Federal Agency for Cartography and Geodesy (BKG), Frankfurt am Main, Germany
e-mail: sadegh.modiri@bkg.bund.de

S. Belda
UAVAC, University of Alicante, Alicante, Spain

cantly improved, the EOP datasets continue to experience delays ranging from a couple of days to weeks. This delay is primarily attributed to the complexity of the data analysis process. Therefore, EOP must be accurately predicted to bridge the gap between the EOP-estimated data and EOP in real time.

Until now, there have been two Earth Orientation Parameters Prediction Comparison Campaigns (EOP PCC), with the first held from 2005 to 2009 and the second initiated in 2021, organized by the International Association of Geodesy (IAG) and the International Earth Rotation and Reference Systems Service (IERS). These campaigns aim to identify the most effective prediction algorithms. According to the first EOP PCC report, prediction techniques can be categorized into two primary groups based on their input data. The first group of methods rely on information that is contained within the EOP time series, using methods such as least squares (LS) collocations (Wlodzimierz 1990), Kriging-based prediction (Michalczak and Ligas 2021) and neural networks (Schuh et al. 2002). Hybrid techniques, combining least squares (LS) with auto-regressive (AR), auto-regressive moving average (ARMA), auto-covariance, and neural network methods (Xu and Zhou 2015; Wu et al. 2021), as well as the combination of dynamic mode decomposition and bivariate autoregressive (Ligas and Michalczak 2024), or the integration of Singular Spectrum Analysis (SSA) and Copula-based analysis (Modiri et al. 2018; Modiri 2021), have shown promise. The second group also incorporates geophysical parameters such as the axial component of effective angular momentum (EAM) (Modiri et al. 2020; Guessoum et al. 2022; Kiani Shahvandi et al. 2022, 2023). These techniques are predominantly designed to utilize multi-technique combined EOP products. The IERS currently provides the official EOP dataset in the form of multi-technique combined EOP time series (available at https://datacenter.iers.org/data/latestVersion/EOP_20_C04_one_file_1962-now.txt). Also, the IERS Rapid Service Prediction Center (IERS-RS/PC) provides weekly predictions of EOP based on multi-technique combined EOP products.

This study explores the prospect of utilizing new technique-specific and multi-technique combined EOP series produced by the Federal Agency of Cartography and Geodesy (BKG) for EOP prediction. The present investigation also establishes the groundwork for an operational EOP prediction service aimed at revolutionizing real-time EOP time series availability at BKG. The study involves a comparison between predictions from the IERS C04 time series and BKG's EOP time series (both Technique-Specific and multi-techniques combined) to illustrate the feasibility of utilizing BKG's EOP time series. Our results indicate that, although BKG's multi-technique combined EOP is likely more accurate than technique-specific EOP, both exhibit high accuracy. Furthermore, our objective is to achieve high precision for the complete set of parameters, including Polar Motion, dUT1, Length of Day (LOD), and Celestial Pole Offset (CPO), surpassing the best results from the 2nd EOP-PCC (Śliwińska-Bronowicz et al. 2024). The final emphasis is on highlighting the robustness and accuracy of BKG's Rapid EOP, characterized by a shorter latency of 1 to 2 days.

2 Methodology

In this study, the combination of the SSA and Copula methods is employed for the precise prediction of Earth rotation parameters (ERP), encompassing Polar Motion in the x-axis (PMx), Polar Motion in the y-axis (PMy), the difference between Universal Time (UT1) and Coordinated Universal Time (UTC) (dUT1), and Length of Day (LOD). The SSA is utilized to model the deterministic part of ERP time series, and the Copula-based analysis method is applied to model the stochastic behavior remaining in the ERP time series after reducing the SSA-reconstructed deterministic terms from the original ERP time series. Additionally, the empirical Free-Core Nutation (FCN) model (B16) proposed by Belda et al. (2016) is employed to forecast Celestial Pole Offsets (CPO) (Belda et al. 2018). The algorithms for the combination of SSA and Copula-based analysis are well described in (Modiri et al. 2018), named Algorithm 1, where only the EOP time series is taken into the model, and (Modiri et al. 2020), named Algorithm 2, where geophysical parameters are introduced as an additional input together with EOP to the SSA+Copula-based analysis model to enhance prediction accuracy.

2.1 Prediction of Earth Rotation Parameters (PMx, PMy, dUT1, and LOD)

2.1.1 Algorithm 1: PMx, PMy, and dUT1 Modeling

In Algorithm 1 (Modiri et al. 2018; Namazi 2022), we focus on the prediction of Polar Motion (PMx and PMy) and the variation in the difference between Coordinated Universal Time (UTC) and Universal Time (UT1), denoted as dUT1. The process can be summarized as follows (Modiri 2021):

1. **Deterministic Component Modeling with SSA**:
   - SSA is employed to extract the deterministic components of PMx, PMy, and dUT1 from the observed data. SSA helps capture underlying trends and periodic behavior.
   - The SSA algorithm involves window length, eigenvalue decomposition of the time series matrix and selection of eigenvectors based on their eigenvalues.

- The deterministic components obtained from SSA are used as the initial estimate for the modeling process.
2. **Copula-Based Analysis for Residual Modeling**:
   - Copula-based analysis is used to model the residual components, capturing variations not accounted for by SSA.
   - Copula selection is guided by goodness-of-fit tests and Akaike Information Criterion (AIC) values to choose the most suitable model for capturing the dependence structure between PMx, PMy, and dUT1.
   - Maximum Likelihood Estimation (MLE) is employed to estimate the parameters of the chosen Copula model, effectively characterizing the interdependencies between these parameters (Modiri 2015).
3. **Extrapolation of Periodic Terms and Anomaly Prediction**:
   - The periodic components of PMx, PMy, and dUT1 are extrapolated based on historical data patterns.
   - Anomaly terms are predicted separately using the Copula-based analysis.
4. **Combining Predicted Periodic and Anomaly Terms**:
   - The predicted periodic and anomaly terms are integrated to obtain the final predictions for PMx, PMy, and dUT1.

2.1.2 Algorithm 2: LOD Modeling

In Algorithm 2, the focus shifts to modeling Length of Day (LOD) with particular attention to Effective Angular Momentum (EAM). The process can be summarized as follows:

1. **Utilizing Effective Angular Momentum (EAM)**:
   - EAM, derived from geophysical parameters, is considered as a key input for modeling LOD.
   - The EAM-based model is used to capture irregularities and variations in Earth's rotation, allowing for precise LOD predictions.
   - The model parameters are estimated through a regression-based approach to align the observed LOD data with the modeled values.
2. **SSA for Residual Extrapolation**:
   - SSA is employed to predict the residual part between the original LOD data and the LOD time series predicted by the Copula-based analysis, which is based on the EAM time series.
3. **Combining Predicted LOD using EAM and extrapolated Residual**:
   - The predicted EAM is then used to forecast the LOD, and the residual part is extrapolated using SSA.
   - The predicted LOD from the Copula-based analysis and the extrapolated residual from SSA are combined to create the final LOD predicted time series.

2.2 Prediction of Celestial Pole Offsets (CPO)

The prediction of Celestial Pole Offsets (CPO) is achieved through the use of the empirical FCN model, specifically the B16 model developed by Belda et al. (2016). This model is well-suited for capturing the irregularities in the Earth's rotation. The B16 model is a function of time and includes empirical parameters that are fitted to observed CPO data. The process involves estimating these parameters to minimize the difference between observed and modeled CPO values. More details can be found here in the paper of Belda et al. (2018).

3 Dataset

This section will provide an overview of the EOP time series data used in the study, highlighting the different sources and techniques employed. The EOP time series were compared with the IERS 14 C04 time series, which served as the official reference (Bizouard et al. 2019). The EOP time series included the following components:

Multi-Technique Combined EOP

BKG Final Solution: This dataset is derived from the combination of different space geodetic observation techniques to obtain EOP (Lengert et al. 2022). It incorporates data from the following sources:

- VLBI Rapid Sessions (R1 and R4) from BKG's IVS (International VLBI Service for Geodesy and Astrometry (IVS)) Analysis Center: These are 24-hour sessions that provide high-precision measurements of Earth's rotation (Lengert et al. 2022; Klemm et al. 2024).
- VLBI Intensive Sessions from BKG's IVS Analysis Center: These are 1-hour sessions that are intended to determine solely dUT1.
- GNSS Rapid Sessions from CODE: CODE (Center for Orbit Determination in Europe) provides precise geodetic products including EOP series based on Global Navigation Satellite System (GNSS) data.

The combination of the above mentioned data sets results in the "BKG Final Solution" which has a latency of about two weeks.

BKG Rapid Solution: This dataset is obtained by combining VLBI and GNSS observation.

- VLBI Intensive Sessions from BKG's IVS Analysis Center
- GNSS Rapid Sessions from CODE

The resulting combined EOP series labelled "BKG Rapid Solution" has a shorter latency, typically ranging from 1 to 2 days. At present, this BKG Final and BKG Rapid dataset are not officially available, but we plan to make it accessible online in the near future. However, interested parties can request access to the data upon inquiry.

IERS EOP 14C04: These series are generated by combining operational EOP series obtained from VLBI, GNSS, Satellite Laser Ranging (SLR), and Doppler Orbitography and Radiopositioning Integrated by Satellite (DORIS).

Technique-Specific EOP Series

The study utilized also the EOP solution provided by the BKG ILRS (International Laser Ranging Service) Analysis Center which is a solution based solely on SLR data. Similar to the BKG Rapid Solution, this EOP series also has a latency of less than one day.

Effective Angular Momentum (EAM): In addition to the previous EOP time series, an additional external input was utilized in this study. This input is derived from the z component of effective angular momentum, provided by the GFZ (German Research Centre for Geosciences) (Dobslaw and Dill 2018). The EAM time series are available from: (http://rz-vm115.gfz-potsdam.de:8080/repository). The EAM, included Atmospheric angular momentum (AAM), Oceanic angular momentum (OAM), and Hydrological Angular Momentum (HAM).

4 Results

In this study, all the EOP time series explained above were used as input to the prediction model. The goal was to generate predictions for EOP for the next ten days. By comparing these predictions with the IERS C04 reference dataset, the study aimed to evaluate the accuracy and performance of the prediction model.

4.1 Polar Motion

For polar motion in the x-axis and y-axis, different datasets were employed, each represented by a distinct color: IERS C04 in blue, BKG ILRS in green, BKG Rapid in yellow, and BKG Final Solution in red (see Figs. 1 and 2).

Six years of data were utilized to train the model, and subsequently, predictions were made for the next ten days, starting from 550 epochs after January 2021. The Singular Spectrum Analysis (SSA) window length was defined as 90 days, and the reconstructed component and Copula parameters were dynamically estimated for each window. The lower panel of Fig. 1 displays the Mean Absolute Errors (MAE) of the predictions. It demonstrates that our prediction model consistently provides errors of less than one milliarcsecond (mas) for the first third of the days, irrespective of the input data. Notably, the BKG Rapid and Final Solution exhibit similar levels of error. The technique-specific Satellite Laser Ranging (SLR) solution also yields a comparable error; however, the MAE is slightly larger than that of the other datasets. This is not surprising, as the Rapid and Final solutions contain GNSS data, which usually determines polar motion more stably than SLR data can do. For polar motion in the y-direction (PM_y), the model configuration is nearly identical to that of PM_x. In comparison to PM_x, the error is consistently less than 1mas for all time series up to the first three days. Furthermore, for the BKG Rapid, Final, and IERS C04 time series, this low error rate extends up to day 6 (see Fig. 2).

4.2 dUT1

The dUT1 time series from both the BKG Rapid and Final solutions were utilized for comparison with the IERS C04 solution. The decision to use four years of data for training the prediction model was based on the need to balance capturing relevant patterns while avoiding over-fitting, considering the specific characteristics of the dUT1 data. This time span differs from PM due to the dominated signals in both time series. In PM, the main signals are Chandler Wobble and annual oscillation, but in the dUT1 time series, the annual oscillation is dominant. Therefore, we aimed to capture at least two cycles of the main signals.

In our study, the prediction model was trained on four years of data, and subsequently, 550 epochs were predicted for the next ten days. The choice of the Singular Spectrum Analysis window length at 120 days aimed to strike a balance between capturing long-term patterns in the dUT1 time series and maintaining responsiveness to shorter-term variations. This window length was determined through iterative testing and experimentation to optimize the model's performance. The MAE in the BKG time series closely approximates that of the IERS C04 counterpart. Nevertheless, examination of the IERS C04 dataset indicates a marginally lower error, as illustrated in Fig. 3. This MAE analysis implies that our findings adhere to the elevated benchmarks established by the second EOP Prediction Comparison Campaign (2nd EOP PCC), as detailed in the evaluation conducted by Kur et al. (2022).

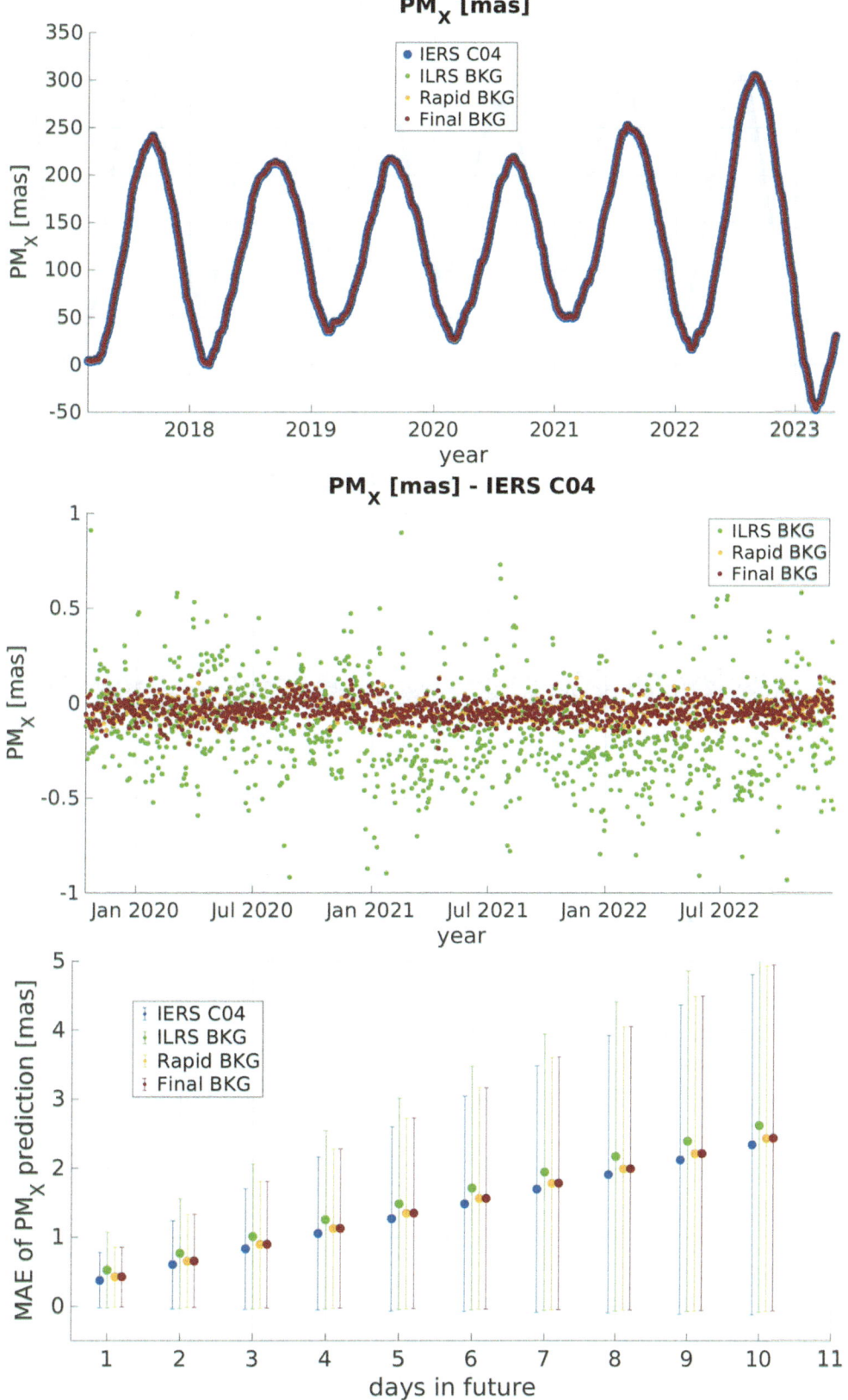

Fig. 1 Upper plot: Polar motion time series in the x-direction. Middle plot: the difference between various BKG's EOP time series and IERS C04. Lower plot: Mean absolute error of the prediction results

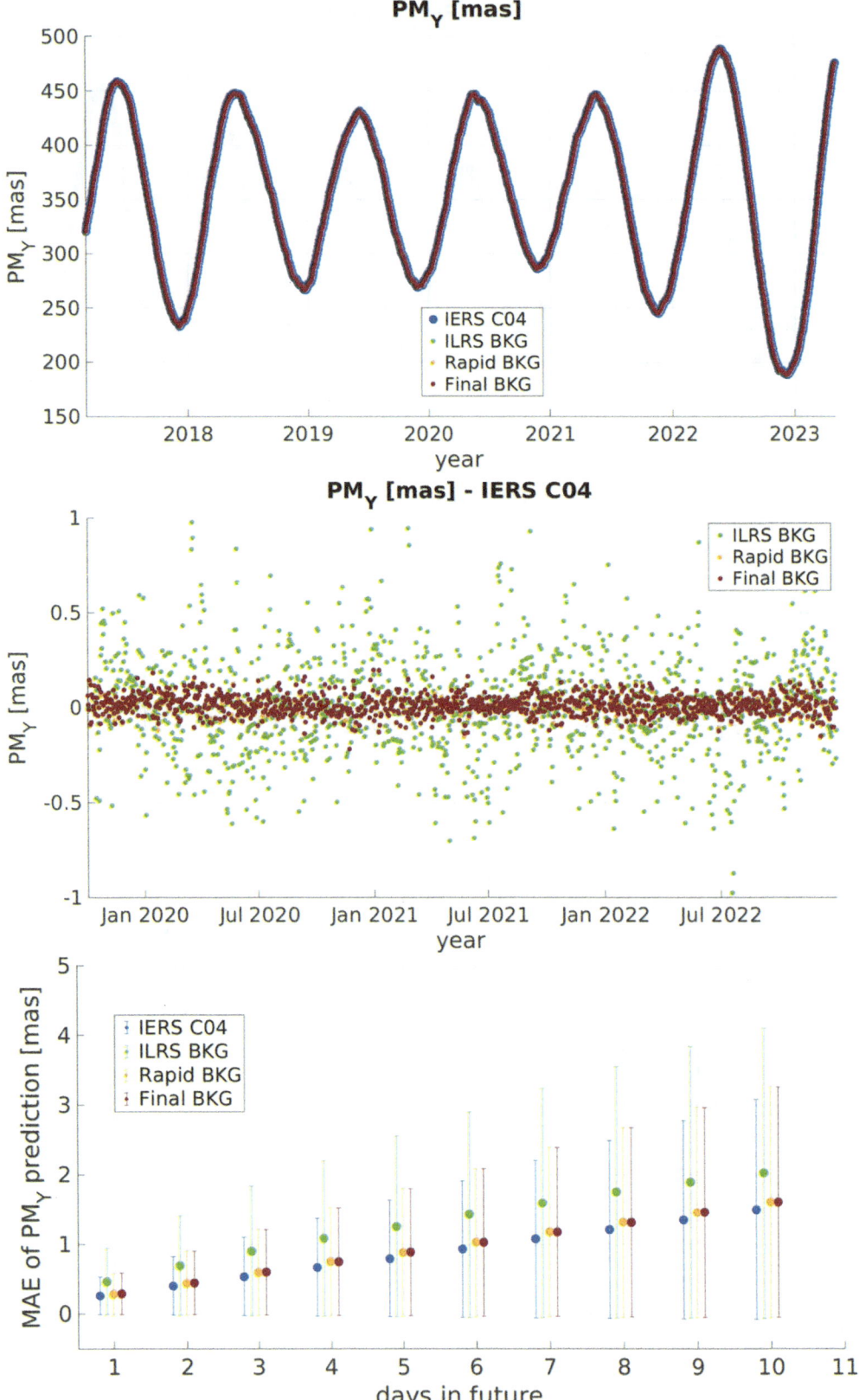

Fig. 2 Upper plot: Polar motion time series in y direction. Middle plot: difference between various BKG's EOP time series and IERS C04. Lower plot: Mean absolute error of the prediction results

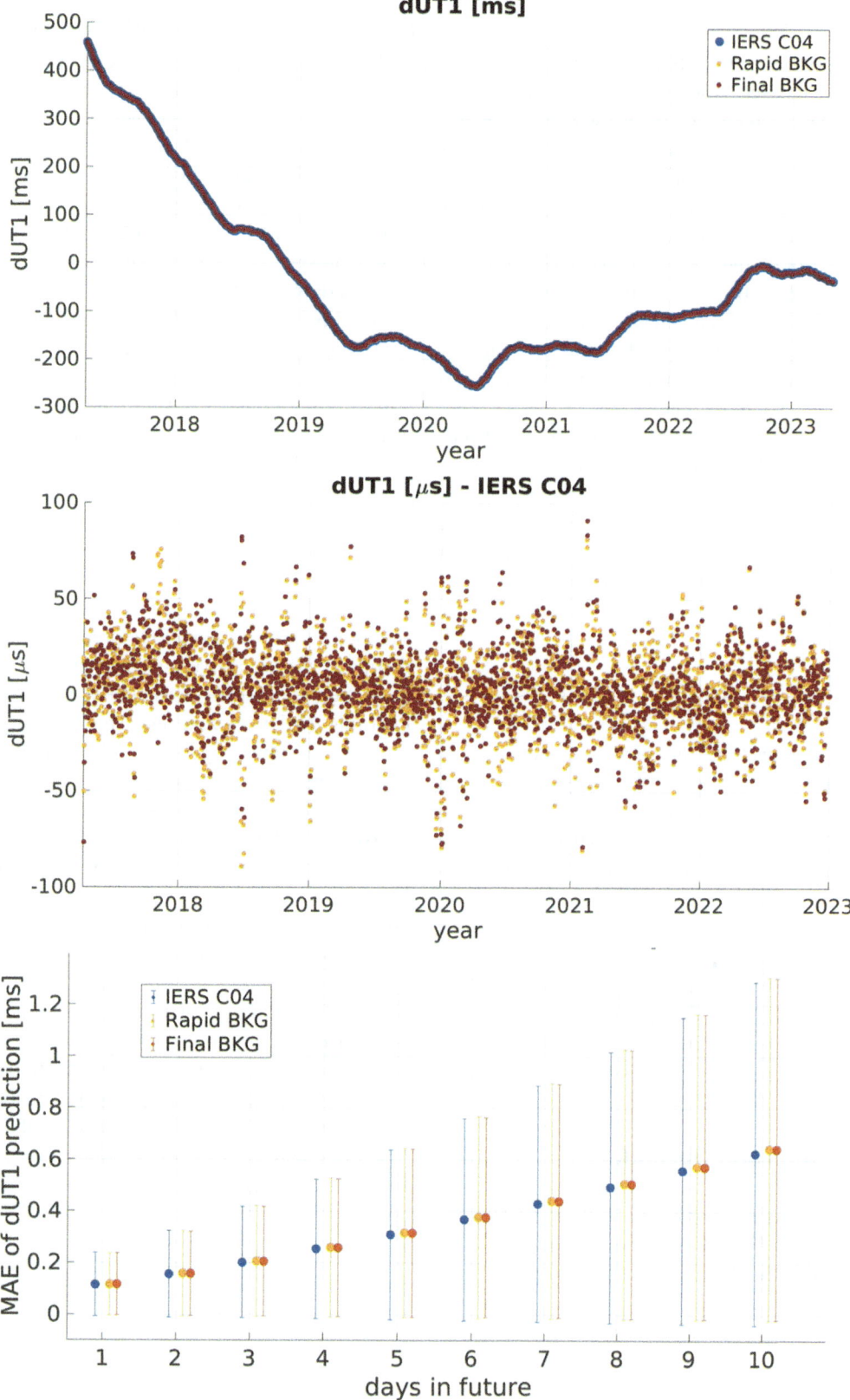

Fig. 3 Upper plot: dUT1 time series. Middle plot: difference between various BKG's EOP time series and IERS C04. Lower plot: Mean absolute error of the prediction results

4.3 LOD

The LOD data was retrieved from BKG's ILRS series as well as from IERS C04 dataset. Additionally, the z-component of Effective Angular Momentum (EAM) was incorporated as described in Sect. 2.1.2. A dataset spanning almost seven years was employed to train the prediction model for LOD. The decision to use a longer duration for LOD data is attributed to the application of a distinct algorithm. This algorithm, designed to capture the intricate dependency structure between LOD and EAM, benefits from a larger dataset. In contrast, PM employed a six-year dataset, and dUT1 utilized four years, each tailored to the specific characteristics of the respective phenomena. Predictions were made for 550 epochs, covering the next 10 days (Fig. 4).

The results of the prediction indicate that the LOD predictions using BKG's SLR series exhibit a level of error comparable to that of IERS C04. The error remains below 0.1 ms/day until the 9th day as can be seen in Fig. 5. Notably, the error on the 10th day is among the best levels reported in the 2nd EOP PCC range, ranging from 0.07 to 0.028 ms/day (Kur et al. 2022).

4.4 CPO

In this study, to demonstrate the predictive potential of the full set of EOP, CPO from VLBI observations were employed as input data for estimating the Free Core Nutation (FCN). Specifically, the B16 (Belda et al. 2016) FCN model predicted CPO values for 365 days. The FCN amplitudes were estimated daily, employing a sliding window size of 400 days and advancing by one day step (Belda et al. 2018). The prediction was conducted from January 2022 to the end of 2022. As depicted in Fig. 5, the MAE consistently remained at the level of μas. This performance places the error among the best levels reported for CPO prediction (2ndEOP PCC 2023).

5 Conclusions and Outlook

In conclusion, this study has explored and demonstrated the potential for advancing Earth Orientation Parameter (EOP) prediction using a combination of BKG's diverse EOP time series based on space geodetic techniques. By harnessing both single-technique and the multi-technique combined EOP time series, we have achieved consistent and precise EOP series predictions with daily temporal resolution.

Our comparison of predictions from the IERS C04 dataset and BKG's EOP data showed high accuracy, with similar MAE levels. Furthermore, our findings underscore the superior performance of the multi-technique combined EOP series over single-technique EOP products, albeit with technique-specific EOP products are also maintaining a commendable level of accuracy.

Notably, our predictions exhibited exceptional accuracy in various key parameters. Specifically, Polar Motion in both the x and y axes demonstrated prediction errors of less than 1 mas for multiple days. Additionally, our dUT1 predictions were on par with the best prediction series generated within the 2nd EOPPCC results (Kur et al. 2022). The LOD predictions achieved remarkable precision, with an error of less than 0.1 milliseconds per day on the 9th day, reaching the pinnacle of 2nd EOP-PCC results on the 10th day.

The reliability and consistency of the Rapid series of BKG's EOP products have been established, affirming its suitability for operational EOP prediction requirements. Furthermore, we are actively testing and implementing various prediction techniques, particularly those rooted in Machine Learning (ML) approaches. This ongoing effort aims to establish a robust operational EOP prediction service at BKG. Furthermore, the variety of EOP series used as input for the prediction methods will be further increased using our VLBI- and GNSS-only EOP series. Investing in continuous research and development remains essential to enhance our EOP estimation methods. This investment holds significant promise for advancing our ability to predict EOP with even greater precision and accuracy.

Moreover, with the refinement of prediction capabilities, we anticipate revolutionizing satellite orbit prediction with far-reaching implications for satellite navigation, space exploration, and other critical applications. In conclusion, our commitment to pushing the boundaries of EOP prediction strengthens our understanding of Earth's rotational dynamics and contributes to advancing a wide range of scientific and practical fields.

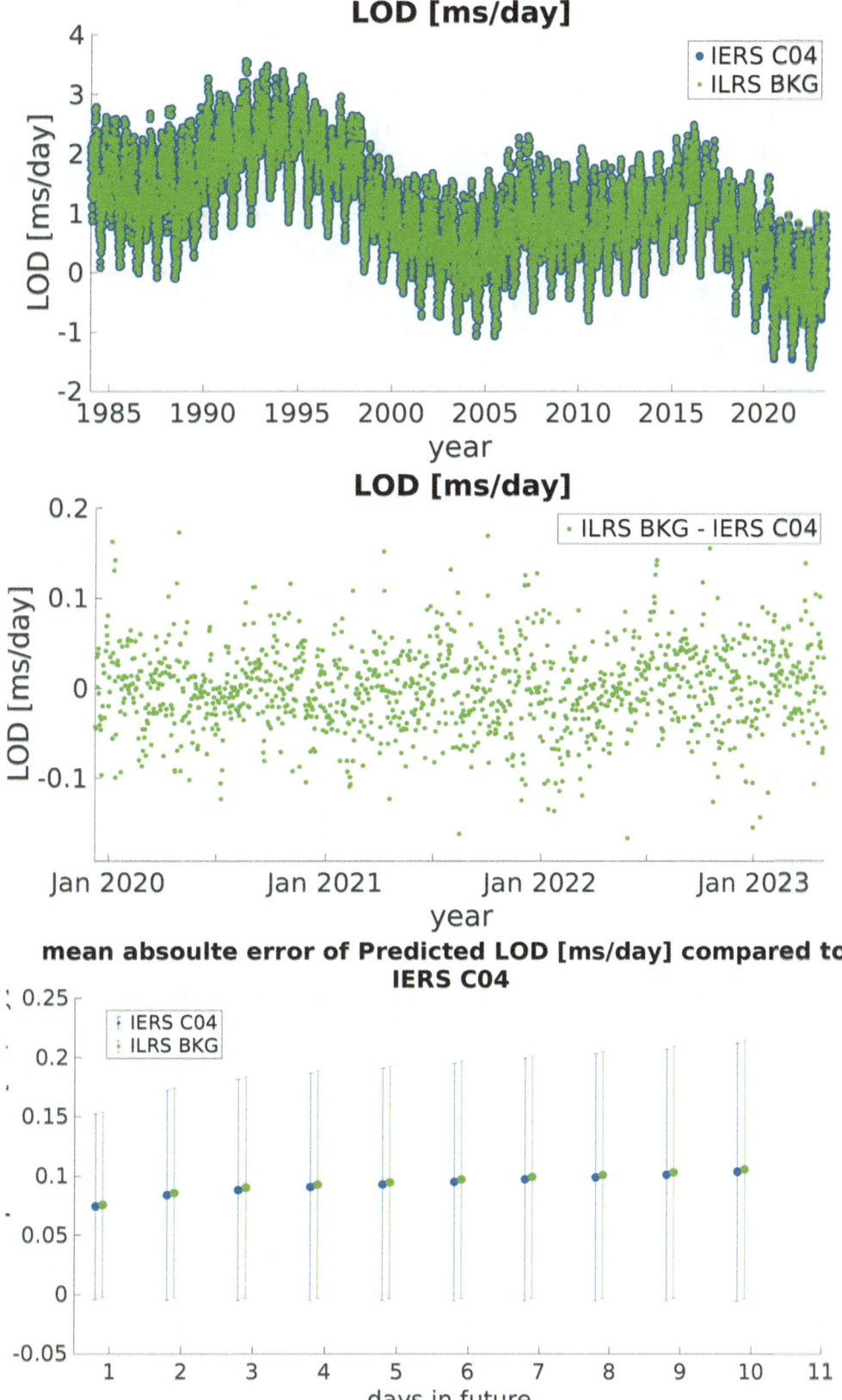

Fig. 4 Upper plot: LOD time series. Middle plot: difference between various BKG's EOP time series and IERS C04. Lower plot: Mean absolute error of the prediction results

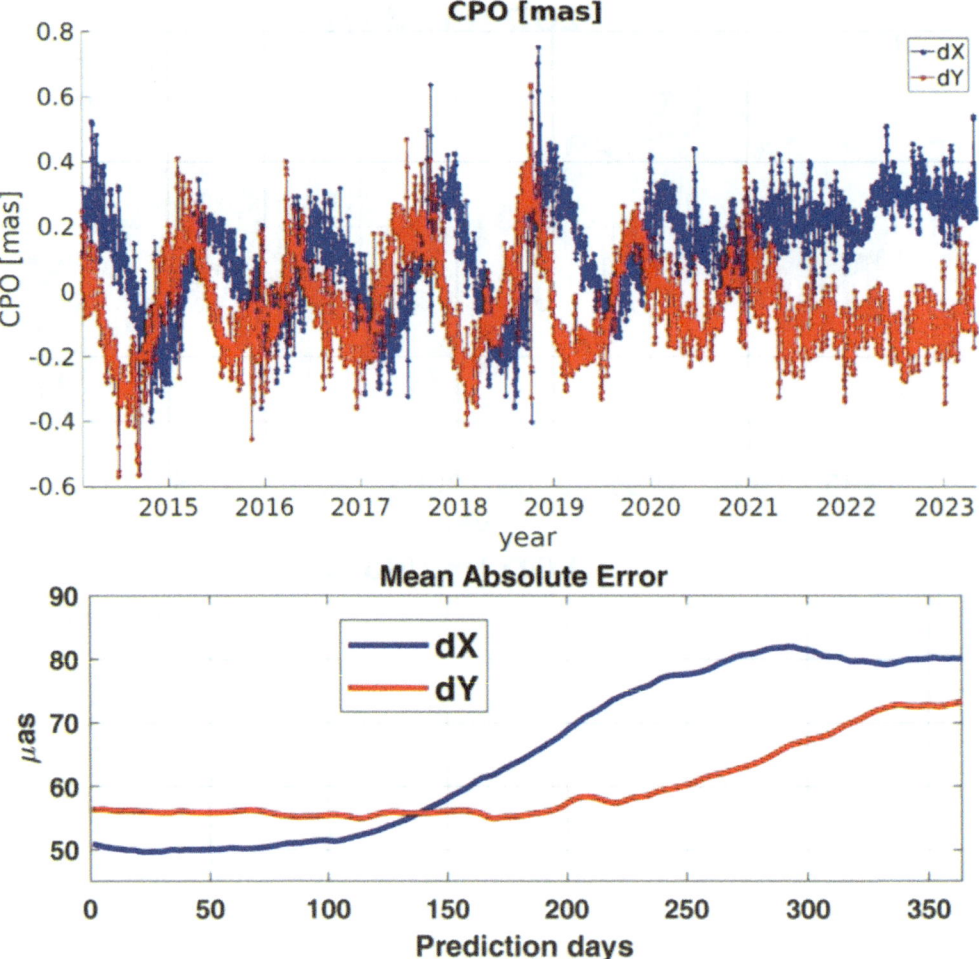

Fig. 5 Upper plot: Daily values of dX and dY. Lower plot: MAE of prediction. Units μas

Acknowledgements SB was partially supported by Generalitat Valenciana (SEJIGENT/2021/001), the European Union–NextGenerationEU (ZAMBRANO 21-04) and Ministerio de Ciencia e Innovación (MCIN/AEI/10.13039/501100011033/).

Conflict of Interest The authors declare that they have no conflict of interest.

References

2ndEOP PCC (2023) Operational phase predictions of the 2nd eop pcc. https://doi.org/10.5880/GFZ.1.3.2023.001

Belda S, Ferrándiz JM, Heinkelmann R, Nilsson T, Schuh H (2016) Testing a new free core nutation empirical model. J Geodynam 94:59–67

Belda S, Ferrándiz JM, Heinkelmann R, Schuh H (2018) A new method to improve the prediction of the celestial pole offsets. Scientific Reports 8(1):13861

Bizouard C, Lambert S, Gattano C, Becker O, Richard JY (2019) The iers eop 14c04 solution for earth orientation parameters consistent with itrf 2014. J Geodesy 93(5):621–633

Dobslaw H, Dill R (2018) Predicting earth orientation changes from global forecasts of atmosphere-hydrosphere dynamics. Adv Space Res 61(4):1047–1054

Ferrándiz JM, Modiri S, Belda S, Barkin M, Bloßfeld M, Heinkelmann R, Schuh H (2020) Drift of the earth's principal axes of inertia from grace and satellite laser ranging data. Remote Sens 12(2):314

Fodor C, Heinkelmann R, Modiri S, Raut S, Schuh H, Varga P (2019) On the mutual interrelation between earth rotation and earthquake activity. In: Proceedings of the Journées Systèmes de Référence Spatio-temporels 2019 "Astrometry, earth rotation and reference system in the Gaia era, pp 85–90

Guessoum S, Belda S, Ferrandiz JM, Modiri S, Raut S, Dhar S, Heinkelmann R, Schuh H (2022) The short-term prediction of length of day using 1d convolutional neural networks (1d cnn). Sensors 22(23):9517

Kalarus M, Kosek W, Akyilmaz O, Bizouard C, Gross R, Jovanović B, Kumakshev S, et al. (2010) Achievements of the earth orientation parameters prediction comparison campaign. J Geodesy 84:587–596

Kiani Shahvandi M, Schartner M, Soja B (2022) Neural ode differential learning and its application in polar motion prediction. J Geophys Res Solid Earth 127(11):e2022JB024775

Kiani Shahvandi M, Dill R, Kehm A, Bloßfeld M, Schartner M, Mishra S, Soja B (2023) Geophysically informed machine learning for improving rapid estimation and short-term prediction of earth orientation parameters. J Geophys Res Solid Earth 128(10)

Klemm L, Thaller D, Flohrer C, Walenta A, Ullrich D, Hellmers H (2024). Intra-technique combination of VLBI intensives and rapid data to improve the temporal regularity and continuity of the UT1-UTC series. In: International Association of Geodesy Symposia. Springer, Berlin, Heidelberg. https://doi.org/10.1007/1345_2023_235

Kur T, Dobslaw H, Śliwińska J, Nastula J, Wińska M, Partyka A (2022) Evaluation of selected short-term predictions of ut1-utc and lod collected in the second earth orientation parameters prediction comparison campaign. Earth Planets Space 74(1):1–9

Lengert L, Thaller D, Flohrer C, Hellmers H, Girdiuk A (2022) On the improvement of combined EOP series by adding 24-h VLBI sessions to VLBI intensives and GNSS data. Springer, Berlin, Heidelberg, pp 1–8. https://doi.org/10.1007/1345_2022_175

Ligas M, Michalczak M (2024) Dynamic mode decomposition and bivariate autoregressive short-term prediction of earth rotation parameters. J Appl Geodesy 18(2):211–221. https://doi.org/10.1515/jag-2023-0030

Malkin Z, Belda S, Modiri S (2022) Detection of a new large free core nutation phase jump. Sensors 22(16):5960

Michalczak M, Ligas M (2021) Kriging-based prediction of the earth's pole coordinates. J Appl Geodesy 15(3):233–241

Modiri S (2015) Copula-based analysis of correlation structures in case of GRACE coefficients. University of Stuttgart, Germany

Modiri S (2021) On the improvement of earth orientation parameters estimation: using modern space geodetic techniques. Technische Universitaet Berlin (Germany)

Modiri S, Belda S, Heinkelmann R, Hoseini M, Ferrándiz JM, Schuh H (2018) Polar motion prediction using the combination of ssa and copula-based analysis. Earth Planets Space 70(1):1–18

Modiri S, Belda S, Hoseini M, Heinkelmann R, Ferrándiz JM, Schuh H (2020) A new hybrid method to improve the ultra-short-term prediction of lod. J Geodesy 94:1–14

Modiri S, Heinkelmann R, Belda S, Malkin Z, Hoseini M, Korte M, Ferrándiz JM, Schuh H (2021) Towards understanding the interconnection between celestial pole motion and earth's magnetic field using space geodetic techniques. Sensors 21(22):7555

Namazi A (2022) On the improvement of heart rate prediction using the combination of singular spectrum analysis and copula-based analysis approach. PeerJ 10:e14601

Raut S, Modiri S, Heinkelmann R, Balidakis K, Belda S, Kitpracha C, Schuh H (2022) Investigating the relationship between length of day and el-nino using wavelet coherence method. In: Geodesy for a sustainable earth: Proceedings of the 2021 scientific assembly of the international association of Geodesy, Beijing, China, June 28–July 2, pp 253–258, 2021. Springer International Publishing, Cham

Schuh H, Ulrich M, Egger D, Müller J, Schwegmann W (2002) Prediction of earth orientation parameters by artificial neural networks. J Geodesy 76:247–258

Seitz F, Schuh H (2010) Earth rotation. Sciences of geodesy-I: advances and future directions, pp 185–227

Śliwińska-Bronowicz J, Kur T, Wińska M, et al. (2024) Assessment of length-of-day and universal time predictions based on the results of the Second Earth Orientation Parameters Prediction Comparison Campaign. J Geod 98:22. https://doi.org/10.1007/s00190-024-01824-7

Wlodzimierz H (1990) Polar motion prediction by the least-squares collocation method. In: Earth rotation and coordinate reference frames: Edinburgh, Scotland, August 10–11, pp 50–57, 1989. Springer, New York

Wu F, Chang G, Deng K (2021) One-step method for predicting lod parameters based on ls+ ar model. J Spat Sci 66(2):317–328

Xu X, Zhou Y (2015) Eop prediction using least square fitting and autoregressive filter over optimized data intervals. Adv Space Res 56(10):2248–2253

Open Access This chapter is licensed under the terms of the Creative Commons Attribution 4.0 International License (http://creativecommons.org/licenses/by/4.0/), which permits use, sharing, adaptation, distribution and reproduction in any medium or format, as long as you give appropriate credit to the original author(s) and the source, provide a link to the Creative Commons license and indicate if changes were made.

The images or other third party material in this chapter are included in the chapter's Creative Commons license, unless indicated otherwise in a credit line to the material. If material is not included in the chapter's Creative Commons license and your intended use is not permitted by statutory regulation or exceeds the permitted use, you will need to obtain permission directly from the copyright holder.

Part III
Gravity Field Modelling and Height Systems

On the Treatment of Static Gravity Field Signal for Time-Variable Gravity Field Recovery

Martin Lasser, Ulrich Meyer, Daniel Arnold, and Adrian Jäggi

Abstract

When estimating time-variable gravity field models from GRACE Follow-On data, a set of a priori given background force models is introduced in the processing to enable the computation of monthly snapshots of spherical harmonic coefficients representing the state of the Earth's gravity field. The to-be-estimated spherical harmonic series has to be truncated at a certain point, for GRACE Follow-On commonly at degree and order 96, and one of the background models is usually a model for the gravity field itself, which is used to reduce higher frequency static gravity field signal to avoid aliasing (contained in degrees above 96).

In this study we take a look on the influence of different strategies to treat the high degree gravity field signal in monthly gravity field solutions from GRACE Follow-On data. We estimate temporal gravity fields with fixed high degrees of different a priori background gravity field models, and opposed to this, we also co-estimate static spherical harmonic coefficients from degree 97 up to degree and order 160 from 51 months of GRACE Follow-On data along with the monthly snapshots to enable a consistent handling of correlations between time-variable and static gravity field coefficients. The observation noise modelling of the data is handled by an empirical covariance estimation for the noise based on post-fit residuals between the final GRACE Follow-On orbits, that are co-estimated together with the gravity field, and the observations. Since the post-fit residuals, amongst other things, depend on the choice of the background force models they are a potential carrier of a priori information into the final solution.

The results show that a formal correlation between the time-variable and static gravity field coefficients is almost non-existent, and also the empirical covariance model has only minor impacts on this correlation. Only a poor choice of the background gravity field requires a prior or co-estimation of the static gravity field.

Keywords

GRACE · GRACE follow-On · Gravity field · Gravity field recovery · Static gravity field · Time-variable gravity field

1 Introduction

The successor mission of the Gravity Recovery and Climate Change Mission (GRACE, Tapley et al. 2004) GRACE Follow-On (Landerer et al. 2020) prolongs the unprecedented insights into global scale mass distribution and their

M. Lasser (✉) · U. Meyer · D. Arnold · A. Jäggi
Astronomical Institute, University of Bern, Bern, Switzerland
e-mail: martin.lasser@unibe.ch; ulrich.meyer@unibe.ch; daniel.arnold@unibe.ch; adrian.jaeggi@unibe.ch

development in time of the dynamic system Earth by precisely measuring variations in satellite orbital trajectories, which are caused by changes in the underlying potential field, i.e., by mass re-distributions, to ultimately recover the latter.

The main observable for tracking the relative orbital trajectory between the satellite pair of GRACE Follow-On are inter-satellite K-band range-rates. These range-rates are sensitive to the full mass distribution of the Earth, and by implication, the relative motion of both GRACE Follow-On satellites. In practise, gravity field modelling from GRACE Follow-On data is implemented by computing monthly snapshots which represent the mean state of the Earth's gravity field in each month. Associated to this, the formulation of the problem has to take into account all gravitational effects which cannot be sufficiently recovered by one month of range-rate observations, and, therefore, would lead to signal aliasing. Typically, the forces acting on the satellites caused by Sun, Moon and planets, as well as relativistic effect, all kinds of tides and atmospheric and oceanic de-aliasing are reduced as to the extent possible by introducing so called *background force models*. Non-gravitational effects are considered to be known from the onboard accelerometer measurements.

The to-be-estimated gravity field is commonly expanded into a spherical harmonic series (Heiskanen and Moritz 1967), which is truncated at a certain point, typically at degree an order (d/o) 96 (see e.g., University of Texas Center for Space Research 2018). Higher degree and order gravity field coefficients are fixed to a set of a priori values and are, therefore, part of the background force models.

In this investigation, we pose the question about the influence of the high degrees of different a priori background gravity field models on the monthly gravity field models recovered from GRACE Follow-On data: Does the a priori chosen gravity field influence our monthly solutions or can we do better by co-estimating monthly solutions (up to d/o 96) together with a static component (degrees > 96)? We compare between two main scenarios:

- Introducing a priori background gravity field information for degrees > 96, and
- rigorously co-estimating high degree (> 96) static gravity field components (from GRACE Follow-On data only) along with monthly snapshots (up to d/o 96) to enable a consistent handling of correlations between time-variable (monthly) and static gravity field coefficients.

In addition, to further quantify the influence of the high degrees of the a priori background gravity field model and account for recent modelling strategies, which reflect the choice of background models, the stochastic noise of the data is modelled with an empirical covariance estimation of the noise based on the post-fit residuals. They are computed between the final GRACE Follow-On orbits, that are co-estimated together with the gravity field, and the observations. Post-fit residuals are expressed in position residuals to the kinematic positions and in K-band range-rate residuals.

2 Gravity Field Recovery from GRACE Follow-on Data with the CMA

The orbit and gravity field recovery process is implemented as one common least-squares adjustment using the Celestial Mechanics Approach (CMA, Beutler et al. 2010), developed at the Astronomical Institute of the University of Bern (AIUB). The processing is based on daily arcs for which a set of orbital elements as well as accelerometer bias and scale factors are estimated. Additionally, constrained piecewise-constant accelerations (pseudo-stochastic parameters), which are set up in regular intervals of 15 min, are jointly estimated to compensate for deficiencies in the background force modelling. The observation noise modelling employs the assumption of white noise for the K-band range-rates and a propagation of white noise for carrier phase observations through a kinematic point positioning process. Gravity field parameters are set up for each arc, however, then stacked to monthly normal equations.

GRACE Follow-On Level-1B data (Wen et al. 2019; NASA Jet Propulsion Laboratory 2019) for the time span of June 2018 through October 2022 (51 months) are included in this study. Data from the K-band ranging system (range-rates) together with kinematic positions (Arnold and Jäggi 2022) are introduced as observations and pseudo-observations, respectively. Non-gravitational perturbing forces are considered error-free as measured by the onboard accelerometer by the GRACE Follow-On-1 satellite (GF1) and from the transplant product of Behzadpour et al. (2021) for GF2. Attitude information stems from the Level-1B data as well. The background force modelling includes the reduction of the effect of gravitational perturbing forces on the satellites and treats the direct influence of Sun, Moon and planets as well as tidal variations for the solid Earth (Petit and Luzum 2010), the poles (Petit and Luzum 2010; Desai 2002), in the oceans (Carrere et al. 2016, introduced up to d/o 100) and atmosphere and non-tidal short term mass variations (Dobslaw et al. 2017, introduced up to d/o 100). The background gravity field model is introduced up to d/o 160, however, the monthly gravity field coefficients are only estimated up to d/o 96. Three different background gravity field models are investigated in this study:

- The Earth Gravitational Model 1996 (EGM96, Lemoine et al. 1998), which can be considered as outdated, since it was published before satellite missions dedicated to sensing the Earth's gravity field were launched,
- the AIUB-GRACE03S model (Jaeggi et al. 2011), representing the mean state of the gravity field in the time

frame 2004–2010 (static model), which was derived from GRACE data. This model is commonly used as background gravity field model in the GRACE and GRACE Follow-On gravity field models published by the AIUB (Darbeheshti et al. 2023; Lasser et al 2020; Meyer et al. 2016) and

- the GOCO06s (Kvas et al. 2021) computed in the GOCO (Gravity Observation Combination) project combining 15 years of data from 19 satellites. This model is composed of a static element and time-variable component, expressed by secular (trend) and annual variations. The full static and time-variable model is employed in this study.

To refine the modelling procedure, the noise in the observations is approximated by estimating a covariance function $\hat{C}$ with the means of serial correlation for lags up to $K = 3\,\mathrm{h}$ of the post-fit residuals $\hat{e}$ while assuming that they follow a stationary process

$$\hat{C}(\Delta_k) = \frac{1}{N} \sum_{n=1}^{N-k} \hat{e}(t_{n+k}) \hat{e}(t_n)$$
$$\text{with} \quad k \in \{0, \ldots, K\} \ , \quad (1)$$

with Δ_k defining the lag between epoch t_n and epoch t_{n+k} and N being the number of data points. A covariance matrix may be compiled $\hat{C}$ in block Toeplitz structure. The inverted covariance matrix is then used in a further orbit and gravity field recovery step as weight matrix for the corresponding observations. The complete process may be found in Ellmer (2018) and was introduced by Lasser (2022) to the gravity field recovery with the CMA. With this approach, an iterative treatment of the problem is necessary, including the task to determine a reliable initial estimate of the gravity field in order not to introduce information from the a priori gravity field via the estimated covariance matrix into the final solution. Note that this approach might amplify the dependency on the background models since the post-fit residuals already depend on the choice of the background force models.

3 Co-estimation of Static and Time-Variable Gravity Fields

In a typical monthly processing scheme, daily Normal Equations (NEQs), implicitly containing orbit, accelerometer and pseudo-stochastic parameters, which were pre-eliminated, and explicitly containing the monthly gravity field parameters (up to d/o 96), are stacked for one month and solved to obtain an estimate for a monthly gravity field model.

When jointly estimating both, the time-variable (monthly, up to d/o 96, in the vector $\hat{y}$) and static components (from degree 97 up to d/o 160, in the vector $\hat{x}$), monthly NEQs containing gravity field coefficients up to d/o 160 are set up. In principle one could now stack all monthly NEQs in a way where the entries related to the time-variable coefficients are not accumulated but only the entries related to the static coefficients. This however, would lead to a NEQ system with 496,167 unknown parameters. For the sake of efficiency, we pre-eliminate (see e.g., Brockmann 1996) the monthly components ($\hat{y}$) of each NEQ by extracting the solution for $\hat{y}$ and assuming that the normal equation matrix related to the to-be-eliminated parameters $\mathbf{N}_{\hat{y}\hat{y}}$ is nondegenerate

$$\hat{y} = \mathbf{N}_{\hat{y}\hat{y}}^{-1} \left(\boldsymbol{b}_{\hat{y}} - \mathbf{N}_{\hat{x}\hat{y}}^{T} \hat{x} \right) \ , \quad (2)$$

where $\mathbf{N}_{(\cdot,\cdot)}$ denotes the normal equations matrix related to the parameters in the subscript and $\boldsymbol{b}_{(\cdot)}$ is the corresponding right-hand side of the normal equations. The full normal equation system may then be re-formulated to a parameter space only containing the static gravity field coefficients explicitly by

$$\left(\mathbf{N}_{\hat{x}\hat{x}} - \mathbf{N}_{\hat{x}\hat{y}} \mathbf{N}_{\hat{y}\hat{y}}^{-1} \mathbf{N}_{\hat{x}\hat{y}}^{T} \right) \hat{x} = \boldsymbol{b}_{\hat{x}} - \mathbf{N}_{\hat{x}\hat{y}} \mathbf{N}_{\hat{y}\hat{y}}^{-1} \boldsymbol{b}_{\hat{y}} \ . \quad (3)$$

Stacking all NEQs, where only the static coefficients are explicitly contained (Eq. 3), and solving yields the solution of the static gravity field part. The time-variable components are available by back-substitution of $\hat{x}$ into Eq. 2. However, for a more efficient handling within the software and direct access to covariances, the back-substitution process is omitted and for each month two NEQs are created: One containing all parameters and one where the time-variable part is pre-eliminated and only the static part remains explicitly. The solution of each month, where the time-variable and static component of the gravity field are co-estimated, is then obtained by stacking all monthly NEQs with only the static coefficients, except the one of the month of interest. For this month the full NEQ (time-variable and static coefficients explicitly available) is stacked to the others. The resulting NEQ is then solved. This procedure is repeated for each month. Figure 1 gives an overview about the structure of the design matrix and parameters involved.

4 Results

In total six major test scenarios, listed in Table 1, were investigated in detail, introducing three different a priori gravity field models and making use of an empirical covariance modelling based on serial correlation of post-fit residuals (see Eq. 1) and jointly estimating static gravity field coefficients together with the time-variable ones. Scenarios #4b and #5b include the empirical covariance model before the co-

Table 1 Test scenarios investigated in this study of background gravity models, applied covariance modelling strategies and co-estimation of time-variable and static gravity field coefficients. The flag 'no/yes' indicates that both solutions were created

Scenario #	Gravity field model	Remark	Covariance model	Co-estimation
1[a,b]	AIUB GRACE03S	Static	None	No [a] / Yes [b]
2[a,b]	GOCO06s	Time-variable	None	No [a] / Yes [b]
3[a,b]	EGM96	Static	None	No [a] / Yes [b]
4[a,b]	AIUB GRACE03S	Static	Empirical	No [a] / Yes [b], after covariance model is applied
5[a,b]	GOCO06s	Time-variable	empirical	No [a] / Yes [b], after covariance model is applied
6	AIUB GRACE03S	Static	Empirical	Yes

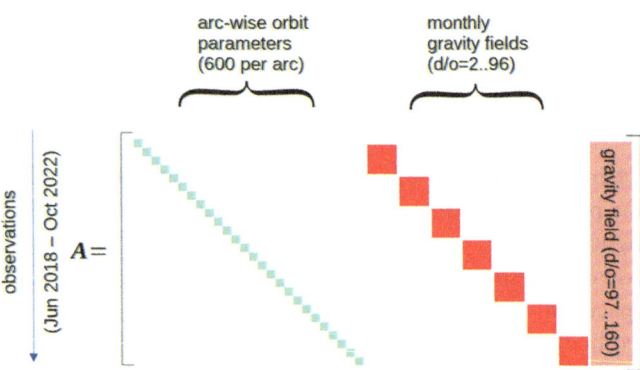

Fig. 1 Sketch of the design matrix with the parameter set up scheme of the co-estimated monthly and static gravity field coefficients

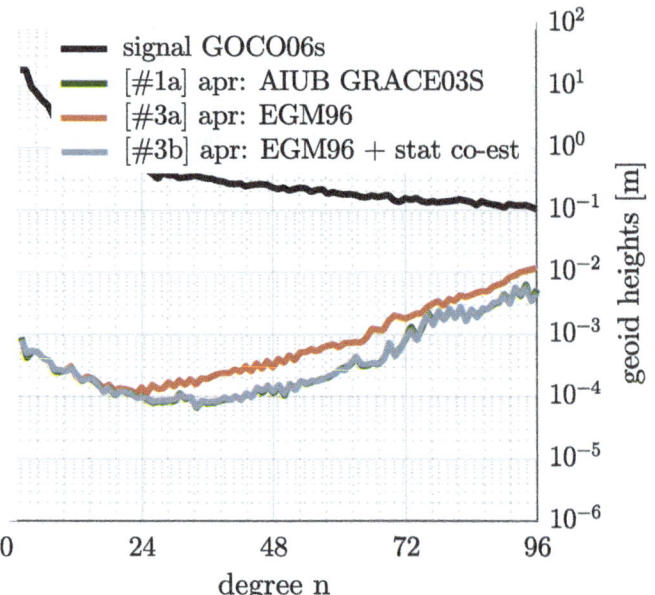

Fig. 2 Difference degree amplitudes for May 2020 w.r.t. GOCO06s when starting with EGM96 as a priori gravity field model. The light blue curve shows the solution when a static component of the gravity field is co-estimated, whereas red depicts the solution without co-estimation. The green curve (overlapped by light blue) was created with the better AIUB GRACE03S a priori gravity field without co-estimation of high degree coefficients

estimation of the static degrees of the gravity field, where, thus, the covariance model depends on the high degrees (> 96) of the a priori background gravity field. Scenario #6, in contrast, is implemented with the co-estimation before the covariance model is derived, then takes a thereof based empirical covariance model into account and again makes use of the co-estimation of the time-variable and static gravity field coefficients. The quality assessment of the co-estimated static gravity field when starting with different a priori gravity field models (AIUB GRACE03S, GOCO06s, EGM96) and applying the different covariance modelling techniques (none, empirical covariance model) shows two groups based on the covariance modelling technique. Referencing to the GOCE TIM R6 model (Brockmann et al. 2019) as independent source of information about the static gravity field, the co-estimated static components feature a RMS of 7.47 ± 0.02 cm in terms of geoid height differences when not applying any dedicated covariance modelling technique (scenario #1–3) and 7.47 ± 0.02 cm when introducing an empirical covariance model (scenario #4–6).

The main impact of co-estimating static gravity field coefficients may be seen by using the outdated EGM96 model, which, by nowadays standards, is considered of rather limited quality. Comparing difference degree amplitudes for May 2020 (Fig. 2) between the GOCO06s model as reference signal (any reasonably up-to-date reference is sufficient here) and taking the EGM96 as a priori gravity field either without or with co-estimating the static components yields an significant improvement for the latter (light blue). However, co-estimation is not the only means to achieve a solution of good quality. Taking a better a priori gravity field (as the AIUB GRACE03S with the green line from Lasser et al (2020)), which was determined beforehand in a separate gravity field recovery process, yields an almost identical solution.

As a quality measure for the estimated solutions the RMS of geoid height differences over the open oceans w.r.t. a mean field, which was derived by estimating from all 51 months for each coefficient a model composed of a mean, a trend and an annual and semi-annual signal, is computed. Results (Fig. 3, with the reference of the monthly solutions from the Center of Space Research of the University of Texas (CSR, Save 2019) and Graz University of Technology (TUG, Mayer-Gürr et al. 2018)) show that the solutions are all of a similar quality, rather independent of the a priori gravity field and independent of co-estimating a static

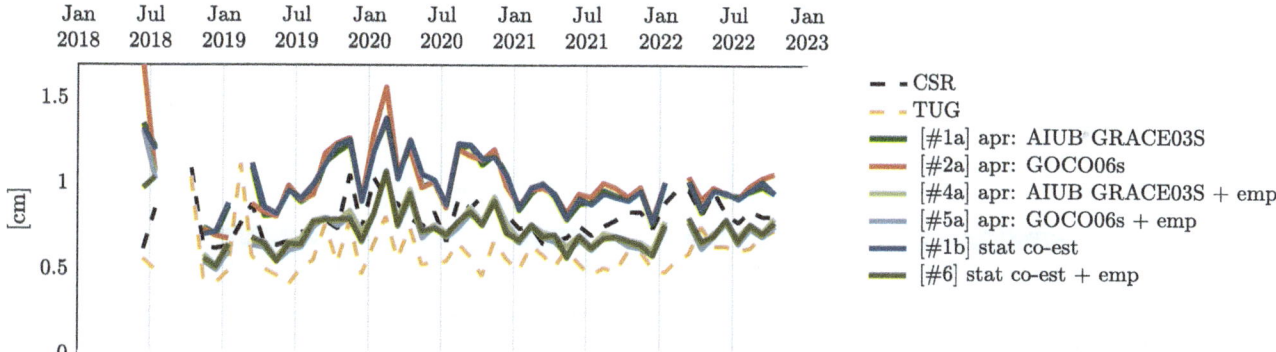

Fig. 3 RMS of geoid height differences w.r.t. a mean model over the open oceans for test cases of different a priori gravity field models, with and without empirical covariance modelling and co-estimation of the static gravity field. The solutions from CSR and TUG are drawn for a general classification

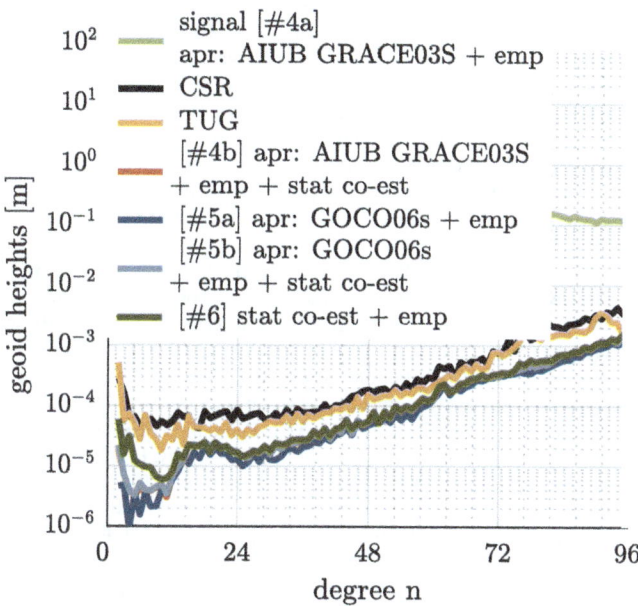

Fig. 4 Difference degree amplitudes for May 2020 w.r.t. the solution using the AIUB GRACE03S model as a priori gravity field with empirical covariance modelling. The difference degree amplitudes of the individual solutions with and without co-estimating the static coefficients while applying empirical covariance modelling scheme are well below the differences w.r.t. the solutions from CSR and TUG

gravity field together with the time-variable components. The main difference emerges from the processing strategies: The adopted covariance modelling technique reduces the RMS over the oceans significantly.

The differences in the spectral domain between individual solutions (only shown for the case of empirical covariance modelling) of May 2020 are depicted in Fig. 4. The solution using the AIUB GRACE03S as a priori gravity field serves as a reference, the solutions of CSR and TUG are given as quality indicator to assess what could be expected from different analysis strategies.

The differences of solutions starting from different a priori gravity field models and applying the covariance modelling are well below the references from CSR and TUG, showing that the chosen background gravity field model and/or co-estimation of static coefficients only have a minor impact.

The process of not straightforwardly obtaining the jointly estimated time-variable gravity field coefficients via back-substitution but by computing a full solution for each month allows to directly access the covariances and thereof derived linear correlation coefficients (Pearson 1895) between the time-variable and static coefficients. Figure 5 shows the mean correlation coefficients between each time-variable spherical harmonics coefficient and all static coefficients for May 2020, chosen as month in the middle of the data period, thus, close to the 'epoch' of the co-estimated static gravity field. The mean correlation is almost zero, however, slightly larger when applying the empirical covariance modelling, which is an indication that the post-fit residuals transport some information about the high degrees as well.

The results are most likely extrapolable to GRACE data as well, however, one has to keep in mind that GRACE was flying lower than GRACE Follow-On in the test years of 2018–2022 for most of its lifetime. Thus, the sensitivity of GRACE to high degrees (>96) was better, implicating also a (slightly) higher dependency on the background force models covering these spectra.

5 Conclusions

In this study we have tested different a priori background gravity field models for their impact in time-variable gravity field recovery from GRACE Follow-On data. We included empirical covariance models derived from post-fit residuals into the investigation to introduce a quantity which additionally reflects background models choices and computed monthly gravity field solutions where high degrees of the gravity field where either fixed to a priori values from different gravity field models or co-estimated together with

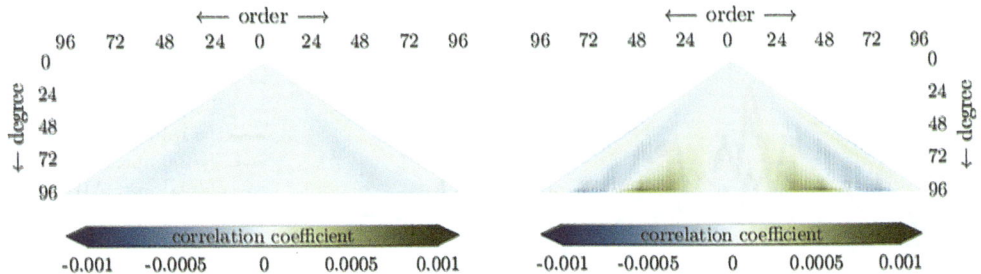

Fig. 5 Formal mean correlation coefficients for May 2020 between each time-variable spherical harmonics coefficient and all static coefficients for a solution without empirical covariance modelling (left) and including covariance modelling based on post-fit residuals (right)

the monthly gravity fields from 51 month of GRACE Follow-on data.

The outcomes show that the dependency between the time-variable and static gravity field coefficients is limited – almost zero when expressed in correlation coefficients between a co-estimated static gravity field solution – and also the empirical covariance model has only minor impacts. In conclusion, based on GRACE Follow-On data introducing a rather recent background gravity field model is sufficient for estimating time-variable monthly gravity field models. The (cumbersome) co- or prior estimation of a static gravity field solution is only indispensable when having poor or limited knowledge about the gravity field beyond the time-variable spectrum.

Supplementary Information GRACE Follow-On Level-1B data is obtained from https://podaac.jpl.nasa.gov/dataset/GRACEFO_L1B_ASCII_GRAV_JPL_RL04. The data may be found and referred to under NASA Jet Propulsion Laboratory (2019) and Wen et al. (2019).

Declarations The authors declare that they have no conflict of interest. This research was supported by the European Research Council under the grant agreement no. 817919 (project SPACE TIE). All views expressed are those of the authors and not of the European Research Council. Calculations were performed on UBELIX (http://www.id.unibe.ch/hpc), the HPC cluster at the University of Bern.

References

Arnold D, Jäggi A (2022) AIUB GRACE-FO kinematic orbits, release 02. Astronomical Institute, University of Bern, dataset. https://doi.org/10.48350/169040

Behzadpour S, Mayer-Gürr T, Krauss S (2021) GRACE follow-on accelerometer data recovery. J Geophys Res Solid Earth 126(5). https://doi.org/10.1029/2020JB021297

Beutler G, Jäggi A, Mervart L, et al (2010) The celestial mechanics approach: theoretical foundations. J Geodesy 84(10):605–624. https://doi.org/10.1007/s00190-010-0401-7

Brockmann E (1996) Combination of solutions for geodetic and geodynamic applications of the Global Positioning System (GPS). PhD thesis, University of Bern, Bern, Switzerland

Brockmann JM, Schubert T, Mayer-Gürr T, et al (2019) The Earth's gravity field as seen by the GOCE satellite – an improved sixth release derived with the time-wise approach (GO_CONS_GCF_2_TIM_R6). GFZ Data Services, dataset. https://doi.org/10.5880/ICGEM.2019.003

Carrere L, Lyard F, Cancet M, et al (2016) FES 2014, a new tidal model – Validation results and perspectives for improvements. In: ESA Living Planet Conference, Prague, Czech Republic

Darbeheshti N, Lasser M, Meyer U, et al (2023) AIUB-G3P GRACE monthly gravity field solutions. GFZ Data Services, dataset. https://doi.org/10.5880/icgem.2023.001

Desai SD (2002) Observing the pole tide with satellite altimetry. J Geophys Res Oceans 107(C11):1–13. https://doi.org/10.1029/2001JC001224

Dobslaw H, Bergmann-Wolf I, Dill R, et al (2017) A new high-resolution model of non-tidal atmosphere and ocean mass variability for de-aliasing of satellite gravity observations: AOD1B RL06. Geophys J Int 211(1):263–269. https://doi.org/10.1093/gji/ggx302

Ellmer M (2018) Contributions to GRACE Gravity Field Recovery: Improvements in Dynamic Orbit Integration Stochastic Modelling of the Antenna Offset Correction, and Co-Estimation of Satellite Orientations. PhD thesis, Graz University of Technology, In: Monographic Series TU Graz, number 1, Verlag der Technischen Universität Graz, Graz, Austria. https://doi.org/10.3217/978-3-85125-646-8

Heiskanen WA, Moritz H (1967) Physical Geodesy. W. H. Freeman and Company, San Francisco, CA, USA

Jaeggi A, Meyer U, Beutler G, et al (2011) AIUB-GRACE03S: A static gravity field model computed with simultaneously solved-for time variations from 6 years of GRACE data using the Celestial Mechanics Approach. Paper in preparation. http://icgem.gfz-potsdam.de/tom_longtime

Kvas A, Brockmann JM, Krauss S, et al (2021) GOCO06s – a satellite-only global gravity field model. Earth Syst Sci Data 13(1):99–118. https://doi.org/10.5194/essd-13-99-2021

Landerer FW, Flechtner FM, Save H, et al (2020) Extending the global mass change data record: GRACE follow-on instrument and science data performance. Geophys Res Lett 47(12). https://doi.org/10.1029/2020GL088306

Lasser M (2022) Noise Modelling for GRACE Follow-On Observables in the Celestial Mechanics Approach. PhD thesis, University of Bern, Bern, Switzerland. https://boristheses.unibe.ch/id/eprint/4127

Lasser M, Meyer U, Arnold D, et al (2020) AIUB-GRACE-FO-operational — Operational GRACE Follow-On monthly gravity field solutions. GFZ Data Services, dataset. https://doi.org/10.5880/icgem.2020.001

Lemoine FG, Kenyon SC, Factor JK, et al (1998) The Development of the Joint NASA GSFC and NIMA Geopotential Model EGM96. NASA Goddard Space Flight Center, NASA Goddard Space Flight Center, Greenbelt, MD, USA

Mayer-Gürr T, Behzadpur S, Ellmer M, et al (2018) ITSG-Grace2018 - Monthly, Daily and Static Gravity Field Solutions from GRACE. GFZ Data Services, dataset. https://doi.org/10.5880/ICGEM.2018.003

Meyer U, Jäggi A, Jean Y, et al (2016) AIUB-RL02: an improved time-series of monthly gravity fields from GRACE data. Geophys J Int 205(2):1196–1207. https://doi.org/10.1093/gji/ggw081

NASA Jet Propulsion Laboratory (2019) GRACE-FO Level-1B Release version 4.0 from JPL in ASCII. NASA Physical Oceanography Distributed Active Archive Center, dataset. https://doi.org/10.5067/GFL1B-ASJ04

Pearson K (1895) Notes on Regression and Inheritance in the case of Two Parents. In: Proceedings of the Royal Society of London, vol 58, pp 240–242. Royal Society of London, London, United Kingdom

Petit G, Luzum B (2010) IERS Conventions (2010), IERS Technical Note No. 36. Verlag des Bundesamts für Kartographie und Geodäsie, Frankfurt am Main, Germany

Save H (2019) CSR Level-2 Processing Standards Document For Level-2 Product Release 06. Processing Standards Document, The University of Texas at Austin, Austin, TX, USA. https://podaac.jpl.nasa.gov/gravity/gracefo-documentation

Tapley BD, Bettadpur S, Watkins MM, et al (2004) The gravity recovery and climate experiment: Mission overview and early results. Geophys Res Lett 31(9). https://doi.org/10.1029/2004GL019920

University of Texas Center for Space Research (2018) GRACE STATIC FIELD GEOPOTENTIAL COEFFICIENTS CSR RELEASE 6.0. NASA Physical Oceanography Distributed Active Archive Center, dataset. https://doi.org/10.5067/GRGSM-20C06

Wen HY, Kruizinga G, Paik M, et al (2019) Gravity Recovery and Climate Experiment Follow-On (GRACE-FO) Level-1 Data Product User Handbook. Technical Report JPL D-56935 (URS270772), NASA Jet Propulsion Laboratory/California Institute of Technology, Pasadena, CA, USA. https://doi.org/doi.org/10.5067/GFL1B-ASJ04

Open Access This chapter is licensed under the terms of the Creative Commons Attribution 4.0 International License (http://creativecommons.org/licenses/by/4.0/), which permits use, sharing, adaptation, distribution and reproduction in any medium or format, as long as you give appropriate credit to the original author(s) and the source, provide a link to the Creative Commons license and indicate if changes were made.

The images or other third party material in this chapter are included in the chapter's Creative Commons license, unless indicated otherwise in a credit line to the material. If material is not included in the chapter's Creative Commons license and your intended use is not permitted by statutory regulation or exceeds the permitted use, you will need to obtain permission directly from the copyright holder.

Analysis of Novel Sensors and Satellite Formation Flights for Future Gravimetry Missions

Alexey Kupriyanov, Arthur Reis, Annike Knabe, Nina Fletling, Alireza HosseiniArani, Mohsen Romeshkani, Manuel Schilling, Vitali Müller, and Jürgen Müller

Abstract

Accelerometers (ACCs) in low-low satellite-to-satellite gravimetry missions measure the non-gravitational forces acting on the spacecraft that have to be taken into account to derive the gravitational contribution in the distance variations. Multiple ACCs form a so-called gradiometer that measure the gravity gradient. In satellite gravimetry up to now, only electrostatic ACCs were used, which are one of the main instrumental limitations due to their error contribution at low frequencies, known as drift.

In this paper, we compare the performance of electrostatic ACCs at low Earth orbits with other sensors, i.e. so-called Optical ACCs based on flight heritage of the LISA-Pathfinder mission, and theoretical ACC concepts, for example Cold Atom Interferometer (CAI) ACCs and hybridized sensors (combination of electrostatic and CAI ACCs) in terms of static gravity field recovery. Under our assumptions, in particular that high-frequency variations of the gravity field can be perfectly modeled and removed during gravity field recovery, the results may be limited in the future by the performance of the LRI.

We also discuss the outcomes from the various novel satellite formation flights (SFF) that utilize two orbits that differ either by right ascension of the ascending node (RAAN) or by inclination in order to acquire ranging information in the cross-track direction. The closed-loop simulations from both scenarios showed significantly lower order of magnitude of the residuals w.r.t. reference gravity field than from the anticipated future performance of the solely in-line GRACE-like satellite pair. Moreover, these triple satellite formations provide better multi-directionality of the retrieved data, avoiding the North-South striping behavior. However, it is worth noting that in such formations significant modifications are needed in the satellite bus, ACC test mass readout, LRI beam steering mechanism, etc. in order to be capable of measuring the cross-track range changes at higher range rates w.r.t.

A. Kupriyanov (✉) · A. Knabe · N. Fletling · A. HosseiniArani · M. Romeshkani · J. Müller
Institute of Geodesy, Leibniz University Hannover, Hannover, Lower Saxony, Germany
e-mail: kupriyanov@ife.uni-hannover.de; knabe@ife.uni-hannover.de; fletling@ife.uni-hannover.de; hosseiniarani@ife.uni-hannover.de; romeshkani@ife.uni-hannover.de; mueller@ife.uni-hannover.de

A. Reis · V. Müller
Max Planck Institute for Gravitational Physics (IGP), Albert Einstein Institute, Hannover, Lower Saxony, Germany

Institute for Gravitational Physics, Leibniz University Hannover, Hannover, Lower Saxony, Germany
e-mail: arthur.reis@aei.mpg.de; vitali.mueller@aei.mpg.de

M. Schilling
Institute for Satellite Geodesy and Inertial Sensing, German Aerospace Center (DLR), Hannover, Lower Saxony, Germany
e-mail: manuel.schilling@dlr.de

© The Author(s) 2024
J. T. Freymueller, L. Sánchez (eds.), *Together Again for Geodesy*,
International Association of Geodesy Symposia 157, https://doi.org/10.1007/1345_2024_279

in-line GRACE-like configuration. In addition, a substantial reduction of costs in building and launching only three satellites rather than four as in double-pair constellations could be an advantage for such formations.

Keywords

Accelerometer · Future gravimetry missions · Gradiometer · Satellite formation flights

1 Introduction

Dedicated satellite gravimetry missions such as CHAMP (Mehta et al. 2017; Torge et al. 2023), GOCE (Flechtner et al. 2021), GRACE (Chen et al. 2022; Panet et al. 2022), its successor GRACE-FO (Peidou et al. 2022) and a Chinese satellite gravimetry pair (Xiao et al. 2023) provide unique data about Earth's gravity field variations at different spatio-temporal scales. However, despite the impressive results from the above-mentioned missions with unprecedented accuracy at the time when they were obtained, there was a common limiting factor on the instrument level. This well-known limitation results from the behaviour of the electrostatic accelerometer (EA), which shows a so-called drift in the low frequency domain (Frommknecht et al. 2003; van Camp et al. 2021), mainly caused by the polarization wire that connects the test mass to the surrounding electrode housing and is a significant source of stiffness (Christophe et al. 2015).

In order to overcome the drawback of EAs and better satisfy the user needs discussed by Pail et al. (2015), various concepts and novel technologies were developed. For example, enhanced EAs with modified test mass parameters were analyzed by the French aerospace lab ONERA (Liorzou et al. 2023). Another technical improvement that has been preliminary evaluated by ONERA (Boulanger et al. 2020) is the use of an EA without a polarization wire. It was substituted by a wireless charge management system utilizing ultraviolet light, which excites and expels extra electrons and keeps the electrostatic noise sources at low frequencies at an acceptable level (Sumner et al. 2020). This technology has flight heritage since it was on-board the LISA-Pathfinder mission (Armano et al. 2021) where optical accelerometers, also known as Gravitational Reference Sensor (GRS), were also tested in space for the first time (Armano et al. 2018). Promising results of the GRS from the LISA-Pathfinder gave a start to various studies that evaluate the performance of a simplified-GRS (SGRS) for low Earth orbits. SGRS is an enhanced EA with a free-floating cubic shaped test mass, with capacitive position readout sensing and without polarization wire proposed by Dávila Álvarez et al. (2022) and LISA-like SGRS by Weber et al. (2022). SGRS with a wider range of parameters, electrostatic and optical readouts of the test mass displacements was modeled and evaluated by Kupriyanov et al. (2024).

A totally different technology that could overcome the drawback of EAs at low frequencies is the Cold Atom Interferometry (CAI) where atom clouds act as test masses (Alonso et al. 2022). In CAI accelerometry, the unknown acceleration is calculated from the phase shift of two interfering atomic states of an atom cloud after it was manipulated with pulses of two counter propagating laser beams. Corresponding studies evaluating CAI accelerometry were done by Knabe et al. (2022) and for the usage in gradiometry by Trimeche et al. (2019). However, the utilization of a CAI accelerometer as a standalone instrument has certain drawbacks due to the long interrogation times in which affected short-term non-gravitational forces could not be observed. A possibility to solve this problem is the application of hybridization. Here, the idea is to combine the EA, which are able to measure in the high frequency domain, with the precise CAI ACC. HosseiniArani et al. (2022) and Zahzam et al. (2022) studied hybrid sensors and different ways of hybridization.

It is important to note, that in contrast to EA or optical accelerometers, CAI has not yet a flight heritage in space. However, the recently started CARIOQA-PMP (Cold Atom Rubidium Interferometer in Orbit for Quantum Accelerometry – Pathfinder Mission Preparation) project, funded by the European Union, aims to increase the Technology Readiness Level and prepare a Quantum Pathfinder Mission for space gravimetry by 2030 (Lévèque et al. 2023).

2 Performance of Sensors

2.1 Comparison of Accelerometer Performance

In Fig. 1 the Amplitude Spectral Density (ASD) of various ACCs types are shown w.r.t. typical non-gravitational accelerations (black curve). The above-mentioned drift of the EA that is used in GRACE-FO (Daras and Pail 2017) (grey curve) is very prominent in the frequencies below 1 mHz. Another EA which was used in GOCE mission has a similar

Analysis of Novel Sensors and Satellite Formation Flights for Future Gravimetry Missions

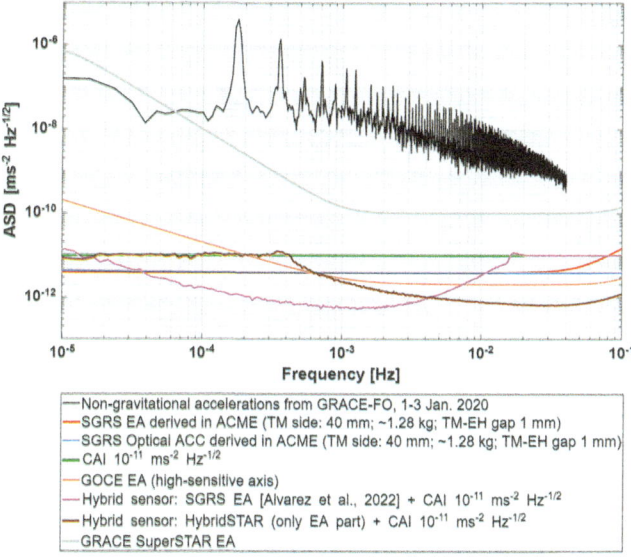

Fig. 1 Comparison of the ASD sensitivities of accelerometers for current instruments and anticipated enhanced concepts w.r.t. non-gravitational accelerations measured by GRACE-FO

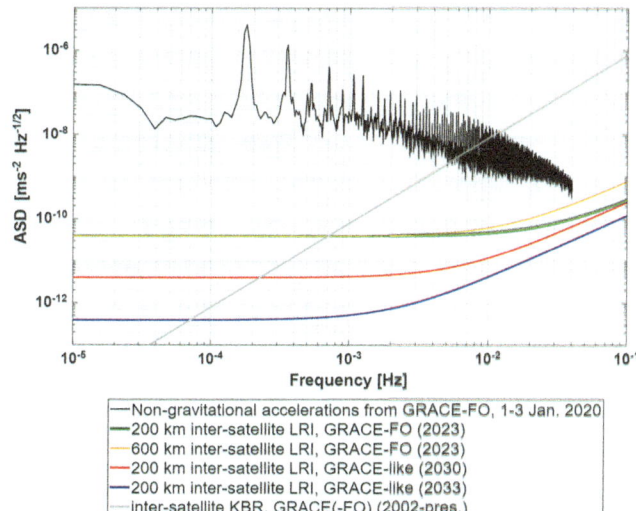

Fig. 2 Comparison of the ASDs of the inter-satellite LRI and KBR range measurement instrument errors used in the studies

drift. The orange curve represents the high-sensitive axis of the accelerometer that forms the gradiometer (Touboul et al. 2016; Marque et al. 2010). The red and blue curves represent the noise levels of SGRS sensors with electrostatic and optical test mass position readout, developed by Kupriyanov et al. (2024). Due to the modeling features of the noise budgets of these SGRS sensors, the difference of their ASDs appears only above 0.01 Hz. The green curve represents the realistic level of accuracy in one degree of freedom of a near-future CAI ACC introduced by white noise at the level of 10^{-11} m/s$^2/\sqrt{\text{Hz}}$. For this sensor the following major parameters were assumed: laser waist 20×10^{-3} m, atomic temperature 10×10^{-12} K, number of atoms 1×10^6, interrogation time 10 s (HosseiniArani et al. 2024). As it was demonstrated by Barrett et al. (2019), achieving a same level of accuracy in all three axes for CAI inertial sensor is quite challenging. Additionally, two curves of hybrid sensors are shown in the same graph. The pink curve corresponds to a hybridized instrument that consists of an SGRS EA from Dávila Álvarez et al. (2022) and the aforementioned CAI ACC. The hybridization was done at the level of ASDs by applying low- and high-pass filters at two cut-off frequencies at 11 mHz and 17 mHz in order to have a smooth transition between the ASDs. A similar procedure was applied for the hybridization of another sensor (brown curve) that consists of an electrostatic part of the HybridSTAR accelerometer (Dalin et al. 2020) and the CAI ACC. Here, only one cut-off frequency at 0.35 mHz was used for both low- and high-pass filters.

2.2 Comparison of Inter-Satellite Range Measurement Instruments

Figure 2 shows the ASDs of the inter-satellite LRI and K-band ranging (KBR) measurement instrument errors. The grey curve corresponds to the KBR instrument that is used in GRACE and GRACE-FO (Frommknecht et al. 2006). All other curves represent our assumptions on the distance-dependent errors of Laser Range Interferometer (LRI) system on-board GRACE-FO (green curve) or the anticipated LRI error noise in the future (red and blue curves) (Kupriyanov et al. 2024).

3 Closed-Loop Simulation

The block diagram of the closed-loop simulation procedure that was applied in the context of GRACE-like, GOCE-like and novel satellite constellations is shown in Fig. 3. The simulation procedure was carried out in multiple steps and using software parts in different language tools and programming platforms. In all simulations the EIGEN-6C4 (Förste et al. 2011) static gravity field model was used as the reference. The time-variable background models were not considered in our simulations in order to focus on the advantages of the novel sensors and concepts. Therefore, this study investigates the retrieval of the static gravity field. The evaluation is carried out on the level of residuals between recovered and reference gravity field models that occur due to considering various error sources, i.e. instrument errors.

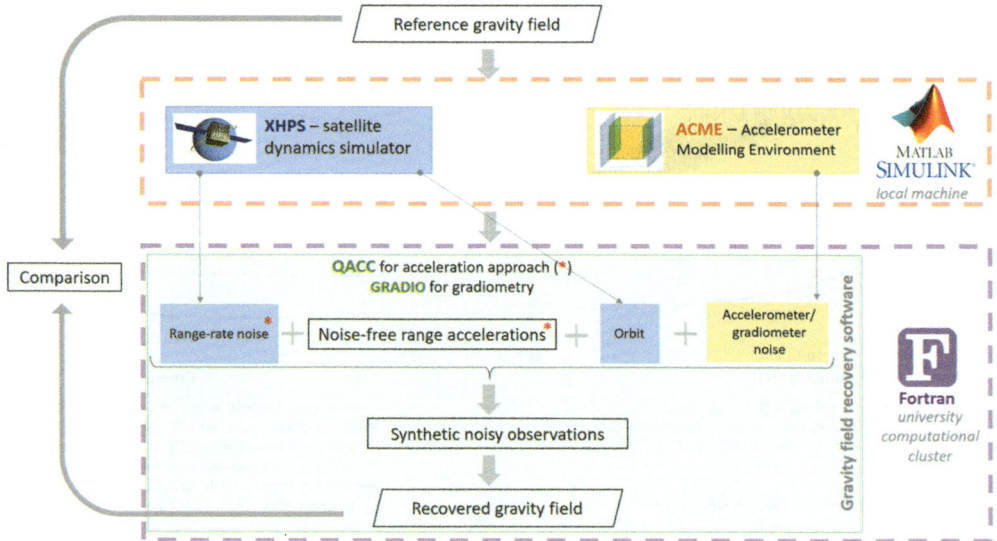

Fig. 3 Block diagram of the closed-loop simulation procedure

The Extended High Performance Satellite Dynamics Simulator (XHPS) developed by ZARM/DLR (Wöske et al. 2016) in Matlab/Simulink is used for simulating the satellite's orbit taking into account various non-gravitational forces and a detailed finite element model of the satellite (Wöske et al. 2018). Accelerometer Modeling Environment (ACME) (Kupriyanov et al. 2024) developed by Max Planck Institute for Gravitational Physics is also implemented in Matlab/Simulink and utilize to simulate different type of ACCs, e.g. electrostatic or optical ones. In this work ACME is operated concurrent to XHPS, running time-domain simulations and generating mock data of various sensors. Two gravity field recovery software tools, named Quantum Accelerometry (QACC) and GRADIO, both developed by IfE, LUH (Wu 2016) written in Fortran are used in the simulations.

For the GRACE-like cases QACC toolbox with acceleration approach is used. Here the noise-free range accelerations time-series are generated and combined afterwards with corresponding instrument noise time-series, i.e. range-rate sensor noise, LRI or KBR (Frommknecht et al. 2006) and with ACC noise. Then the synthesized noisy observations are used for least-squares estimation of the gravity field parameters – spherical harmonic coefficients. At the end the recovered gravity field is compared with the reference gravity field, by computing the difference of the coefficients and plotting them either on a global map in terms of equivalent water height (EWH) or in two dimensional spherical harmonic error spectrum (similar as it was done by Wu (2016)).

For the gravity field recovery from the GOCE-like gradiometry missions GRADIO toolbox is used. There, the satellites' orbit from XHPS synthesised together with the corresponding gradiometer noise time-series. Then following the aforementioned steps the retrieved gravity field are compared with the reference one.

4 Gravity Field Recovery Results from Simulation

For all simulations a one-month mission duration in May 2002 was considered, since this was a period of strong solar activity (SILSO World Data Center 2023). As a result, high solar activity means that the solar radiation pressure and the air-drag on the spacecraft in orbit are larger. The selection of the year/month with the high solar activity was important for the testing accelerometers, developed in ACME, in critical conditions. Furthermore, all gravity field recovery simulations were calculated up to degree and order 180, while the input reference gravity field model was given up to degree and order 2190.

4.1 Low-Low Satellite-to-Satellite

Figure 4 represents the averaged error degree variance per specific degree in terms of geoid height for the low-low satellite-to-satellite (ll-sst) missions on a GRACE-like orbit with different types of ACCs and inter-satellite range instruments. Black dashed line represent the static gravity field signal EIGEN-6C4 (Förste et al. 2011). The mean monthly Hydrology, Ice and solid Earth (HIS) signal (Dobslaw et al. 2015), depicted by the grey dashed curve. This curve is shown here for the understanding that in an idealized case, when the temporal aliasing is sufficiently considered and the high performance of the instruments can be fully exploited,

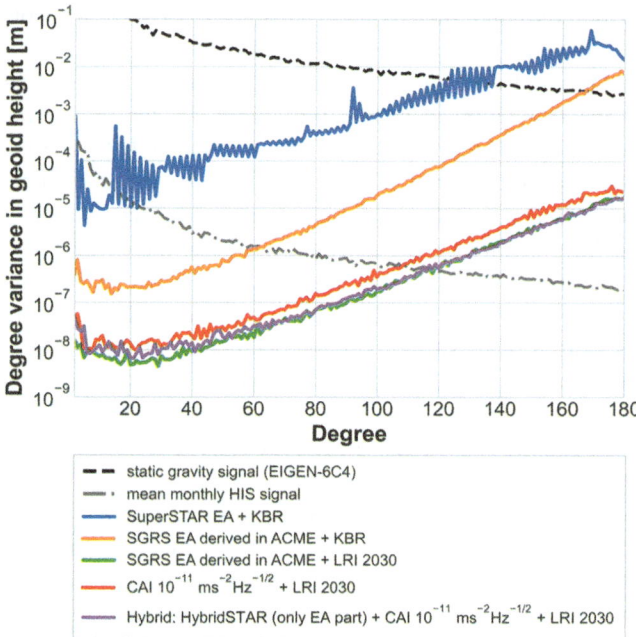

Fig. 4 Averaged error degree variance per specific degree in terms of geoid height of the different on-board ACC and inter-satellite range instruments in the context of GRACE-FO like missions

4.2 Cross-Track Gradiometry

Since the ll-sst type of measurements and satellite gradiometry are both sensitive to different wavelengths of the gravity field signal (Pail et al. 2019), we also checked the performance of novel instruments w.r.t. high-sensitive axis of the GOCE electrostatic gradiometer (Touboul et al. 2016; Marque et al. 2010) arranged in the cross-track direction. Gradiometry simulations were computed for the GOCE-like altitude, near-polar, drag-compensated orbit ($i = 89°$, $h = 246$ km) with a gradiometer arranged in cross-track direction with a baseline length $b = 0.5$ m.

Figure 5 shows the average error degree variance per specific degree for various gradiometer instruments. The blue curve corresponds to the GOCE high-sensitive axis of the electrostatic gradiometer, acquired from the ASD orange curve from Fig. 1. The orange and green curves in Fig. 5 correspond to the gradiometers that consist of SGRS EA or an optical ACC. Simulated CAI and hybridized gradiometers also include the effect of CAI gyroscopes (Savoie et al. 2018) which are required to correct for the rotational movement of the gradiometer axis in cross-track direction. The considered CAI gyroscope has a white-noise level close to 10^{-8} rads^{-1} in terms of angular velocity which corresponds to interrogation time of 10 s. However, the effect of the

potentially time-variable gravity field can be determined up to a much higher degree and order than now. The blue curve represents the GRACE-FO mission with a SuperSTAR EA and KBR as the range measurement instrument. The static gravity field signal could be resolved up to degree 130 by this mission and the mean monthly HIS signal only up to degree 15. The strong oscillations of the blue curve corresponds to the drift of the EA in the low frequency domain. The orange and green curves represent the missions with a SGRS EA, derived in ACME, on-board and KBR or an anticipated LRI of the year 2030 as the inter-satellite instrument. In comparison to the GRACE-FO scenario with a maximum achievable level of a geoid height 10^{-5} m, these mission concepts are able to reach 10^{-7} m and 10^{-8} m respectively. They also can potentially resolve the time-variable HIS signal under the above-mentioned conditions up to degree 60 and 120. The CAI ACC as a standalone accelerometer together with the anticipated LRI 2030 performance (red curve) shows a slightly worse performance w.r.t. the previously discussed SGRS EA with LRI 2030. Hybridizing a CAI ACC with an electrostatic part of the HybridSTAR accelerometer together with the LRI 2030 (purple curve) brings only minor improvements w.r.t. the standalone CAI ACC case (red curve). However, the SGRS EA derived in ACME with LRI 2030 (green curve) shows the best performance up to degree 60 among the considered sensors.

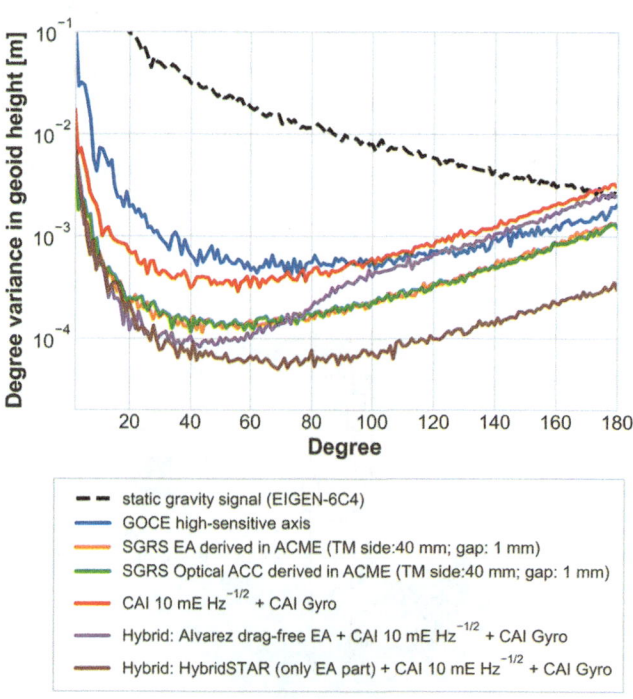

Fig. 5 Averaged error degree variance per specific degree in terms of geoid height of the different gradiometers in cross-track direction in the context of a potential GOCE-FO like mission

CAI gyroscope to the total CAI or hybridized gradiometer performance was negligibly small. As it is depicted in the graph, hybridized gradiometers show the best performance roughly up to degree 70.

4.3 SFF with Different Right Ascension of the Ascending Node

So far, only near-polar satellite orbits are considered in our simulations. This kind of orbits causes resonance effects (Kvas et al. 2019), which together with drift of the EA in the low frequency domain, lead to the North-South striping behavior of the retrieved gravity field models. Hence, the North-South striping effect can be reduced either by utilizing the novel enhanced accelerometers or by providing additional measurements in the cross-track direction. One of the options for getting measurements in the cross-track direction could be by adding another satellite, see Fig. 6. In this section we consider two near-polar orbits ($i = 89°$) with the GRACE-like altitudes ($h = 450$ km) but with different RAAN angles $\Omega_1 = 127°$ and $\Omega_2 = 128°$. This brings a stable orbital configuration, i.e. the maximum distance at the equator between the satellites B and C does not increase over time, since the Ω does not impact the secular variations. In principle, this SFF could be treated as an enhanced version of the Pendulum constellation (Elsaka et al. 2014) with the difference that here, it is assumed that satellite C is always in cross-track direction w.r.t. the in-line along-track formation of the satellites A and B. The satellites A and B form an in-line GRACE-like formation separated by roughly $\rho = 200$ km. In this SFF, the two orbital planes intersect close to the poles and the maximum separation between the orbits is at the equator, with about $l = 120$ km.

Different inter-satellite ranging instruments are assumed in the simulation of such a formation. For the in-line formation (satellites A and B) an anticipated LRI 2033 is assumed (blue curve in Fig. 2), while for the cross-track inter-satellite range measurement, a certain degradation of the laser beam steering mechanism due to increased tilt-to-length noise and the coupling caused by satellites pointing variations (Wegener et al. 2020) is considered. Furthermore, the current level of the pointing accuracies of the GRACE-like satellites, i.e. 2.5 rad in roll axis and 250 µrad in pitch and yaw (Goswami et al. 2021), is taken into account. Hence, a less accurate LRI 2023 (green curve on Fig. 2) is considered for the cross-track SFF.

However, on-board of all three satellites a SGRS Optical accelerometer derived in ACME was assumed (Kupriyanov et al. 2024). The same level of accuracy was taken for all three axes. Figure 7 shows the averaged error degree variance per specific degree in terms of geoid height for the in-line (blue curve), cross-track (orange curve) and combined formation (green curve). Identically to the previous graphs, the static gravity signal is depicted as a black dashed line. The in-line formation has a better performance than the cross-track formation. Combining these solutions at the level of normal equations (Wu 2016) with a weighting based on the posteriori variances results in the green curve.

Figure 8 represents global gravity field maps recovered from the in-line (top), cross-track (middle) and combined (bottom) satellite formations. Geoid height amplitudes represent the difference of the recovered gravity field models with the reference gravity field model EIGEN-6C4. The same order of magnitude of the spatial residuals, $\pm 5 \times 10^{-6}$ m, are used in all global maps for a more explicit difference. The combination of three satellites allow to get

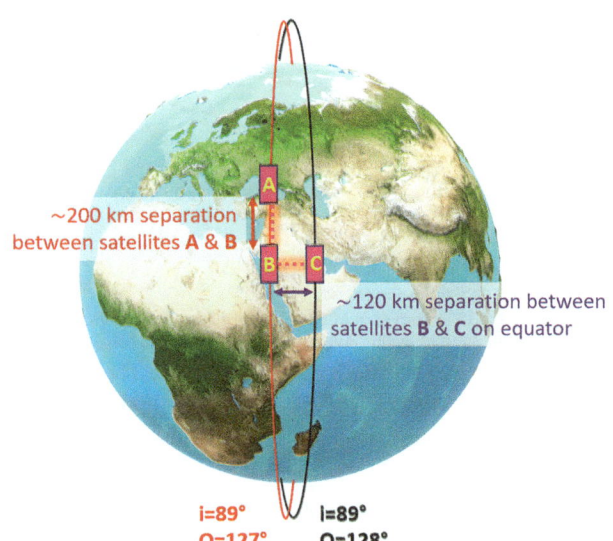

Fig. 6 Scheme of the combination of the in-line and cross-track formation differing by the right ascension of the ascending node (RAAN)

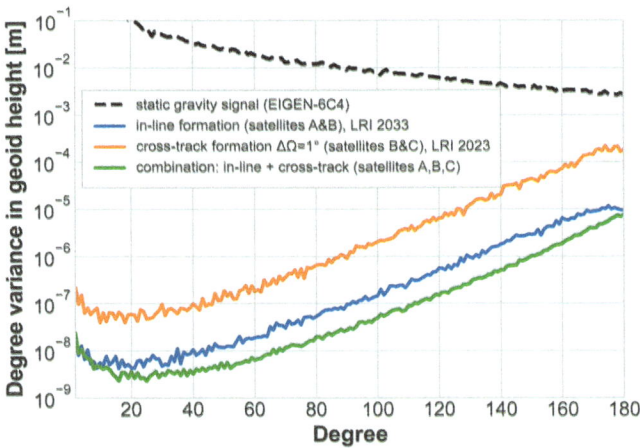

Fig. 7 Averaged error degree variance per specific degree in terms of geoid height of the in-line, cross-track and combined formations

Fig. 8 Global maps with the residuals in geoid heights, raw data, without post-processing and filtering. *Top*: In-line formation (satellites A and B); *Middle*: Cross-track formation (satellites B and C); *Bottom*: Combined formation (satellites A, B and C)

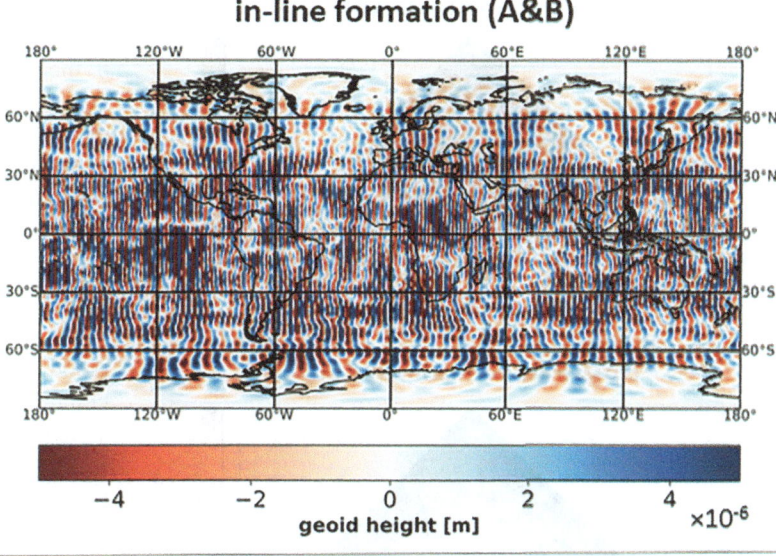
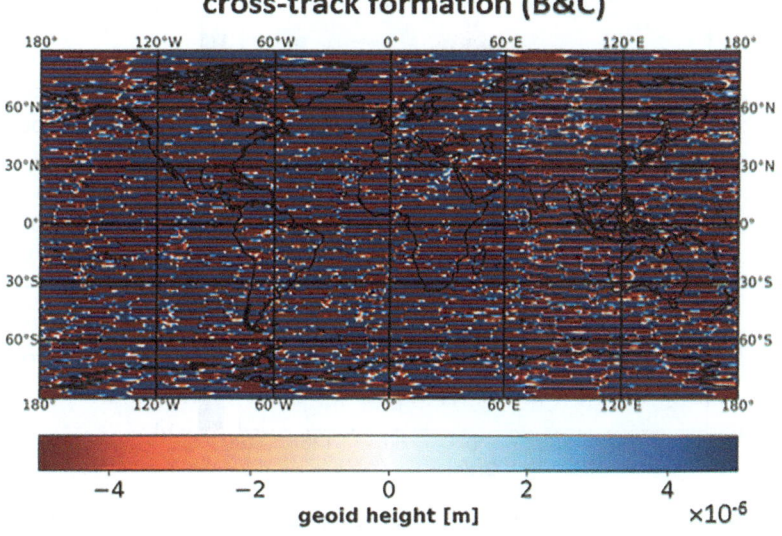
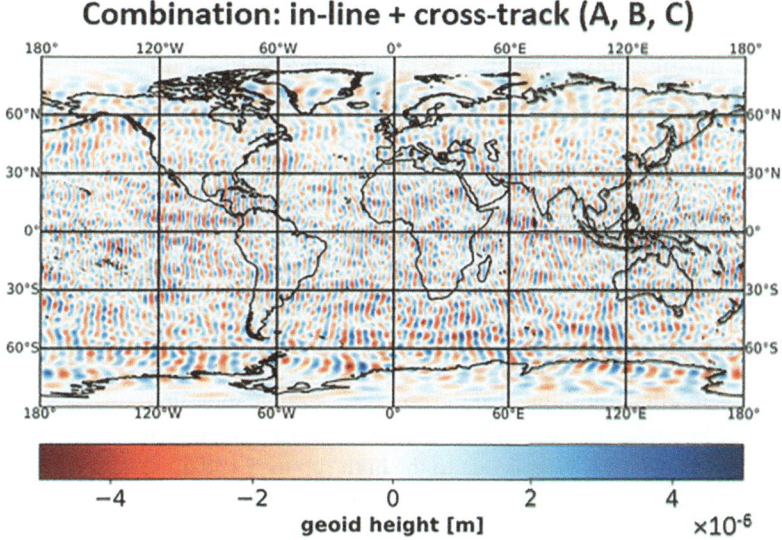

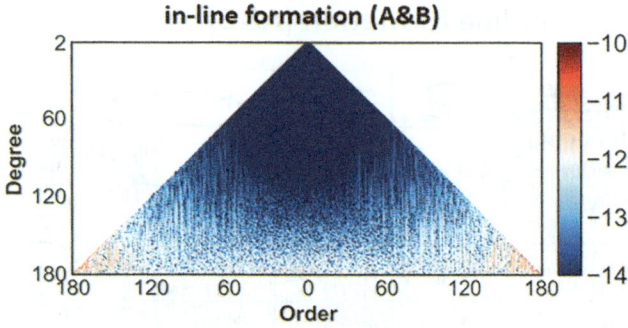

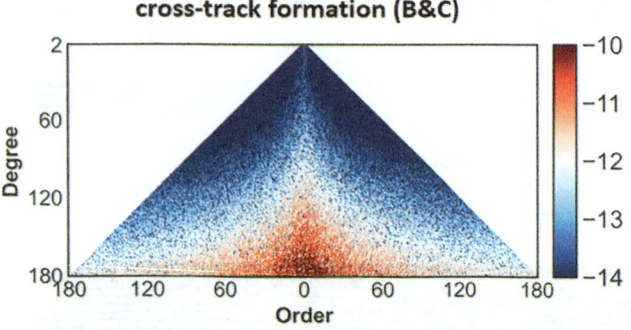

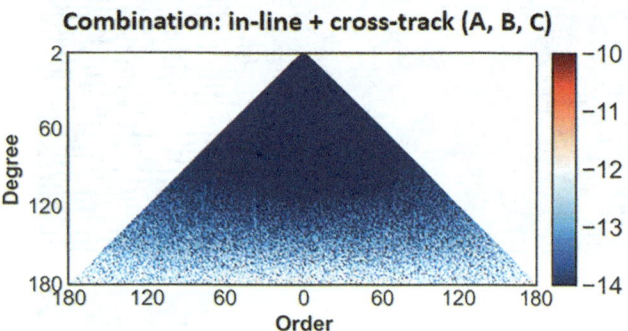

Fig. 9 Comparison of the spherical harmonic error spectra between in-line (top), cross-track (middle) and combined (bottom) formations

multi-directionality of the data by avoiding the North-South striping effect of the in-line formation without any post-processing or filtering.

Figure 9 shows the two dimensional spherical harmonic error spectrum or, in other words, the relative error of each spherical harmonic coefficient w.r.t. the reference gravity field. The color bars are given in logarithmic scale for better representation. On the top, we show the results from the in-line SFF, in the middle from the cross-track and at the bottom combined satellite formations.

It is worth mentioning that the cross-track range measurements in such SFF are technically challenging for current inter-satellite LRI systems due to the high (up to ± 130 m/s) relative range-rates between the satellites B and C. The current feasible range-rate for heterodyne lasers that were used on-board the GRACE-FO mission are ± 10 m/s (Sheard

et al. 2012; Elsaka et al. 2014), which is acceptable and sufficient even for the next upcoming missions to be launched in the time frame of 2028 to 2031 (Haagmans and Tsaoussi 2020; Massotti et al. 2021). However, in order to realize the additional cross-track link at some point, future developments should also focus extending the dynamic range of the LRI in terms of pointing angles and range rates.

4.4 SFF with Different Inclinations

Another satellite formation consisting of three satellites that could utilize information in the cross-track direction, is possible with orbits that differ by inclination, see Fig. 10. Similar to the previous SFF, this constellation assumed to have altitudes around 450 km, but inclinations equal to $i = 89.048°$ for the in-line and $i = 90.048°$ for the second orbital plane. However, a stable formation is not guaranteed (Elsaka 2012) in this case because of the secular perturbations of some Keplerian elements, in particular $\dot{\Omega}$ (Bloßfeld et al. 2014). It was found that after one month the maximum distance close to the poles in the cross-track direction between the satellites B and C increased from 120 to 520 km. In order to take into account adequately this increasing inter-satellite distance, an LRI 2023 model with $\rho = 600$ km baseline (yellow curve Fig. 2) was used for the cross-track formation.

After following the similar procedure as in the previous SFF, i.e. calculating the recovered gravity field from the in-line, cross-track and combined formations, the averaged error

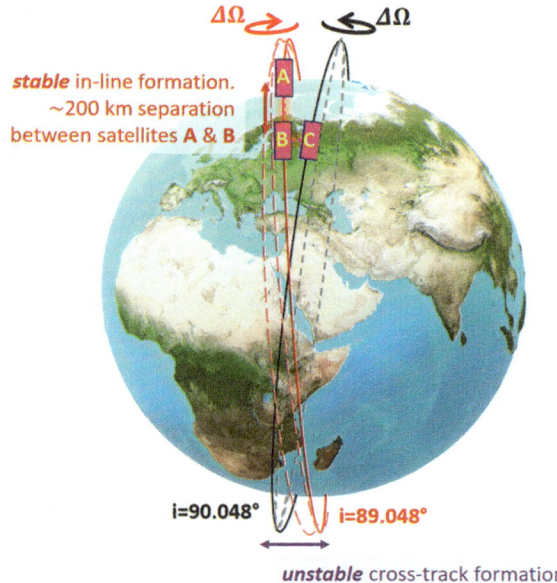

Fig. 10 Scheme of the combination of the in-line and cross-track formation differing by the inclination

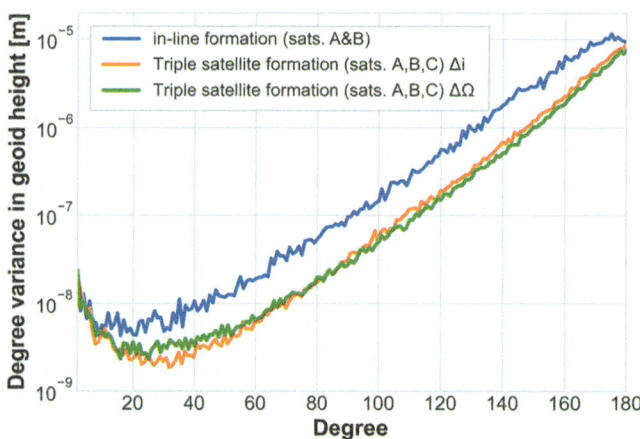

Fig. 11 Averaged error degree variance per specific degree in terms of geoid height of the in-line formation solely (blue), combined satellite formation with different RAAN (green) and combined satellite formation with different inclinations (orange)

degree variances per specific degree in terms of geoid height have been calculated. Figure 11 represents the comparison between the error degree variances from the in-line formation solely or, in other words, satellites A and B (blue curve), combined SFF from three satellites with different RAAN (green curve) and combined SFF with different inclinations (orange curve). In principle, combined solutions (green and orange curves) from both SFFs have a similar level of geoid accuracy. Up to degree 80, SFF with different inclinations show superior performance, while at higher degrees SFF with different RAAN performs better.

From technical point of view, SFF with different inclinations is even more technologically challenging than the one with different RAAN, because the relative range-rates between the satellites B and C are up to ±600 m/s here. This is 60 times larger than the current range-rate required for heterodyne lasers (Sheard et al. 2012; Elsaka et al. 2014). Furthermore, due to the unstable formation, a certain orbit maintenance has to be done in order to avoid drifting the orbital planes apart.

5 Conclusions

In this study, closed-loop simulations were performed in order to investigate the benefits of novel and enhanced instruments, i.e. accelerometers and gradiometers w.r.t. 'classical' electrostatic sensors. Moreover, static gravity field recovery from two novel triple satellite formations was carried out. To quantify the advantages of the novel instruments on the gravity field retrieval, the impact of temporal aliasing due to the insufficient background models was neglected.

Time-variable background models and associated aliasing errors are a major limiting factors in current satellite gravimetry. However, the background models will certainly be improved in future. It is anticipated that the processing strategies of gravitational data also enhance, for example, different parametrization, co-estimation of certain signals, etc. Therefore, one would be able to re-process data somewhen in future, as it was done with the GRACE or satellite altimetry data. There the observations from the 80s and 90s were re-processed benefiting from better orbits due to improved gravity field and reduction models. On the other hand, if sensors with poor performance are used in a mission, one can merely improve anything in post-processing. Therefore, it is necessary to consider the best possible sensors of a time on a mission and this study underlines what might be achieved in this respect. Moreover, the upcoming MAGIC mission, consisting of two satellite pairs, will also contribute to reducing the temporal aliasing effect (Purkhauser et al. 2020).

It was demonstrated that in low-low satellite-to-satellite configurations the static gravity field solutions could be improved up to three orders of magnitude with enhanced electrostatic, CAI and hybridized instruments

In single-axis cross-track gradiometry simulations, novel instruments also significantly improved the gravity field retrieval by one order of magnitude w.r.t. a GOCE-like electrostatic gradiometer. This would enable the geoid determination with a higher accuracy.

In addition, gravity field recovery results were investigated for two new satellite formation flights each with two orbital planes varying in RAAN or inclination relatively to each other. Both configurations showed an improvement in recovering the gravity field models from the combined solutions w.r.t. in-line and cross-track formations individually. However, the technical constraints, like formation maintenance with associated propellant consumption and maximum LRI range rates, make the option with RAAN separation more likely to be realized. Nonetheless, both alternative satellite formations would require significant modifications of the current GRACE-like platforms.

Further research is required to obtain more realistic simulations. This could be done in multiple ways, in particular with an improvement of the instrument modeling, e.g. analysing the noise budget of novel accelerometers, or including ancillary instruments in the gradiometry simulations, e.g. by providing additional angular velocity noise. Another to include time-variable gravity signals and related background models into the simulations. They are the source of the short-term mass variations from the atmosphere and oceans and produce aliasing errors which are typically the main contributors to inaccuracies in temporal gravity solutions.

Acknowledgements The authors acknowledge funding by:
1. Deutsche Forschungsgemeinschaft (DFG) – TerraQ (Project-ID 434617780 – SFB 1464)

2. European Union - CARIOQA-PMP (Project-ID 101081775)
3. German Aerospace Center (DLR) – Q-BAGS (Project-ID 50WM2181)
4. German Aerospace Center (DLR) – QUANTGRAV (Project-ID 50EE2220B)

References

Alonso I, Alpigiani C, Altschul B, et al (2022) Cold atoms in space: community workshop summary and proposed road-map. EPJ Quantum Technol 9(1):1–55. https://doi.org/10.1140/epjqt/s40507-022-00147-w

Armano M, Audley H, Baird J, et al (2018) Beyond the required LISA free-fall performance: New LISA pathfinder results down to 20 mHz. Phys Rev Lett 120(6):061101. https://doi.org/10.1103/PhysRevLett.120.061101

Armano M, Audley H, Baird J, et al (2021) Sensor noise in LISA pathfinder : In-flight performance of the optical test mass readout. Phys Rev Lett 126(13). https://doi.org/10.1103/PhysRevLett.126.131103

Barrett B, Cheiney P, Battelier B, Napolitano F, Bouyer P (2019) Multidimensional atom optics and interferometry. Phys Rev Lett 122(4):043604. https://doi.org/10.1103/PhysRevLett.122.043604

Bloßfeld M, Gerstl M, Hugentobler U, et al (2014) Systematic effects in LOD from SLR observations. Adv Space Res 54(6):1049–1063. https://doi.org/10.1016/j.asr.2014.06.009

Boulanger D, Christophe B, Rodrigues M, et al (2020) Future electrostatic accelerometer without polarization wire. European Geosciences Union General Assembly 2020, Online, 4–8 May 2020. https://doi.org/10.5194/egusphere-egu2020-4640

Chen J, Cazenave A, Dahle C, et al (2022) Applications and challenges of GRACE and GRACE Follow-On satellite gravimetry. Surv Geophys 43(1):305–345. https://doi.org/10.1007/s10712-021-09685-x

Christophe B, Boulanger D, Foulon B, et al (2015) A new generation of ultra-sensitive electrostatic accelerometers for GRACE Follow-on and towards the next generation gravity missions. Acta Astronautica 117:1–7. https://doi.org/10.1016/j.actaastro.2015.06.021

Dalin M, Christophe B, Liorzou F, et al (2020) ONERA accelerometers for future gravity mission. European Geosciences Union General Assembly 2020, online poster

Daras I, Pail R (2017) Treatment of temporal aliasing effects in the context of Next Generation satellite Gravimetry Missions. J Geophys Res Solid Earth 122(9):7343–7362. https://doi.org/10.1002/2017JB014250

Dávila Álvarez A, Knudtson A, Patel U, et al (2022) A simplified gravitational reference sensor for satellite geodesy. J Geodesy 96(10):70. https://doi.org/10.1007/s00190-022-01659-0

Dobslaw H, Bergmann-Wolf I, Dill R, et al (2015) The updated ESA Earth System Model for future gravity mission simulation studies. J Geodesy 89:505–513. https://doi.org/10.1007/s00190-014-0787-8

Elsaka B (2012) Simulated satellite formation flights for detecting the temporal variations of the earth's gravity field. PhD thesis, Rheinische Friedrich-Wilhelms-Universität Bonn. https://hdl.handle.net/20.500.11811/4203

Elsaka B, Raimondo J-C, Brieden P, et al (2014) Comparing seven candidate mission configurations for temporal gravity field retrieval through full-scale numerical simulation. J Geodesy 88:31–43. https://doi.org/10.1007/s00190-013-0665-9

Flechtner F, Reigber C, Rummel R, et al (2021) Satellite gravimetry: A review of its realization. Surv Geophys 42(5):1029–1074. https://doi.org/10.1007/s10712-021-09658-0

Förste C, Bruinsma S, Flechtner F, et al (2011) EIGEN-6C3 - The latest Combined Global Gravity Field Model including GOCE data up to degree and order 1949 of GFZ Potsdam and GRGS Toulouse. In: AGU Fall Meeting Abstracts, vol 2011, pp G51A–0860. Harvard ASD: 2011AGUFM.G51A0860F

Frommknecht B, Oberndorfer H, Flechtner F, et al (2003) Integrated sensor analysis for GRACE–development and validation. Adv Geosci 1:57–63

Frommknecht B, Fackler U, Flury J (2006) Integrated sensor analysis GRACE. In: Flury J, et al (eds), Observation of the Earth system from space. Springer, pp 99–113. doi: 10.1007/3-540-29522-4_8

Goswami S, Francis SP, Bandikova T, et al (2021) Analysis of GRACE Follow-On Laser Ranging Interferometer derived inter-satellite pointing angles. IEEE Sensors J, 1. https://doi.org/10.1109/JSEN.2021.3090790

Haagmans R, Tsaoussi L (2020) Next generation gravity mission as a Mass-change And Geosciences International Constellation (MAGIC) mission requirements document. Earth and Mission Science Division, European Space Agency; NASA Earth Science Division. https://doi.org/10.5270/esa.nasa.magic-mrd.2020

HosseiniArani A, Tennstedt B, Schilling M, et al (2022) Kalman-filter based hybridization of classic and cold atom interferometry accelerometers for future satellite gravity missions. In: International Association of Geodesy Symposia. https://doi.org/10.1007/1345_2022_172

HosseiniArani A, Schiling M, Beaufils Q, et al (2024) Advances in atom interferometry and their impacts on the performance of quantum accelerometers on-board future satellite gravity missions. Adv Space Res. https://doi.org/10.1016/j.asr.2024.06.055

Knabe A, Schilling M, Wu H, et al (2022) The benefit of accelerometers based on cold atom interferometry for future satellite gravity missions. International Association of Geodesy Symposia. Springer, Berlin, Heidelberg. https://doi.org/10.1007/1345_2022_151

Kupriyanov A, Reis A, Schilling M, et al (2024) Benefit of enhanced electrostatic and optical accelerometry for future gravimetry missions. Adv Space Res, 73. https://doi.org/10.1016/j.asr.2023.12.067

Kvas A, Behzadpour S, Ellmer M, et al (2019) ITSG–Grace2018: overview and evaluation of a new GRACE–only gravity field time series. J Geophys Res Solid Earth 124(8):9332–9344. https://doi.org/10.1029/2019JB017415

Lévèque T, Fallet C, Lefebve J, et al (2023) CARIOQA: Definition of a quantum pathfinder mission. In: International Conference on Space Optics–ICSO 2022, vol 12777, pp 1536–1545. SPIE. arXiv:2211.01215

Liorzou F, Lebat V, Christophe B, et al (2023) ONERA accelerometers for future gravity mission. In: EGU General Assembly, Vienna, Austria. https://doi.org/10.5194/egusphere-egu23-8155

Marque J-P, Christophe B, Foulon B (2010) Accelerometers of the GOCE mission: return of experience from one year of in-orbit. In: Proceedings of the ESA Living Planet Symposium, Bergen, Norway. Harvard ADS: 2010ESASP.686E..57M

Massotti L, Siemes C, March G, et al (2021) Next generation gravity mission elements of the mass change and geoscience international constellation: from orbit selection to instrument and mission design. Remote Sensing 13(19):3935. https://doi.org/10.3390/rs13193935

Mehta PM, Walker AC, Sutton EK, et al (2017) New density estimates derived using accelerometers on board the CHAMP and GRACE satellites. Space Weather 15(4):558–576

Pail R, Bingham R, Braitenberg C, et al (2015) Science and user needs for observing global mass transport to understand global change and to benefit society. Surv Geophys 36(6):743–772. https://doi.org/10.1007/s10712-015-9348-9

Pail R, Bamber J, Biancale R, et al (2019) Mass variation observing system by high low inter-satellite links (MOBILE)—a new concept for sustained observation of mass transport from space. J Geodetic Sci 9(1):48–58. https://doi.org/10.1515/jogs-2019-0006

Panet I, Narteau C, Lemoine J-M, et al (2022) Detecting preseismic signals in GRACE gravity solutions: Application to the 2011 Tohoku

M w 9.0 earthquake. J Geophys Res Solid Earth 127(8). https://doi.org/10.1029/2022JB024542

Peidou A, Landerer F, Wiese D, et al (2022) Spatiotemporal characterization of geophysical signal detection capabilities of GRACE–FO. Geophys Res Lett 49(1). https://doi.org/10.1029/2021GL095157

Purkhauser AF, Siemes C, Pail R (2020) Consistent quantification of the impact of key mission design parameters on the performance of next-generation gravity missions. Geophysl J Int 221(2):1190–1210. https://doi.org/10.1093/gji/ggaa070

Savoie D, Altorio M, Fang B, et al (2018) Interleaved atom interferometry for high-sensitivity inertial measurements. Sci Adv 4(12):eaau7948. https://doi.org/10.1126/sciadv.aau7948

Sheard B, Heinzel G, Danzmann K, et al (2012) Intersatellite laser ranging instrument for the GRACE follow-on mission. J Geodesy 86:1083–1095. https://doi.org/10.1007/s00190-012-0566-3

SILSO World Data Center (2023) International Sunspot Number Monthly Bulletin and online catalogue. http://www.sidc.be/silso/

Sumner TJ, Mueller G, Conklin JW, et al (2020) Charge induced acceleration noise in the LISA gravitational reference sensor. Classical Quantum Gravity 37(4):045010. https://doi.org/10.1088/1361-6382/ab5f6e

Torge W, Müller J, Pail R (2023) Geodesy. de Gruyter. ISBN 3110723301

Touboul P, Metris S, Le Traon O, et al (2016) Gravitation and geodesy with inertial sensors, from ground to space. AerospaceLab 12. https://doi.org/10.12762/2016.AL12-11

Trimeche A, Battelier B, Becker D, et al (2019) Concept study and preliminary design of a cold atom interferometer for space gravity gradiometry. Classical Quantum Gravity 36(21):215004. https://doi.org/10.1088/1361-6382/ab4548

van Camp M, Pereira dos Santos F, Murböck M, et al (2021) Lasers and ultracold atoms for a changing Earth. Eos Trans Am Geophys Union 102. https://doi.org/10.1029/2021EO210673f

Weber WJ, Bortoluzzi D, Bosetti P, et al (2022) Application of LISA gravitational reference sensor hardware to future intersatellite geodesy missions. Remote Sensing 14(13):3092. https://doi.org/10.3390/rs14133092

Wegener H, Müller V, Heinzel G, et al (2020) Tilt-to-length coupling in the GRACE Follow-On laser ranging interferometer. 0022-4650, pp 1–10. https://doi.org/10.2514/1.A34790

Wöske F, Kato T, List M, et al (2016) Development of a high precision simulation tool for gravity recovery missions like GRACE. In: Proceedings of the 26th AAS/AIAA Space Flight Mechanics Meeting, 14.-18.02.2016, Napa, Califonia, USA, pp 14–18

Wöske F, Kato T, Rievers B, et al (2018) GRACE accelerometer calibration by high precision non-gravitational force modeling. Adv Space Res 63(3):1318–1335. https://doi.org/10.1016/j.asr.2018.10.025

Wu H (2016) Gravity field recovery from GOCE observations. PhD thesis, Fakultät für Bauingenieurwesen und Geodäsie - Leibniz Universität Hannover

Xiao Y, Yang Y, Pan Z, et al (2023) Performance and application of the Chinese satellite-to-satellite tracking gravimetry system. Chinese Sci Bull 68:2655–2664. https://doi.org/10.1360/TB-2022-1057

Zahzam N, Christophe B, Lebat V, et al (2022) Hybrid electrostatic–atomic accelerometer for future space gravity missions. Remote Sensing 14(14):3273. https://doi.org/10.3390/rs14143273

Open Access This chapter is licensed under the terms of the Creative Commons Attribution 4.0 International License (http://creativecommons.org/licenses/by/4.0/), which permits use, sharing, adaptation, distribution and reproduction in any medium or format, as long as you give appropriate credit to the original author(s) and the source, provide a link to the Creative Commons license and indicate if changes were made.

The images or other third party material in this chapter are included in the chapter's Creative Commons license, unless indicated otherwise in a credit line to the material. If material is not included in the chapter's Creative Commons license and your intended use is not permitted by statutory regulation or exceeds the permitted use, you will need to obtain permission directly from the copyright holder.

Automated Anomaly and Outlier Detection in GRACE and GRACE Follow-On Post-Fit Residuals Using Machine Learning

Martin Lasser, Jonas Zbinden, Ulrich Meyer, Brandon Panos, Daniel Arnold, and Adrian Jäggi

Abstract

GRACE and GRACE Follow-On inter-satellite ranges and thereof derived range-rates are the main observables for the determination of monthly snapshots of the Earth's gravity field. These observations are sensitive to the mass distribution on the Earth, and as a consequence, the relative motion between the missions' satellite pairs. The range-rate observations exhibit a number of difficult-to-identify error sources and efficient screening of the data is not trivial. Therefore, we apply machine learning based outlier detection methods such as isolation forests, to flag outliers in an unsupervised fully automated way. We apply the technique to post-fit residuals of monthly, joint orbit and gravity field determination processes, combined with the geographical position of each observation. The flagged outliers are investigated for local geographical correlations to distinguish between unfitted signal from gravitational sources and artefacts caused by the satellites' instrumentation. For that purpose we train a mutual information neural network, learning the mutual information between the post-fit residuals and the geographical location. Outliers flagged as artefacts are removed from the original inter-satellite range-rate data and the orbit and gravity field determination process is repeated to investigate for improvements. The automated outlier detection with isolation forests performs similar, at times slightly better, to empirical screenings by visual inspection of post-fit residuals. In addition, the mutual information can be taken as a valuable source to detect geophysical signal remaining in the post-fit residuals.

Keywords

GRACE · GRACE Follow-On · Gravity field · Isolation forest · Machine learning · Mutual information · Outlier detection

1 Introduction

Measuring and quantifying mass re-distributions in the dynamic system Earth on a global scale defines the mission goal of the Gravity Recovery and Climate Change Mission (GRACE, Tapley et al. 2004) and its successor mission GRACE Follow-On (Landerer et al. 2020). The continuous observation of the Earth's time-variable gravity field provides insights into the continental water cycle or the melting of the ice sheets. One way to observe these changes is to precisely measure variations in the inter-

Authors Martin Lasser and Jonas Zbinden have contributed equally to this work.

M. Lasser (✉) · J. Zbinden · U. Meyer · D. Arnold · A. Jäggi
Astronomical Institute, University of Bern, Bern, Switzerland
e-mail: martin.lasser@unibe.ch

B. Panos
Institute for Data Science, University of Applied Sciences and Arts Northwestern Switzerland, Aargau, Switzerland

satellite distance between two spacecraft, which are caused by changes in the underlying Earth's gravity potential field, i.e., by mass re-distributions, to ultimately recover the latter. The typical time resolution of GRACE and GRACE Follow-On gravity fields is one month, for which the observations collected are accumulated to estimate monthly snapshots of the Earth's gravity field, providing a resolution of about 63,000 km^2 on the Earth's surface for hydrological applications at a noise level of 2 cm equivalent water height (Vishwkarma et al. 2018).

Inter-satellite K-band range-rates are the main observable for tracking the relative orbital trajectory between the satellite pairs of GRACE or GRACE Follow-On. These range-rates are sensitive not only to the mass distribution of the Earth, and as a consequence, the relative motion of both GRACE and GRACE Follow-On satellites respectively, but also to all kinds of anomalous behaviour of the instruments involved in providing a stable platform in space. Potential sources for outliers and mis-modellings are unfitted signal leaking into the post-fit residuals, manoeuvrers, the K-band instrument system, the accelerometer data, which measures a total of drag and radiation effects, and attitude data, as well as effects of the geomagnetic equator on the instruments. Therefore, an efficient screening of the range-rate observations to remove outliers or spurious patterns is not trivial. In this study, we apply machine learning to flag outliers based on post-fit residuals of a first first gravity field estimate in an unsupervised and fully automated way and compare it to empirical screening methods. Additionally, we search for geographically correlated patterns by estimating the mutual information between the geographical location and the post-fit residuals.

2 Data

The goal of this investigation is to quantify the impact of different outlier detection methods on gravity field recovery from GRACE and GRACE Follow-On data. All methods were applied to one year of observations of each mission, 2004 for GRACE and 2019 for GRACE Follow-On. Both years exhibit particularly challenging data samples with sub-monthly repeat orbit cycles in summer 2004 and a significant amount of data not available for February 2019 due to a malfunctioning of the onboard computer of the GRACE Follow-On-2 spacecraft.[1] On top of this large scale patterns in the data outliers due to the instruments' performances are sought-after and, if found, removed from the data.

The orbit and gravity field determination is executed in one common least-squares adjustment using the Celestial Mechanics Approach (CMA, Beutler et al. 2010) developed

---
[1] See the spacecraft events log TN-01a_SCE.txt of the operator.

at the Astronomical Institute of the University of Bern (AIUB). GRACE and GRACE Follow-On Level-1B data (Case et al. 2010; Wen et al. 2019) from the K-band ranging system (range-rates) together with kinematic positions (Arnold 2022) are used as observations and pseudo-observations, respectively. Non-gravitational perturbing forces are introduced error-free as measured by the onboard accelerometer by GRACE and from the transplant product of Behzadpour et al. (2021) for GRACE Follow-On. Additionally, attitude information is taken from the Level-1B data. Background force modelling includes the reduction of the effect of gravitational perturbing forces on the satellites and covers the direct influence of Sun, Moon and planets as well as tidal variations for the solid Earth (Petit and Luzum 2010), the poles (Petit and Luzum 2010; Desai 2002), in the oceans (Carrere et al. 2016) and atmosphere and non-tidal short term mass variations (Dobslaw et al. 2017). The processing is done in daily arcs for which a set of orbital elements as well as accelerometer bias and scale factors are estimated. In addition, constrained piecewise-constant accelerations (pseudo-stochastic parameters), set up in regular intervals of 15 min, are co-estimated to compensate for deficiencies in the background force modelling and the observation noise modelling, which employs the assumption of white noise for the K-band range-rates and a propagation of white noise for carrier phase observations through the kinematic point positioning process. Gravity field parameters are set up for each arc, however, then stacked to monthly normal equations, and solved on monthly level together with the daily orbit, accelerometer and pseudo-stochastic parameters.

The data analysed in this study for outliers are K-band range-rate post-fit residuals of a first joint orbit and gravity field determination process which are the difference between the range-rate observations and computed observations based on the just-estimated orbit and gravity field model. We use post-fit residuals instead of pre-fit residuals (a purely model based fully controlled approach), for several reasons: We intend to diminish the dependency on a priori models of the gravity field in the screening process. In addition, the empirical screening which is taken as a reference was based on post-fit residuals. Furthermore, in the context of time-variable gravity field determination the computational burden is not too heavy and even several iterations with post-fit residuals are feasible.

3 Methods of Outlier Detection

In this study three methods of outlier and anomaly detection are investigated. An empirical screening by visual inspection of time series of post-fit residuals, which has been used for two releases of AIUB GRACE gravity field solutions - this will also serve as reference - and two novel methods using machine learning. In addition, the latter two are combined.

3.1 Empirical Screening

The empirical screening is done by manually looking for anomalous patterns in the time series of post-fit residuals. It was applied for the processing of GRACE gravity fields from AIUB (Meyer et al. 2012, 2016). Data in an interval of $[-10,10]$ min around a suspiciously large oscillation, especially after IPU reboots, is removed. However, a cross-check is made not to exclude variations induced by flying over geophysical features, such as deep sea trenches. Additionally, an empirical screening is performed to detect bad data that are not obvious from the inspection of the time series of post-fit residuals. Each daily arc is removed once from the computation of the gravity field and it is examined whether the estimation of the gravity field improves without this arc. The assessment is performed by referencing to gravity field solution from other analysis centres and by validating the remaining signal left over the open ocean, which ought to be as low as possible since the oceanic mass change and the atmospheric mass change over the ocean is already reduced by the background force modelling.

3.2 Isolation Forest

The anomaly detection performed with the method of an Isolation Forest (IF, Liu et al. 2008) is also applied to the post-fit residuals of the first joint orbit and gravity field determination process. In general, an IF provides a fast and easy way to detect outliers within a high-dimensional dataset. The procedure was developed to compensate the disadvantages of conventional anomaly detection methods based on statistical approaches that represent normal instances through a normal profile and identifying anomalies by instances deviating from the normal profile. Examples for normal profiles are clusters found with unsupervised clustering methods, or a self organizing map automatically sorting the data in a smooth way. Such methods suffer from a high rate of false positives and computational complexity. An IF approaches the problem of anomaly detection with the assumption of anomalous points being unique and rare in a dataset. Both properties allow to detect anomalies by isolating points with a tree approach. Thereby, a dataset is step by step randomly separated, where each split is orthogonal to the previous one, which is shown in Fig. 1 for a two dimensional dataset. In our case along longitude, latitude, post-fit residual (orthogonal to the longitude-latitude plane) and time (orthogonal to the longitude-latitude-post-fit-residual-cube).

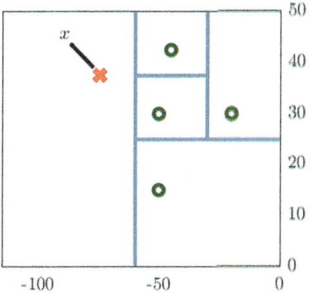

Fig. 1 Example of how an IF isolates points with cuts in the x and y directions when the data is in a two-dimensional plane. The fewer cuts necessary to completely isolate a point, the more likely it is an outlier (one vertical cut for the red x, in contrast to the green circles). In this way outliers may also be found globally embedded in the data. The procedure is repeated many times to get an ensemble of isolation trees and reduce false positives

Each separation represents an edge in the tree

$$s(x,n) = 2^{-\frac{\mathbb{E}[h(x)]}{c(n)}}, \quad (1)$$

where $h(x)$ is the path length to isolate x in a tree, and expectancy value $\mathbb{E}[h(x)]$ represents the average of $h(x)$ over an ensemble of trees. $\mathbb{E}[h(x)]$ is normalised by

$$c(n) = 2\,H(n-1) - 2\frac{n-1}{n}, \quad (2)$$

with n trees in an ensemble, where $H(n)$ is the harmonic number and can be estimated by

$$H(n) = \ln(n) + 0.5772156649. \quad (3)$$

The isolation score $s(x,n)$ is monotonically increasing with x and lies in the interval of $[0, 1]$. A high score indicates easy separability of x from the rest of the dataset, pointing more likely to an anomaly. In practise, not only one IF is computed but an ensemble, in which a threshold of $s > 0.9$ is taken to indicate anomalies. Inliers are flagged with the binary score of 1, outliers with -1. The experiment is repeated several times with a range of hyper-parameters and the average over the binary scores is build to finally separate outliers at a threshold of -0.9. In this way the certainty for an outlier to be a true outlier is increased, independent of our choice of hyper-parameters. The threshold was chosen by looking at the spread of the final averaged outlier scores and trial and error of different thresholds.

3.3 Mutual Information

Based on the post-fit residuals of the first joint orbit and gravity field determination process, and their geographical location, which is defined as the projection of the mid-point between the two satellites, Mutual Information (MI) tries to learn about dependencies between the post-fit residuals and the location.

Two random variables X and Y are independent if and only if the probability $P(X = x, Y = y) = P(X = x)P(Y = y)$. To quantify the linear dependency of two dependent random variables X and Y, a typical measure is the correlation between X and Y (Pearson 1895). However, this kind of correlation only captures the linear dependence between two random variables. MI, in contrast, captures the higher order dependencies as well as the linear dependence between two random variables, and is therefore a generalisation of the correlation measure. The MI can be derived from the Kullback-Leibler (KL) divergence:

$$D_{\text{KL}}(p\|q) = \sum_x p(x) \log \frac{p(x)}{q(x)} \quad (4)$$

may be a distance metric for the distance between the two distributions p and q. Defining the probability $p(x, y)$ as the true distribution of X and Y and $p(x)p(y)$ as the marginal probability distribution we get the MI by computing the KL-divergence

$$MI(X; Y) = \sum_x \sum_y p(x, y) \log \frac{p(x, y)}{p(x)p(y)} \quad . \quad (5)$$

To compute the MI between the positions of the satellites and the post-fit residuals corresponding to these positions we follow the same procedure as Panos et al. (2021). Equation 4 may be re-written with a function T in the so-called dual representation (Donsker and Varadhan 1983), where

$$D_{\text{KL}}(p\|q) = \sup_{T:\Omega \to \mathbb{R}} \left\{ \mathbb{E}_p[T] - \log\left(\mathbb{E}_q\left[e^T\right]\right) \right\} \quad . \quad (6)$$

The function T must be found such that the argument of the supremum holds. This function can be approximated by function approximators such as Neural Networks (NN). Thus, a maximisation problem in T arises and one can estimate the MI by learning the function T which maximises

$$MI(X; Y)_n \geq \sup_{\theta \in \Theta} \left\{ \mathbb{E}_{P_n}[T_\theta] - \log\left(\mathbb{E}_{P_n^\dagger}\left[e^{T_\theta}\right]\right) \right\} \quad , \quad (7)$$

where P_n stands for the joint probability distribution and is sampled by picking the position and post-fit residual from the same pixel, while $P_n^\dagger$ is the independent probability distribution, and is sampled by picking the position data and post-fit residual from two random pixels. The idea is depicted in Fig. 2. θ are the parameters of the NN which are optimized through back-propagation, and the subscript n denotes the batch number which determines the subset we sample over and therefore provides a unique estimate of MI. Since neural networks are universal function approximators (Cybenko 1989) it is guaranteed to find a function T_θ close to the real function T with arbitrary accuracy.

While training, the estimate of the MI converges with the progress of the NN capturing the dependencies between the variables to the true MI. This approach is called a Mutual Information Neural Estimator (MINE) and was developed by Belghazi et al. (2018). However, McAllester and Stratos (2020) have found in recent investigations that the lower bound cannot be greater than $\mathbb{O}(\log n)$ with n the number of samples. Therefore, the lower bound can be greatly underestimated.

The MI provides an estimate of how related an individual residual is with its location. The estimate of the MI is defined in the interval of $[0, 1]$ with $MI = 0$ meaning no correlation with the geographical location. The MINE, on the other hand, is in principle unbound retaining the property of $MI = 0$ indicating no dependency.

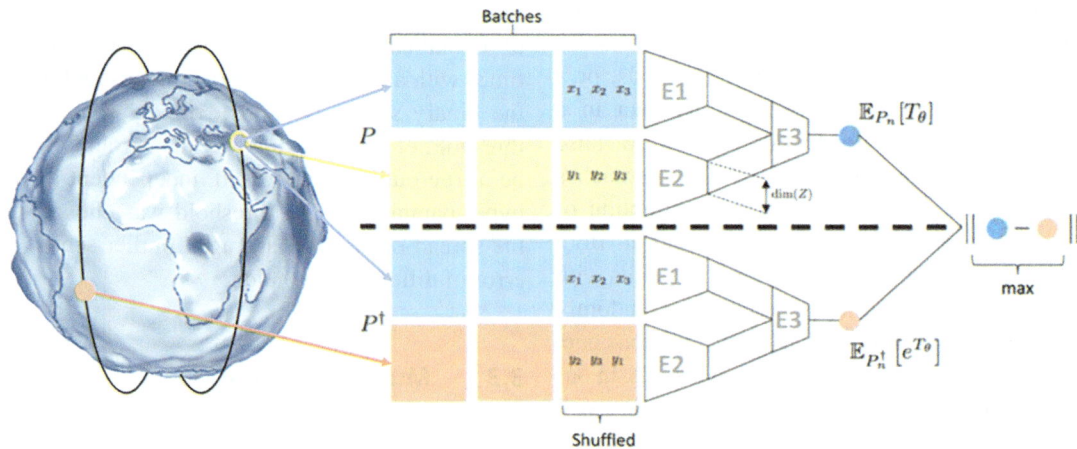

Fig. 2 Graphical depiction how the two probability distributions P and $P^\dagger$ are sampled, and processed by the neural network. (Figure adapted from Panos et al. (2021) under © American Astronomical Society (AAS))

4 Results

All methods of outlier and anomaly detection were applied to the GRACE and GRACE Follow-On data set of 2004 and 2019, respectively. Observations corresponding to post-fit residuals which have been characterised as outliers are removed and the process of orbit and gravity field recovery is repeated. The results give a qualitative assessment about the performance of the outlier detection methods.

Outlier detection is applied differently to the GRACE 2004 dataset than to the GRACE Follow-On 2019 dataset. This is due to the inherent difference in the orbital coverage of GRACE compared to GRACE Follow-on (repeat cycles). In the first part we have only applied IFs on complete orbits, where each segment goes from North pole–North pole. For comparison, we take the empirical screening as a reference which is also based on complete orbits. In the second part we employed IFs together with the MINE to the GRACE Follow-On dataset, where the method is applied point-wise. This means each data point is evaluated individually and removed from the dataset if it is deemed an outlier. For this dataset the empirical screening does not indicate outliers and the only quantitative and qualitative measure to investigate if IF found real outliers is the improvement of the gravity field determination.

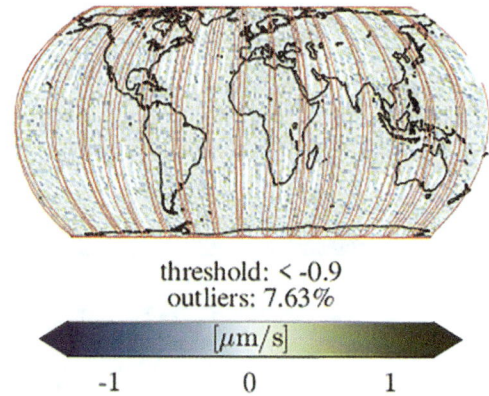

Fig. 3 Post-fit residuals for May 2004 binned on a $2° \times 2°$ grid and outliers (red) as marked by the IF with a threshold score of -0.9

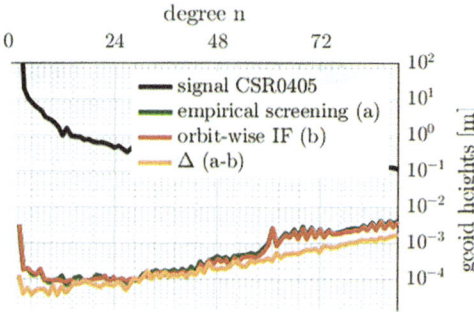

Fig. 4 Difference degree amplitudes w.r.t. a solution from CSR for the empirical screening and the IF based detection

4.1 GRACE—Isolation Forest

The IF algorithm is employed on the residuals of each orbital segment individually. This allows to find anomalous signals in time, since each segment is compared to every other segment in its entirety. It also helps reducing the bias induced by the similar geographical distribution of the orbits, i.e., to avoid rejecting data just because GRACE covered some geographical locations with only few passes. As a consequence outliers are removed as entire orbital segments, even if only one point along the segment was anomalous. Since we are not using MINE on GRACE data, the outliers found cannot be distinguished from unfitted signal. Figure 3 shows the outliers (in red), which make about 7.63% of the data, found for May 2004 with binned post-fit residuals in the background. For the whole test year of 2004 the number of outliers ranges from 0 to 15.3% of all K-band range-rates or in average over the year 4.62%. The empirical screening rejects slightly less observations with an annual average of 3.3%. The overlap between the empirically found outliers and the automatically found outliers is 88.62%. If one wishes to increase this overlap, a hyper-parameter search could be performed optimising for the overlap factor or train some other machine learning model in a supervised fashion, meaning the empirically flagged outliers are introduced as positive instances. However, it is already remarkable to see that the overlap with a completely unsupervised method and no hyper-parameter tuning reaches almost 90%.

The gravity field uncertainty estimation yields minor improvements for some months with differences up to roughly one order of magnitude as shown in Fig. 4, assessed by computing difference degree amplitudes w.r.t a solution (RL06 version 02) from the Center of Space Research at the University of Texas at Austin (CSR, University of Texas Center for Space Research 2018), compared to the empirical removal of outliers from visual inspection of post-fit residuals time series. Other months look even more similar to the empirical screening.

The MINE could not be applied to most of the GRACE data from 2004 due to the inhomogeneous coverage and large data gaps. Further training months would be necessary to stabilise the set up of the MINE. GRACE Follow-On post-fit residuals, however, allowed to apply the MINE, as shown in the next section.

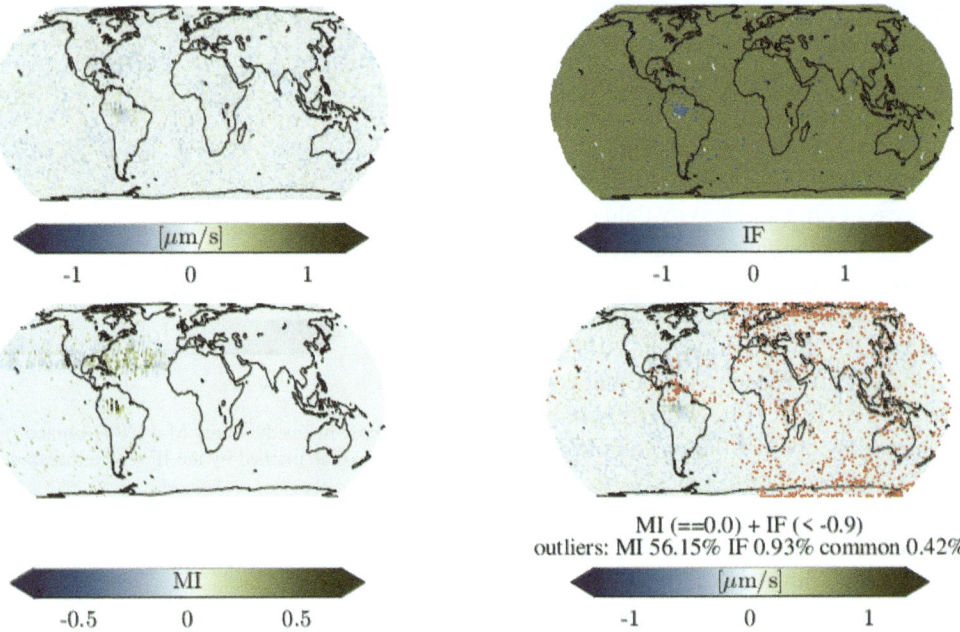

Fig. 5 Post-fit residuals for May 2019 binned on a 2° × 2° grid, which feature a prominent pattern around the Amazon basin (top left) and the corresponding binned IF scores for each residual (top right), where the Amazon area shows lowest scores. Associated MI estimates (bottom left) also exhibit the same prominent pattern for high scores. Applying MI+IF such that only geographically uncorrelated outliers remain yields a few outliers (red) mostly in polar regions (bottom right)

4.2 GRACE Follow-On—Isolation Forest and Mutual Information

The post-fit residuals data of 2019 (shown for May 2019 in Fig. 5, top left) was used to set up the unsupervised ensemble of IFs (not based on orbital segments but on each individual residual) and MI to separate unfitted signals from true outliers. The result is depicted in Fig. 5, top right, where strong negative values would indicate an outlier (blue) in contrast to inliers (brown). It exhibits that the IF based screening flags observations as outliers clustered over the Amazon basin: This, however, may not naturally point to outliers but to mis-modellings in the orbit and gravity field recovery process (unfitted gravity signal, which is then present in the post-fit residuals). Hence, we make use of a MINE to find potential (cor)relations between the geographical position and the post-fit residuals. The MI is able to resolve more complex relations and dependencies between the geographic position and the post-fit residuals. If there is any geographically (cor)related signal in the post-fit residuals, then $MI \neq 0$, which is the case over the Amazon area (Fig. 5, bottom left). In order to separate such anomalies from true outliers we combine the MI with the IF, thus only if $MI = 0$ and $IF < -0.9$ an outlier is assumed and the corresponding observation is removed from the data, see Fig. 5, bottom right with the outliers marked in red. For 2019 the number of outliers (as detected by the IF) which are not correlated with their locations (as found by the MINE) ranges from 0 to 1.2%. Removing those few, usually not gross, more or less randomly distributed outliers from the range-rate data does not lead to significant improvements in the gravity field recovery.

5 Conclusions

In this study we have shown how anomalies in GRACE and GRACE Follow-On post-fit residuals can be identified. We find individual outliers or geographically correlated anomalies, most likely from unfitted signal leaking into the post-fit residuals, using IF combined with MI. Additionally, we have shown that the post-fit uncertainty estimation improves for GRACE, when removing outliers on an orbit-segment basis with IFs. In case of GRACE, the model fit errors improve compared to a (tedious) manual removal of outliers by visual inspection. For GRACE Follow-On the unsupervised method of IF was applied in a combination with a MINE estimating the MI between post-fit residuals and their geographical location. This enables the separation of single outliers from geographically correlated anomalies. While this process could not improve the monthly Earth gravity field solution so far, it provides an automated detection of deficiencies in the underlying functional model of the orbit and gravity field recovery process. In future the stability of the MINE needs to be investigated and the training set expanded, especially for areas with $MI = 0$ and $IF < -0.9$ but no clear explanation for the geographic distribution of the outliers, as seen in Fig. 5. Furthermore, a parameter grid search could be of interest, if there is an optimal set of parameters for the outlier selection to improve the model fits. The next step after the investigation of K-band range-rate based gravity fields is to make use of the Laser Ranging Interferometer (LRI) instrument of GRACE Follow-On, which offers a unique data quality together with a higher sampling rate. Outliers and in particular searching for geographically correlated anomalies from LRI post-fit residuals will be a proper task for the machine learning tools.

Supplementary Information GRACE and GRACE Follow-On Level-1B data is obtained from http://grace.jpl.nasa.gov (NASA Jet Propulsion Laboratory 2018, 2019).

Declarations The authors declare that they have no conflict of interest. This research was supported by the European Research Council under the grant agreement no. 817919 (project SPACE TIE). All views expressed are those of the authors and not of the European Research Council. Calculations were performed on UBELIX (http://www.id.unibe.ch/hpc), the HPC cluster at the University of Bern.

References

Arnold D, Jäggi A (2022) AIUB GRACE-FO kinematic orbits, release 02. Astronomical Institute, University of Bern, dataset. https://doi.org/10.48350/169040

Behzadpour S, Mayer-Gürr T, Krauss S (2021) GRACE Follow-On accelerometer data recovery. J Geophys Res Solid Earth 126(5). https://doi.org/10.1029/2020JB021297

Belghazi MI, Baratin A, Rajeshwar S, et al (2018) Mutual information neural estimation. In: Dy J, Krause A (eds) Proceedings of the 35th International Conference on Machine Learning, vol 80. PMLR, Stockholm, Sweden, pp 531–540

Beutler G, Jäggi A, Mervart L, et al (2010) The celestial mechanics approach: theoretical foundations. J Geodesy 84(10):605–624. https://doi.org/10.1007/s00190-010-0401-7

Carrere L, Lyard F, Cancet M, et al (2016) FES 2014, a new tidal model – Validation results and perspectives for improvements. In: ESA Living Planet Conference, Prague, Czech Republic

Case K, Kruizinga G, Wu SC (2010) Gravity Recovery and Climate Experiment (GRACE) Level-1 Data Product User Handbook. Technical Report JPL D-22027, NASA Jet Propulsion Laboratory/California Institute of Technology, Pasadena, CA, USA. https://doi.org/10.5067/GRJPL-L1B03

Cybenko G (1989) Approximation by superpositions of a sigmoidal function. Math Control Signals Syst 2(4):303–314. https://doi.org/10.1007/BF02551274

Desai SD (2002) Observing the pole tide with satellite altimetry. J Geophys Res Oceans 107(C11):1–13. https://doi.org/10.1029/2001JC001224

Dobslaw H, Bergmann-Wolf I, Dill R, et al (2017) A new high-resolution model of non-tidal atmosphere and ocean mass variability for de-aliasing of satellite gravity observations: AOD1B RL06. Geophys J Int 211(1):263–269. https://doi.org/10.1093/gji/ggx302

Donsker MD, Varadhan SRS (1983) Asymptotic evaluation of certain markov process expectations for large time. IV. Commun Pure Appl Math 36(2):183–212. https://doi.org/10.1002/cpa.3160360204

Landerer FW, Flechtner FM, Save H, et al (2020) Extending the Global Mass Change Data Record: GRACE Follow-On Instrument and Science Data Performance. Geophys Res Lett 47(12). https://doi.org/10.1029/2020GL088306

Liu FT, Ting KM, Zhou ZH (2008) Isolation Forest. In: 2008 Eighth IEEE International Conference on Data Mining, pp 413–422. https://doi.org/10.1109/ICDM.2008.17

McAllester D, Stratos K (2020) Formal Limitations on the Measurement of Mutual Information. In: Chiappa S, Calandra R (eds) Proceedings of the 23rd International Conference on Artificial Intelligence and Statistics, vol 108. PMLR, Stockholm, Sweden, pp 875–884

Meyer U, Jäggi A, Beutler G (2012) Monthly gravity field solutions based on GRACE observations generated with the Celestial Mechanics Approach. Earth Planet Sci Lett 345–348:72–80. https://doi.org/10.1016/j.epsl.2012.06.026

Meyer U, Jäggi A, Jean Y, et al (2016) AIUBRL02: an improved time-series of monthly gravity fields from GRACE data. Geophys J Int 205(2):1196–1207. https://doi.org/10.1093/gji/ggw081

NASA Jet Propulsion Laboratory (2018) GRACE LEVEL 1B JPL RELEASE 3.0. NASA Physical Oceanography Distributed Active Archive Center, dataset. https://doi.org/10.5067/GRJPL-L1B03

NASA Jet Propulsion Laboratory (2019) GRACEFO Level-1B Release version 4.0 from JPL in ASCII. NASA Physical Oceanography Distributed Active Archive Center, dataset. https://doi.org/10.5067/GFL1B-ASJ04

Panos B, Kleint L, Voloshynovskiy S (2021) Exploring Mutual Information between IRIS Spectral Lines. I. Correlations between Spectral Lines during Solar Flares and within the Quiet Sun. Astrophys J 912(2). https://doi.org/10.3847/1538-4357/abf11b

Pearson K (1895) Notes on regression and inheritance in the case of two parents. In: Proceedings of the Royal Society of London, vol 58. Royal Society of London, London, UK, pp 240–242

Petit G, Luzum B (2010) IERS Conventions (2010) IERS technical note no. 36. Verlag des Bundesamts für Kartographie und Geodäsie, Frankfurt am Main, Germany

Tapley BD, Bettadpur S, Watkins MM, et al (2004) The gravity recovery and climate experiment: Mission overview and early results. Geophys Res Lett 31(9). https://doi.org/10.1029/2004GL019920

University of Texas Center for Space Research (2018) GRACE STATIC FIELD GEOPOTENTIAL COEFFICIENTS CSR RELEASE 6.0. NASA Physical Oceanography Distributed Active Archive Center, dataset. https://doi.org/10.5067/GRGSM-20C06

Vishwakarma BD, Devaraju B, Sneeuw N (2018) What is the spatial resolution of grace satellite products for hydrology? Remote Sens 10(6). https://doi.org/10.3390/rs10060852

Wen HY, Kruizinga G, Paik M, et al (2019) Gravity Recovery and Climate Experiment Follow-On (GRACE-FO) Level-1 data product user handbook. Technical report JPL D-56935 (URS270772), NASA Jet Propulsion Laboratory/California Institute of Technology, Pasadena, CA, USA. https://doi.org/doi.org/10.5067/GFL1B-ASJ04

Open Access This chapter is licensed under the terms of the Creative Commons Attribution 4.0 International License (http://creativecommons.org/licenses/by/4.0/), which permits use, sharing, adaptation, distribution and reproduction in any medium or format, as long as you give appropriate credit to the original author(s) and the source, provide a link to the Creative Commons license and indicate if changes were made.

The images or other third party material in this chapter are included in the chapter's Creative Commons license, unless indicated otherwise in a credit line to the material. If material is not included in the chapter's Creative Commons license and your intended use is not permitted by statutory regulation or exceeds the permitted use, you will need to obtain permission directly from the copyright holder.

Impact of a Priori Gravity Field Models on SLR Data Processing

Linda Geisser, Ulrich Meyer, Daniel Arnold, and Adrian Jäggi

Abstract

Satellite Laser Ranging (SLR) is essential for the geodetic parameter determination, e.g., geocenter and station coordinates, and, therefore, for long-term stable reference frame realizations. However, the orbit modeling and the quality of the parameter estimation partially depend on the background models. This study analyses the impact of static and time-variable a priori gravity field models provided by the Center for Space Research (CSR), the International Laser Ranging Service (ILRS), and the Combination Service for Time-variable Gravity Fields (COST-G) on the SLR data processing of spherical geodetic SLR satellites (LAGEOS-1/2 and LARES) at different orbital altitudes, by comparing the estimates of Earth rotation parameters, station coordinates and observation residuals. The COST-G model is further used to examine the impact of the mean pole model and the replacement of the spherical harmonic coefficients C_{21}/S_{21} according to convention provided by the International Earth Rotation Service (IERS).

While for LAGEOS-1/2 SLR data processing the a priori gravity field model has only a minor impact, the lower flying LARES satellite is more sensitive to the Earth's gravity field and requires a more sophisticated gravity field modeling, e.g., COST-G Fitted Signal Model (FSM). In order to achieve higher consistency and thus improved solutions, the same mean pole model should be used in the SLR data processing as for the generation of the used a priori gravity field model.

This study confirms the high quality of the COST-G FSM and demonstrates its suitability for potential use in the ILRS operational SLR processing.

Keywords

COST-G FSM · CSR · Earth rotation parameters · Gravity field models · ILRS · LAGEOS-1/2 · LARES · SLR

1 Introduction

Satellite Laser Ranging (SLR) makes an indispensable contribution to the determination of geodetic parameters, e.g., geocenter coordinates and terrestrial scale, which are used for the realization of long-term stable terrestrial reference frames (Altamimi et al. 2023). The reliability of the estimated parameters partially depends on the gravitational and non-gravitational background models. While some of the mis-modelings are absorbed by additional orbit parameters, e.g., empirical accelerations or pseudo-stochastic pulses (see Sect. 2), others may lead to systematic errors in the parameter estimation.

This study focuses on the impact of a priori gravity field models on the parameter estimation by SLR. In addition to a static gravity field model provided by the Center for

L. Geisser (✉) · U. Meyer · D. Arnold · A. Jäggi
Astronomical Institute, University of Bern, Bern, Switzerland
e-mail: linda.geisser@unibe.ch; ulrich.meyer@unibe.ch; daniel.arnold@unibe.ch; adrian.jaeggi@unibe.ch

Table 1 Orbit and geodetic parameterization for SLR data processing

Parameters	LAGEOS-1/2, LARES	
Osculating elements	$a, e, i, \Omega, \omega, u_0$	One set per 7-day arc
Dynamic orbit parameters	S_0, S_S, S_C, W_S, W_C	One set per 7-day arc
Pseudo-stochastic pulses in along-track	For LAGEOS-1/2	None
	For LARES	Twice per day
Station coordinates	NNR/NNT minimal constraint	One set per 7-day arc
Geocenter coordinates		One set per 7-day arc
ERP	Piecewise linear	One set per day
Range biases[a]	For LAGEOS-1/2: selected stations	One set per 7-day arc
	For LARES: all stations	One set per 7-day arc

[a] https://ilrs.dgfi.tum.de/fileadmin/data_handling/ILRS_Data_Handling_File.snx (Version: 2015/07/09)

Space Research (CSR), also time-variable models provided by the International Laser Ranging Service (ILRS, Pearlman et al. 2019) and the Combination Service for Time-variable Gravity Fields (COST-G, Jäggi et al. 2020) are analyzed. The sensitivity of a satellite w.r.t. time-variable mass changes on Earth depends on the orbital altitude of the satellite. Consequently, the LAGEOS-1/2 SLR combinations are extended by SLR observations to the low-flying LARES satellite, which has an orbital altitude of 1450 km. Furthermore, the influence of the old and updated mean pole model provided by the International Earth Rotation Service (IERS, Petit and Luzum 2010) is examined, when using the COST-G Fitted Signal Model (FSM) as a priori. The mean pole model is essential for computation of pole tides on the one hand, and for a possible replacement of the Spherical Harmonic (SH) geopotential coefficients C_{21}/S_{21} (see Sect. 3) on the other hand.

The impact of the a priori gravity field and mean pole models is validated by comparing a subset of the estimated parameters with external references and by inspecting the observation residuals.

2 SLR Data Processing at AIUB

The SLR data are analyzed in a development version of the Bernese GNSS Software (Dach et al. 2015), which is developed and maintained at the Astronomical Institute of the University of Bern (AIUB).

The satellite 7-day orbits (see Table 1) are represented by the six osculating Keplerian elements referring to the beginning of the arc and five dynamic orbit parameters, i.e., a constant acceleration S_0 in along-track (S) and Once-Per-Revolution (OPR) sine and cosine accelerations (S_S, S_C) in along-track and (W_S, W_C) in cross-track (W), respectively. To partially account for the more variable environment, e.g., increased air density or sensitivity to the Earth's gravity field of the low-flying satellite, additional pseudo-stochastic pulses, i.e., instantaneous velocity changes, in S are introduced for LARES with a spacing of 12h.

The satellite orbits are estimated together with the geodetic parameters of interest, i.e., station coordinates, geocenter coordinates and Earth Rotation Parameters (ERPs), as well as range biases (see Table 1) by means of a least-squares adjustment.

Table 2 lists the background models used in the SLR data processing. The different a priori gravity field models are described in more detail in the next section.

3 A Priori Gravity Field and Mean Pole Models

If the SH geopotential coefficients are not co-estimated, they will be fixed to the a priori values defined by the background model. Consequently, errors in the background model may lead to systematics in the parameter estimation.

The impact of the a priori gravity field model on SLR data processing and the resulting estimates are analyzed based on the following three gravity field models (see Fig. 1):

- **GGM05S** (Tapley et al. 2013) is a static gravity field model provided by the Center of Space Research, University of Texas at Austin (CSR). It is based on ten years of Gravity Recovery And Climate Experiment (GRACE, Tapley et al. 2004) data. Since, however, C_{20} is better determined by SLR (Cheng and Ries 2017), its time series is replaced by analysis of SLR observations.
- **ILRS model** is a partial time-variable gravity field provided by the ILRS for the generation of daily and weekly operational products (Pavlis 2021, pers. communication). The static GGM05S model forms the basis, where C_{20} and C_{21}/S_{21} are modified by time-variable corrections derived from CSR 15-day SLR solutions and after 20.09.2017 by a forecast model of Joint Center for Earth Systems Technology (JCET). Furthermore, the zonal SH geopotential coefficients of degrees 3-6 are replaced by time series provided by the ILRS. The ILRS coefficients already contain the AOD corrections.

Table 2 Background models used in the SLR data processing

Models	Description
Reference frame	SLRF2014[a]
ERPs	IERS-14-C04[b] (Bizouard et al. 2019)
Nutation model	IAU2000 (Mathews et al. 2002)
Subdaily pole model	DESAI (Desai and Sibois 2016)
Ocean tide model	FES2014b: d/o 30 (Lyard et al. 2021) + admittances
Earth Tides	Solid Earth tides, pole tides and
	Ocean pole tides: IERS 2010 (Petit and Luzum 2010)
Loading corrections	Ocean tidal loading: FES2014 (Lyard et al. 2021)
	Atmospheric tidal loading: Ray and Ponte (Ray and Ponte 2003)
De-aliasing products	Atmosphere + Ocean RL06: d/o 30
	incl. S1- and S2-atmosphere tides (Dobslaw et al. 2017)
Air drag	NRLMSISE-00 (Picone et al. 2002) (only for LARES)
Earth gravity field	GGM05S: d/o 90 (Tapley et al. 2013)
	ILRS model: d/o 90 provided by ILRS (Pavlis 2021)
	COST-G FSM: d/o 90 (Peter et al. 2022)

[a] https://cddis.nasa.gov/archive/slr/products/resource/SLRF2014_POS+VEL_2030.0_200325.snx (Accessed: 22/12/2022)
[b] https://hpiers.obspm.fr/eoppc/eop/eopc04/ (Accessed: 22/12/2022)

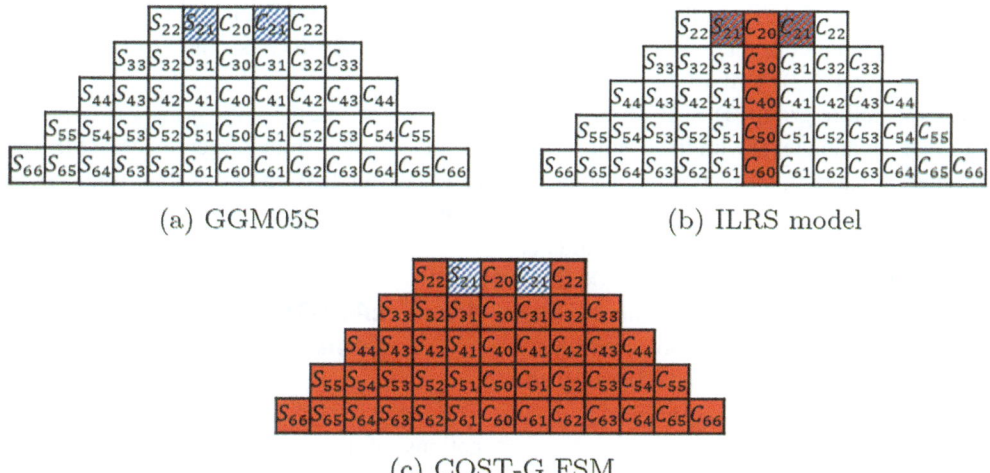

Fig. 1 A priori gravity field models: GGM05S, ILRS and COST-G FSM. Static and time-variable SH coefficients are marked in white and red, respectively. C_{21} and S_{21} (marked with blue stripes) can be replaced with the mean pole model according to Eq. (1)

- **COST-G FSM** (Peter et al. 2022) is a Fitted Signal Model (FSM) derived from monthly gravity field models resulting from combinations of monthly solutions provided by various analysis centers of COST-G. The solutions are based on the data of the GRACE and GRACE Follow-On (Landerer et al. 2020) satellite missions.

Since C_{21} and S_{21} have a geometrical meaning by describing the position of the Earth's figure axis, the conventional values can be expressed according to the IERS conventions (Petit and Luzum 2010) with a time-dependent mean pole (x_p, y_p) by

$$C_{21}(t) = \sqrt{3}x_p(t)C_{20} - x_p(t)C_{22} + y_p(t)S_{22},$$
$$S_{21}(t) = -\sqrt{3}y_p(t)C_{20} - y_p(t)C_{22} + x_p(t)S_{22}.$$
(1)

This realisation of C_{21}/S_{21} ensures that the geopotential coefficients take into account the time-variable effect of polar motion.

The mean pole model given in the IERS conventions (Petit and Luzum 2010) was updated in the year 2018. In this study, these two different mean pole models, the old and updated version, are labeled as MP-1 and MP-2, respectively.

The impact of replacing C_{21}/S_{21} according to Eq. (1) when either using MP-1 or MP-2 is examined based on the COST-G FSM gravity field model.

4 Validation of a Priori Gravity Field Models Used in SLR Combinations

The impact of the different a priori gravity field models are analyzed based on LAGEOS-1/2 SLR combinations (see

Table 3 Estimated ERP corrections w.r.t. IERS-14-C04 reference series of LAGEOS-1/2 (A) SLR combinations, when different a priori gravity field and mean pole models are used (for the years 2015–2020)

Sats	Grav. field model	Replace C_{21}/S_{21}	X-pole [μas] Bias	X-pole [μas] WRMS	Y-pole [μas] Bias	Y-pole [μas] WRMS	UT1-UTC [μs] Bias	UT1-UTC [μs] WRMS
A	GGM05S	No (MP-1)	86.4	152.3	28.3	129.0	−1.5	82.5
A	ILRS	No (MP-1)	85.6	142.1	31.9	118.1	−1.7	64.3
A	COST-G	No (MP-1)	53.5	126.0	36.8	124.8	−2.1	71.4
A	COST-G	No (MP-2)	69.2	129.2	23.8	117.4	−6.5	70.9
A	COST-G	Yes (MP-1)	61.8	127.0	21.9	114.4	−9.2	76.6
A	COST-G	Yes (MP-2)	66.0	130.7	28.2	122.4	−4.3	71.6

Table 4 Weighted (W) mean RMS of the Helmert transformation residuals of the station coordinates and the mean RMS of observation residuals for LAGEOS-1/2 (A) SLR combinations

Sats	Grav. field model	Replace C_{21}/S_{21}	W mean RMS of Helmert transf. of station coordinates [mm] North	W mean RMS of Helmert transf. of station coordinates [mm] East	W mean RMS of Helmert transf. of station coordinates [mm] Up	Mean RMS of obs. res. for sat. group [mm] A
A	GGM05S	No (MP-1)	6.18	5.12	11.82	9.91±1.41
A	ILRS	No (MP-1)	5.37	4.22	11.34	8.90±1.60
A	COST-G	No (MP-1)	5.46	4.36	11.48	8.95±1.48
A	COST-G	No (MP-2)	5.36	4.10	11.43	8.70±1.60
A	COST-G	Yes (MP-1)	5.32	4.21	11.28	8.91±1.60
A	COST-G	Yes (MP-2)	5.55	4.31	11.53	8.84±1.49

Sect. 4.1) for the years 2015-2020. To further increase the sensitivity to the gravity field models, the low-flying LARES satellite is included (see Sect. 4.2). LAGEOS-1/2 (satellite group A) and LARES (satellite group B) SLR observations are combined on normal equation level with fixed empirically determined weightings of 1 and 0.45, respectively (Geisser 2023).

The impact of the a priori gravity field models is validated by comparing the estimated ERPs to the IERS-14-C04 reference series (Bizouard et al. 2019), the RMS of the Helmert transformation residuals of the estimated station coordinates w.r.t. SLRF2014[a] reference frame, and by analyzing the observation residuals.

4.1 LAGEOS-1/2 SLR Combinations

When different gravity field models are used without replacing C_{21}/S_{21} and with MP-1, the biases of the estimated ERP corrections w.r.t. IERS-14-C04 reference series agree within 33 μas in X-pole, 8.5 μas in Y-pole and 0.6 μs in UT1-UTC, respectively (see Table 3). The comparison between solutions using the same gravity field model (COST-G FSM), which differ in mean pole models and C_{21}/S_{21} handling, reveals that except for the case, where C_{21}/S_{21} are not replaced and MP-1 is used, the biases are comparable.

The cluster of SLR stations in Europe leads to a better determination of the meridian plane passing through Greenwich and, therefore, of the Y-pole. Consequently, the bias of the polar motion in y-direction is half the size of the bias in x-direction and shows a higher stability in terms of Weighted Mean RMS (WRMS).

Except for the use of the static gravity field model GGM05S, the weighted mean RMS of the Helmert transformation residuals of station coordinates agree within 0.3 mm in North, East and Up direction, respectively (see Table 4).

Furthermore, the static gravity field model GGM05S leads to an almost 1 mm larger mean RMS of the observation residuals for LAGEOS-1/2 than time-variable gravity fields from ILRS and COST-G FSM (see Table 4).

4.2 LAGEOS-1/2 and LARES SLR Combinations

The inclusion of LARES causes an imbalance in the inclinations of the satellites, i.e., retrograde (LAGEOS-1) or prograde (LAGEOS-2, LARES) motion, and introduces systematic biases in the estimated ERP corrections compared to LAGEOS-1/2 SLR only combinations (Geisser 2023). However, in contrast to the LAGEOS-1/2 SLR combination, the gravity field model has a remarkable impact on the biases of the estimated ERP corrections, which differ by 220 μas, 73 μas and 27 μs in polar motion and UT1-UTC, respectively (see Table 5). The COST-G FSM leads to the most stable ERPs in terms of WRMS. Apart from the gravity field model, also the mean pole model and the replacement of C_{21}/S_{21} have a significant impact on polar motion. The bias of the X- and Y-pole differ by 124.5 μas and 43.2 μas when using different mean pole models, i.e., MP-1 and MP-2, for computing the pole tides. The COST-G FSM is based on MP-2 such that C_{21}/S_{21} are comparable with the computed values according to Eq. (1).

While the weighted mean RMS of the Helmert transformation residuals of the station coordinates differ on the mm-level for different gravity field models, they only vary in the sub-millimeter level for different mean pole models or the replacement of C_{21}/S_{21} (see Table 6).

Table 5 Estimated ERP corrections w.r.t. IERS-14-C04 reference series of LAGEOS-1/2 (A) and LARES (B) SLR combinations, when different a priori gravity field and mean pole models are used (for the years 2015-2020)

Sats	Grav. field model	Replace C_{21}/S_{21}	X-pole [µas] Bias	X-pole [µas] WRMS	Y-pole [µas] Bias	Y-pole [µas] WRMS	UT1-UTC [µs] Bias	UT1-UTC [µs] WRMS
A+B	GGM05S	No (MP-1)	−295.1	350.8	−12.5	158.7	−36.2	133.7
A+B	ILRS	No (MP-1)	−86.8	166.5	38.3	146.1	−16.3	87.0
A+B	COST-G	No (MP-1)	−74.8	148.7	60.4	136.8	−8.9	79.1
A+B	COST-G	No (MP-2)	49.7	126.5	17.2	116.3	−8.4	71.5
A+B	COST-G	Yes (MP-1)	126.6	177.3	−16.0	117.6	−10.2	75.2
A+B	COST-G	Yes (MP-2)	−46.4	135.6	41.1	128.9	−9.2	78.5

Table 6 Weighted (W) mean RMS of the Helmert transformation residuals of the station coordinates and the mean RMS of observation residuals for LAGEOS-1/2 (A) and LARES (B) SLR combinations

Sats	Grav. field model	Replace C_{21}/S_{21}	W mean RMS of Helmert transf. of station coordinates [mm] North	East	Up	Mean RMS of obs. res. for sat. groups [mm] A	B
A+B	GGM05S	No (MP-1)	8.91	6.46	12.43	11.90±1.88	24.7±3.08
A+B	ILRS	No (MP-1)	6.74	5.02	11.71	9.85±1.53	18.7±2.60
A+B	COST-G	No (MP-1)	5.98	4.68	11.63	9.49±1.41	15.7±2.82
A+B	COST-G	No (MP-2)	5.52	4.28	11.50	9.03±1.53	14.3±3.16
A+B	COST-G	Yes (MP-1)	5.59	4.43	11.37	9.24±1.52	14.9±2.97
A+B	COST-G	Yes (MP-2)	5.91	4.55	11.72	9.33±1.42	15.4±2.82

The RMS of observation residuals for LAGEOS-1/2 and LARES are smallest for the COST-G FSM without replacing C_{21}/S_{21} and using MP-2 (see Table 6).

5 Conclusions

SLR data processing based on satellites orbiting at high altitudes, e.g., LAGEOS-1/2, is marginally affected by the a priori gravity field and mean pole models. However, as soon as lower flying satellites such as LARES are included, a more sophisticated a priori gravity field model is required. Furthermore, the mean pole model has also a significant impact on the estimates and the observation residuals. If the same mean pole model as in the computation of the time-variable gravity field model is used, it is more consistent to not replace C_{21}/S_{21} according to Eq. (1).

This study confirms the high quality of the COST-G FSM. Since it is a fitted signal model, it enables a few months prediction and therefore offers a potential use in the ILRS operational daily and weekly SLR processing, especially when low-flying satellites will be included in the future.

Acknowledgements This research was supported by the European Research Council under the grant agreement no. 817919 (project SPACE TIE). All views expressed are those of the authors and not of the European Research Council. Calculations were performed on UBELIX (http://www.id.unibe.ch/hpc), the HPC cluster at the University of Bern.

References

Altamimi Z, Rebischung P, Collilieux X, et al (2023) ITRF2020: an augmented reference frame refining the modeling of nonlinear station motions. J Geodesy 97(5):47. https://doi.org/10.1007/s00190-023-01738-w

Bizouard C, Lambert S, Gattano C, et al (2019) The IERS EOP 14c04 solution for earth orientation parameters consistent with ITRF 2014. J Geodesy 93(5):621–633. https://doi.org/10.1007/s00190-018-1186-3

Cheng M, Ries J (2017) The unexpected signal in GRACE estimates of c20. J Geodesy 91(8):897–914. https://doi.org/10.1007/s00190-016-0995-5

Dach R, Lutz S, Walser P, et al (2015) Bernese GNSS software version 5.2. https://doi.org/10.7892/BORIS.72297

Desai SD, Sibois AE (2016) Evaluating predicted diurnal and semidiurnal tidal variations in polar motion with GPS-based observations. J Geophys Res Solid Earth 121(7):5237–5256. https://doi.org/10.1002/2016JB013125

Dobslaw H, Bergmann-Wolf I, Dill R, et al (2017) A new high-resolution model of non-tidal atmosphere and ocean mass variability for de-aliasing of satellite gravity observations: AOD1b RL06. Geophys J Int 211(1):263–269. https://doi.org/10.1093/gji/ggx302

Geisser L (2023) Generation and analysis of satellite laser ranging normal points for geodetic parameter estimation. (Thesis). University of Bern. https://boristheses.unibe.ch/4855/

Jäggi A, Meyer U, Lasser M, et al (2020) International combination service for time-variable gravity fields (COST-G): start of operational phase and future perspectives. In: Freymueller JT, Sánchez L (eds) Beyond 100: The next century in geodesy, vol 152. Springer International Publishing, pp 57–65. https://doi.org/10.1007/13452020109

Landerer FW, Flechtner FM, Save H, et al (2020) Extending the global mass change data record: GRACE follow-on instrument and science data performance. Geophys Res Lett 47(12). https://doi.org/10.1029/2020GL088306

Lyard FH, Allain DJ, Cancet M, et al (2021) FES2014 global ocean tide atlas: design and performance. Ocean Sci 17(3):615–649. https://doi.org/10.5194/os-17-615-2021

Mathews PM, Herring TA, Buffett BA (2002) Modeling of nutation and precession: New nutation series for nonrigid earth and insights into the earth's interior: New nutation series and the earth's interior. J Geophys Res Solid Earth 107:ETG 3-1–ETG 3-26. https://doi.org/10.1029/2001JB000390

Pavlis EC (2021) Updated/extended gravity coefficients series for OPERATIONAL series ONLY (e-mail)

Pearlman MR, Noll CE, Pavlis EC, et al (2019) The ILRS: approaching 20 years and planning for the future. J Geodesy 93(11):2161–2180. https://doi.org/10.1007/s00190-019-01241-1

Peter H, Meyer U, Lasser M, et al (2022) COST-g gravity field models for precise orbit determination of low earth orbiting satellites. Adv Space Res 69(12):4155–4168. https://doi.org/10.1016/j.asr.2022.04.005

Petit G, Luzum B (2010) IERS conventions (2010). IERS Tech Note (36):179

Picone JM, Hedin AE, Drob DP, et al (2002) NRLMSISE-00 empirical model of the atmosphere: Statistical comparisons and scientific issues: TECHNIQUES. J Geophys Res Space Phys 107:SIA 15-1–SIA 15-16. https://doi.org/10.1029/2002JA009430

Ray RD, Ponte RM (2003) Barometric tides from ECMWF operational analyses. Ann Geophys 21(8):1897–1910. https://doi.org/10.5194/angeo-21-1897-2003

Tapley BD, Bettadpur S, Ries JC, et al (2004) GRACE measurements of mass variability in the earth system. Science 305(5683):503–505. https://doi.org/10.1126/science.1099192

Tapley BD, Flechtner F, Bettadpur SV, et al (2013) The status and future prospect for GRACE after the first decade. In: AGU Fall Meeting Abstracts 2013:G32A–01. Conference Name: AGU Fall Meeting Abstracts ADS Bibcode: 2013AGUFM.G32A..01T

Open Access This chapter is licensed under the terms of the Creative Commons Attribution 4.0 International License (http://creativecommons.org/licenses/by/4.0/), which permits use, sharing, adaptation, distribution and reproduction in any medium or format, as long as you give appropriate credit to the original author(s) and the source, provide a link to the Creative Commons license and indicate if changes were made.

The images or other third party material in this chapter are included in the chapter's Creative Commons license, unless indicated otherwise in a credit line to the material. If material is not included in the chapter's Creative Commons license and your intended use is not permitted by statutory regulation or exceeds the permitted use, you will need to obtain permission directly from the copyright holder.

Dynamical Evaluation of Gravity Spherical Harmonic Coefficients due to Generally Shaped Polyhedra

Georgia Gavriilidou and Dimitrios Tsoulis

Abstract

The gravitational potential uncertainty process arising from the stochastic consideration of generally shaped polyhedra is outlined and tested on the real shape model of asteroid Psyche. The examined method is based on the computation of partial derivatives of spherical harmonic coefficients as implied by corresponding coordinate changes of the polyhedron's vertices, while the derived results are compared with gravity signal differences induced by the shape's variations using the line integral analytical approach. For the numerical tests, 3 regular grids of points with dimensions 600 km^2 were considered. The differences of the obtained results between the two approaches range from 85 m^2/s^2 to 300 m^2/s^2 for the gravitational potential uncertainties and from 2% to 2.4% for the normalized gravitational potential uncertainties. Additional tests were carried out on different points with increasing distance from the asteroid's surface to correlate the computed uncertainties with the spherical harmonic coefficients' maximum degree of expansion. As seen, inside the uncertainty region defined by the boundary of Brillouin sphere, the computed normalized gravitational potential uncertainties differ at the level of 0.04% for solutions of maximum degree of expansion {5, 10, 15, 20} while outside they gradually become identical. Therefore, the position of the computation points as well as the morphology of the examined mass distribution that defines the Brillouin sphere seem to strongly affect the derived results.

Keywords

Analytical gravity signal · Gravitational potential uncertainties · Polyhedral modelling · Spherical harmonic coefficients

1 Introduction

Gravity field forward modelling and interpretation define an active interdisciplinary research topic with specific tasks including terrain modelling, mass balance studies and satellite data analysis. Gravity field evaluation is directly linked with the geometrical shape selected to represent the examined source mass, e.g., prisms, polyhedra, tesseroids or mascons. Based on the adopted modelling approach, various algorithms exist for the computation of the induced gravity signal, which differ in terms of the used mathematical tools. The solutions are characterized as numerical or analytical, where the former give approximate results, are easier to handle and are computationally faster, while the latter give accurate results but are mathematically demanding and require more CPU time. Numerical methods solve the polyhedral gravitational potential expression by means of numerical integration or by formulating it into an infinite series of spherical harmonic coefficients (Balmino 1994; Colombo 1981; Cunningham 1970; Gottlieb 1993; Chen et al. 2019; Jamet and Tsoulis 2020). Analytical methods on the other hand, compute the gravitational potential straightforwardly

G. Gavriilidou (✉) · D. Tsoulis
Department of Geodesy and Surveying, Aristotle University of Thessaloniki, Thessaloniki, Greece
e-mail: georgiaga@topo.auth.gr

using different mathematical algorithms (D'Urso 2014a, b; Singh and Guptasarma 2001; Fukushima 2017; Saraswati et al. 2019; Tsoulis 2012; Holstein 2003).

Most of the available methodologies and investigations represent the examined distributions as polyhedra with deterministic geometry, which is not a realistic contemplation especially for time varying mass distributions. In fact, all celestial bodies are characterized by a continuous variation of their shape due to dynamical internal mass redistributions and physical processes. Considering a polyhedral mass distribution as stochastic or otherwise as a 3D body whose vertices have not a standard position, leads to the definition of an uncertain gravity field with respect to the shape's variations. From a statistical point of view, in a linear framework, when the distribution is considered Gaussian and the partial derivatives of spherical harmonic coefficients, resulting from corresponding coordinate changes of the polyhedron's vertices, are considered stochastic, their uncertainty quantification is feasible, and it remains Gaussian. Thus, the extracted gravity field variations constitute an estimation and not a straightforward deterministic computation. The gravitational signal variations can be useful for investigations such as deep-space exploration, satellite orbit determination and internal mass redistribution monitoring. One of the first attempts to quantify the gravity signal uncertainty considering the examined distribution as Gaussian is that of Muinonen (1998). Other papers that investigate the uncertainties but do not propose a specific mathematical formulation for its straightforward computation are those of Melman et al. (2013) and Santos et al. (2020). Some recent contributions on gravity signal uncertainty evaluation of stochastic polyhedra are those of Panicucci et al. (2020) and Bercovici et al. (2020), as well as Bercovici and McMahon (2019).

This paper aims at the evaluation of gravitational potential uncertainties derived from the stochastic consideration of asteroid Psyche. The implemented algorithm for the uncertainty quantification follows Panicucci et al. (2020) and is based on the method proposed by Werner (1997). The uncertainty results are compared with the gravity signal differences derived from stochastic bodies using the line integral analytical methodology (Tsoulis 2012). The first approach expresses the gravitational potential as an infinite series of spherical harmonic coefficients, while the uncertainties are computed by the harmonic synthesis of the partial derivatives of the coefficients that are defined by coordinate changes of the polyhedral vertices. The second approach computes straightforwardly the gravity signal from different stochastic representations of the same 3D mass distribution, while the differences between the solutions captures the range of the uncertainties induced by this shape variation.

2 Theory

The fundamental expression of gravitational potential induced by a 3D mass distribution reads

$$V = G \iiint_U \rho \frac{dU(Q)}{\ell}, \quad (1)$$

where ρ is the used density, G the Newtonian gravitational constant, Q the integration point, U the total volume of the distribution and ℓ the distance between the computation point and the infinitesimal volume element dU. The volume integral shown in Eq. (1) can be solved either analytically or numerically by replacing it with an infinite series of normalized spherical harmonic coefficients $\overline{C}_{n,m}, \overline{S}_{n,m}$ (Heiskanen and Moritz 1967) as follows

$$V(r, \theta, \lambda) = \frac{GM}{r} \left\{ 1 + \sum_{n=1}^{\infty} \left(\frac{a}{r}\right)^n \sum_{m=0}^{n} \overline{P}_{n,m}(\cos \theta) \times \left(\overline{C}_{n,m} \cos(m\lambda) + \overline{S}_{n,m} \sin(m\lambda)\right) \right\}, \quad (2)$$

where (r, θ, λ) are the spherical coordinates of the computation point, $\overline{P}_{n,m}$ the normalized associated Legendre functions, M the total mass of the distribution and a the minimum radius of a sphere that can enclose the whole mass distribution, also known as Brillouin sphere. The convergence of the harmonic series of Eq. (2) is guaranteed only outside the Brillouin sphere.

2.1 Werner's Approach

Werner's method (Werner 1997) derives recursive formulas for the computation of the spherical harmonic coefficients of Eq. (2), when the distribution is described by a polyhedral shape. First, the considered polyhedron is divided into smaller tetrahedra, each defined by one triangular polyhedral face and the origin of the coordinate system. Thus, the spherical harmonic coefficients are calculated for each tetrahedral division separately and then added over the extended polyhedral domain. The final expression of the fully normalized spherical harmonic coefficients, using the integral solution by Lien and Kajiya (1984), read

$$\begin{bmatrix} \overline{C}_{n,m} \\ \overline{S}_{n,m} \end{bmatrix} = \rho \sum_{tetrahedra} \left(\frac{\det[\mathbf{J}]}{(n+3)!} \sum_{i+j+k=n} i!j!k! \begin{bmatrix} \overline{\alpha}_{i,j,k} \\ \overline{\beta}_{i,j,k} \end{bmatrix} \right) \quad (3)$$

where **J** is the Jacobian matrix of the coordinates defining one tetrahedron and $\overline{\alpha}_{i,j,k}, \overline{\beta}_{i,j,k}$ are the trinomial coefficients describing the spherical harmonics for the respective tetrahedron, which are computed recursively as in (Werner 1997).

2.2 Line Integral Approach

The line integral analytical method (Petrović 1996; Tsoulis and Petrović 2001; Tsoulis 2012) for the computation of the gravity signal induced by a generally shaped polyhedron is based on a subsequent application of the Gauss divergence theorem to Eq. (1) which produces closed exact formulas. It comprises a constant monitoring of the relative position of the computation point and its projections onto the planes bearing the polyhedral faces and segments relative to the polyhedron, in order to evaluate the corresponding numerical singularities. Thus, the computation point can be placed anywhere in 3D space outside the polyhedral hull, or even on its boundary. There is no prerequisite for the present method, as each face can be built by a varying number of vertices. The final equation for the gravitational potential reads

$$V = \frac{G\rho}{2} \sum_{p=1}^{nf} \sigma_p h_p \left[\sum_{q=1}^{ns} \sigma_{pq} h_{pq} LN_{pq} + h_p \sum_{q=1}^{ns} \sigma_{pq} AN_{pq} + SING_{A_p} \right], \quad (4)$$

where $SING_{A_p}$ is the aforementioned singular term, nf the number of polyhedral faces, ns the number of segments of each face, while the rest of the parameters being linked to the distribution's geometry are explained in Tsoulis and Gavriilidou (2021).

2.3 Uncertainty Methodology

The algorithm for the quantification of gravitational potential uncertainties induced by stochastic generally shaped polyhedra proposed by Panicucci et al. (2020) is based on Werner's recursive formulas (c.f. Sect. 2.1). The only additional input data required is the polyhedron's vertices coordinate covariance matrix that defines its stochastic shape. The algorithm comprises two basic steps. First, it derives the partial derivatives of Werner's recursive formulas with respect to the polyhedron's vertex coordinates. Second, inserting these partial derivatives in the spherical harmonic synthesis, the partial derivative of the gravitational potential resulting by the corresponding changes in the polyhedral shape is obtained. Thus, the initial variances Δx, Δy, Δz lead to the definition of first derivative for the spherical harmonic coefficients $\delta \overline{C}_{nm}, \delta \overline{S}_{nm}$ over the vertex coordinates which subsequently result in the gravitational potential uncertainty estimation δV.

The algorithm's steps start with the reformulation of Werner's method into a matrix-oriented representation, where $\overline{\Phi}_{n,m} = (\overline{C}_{nm}, \overline{S}_{nm})^T$ is the vector of normalized spherical harmonic coefficients for a single combination of (n,m) and $\overline{\Phi} = (\overline{\Phi}_{0,0}, \overline{\Phi}_{1,0}, \overline{\Phi}_{1,1}, \ldots \overline{\Phi}_{N,N-1}, \overline{\Phi}_{N,N})^T$ is the vector containing all the coefficients up to expansion degree N. The partial derivatives of the recursive formulas are computed as the normalized spherical harmonic coefficients partial derivatives for a single tetrahedron with respect to the polyhedron's vertex coordinates **C** according to $\left[\frac{\partial \overline{\Phi}^t_{n,m}}{\partial \mathbf{C}}\right]$ (Panicucci et al. 2020). Adding the contribution of all the individual tetrahedra, the partial derivatives of the spherical harmonic coefficients emerging from the corresponding coordinate changes of the polyhedral vertices read

$$\left[\frac{\partial \overline{\Phi}_{n,m}}{\partial \mathbf{C}}\right] = \frac{1}{U} \left(\sum_{tetrahedra} \left[\frac{\partial \overline{\Phi}^t_{n,m}}{\partial \mathbf{C}}\right] - \overline{\Phi}_{n,m} \left[\frac{\partial U}{\partial \mathbf{C}}\right] \right), \quad (5)$$

where U is the total volume and $\left[\frac{\partial U}{\partial \mathbf{C}}\right]$ its partial derivative with respect to the polyhedron's vertices. Using the aforementioned abbreviations, the fundamental equation of the gravitational potential Eq. (2) is transformed to

$$V(r, \theta, \lambda) = \frac{G\rho U}{r} \overline{\Pi}^T \overline{\Phi}, \quad (6)$$

where $\overline{\Pi}$ is the vector containing the associated Legendre functions. Furthermore, the partial derivative of the gravitational potential read

$$\left[\frac{\partial V}{\partial \mathbf{C}}\right] = \frac{G\rho}{r} \left[\overline{\Pi}^T \overline{\Phi} \frac{\partial U}{\partial \mathbf{C}} + U \overline{\Pi}^T \frac{\partial \overline{\Phi}}{\partial \mathbf{C}} \right]. \quad (7)$$

Finally, the gravitational potential uncertainty P_V becomes

$$P_V(r, \theta, \lambda) = \left[\frac{\partial V}{\partial \mathbf{C}}\right] [\mathbf{P_c}] \left[\frac{\partial V}{\partial \mathbf{C}}\right]^T, \quad (8)$$

where $[\mathbf{P_c}]$ is the vertices coordinate covariance matrix, while the normalized uncertainty with respect to the gravitational potential value

$$\sigma_V(r, \theta, \lambda) = \frac{\sqrt{P_V(r, \theta, \lambda)}}{V(r, \theta, \lambda)}. \quad (9)$$

[$\mathbf{P_c}$] is manually constructed. Here the gaussian correlation function was implemented (Muinonen and Pieniluoma 2011; Panicucci et al. 2020) which uses two stochastic parameters, namely the standard deviation σ of each vertex and the correlation length l.

3 Numerical Implementation

The implemented algorithms are applied on the real shape data of asteroid Psyche (Vernazza et al. 2021; Shepard et al. 2017; Shepard et al. 2021). The shape model consists of 3234 vertices that define 6464 triangular faces, while the constant density value is set to 4500 kg/m^3 and the radius of Brillouin sphere equals 140 km. For the uncertainty algorithm, the spherical harmonic coefficients and their corresponding partial derivatives as implied by changes of the polyhedral shape were computed up to degree 20 while the needed CPU time was 64 h at a standard PC (16 Gbytes of RAM and a 10-core 1.3 GHz Intel Core i5 processor). For the analytical approach, based on the vertices coordinate covariance matrix [$\mathbf{P_c}$], the shapes of the largest and smallest possible considerations of the asteroid were created. Thus, the differences on the derived gravitational potential between these two considerations with the original shape should capture the range of the uncertainties. The comparisons took place on points distributed at the three Cartesian planes ($X = 0$, $Y = 0$ and $Z = 0$) outside the non-convergence area defined by the Brillouin sphere. The 3 created planes consist of 61×61 grid points with coordinates vary between -300 km and 300 km.

The uncertainties defined by the square root of Eq. (8) were computed, using the stochastic parameters $\sigma = 2.5144$ km and $l = 100$ km, as an optimal choice indicated by Panicucci et al. (2020). The fundamental statistical indicators of maximum, minimum, mean and standard deviation for the differences with the analytical approach are shown in Table 1. The "Largest polyhedron" columns show the differences between the uncertainty algorithm results and the gravity signal differences derived from the analytical approach concerning the largest possible asteroid consideration minus the original shape, while the "Smallest polyhedron" columns give the differences of the original shape minus the smallest polyhedron. The differences between the two aspects of the analytical approach range from 0 to 50 m^2/s^2, while the differences between the uncertainty algorithm and the analytical approach range between 85 and 300 m^2/s^2. The mean value is at the level of 150 m^2/s^2 and the std. at the 40 m^2/s^2 level.

Furthermore, the normalized uncertainties were also computed using Eq. (9), while the results only for the $Z = 0$ Cartesian plane are shown in Fig. 1. The first row represents the uncertainty algorithm results, the second row the normalized uncertainties derived from the analytical solution, specifically differences of the original shape minus the smallest possible polyhedron, while the third row shows the uncertainties from the analytical solution for the largest polyhedron minus the original shape. The computed statistics of the differences between the uncertainty algorithm and the two aspects of the analytical approach are shown in Table 1. For all computed planes, both approaches lead to the same magnitudes of highest and lowest normalized uncertainties, while the numerical differences between the uncertainty algorithm and the analytical approach are at the level of 2%. The corresponding mean values are at the 2.2% level, while the standard deviation is 0.03%.

Additional tests were made to address the influence of the spherical harmonics' maximum degree of expansion to the uncertainty quantification. Here, the gravitational potential normalized uncertainties were re-evaluated four times for maximum expansion degrees $n = \{5, 10, 15, 20\}$. The comparisons took place on a set of 116 3D points located inside the theoretical box defined by the Cartesian plane limits of the first test with increasing distance from the asteroid's boundary. The results for points outside the Brillouin sphere are shown in Fig. 2. There are differences between the four solutions in the range of 0.04% at the closest point on the asteroid, 0.005% on the Brillouin sphere boundary and 2E-5% on the last distant point. Thus, as the distance between the asteroid and the computation point increases so does the convergence between the solutions. Additionally, at the point of radius 200 km approximately, the solutions with $n = \{10, 15, 20\}$ start to coincide and differ from the solution of $n = 5$ at the level of 4E-4%.

Table 1 Statistics for the gravitational potential uncertainty differences between the uncertainty algorithm results and the analytical solution gravity signal differences

	Uncertainty Differences [m^2/s^2]		Normalized Uncertainty Differences [%]	
	Largest polyhedron	Smallest polyhedron	Largest polyhedron	Smallest polyhedron
Min	94.26	85.39	2.17	1.98
Max	300.97	268.06	2.37	2.14
Mean	161.87	146.23	2.32	2.10
Std	42.04	37.61	0.03	0.03

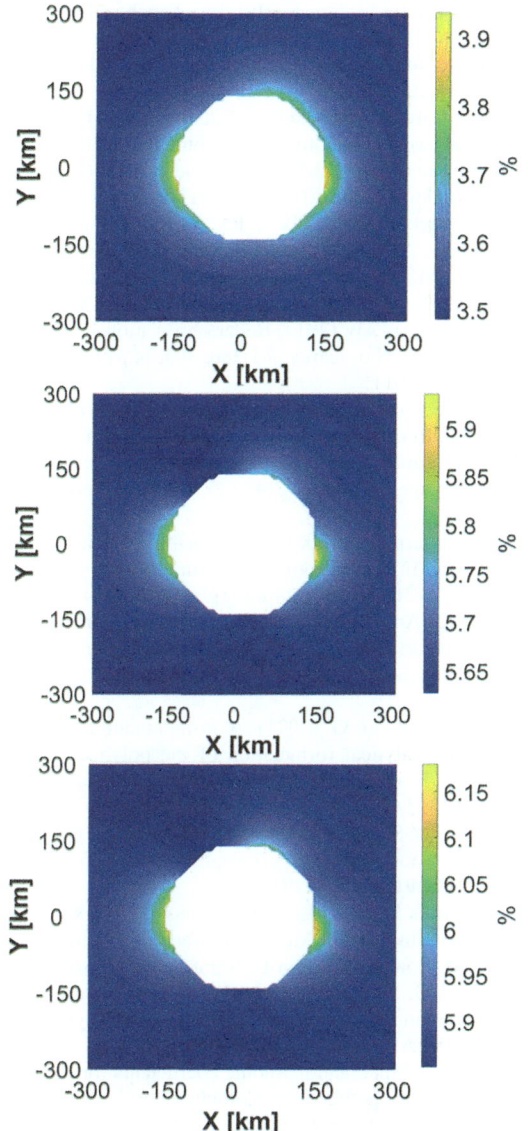

Fig. 1 Normalized uncertainties in the Z Cartesian plane

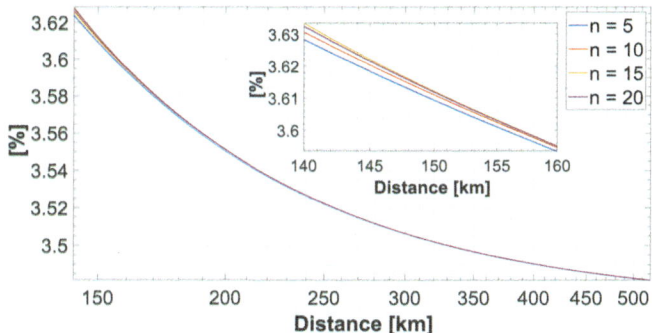

Fig. 2 Gravitational potential normalized uncertainties in logarithmic scale for points outside the Brillouin sphere

4 Discussion

The basic steps for the estimation of gravitational potential uncertainties induced by stochastic polyhedra were outlined. For the implementation of the presented algorithm, some basic cases of computation points with respect to asteroid Psyche were considered. The results derived from this stochastic representation seem to be acceptable but raise some additional questions that need to be addressed in future investigations. First, a wider range of mass distribution cases with irregular shapes must be considered, to show a connection between the morphology of the examined polyhedron and its gravity field uncertainty quantification. Second, the convergence of uncertainty results within the Brillouin sphere needs to be addressed, especially for the cases of irregular mass distributions and low spherical harmonics maximum degree of expansion, by running scenarios with special computation point locations. Furthermore, the relation of the maximum degree of expansion on the obtained uncertainties needs to be evaluated, for points not only near the examined polyhedron's hull but also at a large distance from it. Finally, a comprehensive statistical review is of importance to link the arbitrary definition of the statistical parameters that are selected to construct the covariance matrix $[\mathbf{P}_c]$ in the induced uncertainties.

Acknowledgments The first author (GG) was supported by the Hellenic Foundation for Research and Innovation (HFRI) under the 4th Call for HFRI PhD Fellowships (Fellowship Number: 10834). Thanks are due to the two anonymous reviewers for their constructive comments.

References

Balmino G (1994) Gravitational potential harmonics from the shape of an homogeneous body. Celest Mech Dyn Astron 60(3):331–364. https://doi.org/10.1007/BF00691901

Bercovici B, McMahon JW (2019) Inertia parameter statistics of an uncertain small body shape. Icarus 328:32–44. https://doi.org/10.1016/j.icarus.2019.02.016

Bercovici B, Panicucci P, McMahon J (2020) Analytical shape uncertainties in the polyhedron gravity model. Celest Mech Dyn Astron 132(5):29. https://doi.org/10.1007/s10569-020-09967-3

Chen C, Chen Y, Bian S (2019) Evaluation of the spherical harmonic coefficients for the external potential of a polyhedral body with linearly varying density. Celest Mech Dyn Astron 131(2):8. https://doi.org/10.1007/s10569-019-9885-5

Colombo OL (1981) Numerical methods for harmonic analysis on the sphere

Cunningham LE (1970) On the computation of the spherical harmonic terms needed during the numerical integration of the orbital motion of an artificial satellite. Celest Mech 2(2):207–216. https://doi.org/10.1007/BF01229495

D'Urso MG (2014a) Analytical computation of gravity effects for polyhedral bodies. J Geod 88(1):13–29. https://doi.org/10.1007/s00190-013-0664-x

D'Urso MG (2014b) Gravity effects of polyhedral bodies with linearly varying density. Celest Mech Dyn Astron 120(4):349–372. https://doi.org/10.1007/s10569-014-9578-z

Fukushima T (2017) Precise and fast computation of the gravitational field of a general finite body and its application to the gravitational study of asteroid Eros. AJ 154(4):145. https://doi.org/10.3847/1538-3881/aa88b8

Gottlieb RG (1993) Fast gravity, gravity partials, normalized gravity, gravity gradient torque and magnetic field: derivation, code and data. In: NASA contractor report 188243, National Aeronautics and Space Administration Lyndon B. Johnson Space Center Houston, Texas

Heiskanen WA, Moritz H (1967) Physical geodesy. W.H. Freeman and Company, San Francisco

Holstein H (2003) Gravimagnetic anomaly formulas for polyhedra of spatially linear media. Geophysics 68(1):157–167. https://doi.org/10.1190/1.1543203

Jamet O, Tsoulis D (2020) A line integral approach for the computation of the potential harmonic coefficients of a constant density polyhedron. J Geod 94(3):30. https://doi.org/10.1007/s00190-020-01358-8

Lien SL, Kajiya JT (1984) A symbolic method for calculating the integral properties of arbitrary nonconvex polyhedra. IEEE Computer Graphics and Applications 4:35–42. https://doi.org/10.1109/MCG.1984.6429334

Melman JCP, Mooij E, Noomen R (2013) State propagation in an uncertain asteroid gravity field. Acta Astronaut 91:8–19. https://doi.org/10.1016/j.actaastro.2013.04.027

Muinonen K (1998) Introducing the Gaussian shape hypothesis for asteroids and comets. Astron Astrophys 332:1087–1098

Muinonen K, Pieniluoma T (2011) Light scattering by Gaussian random ellipsoid particles: first results with discrete-dipole approximation. J Quant Spectrosc Radiat Transf 112(11):1747–1752. https://doi.org/10.1016/j.jqsrt.2011.02.013

Panicucci P, Bercovici B, Zenou E, McMahon J, Delpech M, Lebreton J, Kanani K (2020) Uncertainties in the gravity spherical harmonics coefficients arising from a stochastic polyhedral shape. Celest Mech Dyn Astron 132(4):23. https://doi.org/10.1007/s10569-020-09962-8

Petrović S (1996) Determination of the potential of homogeneous polyhedral bodies using line integrals. J Geod 71(1):44–52. https://doi.org/10.1007/s001900050074

Santos LBT, Marchi L, Sousa-Silva PA, Sanchez DM, Aljbaae S, Prado AFBA (2020) Dynamics around an asteroid modeled as a mass tripole. RevMexAA 56(2):269–286. https://doi.org/10.22201/ia.01851101p.2020.56.02.09

Saraswati AT, Cattin R, Mazzotti S, Cadio C (2019) New analytical solution and associated software for computing full-tensor gravitational field due to irregularly shaped bodies. J Geod 93(12):2481–2497. https://doi.org/10.1007/s00190-019-01309-y

Shepard MK, Richardson J, Taylor PA, Rodriguez-Ford LA, Conrad A, de Pater I, Adamkovics M, de Kleer K, Males JR, Morzinski KM, Close LM, Kaasalainen M, Viikinkoski M, Timerson B, Reddy V, Magri C, Nolan MC, Howell ES, Benner LAM, Giorgini JD, Warner BD, Harris AW (2017) Radar observations and shape model of asteroid 16 psyche. Icarus 281:388–403. https://doi.org/10.1016/j.icarus.2016.08.011

Shepard MK, de Kleer K, Cambioni S, Taylor PA, Virkki AK, Rívera-Valentin EG, Rodriguez S-VC, Fernanda Z-ML, Magri C, Dunham D, Moore J, Camarca M (2021) Asteroid 16 psyche: shape, features, and global map. Planet Sci 2(4):125. https://doi.org/10.3847/PSJ/abfdba

Singh B, Guptasarma D (2001) New method for fast computation of gravity and magnetic anomalies from arbitrary polyhedra. Geophysics 66(2):521–526. https://doi.org/10.1190/1.1444942

Tsoulis D (2012) Analytical computation of the full gravity tensor of a homogeneous arbitrarily shaped polyhedral source using line integrals. Geophysics 77(2):F1–F11. https://doi.org/10.1190/geo2010-0334.1

Tsoulis D, Gavriilidou G (2021) A computational review of the line integral analytical formulation of the polyhedral gravity signal. Geophys Prospect 69(8–9):1745–1760. https://doi.org/10.1111/1365-2478.13134

Tsoulis D, Petrović S (2001) On the singularities of the gravity field of a homogeneous polyhedral body. Geophysics 66(2):535–539. https://doi.org/10.1190/1.1444944

Vernazza P, Ferrais M, Jorda L, Hanuš J, Carry B, Marsset M, Brož M, Fetick R, Viikinkoski M, Marchis F, Vachier F, Drouard A, Fusco T, Birlan M, Podlewska-Gaca E, Rambaux N, Neveu M, Bartczak P, Dudziński G, Socha L (2021) VLT/SPHERE imaging survey of the largest main-belt asteroids: final results and synthesis. Astron Astrophys 654:A56. https://doi.org/10.1051/0004-6361/202141781

Werner RA (1997) Spherical harmonic coefficients for the potential of a constant-density polyhedron. Comput Geosci 23(10):1071–1077. https://doi.org/10.1016/S0098-3004(97)00110-6

Open Access This chapter is licensed under the terms of the Creative Commons Attribution 4.0 International License (http://creativecommons.org/licenses/by/4.0/), which permits use, sharing, adaptation, distribution and reproduction in any medium or format, as long as you give appropriate credit to the original author(s) and the source, provide a link to the Creative Commons license and indicate if changes were made.

The images or other third party material in this chapter are included in the chapter's Creative Commons license, unless indicated otherwise in a credit line to the material. If material is not included in the chapter's Creative Commons license and your intended use is not permitted by statutory regulation or exceeds the permitted use, you will need to obtain permission directly from the copyright holder.

Optimizing Airborne Flight Line Spacing for Geoid Determination with Full Gravity Vectors

Ismael Foroughi, Mehdi Goli, Stephen Ferguson, and Spiros Pagiatakis

Abstract

The horizontal components of the airborne gravity vector are equivalent to the deflection of the vertical at the flight level and contain signals of the slope of Earth's gravity field. We test the contribution of such components in finding the optimum flight line spacing for geoid modelling. We use the one-step integration method and create a system of linear equations containing the three components of the airborne gravity vector as observations and solve the geodetic boundary value problem on the reference ellipsoid as an overdetermined weighted least-squares problem. We test our methodology in the Colorado region in the USA given that it is one of the most challenging areas for geoid modelling. We show that by incorporating the horizontal components at the flight level, one can increase the flight line spacing by almost 40%, thereby significantly reducing the cost of airborne surveys while maintaining the same accuracy in the estimated geoid heights as when the scalar value of gravity is used.

Keywords

Airborne gravimetry · Geoid determination · Inverse problems · Survey flight design

1 Introduction

Precise local geoid determination requires global integration of gravity observations with homogenous accuracy and resolution (Heiskanen and Moritz 1967). Such a global dataset is not yet available, therefore, the determination of the local geoid mostly depends on the accuracy and resolution of the regional gravity observations, i.e., the near-zone contribution (Fotopoulos 2005; McCubbine et al. 2017; Foroughi et al. 2019; Abd-Elmotaal and Kühtreiber 2019; Ophaug and Gerlach 2020; Goyal et al. 2023). The effect of observation points far from the area of geoid calculation, i.e., the far-zone contribution, is computed using one of the Global Gravity Models (GGMs). Regional gravity observations are provided by gravity measurements on and/or above the Earth's surface, e.g., land, airborne, and satellite gravity observations. Most present-day geodetic applications use the geoid as the reference surface for heights with the requirement of a sub-centimetre level of accuracy. However, the satellite gravity missions alone cannot meet this requirement even though they provide a homogenous figure of Earth's gravity field (Huang et al. 2022). Land-based gravity observations yield data of the highest frequency content on the Earth's gravity field, yet they come with the drawback of being expensive to acquire and they are often of lower density

I. Foroughi (✉)
Department of Earth and Space Science and Engineering, York University, Toronto, ON, Canada

Canadian Geodetic Survey, Surveyor General Branch, Natural Resources Canada, Ottawa, ON, Canada
e-mail: i.foroughi@unb.ca

M. Goli
Faculty of Civil Engineering, Shahrood University of Technology, Shahrood, Iran

S. Ferguson
Sander Geophysics Ltd., Ottawa, ON, Canada

S. Pagiatakis
Department of Earth and Space Science and Engineering, York University, Toronto, ON, Canada

in rough topography and coastal regions. Airborne gravity surveys offer consistent regional gravity observations for densifying the ground-based gravity observations. They also serve as an intermediate data source, bridging the frequency gap between satellite and land-based gravity measurements, see, e.g., Gravity for Redefinition of the American Vertical Datum (GRAV-D) (Smith 2007). The power of airborne gravity observations in improving the accuracy of the local geoid models has been reported in the last two decades, please see e.g. (Kern et al. 2003; Huang et al. 2019; Jiang et al. 2020; Li et al. 2022; Foroughi et al. 2023).

The standard practice of using scattered airborne gravity observations for geoid modelling is to interpolate them to grid points, downward continue, and combine them with terrestrial gravity observations (Novák et al. 2003). In such cases, the scalar value of the airborne gravity vector (or its magnitude) is used (Forsberg et al. 2000). This is equivalent to utilizing only the vertical component of the full three-dimensional (3D) gravity vector. This limitation arises either from the lack of observations of the horizontal components (HC) or from the large long-wavelength errors of such measurements from most of the airborne gravimeter platforms used (Serpas 2003). With the recent progress in instrumentation, and updates in postprocessing methods, the accuracy of the measurements of the horizontal components of the airborne gravity vector has improved (Senobari 2010; Deng et al. 2020), therefore, their impact on local geoid modelling needs to be investigated. Only a few studies have considered this impact. Serpas and Jekeli (2005) employed the Hotine and line integral methods for geoid determination in the Canadian Rocky Mountains using test airborne gravity vector observations adjusted by simulated crossover profiles. They showed that geoid estimates using only the HC provide a better accuracy compared to the use of the vertical component (VC), however, the combination of the three (the focus of this study) requires more accurate horizontal components and an adequate covariance model. There have been many improvements in the accuracy of airborne gravimetry since this study, see, e.g., (Sander et al. 2011; Mayunga 2016; Ferguson et al. 2019; Vyazmin and Golovan 2023). Ferguson et al. (2019) combined the airborne gravity vector and land gravity observations for local geoid modelling using Least Squares Collocation (LSC) in Kimberley, Australia. Deng et al. (2020) used a simulated airborne gravity vector at a constant flight altitude to confirm the improvement in the geoid determination by using only the horizontal components of the airborne gravity vector.

The main question of this study is that by using the full airborne gravity vector, specifically by including the HC, we can increase the necessary separation between flight lines to achieve certain accuracy in geoid modelling as HCs can fill in the gravity gaps between flight lines. What follows is the theoretical background used in Sect. 2, with the numerical results in Sect. 3 and the concluding remarks in Sect. 4.

2 Theory

The one-step integration method is used to convert the airborne gravity vector observations at the flight level to disturbing potential at the reference ellipsoid (and geoid heights using the Bruns formula). What is considered standard practice is using the magnitude of the airborne gravity vector or its VC only. The theory of the one-step integration method using scalar-valued gravity observations was fully developed by Novák (2003) and used in practice by Goli et al. (2019). We follow the same approach but use vector gravity observations instead. An implementation flowchart of this method is also provided in Foroughi et al. (2023).

In spherical approximation, the one-step integration method in vector mode reads (bold variables indicate vector):

$$\delta \mathbf{g}(r, \Omega) = \frac{1}{4\pi R} \int_{\Omega_G} T(R, \Omega') \, \mathbf{P}(r, \psi, R) \, d\Omega', \quad (1)$$

where R is the mean radius of the Earth and $r = R + h$ is the geocentric radius of computation points with h being the geodetic height of the airborne gravity observations. $\Omega = (\varphi, \lambda)$ and $\Omega' = (\varphi', \lambda')$ indicate the latitude and longitude of the computation and integration points respectively and ψ represents their spherical distance. Ω_G represents the integration domain and T is the (sought) disturbing potential computed at the reference ellipsoid approximated by a sphere with radius R. In Eq. (1), $\delta \mathbf{g}$ is the gravity disturbance vector (measured at flight level) in the local right-handed Cartesian coordinate system where the z-axis observation (δg_Z) is upward and gravity observations towards North and West are indicated by the δg_X and δg_Y:

$$\delta \mathbf{g} = \begin{bmatrix} \delta g_X \\ \delta g_Y \\ \delta g_Z \end{bmatrix}. \quad (2)$$

The kernel function in Eq. (1) is the first-order gradient of the spherical Poisson function:

$$\mathbf{P}(r, \psi, R) = R \begin{bmatrix} P_X \\ P_Y \\ P_Z \end{bmatrix} = R \begin{bmatrix} \frac{1}{r} \frac{\partial P}{\partial \varphi} \\ \frac{1}{r \cos \varphi} \frac{\partial P}{\partial \lambda} \\ -\frac{\partial P}{\partial r} \end{bmatrix}, \quad (3)$$

and the spatial form of the Poisson kernel reads:

$$P(r, \psi, R) = \frac{R(r^2 - R^2)}{L(r, \psi, R)^3}, \quad (4)$$

where L is the spatial distance between the computation and integration points.

Numerical evaluation of Eq. (1) requires global integration of the gravity vector disturbances, although such data are not available. For this case, the vector components of the Poisson kernel in Eq. (3) are modified so they are more sensitive to the near-zone points and the effect of the far zone is computed by an EGM.

Using a discretized form of Eq. (1), one can create a system of linear equations as:

$$\mathbf{A} = \begin{bmatrix} \mathbf{A}_X \\ \mathbf{A}_Y \\ \mathbf{A}_Z \end{bmatrix} \text{ and } \mathbf{Ax} = \begin{bmatrix} \mathbf{b}_X \\ \mathbf{b}_Y \\ \mathbf{b}_Z \end{bmatrix} + \begin{bmatrix} \mathbf{r}_X \\ \mathbf{r}_Y \\ \mathbf{r}_Z \end{bmatrix}, \quad (5)$$

where $\mathbf{x}$ is the vector of unknown quantities, which here are the disturbing potential (or geoid heights), and $\mathbf{b}_X, \mathbf{b}_Y, \mathbf{b}_Z$ are gravity disturbance vector components in which the effect of the far zone is removed. The $\mathbf{r}_X, \mathbf{r}_Y, \mathbf{r}_Z$ are the components of the observation error vector. The weighted least-squares (WLS) solution of Eq. (5) reads:

$$\mathbf{x} = \left(\mathbf{A}^T \mathbf{W} \mathbf{A} \right)^{-1} \mathbf{A}^T \mathbf{W} \, \mathbf{b}. \quad (6)$$

where $\mathbf{W}$ is the weight matrix of the gravity observations determined from the observation error. For numerical computations, we used the modified band-limited kernels corresponding to the frequency band of the input data, see Sect. 3. Please note that when $\psi = 0$, the P_X and P_Y kernels are zero, and the P_Z has larger values but it is not singular when $h > 0$ (Novák 2003).

3 Numerical Test

We select our test area in the Colorado region of the USA, one of the most challenging environments for geoid modelling in terms of its rough topography. We select the data coverage area between $-109° < \lambda < -102°$ and $35° < \varphi < 39°$ and geoid computation area between $-108° < \lambda < -104°$ and $35.5° < \varphi < 38.5°$ where we have most of the historical GPS/Levelling points (cf., Wang et al. 2021) and also to avoid the edge effect when computing the near-zone contribution. To accurately test our methodology we use synthetic data observations computed from XGM2019 (Zingerle et al. 2020) whose highest degree of expansion is 5,540, corresponding to 2 arc-min spatial resolution (~ 4 km). This allows us to compare our estimated geoid heights with the "true" geoid heights synthesized from the same model. To simulate a realistic airborne survey over this area we:

- Develop a gentle (smooth) drape flight surface over our data coverage area. Such a surface provides a semi-constant flight height that is very common for geophysical airborne surveys, please see (Sander 1998; Dumont and Bardossy 2014).
- Extract five different test flight lines with 2, 5, 7, 10, and 16 km spacing and 100 km cross profiles from the gentle drape surface. As an example, please see Fig. 1 for the 16 km line spacing case.
- Compute the synthetic gravity disturbance vector (3D) using XGM2019 up to degree/order 5,540 at the scattered points with 500 m spacing on the test flight lines.

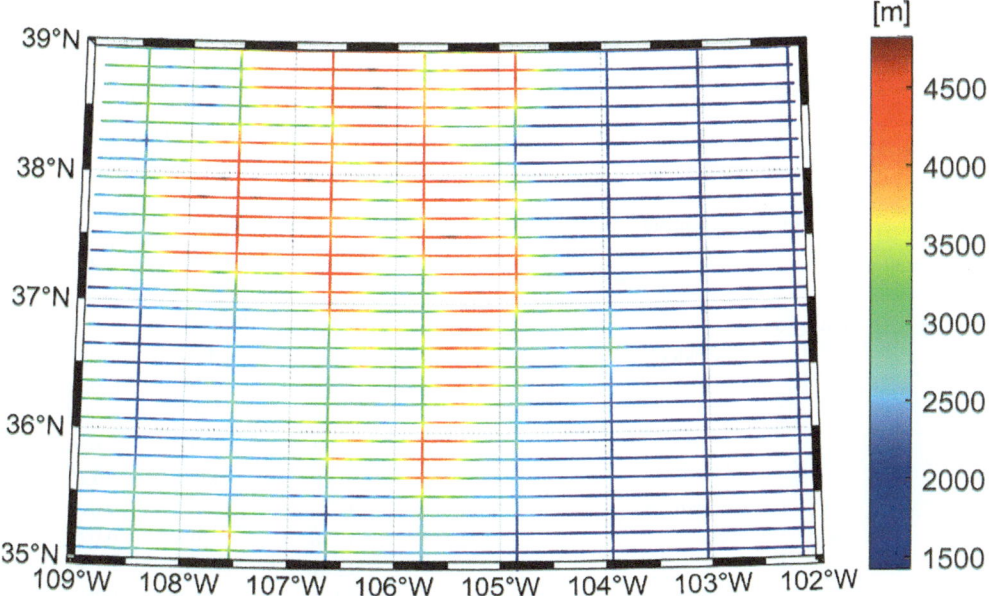

Fig. 1 Flight altitude at the gentle drape surface and with 16 km line spacing

- Subtract the reference gravity disturbance vector computed up to degree/order 360 from the full synthetic observations to work with the residual gravity disturbance vector corresponding to degree/order 361-5,540.
- Add coloured noise to the residual synthetic gravity disturbance vector as the sum of low-pass and band-pass filtered white noise with a standard deviation of 1.2 mGal. This level of noise is compatible with the noise level of the state-of-the-art airborne gravimetry, e.g., (Ferguson et al. 2017; Deng et al. 2020; Bidel et al. 2023)
- Interpolate the VC of the scattered residual gravity disturbance vectors to grid points with a resolution of $1' \times 1'$. These grid values are used to test the standard practice of using airborne gravity observations for geoid modelling.
- Compute the synthetic residual geoid heights using XGM2019 between degree/order 361-5,540 on a $1' \times 1'$ spatial grid to use as a "true" solution when testing our geoid solutions for different airborne survey scenarios. We also computed the same residual geoid heights on the locations of the available historical GPS-on-BMs to use as a stopping condition in our iterative process of solving the WLS in Eq. (6).

To find the solution to Eq. (6), we use different configurations for our input gravity observations and also the spatial resolution of our solution. We use the residual (3D) full gravity vector disturbances at the flight level and in their position of observation (scattered mode) and invert them to geoid heights, see case B of Table 1. The estimated geoid heights, in this case, are the solution to an overdetermined weighted least-squares problem where HCs of the gravity vector are combined with the VC. We consider a diagonal weight matrix filled with the inverse of the variances of the considered observation noises. To test the standard approach, we use the (interpolated) VC grid points of $1' \times 1'$ resolution and invert them to geoid heights, see case A of Table 1. The geoid heights in this part are the solutions of the standard approach.

For all the solutions above, we use a $30'$ integration cap size; compute the far-zone contribution using XMG2019 up to degree/order 720; and use the band-limited spectral form of the kernel in Eq. (3) within the degrees of 361-5,540 in the integration cap. The solution in all cases is obtained iteratively using the LSQR method (Paige and Saunders 1982; Fürst 2008) until the smallest RMS is achieved when compared to the synthetic (residual) geoid heights at the GPS-on-BMs.

By comparing the statistics in Table 1, one can confirm that adding the HC and using the airborne gravity vector in the scattered mode, not only removes the unnecessary step of gridding but also improves the fit to the true data between 15% to 50% for different line spacing cases. For a better evaluation of the effect of the HC on the estimation of the geoid heights, we plot the difference between the estimated geoid heights for 5 km line spacing in cases A and B of Table 1 in Fig. 2.

Please note that we choose a 500 m spacing between the point observations along the lines to optimize computational efficiency. Our internal results for a smaller computation area, using 100 m point observations (closer to real processed airborne gravity observations), show better accuracy with a slightly smaller standard deviation (a few millimetres) compared to what is provided in Table 1. However, for the case of our study, which involves comparisons of the geoid accuracy across different flight line spacings, it is not feasible to densify simulated observations along the lines. To account for the effect of different noise levels, we considered five different standard deviations of noise between $0.5 mGal$ and $5 mGal$ and repeated our analysis. In all circumstances, still using full vector gravity allowed us for sparser flight lines compared to using the vertical component only while preserving the same level of geoid accuracy. Nevertheless, it is worth noting that employing a higher noise level resulted in a lower final accuracy of the geoid, which is not surprising. Due to space constraints and the similarity of results, the tables for this test are not included in this study.

Table 1 Statistics of the inversion errors in different survey designs

Input gravity data configurations				Statistics of discrepancies of the geoid solutions with "true" geoid			
				$1' \times 1'$			
Cases	Line spacing	Gravity data		Min	Max	Mean	STD [m]
A	2 km	$[\delta g_z^{res}]$	Gridded to $1' \times 1'$	−0.099	0.107	0.000	0.017
	5 km			−0.548	0.530	0.008	0.052
	7 km			−0.753	0.754	0.010	0.069
	10 km			−0.872	0.961	0.009	0.086
	16 km			−0.936	1.069	0.012	0.126
B	2 km	$\begin{bmatrix} \delta g_x^{res} \\ \delta g_y^{res} \\ \delta g_z^{res} \end{bmatrix}$	Scattered	−0.154	0.213	0.000	0.011
	5 km			−0.235	0.258	0.004	0.025
	7 km			−0.319	0.381	0.006	0.042
	10 km			−0.563	0.460	−0.002	0.074
	16 km			−0.719	0.709	−0.010	0.106

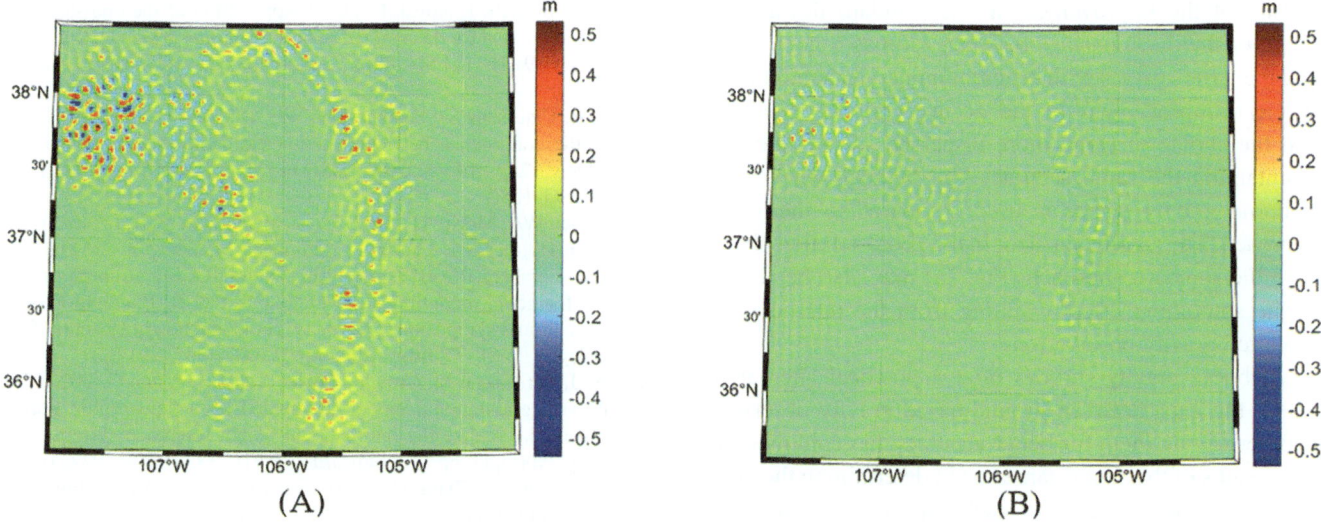

Fig. 2 The difference between estimated geoid heights and the "true" values when a gridded vertical component is used (**a**) and when the scattered full gravity vector is used for inversion (**b**) units in [m]

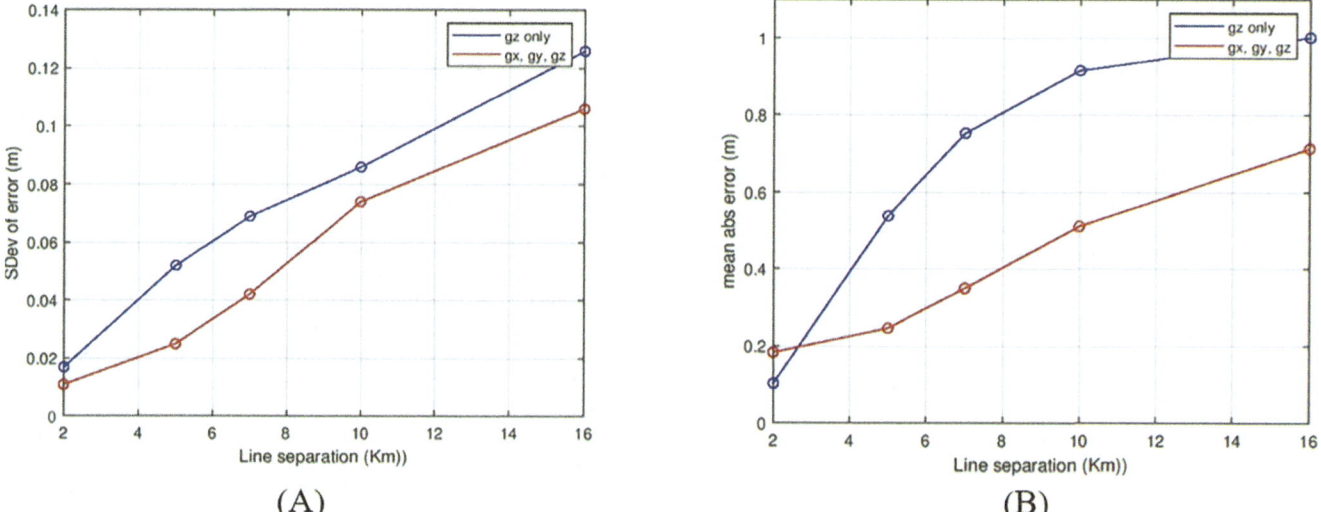

Fig. 3 STD (**a**) and Range (**b**) of discrepancies between estimated geoid heights and the "true" values with different input data configurations

4 Conclusion

In this study, we posed a fundamental question: Can the full 3D airborne gravity observations allow us to space the airborne survey lines farther apart while maintaining the same level of accuracy in geoid models, as compared to only using the vertical component? The answer is yes. Our numerical findings confirmed that the horizontal components of the airborne gravity vector provide gravity signals between the flying lines to increase flight line spacing by up to 40% depending on the required level of geoid accuracy. For instance, geoid heights obtained from an airborne survey at 7 km line spacing and complete (3D) airborne gravity vector are equivalent to, and even slightly better than the geoid heights obtained from surveys at 5 km line spacing and only the scalar-valued or vertical component. Compared with the standard approach in geoid modelling involving interpolation of airborne gravity data at grid points with a specific resolution, we also showed that incorporating the full gravity vector can improve the accuracy of the geoid heights by between 15% to 50% at the same line spacing while also skipping the unnecessary step of gridding.

We used synthetic vector gravity from the XGM2019 model up to full degree/order burdened with realistic systematic noise, and compared our geoid height estimates with the reference geoid heights computed from the same model. We used the one-step integration method for geoid modelling and expanded it to include the use of the HCs of the airborne gravity vector. Figure 3 provides an overall picture of the improvement in the standard deviation and range of the discrepancies between estimated geoid heights and the "true" values with different input data configurations. A clear improvement in the accuracy of the final geoid and the choice of using more dispersed survey lines while maintaining the same accuracy can be seen when the full airborne gravity vector is used compared to the standard approach. The significant error in the geoid solution observed with the standard approach (Fig. 2a) may also be attributed to the smoothing effect resulting from the interpolation of scattered gravity observations.

Airborne gravity surveys offer a practical alternative to filling the gaps between ground-based gravity observations and densifying them for better accuracy in geoid modelling. The results of this study can assist in determining the optimal flight line spacing needed to reach a specific accuracy in local geoid models.

References

Abd-Elmotaal HA, Kühtreiber N (2019) Suitable gravity interpolation technique for large data gaps in Africa. Stud Geophys Geod 63(3):418–435. https://doi.org/10.1007/s11200-017-0545-5

Bidel Y, Zahzam N, Bresson A, Blanchard C, Bonnin A, Bernard J, Cadoret M, Jensen TE, Forsberg R, Salaun C, Lucas S, Lequentrec-Lalancette MF, Rouxel D, Gabalda G, Seoane L, Vu DT, Bruinsma S, Bonvalot S (2023) Airborne absolute gravimetry with a quantum sensor, comparison with classical technologies. J Geophys Res Solid Earth 128(4):e2022JB025921. https://doi.org/10.1029/2022JB025921

Deng K, Chang G, Huang M, Zeng H, Chen X (2020) Geoid determination using band limited airborne horizontal gravimetric data. J Spat Sci 0(0):1–14. https://doi.org/10.1080/14498596.2020.1746703

Dumont R, Bardossy Z (2014) Drape DTM 2.0: Software to calculate a smooth drape surface for an airborne geophysical survey / [by] R. Dumont and Z. Bardossy.: M183-2/7690E-PDF - Government of Canada Publications - Canada.ca. https://publications.gc.ca/site/eng/9.819766/publication.html

Ferguson S, Wang Y, Ellief S, Li X, Holmes S, Ahlgren K, Xi R (2017) A comparison of airborne vector gravimeter measurements with the Geoid Slope Validation Survey 2014. IAG-IASPEI, Kobe

Ferguson S, Forsberg R, Brown N, Elieff S (2019) Geoid models using combinations of airborne and ground gravity data—a case study from the Kimberley region, Australia 2019, G43A-08

Foroughi I, Vaníček P, Kingdon RW, Goli M, Sheng M, Afrasteh Y, Novák P, Santos MC (2019) Sub-centimetre geoid. J Geod 93(6):849–868. https://doi.org/10.1007/s00190-018-1208-1

Foroughi I, Goli M, Pagiatakis S, Ferguson S, Novák P (2023) Data requirements for the determination of a sub-centimetre geoid. Earth Sci Rev 239:104326. https://doi.org/10.1016/j.earscirev.2023.104326

Forsberg R, Olesen A, Bastos L, Gidskehaug A, Meyer U, Timmen L (2000) Airborne geoid determination. Earth Planets Space 52(10):863–866. https://doi.org/10.1186/BF03352296

Fotopoulos G (2005) Calibration of geoid error models via a combined adjustment of ellipsoidal, orthometric and gravimetric geoid height data. J Geod 79(1):111–123. https://doi.org/10.1007/s00190-005-0449-y

Fürst J (2008) The WLSQR Scheme for the flows in complex geometry. In: 5th international symposium on finite volumes for complex applications

Goli M, Foroughi I, Novák P (2019) Application of the one-step integration method for determination of the regional gravimetric geoid. J Geod 93(9):1631–1644. https://doi.org/10.1007/s00190-019-01272-8

Goyal R, Claessens SJ, Featherstone WE, Dikshit O (2023) Investigating the congruence between gravimetric geoid models over India. J Surv Eng 149(3):04023005. https://doi.org/10.1061/JSUED2.SUENG-1382

Heiskanen WA, Moritz H (1967) Physical geodesy. W.H. Freeman

Huang J, Holmes SA, Zhong D, Véronneau M, Wang Y, Crowley JW, Li X, Forsberg R (2019) Analysis of the GRAV-D airborne gravity data for geoid modelling. In: Vergos GS, Pail R, Barzaghi R (eds) International symposium on gravity, geoid and height systems 2016. Springer, pp 61–77. https://doi.org/10.1007/1345_2017_23

Huang J, Véronneau M, Crowley JW, D'Aoust B, Pavlic G (2022) Can an earth gravitational model augmented by a topographic gravity field model realize the international height reference system accurately? Springer, pp 1–7. https://doi.org/10.1007/1345_2022_162

Jiang T, Dang Y, Zhang C (2020) Gravimetric geoid modeling from the combination of satellite gravity model, terrestrial and airborne gravity data: a case study in the mountainous area, Colorado. Earth Planets Space 72(1):189. https://doi.org/10.1186/s40623-020-01287-y

Kern M, Schwarz KKPP, Sneeuw N (2003) A study on the combination of satellite, airborne, and terrestrial gravity data. J Geod 77(3):217–225. https://doi.org/10.1007/s00190-003-0313-x

Li X, Huang J, Klees R, Forsberg R, Willberg M, Slobbe DC, Hwang C, Pail R (2022) Characterization and stabilization of the downward continuation problem for airborne gravity data. J Geod 96(4):18. https://doi.org/10.1007/s00190-022-01607-y

Mayunga SD (2016) Towards a new geoid model of Tanzania using precise gravity data. J Environ Sci Eng A 5(5) https://doi.org/10.17265/2162-5298/2016.05.005

McCubbine JC, Featherstone WE, Kirby JF (2017) Fast-Fourier-based error propagation for the gravimetric terrain correction. Geophysics 82(4):G71–G76. https://doi.org/10.1190/geo2016-0627.1

Novák P (2003) Geoid determination using one-step integration. J Geod 77(3):193–206. https://doi.org/10.1007/s00190-003-0314-9

Novák P, Kern M, Schwarz K-P, Sideris MG, Heck B, Ferguson S, Hammada Y, Wei M (2003) On geoid determination from airborne gravity. J Geod 76(9):510–522. https://doi.org/10.1007/s00190-002-0284-3

Ophaug V, Gerlach C (2020) Error propagation in regional geoid computation using spherical splines, least-squares collocation, and Stokes's formula. J Geod 94(12):120. https://doi.org/10.1007/s00190-020-01443-y

Paige CC, Saunders MA (1982) LSQR: an algorithm for sparse linear equations and sparse least squares. ACM Trans Math Softw 8(1):43–71. https://doi.org/10.1145/355984.355989

Sander L (1998) Pre-planned drape surfaces: a new survey planning tool. Can J Explor Geophys 34(1, 2):4–8

Sander L, Ferguson S, Geophysics S (2011) Advances in SGL AIRGrav acquisition and processing. https://www.semanticscholar.org/paper/Advances-in-SGL-AIRGrav-Acquisition-and-Processing-Sander-Ferguson/942505891499f80fcb3c2c8688842c9070c50beb#citing-papers

Senobari MS (2010) New results in airborne vector gravimetry using strapdown INS/DGPS. J Geod 84(5):277–291. https://doi.org/10.1007/s00190-010-0366-6

Serpas JG (2003) Local and regional geoid determination from vector airborne gravimetry, The Ohio State University. https://etd.ohiolink.edu/apexprod/rws_olink/r/1501/10?clear=10&p10_accession_num=osu1066757143

Serpas JG, Jekeli C (2005) Local geoid determination from airborne vector gravimetry. J Geod 78(10):577–587. https://doi.org/10.1007/s00190-004-0416-z

Smith D (2007) The GRAV-D project gravity for the redefinition of the American vertical datum (NOAA Special Publication NOS NGS 55454; 7)

Vyazmin V, Golovan A (2023) Scalar and vector strapdown airborne gravimetry on aircraft and UAV: methodology of surveying and data processing. In: 2023 30th Saint Petersburg international conference on integrated navigation systems (ICINS), pp 1–6. https://doi.org/10.23919/ICINS51816.2023.10168453

Wang YM, Sánchez L, Ågren J, Huang J, Forsberg R, Abd-Elmotaal HA, Ahlgren K, Barzaghi R, Bašić T, Carrion D, Claessens S, Erol B, Erol S, Filmer M, Grigoriadis VN, Isik MS, Jiang T, Koç Ö, Krcmaric J et al (2021) Colorado geoid computation experiment: overview and summary. J Geod 95(12):127. https://doi.org/10.1007/s00190-021-01567-9

Zingerle P, Pail R, Gruber T, Oikonomidou X (2020) The combined global gravity field model XGM2019e. J Geod 94(7):66. https://doi.org/10.1007/s00190-020-01398-0

Open Access This chapter is licensed under the terms of the Creative Commons Attribution 4.0 International License (http://creativecommons.org/licenses/by/4.0/), which permits use, sharing, adaptation, distribution and reproduction in any medium or format, as long as you give appropriate credit to the original author(s) and the source, provide a link to the Creative Commons license and indicate if changes were made.

The images or other third party material in this chapter are included in the chapter's Creative Commons license, unless indicated otherwise in a credit line to the material. If material is not included in the chapter's Creative Commons license and your intended use is not permitted by statutory regulation or exceeds the permitted use, you will need to obtain permission directly from the copyright holder.

Update of the Atmospheric Attraction Computation Service (Atmacs) for High-Precision Terrestrial Gravity Observations

Ezequiel D. Antokoletz, Hartmut Wziontek, Thomas Klügel, Kyriakos Balidakis, and Henryk Dobslaw

Abstract

The Atmospheric attraction computation service (Atmacs) of BKG provides atmospheric corrections for terrestrial high-precision gravity measurements based on operational weather models of the German Weather Service (DWD). In Atmacs, Newtonian attraction and deformation contributions to atmospheric loading are computed separately. The attraction component benefits from the discrete 3D distribution of air masses around the station, while deformation effects are derived from surface atmospheric pressure changes assuming that the oceans respond to atmospheric forcing as an Inverse Barometer (IB). Several improvements in the modelling approach of Atmacs are presented. A revision of the IB hypothesis implementation revealed that the attraction component over oceans was overestimated. A modification of the IB implementation not only resolves this issue but further enhances the compatibility between the atmospheric modelling and ocean models. This allows to complement Atmacs with non-tidal ocean loading effects, here based on the Max-Plank-Institute for Meteorology Ocean Model (MPIOM). These updates allow for a consistent combination of atmospheric and ocean models and a more efficient reduction of the signal recorded by high-precision terrestrial gravimeters.

Keywords

Atmospheric loading · Non-tidal ocean loading · Terrestrial gravity time series

1 Introduction

Atmospheric mass changes are one of the major contributions to gravity variations as measured by high-precision terrestrial gravimeters. Corrections for such effects are classically based on an admittance factor for the local air pressure recorded at the station but this simple approach does not account for the spatial distribution of air masses around the station. Improved atmospheric corrections have been successfully accomplished through considering numerical weather models. Merriam (1992) considered for the first-time global surface pressure fields to compute atmospheric effects on gravity, including the vertical air mass distribution based on a standard atmosphere. More sophisticated procedures benefit from operational weather models by accounting for the 3D air mass distribution around the computation point (e.g., Neumeyer et al. 2004; Klügel and Wziontek 2009).

Superconducting gravimeters (SGs) are recognised to provide the most reliable, precise and stable gravity signal which allows the study of a wide range of effects (e.g., Neumeyer 2010). Since gravity measurements always contain the sum of all physically relevant phenomena, those effects of larger amplitudes have to be precisely modelled and removed to facilitate the separation and study of smaller signals. Thus, atmospheric corrections have to be as precise as possible to

E. D. Antokoletz (✉) · H. Wziontek · T. Klügel
Federal Agency for Cartography and Geodesy (BKG), Frankfurt am Main, Germany
e-mail: Ezequiel.Antokoletz@bkg.bund.de

K. Balidakis · H. Dobslaw
German Research Centre for Geosciences (GFZ), Potsdam, Germany

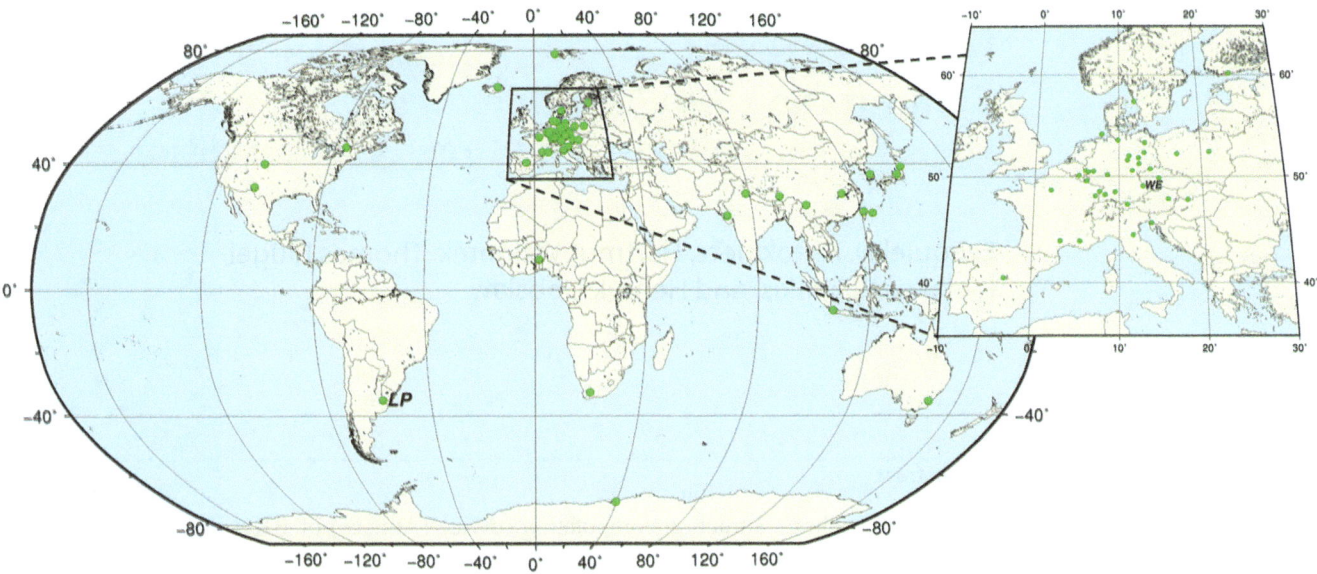

Fig. 1 Distribution of stations where atmospheric effects are routinely computed

support further interpretation of the remaining signal in terms of e.g., hydrology (Güntner et al. 2017).

The oceans are well known to be coupled and highly-anticorrelated with the atmosphere (Dobslaw et al. 2017a). Generally, the ocean response to atmospheric pressure forcing is supposed to follow, as a first approximation, the Inverse Barometer (IB; Wunsch and Stammer 1997) assumption. However, for periods shorter than a few weeks, the dynamic response of the ocean to atmospheric forcing cannot be neglected (Boy et al. 2009). Therefore, a complete and precise modelling of atmospheric effects should also account for the dynamics of the ocean. Non-tidal ocean loading effects turned out to provide significant contribution to gravity variations especially for stations located close to the coast (e.g. Kroner et al. 2009; Boy and Lyard 2008; Oreiro et al. 2018).

Nowadays, the Atmospheric attraction computation service (Atmacs) is one of a few services that provide atmospheric corrections for high-precision terrestrial gravimeters and is based on operational weather models from the German Weather Service (DWD), assuming the IB hypothesis to be valid over oceans. In the present study, some improvements in the modelling scheme of Atmacs are presented. In order to complement Atmacs, non-tidal ocean loading effects are evaluated based on the Max-Planck-Institute for Meteorology Ocean Model (MPIOM; Jungclaus et al. 2013).

2 The Atmospheric Attraction Computation Service (Atmacs)

Atmacs routinely provides atmospheric gravity corrections for about 60 stations (Fig. 1) that mostly contribute to the International Geodynamics and Earth Tides Service (IGETS;

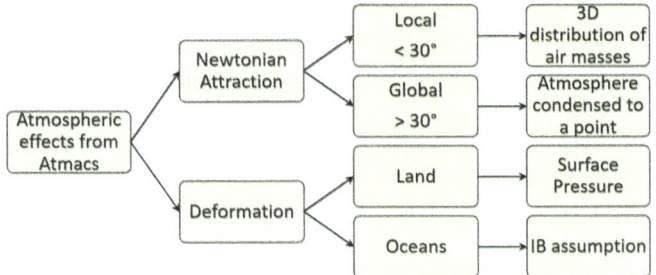

Fig. 2 Computation scheme of atmospheric effects in Atmacs

Boy et al. 2020). Atmospheric density data together with surface pressure is provided by DWD on a daily basis for the global and the European solutions of the Icosahedral Nonhydrostatic (ICON; Zängl et al. 2015) model with a temporal resolution of 3 hours. The global ICON model has a triangular grid of about 13 km spatial resolution and it vertically extends up to 75 km divided into 120 vertical layers where the thickness of each layer is constant. ICON-EU is given on a latitude-longitude grid of about 7 km spatial resolution and a vertical extension of 22 km divided into 60 vertical layers also with constant thickness. This model is nested in the global ICON for a complete representation of the atmosphere. Consequently, the solution based on ICON-EU was only provided for stations located in Europe. All timeseries are provided as ASCII-files and available for any user through a web site.[1]

In Atmacs, attraction and deformation contributions to gravity are computed separately (Fig. 2). The calculation of Newtonian attraction effects is divided into a region

---

[1] https://atmacs.bkg.bund.de, last access: 27 December 2023.

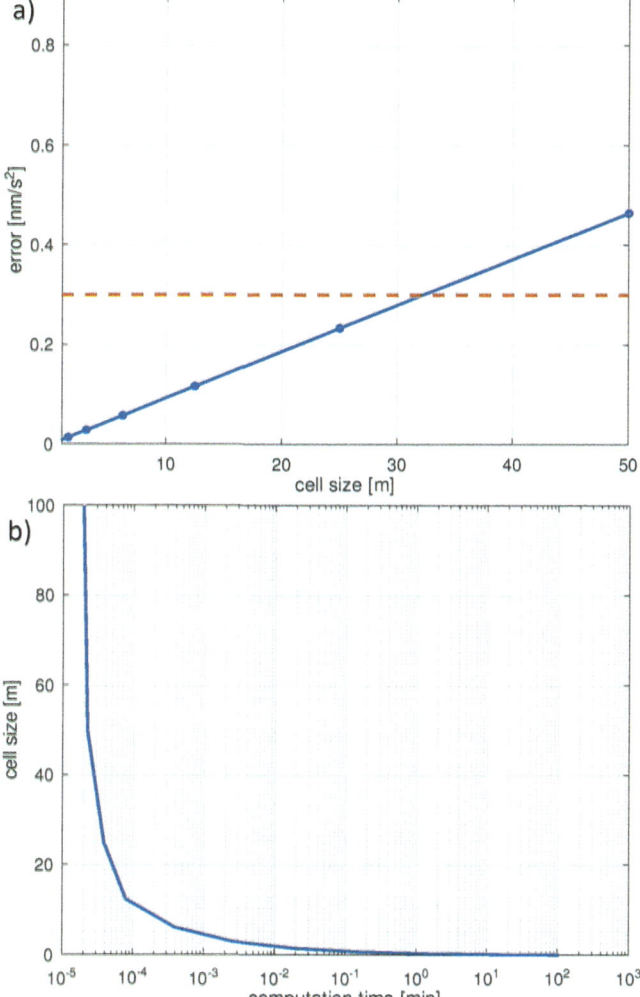

Fig. 3 (a) Approximation error in the atmospheric attraction computation as a function of the cell size for the densification of point masses. In orange a threshold of $0.3\,\text{nm/s}^2$ is depicted. (b) Computation time required for different cell sizes

Table 1 Densification of point masses for the ICON model with 13 km spatial resolution

Range	Densification factor
Innermost cell	500
Up to 25 km	100
Up to 200 km	20
Up to 30°	1

error approximation (orange line in Fig. 3a), a cell size of 30 m was found to be the optimum size that allows also a fast computation time. The densification is then applied by linear interpolation, depending on the distance from the computation point. For the recent global ICON model with a spatial resolution of about 13 km, different densifications factors are applied depending on the distance of the cell to the computation point, as shown in Table 1. Up to 30°, the vertical air mass distribution is accounted for, while outside those regions, air masses are condensed to a single point mass per column located at its centre of mass (about 5.4 km height) and derived from surface pressure assuming hydrostatic equilibrium. By this, the triangular icosahedral mesh of ICON can be used, providing an almost equal cell size globally and avoiding artificial densification towards (and a singularity at) the poles of classical latitude-longitude grids.

Prior to the Global Model (GME) 384 of DWD, the height of the European and the global models released until February, 2012 were limited to about 22 and 30 km, respectively and with a time resolution of 6 hours (Majewski et al. 2002). It turned out that these models had no closed mass balance, causing an artificial seasonal variation. This was avoided by a model extension up to 80 km by continuation with the ISO standard atmosphere based on temperature and pressure at the uppermost boundary. This is not necessary for GME 384 and the global ICON anymore which have a sufficient height of 75 km.

Elastic deformation effects are computed following the classical approach of Farrell (1972), applying load Love numbers from the Preliminary Earth Model (PREM) given by Jentzsch (1997). The mass load is derived from surface atmospheric pressure assuming hydrostatic equilibrium. Over the oceans, the IB hypothesis was assumed to be valid, meaning no net deformations on the seafloor due to atmospheric pressure changes occur. Practically, atmospheric pressure variations over oceans were neglected and load mass variations were considered only over continents. The discrimination between continents and oceans was based on the model orography: cells with height equal or below 0 m were simply excluded from the deformation contribution. The dynamic response of the ocean to atmospheric forcing was missing in Atmacs as an ocean dynamic model was not included at all.

where the vertical air mass distribution is considered and a remaining global part. Since point masses are not valid close to the computation point, in the early version of Atmacs Klügel and Wziontek (2009) replaced the cells in a local zone of about 10 km distance by a pile of disks of equal density which was integrated analytically. The height of each disk corresponded to the height of each layer of the model. This first solution was only available for European stations as it could not be applied to the global models of DWD because of the distinct lower spatial resolution and the triangular grid structure. For worldwide computations introduced in 2013, the local cylinder was replaced by densified point masses. To find the appropriate cell size, several tests were performed. Figure 3a depicts the approximation error of the point mass approach as a function of the cell size and Fig. 3b its computation time. With a threshold of $0.3\,\text{nm/s}^2$ in the

3 Updates in the Atmospheric Loading Computation

By a careful revision of the implementation of the IB assumption in Atmacs, several modifications to the computation scheme have been introduced. Although the IB assumption implies that no net deformation of the seafloor occurs due to surface atmospheric pressure changes, a redistribution of air and water masses takes place which has to be accounted for the Newtonian attraction computation, as pointed out by Antokoletz et al. (2023). While air mass variations over the oceans were considered in Atmacs, the compensation effect of water mass redistribution due to the IB was missing, causing an overestimation of the ocean contribution.

In order to account for the compensating contribution due to the IB response of the ocean, the attraction effect of the compensating water mass must be included. This mass (dm) is derived from the atmospheric surface pressure (P) following

$$dm = -\frac{(P - \bar{P})}{g_0} dA \quad (1)$$

where dA is the area of the respective ocean cell, $g_0 = 9.80665 \, \text{m/s}^2$ is the gravity reference value given by WMO (2008) and $\bar{P} = \frac{1}{A_{Oc}} \iint_{Oc} P \, dA$ represents a mean surface pressure value over the whole ocean area (A_{Oc}). Attraction effects due to this term are computed based on point masses located at the sea surface.

This contribution was found to be significant and allows to further reduce the variance of gravity residuals (Antokoletz et al. 2023). Significant amplitudes were found for stations located close to the coast or more exposed to ocean mass variability. For the set of stations currently included in Atmacs, the largest Root Mean Square (RMS) variability of this contribution was found at Ny Ålseund (NY, Norway), Theistareykir (TH, Iceland) and Ishigakijima (IG, Japan) with 4.6, 4.1 and 3.4 nm/s², respectively. Even for stations far away from the coast such as Apache Point (AP, USA), Djougou (DJ, Benin) or Wettzell (WE, Germany), the RMS still reaches the nm/s² level.

In order to account for the missing ocean mass conservation, a small modification in the deformation computation is also implemented. Instead of setting atmospheric surface pressure variations to zero over oceans, now $P = \bar{P}$ is introduced over oceans. Besides temporal variations due to $\bar{P}$ provide a small contribution to loading, it enhances the compatibility with ocean dynamic models and ensures the conservation of the total ocean mass along time.

Both modifications in the atmospheric attraction and deformation rely on a precise definition of the coastlines. The simple definition applied before may lead to artefacts in case a station is located close to the coast, continental areas are below the sea level or limited representation of coastal areas by the orography. Therefore, a new coastline definition is now implemented based on the spatial resolution of the global ICON model and considering large catchments such as the Caspian Sea which behaves as a lake as it is not connected to the global ocean, and the major lakes as the Great Lakes in North America or lakes Victoria and Tanganyika in Africa. Local IB effects are also accounted for these areas.

Finally, for all European stations, two atmospheric corrections were provided in Atmacs, based on the European and the global ICON models. Comparisons between both results showed no significant differences in the total effect. Therefore, the ICON-EU solution was interrupted since November 2022 and only the global solution is continued.

4 Implementation of Non-Tidal Ocean Loading Effects

As mentioned before, the IB response of the ocean reflects neither its dynamic response to atmospheric pressure and wind forcing nor ocean circulation effects or temperature and salinity changes. These effects can only be considered by global ocean general circulation models.

Non-tidal ocean loading effects are principally available from the surface loading products routinely calculated by the German Research Centre for Geosciences (GFZ; Dill and Dobslaw 2013). These are based on an ocean simulation of MPIOM, forced by atmospheric fields from the Operational European Centre for Medium-Range Weather Forecasts (ECMWF) model (Dobslaw et al. 2017b). Global loading effects are given on a latitude-longitude grid of $0.5° \times 0.5°$ resolution and a temporal resolution of 3 hours. From the gridded data, the effect for a given station can be obtained through interpolation. For all routine products maintained by GFZ, calculations typically extend back to the year 1976 when the first geodetic satellites were launched. The data products are routinely updated daily for all timesteps of the previous day and is thus technically very well suited to complement Atmacs.

For additional comparison, the ocean mass grids from MPIOM were also considered in this study with the same spatial and temporal resolution as of the loading grids in order to analyse the performance of the latter to derive gravity loading effects. In this case, non-tidal ocean loading effects were computed for specific stations employing SPOTL (Agnew 2012) and compared with the interpolated grid-data.

In order to assess a possible dependency on the distance of the stations to the coast, two particular stations were selected (Fig. 1): Wettzell (WE, Germany) and La Plata (LP,

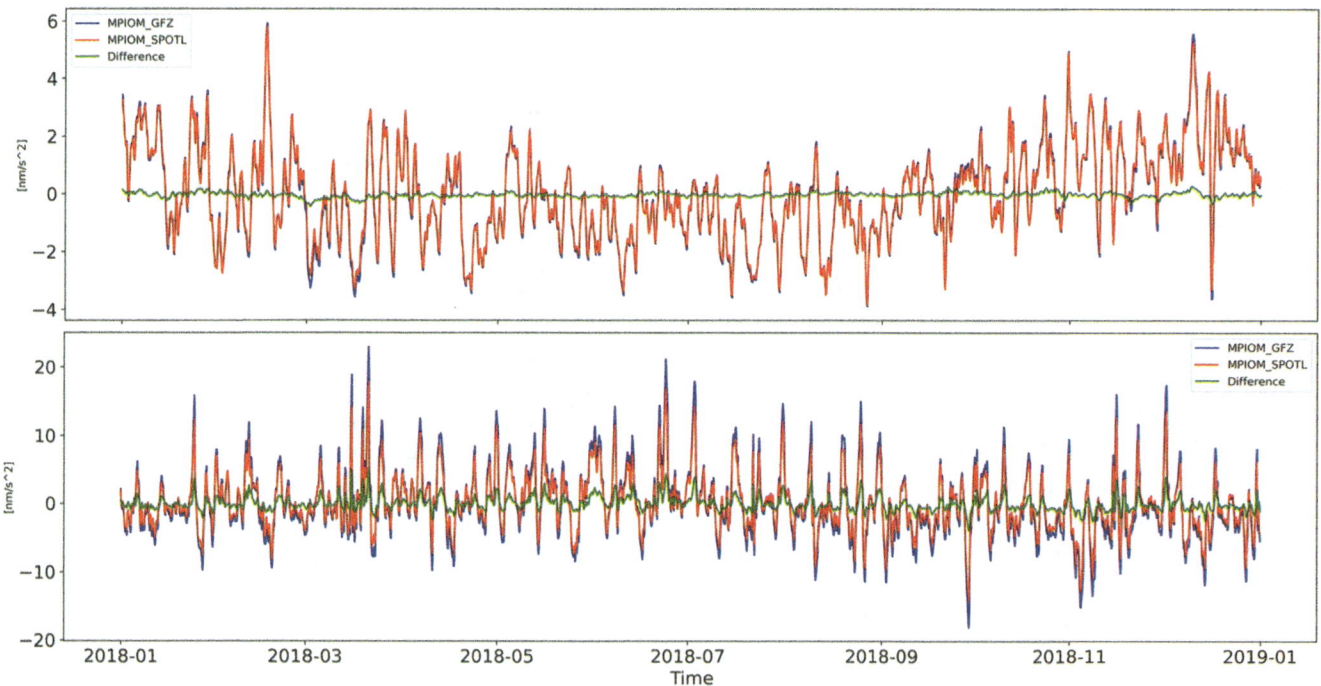

Fig. 4 Non-tidal ocean loading effects as interpolated from the GFZ gridded loading products (blue) and as computed with SPOTL (red). In green the difference is depicted. Top: station WE (Germany). Bottom: station LP (Argentina)

Argentina), located at approximately 380 and 15 km from the coast, respectively. Further discrepancies between loading computations, e.g., the use of different Green's functions, are not discussed here as they are expected to have only a minor impact on the computed effect (e.g., Bos and Baker 2005).

Figure 4 depicts non-tidal ocean loading effects from both solutions and their differences. For station WE, the differences are insignificant, meaning that for inland stations, almost equivalent solutions are obtained from both approaches. However, for station LP, the interpolated grid solution shows larger effects compared to the point wise computation from SPOTL with a RMS variability of the differences exceeding the nm/s^2 level. Such differences have been traced back to the coarse spatial resolution of the coastline included in the grid product. This coastline has a spatial resolution of $0.5° \times 0.5°$ and causes a misrepresentation of the total loading effect. Loading computations in SPOTL include a detailed coastline and a further refinement of the integration mesh for a more precise representation of Newtonian attraction effects. Figure 5 depicts the simple coastline superimposed with the one implemented in SPOTL for the vicinity of LP station. As it becomes apparent from the Fig. 5, the simple coastline includes a few ocean cells that are actually over land.

In conclusion, while the surface loading grid readily allows to derive non-tidal ocean loading effects everywhere

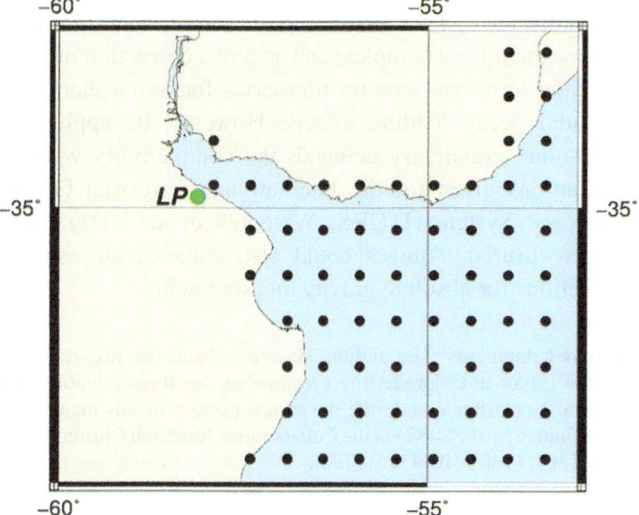

Fig. 5 Coastline from the GFZ gridded loading products (black dots) over the SPOTL coastline definition in the vicinity of station LP (green dot)

on our planet, stations close to the coast require a more detailed representation of the coastline and direct Newtonian effects caused by the water mass redistribution. For this reason, the implementation for Atmacs is based on the SPOTL computation on an operational basis at GFZ.

5 Conclusions

Several improvements of the Atmospheric attraction computation service (Atmacs) have been presented. A revision of the Inverse Barometer (IB) assumption allowed to consistently consider all Newtonian attraction effects over oceans, which were previously overestimated. Moreover, the ocean mass conservation was included and a new detailed coastline definition was implemented also accounting for local IB effects over large catchments. These modifications ensure the compatibility with ocean dynamic models and have been operationally included since July 1st, 2023 and available through the web page of Atmacs. Further, the European solution based on ICON-EU has been discontinued.

Non-tidal ocean loading effects as provided through the surface loading products of GFZ have been evaluated and compared with point wise loading computations with SPOTL, based on the same ocean dynamic model. Although the gridded data gives direct access to the loading effects, the coarse coastline definition included leads to a misrepresentation of mass anomalies and, consequently, a misestimation of the loading effects especially for stations close to the coast. Therefore, a point wise computation for each station will be included in Atmacs to complement atmospheric effects.

Finally, the updates introduced here represent a significant improvement for a complete and accurate correction of high-precision terrestrial gravity timeseries for atmospheric and non-tidal ocean loading effects. However, its application in absolute gravimetry demands the compatibility with the conventions given for the International Terrestrial Gravity Reference System (ITGRS; Wziontek et al. 2021). Once this is ensured, Atmacs could also enhance atmospheric corrections for absolute gravity measurements.

Acknowledgements The authors sincerely thank the two reviewers and the Editor-in-Chief, Jeffrey Freymueller, for their valuable comments and contributions during the review process of this manuscript. KB is funded by the DFG via the Collaborative Research Cluster TerraQ (SFB 1464, Project-ID 434617780).

Data Availability Atmospheric corrections are available through the Atmospheric attraction computation service (Atmacs): http://atmacs.bkg.bund.de/. Geodetic data products based on MPIOM global ocean simulations are publicly available from GFZ's Information System and Data Center (https://doi.org/10.5880/GFZ.1.3.2022.003).

Disclosure of Interests The authors declare that they have no conflict of interest.

References

Agnew DC (2012) SPOTL: Some programs for oceantide loading. SIO Technical Report. http://escholarship.org/uc/item/954322pg

Antokoletz ED, Wziontek H, Dobslaw H, Balidakis K, Klügel T, Oreiro FA, Tocho CN (2023) Combining atmospheric and non-tidal ocean loading effects to correct high precision gravity time-series. Geophys J Int 236(1):88–98. https://doi.org/10.1093/gji/ggad371

Bos M, Baker T (2005) An estimate of the errors in gravity ocean tide loading computations. J Geodesy 79(1):50–63. https://doi.org/10.1007/s00190-005-0442-5

Boy JP, Lyard F (2008) High-frequency non-tidal ocean loading effects on surface gravity measurements. Geophys J Int 175(1):35–45. https://doi.org/10.1111/j.1365-246X.2008.03895.x

Boy JP, Longuevergne L, Boudin F, Jacob T, Lyard F, Llubes M, Florsch N, Esnoult MF (2009) Modelling atmospheric and induced non-tidal oceanic loading contributions to surface gravity and tilt measurements. J Geodyn 48(3):182–188. https://doi.org/10.1016/j.jog.2009.09.022.

Boy JP, Barriot JP, Förste C, Voigt C, Wziontek H (2020) Achievements of the first 4 years of the international geodynamics and earth tide service (IGETS) 2015–2019. Springer, Berlin, Heidelberg, pp 1–6. https://doi.org/10.1007/1345_2020_94

Dill R, Dobslaw H (2013) Numerical simulations of global-scale high-resolution hydrological crustal deformations. J Geophys Res Solid Earth 118(9):5008–5017. https://doi.org/10.1002/jgrb.50353.

Dobslaw H, Bergmann-Wolf I, Dill R, Poropat L, Flechtner F (2017a) Product description document for AOD1B release 06. GFZ German Research Centre for Geosciences Department 1. https://gfzpublic.gfz-potsdam.de/pubman/item/item_5006868

Dobslaw H, Bergmann-Wolf I, Dill R, Poropat L, Thomas M, Dahle C, Esselborn S, König R, Flechtner F (2017b) A new high-resolution model of non-tidal atmosphere and ocean mass variability for de-aliasing of satellite gravity observations: AOD1B RL06. Geophys J Int 211(1):263–269. https://doi.org/10.1093/gji/ggx302

Farrell W (1972) Deformation of the Earth by surface loads. Rev Geophys 10(3):761–797. https://doi.org/10.1029/RG010i003p00761

Güntner A, Reich M, Mikolaj M, Creutzfeldt B, Schroeder S, Wziontek H (2017) Landscape-scale water balance monitoring with an iGrav superconducting gravimeter in a field enclosure. Hydrol Earth Syst Sci 21(6):3167–3182. https://doi.org/10.5194/hess-21-3167-2017.

Jentzsch G (1997) Earth tides and ocean tidal loading. In: Wilhelm H, Zürn W, Wenzel HG. (eds) Tidal Phenomena. Lecture Notes in Earth Sciences, vol 66. Springer, Berlin, Heidelberg. https://doi.org/10.1007/BFb0011461

Jungclaus J, Fischer N, Haak H, Lohmann K, Marotzke J, Matei D, Mikolajewicz U, Notz D, Von Storch J (2013) Characteristics of the ocean simulations in the Max Planck Institute Ocean Model (MPIOM) the ocean component of the MPI-Earth system model. J Adv Model Earth Syst 5(2):422–446. https://doi.org/10.1002/jame.20023

Klügel T, Wziontek H (2009) Correcting gravimeters and tiltmeters for atmospheric mass attraction using operational weather models. J Geodynam 48(3):204–210. https://doi.org/10.1016/j.jog.2009.09.010

Kroner C, Thomas M, Dobslaw H, Abe M, Weise A (2009) Seasonal effects of non-tidal oceanic mass shifts in observations with superconducting gravimeters. J Geodyn 48(3):354–359. https://doi.org/10.1016/j.jog.2009.09.009

Majewski D, Liermann D, Prohl P, Ritter B, Buchhold M, Hanisch T, Paul G, Wergen W, Baumgardner J (2002) The operational global icosahedral–hexagonal gridpoint model GME: Description and highresolution tests. Month Weather Rev 130(2):319–338. https://doi.org/10.1175/1520-0493(2002)130%3C0319:TOGIHG%3E2.0.CO;2

Merriam JB (1992) Atmospheric pressure and gravity. Geophys J Int 109(3):488–500. https://doi.org/10.1111/j.1365-246X.1992.tb00112.x

Neumeyer J (2010) Superconducting gravimetry. In: Sciences of Geodesy-I. Springer, New York, pp 339–413. https://doi.org/10.1007/978-3-642-11741-1_10

Neumeyer J, Hagedoorn J, Leitloff J, Schmidt T (2004) Gravity reduction with three-dimensional atmospheric pressure data for precise ground gravity measurements. J Geodyn 38(3):437–450. https://doi.org/10.1016/j.jog.2004.07.006

Oreiro F, Wziontek H, Fiore M, D'Onofrio E, Brunini C (2018) Nontidal ocean loading correction for the Argentinean-German Geodetic Observatory using an empirical model of storm surge for the Río de La Plata. Pure Appl Geophys 175(5):1739–1753. https://doi.org/10.1007/s00024-017-1651-6

WMO (2008) Guide to meteorological instruments and methods of observation (WMO-No. 8). World Meteorological Organisation, Geneva, Switzerland, 29

Wunsch C, Stammer D (1997) Atmospheric loading and the oceanic "inverted barometer" effect. Rev Geophys 35(1):79–107. https://doi.org/10.1029/96RG03037

Wziontek H, Bonvalot S, Falk R, Gabalda G, Mäkinen J, Pálinkás V, Rülke A, Vitushkin L (2021) Status of the international gravity reference system and frame. J Geodesy 95(1):1–9. https://doi.org/10.1007/s00190-020-01438-9

Zängl G, Reinert D, Rípodas P, Baldauf M (2015) The ICON (ICOsahedral Non-hydrostatic) modelling framework of DWD and MPI-M: Description of the non-hydrostatic dynamical core. Q J Roy Meteorol Soc 141(687):563–579. https://doi.org/10.1002/qj.2378.

Open Access This chapter is licensed under the terms of the Creative Commons Attribution 4.0 International License (http://creativecommons.org/licenses/by/4.0/), which permits use, sharing, adaptation, distribution and reproduction in any medium or format, as long as you give appropriate credit to the original author(s) and the source, provide a link to the Creative Commons license and indicate if changes were made.

The images or other third party material in this chapter are included in the chapter's Creative Commons license, unless indicated otherwise in a credit line to the material. If material is not included in the chapter's Creative Commons license and your intended use is not permitted by statutory regulation or exceeds the permitted use, you will need to obtain permission directly from the copyright holder.

Geoid Computation for the Future Circular Collider at CERN

Julia Azumi Koch, Urs Marti, Iván Darío Herrera Pinzón, Daniel Willi, Benedikt Soja, and Markus Rothacher

Abstract

In the scope of initial studies for a post-LHC (Large Hadron Collider) particle accelerator, the design and feasibility of the Future Circular Collider (FCC) at CERN are studied. In the FCC Geodesy project, which belongs to these initial studies, a high-precision geoid model for the significantly larger FCC area will be computed. The pre-alignment of the magnets and thrusters in the tunnel requires a very accurate high-resolution local geoid model with a precision of up to 30 μm over 225 m.

The data sets containing gravity and deflections of the vertical measurements are described. Then, an overview of the existing (quasi)geoid models in the area is shown and representative models are compared to a newly calculated quasigeoid model. This model is the first gravitational (quasi)geoid model that was estimated specifically for this region in a least-squares adjustment. The highest correspondence of the preliminary FCC Quasigeoid was found with CHQua04 published by Switzerland and RAF20, which is a hybrid regional quasigeoid for France. The smallest offset between two models was found between the FCC Quasigeoid and XGM2019. A comparison of the new solution with GNSS-levelling datasets showed a good agreement with a standard deviations of < 1 cm in the region of interest.

This publication aims to provide an overview of the objectives of the FCC Geodesy project, to show the available data sets of gravity and deflections of the vertical measurements and the results of a first comparison of existing and the newly calculated FCC (quasi)geoid with GNSS-levelling points.

Keywords

CERN · Deflections of the vertical · FCC · Gravity · Regional geoid

1 Introduction

The Large Hadron Collider (LHC) at CERN (Conseil Européen pour la Recherche Nucléaire) is currently the biggest particle accelerator of its type with a circumference of 27 km and is located at an average depth of 100 m under the Earth's surface. The installation of the LHC was finished in 2008 and after six realignment and reconstruction phases, it will reach its conclusion by the year 2040. Therefore, CERN is planning a post-LHC particle accelerator, the so-called Future Circular Collider (FCC) with a circumference of around 100 km and situated 300 m below the surface.

J. A. Koch (✉) · B. Soja · M. Rothacher
Institute of Geodesy and Photogrammetry, ETH Zurich, Zurich, Switzerland
e-mail: jukoch@ethz.ch

U. Marti · D. Willi
swisstopo, Federal Office of Topography, Wabern, Switzerland

I. D. Herrera Pinzón
Institute of Geodesy and Photogrammetry, ETH Zurich, Zurich, Switzerland

Astronomical Institute (AIUB), University of Bern, Bern, Switzerland

© The Author(s) 2024
J. T. Freymueller, L. Sánchez (eds.), *Together Again for Geodesy*,
International Association of Geodesy Symposia 157, https://doi.org/10.1007/1345_2024_275

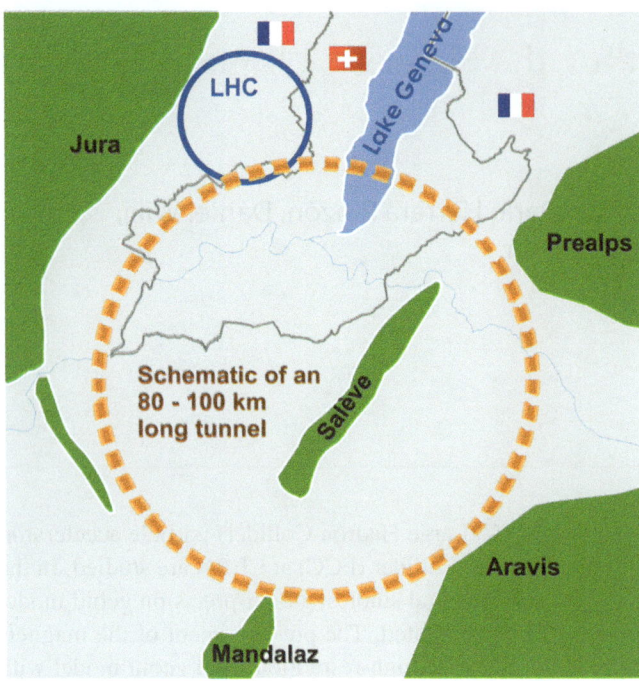

Fig. 1 Location of LHC and FCC with the main topographic features in the region. (© CERN 2014)

In the scope of initial studies, the design and feasibility of the FCC are studied until the next update of the European Strategy for Particle Physics in 2027.

From the geodetic point of view, the area of the FCC bears several challenges. As depicted in Fig. 1, firstly, the area of the LHC and FCC crosses the border between France and Switzerland. Due to different national height systems and political constitutions, unified models and reference systems need to be established specifically for the CERN region. Secondly, even though the region of the FCC is topographically relatively flat, its planned trajectory circles Mont Salève and passes below Lake Geneva covering an area with several tectonic faults. In addition, the FCC is enclosed by the Jura Mountains, Vuache mountain range and (Pre)Alps. The significantly larger size of the FCC and the topographic features made it necessary to initiate the FCC Geodesy project as a feasibility study funded by CHART to develop the reference network on the surface as well as in the tunnel and to compute new high-precision geoid models for this specific area.

The first part of the FCC Geodesy project established a conceptual design of the surface geodetic network including control baselines needed for the construction of the FCC tunnel. The surface geodetic network will be used for the civil engineering and surveying works during the construction and pre-alignment phase. In addition, it provides the reference for the geo-kinematic monitoring in the FCC area.

The second part of the project plans the vertical reference on the surface as well as in the tunnel. Therefore, several geoid models with different levels of precision will be computed. For the civil engineering site investigations and construction a 1-cm geoid will be established. In the first stage of the FCC, the so-called FCC-ee will be constructed, which will perform electron-to-positron (e^+e^-) collisions and will act as "Higgs factory". For this collider, the geoid must be determined with a precision of 30 μm over 225 m along the nominal location of the collider. In the second stage, the collider will be transformed into the FCC-hh, in which hadrons (h) are collided at extremely high energies. For this type of collider, the geoid precision is 100 μm over 225 m. These geoid precisions have not been realized so far and in the future scope of this project comprehensive numerical simulations will be performed to find out what kind of and where observations are needed as well as which auxiliary data sets are required in order to improve the 1-cm geoid to the level of a sub-mm geoid along the trajectory of the collider. To be able to assess the internal accuracy of existing geoid models as well as the geoid models, which will be calculated in the scope of the FCC Geodesy project, a high-precision geoid profile was established in the area of the FCC and will be used in addition to GNSS-levelling points.

This publication consists of five sections. The alignment methodology at CERN is introduced in Sect. 2. The data available in the area of the FCC for regional geoid modeling is listed in Sect. 3. The existing geoid models as well as a preliminary geoid specifically computed for the FCC are introduced in Sect. 4. A comparison between the models and GNSS-levelling data is provided in Sect. 5 and the study is concluded in Sect. 6.

2 Pre-alignment of Accelerator Components at CERN

A distinctive difference between the needs of a geoid model for classical tunnel construction and the tunnel for the FCC is that the geoid model is not only needed for the construction of the tunnel but also for the monitoring and pre-alignment of the accelerator components during the lifetime of the FCC. The following part is based on the knowledge from and realization of the current LHC, since the exact structure of the facilities in the FCC tunnel is still in the planning phase.

Along the tunnel, the particles are accelerated to close the speed of light and are then brought to collision at interaction points (IP) located in several test facilities within the tunnel. In these experimental areas, particle physicists track the particle collisions and reconstruct the three-dimensional particle trajectory with respect to the nominal beam line (Herty 2009).

Particle beams are considered not to be influenced by gravity, therefore the beam line is planned in a Euclidean plane. Moreover, in order to divert hadrons from their straight path, high energy is needed because of their comparatively high mass. Charged particles such as electrons and positrons emit synchrotron radiation when they are accelerated and radiate beamstrahlung photons when they are deflected from their straight path. All effects lead to an energy loss and, therefore, all unnecessary deviations from the nominal path must be avoided. Additionally, the beam oscillates around its nominal trajectory, therefore small steps in the alignment can trigger an excitation of the amplitude and lead to strong deviations from the nominal path and a strong increase in the beam size at a location far away from the location of the misalignment.

For the surveying work in the LHC, this means that once all accelerator components are correctly placed in the reference frame of the project by using classical surveying methods like geodetic levelling, total station and laser tracking measurements, a second survey is conducted using digital levels and the so-called ecartometer, which measures the offset with respect to a stretched wire that is attached to each reference point on the magnets (Myers and Schopper 2020). Finally, on each side of an experimental area, so-called low beta magnet triplets are used to reduce the diameter of the particle clouds and adjust their flight direction before they collide at the nominal interaction point (Herty 2009). In these areas, an additional survey is conducted to obtain the local best fit of the three low beta quadrupoles Q1, Q2 and Q3, which build the low beta magnet triplet. For this purpose, an alignment system consisting of a Wire Positioning System (WPS), a Hydrostatic Levelling System (HLS), and a Distance Offset Measurement Sensor (DOMS) is installed at each low beta magnet triplet (see Fig. 2). After these pre-alignment steps, requiring a precision of 30 μm over 225 m for the vertical component, at last the beam-based alignment can begin, which then realizes a vertical beam size of only about 30 nm.

Table 1 gives an overview of a selection of parameters that are relevant for the determination of the error budget of 30 μm over a sliding window of 225 m for the geoid computation. Luminosity L is, apart from the energy level, the most important target value of a particle collider. It is the relation of the number of collisions per second dR/dt

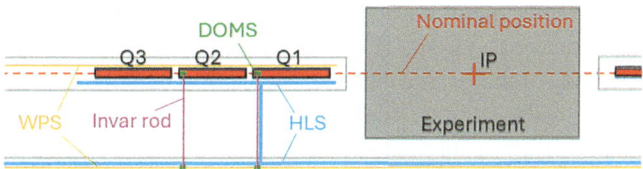

Fig. 2 Top View of WPS, HLS, and DOMS, which position the low beta quadrupoles on each side of the experimental areas in the accelerator tunnel. The quadrupoles in the LHC experiment are 6.3 m (Q1, Q3) and 5.5 m (Q2) long. The ATLAS experiment has a total length of 43 m (Rude et al. 2022)

and the IP beam size σ: $dR/dt = L \cdot \sigma$. For the FCC-ee and the Compact Linear Collider (CLIC, another planned accelerator at CERN), flat beams with a strong difference in the focusing of the beams in the vertical and horizontal direction are designed to overcome the increase in beam size caused by synchrotron radiation (Myers and Schopper 2020). Comparing the vertical beam sizes, it becomes apparent that the beam size of the FCC-ee ($\sigma_y = 0.028$ μm) is more than 250 times smaller than the beam size of the improved LHC, the High-Luminosity LHC (HL-LHC; $\sigma_x = \sigma_y = 7.1$ μm). In addition, the FCC is approximately four times longer than the LHC, which means that the realignment process (iterative loop of measuring - analyzing - readjusting) during shutdown phases will take at least four times longer because it is not possible to significantly increase the number of trained experts. In addition, seismic ground movement, vibrations from components and cultural noise will move the components out of the (pre-)alignment tolerance and therefore, it will be necessary to carry out even more realignments during its lifetime. In order to overcome the limit in human and time resources in the shut-down phases for the realignment, one option is to install the alignment and monitoring system consisting of WPS, HLS, and DOMS on all components along the collider. This would make it possible to analyze the changes in the alignment already before the shut-down phase or even to adjust each component remotely (Mainaud 2022).

Other alignment systems based on optical or mechanical reference lines are also considered, however, they suffer from technical difficulties, cannot operate in radioactive environments or are taking up too much space in the FCC tunnel (Stern 2016). Therefore, the vertical pre-alignment

Table 1 Comparison of selected parameters of collider projects at CERN

Parameter	Unit	LHC (Abada et al. 2019a)	HL-LHC (Abada et al. 2019a)	FCC-hh (nominal) (Abada et al. 2019a)	FCC-ee (Z-pole) (Abada et al. 2019b)	CLIC (Linssen et al. 2012)
Length	km	26.7	26.7	97.8	97.8	48.4
Luminosity target	10^{34} cm^{-2}s^{-1}	1.0	5.0	< 30.0	193.0	5.9
IP beam size σ	μm	16.7	7.1	6.8	$\sigma_x = 6.4$	$\sigma_x = 0.045$
					$\sigma_y = 0.028$	$\sigma_y = 0.001$

and monitoring system consisting of WPS and HLS is the most favored system. However, to benefit from the high precision of the HLS, a high-precision geoid model is needed to relate the equipotential surface created by the HLS to the Euclidean plane, which is needed for the collider pre-alignment.

3 Available Data for Regional Geoid Modeling and Validation

3.1 Gravity Measurements

In the area of the FCC, over 16,000 gravity measurements are available. Figure 3 shows their location with different colors indicating the variety of sources providing the data. These heterogeneous data sources and qualities require a thorough pre-processing and assessment of each measurement before it is used for the geoid calculation.

In Switzerland, the largest data set is given by the Gravimetric Atlas of Switzerland. These measurements were conducted between 1968 and 2000 (Olivier et al. 2010). Another data set consists of several geodetic leveling campaigns, which were conducted either by the Canton of Geneva or the Federal Office of Topography (swisstopo). In France, the main contributions come from IGN (Institut national de l'information géographique et forestière) and BRGM (Bureau de Recherches Géologiques et Minières), the bureau for geological services in France.

In the last 70 years, in which these data sets were growing and merged, several coordinate system changes were implemented, which may lead to wrong coordinates, if the transformation parameters are not clearly defined. In addition, the geolocalization of these measurements was difficult in the past, when access to high-precision GNSS solutions was non-existing or limited. These circumstances led to points that are duplicates but with different horizontal as well as vertical coordinates. The erroneous points must be excluded to avoid firstly an overweight in the geoid calculation and secondly the derivation of an inaccurate geoid model.

The distribution of the gravity measurements in the region of the FCC is very heterogeneous. Especially along the border of the region of interest (RoI), which is shown with a grey polygon in Fig. 3, not a lot of data is available. The RoI defined for the FCC Geodesy project includes all current and planned facilities of CERN. Around the eight vertical shafts currently planned for the FCC, several facilities will be set up on the surface in a radius of 5 km. They are used to define the extent of the RoI in addition to the existing CERN infrastructure in the North.

In order to fill these gaps for the 1-cm geoid needed already in the construction phase, a first measurement campaign was conducted in 2022, which densified the area located in the Northwest of Lake Annecy with gravity measurements in a grid-like pattern with an average distance of 2 km by 2 km (see purple dots in Fig. 3). The FCC area is also included in the European Alps Geoid (EAlpsG) project area (Baue and Schwabe 2023), favoring a collaboration to conduct more measurement campaigns benefiting both parties.

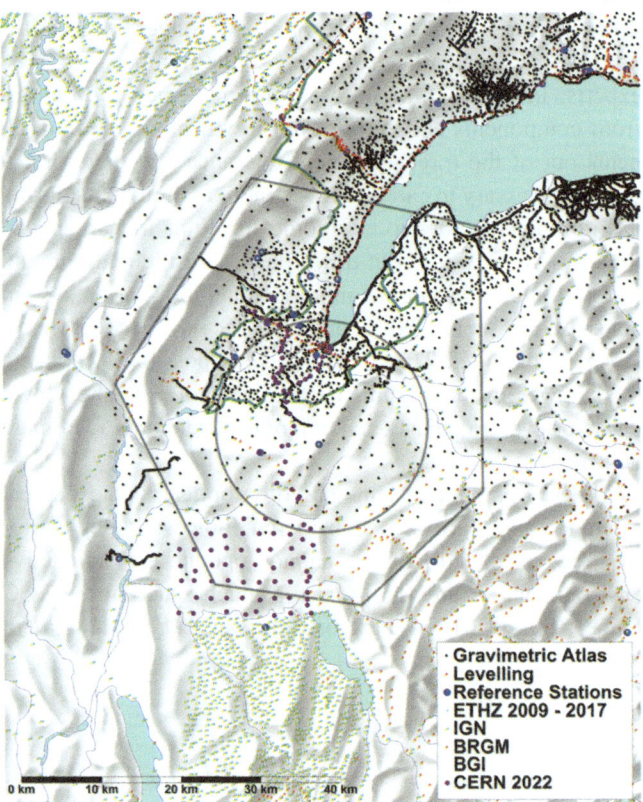

Fig. 3 Gravity measurement points in the area of the FCC. Different data sources are shown with different colors. The grey polygon shows the approximate region of interest for the FCC Geodesy project and the grey circle the approximate location of the FCC tunnel. © swisstopo

3.2 Deflections of the Vertical Measurements

In the area of the FCC, over 120 deflection of the vertical (DoV) measurement points are available. The deviation from the normal on the reference ellipsoid is divided into the North-South component ξ and the East-West component η. They were derived from zenith camera observations and, due to the lack of precise clocks in the past, the older measurements show lower accuracies in the η component.

The zenith camera systems were mainly developed at University of Hannover (TZK2-D) and ETH Zurich (DIADEM replaced by CODIAC). Therefore, almost all available DoV measurements were made with these systems. Consequently, most of the measurement sites are in the Swiss part of the FCC region. During an observation period of 20 minutes, 50 individual measurements are conducted and the precision of the solution for a full session is between 0.08" and 0.10" (Hirt et al. 2010).

Studies in the past have shown that DoV measurements become especially useful in geoid modeling to bridge the gap between the geoid's longer wavelengths provided by global geoid models and the very short wavelengths captured with gravity measurements (Guillaume 2015).

4 Regional Geoid and Quasigeoid Models in the FCC Area

4.1 Existing Geoid and Quasigeoid Models

In the FCC region, several regional geoid models are already available. In Table 2, a selection is listed.

CERN determined two geoid models. The CG1985 was introduced for the reference system of the Large Electron-Positron Collider (LEP), but since then the facilities of CERN have grown and the CG2000 was established. Since the FCC area is around 100 times larger than the current extent of the CERN facilities and infrastructure, a new solution is calculated in this project. The most recent geoid of Switzerland CHGeo2004 was published in 2004. It covers the area of Switzerland including the RoI of the FCC. The solution is based on the gravity and DoV measurements, which were introduced in Sects. 3.1 and 3.2 that were already available at the time of calculation. France has published quasigeoid models since 1996 along with hybrid quasigeoid models, which were adjusted to GNSS-levelling points. Currently, the D-A-CH-Geoid and its extension to the European Alps Geoid (EAlpG) are in the making. In a joint effort of several national groups, a geoid solution is derived that captures the high-frequency changes in the geoid caused by large height changes and faults in the geological layers. The largest region is covered by EGG2015, which covers the full extent from Iceland to the area of Uzbekistan. This is possible by deriving additional gravity values from altimetry and global geoid models.

4.2 An Initial Gravimetric Geoid for the FCC Region

For the calculation of the FCC Geoid existing software packages are used, which are going to be adapted based on the available data and needs in this research project. Since former studies have shown that the incorporation of deflection of the vertical measurements is key to reaching the high precisions needed for the FCC project (Guillaume 2015), software packages that can already process DoV data are selected. One of these software toolkits is GROOPS (Mayer-Gürr et al. 2021). In Fig. 4 the preliminary gravimetric quasigeoid, which was calculated with the GROOPS software, is shown.

The GROOPS software uses the Remove-Compute-Restore approach and the residual geoid estimation uses a global geoid model as reference surface to minimize the linearization uncertainties. In the depicted solution, a combination of GOCO06s and EGM2008 was used. A strength of this software is the calculation of the residual terrain correction for the reduction of the observations on the surface by calculating the difference in the spherical harmonic expansion of the DEM and the global geoid model. The computation uses a least squares approach with spherical radial basis functions (SRBF) for parameterization. SRBFs are based on the spherical harmonic functions, which are well known from global geoid modeling, but they have the advantage that they can achieve high spatial resolution for a regional solution. They have been gaining more popularity in recent years (Pock 2017; Yu et al. 2023).

Since this quasigeoid solution is still in its preliminary phase, it will mainly be used to detect critical areas and to define future approaches. In the RoI, the quasigeoid varies from 49.418 m to 50.469 m and the average geoid height

Table 2 Selected list of regional geoid and quasigeoid models that cover the area of the FCC

Region	Name	Type of Geoid	Reference
CERN	CG1985	Geoid	Jones et al. (2002)
CERN	CG2000	Geoid	Jones (2003)
Switzerland	CHGeo98	Geoid	Marti (2002)
Switzerland	CHGeo2004	Hybrid Geoid	Marti (2007)
Switzerland	CHQua04	Hybrid Quasigeoid	Marti (2007)
France	QGF98	Quasigeoid	Duquenne (1998b)
France	RAF98	Hybrid Quasigeoid	Duquenne (1998a)
France	QGF2016	Quasigeoid	L'Ecu (2017)
France	RAF2018b	Hybrid Quasigeoid	IGN (2019)
France	RAF20	Hybrid Quasigeoid	L'Ecu (2021)
Europe	D-A-CH-Geoid	Geoid	Baue and Schwabe (2023)
Europe	EGG2015	Quasigeoid	Denker (2015)

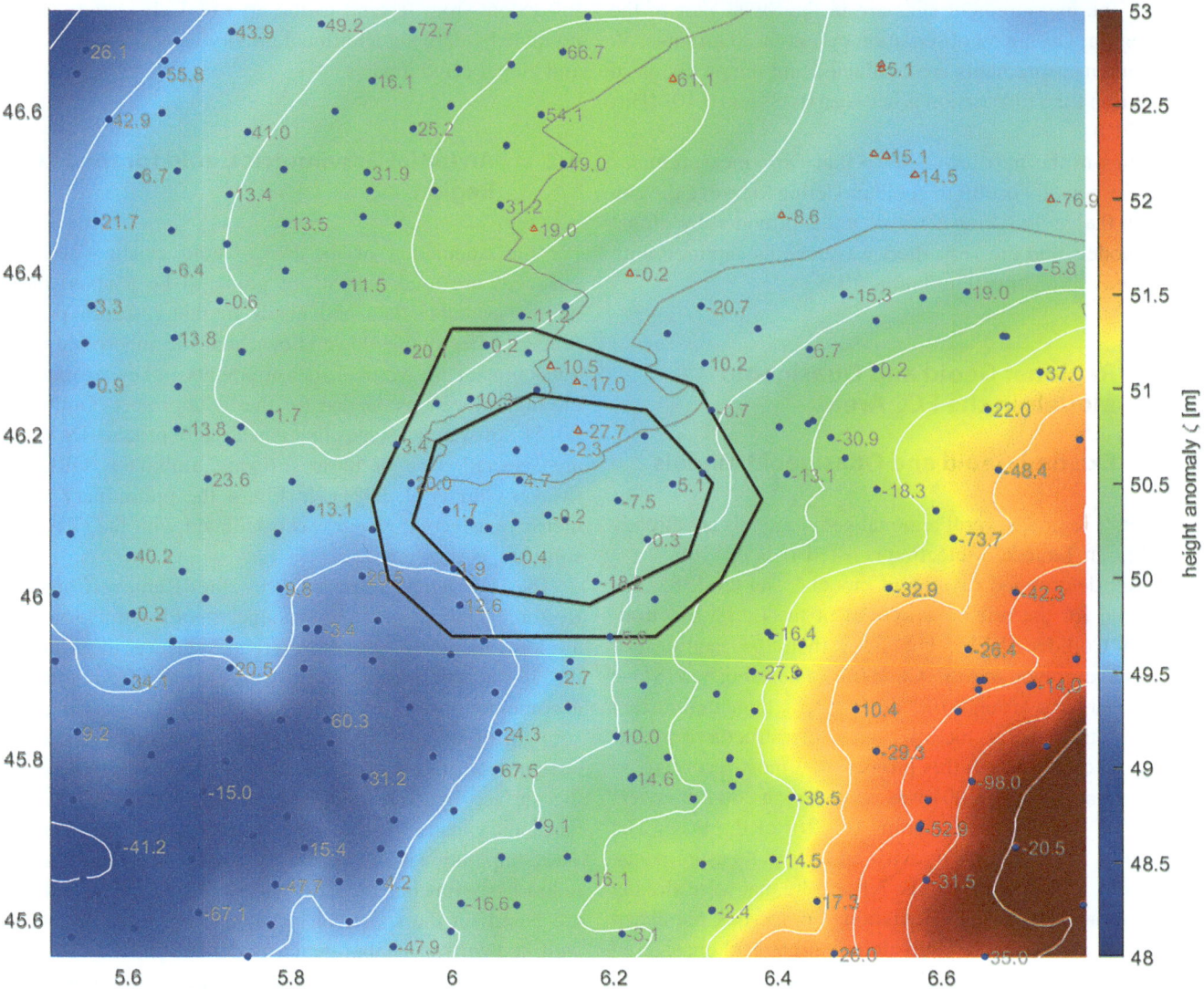

Fig. 4 A preliminary quasigeoid for the FCC region based on the hybrid global geoid GOCO06EGM08 and gravity measurements calculated with the GROOPS software. The region of interest and approximate location of the FCC tunnel are shown as black polygons. Blue filled circles and red triangles depict the location of GNSS-levelling points. For a selection of locations, the deviation between the FCC-QG and the GNSS-levelling datasets are depicted in grey. The values are given in millimeters and the average offset of 34 cm (swisstopo) and −13.2 cm (IGN) is reduced

is 49.777 m. The northwestern and southeastern borders of the RoI show the highest values, which are the impact of the Jura Mountains and the Alps, respectively. In the central part, slightly higher values are visible, which are caused by the Salève mountain. Overall, the RoI is dominated by the impact of the Molasse basin. In order to fully capture the influence of Earth's gravity field on the HLS measurements, for the FCC, an orthometric height system based on the geoid will be used. Therefore, the planned inclusion of the density values derived from the detailed geological models available will likely affect the geoid heights in this region the most.

5 Comparison of Quasigeoid Solutions

5.1 Comparison Between Quasigeoid Models

In this section, a selection of quasigeoid models that cover the area of the FCC are compared and the statistics are shown in Table 3. This comparison includes the global models GOCO06s (d/o 300), EGM2008 (d/o 2190) and XGM2019 (d/o 2190) as well as the new quasigeoid calculated with the GROOPS software.

Table 3 Comparison of quasigeoid solutions available in the region of the FCC. For each comparison, the offset and the standard deviation are given in millimeters

		CHQua04	QGF16	RAF20	EGG2015	GOCO06s	EGM2008	XGM2019
FCC-QG	Offset	227	−121	−131	333	111	95	14
	Std	27	49	14	41	147	40	46
CHQua04	Offset		−348	−359	106	−116	−132	−213
	Std		61	21	57	154	33	50
QGF16	Offset			−10	454	233	216	136
	Std			53	23	115	61	48
RAF20	Offset				464	243	226	146
	Std				46	146	34	47
EGG2015	Offset					−222	−238	−319
	Std					125	57	45
GOCO06s	Offset						−17	−97
	Std						147	139
EGM2008	Offset							−81
	Std							37

As expected, the FCC Quasigeoid (FCC-QG) shows the smallest standard deviations with 14 mm and 27 mm, respectively, with respect to the geoid models RAF20 and CHQua04, since these three models are based on very similar data sets. For the same reason, EGG2015 and EGM2008 also have a good agreement with the FCC-QG. However, the smallest offset to the FCC-QG is found for XGM2019. This might be because the FCC-QG uses as a reference surface a global geoid model that was derived by combining GOCO06s and EGM2008, which is very similar to the signal content of XGM2019. This hypothesis is supported by the small standard deviation for the comparison of the global models EGM2008 and XGM2019.

In order to estimate the current precision along the sliding window of 225 m of the target tolerance, the difference between the FCC-QG and RAF20 and CHQua04, respectively, was calculated for each grid point. Then the gradients of these differences to neighboring grid points were determined. The results were then rescaled from the current grid spacing of approximately 1 km to the 225 m of the tolerance. For the comparison, the neighboring grid points along the FCC tunnel depicted in Fig. 4 were used, which resulted in values of $RMSE_{RAF20,225m,FCCline} = 1.1$ mm and $RMSE_{CHQua04,225m,FCCline} = 0.7$ mm. In addition, the analysis of the differences between several (quasi)geoid models showed the strongest deviations between the models in the area of the Salève mountain in East-West as well as North-South direction, at the most Northern and most Southern shaft of the FCC tunnel in North-South direction and at the Southeastern shaft in diagonal direction towards the Alps. In these areas the highest deviations between the neighboring grid points are expected, therefore along 5 lines of 7 consecutive grid points the same analysis was repeated. In these areas the values increase to $RMSE_{RAF20,225m,extreme} = 1.5$ mm and $RMSE_{CHQua04,225m,extreme} = 1.2$ mm. These results show that under the assumption that CHQua04 and RAF20 would be error-free quasigeoids, the current precision of the FCC-QG solution in a sliding window of 225 m is approximately $RMSE_{225m,RoI} = 1.1$ mm in the RoI. To reach the required tolerances of 30 μm and 100 μm the precision for the target geoid must be improved by a factor of 10 to 30 in the worst case. In the studies for the CLIC linear collider, it was shown that a difference of < 20 μm can be achieved with respect to a line of 200 m (Guillaume 2015), which makes us confident that it is possible to calculate a geoid model in the tolerance specifications for the FCC.

5.2 Comparison with GNSS-Levelling Points

The most common validation method for geoid modelling is the comparison with GNSS-levelling points. In this section, the quasigeoid models are thus compared to GNSS-levelling points (see Table 4). The dataset of swisstopo contains 4 points in the RoI, the dataset of IGN 30 points. The location of the points are shown in Fig. 4.

It is encouraging to see that the comparison of the GNSS-levelling data with the newly calculated FCC-QG within the RoI resulted in the best standard deviation (STD) of 9 mm for the IGN dataset and in 7 mm for the small swisstopo dataset, respectively. When looking at the next-best STDs for the statistically more relevant IGN dataset stemming from the hybrid quasigeoid solutions CHQua04 and RAF20, one should keep in mind that these were created by including the GNSS-levelling data in the calculation process. A more realistic estimate for the precision of the French geoid solution can be drawn from the gravimetric quasigeoid QGF16 with an STD of 53 mm (IGN) and 28 mm (swisstopo).

For the full region, which will be used in the future only to analyze the gradual deterioration of the FCC geoid, the STD of the FCC-QG increases to approximately 3 cm. In Fig. 4 the deviations to a selection of the GNSS-levelling points are shown. It becomes apparent that the highest differences are found at the border of the displayed region and regions with

Table 4 Comparison of quasigeoid solutions available in the region of the FCC with GNSS-levelling points provided by IGN and swisstopo. The offset and standard deviation are calculated for the full region depicted in Fig. 4 and for the region of interest. The parameters are given in millimeters

			FCC-QG	CHQua04	QGF16	RAF20	EGG2015	GOCO06s	EGM2008	XGM2019
IGN	Full region	Offset	−132	−343	−35	−2	−493	−141	−185	−149
		Std	31	56	71	10	63	397	54	62
	RoI	Offset	−129	−358	2	0	−454	−200	−234	−150
		Std	9	21	53	10	41	174	35	46
swisstopo	Full region	Offset	340	68	454	479	3	321	228	323
		Std	30	10	53	24	37	383	38	49
	RoI	Offset	322	68	460	449	15	197	224	348
		Std	7	2	28	5	29	98	16	22

distinctive topographic features, such as the Jura mountain range in the North as well as the Prealps in the South to Southeast direction.

Unfortunately, the dataset of swisstopo is rather sparse and in the RoI only 4 points are available because it is a set of GNSS-levelling points carefully selected and rechecked for errors. In contrast to the swisstopo dataset, the dataset of IGN includes 237 points in the full region and even in the RoI 30 points are available for comparison. However, for this dataset no metadata exist, which makes it difficult to assess whether the deviations are caused by erroneous heights in the GNSS-levelling datasets or inaccuracies in the FCC-QG due to simplifications and approximations in the GROOPS software or the lack of data for the calculation. Therefore, in the next steps of the project, the overall accuracy of the points on the validation profile will be determined and can then be used to better estimate the errors in the geoid solution.

6   Summary and Outlook

The FCC region bears several geological features that pose considerable challenges for regional geoid modeling. The amount and variety of available data ranging from measurements of gravity and deflections of the vertical to three-dimensional geological models give a unique opportunity to study the current limits of regional geoid modeling. The external accuracy of the new preliminary geoid model FCC-QG was determined by a comparison with GNSS-levelling data to < 1 cm in the RoI and approximately 3 cm in the full region. The comparison of the FCC-QG with existing solutions showed that the FCC-QG has a good concordance with CHQua04 and RAF20, with an STD of approximately 1.5 to 3 cm. Analyzing the changes in the differences of the FCC-QG compared to CHQua04 and RAF20 between neighboring grid points and rescaling them to the length of the sliding window of 225 m, resulted in an RMSE of 1.1 mm in the RoI. In order to reach the target tolerance of 30 μm the precision must be improved by a factor of 10 to 30. For this, as the CLIC study results indicate, the addition of a sufficient number of DoV measurements will be crucial.

Currently, a geoid derived from gravity and DoV measurements is in preparation and the already available solutions will be compared to the validation profile with realistic error estimates for the validation data. The better DEMs provided by swisstopo and IGN are going to be merged and will replace the DEM currently in use. This will be followed by a second thorough examination of the gravity data, which can hopefully resolve the still-existing problems in the data sets.

The next step will be the calculation of a density model based on the layers given by geological models and the estimation of a combined uncertainty caused by the uncertainties of the exact location of the boundaries of the geological layers and the uncertainty in the rock densities for each layer. By analyzing the changes in the geoid depending on whether a constant density, a surface density model or a 3D model is used, the range of error induced by presumptions on the density in geoid modeling can be estimated.

In addition, the variety of input data as well as the very high resolution and precision of the targeted geoid solution require the thorough testing of the GROOPS software for numerical accuracy at these extreme scales.

In the end, the most limiting factor will be the quality of density information and the distribution and resolution of the gravity and DoV observations. According to the Kaula rule (Kaula 1966) the average RMS variation of geoid heights below 10 m spatial resolution is 30 μm, therefore, numerical simulations will be performed to determine the areas and the densities of the measurements that are needed to reach the sub-mm level precision for the geoid along the tunnel of the FCC. In the studies for the CLIC linear collider, it was shown that a difference of < 20 μm can be achieved between the predicted and measured variations of the equipotential line determined with gravimetric and astro-geodetic observations along a line of 650 m (Guillaume 2015). These results based on real observations make us confident that it is possible to calculate a geoid model in the tolerance specifications for the FCC.

Acknowledgements This work was performed under the auspices and with support from the Swiss Accelerator Research and Technology (CHART) program (www.chart.ch).

References

Abada A, Abbrescia M, AbdusSalam SS, et al (2019a) FCC-hh: The Hadron collider. Eur Phys J Spec Top 228(4):755–1107. https://doi.org/10.1140/epjst/e2019-900087-0

Abada A, Abbrescia M, AbdusSalam SS, et al (2019b) FCC-ee: The Lepton collider. Eur Phys J Spec Top 228(2):261–623. https://doi.org/10.1140/epjst/e2019-900045-4

Bauer T, Schwabe J, The European Alps Geoid group (2023) The European Alps Geoid (EAlpG) Project – a joint initiative for improved cross-border regional geoid modelling and height transformation. In: EGU General Assembly 2023–7431. https://doi.org/10.5194/egusphere-egu23-7431

Denker H (2015) A new European Gravimetric (Quasi)Geoid EGG2015. Poster Presentation at the XXVI General Assembly of the International Union of Geodesy and Geophysics (IUGG), Earth and Environmental Sciences for Future Generations

Duquenne H (1998a) Grille de correction pour effectuer du nivellement par GPS. Revue Geometre

Duquenne H (1998b) QGF98, a new solution for the quasi-geoid in France. In: Report of the Finnish Geodetic Institute, vol 98, 4th edn. Finnish Geodetic Institute, p 251–255

Guillaume S (2015) Determination of a precise gravity field for the CLIC feasibility studies. PhD thesis, ETH Zurich, Zurich. https://doi.org/10.3929/ethz-a-010549038

Herty A (2009) Micron precision calibration methods for alignment sensors in particle accelerators. PhD thesis, School of Architecture, Design and the Built Environment, Nottingham Trent University. https://irep.ntu.ac.uk/id/eprint/364/

Hirt C, Bürki B, Somieski A, et al (2010) Modern determination of vertical deflections using digital Zenith cameras. J Surv Eng 136(1):1–12. https://doi.org/10.1061/(ASCE)SU.1943-5428.0000009. https://ascelibrary.org/doi/10.1061/%28ASCE%29SU.1943-5428.0000009

IGN (2019) National Report - France. In: EUREF Symposium 2019. http://www.euref.eu/symposia/2019Tallinn/05-08-France.pdf

Jones M (2003) Determination of updated vertical geodetic reference surfaces for the CERN site. CERN EDMS Document N°411060

Jones M, Mayoud M, Wiart A (2002) Geodetic Parameterization of the CNGS Project. In: Proceedings of the 7th International Workshop on Accelerator Alignment (IWAA 2002), pp 172–184. https://proj-cngs.web.cern.ch/PDFfiles/MJone02.pdf

Kaula WM (1966) Global harmonic and statistical analysis of gravimetry. In: Gravity Anomalies: Unsurveyes Areas, vol 9. Blaisdell Publishing Company, Waltham, MA, pp 58–67. https://doi.org/10.1029/GM009p0058

L'Ecu F (2017) Calcul du quasi-géoïde QGF16 et de la grille de conversion altimétrique RAF16 Aétat d'avancement et perspectives. Revue XYZ 150:49–51. https://documentation.ensg.eu/index.php?lvl=noticedisplay&id=84433

Linssen L, Miyamoto A, Stanitzki M, et al (eds) (2012) Physics and detectors at CLIC, vol 2. CERN, Geneva. https://doi.org/10.5170/CERN-2012-003

L'Ecu F (2021) Description technique de la surface de conversion altimétrique RAF20. Tech. rep., IGN, Service de Géodésie et de Métrologie

Mainaud Durand H (2022) Alignment options for the FC-ee. In: FCCIS Workshop. https://indico.cern.ch/event/1203316/contributions/5125322/attachments/2561632/4415400/FCC-alignmentoptionsdec22.pdf

Marti U (2002) Das Geoid der Schweiz 1998 "CHGEO98". Swisstopo Doku 16(10). https://shop.swisstopo.admin.ch/de/buecher-und-publikationen/geodaesie/swisstopo-doku

Marti U (2007) Comparison of high precision geoid models in Switzerland. In: Dynamic planet. Springer, Berlin, Heidelberg, pp 377–382. https://doi.org/10.1007/978-3-540-49350-155

Mayer-Gürr T, Behzadpour S, Eicker A, et al (2021) GROOPS: A software toolkit for gravity field recovery and GNSS processing. Comput Geosci 155:104864. https://doi.org/10.1016/j.cageo.2021.104864

Myers S, Schopper H (eds) (2020) Particle physics reference library, vol 3: Accelerators and colliders. Springer International Publishing, Cham. https://doi.org/10.1007/978-3-030-34245-6

Olivier R, Dumont B, Klingelé E (2010) L'Atlas gravimetrique de la Suisse. Geophysique 43

Pock C (2017) Consistent combination of satellite and terrestrial gravity field observations in regional geoid modeling. PhD thesis, Graz University of Technology. https://online.tugraz.at/tugonline/wbAbs.showThesis?pThesisNr=61642&pOrgNr=34102

Rude V, Kautzmann G, Mainaud Durand H, et al (2022) 3D calculation for the alignment of LHC low-beta quadrupoles. In: 16th International Workshop on Accelerator Alignment (IWAA 2022), Ferney-Voltaire, France. https://inspirehep.net/literature/2635645

Stern G (2016) Study and development of a laser based alignment system for the compact linear collider. PhD thesis, Institute of Geodesy and Photogrammetry, ZĄNurich. https://doi.org/10.3929/ethz-a-010621412

Yu H, Chang G, Zhang S, et al (2023) Application of sparse regularization in spherical radial basis functions-based regional geoid modeling in Colorado. Remote Sens 15(19):4870. https://doi.org/10.3390/rs15194870

Open Access This chapter is licensed under the terms of the Creative Commons Attribution 4.0 International License (http://creativecommons.org/licenses/by/4.0/), which permits use, sharing, adaptation, distribution and reproduction in any medium or format, as long as you give appropriate credit to the original author(s) and the source, provide a link to the Creative Commons license and indicate if changes were made.

The images or other third party material in this chapter are included in the chapter's Creative Commons license, unless indicated otherwise in a credit line to the material. If material is not included in the chapter's Creative Commons license and your intended use is not permitted by statutory regulation or exceeds the permitted use, you will need to obtain permission directly from the copyright holder.

Meteorite Impact Origin of Yangju Circular Structure in the Middle Part of the Korean Peninsula Estimated by Gravity Field Interpretation

Sungchan Choi, Sung-Wook Kim, Younghong Shin, and Eun-Kyeong Choi

Abstract

To ascertain the origin of the Yangju-Circular-Structure (YCS), located north of Seoul, South Korea, extensive gravity surveys were conducted. A collaborative analysis, integrating geology, geomorphology, gravity field interpretation, and density modeling, yielded significant insights:

(1) The eastern region of the YCS, adjacent to the Dongducheon fault line, has been displaced approximately 3,000 m southward. This substantial shift provides crucial evidence of significant tectonic activity affecting the area. (2) Our analyses indicate that the formation of the YCS is unlikely to be a result of differential weathering processes. (3) The YCS features three distinct concentric circular structures with varying in diameter. (4) The subsurface structure beneath the YCS appears to be symmetrical, likely resulting from concentric energy waves caused by an external impact, such as a meteorite.

Keywords

Density modelling · Gravity field interpretation · Meteorite impact · Yangju-circular-structure

1 Introduction

The Yangju Circle Structure (YCS), a significant geomorphological circular feature marked in the red rectangle of Fig. 1b, is located north of Seoul, the capital of South Korea, a densely populated area with over 20 million residents.

This structure is bisected by the NS-trending Dongducheon fault (DF, Fig. 1b), a key component of the Chugaryung-Fault-System (CFS in Fig. 1a). This structure is bisected by the north-south trending Dongducheon Fault (DF, Fig. 1b), a key component of the Chugaryung Fault System (CFS, Fig. 1a). The main rock type, diameter, and average height of the eastern half-circle of the YCS are as follows: it comprises Jurassic garnet-bearing biotite Granite (Jgbgr, Fig. 1b), spans 11–13 km in diameter (e1 and e2, Fig. 1e), and rises approximately 200 m in height (Fig. 1e) (Yun 1995, 1997; Jung 1998; Kwon and Sagong 1998). Initially, the YCS was hypothesized as a cauldron subsidence structure linked to volcanic activities (Yang et al. 2008). Contrary to expectations of volcanic rock composition, the outer rim of this crater-like structure primarily consists of plutonic rocks, notably the Jgbgr (Fig. 1b). An alternative explanation for the YCS's origin involves differential weathering between the Jgbgr, which forms the outer ring mountains, and the Jurassic biotite Granite (Jbgr, Fig. 1b) found in the central flatlands of the YSC. The Jbgr is distinguished by a relatively higher plagioclase to K-feldspar ratio and more mafic minerals than the Jgbgr (Yun 1995). Although mafic minerals and plagioclase generally undergo faster chemical weathering compared to K-feldspar and quartz (Goldich 1938; Yun 1995, 1997; Kwon and Sagong 1998), the Jgbgr displays a greater tendency for microfractures, porosity, and absorption,

S. Choi · S.-W. Kim (✉) · E.-K. Choi
Geo-Information Research Group, GI Co. LTD, Busan, South Korea
e-mail: suwokim@naver.com

Y. Shin
Korea Institute of Geoscience and Mineral Resources, Daejeon, South Korea

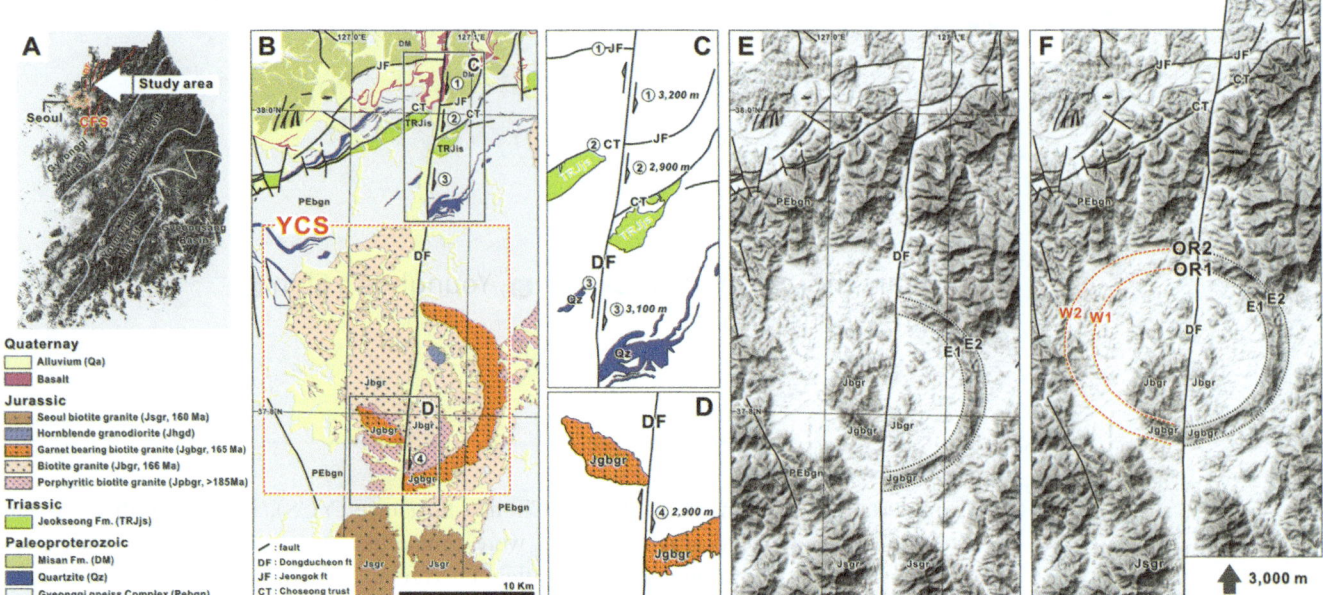

Fig. 1 (**a**) displays the location of the study area, identifying the YCS within the central region of the Korean peninsula. (**b**) presents a geological map highlighting the division of the YCS by North-South trending Dongducheon fault (DF). The eastern half of the YCS is notable for its ring-shaped structure composed of garnet-bearing biotite Granite (Jgbgr). The dextral (right-lateral) displacement along the DF is estimated to be approximately 3,000 m inferred from the comparative positions of fault lines (e.g., Jeongok fault, Choseong Trust in **c**) and rock units (e.g., Quartzite and Jgbgr in **d**) on either side of the DF. (**e**) A topographic map showcases the physical landscape of the YCS region, particularly emphasizing the ring-form structures (*e1* and *e2*) located east of the DF. (**f**) Illustrates the hypothesized initial structure of the YCS, which is reconstructed by virtually moving the eastern section of the DF approximately 3,000 m to the north. This reconstruction reveals an outermost ring (OR2) and a smaller outer ring (OR1), possibly providing insight into the historical geological activity in the region

suggesting it is more prone to both mechanical and chemical weathering than the Jbgr. This raises uncertainties about whether differential weathering between these granitic types fully accounts for the YCS's formation.

Another hypothesis considers the possibility of a meteorite impact, characterized by stages of contact and compression, excavation, and modification (Melosh 1989; Collins et al. 2020; Choi et al. 2022). During the excavation phase, a shock wave radiates through the target rock and diminishes outward, while the modification phase produces bowl-shaped ring structures of varying diameters surrounding the crater (Collins et al. 2020; Choi et al. 2022; Hildebrand et al. 1991; Pilkington et al. 1994; Pilkington and Hildebrand 2000). Geophysical methods such as gravimetry have long been employed to image the impact-derived concentric circle structure (Choi et al. 2022; Hildebrand et al. 1991; Pilkington et al. 1994; Pilkington and Hildebrand 2000; Melosh 1982; Ernstson 1984). Our recent gravity survey around the YCS, conducted in 2022 with over 1,000 stations, yielded substantial data that will be crucial in elucidating the origin of the YCS.

2 Geologic and Topographic Overviews

The YCS, delineated by the red rectangle in Fig. 1b, is a prominent feature within the central part of the Seoul Granitic Batholith. This batholith is a north-northeast trending Jurassic (185 160 Ma) intrusive complex located in the Precambrian Gyeonggi Massif (as shown in Figs. 1a, 1–4). The YCS is predominantly composed of two major granitic rock types: the biotite Granite (Jbgr, Fig. 1b) and the garnet-bearing biotite Granite (Jgbgr, Fig. 1b), both of which intrude the Precambrian Gneiss (PEbgn, Fig. 1b). A comparison of the geological (Fig. 1b) and topographic maps (Fig. 1e) reveals that the Jgbgr is mainly outcropped in the eastern half-circle, forming mountains with an average height of about 200 m. In contrast, the Jbgr spans the central region of the YCS with elevations ranging from 50 to 250 m. Although there is little age difference between the Jbgr (approx. 166 Ma) and the Jgbgr (approx. 165 Ma) as determined by K-Ar dating, there is a notable difference in magnetic susceptibility between the two, with the Jbgr showing 332 μSI and the Jgbgr at 2.3 μSI (Yun 1995). Multiple theories have been proposed regarding the origin of the YCS:

1. Kwon and Sagong (1998) suggested that the YCS's formation could be due to differential weathering between the more resistant Jgbgr and the more easily weathered Jbgr. Contrarily, other petrophysical analyses indicate that the Jgbgr is more susceptible to weathering than the Jbgr (Yun 1995, 1997).
2. The association with a volcanic cauldron subsidence has been speculated due to the shallow emplacement depth of the Jgbgr and its rapid cooling rate of about 100 °C/Ma, although no volcanic rocks have been found in the vicinity of the YCS (Kwon and Sagong 1998).
3. Another compelling theory considers the possibility of a meteorite impact. The typical impact process involves contact and compression, excavation, and modification phases, which could explain the formation of concentric circular structures (Melosh 1982; Ernstson 1984; Choi et al. 2022). This hypothesis is supported if similar circular structures are identifiable in gravity field maps.

Significantly, half of the YCS has been displaced due to the dextral tectonic movement along the DF. The displacement measurements along the DF, such as those of the Jeongok fault line (JF), Choseong-Trust (CT), and Jgbgr, suggest an average dextral strike-slip movement of about 3,000 m. By projecting the eastern section of the DF approximately 3,000 m northward from its present position (Fig. 1f), it appears that the geological boundary and the largest outer rings of the structure connect to form a coherent geological feature spanning approximately 14 km in diameter. The reformation of the topographic and geologic maps (Fig. 1f) based on this movement illustrates a potential connection between various geological features across the DF, supporting theories of both tectonic activity and impact origins. This comprehensive analysis sets the stage for further investigations that may conclusively determine the genesis of the YCS.

3 Gravity Field Interpretation and Modeling

3.1 Gravity Field Interpretation

As a part of the regional geophysical mapping project, the Korean Institute of Geoscience and Mineral resources (KIGAM) and the Geo-Information Research Group conducted gravimetric surveys in the area around the YCS. Approximately 1,100 gravity values, depicted as white point in Fig. 2a, were extracted from the KIGAM national database (2014). Topographical data, ranging from 20 to 700 m in elevation, were sourced from the National Geographic Information Institute of Korea (NGII) dataset, with a resolution of 30 m. These data were essential for calculating the complete Bouguer anomaly after accounting for a constant rock density of 2,670 kg/m^3, which aligns with the measured mean bulk density across the Korean peninsula (Choi et al. 2019, 2020) and typical continental crust values (Christensen and Moony 1995). The terrain effect, calculated using proprietary software, showed variations from 0 to 5.5 mGal (10^{-5} m/s^2), with most of the study area requiring corrections of less than 3.0 mGal. Notably, regions with more than 300 m of relief exhibited values exceeding 2.5 mGal, whereas flatlands showed negligible terrain effects.

The complete Bouguer anomaly map (Fig. 2a) displayed a mean anomaly of about −10 mGal, ranging from −20 to 16 mGal. Higher Bouguer gravity anomalies, above the mean value, were predominantly observed in the western and northwestern parts of the YCS, likely reflecting the denser Precambrian Gneiss (PEbgn) with a mean density of approximately 2,700 kg/m^3, higher than the Bouguer slab's density. In contrast, the middle part of the YCS shows NW-SE directing very-low anomalies, which are probably caused by the spatial distribution of the Jgbgr and the Jbgr. The both granites have same average densities of about 2,600 kg/m^3 (Park et al. 2009) that is less than the Bouguer slab. The NW-SE direction of these very low anomalies should be caused by NS striking dextral movement of the DF. If the YCS was originated either from a meteorite impact or caused by differential weathering between the Jbgr and the Jgbgr (Kwon and Sagong 1998), it makes sense that the traces such as the impact derived ejecta deposits or weathered structures remain close to the subsurface rather than in the deep crustal area. From that reason, we calculated the residual anomaly field, which displays the gravity effects of the near surface density structure much more clearly than the Bouguer anomalies. The core of our analysis focused on the residual anomalies, which provide a clearer indication of near-surface density structures.

By employing a high-pass filter with a cutoff at a wavelength of approximately 3,000 m, you have effectively isolated the near-surface density variations in the Bouguer gravity field. The resulting residual anomaly map (Fig. 2b) ranged from −3 to 3 mGal, highlighting several key features: (1) Positive residual gravity anomalies (>1.0 mGal) were primarily associated with areas underlain by the PEbgn. (2) Prominent negative residual anomalies were located along the Jeongok (JF in Fig. 2b) and Choseong Trust lines (CT in Fig. 2b). (3) Negative anomalies were also observed along the ring-shaped geological structure marked by e1 and e2. (4) Circular-wise negative anomalies in the northwestern part of the YCS delineated the density boundary between the PEbgn and the Jbgr. (5) Positive anomalies were observed in the central part of the YCS, mainly over areas dominated by the Jbgr.

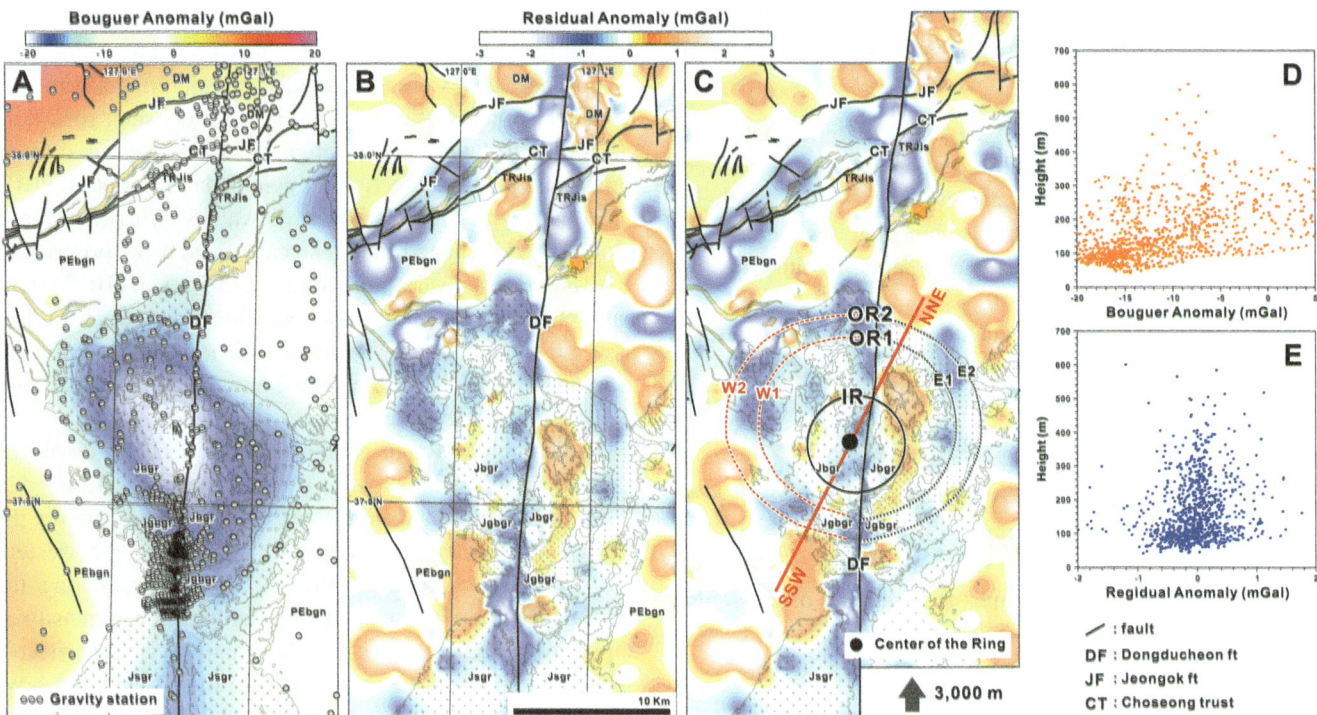

Fig. 2 (**a**) shows the Bouguer anomaly map for the YCS area. (**b**) illustrates the residual gravity field map. Residual anomalies provide more detail about the local variations by removing the regional gravity effect. This helps to highlight smaller-scale features like faults, cavities, or denser bodies that might not be visible in the broader Bouguer anomaly data. (**c**) depicts the residual anomaly field after it has been adjusted to reflect the theorized original position prior to the dextral movement along the DF. By moving the eastern side of the DF approximately 3,000 m to the north. This reveals different circular gravity boundaries, such as OR2, OR1, and IR, which are hypothesized to be related to the structure created by an impact event. (**d**) and (**e**) are scatter plots comparing the gravity anomaly values with topography. The lack of correlation between the gravity field and topography in these plots suggests that topographical features are not closely related to the variations in gravity. This finding implies that the YCS is less likely to have formed solely through differential weathering

The residual anomaly field was further analyzed by retroactively adjusting the eastern side of the DF northward by 3,000 m, approximating the original pre-tectonic movement position. This recalculated map (Fig. 2c) revealed several circular gravity boundaries correlating with topographic features (as per Fig. 1f). Intriguingly, an inner ring (IR in Fig. 2c), approximately 7 km in diameter and characterized by positive local anomalies, was identified. This feature, not visible in the topographic map, aligns with peak rings typically observed in meteorite impact structures, supporting the impact origin hypothesis for the YCS (Collins et al. 2020; Choi et al. 2022; Pilkington and Hildebrand 2000).

3.2 Gravity Field Modeling

To further validate the geological interpretations of the YCS, advanced gravity modeling was conducted along the NE-SW model line depicted by the red solid line in Fig. 2c. For this purpose, the IGMAS+ (3D gravity modelling system) was employed, which utilizes a sophisticated approach to approximate subsurface mass distributions. IGMAS+ operates on a kernel that employs polyhedral shapes with triangulated surfaces, a method well-suited for detailed geophysical analysis. These shapes are arranged in a series of parallel vertical modeling sections, enabling precise representation of complex geological structures within the Earth's crust and mantle. This methodology, detailed in foundational studies by Götze and Lahmeyer (1988) and Götze and Schmidt (2002), allows for an three-dimensional approximation of mass distributions and aids in the resolution of density variations across different geological layers. The primary objective of using this gravity modeling approach was to establish a clear picture of the underground structures and the density distribution within the crust along the selected NE-SW model line. This direction was chosen to align closely with the geological features of interest, particularly the configuration and orientation of the YCS as indicated by previous surface and subsurface observations. By implementing this detailed gravity modeling technique, we aim to enhance the resolution and reliability of the geological interpretations, providing a robust scientific basis for understanding the formation and characteristics of the YCS.

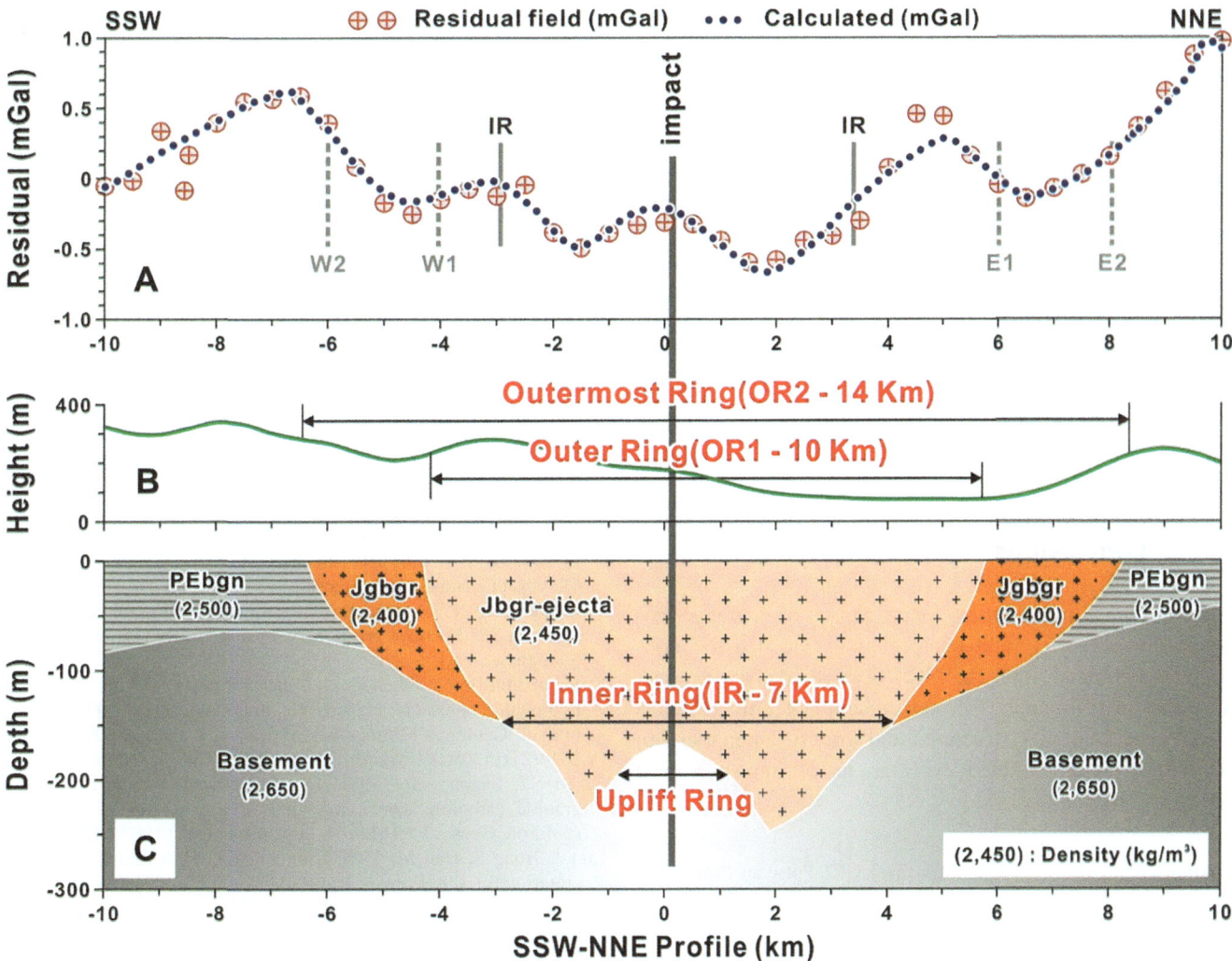

Fig. 3 (**a**) shows the residual gravity anomalies along a SW-NE profile (aligned with the black dotted line). Key features labeled include the concentric rings identified as "W1," "W2," and "IR," suggesting variations in subsurface densities that could be interpreted as the effects of an external impact. (**b**) illustrates changes in the topography along the profile. The geological model (**c**) shows a cross-sectional view of the subsurface structure along the profile. The model identifies three distinct geological features: Outermost ring (OR2) and outer ring (OR1) are interpreted as potential impact crater rims. Inner ring (IB) is postulated to be the peak ring with a diameter of around 7 km. An uplifted basement beneath the central part of the structure suggests a uplift ring commonly observed in complex craters

The gravity modeling conducted on the Yangju-Circle-Structure (YCS) offers significant insights into its subsurface structure, suggesting it was shaped by symmetrical, concentric energy waves typically associated with external impacts. The detailed analysis using the SW-NE density model facilitated by IGMAS+ revealed several key features indicative of such an event: (1) Notable features identified include the outermost ring (OR2 in Figs. 2c and 3c) and an outer ring (OR1 in Figs. 2c and 3c), which likely represent impact crater rims. (2) The inner ring (IB in Figs. 2c and 3c), with a diameter of about 7 km, is presumed to be the peak ring. (3) While the uplift ring is not overtly evident in the gravity field interpretation (Fig. 2c), the density model suggests the presence of an uplifted basement approximately 2 km in diameter beneath the central part of the ejecta.

4 Discussion

The relationship between weathering, density, and topography is well documented under similar environmental and petrophysical conditions (Yun 1995, 1997; Jung 1998; Kwon and Sagong 1998; Choi et al. 2020, 2022). Typically, regions that are more weathered show more fracturing and higher porosities, resulting in lower densities and reduced topographical elevations compared to less weathered areas. Initially, it was posited that the YCS's formation might be attributable to differential weathering between the Jgbgr and the Jbgr, both of which have a laboratory-measured density of about 2,600 kg/m^3 (Park et al. 2009). This theory suggested that lower residual gravity anomalies should correlate with lower elevations due to increased weathering. However, our analysis comparing topographic elevation changes with Bouguer and residual anomalies (Fig. 2d, e) showed no correlation, indicating that differential weathering does not explain the origin of the YCS.

The possibility of a meteorite impact forming the YCS has been a topic of considerable debate. Unlike previous studies which did not find physical evidence such as ring form circular geomorphology or impact-driven metamorphic features (e.g., Choi et al. 2022; Lim et al. 2021), our investigation marks the first geophysical approach to addressing this hypothesis. Our gravity field interpretation (Fig. 2c) revealed at least three circular geological and geomorphological structures within the YCS that could be indicative of meteorite impacts.

5 Conclusions

Based on the comprehensive joint analysis of geology, geomorphology, gravity field interpretation, and density modeling, we conclude that the YCS is most likely a product of a meteorite impact. This conclusion not only explains the observed circular structures and gravity anomalies but also aligns with modeled expectations of impact-derived geological formations.

Acknowledgments We would like to thank GFZ-Potsdam, Prof. Götze and Dr. Schmidt who provided the software package (IGMAS+) for the gravity modelling.

Funding This work was supported by the National Research Foundation of Korea (NRF) grant, funded by the Korean government (MSIT) [RS-2023-00250472]. This study was also supported in part by the Korea Meteorological Administration Research and Development Program (KMI2022-00710).

References

Choi S, Ryu I, Lee Y (2019) An analysis of the intraplate earthquake (2016M5.8_GY) that occurred in the Gyeongsang Basin in the SE of the Korean Peninsula, based on 3D modelling of the gravity and magnetic field. Geophys J Int 217:90–107. https://doi.org/10.1093/gji/ggz001

Choi S, Ryu I, Lee Y, Son Y (2020) Gravity and magnetic field interpretation to detect deep buried paleobasinal fault lines contributing to intraplate earthquakes: a case study from Pohang Basin, SE Korea. Geophys J Int 220:490–500. https://doi.org/10.1093/gji/ggz464

Choi S, Kim SW, Choi EK, Lee Y (2022) Estimating the impact process of the Jeokjung-Chogye Basin in Korea from gravity field interpretation. Geophys J Int 228:1457–1463. https://doi.org/10.1093/gji/ggab410

Christensen NI, Moony WD (1995) Seismic structure and composition of the continental crust: a global review. J Geophys Res 100:9761–9788. https://doi.org/10.1029/95JB00259

Collins GS, Patel N, Davison TM, Rae ASP, Morgan JV, Gulick SPS, IODP-ICDP Expedition 364 Science Party (2020) A steeply-inclined trajectory for the Chicxulub impact. Nat Commun 11:1480. https://doi.org/10.1038/s41467-020-15269-x

Ernstson K (1984) A gravity derived model for the Steinheim impact crater. Int J Earth Sci 73:483–498. https://doi.org/10.1007/BF01824969

Goldich SS (1938) A study in rock-weathering. J Geol 46:17–58. https://doi.org/10.1086/624619

Götze H-J, Lahmeyer B (1988) Application of three-dimensional interactive modelling in gravity magnetics. Geophysics 53(8):1096–1108

Götze H-J, Schmidt S (2002) Geophysical 3D modelling using GIS-functions. In: Proceedings of the 8th annual conference of the International Association for Mathematical Geology, Terra Nostra, pp 87–92, ISSN: 0946-8978

Hildebrand AR, Penfield GT, Kring DA, Pilkington M, Camargo ZA, Jacobsen SB (1991) Chicxulub Crater: a possible Cretaceous/Tertiary boundary impact crater on the Yucatan Peninsula, Mexico. Geology 19:867–871. https://doi.org/10.1130/0091-7613(1991)019{\mathsurround=\opskip$<$}0867:CCAPCT{\mathsurround=\opskip$>$}2.3.CO;2

Jung HS (1998) Paleomagnetic study for the Daebo granite in Dongducheon and Pocheon areas. M.S. thesis, Yonsei University, Seoul [GoogleScholar]

Korea Institute of Geoscience and Mineral Resources (KIGAM) (2014) Regional geophysical anomaly mapping, published by Korea Institute of Geology, Mining and Materials, Daejeon, GP2013-020-2014(2) (in Korean with English abstract) [GoogleScholar]

Kwon ST, Sagong H (1998) Uijongbu circular structure of Seoul granitic batholith, Korea: ring dike origin of a Jurassic volcanic cauldron. Geosci J 2:161–164. https://doi.org/10.1007/BF02910161

Lim J, Hong S, Han M, Yi S, Kim S (2021) First finding of impact cratering in the Korean Peninsula. Gondwana Res 91:121–128. https://doi.org/10.1016/j.gr.2020.12.004

Melosh HJ (1982) A schematic model of crater modification by gravity. J Geophys Res 87:371–380. https://doi.org/10.1029/JB087iB01p00371

Melosh HJ (1989) Impact cratering: a geologic process. Oxford University Press, U.K. Paperback, 245pp. ISSN 0952-7028

Park J, Kim HC, Lee Y, Shim BO, Song MY (2009) Thermal properties of rocks in the republic of Korea. Econ Environ Geol 42:591–598 (in Korean with English abstract) [GoogleScholar]

Pilkington M, Hildebrand AR (2000) Three-dimensional magnetic imaging of the Chicxulub. J Geophys Res 105:23479–23491. doi: https://doi.org/10.1029/2000JB900222

Pilkington M, Hildebrand AR, Ortiz-Aleman C (1994) Gravity and magnetic field modeling and structure of the Chicxulub Crater, Mexico. J Geophys Res 99:13147–13162. https://doi.org/10.1029/94JE01089

Yang J, Kwon BD, Cho IK, Lee H, Park G, Um J (2008) Geophysical study on the geoelectrical structure of the Hwasan Caldera in the Euisung sub-basin using magnetotelluric survey. Geophys Geophys Explor 11:99–108 (in Korean with English abstract) [GoogleScholar]

Yun HS (1995) Occurrence and petrochemistry of the granites in the Pocheon - Euijongbu area. J Petrol Soc 4:91–103 (in Korean with English abstract) [GoogleScholar]

Yun HS (1997) Petrological characteristics on stone resources of granites in the Pocheon - Euijongbu area. J Petrol Soc 6:34–44 (in Korean with English abstract) [GoogleScholar]

Open Access This chapter is licensed under the terms of the Creative Commons Attribution 4.0 International License (http://creativecommons.org/licenses/by/4.0/), which permits use, sharing, adaptation, distribution and reproduction in any medium or format, as long as you give appropriate credit to the original author(s) and the source, provide a link to the Creative Commons license and indicate if changes were made.

The images or other third party material in this chapter are included in the chapter's Creative Commons license, unless indicated otherwise in a credit line to the material. If material is not included in the chapter's Creative Commons license and your intended use is not permitted by statutory regulation or exceeds the permitted use, you will need to obtain permission directly from the copyright holder.

Achievements of the GGOS Focus Area Unified Height System

Laura Sanchez and Riccardo Barzaghi

Abstract

The Global Geodetic Observing System (GGOS) of the International Association of Geodesy (IAG) promotes the standardisation of height systems worldwide. The GGOS Focus Area Unified Height System (GGOS-FA-UHS) was established to lead and coordinate the efforts needed towards the establishment of a global standard for the precise determination of physical heights. During the 2011–2015 term, various discussions focused on the best possible definition of a global unified vertical reference system, resulting in the IAG Resolution for the *Definition and Realisation of an International Height Reference System* (IHRS), which was adopted at the 2015 General Assembly of the International Union of Geodesy and Geophysics (IUGG) in Prague, Czech Republic. During the period 2015–2019, activities were undertaken to investigate the best strategy for the implementation of the IHRS; i.e., the establishment of the *International Height Reference Frame* (IHRF). A preliminary selection of stations for the IHRF reference network was made and different calculation methods for the determination of potential values as IHRF coordinates were evaluated. For the period 2019–2023, the objectives of the GGOS-FA-UHS focused on (*i*) compiling detailed standards, conventions and guidelines to support a consistent determination of the IHRF at global, regional and national levels; (*ii*) coordinating with regional/national experts in gravity field modelling the computation of a first IHRF solution; and (*iii*) designing an operational infrastructure that will ensure the long-term sustainability and reliability of the IHRS/IHRF. This infrastructure was approved by the IAG Executive Committee in December 2023 and will operate under the responsibility of the International Gravity Field Service (IGFS). With these objectives achieved, the GGOS-FA-UHS completed its goals and was closed during the IUGG 2023 General Assembly in Berlin, Germany. This paper presents a comprehensive report on the activities and achievements of the GGOS-FA-UHS.

Keywords

Global unified height system · IHRF Coordination Centre · International Height Reference System (IHRS) and Frame (IHRF) · Reference level W_0 · World height system

L. Sanchez (✉)
Deutsches Geodätisches Forschungsinstitut, Technische Universität München (DGFI-TUM), Munich, Germany
e-mail: lm.sanchez@tum.de

R. Barzaghi
Department of Civil and Environmental Engineering, Politecnico di Milano, Milan, Italy
e-mail: riccardo.barzaghi@polimi.it

© The Author(s) 2024
J. T. Freymueller, L. Sánchez (eds.), *Together Again for Geodesy*,
International Association of Geodesy Symposia 157, https://doi.org/10.1007/1345_2024_249

1 Introduction

Most existing physical height systems do not meet the accuracy requirements of modern geodesy. They refer to local sea surface levels, are stationary (do not consider time variations), realise different types of physical heights (orthometric, normal, normal-orthometric heights, etc.), and their combination in a global frame leads to uncertainties on the metre scale. The International Association of Geodesy (IAG), as the organisation responsible for the advancement of the science of Geodesy, promotes the definition and implementation of geodetic reference systems that respond to the increased precision of modern observation techniques and can support the current needs of science and society for high-resolution georeferenced data (IAG 2017). In recent decades, enormous progress has been made in the *International Celestial Reference System* (ICRS, Ma and Feissel 1997) and the *International Terrestrial Reference System* (ITRS, Petit and Luzum 2010) as well as in their realisations the *International Celestial Reference Frame* (ICRF) and the *International Terrestrial Reference Frame* (ITRF), respectively. The definition, implementation, and maintenance of the ICRS/ICRF and ITRS/ITRF guarantee a globally unified geometric reference frame with cm-level reliability. An equivalent high accuracy global physical reference system needs to be implemented. There is no doubt that the existing height systems follow the best conditions offered by the state of the art at the time they were established. However, they have been realised individually, generally using non-standardised procedures. As a result, there are currently about hundred local and regional physical height systems in use, with discrepancies between them of up to ± 2 m. The geodetic data that depend on them (e.g., physical heights, gravity anomalies, geoid models, digital terrain models, etc.) are usable only in limited geographical areas; their global combination or with satellite-based data (in particular Global Navigation Satellite System (GNSS) positioning) show discrepancies of much greater magnitude than the accuracy required today.

The GGOS Focus Area Unified Height System (GGOS-FA-UHS, formerly GGOS Theme 1) was established during the 2010 GGOS Planning Meeting with the aim of bringing together existing initiatives for the establishment of a global unified vertical reference system and to address the activities to be undertaken. The starting point was the results of the IAG Inter-Commission Project 1.2 Vertical Reference Frames (IAG-ICP1.2-VRF; Ihde 2007), summarised in the document *Conventions for the Definition and Realisation of a Conventional Vertical Reference System—CVRS* (Ihde et al. 2007). Based on this document, the initial objectives of the GGOS-FA-UHS were defined as (Sideris and Ihde 2012; Sánchez 2016):

- Refine standards and conventions for the definition and realisation of a global unified vertical reference system, including unification/harmonisation of standards and conventions that are used by the geometric and gravity Scientific Services of the IAG.
- Make a recommendation about the reference value W_0 to be adopted as the conventional reference level for the global vertical reference system.
- Coordinate the generation of a set of consistent geodetic products for the realisation of the global vertical reference system, including a global vertical reference frame with regional and national densifications and a catalogue and guidelines for height system unification.
- Design of strategies for the appropriate maintenance and use in practice of the global vertical reference system considering determination of time-dependent changes and the alignment/update of the definition and its realisation with future improvements in geodetic observations, data analysis, and modelling.
- Servicing the vertical datum needs of other geosciences such as, e.g., hydrography and oceanography.

In line with these objectives, the following is a summary of the progress made towards the establishment of a unified global vertical reference system.

2 Inventory of Standards Presently Used in the Vertical Coordinate Determination

GGOS, through its Bureau of Products and Standards (GGOS-BPS), started in 2012 to compile an inventory of standards, constants, resolutions, and conventions adopted and used by IAG and its components for the generation of IAG products. The aims are to contrast adopted and applied standards and conventions, to identify gaps, inconsistencies, and deficiencies, and to propose new standards where appropriate. The first version of the GGOS-BPS inventory was published in 2016 (Angermann et al. 2016) and an updated version, including recent innovations in geodetic data analysis and modelling, was published in 2020 (Angermann et al. 2020). The GGOS-FA-UHS supported this activity by compiling and updating *Chapter 4.6 Height Systems and their Realisations*. This document describes in detail the discrepancies of the local physical height systems and their combination with geometric (ellipsoidal) heights and geoid undulations or height anomalies. In the 2016 version of the inventory, particular care was taken to provide a detailed list of corrections or reductions applied to the various vertical coordinates to remove or preserve geophysical effects that affect vertical positioning. In the 2020 update, a description of the standards outlined for the implementation of a global unified height system has been included (see next section).

3 Conventions for the Definition of a Global Vertical Reference System

In 2014, an ad-hoc group was established with the objective to outline the minimal requirements for the definition and realisation of a global unified physical vertical reference system (Ihde et al. 2015). The first recommendation of the ad-hoc group was to introduce a univocal name for the new system. During the last four decades, various names have been used to identify a global vertical reference system; e.g., world height system, global vertical datum, world vertical datum, global vertical network, global height datum unification, global unification of height systems, global unified height reference system, etc. To avoid this multiplicity of names, the ad-hoc group recommended the name *International Height Reference System* (IHRS) with the realisation *International Height Reference Frame* (IHRF). This name is consistent with other reference systems and frames used in Geodesy: ICRS/ICRF, ITRS/ITRF and ITGRS/ITGRF. The latter refers to the *International Terrestrial Gravity Reference System and Frame* (Wziontek et al. 2021).

The ad-hoc group focused on discussing the basic requirements for the establishment of a physical reference system, including a reference for gravity field dependent heights and a reference for gravimetry. The recommendations of this group were discussed during the 2015 General Assembly of the International Union of Geodesy and Geophysics (IUGG) and were presented and officially adopted by two IAG resolutions (Drewes et al. 2016): The first resolution is dedicated to the definition and implementation of an *International Height Reference System*. The second resolution focuses on the establishment of a *Global Absolute Gravity Reference System*.

The foundations for the definition and realisation of the IHRS are extensively discussed in Ihde et al. (2017). This publication is the scientific basis for the IAG Resolution No. 1, 2015 and provides the framework for the realisation of the IHRS. The fundamental conventions for the definition of the IHRS are:

– The vertical reference level is an equipotential surface of the Earth gravity field with the geopotential value $W_0 = 62{,}636{,}853.4$ m^2 s^{-2}. W_0 is understood to be the potential of the geoid or the geoidal potential value.
– Parameters, observations, and data shall be related to the mean tidal system/mean crust.
– The unit of length is the metre, and the unit of time is the second of the International System of Units (SI).
– The vertical coordinates are the differences $-\Delta W(P)$ between the potential $W(P)$ of the Earth's gravity field at the considered point P, and the geoidal potential value W_0; the potential difference $-\Delta W(P)$ is also designated as geopotential number $C(P) = -\Delta W(P) = W_0 - W(P)$.
– The spatial reference of the position P for the potential $W(P) = W(\mathbf{X}_P)$ is given by the coordinate vector $\mathbf{X}_P = \mathbf{X}(P)$ in the ITRS/ITRF.

The estimation of the coordinates $\mathbf{X}(P)$, $W(P)$ (or $C(P)$) includes their variation with time; i.e., $d\mathbf{X}(P)/dt$, $dW(P)/dt$ (or $dC(P)/dt$). For practical purposes, positions $\mathbf{X}(P)$ may be transformed to ellipsoidal coordinates to get the geometric (or ellipsoidal) heights $h(P)$, and $C(P)$ may be transformed to a physical height (orthometric H, dynamic H^d or normal height H^*).

4 Conventional Reference Value W_0

W_0 is defined as the potential value of a particular level surface of the Earth's gravity field called the geoid. Since the most accepted definition of the geoid is understood to be the equipotential surface that coincides with the worldwide mean ocean surface, a usual empirical approximation to W_0 is the averaged potential value W_S at the mean sea surface. In this way, W_0 depends not only on the Earth's gravity field modelling, but also on the mean sea surface modelling. Consequently, like any reference parameter, W_0 should be based on adopted conventions that guarantee its uniqueness, reliability, and reproducibility; otherwise, there would be as many W_0 reference values as computations. During the 2011 IUGG General Assembly, the GGOS-FA-UHS, the IAG Commissions 1 (Reference Frames) and 2 (Gravity Field) and the IGFS established a joint working group devoted to the Vertical Datum Standardisation (Sánchez 2012; Sánchez et al. 2014). The main objective was to recommend a convention for the geopotential value W_0 to be introduced as the reference level for the realisation of the IHRS. At that time, the most used W_0 value was the one included as a conventional constant in the conventions of the International Earth Rotation and Reference Systems' Service (IERS; Petit and Luzum 2010). This so-called IERS W_0 value corresponded to a best estimate available in 1998 (Burša et al. 1998; Groten 1999, 2004). It presents discrepancies of about -2.6 m^2 s^{-2} (corresponding to a level difference of around $+27$ cm) with respect to newest computations based on the latest Earth's surface and gravity field models (e.g. Čunderlík and Mikula 2009; Čunderlík et al. 2014; Dayoub et al. 2012; Sánchez 2008). This working group convened the different groups working on the determination of a global W_0 to coordinate a unified computation (cf. Sánchez et al. 2014). Following aspects were analysed:

– Sensitivity of the W_0 estimation to the Earth's gravity field model (especially omission and commission errors and time-dependent Earth's gravity field changes).
– Sensitivity of the W_0 estimation to the mean sea surface model (e.g., geographical coverage, time-dependent

sea surface variations, accuracy of the mean sea surface heights).
- Weighted computation of the W_0 value based on the input data quality.

Different methodologies, different global gravity models, different mean sea surface models, different reference epochs, and different weights for the input data were evaluated. Based on the results, detailed conventions to ensure the reproducibility of a reference W_0 value were outlined. As the usual approximation of W_0 is the averaged potential value W_S at the mean sea surface; it is expected that W_0 changes in the same way as W_S changes. However, W_0 as a reference parameter should be defined as time-independent, and it should be necessary to decouple it from the Earth's gravity field and sea surface variations. Thus, it was recommended to adopt the potential value valid at a certain epoch and to keep it fixed for a long-term period (e.g., 30 years). If desired, it is possible to monitor the changes of the potential value W_S at the sea surface and to compare it with the adopted W_0 value. When large differences appear (e.g., $> \pm 2$ m^2 s^{-2}, equivalent to a mean sea level change of ± 20 cm) the adopted W_0 may be replaced by an updated value. In conclusion, the working group members recommended the potential value obtained for the epoch 2010.0 (62,636,853.4 m^2 s^{-2}) as the present best estimate for the W_0 value. IAG accepted this recommendation and adopted this value as the conventional reference level for the realisation of the IHRS, see IAG Resolution 1, 2015. A detailed description of the W_0 computation strategy, conventions, and results is given by Sánchez et al. (2016).

During the GGOS/IERS Unified Analysis Workshop held in Paris, France, July 10–12, 2017, the GGOS-BPS pointed out the necessity of consistency between the IERS Conventions and the IAG Resolution No. 1, 2015 and recommended the use of the IAG conventional W_0 value whenever a reference potential is needed in geodetic work. IERS followed this recommendation and in Nov 17, 2017, the old W_0 value from 1998 was replaced with the new one in the IERS Conventions (see IERS Convention 2010, version 1.1.0, available at http://iers-conventions.obspm.fr/conventions_versions.php#official_target).

5 Reference Network for the Establishment of the International Height Reference Frame (IHRF)

It is proposed that the IHRF follows the same structure as the ITRF: a global network with regional and national densifications, whose geopotential numbers referring to the global IHRF are known. To advance in this goal, the GGOS-FA-UHS installed the joint working group *Strategy for the Realisation of the IHRS* for the term 2015–2019 (Sánchez 2019). This working group was supported by the IGFS, the IAG Commissions 1 and 2 (*Reference Frames* and *Gravity field*), the *Inter-commission Committee on Theory* (ICCT), the *regional sub-commissions for reference frames and geoid modelling*, and both *GGOS Bureaus* (*Networks and Observations* and *Products and Standards*). In particular, there was a strong cooperation with the IAG joint working group 2.2.2: *The 1-cm geoid experiment* (Wang and Forsberg 2019); the IAG Sub-Commission 2.2: *Methodology for geoid and physical height systems* (Ågren and Ellmann 2019), the ICCT joint study group 0.15: *Regional geoid/quasi-geoid modelling - Theoretical framework for the sub-centimetre accuracy* (Huang and Wang 2019) and the IAG joint working group 2.1.1: *Establishment of a global absolute gravity reference system* (Wziontek and Bonvalot 2019).

A brainstorming and definition of action items took place at a working group meeting carried out during the *International Symposium on Gravity, Geoid and Height Systems 2016* (GGHS2016) in Thessaloniki (Greece) in Sep 2016. This meeting was attended by 70 colleagues and allowed us to identify the activities to be faced immediately. A main output are the criteria for the selection of IHRF reference stations:

- GNSS continuously operating reference stations to detect reference frame deformations.
- Co-location with fundamental geodetic observatories to ensure a consistent connection between geometric coordinates, potential and gravity values, and reference clocks.
- Co-location with reference stations of the ITGRF.
- Preference of stations belonging to the ITRF and the regional reference frames (like SIRGAS, EPN, APREF, etc.).
- Co-location with reference tide gauges and connection to the national levelling networks to facilitate the vertical datum unification.
- Availability of terrestrial gravity data around the IHRF reference stations as main requirement for high-resolution gravity field modelling (i.e., precise estimation of potential values).

Based on these criteria, a preliminary station selection for the IHRF was initiated in 2016. This selection was based on a global network with worldwide distribution, including a core network (to ensure sustainability and long-term stability of the reference frame) and regional/national densifications (to provide local accessibility to the global frame). The core network includes fundamental geodetic observatories, ITRF sites with more than two space geodetic techniques, ITGRF reference stations and selected IGS reference stations to ensure a global coverage as homogeneous as possible. During 2017–2018, regional and national experts were asked to evaluate whether the preliminary selected sites are suitable to be included in the IHRF (availability of gravity data or possibilities to survey them); and to propose additional geodetic sites to improve the density and distribution of the IHRF stations in their regions/countries. After the feedback from the regional/national experts, the first approximation to the IHRF reference network was completed in 2019. This network comprises about 170 stations and currently, it is regularly refined in agreement with changes/updates of other geodetic reference frames (ITRF and ITGRF and their densifications).

6 Determination of Potential Values as IHRS Coordinates

After the preliminary station selection for the IHRF reference network, efforts concentrated on the computation of station potential values and the assessment of their accuracy. Different approaches were evaluated:

- As some national/regional experts provided us with terrestrial gravity data around some IHRF sites, a direct computation of potential values was performed using a combination of terrestrial gravity data and different global gravity models (GGM) as well as different mathematical formulations (least-squares collocation, Fast Fourier Transformation, radial basis functions, etc.).
- Computation of potential values by national/regional experts responsible for the geoid modelling using their own data and methodologies.
- Computation of potential values based on GGM of high-resolution (such as XGM2016 (Pail et al. 2018), EIGEN-6C4 (Förste et al. 2014), EGM2008 (Pavlis et al. 2012), etc.).
- Recovering potential values from existing local models of geoid undulations or height anomalies.

The comparison of the results showed discrepancies up to the dm-level. The main conclusions of this experiment were:

- The use of only GGMs is (at present) not suitable for the estimation of precise potential values. GGMs may be used if there is *no other way* to determine potential values (e.g., Sánchez et al. 2021; Wang et al. 2021)
- A *standard* procedure for the computation of potential values may be not appropriate as different data availability and different data quality exist around the world and regions with different characteristics require particular approaches (e.g., modification of kernel functions, size of integration caps, geophysical reductions like the global isostatic adjustment, etc.).
- A *centralised* computation (like in the ITRF) is complicated due to the restricted accessibility to terrestrial gravity data.

To overcome these inconveniences, during the *IAG-IASPEI Joint Scientific Assembly* (Kobe, Japan, Aug 2017) was agreed to initiate a new experiment towards:

- The computation of IHRF coordinates using exactly the same input data and the own methodologies (software) of colleagues involved in the gravity field modelling, and
- The comparison of the results, to identify a set of standards that allow to get as similar and compatible results as possible.

In the same IAG-IASPEI 2017 Assembly, J. Ågren (Chair of IAG SC 2.2; Ågren and Ellmann (2019)) and J. Huang (Chair of ICCT JSG 0.15, Huang and Wang (2019)) proposed to establish an interaction with the JWG 2.2.2 (chaired by Y.M. Wang, Wang and Forsberg (2019)). Aim of JWG 2.2.2 was the computation and comparison of geoid undulations using the same input data and the own methodologies/software of colleagues involved in the geoid computation. The comparison of the results should highlight the differences caused by disparities in the computation methodologies. In this frame, it was decided to extend the *geoid experiment* to the computation of station potential values as IHRF coordinates. With this proposal, the US NGS/NOAA agreed to provide terrestrial and airborne gravity data and a digital terrain model for an area of about 730 km × 560 km with height variations up to 3,000 m in Colorado (USA). With the NGS/NOAA data, different groups working on the determination of IHRF coordinates computed potential values for some virtual geodetic stations located in that region. Afterwards, the results of the individual groups were compared with the *Geoid Slope Validation Survey 2017* (GSVS17, Van Westrum et al. 2021), which provides potential differences inferred from first order levelling measurements and gravity corrections along a validation line. The Colorado data were distributed in Feb 2018, together with a document summarising a minimum set of basic requirements (standards) for the computations in order to get as similar and compatible results as possible (Sánchez et al. 2018).

Fourteen solutions contributed to this experiment (Wang et al. 2021; Sánchez et al. 2021). When evaluating them to the independent GSVS17 GNSS/levelling data, it was proved that all methods and processing approaches provide results that agree to each other at the 2-cm level in terms of standard deviation from the mean value. The overall discrepancies range from −9 cm to +8 cm. These discrepancies mainly reflect the disagreement between the data preprocessing and computation methods as the input data are assumed free of error and a proper error propagation analysis is not performed yet. However, it is evident that the discrepancies between the different solutions are highly correlated with the topography, suggesting further investigations on the handling of terrain gravity effects (model and strategy). Wang et al. (2021) and Sánchez et al. (2021) summarise a detailed comparison of the 14 solutions that contributed to the Colorado experiment. Van Westrum et al. (2021) provide a detailed description of the measurement and data analysis of the reference GNSS/levelling validation data along the GSVS17 profile. The input gravity and topographic data, the GNSS/levelling validation data, and the 14 geoid and quasi-geoid models produced within the Colorado experiment are available from the International Service for the Geoid (ISG, Reguzzoni et al. 2021) and can be used as a basis to evaluate any geoid computation method or software anywhere. Based on the results of the Colorado experiment, Sánchez et al. (2021) present a detailed roadmap for the realisation of the IHRS, including:

- Strategy for the determination and evaluation of IHRF coordinates depending on the data availability (especially surface gravity data and topography models),
- Strategy to improve the input data required for the determination of IHRF coordinates,
- Strategy for the IHRF implementation at the regional and national level,
- Strategy to ensure the usability and long-term sustainability of the IHRF.

Following this, during the 2019 IUGG General Assembly in Montreal, Canada, the IAG released a new resolution promoting the implementation of the IHRS at regional and national levels; see IAG Resolution No. 3, 2019 in Poutanen and Rózsa (2020). Additionally, the GGOS-FA-UHS coordinated the publication of a Journal of Geodesy special issue on *Reference Systems in Physical Geodesy* including most of the solutions contributing to the Colorado experiment. This special issue also contains papers facing important issues related to the establishment of the IHRF and ITGRF as well as to the improvement of accurate geoid modelling and the long-term stability of absolute gravity observations. (Sánchez et al. 2023).

7 Vertical Datum Unification for the International Height Reference System (IHRS)

A main component of the IHRS realisation is the integration of the existing height systems into the global one; i.e., existing physical heights (or geopotential numbers) should be referred to one and the same reference level realised by the conventional W_0. This procedure is known as vertical datum unification and its main result are the potential differences (called vertical datum parameters) between the local and the global reference levels. The motivation for the vertical datum unification rises from the fact that the local physical height systems have been the reference for height determination during the last 150 years and they provide a higher accuracy in contiguous areas than the combination of ellipsoidal heights with geoid undulations or height anomalies. If the local height systems are appropriately integrated into the IHRS, the existing vertical data can be modernised and be useful for geodetic applications of global context.

Sánchez and Sideris (2017) rigorously derive the observation equations for the vertical datum unification in terms of potential quantities based on the geodetic boundary value problem (GBVP) approach. Those observation equations are then empirically evaluated for the vertical datum unification of the North American and South American height systems. In the first case, simulations performed in North America provide numerical estimates about the impact of omission errors and direct and indirect effects on the vertical datum parameters. In the second case, a combination of local geopotential numbers, ITRF coordinates, satellite altimetry observations, tide gauge registrations, and high-resolution gravity field models is performed to estimate the level differences between the South American height systems and the global level W_0. Results show that indirect effects vanish when a satellite-only gravity field model with a degree $n \geq 180$ is used for the solution of the GBVP. However, the component derived from satellite-only global gravity models has to be refined with terrestrial gravity data to minimise the omission error and its effect on the vertical datum parameter estimation. The empirical evaluations demonstrate that the vertical datum unification should be based on geodetic stations of highest quality and standardised geodetic data; for example, geometric coordinates should refer to the same ITRF and be given in the same tide system and reference epoch as the geopotential numbers and gravity field model. After a standardisation of the input data used in the unification of the South American height systems and a rigorous error propagation analysis, it is evident that the vertical datum parameters can be estimated with accuracy better than ±5 cm in well-surveyed regions and some decimetres (± 40 cm) in sparsely surveyed regions. Sánchez and Sideris (2017) also provide detailed guidelines for the appropriate data treatment when the integration of a local vertical datum into the IHRS is desired.

8 A First Solution for the IHRF

Based on the outcomes of the Colorado experiment, we classified the computation of potential values in three main scenarios:

(a) Regions without (or with very few) surface gravity data,
  - The only option to determine potential values is the use of GGM of high resolution (GGM-HR).

- Expected mean accuracy values around the ± 4.0 m^2 s^{-2} (± 40.0 cm in terms of height) level or even worse in regions with strong topography gradients.
- It could be improved for instance to the ± 1.0 m^2 s^{-2} (± 10.0 cm) level if new and better surface gravity data are included in the GGMs.
- To avoid multiple potential values provided by different GGM-HRs at the same point, it is necessary to select one GGM-HR as reference model.

(b) Regions with some surface gravity data, but with poor data coverage or unknown data quality,
- The reliability of the existing (quasi-)geoid models is poor.
- Additional gravity surveys around the IHRF stations would help to increase the accuracy of the geopotential numbers computed at those specific stations.

(c) Regions with good surface gravity data coverage and quality.
- Potential values may be inferred from precise geoid/quasi-geoid regional models.

Using this classification, we started in the beginning of 2021 the computation of a first solution for the IHRF. As an initial action, a short description of the "step by step" to infer IHRF potential values from local/regional geoid/quasi-geoid models was prepared. It is based on the IHRS paper published by Sánchez et al. (2021) and was distributed to the members of the working group *Implementation of the International Height Reference Frame* (Sánchez 2023), so that they can compute potential values at the IHRF stations located in their countries using their present/latest geoid/quasi-geoid models. This activity is supported by about 40 colleagues from Canada, Mexico, USA, Germany, Italy, Switzerland, Austria, Sweden, Finland, Australia, Japan, China, South America, Russia, and Africa. Complementary, the ISG and the IGFS are evaluating the quality and documentation of the different regional models available at the Geoid Repository of ISG in order to identify which models can be used to infer potential values. This action is useful for the IHRF computation in areas underrepresented in the working group. Simultaneously, we are computing potential values for all the IHRF stations using GGM extended with topography-based synthetic gravity signals, reaching resolutions up to degree 80,000 ... 90,000. As mentioned, this would be the only option available in those regions where no geoid/quasi-geoid models are available. At the end, we have different potential values for the same points. The agreement of the different GGM and the models stored by ISG with the own computations performed by the colleagues of the working group will allow us to decide which GGM + topography models perform better. The results of these computations were presented at the IUGG2023 General Assembly in Berlin, Germany and are being compiled in a paper to be published in the near future.

9 Operational Infrastructure for the Long-Term Sustainability of the IHRF

An IHRS/IHRF objective is to support the monitoring and analysis of Earth's system changes. The more accurate the IHRS/IHRF is, the more phenomena can be identified and modelled. Thus, the IHRS/IHRF must provide vertical coordinates and their changes with time as accurately as possible. As many global change phenomena occur at different scales, the global frame should be extended to regional and local levels to guarantee consistency in the observation, detection, and modelling of their effects. From this perspective, we are proposing the establishment of an operational infrastructure within the IGFS that takes care of

(a) Maintenance of the IHRF reference network in accordance with the GGOS Bureau of Networks and Observations (Pearlman et al. 2019) and the coordinators of the reference networks for the ITRF, ITGRF and their regional densifications. This activity should be faced by the *IHRF Reference Network Coordination*.

(b) Maintenance of a catalogue with the conventions and standards needed for the IHRF. This should consider a harmonisation with the conventions and standards kept by the GGOS-BPS, the IERS Conventions (for the determination of the ITRF), and the standards applied in the ITGRF and the global gravity field modelling. This task should be carried out by the *IHRF Conventions' Coordination*.

(c) The national/regional agencies/entities contributing to the realisation of the IHRF in their regions may be considered as *IHRF Associate Analysis Centres*. The input data would then be provided by existing IAG gravity field services and local data centres; e.g., GGM are provided by International Centre for Global Earth Models (ICGEM, Ince et al. 2019) and surface gravity data are provided by the Bureau Gravimétrique International (BGI) and refined/complemented with gravity data available at local data centres. In a similar way, one can proceed with digital elevation models.

(d) The combination and quality assessment of the regional/national solutions as well as the release of the final (official) IHRF solution will be faced by the *IHRF Combination Coordination*.

(e) Finally, the IHRF Reference Network Coordination, Conventions' Coordination, Associate Analysis Centres and Combination Coordination will report to the *IHRF Coordination Centre*, which, in turn, would report directly to the IGFS Central Bureau

The IGFS presented this proposal to the IAG Executive Committee at its meeting on 10 December 2023 and it was unanimously approved. Thus, a new component of the IGFS dedicated to the IHRF has been created and will ensure the long-term availability and reliability of the IHRF. More details about this operational infrastructure are presented by Sánchez et al. (2024).

10 Closing Remarks

The implementation of a global reference system for physical heights such as the IHRS is a major challenge and requires the support of a broad scientific community. Therefore, the establishment of the IHRS/IHRF is only possible within a global and structured organisation such as the IAG. The IHRS/IHRF provides a unified frame for height determination around the world, ensuring that different national and regional height systems can be related and compared in a consistent manner. However, strong international cooperation on a voluntary basis is essential to ensure its long-term stability and availability. The GGOS-FA-UHS has motivated this cooperation for 12 years. From now on, this cooperation will be facilitated by the IGFS, which is establishing an appropriate organisational infrastructure to provide a framework for countries to work together towards common goals related to the maintenance and continuous development of the IHRS/IHRF.

Acknowledgements The progress described in this report has been made possible thanks to the support of more than 70 colleagues over 12 years. Their contribution is deeply appreciated.

References

Ågren J, Ellmann A (2019) Report of the sub-commission 2.2: methodology for geoid and physical height systems, reports 2015-2019 of the International Association of Geodesy (IAG), Travaux de l'AIG, vol 41, pp 155–160

Angermann D, Gruber T, Gerstl M, Heinkelmann R, Hugentobler U, Sánchez L, Steigenberger P (2016) GGOS Bureau of Products and Standards: inventory of standards and conventions used for the generation of IAG products. In: Drewes H, Kuglitsch F, Adám J, Rózsa S (eds) The geodesist's handbook 2016, J Geod 90:1095–1156. https://doi.org/10.1007/s00190-016-0948-z

Angermann D, Gruber T, Gerstl M, Heinkelmann R, Hugentobler U, Sánchez L, Steigenberger P (2020) Bureau of Products and Standards: inventory of standards and conventions used for the generation of IAG products. In: The geodesists' handbook 2020, J Geod 94(11):221–292. https://doi.org/10.1007/s00190-020-01434-z

Burša M, Kouba J, Raděj K, True S, Vatrt V, Vojtíšková M (1998) Mean Earth's equipotential surface from TOPEX/Poseidon altimetry. Studia geoph et geod 42:456–466. https://doi.org/10.1023/A:1023356803773

Čunderlík R, Mikula K (2009) Numerical solution of the fixed altimetry–gravimetry BVP using the direct BEM formulation. International Association of Geodesy Symposia Series, vol 133, pp 229–236. https://doi.org/10.1007/978-3-540-85426-5_27

Čunderlík R, Minarechová Z, Mikula K (2014) Realization of WHS based on the static gravity field observed by GOCE. International Association of Geodesy Symposia Series, vol 141, pp 211–220. https://doi.org/10.1007/978-3-319-10837-7_27

Dayoub N, Edwards SJ, Moore P (2012) The Gauss–Listing potential value Wo and its rate from altimetric mean sea level and GRACE. J Geod 86(9):681–694. https://doi.org/10.1007/s00190-012-1547-6

Drewes H, Kuglitsch F, Ádám J, Rózsa S (2016) Geodesist's handbook 2016. J Geod 90:907. https://doi.org/10.1007/s00190-016-0948-z

Förste C, Bruinsma SL, Abrikosov O, Lemoine JM, Marty JC, Flechtner F, Balmino G, Barthelmes F, Biancale R (2014) EIGEN-6C4 the latest combined global gravity field model including GOCE data up to degree and order 2190 of GFZ Potsdam and GRGS Toulouse. GFZ Data Services. https://doi.org/10.5880/icgem.2015.1

Groten E (1999) Report of the International Association of Geodesy Special Commission SC3: fundamental constants. XXII IAG General Assembly, Birmingham

Groten E (2004) Fundamental parameters and current (2004) best estimates of the parameters of common relevance to astronomy, geodesy and geodynamics. The geodesist's handbook 2004. J Geod 77:724–731. https://doi.org/10.1007/s00190-003-0373-y

Huang J, Wang YM (2019). Report of Joint Study Group 0.15: regional geoid/quasigeoid modelling—theoretical framework for the sub-centimetre, reports 2015-2019 of the International Association of Geodesy (IAG), Travaux de l'AIG, vol 41, pp 495–499

IAG (2017) Description of the global geodetic reference frame. Position paper adopted by the IAG Executive Committee in April 2016. J Geod 91:113–116. https://doi.org/10.1007/s00190-016-0994-6

Ihde J (2007) Inter-Commission project 1.2: Vertical Reference Frames. Final report for the period 2003-2007. In: Drewes H, Hornik H (eds) IAG Commission 1—Reference Frames, Report 2003–2007. DGFI, Munich Bulletin No. 20, pp 57–59

Ihde J, Amos M, Heck B, Kersley B, Schöne T, Sánchez L Drewes H (2007) Conventions for the definitions and realization of a conventional vertical reference system (CVRS), unpublished

Ihde J, Barzaghi R, Marti U, Sánchez L, Sideris M, Drewes H, Förste C, Gruber T, Liebsch G, Pail R (2015) Report of the Ad-hoc Group on an International Height Reference System (IHRS). In: Drewes H, Hornik H (eds) Reports 2011-2015 of the International Association of Geodesy (IAG), Travaux de l'AIG, vol 39, pp 549–557

Ihde J, Sánchez L, Barzaghi R, Drewes H, Förste C, Gruber T, Liebsch G, Marti U, Pail R, Sideris M (2017) Definition and proposed realization of the International Height Reference System (IHRS). Surv Geophys 38(3):549–570. https://doi.org/10.1007/s10712-017-9409-3

Ince ES, Barthelmes F, Reißland S, Elger K, Förste C, Flechtner F, Schuh H (2019) ICGEM—15 years of successful collection and distribution of global gravitational models, associated services, and future plans. Earth Syst Sci Data 11(2):647–674. https://doi.org/10.5194/essd-11-647-2019

Ma C, Feissel M (1997) Definition and realization of the International Celestial Reference System by VLBI astrometry of extragalactic objects. IERS technical note 23, Paris, Central Bureau of IERS - Observatoire de Paris, 282 p

Pail R, Fecher T, Barnes D, Factor JF, Holmes SA, Gruber T, Zingerle P (2018) Short note: the experimental geopotential model XGM2016. J Geod 92:443. https://doi.org/10.1007/s00190-017-1070-6

Pavlis N-K, Holmes SA, Kenyon SC, Factor JK (2012) The development of the Earth Gravitational Model 2008 (EGM2008). J Geophys Res 117:B04406. https://doi.org/10.1029/2011JB008916

Pearlman M, Behrend D, Craddock A, Noll C, Pavlis E, Saunier J, Matthews A, Barzaghi R, Thaller D, Maennel B, Bergstrand S, Müller J (2019) GGOS: Current Activities and Plans of the Bureau of Networks and Observations, Abstract No. EGU2019-6181, presented at the European Geosciences Union General Assembly, Vienna, Austria, April 07-12, 2019

Petit G, Luzum B (2010). IERS Conventions (2010). IERS technical note 36, Frankfurt am Main: Verlag des Bundesamts für Kartographie und Geodäsie, 179 pp., ISBN 3-89888-989-6

Poutanen M, Rózsa S (2020) The geodesist's handbook 2020. J Geod 94:109. https://doi.org/10.1007/s00190-020-01434-z

Reguzzoni M, Carrion D, De Gaetani CI, Albertella A, Rossi L, Sona G, Batsukh K, Toro Herrera JF, Elger K, Barzaghi R, Sansó F (2021) Open access to regional geoid models: the International Service for the Geoid. Earth Syst Sci Data 13:1653–1666. https://doi.org/10.5194/essd-13-1653-2021

Sánchez L (2008) Approach for the establishment of a global vertical reference level. International Association of Geodesy Symposia Series, 132, pp 119–125. https://doi.org/10.1007/978-3-540-74584-6_18

Sánchez L (2012) Towards a vertical datum standardisation under the umbrella of Global Geodetic Observing System. J Geod Sci 2(4):325–342. https://doi.org/10.2478/v10156-012-0002-x

Sánchez L (2016) GGOS Focus Area 1: Unified Height System—Terms of reference. In: Drewes H, Kuglitsch F, Ádám J, Rózsa S (eds) Geodesist's handbook 2016, J Geod 90: 1091. https://doi.org/10.1007/s00190-016-0948-z

Sánchez L (2019) Report of the GGOS focus area "unified height system" and the Joint Working Group 0.1.2: strategy for the realization of the International Height Reference System (IHRS), reports 2015-2019 of the International Association of Geodesy (IAG), Travaux de l'AIG, vol 41, pp 583–592

Sánchez L (2023) Report of the GGOS focus area "unified height system", reports 2019-2023 of the International Association of Geodesy (IAG), Travaux de l'AIG, vol 43, pp 710–717

Sánchez L, Sideris MG (2017) Vertical datum unification for the International Height Reference System (IHRS). Geophys J Int 209(2):570–586. https://doi.org/10.1093/gji/ggx025

Sánchez L, Dayoub N, Čunderlík R, Minarechová Z, Mikula K, Vatrt V, Vojtíšková M, Šíma Z (2014) W_0 estimates in the frame of the GGOS working group on vertical datum standardisation. International Association of Geodesy Symposia Series, 141: 203–210. https://doi.org/10.1007/978-3-319-10837-7_26

Sánchez L, Čunderlík R, Dayoub N, Mikula K, Minarechová Z, Šíma Z, Vatrt V, Vojtíšková M (2016) A conventional value for the geoid reference potential W_0. J Geod 90(9):815–835. https://doi.org/10.1007/s00190-016-0913-x

Sánchez L, Ågren J, Huang J, Wang YM, Forsberg R (2018) Basic agreements for the computation of station potential values as IHRS coordinates, geoid undulations and height anomalies within the Colorado 1 cm geoid experiment, unpublished

Sánchez L, Ågren J, Huang J, Wang YM, Mäkinen J, Pail R, Barzaghi R, Vergos GS, Ahlgren K, Liu Q (2021) Strategy for the realisation of the International Height Reference System (IHRS). J Geod 95:3. https://doi.org/10.1007/s00190-021-01481-0

Sánchez L, Wziontek H, Wang YM, Vergos G, Timmen L (2023) Towards an integrated global geodetic reference frame: preface to the special issue on reference systems in physical geodesy. J Geod 97(6). https://doi.org/10.1007/s00190-023-01758-6

Sánchez L, Barzaghi R, Vergos G (2024) Operational infrastructure to ensure the long-term sustainability of the International Height Reference System and Frame—IHRS/IHRF. International Association of Geodesy Symposia Series, IUGG2023, vol 157. https://doi.org/10.1007/1345_2024_250

Sideris MG, Ihde J (2012) GGOS theme 1: unified height system—terms of reference. In: Drewes H (ed) Geodesist's handbook 2012, J Geod 86: 923–924. https://doi.org/10.1007/s00190-012-0584-1

Van Westrum D, Ahlgren K, Hirt C, Guillaume S (2021) A Geoid Slope Validation Survey (2017) in the rugged terrain of Colorado, USA. J Geod 95:9. https://doi.org/10.1007/s00190-020-01463-8

Wang YM, Forsberg R (2019) Report of the Joint Working Group 2.2.2: the 1 cm geoid experiment, reports 2015-2019 of the International Association of Geodesy (IAG), Travaux de l'AIG, vol 41, pp 178–179

Wang YM, Sánchez L, Ågren J, Huang J, Forsberg R, Abd-Elmotaal HA, Barzaghi R, Bašić T, Carrion D, Claessens S, Erol B, Erol S, Filmer M, Grigoriadis VN, Isik MS, Jiang T, Koç Ö, Li X, Ahlgren K, Krcmaric J, Liu Q, Matsuo K, Natsiopoulos DA, Novák P, Pail R, Pitoňák M, Schmidt M, Varga M, Vergos GS, Véronneau M, Willberg M, Zingerle P (2021) Colorado geoid computation experiment—overview and summary. J Geod 95:12. https://doi.org/10.1007/s00190-021-01567-9

Wziontek H, Bonvalot S (2019) Report of the Joint Working Group 2.1.1: establishment of a global absolute gravity reference system, reports 2015-2019 of the International Association of Geodesy (IAG), Travaux de l'AIG, vol 41, pp 147–152

Wziontek H, Bonvalot S, Falk R, Gabalda G, Mäkinen J, Pálinkáš V, Rülke A, Vitushkin L (2021) Status of the international gravity reference system and frame. J Geod 95:7. https://doi.org/10.1007/s00190-020-01438-9

Open Access This chapter is licensed under the terms of the Creative Commons Attribution 4.0 International License (http://creativecommons.org/licenses/by/4.0/), which permits use, sharing, adaptation, distribution and reproduction in any medium or format, as long as you give appropriate credit to the original author(s) and the source, provide a link to the Creative Commons license and indicate if changes were made.

The images or other third party material in this chapter are included in the chapter's Creative Commons license, unless indicated otherwise in a credit line to the material. If material is not included in the chapter's Creative Commons license and your intended use is not permitted by statutory regulation or exceeds the permitted use, you will need to obtain permission directly from the copyright holder.

Operational Infrastructure to Ensure the Long-Term Sustainability of the International Height Reference System and Frame (IHRS/IHRF)

Laura Sánchez, Riccardo Barzaghi, and George Vergos

Abstract

The International Association of Geodesy (IAG) introduced the International Height Reference System (IHRS) in 2015 as an international standard for the accurate determination of physical heights worldwide. Primary vertical coordinates are geopotential numbers referenced to a conventional W_0 value. The realisation of the IHRS is the International Height Reference Frame (IHRF), which corresponds to a global network of reference stations with precise reference coordinates specified in the IHRS. The spatial position of the stations, at which the geopotential numbers are calculated, is defined by their respective coordinates (X, Y, Z) in the International Terrestrial Reference Frame (ITRF). The realisation of the IHRS is thus based on the combination of a geometric component, given by the positions of the stations in the ITRF, and a physical component, given by the determination of the potential values W at these positions. Through a strong international collaboration, framed by the IAG, it has been possible in recent years to pave the scientific foundations of the IHRS, to compute a first solution of the IHRF, and to identify the key requirements for a long-term sustainability of the IHRF. Much progress has been made and continuity is needed to ensure the maintenance and availability of the IHRF in the future. Following IAG practice, the development of theory and methods for the continuous improvement of the IHRS/IHRF should be promoted by the IAG Commissions and the Inter-Commission Committee on Theory (ICCT), while the operational performance should be ensured by the IAG Services. In this paper, we highlight the organisational challenges in maintaining the IHRS/IHRF, discuss how the existing gravity field related IAG Services could contribute to the IHRS/IHRF, and identify the elements needed to establish an operational infrastructure for the IHRS/IHRF that addresses the organisational challenges. Our proposal is to establish a central coordinating body under the responsibility of the International Gravity Field Service (IGFS), composed of individual modules taking care of the main components of the IHRS/IHRF. The central management body is the IHRF Coordination Centre and its modules are the IHRF Reference Network Coordination, the IHRF Conventions' Coordination, the IHRF Associate Analysis Centres, and the IHRF Combination Coordination. The IGFS presented this proposal to the IAG Executive

L. Sánchez (✉)
Technical University of Munich, Deutsches Geodätisches Forschungsinstitut (DGFI-TUM), Munich, Germany
e-mail: lm.sanchez@tum.de

R. Barzaghi
Politecnico di Milano, Department of Civil and Environmental Engineering, Milan, MI, Italy
e-mail: riccardo.barzaghi@polimi.it

G. Vergos
Aristotle University of Thessaloniki, Department of Geodesy and Surveying, Laboratory of Gravity Field Research and Applications, Thessaloniki, Greece
e-mail: vergos@topo.auth.gr

Committee at its meeting on 10 December 2023 and it was unanimously approved. Thus, a new component of the IGFS dedicated to the IHRF has been created and will ensure the long-term availability and reliability of the IHRF.

Keywords

IHRF Coordination Centre · International Gravity Field Service (IGFS) · International Height Reference Frame (IHRF) · International Height Reference System (IHRS)

1 Introduction

The International Association of Geodesy (IAG) introduced the International Height Reference System (IHRS) in 2015 as the international standard for the precise determination of physical heights worldwide (see IAG Resolution 1 (2015) in Drewes et al. 2016). The IHRS is a geopotential-based reference system co-rotating with the Earth. The primary coordinates are potential differences (i.e., geopotential numbers C_P) between the potential W_P of the Earth's gravity field at a point P, and the geoidal potential value W_0 (Ihde et al. 2017; Sánchez et al. 2021)

$$C_P = W_0 - W_P. \tag{1}$$

The realisation of the IHRS is the International Height Reference Frame (IHRF), which is composed of a global network of reference stations with precise reference coordinates (X, Y, Z, C) specified in the IHRS as well as in the International Terrestrial Reference Frame (ITRF, e.g. Altamimi et al. 2023). Thus, the realisation of the IHRS is based on the combination of a geometric component, given by the station positions in the ITRF, and a physical component, given by the determination of the potential values W at these positions.

The definition and initial realisation of the IHRS were implemented under the coordination of the Focus Area Unified Height System (FA-UHS) of the IAG's Global Geodetic Observing System (GGOS) (Plag and Pearlman 2009; Sánchez and Barzaghi 2024). The establishment of a global unified height system was one of the earliest priorities of GGOS, as this topic had been discussed for many decades and required the concurrence of several IAG components, namely, Commission 1 (Reference Frames), Commission 2 (Gravity Field), the Inter-Commission Committee on Theory (ICCT), the International Gravity Field Service (IGFS), and the International Earth's Rotation and Reference Systems Service (IERS). Since 2011, the GGOS FA-UHS has convened representatives from these IAG entities to implement a world height system and these joint efforts have enabled the definition of the IHRS, an initial solution of the IHRF, and the compilation of a cookbook for the IHRF. This cookbook comprises (Sánchez et al. 2021):

- A catalogue of basic standards and conventions including numerical constants, reference ellipsoid, zero degree and mass centre convention, handling of permanent tide effects, etc.,
- guidelines for the determination and evaluation of IHRF coordinates depending on the data availability and quality, and
- strategies for improving the input data required for the determination of IHRF coordinates and for the densification of the global IHRF at the regional and national levels.

With these foundations in place, the next step is to ensure the long-term maintenance and service of the IHRF. Following IAG practice, the operational activities needed for the IHRF should be undertaken by an IAG Service, in this case the IGFS, while future developments in theory and data analysis should be supported by the IAG Commissions and other components, in this case, Commissions 1 and 2, the ICCT and the IAG Project Novel Sensors and Quantum Technology for Geodesy (QuGe). The latter is of particular importance in view of the near future possibility of determining potential differences with precise optical clocks (see e.g., Müller et al. 2018; Wu et al. 2019; Wu and Müller 2020) and quantum satellite gradiometry (see e.g., Migliaccio et al. 2023; Mu et al. 2023).

While the scientific foundations and challenges in the determination of the IHRF are widely discussed in Ihde et al. (2017), Sánchez et al. (2016, 2021), Sánchez and Sideris (2017), Wang et al. (2021), this contribution summarises our findings on the operational and organisational elements required to ensure adequate long-term availability of the IHRS/IHRF. In Sect. 2, we describe the constituent elements of the IHRS/IHRF. In Sect. 3, we highlight the challenges in sustaining the IHRS/IHRF. In Sect. 4, we summarise how the existing gravity field-related IAG Services could contribute to the IHRS/IHRF and vice versa. In Sect. 5, we identify the missing elements and interfaces that need to be established within the IGFS for the maintenance of the IHRS/IHRF.

2 Constituent Elements of the IHRS/IHRF

As with any reference system, the definition of the IHRS outlines the fundamental conventions of the system: the type of coordinates, what they refer to, the units to express the coordinates, the permanent tide concept in which the coordinates must be expressed, etc. The materialisation of the definition is given by a reference frame that realises the system physically and mathematically. The physical realisation establishes a set of stations or continuously operating instruments that allow tangible access to the system. The mathematical realisation determines the reference coordinates at the reference stations following the definition of the system precisely. The latter is achieved through the use of specific constants, standards and procedures that ensure the exact observance of the conventions given in the definition. For example, in the case of the IHRS, two defining conventions state that the coordinates (geopotential numbers) should be given with respect to a specific W_0 reference value (Sánchez et al. 2016) and in the mean permanent tide concept (Mäkinen 2021). Thus, the IHRF coordinates (mathematical realisation) cannot be given with respect to another W_0 value or in another permanent tide concept, because this would not realise the IHRS defined. Continuing with this example, since the potential of the Earth's gravity field cannot be calculated in the mean permanent tide concept (e.g., Sánchez et al. 2021; Wang et al. 2021), common standards are required to remove the direct (and indirect) effects of the permanent tide before the potential values are calculated and then to restore these effects to make the IHRF coordinates consistent with the definition. Definition, realisation, and standards are the first three constituents of the IHRS/IHRF. They represent the state-of-the-art at the time when they were introduced. When theoretical or technological developments in the gravity potential analysis are implemented, the definition of the system, the mathematical realisation, and the standards (procedures) have to be updated accordingly, with rigorous assessment of the changes and implications that these updates bring to IHRS and its realisation. When stations belonging to the realisation are decommissioned or new stations are integrated into the IHRF, the physical realisation has to be updated accordingly. These updates should be timely to ensure that the IHRS and IHRF reflect and are able to support the latest developments in geodesy and that the IHRF is delivered to users in a timely and reliable manner. Thus, the fourth constituent of the IHRS/IHRF is their maintenance, evolution, and long-term availability. Finally, the fifth constituent, as we understand it, is the people who do the work, also the organisational structure that runs the stations, archives and analyses the data, maintains standards and conventions, validates and publishes the results. Figure 1 summarises the IHRS/IHRF constituents and how they interact.

3 Organisational Challenges in the Maintenance of IHRS/IHRF

The determination of the ITRF coordinates follows the IERS Conventions (Petit and Luzum 2010) and is supported by a fully operational infrastructure (reference stations, data centres, analysis centres, combination centres, product centres, etc.) acting under the responsibility of the IAG geometric services and the IERS. Based on more than 30 years of experience (the first ITRF solution was released in 1989: ITRF0; Boucher and Altamimi 1989), the maintenance and routine calculation of the ITRF, including the quality assessment of station positions and velocities, is well established in practice. Such an integrated infrastructure for physical heights is required for the IHRF.

The determination of the IHRF geopotential values C relies on the calculation of the gravity potential W, which is only possible by means of gravity field modelling. As W is non-linear and non-harmonic, a series of approximations are required:

- As W can be determined up to a constant only, the gravitational potential V is assumed to vanish at ∞
- To deal with the non-linearity and non-harmonicity of W, it is linearised by subtracting the gravity potential U generated by a reference level ellipsoid, currently the GRS80 (Geodetic Reference System 1980; Moritz 2000), obtaining the disturbing potential T:

$$T = W - U. \quad (2)$$

Since U is analytically known (Hofmann-Wellenhof and Moritz 2005), if T is known, the potential W and thus, the IHRF geopotential numbers (Eq. 1) can be easily derived. T must be known at the surface of the Earth (at the point P). When calculating the geoid (N), T is determined at the geoid surface. As a result, for the IHRF, T must be continued upwards to the surface of the Earth. If the height anomaly (ζ) is estimated, T can be directly used in Eq. (2) without further modifications (see e.g., Ihde et al. 2017; Sánchez et al. 2021).

- The solution of T demands a *mass-free* external space. Thus, the gravitational effects due to the Sun, Moon, topography, and other disturbing masses must be removed before the calculations and then restored to the final results to obtain the IHRF coordinates in the mean permanent tide concept.
- The determination of T at any point requires the integration of all the Earth masses. Therefore, the combination of

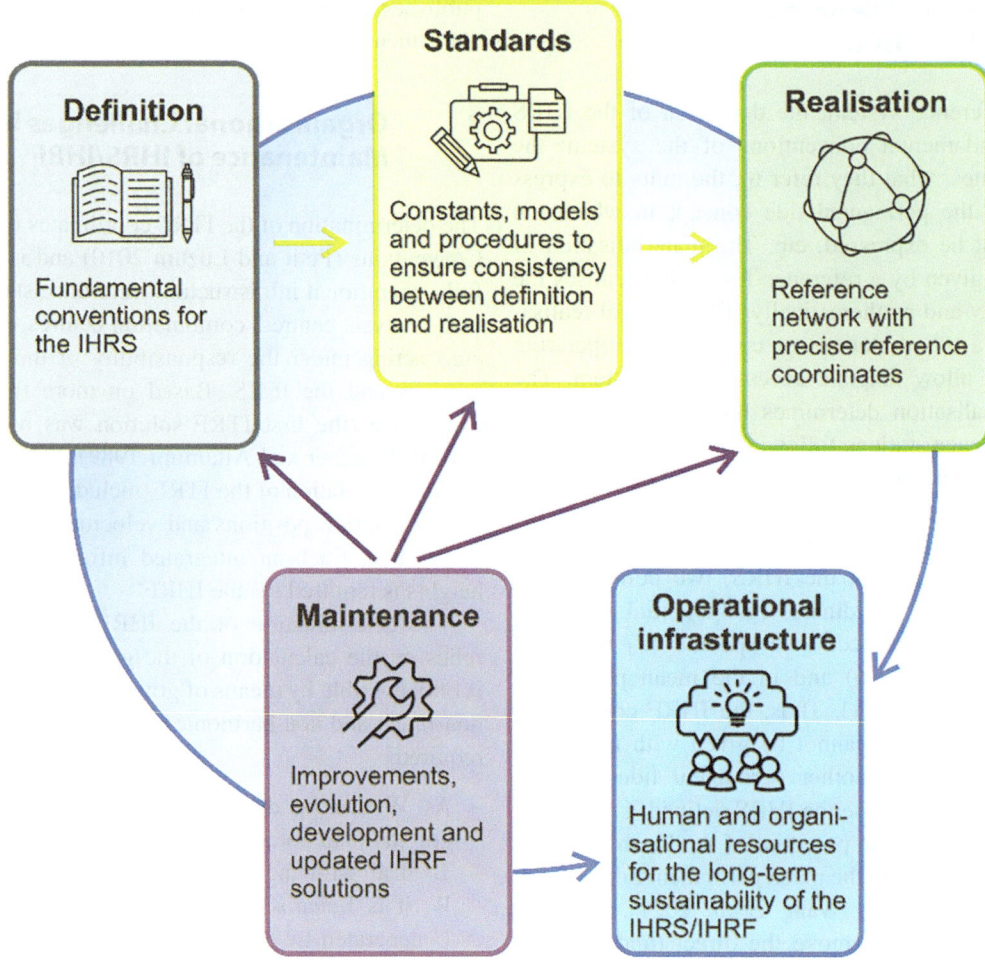

Fig. 1 Constituent elements of IHRS/IHRF

global gravity models with in situ surface (terrestrial, airborne or marine) gravity data is needed. This combination has to be adapted to the available surface gravity data coverage and quality, resulting in integral formulas with different analytical approximations.
- The mathematical evaluation of T can also be performed with different formulations (e.g., least-squares collocation, spherical basis functions, etc.).

Depending on the approximations and the analysis approaches, different results (T-values at the same point) are obtained as there is a high sensitivity to small changes in data handling and processing. As an example, Fig. 2 shows the discrepancies along a levelling profile between the geopotential numbers (Eq. 1) obtained from 13 different gravity field modelling approaches using exactly the same input gravity data to determine W ($=U + T$). The individual solutions agree within ±0.09 m^2 s^{-2} and ± 0.23 m^2 s^{-2} (equivalent to a physical height measure of ±0.009 m and ±0.023 m) in terms of standard deviation from the mean value. The overall discrepancies range from −0.86 m^2 s^{-2} (−0.088 m) for solution 2 to +0.77 m^2 s^{-2} (+0.079 m) for solution 12. The discrepancies generally show a high correlation with the topography, suggesting different responses to different approaches when dealing with the relief effects. The results presented in Fig. 2 were performed within the so-called Colorado Experiment, which represents a milestone in the comparison and calibration of different methods for the calculation of T. Further details about the Colorado Experiment can be found in Wang et al. (2021), Sánchez et al. (2021, 2023), and Van Westrum et al. (2021).

One possibility to minimise discrepancies between different computations would be to define a *standard* set of approximations, but this may not be appropriate as there are large differences in data availability, data density and quality around the world, hence it would not be pragmatic to define a common approach for all cases. Additionally, regions with different geographical (e.g., land only, marine and land, dominated by sea) and topography (smooth versus highly varying terrain) characteristics, which imply totally different spectral properties of the gravity field signal, require

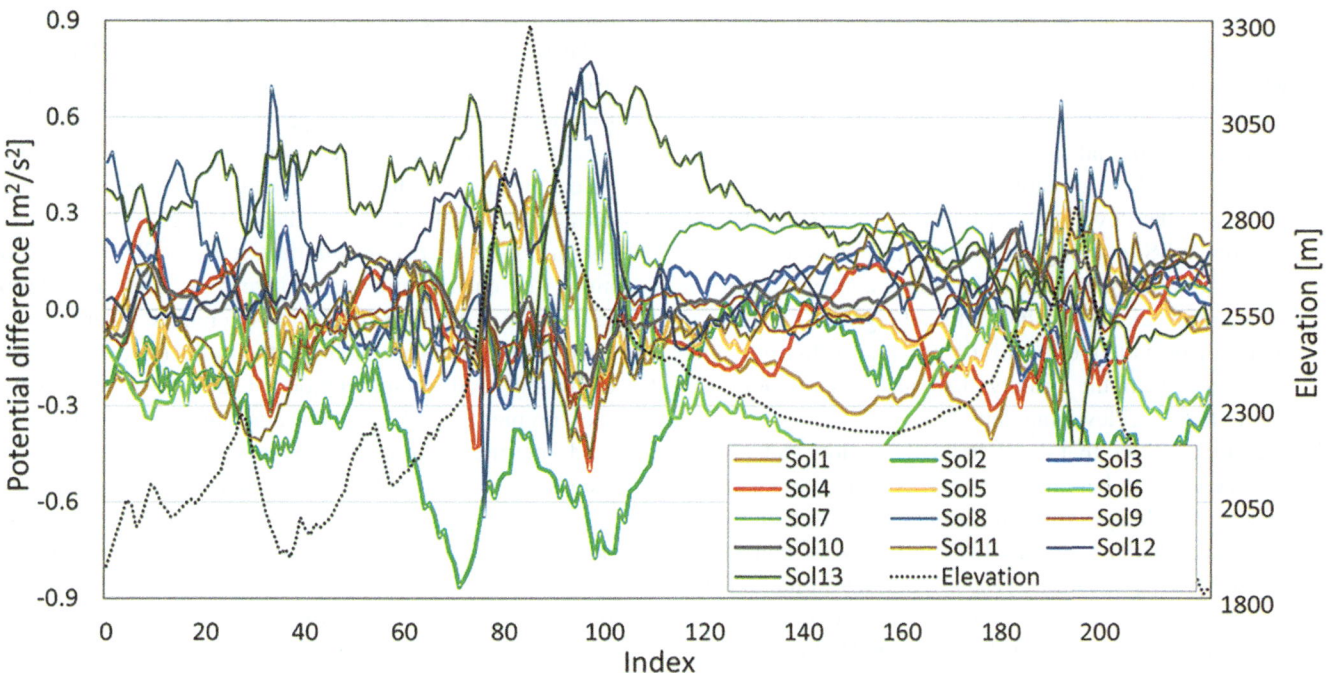

Fig. 2 Comparison of geopotential numbers obtained along a levelling profile from 13 different gravity field modelling approaches using exactly the same input gravity data to determine W

special approaches (e.g., modification of kernel functions, size of integration caps, geophysical reductions, etc.). Given the above, gravity field modelling is still to this day an *art*, meaning that the determination of potential values for each station require dedicated modelling, experience and knowledge of the area under study. Thus, it may be only possible to provide a cookbook rather than stringed guidelines for the estimation of potential values, contrary to what is the practice with the geometric component. A *centralised* calculation (as with the ITRF) is also quite complicated due to the restricted accessibility to surface gravity data. From this point of view, to take advantage of the most available gravity data for the determination of the IHRF coordinates, the potential values W required in Eq. (1) should be calculated by the regional/national experts in the geoid computation. They are knowledgeable about the particular conditions in their regions (i.e., which approximations are best suited to calculate T in their territories) and have access to unpublished surface gravity data and reference validation data (usually levelling networks co-located with GNSS (Global Navigation Satellite Systems) positioning and in some cases deflections of the vertical). In this context, the determination of the IHRF can be understood as a globally distributed calculation under the responsibility of various analysis centres, following a set of basic standards (see Sánchez et al. 2021) but with different processing schemes. This condition poses a number of organisational challenges, the most important of which are as follows:

- It is necessary to rely on colleagues who take over the responsibility for calculating the potential at each IHRF station. This could be managed through the active cooperation of IAG Sub-Commissions on *Physical Heights and Vertical Datum Unification* and the ones for *Regional Geoid Determination*, but not all the regions/countries of the world are represented. In addition, any action oriented to support the determination of the IHRF is voluntary and based on best efforts. Any resources needed have to be provided by the contributing colleagues through their home institutions (universities, agencies, research institutes, etc.). The availability of external funding is not foreseen, and everything depends on the commitment of the colleagues involved.
- Regions with less developed geodetic infrastructure lack not only surface gravity data, but also reliable know-how and computing capabilities for regional gravity field modelling. It is necessary to establish validation mechanisms that allow the quality of their results to be assessed. This is challenging because redundancy of computations is practically impossible.
- The disturbing potential T is the key quantity needed for the geoid (or height anomaly computation). Thus, the potential values W required for the IHRF geopotential numbers (Eq. 1) can be recovered from existing geoid/quasi-geoid models. However, the lack of metadata and detailed calculation reports prevents us from obtaining reliable results, while local/regional solutions are in many cases outdated and do not represent either the cur-

rent state-of-the-art in gravity field modelling and available gravity data holdings. This is particularly problematic when pure gravimetric geoid/quasi-geoid models are fitted to the existing GNSS/levelling data. This practice introduces artificial artefacts into these models in such a way that they are able to reproduce the systematic errors of the levelling networks by applying the so-called GNSS-levelling approach. It is therefore of the utmost importance to have the direct cooperation of the colleagues calculating the national/regional models, so that they provide the disturbing potential values obtained primarily from the gravity data analysis, without any fitting to GNSS/levelling data.

- It is not expected that the IHRF will be widely adopted to replace existing height systems. It is more likely that the existing height systems will continue to be used on a daily basis and that the IHRS/IHRF will only be considered for selected applications. Thus, local ties to the existing height systems will be available. In order to achieve widespread acceptance of the IHRS/IHRF, the reliability of the potential values must be drastically improved along with the adoption of geopotential-based vertical datums as best practice. Ultimately, the essential goal of the IHRF is to compute exactly the same global absolute geoid (independent of local vertical datums) with high accuracy everywhere in the world.

With this said, the next two sections are devoted to describing how the existing gravity field related IAG services could contribute to the IHRS/IHRF, and to identifying the elements needed to establish an operational infrastructure for the IHRS/IHRF that addresses the challenges described above.

4 Existing Gravity Field Related IAG Services and the IHRS/IHRF

Gravity field and geoid related data and products are collected, processed and distributed by several IAG Services, all organised under the IGFS. The IGFS acts as a clearinghouse for these Services to coordinate the standardisation, interoperability and publication of well-documented gravity field-related data and information. The following services are of particular interest to the IHRS/IHRF:

- *International Centre for Global Earth Models* (ICGEM, Ince et al. 2019): It collects, archives and publishes all existing (static and time-dependent) global gravity field models and provides web interfaces for the calculation of various gravity field functionals from the models. The monthly global gravity models calculated by the *International Combination Service for Time-variable Gravity Fields* (COST-G, Peter et al. 2022) based on the combination of GRACE/GRACE-FO data, low-low satellite-to-satellite tracking, high-low satellite-to-satellite tracking, and Satellite Laser Ranging (SLR) are available at the ICGEM as well. The ICGEM also provides global topographic models in terms of spherical harmonics. The gravity and topography models provided by the ICGEM are essential input data for the determination of the potential values needed for the IHRF coordinates, especially over areas with poor local gravity data coverage.
- *Bureau Gravimétrique International* (BGI): The BGI aims to ensure the data inventory and long-term availability of the gravity measurements taken at the Earth's surface. This includes the collection, validation, archiving, and dissemination of all types of gravity measurements (relative or absolute) acquired from land, marine or airborne surveys. The surface gravity data stored at the BGI is essential to improve the spatial resolution of the global gravity models in the determination of the potential values needed for the IHRF coordinates.
- *International Digital Elevation Model Service* (IDEMS): It maintains a catalogue of links to Digital Elevation Models (DEMs), relevant software, and related datasets (including representation of inland water in DEMs). The determination of potential values demands the appropriate handling of topographic effects, which are inferred from topography models. In addition to the global topography models provided by ICGEM, IDEMS offers an updated registry of local and regional digital elevation models of higher resolution.
- *International Service for the Geoid* (ISG, Reguzzoni et al. 2021): It collects, archives and provides regional geoid and quasi-geoid models, collects and distributes geoid determination software, and organises capacity building activities through the International School for the Determination and Use of the Geoid.

Figure 3 summarises the interaction of these Services with the IHRS/IHRF. While the first three Services (ICGEM, BGI and IDEMS) can be understood as input data providers for the determination of IHRF geopotential numbers, the ISG serves in two different ways: as a repository of the geoid or quasi-geoid models that can be computed as a *by-product* when determining T for the IHRF, and as an input data provider when geoid or quasi-geoid models are available in regions or countries that do not participate in the globally distributed IHRS/IHRF calculation. The IHRS/IHRF can also contribute additional by-products to the other Services, such as mean gravity anomaly values in regions with poor gravity data coverage in the BGI databank, or updated GNSS/levelling data sets for the evaluation of global models (ICGEM) and regional models (ISG). Similarly, for IDEMS, elevation models employed for the determination of the IHRF that are not listed in the IDEMS catalogue should be reported to IDEMS in order to extend its catalogue.

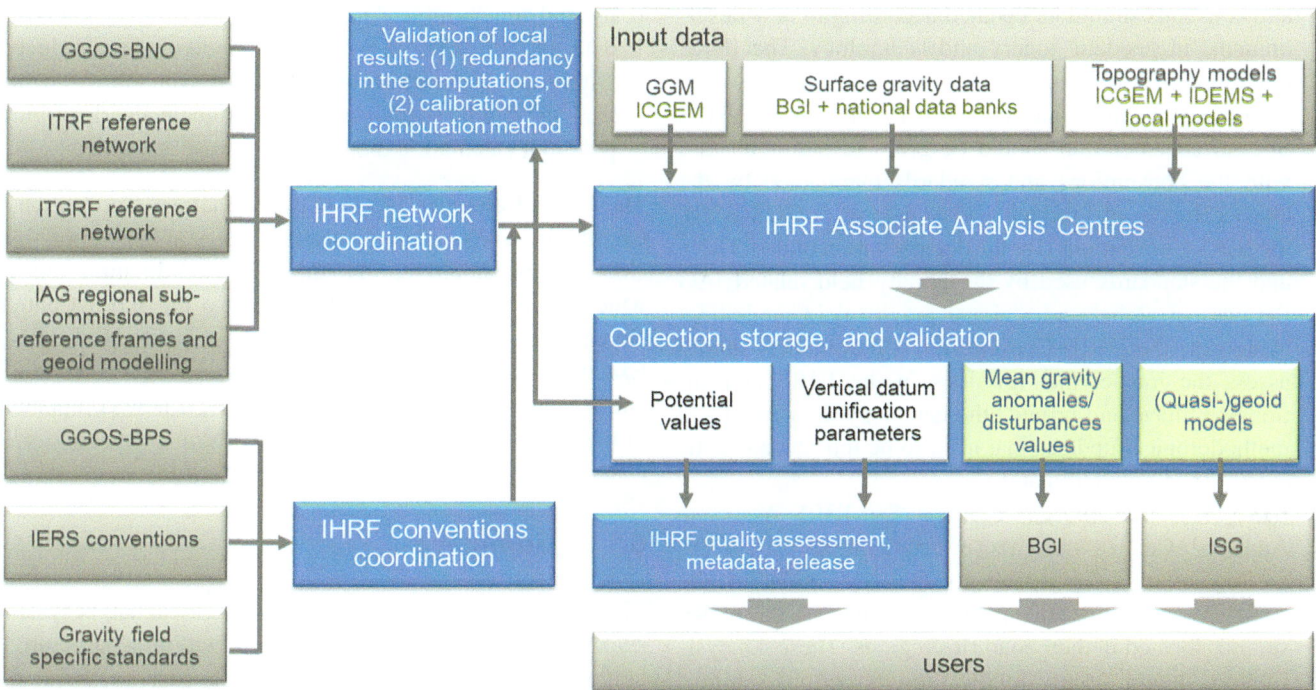

Fig. 3 Data flow between existing gravity field related IAG Services (grey boxes) and new organisational elements for the IHRS/IHRF (blue boxes)

5 Operational Infrastructure for the Long-Term Availability of the IHRS/IHRF

The operational and organisational infrastructure for the IHRS/IHRF should take care of each of the constituent elements described in Sect. 2. Our proposal is to establish a central coordinating body composed of individual modules, which will take care of specific activities as follows (see Fig. 3):

- *IHRF Coordination Centre*: It will be the central management body for the maintenance of the IHRS/IHRF. It will be responsible for the general coordination of activities required for the IHRS/IHRF and for the storage, publication, and service of the IHRF. This includes not only related documentation, products, and relevant information, but also the IHRF coordinates (X, Y, Z, C) at the IHRF reference stations and a catalogue of the vertical datum parameters; i.e., the transformation parameters between the existing local height systems and the IHRF. As in the case of the ITRF, it is foreseen that the IHRF solutions will be regularly updated to take into account new technological developments, and new and improved observation data. Our proposal is to synchronise the release of updated IHRF solutions with the release of updated ITRF solutions. This process should be also coordinated by the IHRF Coordination Centre.

- *IHRF Reference Network Coordination*: According to Ihde et al. (2017) and Sánchez et al. (2021), the IHRF reference stations should be materialised by continuously operating GNSS stations co-located with the GGOS core sites (Appleby et al. 2015), the global ITRF or its regional densifications (such as EUREF, SIRGAS, etc.), the International Terrestrial Gravity Reference Frame (ITGRF, Wziontek et al. 2021) and, if possible, with the national levelling networks. The IHRF Reference Network Coordination should implement and keep updated a catalogue of the IHRF global reference stations. This includes the decommissioning of destroyed stations and the addition of new stations to replace removed stations or improve the geographical distribution. Changes in the IHRF station distribution require interaction with the bodies responsible for the other reference frames: the ITRF (IERS), the ITGRF (IGFS), the GGOS core sites (GGOS Bureau of Networks and Observations) and the IAG Sub-Commissions for the regional reference frames and the regional geoid models. The IHRF Reference Network Coordination should also prepare and provide the set of ITRF coordinates to be used for the determination of updated IHRF solutions.

- *IHRF Conventions' Coordination*: The initial conventions, standards, and constants for the definition and realisation of the IHRS are given in the IAG Resolution 1 (2015), and further commented by Ihde et al. (2017) and Sánchez et al. (2021). These conventions, standards and constants should be updated according to new developments in geodetic theory and technology. The IHRF Conventions' Coordination should maintain a document with the conventions and standards needed for the IHRF. Special consideration should be given to harmonisation with the conventions and standards maintained by the GGOS Bureau of Products and Standards (GGOS-BPS), the IERS Conventions (for the determination of the ITRF), and the standards used by the gravity field related IAG Services in global and regional gravity field modelling. Moreover, the IHRF Conventions' Coordination should assess the impact that revisions in the IHRF conventions will have and provide the necessary theoretical and methodological updates that need to be introduced to the existing station coordinates.

- *IHRF Associate Analysis Centres*: The IHRF Associate Analysis Centres are those national/regional agencies/bodies that contribute to the realisation of the IHRF by providing the potential values at the IHRF stations located in their countries/regions. These Analysis Centres should strictly follow the conventions outlined by the IHRF Conventions Coordination, use the ITRF input coordinates provided by the IHRF Reference Network Coordination, and provide detailed descriptions about their calculations. In an ideal data flow scheme, the IHRF Associate Analysis Centres would provide the IHRF Coordination Centre with the following products: potential values at the IHRF reference stations; vertical datum parameters; mean gravity anomalies or disturbances; and regional geoid or quasi-geoid models of high resolution. The mean gravity anomalies (or disturbances) and the geoid/quasi-geoid models would then be managed by the BGI and ISG, respectively.
- *IHRF Combination Coordination*: The IHRF Combination Coordination will be responsible for the combination and quality assessment of the regional/national solutions and for releasing the final (official) IHRF solution. The quality assessment can be based on redundant calculations or by calibration of computation methods. In the first case, at least two Associate Analysis Centres independently determine the potential values for the same stations. In the second case, IHRF Associate Analysis Centres should determine potential values using a certain set of input data and compare their results with those obtained by other processing approaches. For this purpose, the input gravity and topography data, the GNSS/levelling validation data, and the different geoid/quasi-geoid models produced within the Colorado Experiment are available from the ISG and can be used as a basis to evaluate any disturbing potential calculation method or software anywhere.

The IHRF Reference Network Coordination, Conventions' Coordination, Associate Analysis Centres and Combination Coordination should report to the IHRF Coordination Centre, which, in turn, would report directly to the IGFS Central Bureau (Fig. 4). The IGFS presented this proposal to the IAG Executive Committee at its meeting on 10 December 2023 and it was unanimously approved. Thus, a new component of the IGFS dedicated to the IHRF has been created and will ensure the long-term availability and reliability of the IHRF.

6 Summary

The establishment and maintenance of a global height system such as the IHRS is only possible within a structured, comprehensive and global organisation such as the IAG. The efforts of the GGOS FA-UHS, IAG Commissions 1 and 2, ICCT and IGFS over the last 12 years have made it possible to pave the scientific foundations of the IHRS, to compile a cookbook for the implementation of the IHRF, and to generate a first solution for the IHRF. In particular, the availability of the input data and the results of the Colorado Experiment provide an unprecedented basis for the evaluation of gravity potential (geoid and quasi-geoid) calculation methods. Much progress has been made and continuity must be ensured to guarantee the maintenance and availability of the IHRF in the long term. Following IAG practice, the IAG Commissions, ICCT, working and study groups, etc., are responsible for the development of theory and methods, and the IAG Services operates as centres for standards, networks, data, analysis/combination and products. Our proposal is to establish an IHRF Coordination Centre reporting directly to the IGFS Central Bureau, to coordinate the operational activities necessary for the long-term availability and sustainability of the IHRF. The IGFS presented this proposal to the IAG Executive Committee at its meeting on 10 December 2023 and it was unanimously approved. Thus, a new component of the IGFS dedicated to the IHRF has been created and will ensure the long-term availability and reliability of the IHRF.

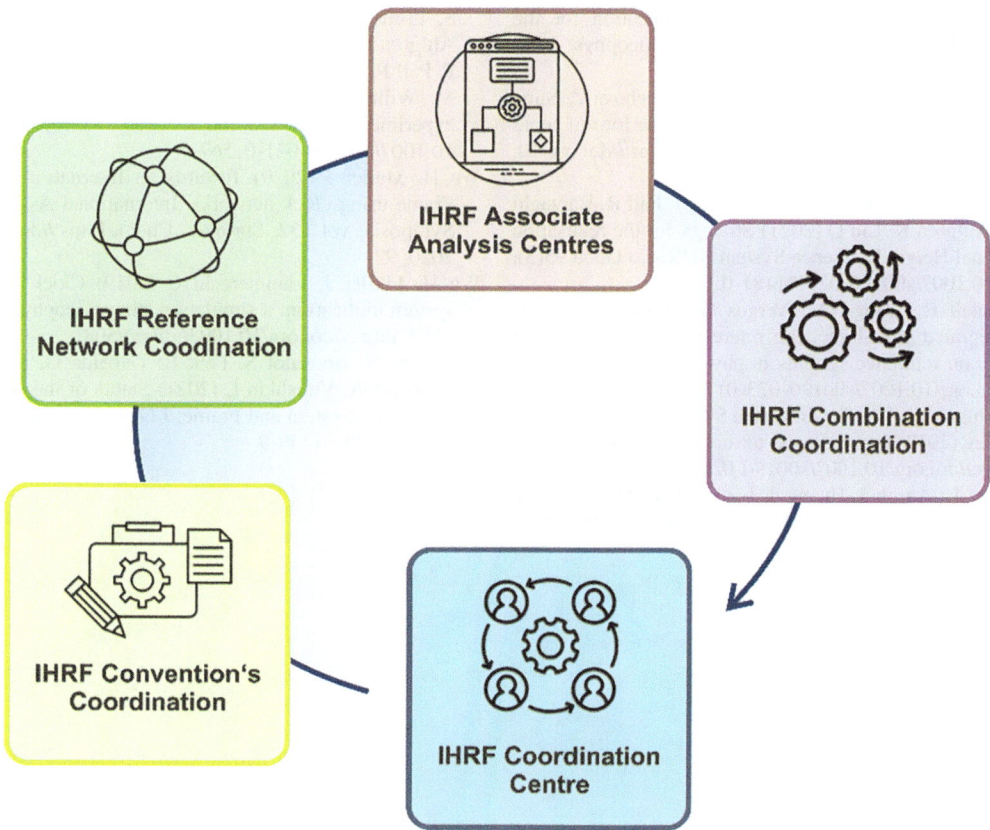

Fig. 4 Data flow between the IHRF Coordination Centre and its modules

References

Altamimi Z, Rebischung P, Collilieux X, Métivier L, Chanard K (2023) ITRF2020: an augmented reference frame refining the modeling of nonlinear station motions. J Geod 97:47. https://doi.org/10.1007/s00190-023-01738-w

Appleby G, Behrend D, Bergstrand S, Donovan H, Emerson C, Esper J, Hase H, Long J, Ma C, McCormick D, Noll C, Pavlis E, Ferrage P, Pearlman M, Saunier J, Stowers D, Wetze S (2015) GGOS requirements for core sites, global geodetic observing system (GGOS), 24p

Boucher C, Altamimi Z (1989) The initial IERS terrestrial reference frame, IERS technical note 1, Central Bureau of IERS - Observatoire de Paris, Paris, 98 p

Drewes H, Kuglitsch F, Ádám J, Rózsa S (2016) Geodesist's handbook 2016. J Geod 90:907. https://doi.org/10.1007/s00190-016-0948-z

Hofmann-Wellenhof B, Moritz H (2005) Physical geodesy, Springer, Wien New York

Ihde J, Sánchez L, Barzaghi R, Drewes H, Foerste C, Gruber T, Liebsch G, Marti U, Pail R, Sideris M (2017) Definition and proposed realization of the International Height Reference System (IHRS). Surv Geophys 38(3):549–570. https://doi.org/10.1007/s10712-017-9409-3

Ince ES, Barthelmes F, Reißland S, Elger K, Förste C, Flechtner F, Schuh H (2019) ICGEM—15 years of successful collection and distribution of global gravitational models, associated services, and future plans. Earth Syst Sci Data 11(2):647–674. https://doi.org/10.5194/essd-11-647-2019

Mäkinen J (2021) The permanent tide and the International Height Reference Frame IHRF. J Geod 95:106. https://doi.org/10.1007/s00190-021-01541-5

Migliaccio F, Reguzzoni M, Rosi G et al (2023) The MOCAST+ study on a quantum gradiometry satellite mission with atomic clocks. Surv Geophys 44:665–703. https://doi.org/10.1007/s10712-022-09760-x

Moritz H (2000) Geodetic reference system 1980. J Geod 74:128–133. https://doi.org/10.1007/s001900050278

Mu Q, Müller J, Wu H, Knabe A, Zhong M (2023) Satellite gradiometry based on a new generation of accelerometers and its potential contribution to earth gravity field determination. Adv Space Res. https://doi.org/10.1016/j.asr.2023.08.023

Müller J, Dirkx D, Kopeikin SM, Lion G, Panet I, Petit G, Visser PNAM (2018) High performance clocks and gravity field determination. Space Sci Rev 214:5. https://doi.org/10.1007/s11214-017-0431-z

Peter H, Meyer U, Lasser M, Jäggi A (2022) COST-G gravity field models for precise orbit determination of low earth orbiting satellites. Adv Space Res 69(12):4155–4168. https://doi.org/10.1016/j.asr.2022.04.005

Petit G, Luzum B (2010) IERS Conventions (2010). IERS technical note 36, Verlag des Bundesamts für Kartographie und Geodäsie, Frankfurt am Main, 179 pp., ISBN 3-89888-989-6

Plag H-P, Pearlman M (2009) Global geodetic observing system: meeting the requirements of a global society. Springer, Berlin

Reguzzoni M, Carrion D, De Gaetani CI, Albertella A, Rossi L, Sona G, Batsukh K, Toro Herrera JF, Elger K, Barzaghi R, Sansó F (2021) Open access to regional geoid models: the International Service for the Geoid. Earth Syst Sci Data 13:1653–1666. https://doi.org/10.5194/essd-13-1653-2021

Sánchez L, Barzaghi R (2024) Achievements of the GGOS Focus Area Unified Height System. In: International Association of Geodesy Symposia, vol 157. Springer, Cham. https://doi.org/10.1007/1345_2024_249

Sánchez L, Sideris MG (2017) Vertical datum unification for the International Height Reference System (IHRS). Geophys J Int 209(2):570–586. https://doi.org/10.1093/gji/ggx025

Sánchez L, Čunderlík R, Dayoub N, Mikula K, Minarechová Z, Šíma Z, Vatrt V, Vojtíšková M (2016) A conventional value for the geoid reference potential W0. J Geod 90(9):815–835. https://doi.org/10.1007/s00190-016-0913-x

Sánchez L, Ågren J, Huang J, Wang YM, Mäkinen J, Pail R, Barzaghi R, Vergos GS, Ahlgren K, Liu Q (2021) Strategy for the realisation of the International Height Reference System (IHRS). J Geod 95(3). https://doi.org/10.1007/s00190-021-01481-0

Sánchez L, Wziontek H, Wang YM, Vergos G, Timmen L (2023) Towards an integrated global geodetic reference frame: preface to the special issue on reference systems in physical geodesy. J Geod 97(6). https://doi.org/10.1007/s00190-023-01758-6

Van Westrum D, Ahlgren K, Hirt C, Guillaume S (2021) A Geoid Slope Validation Survey (2017) in the rugged terrain of Colorado, USA. J Geod 95:9. https://doi.org/10.1007/s00190-020-01463-8

Wang YM, Sánchez L, Ågren J, Huang J, Forsberg R, Abd-Elmotaal HA, Barzaghi R, Bašić T, Carrion D, Claessens S, Erol B, Erol S, Filmer M, Grigoriadis VN, Isik MS, Jiang T, Koç Ö, Li X, Ahlgren K, Krcmaric J, Liu Q, Matsuo K, Natsiopoulos DA, Novák P, Pail R, Pitoňák M, Schmidt M, Varga M, Vergos GS, Véronneau M, Willberg M, Zingerle P (2021) Colorado geoid computation experiment—overview and summary. J Geod 95:12. https://doi.org/10.1007/s00190-021-01567-9

Wu H, Müller J (2020) Towards an International Height Reference Frame using clock networks. International Association of Geodesy Symposia, vol 152. Springer, Cham. https://doi.org/10.1007/1345_2020_97

Wu H, Müller J, Lämmerzahl C (2019) Clock networks for height system unification: a simulation study. Geophys J Int 216(3):1594–1607. https://doi.org/10.1093/gji/ggy508

Wziontek H, Bonvalot S, Falk R, Gabalda G, Mäkinen J, Pálinkáš V, Rülke A, Vitushkin L (2021) Status of the International Gravity Reference System and Frame. J Geod 95:7. https://doi.org/10.1007/s00190-020-01438-9

Open Access This chapter is licensed under the terms of the Creative Commons Attribution 4.0 International License (http://creativecommons.org/licenses/by/4.0/), which permits use, sharing, adaptation, distribution and reproduction in any medium or format, as long as you give appropriate credit to the original author(s) and the source, provide a link to the Creative Commons license and indicate if changes were made.

The images or other third party material in this chapter are included in the chapter's Creative Commons license, unless indicated otherwise in a credit line to the material. If material is not included in the chapter's Creative Commons license and your intended use is not permitted by statutory regulation or exceeds the permitted use, you will need to obtain permission directly from the copyright holder.

Estimation of the Argentinean Vertical Datum Parameter with Respect to the International Height Reference Frame (IHRF)

Agustín R. Gómez, Claudia N. Tocho, Ezequiel D. Antokoletz, Hernán J. Guagni, and Diego A. Piñón

Abstract

One of the current goals of the International Association of Geodesy (IAG) through its Global Geodetic Observing System (GGOS) is the unification of the existing local vertical datums towards the realization of the International Height Reference System (IHRS), i.e. the International Height Reference Frame (IHRF). To achieve this goal, one possible solution is to compute the offset between the equipotential surface of the Earth's gravity field realized by the conventional W_0^{IHRF} value of the IHRS and the unknown geopotential value of the local vertical datum. This offset is known as vertical datum parameter. In this study, the determination of the vertical datum parameter of the Argentinean National Vertical Reference System 2016 (SRVN16) using two approaches is presented. The first approach is based on the Geodetic Boundary Value Problem (GBVP). The second approach combines geopotential numbers obtained with levelling and gravity with geopotential numbers derived from a quasigeoid model. Both methods require GNSS/Levelling data and a high-precision gravimetric quasigeoid model. The quasigeoid model was computed using the remove-compute-restore technique and applying a Fourier representation of Molodensky's integral formula. The vertical datum parameter estimation was carried out in a flat area in the Buenos Aires province due to the availability of high-quality gravity observations and benchmarks with GNSS/Levelling-derived height anomalies, all located near the tide gauge station used to define the Argentinean vertical datum. Estimation results with the first and second approach were -0.46 ± 1.78 m^{2}s^{-2} and -0.46 ± 1.37 m^{2}s^{-2}, respectively. The vertical datum parameter can be further used to integrate SRVN16 into the IHRF.

Keywords

Argentinean vertical reference system · Gravimetric quasigeoid model · International Height Reference Frame (IHRF) · Vertical datum parameter · Vertical datum unification

A. R. Gómez (✉)
Facultad de Ciencias Astronómicas y Geofísicas, Universidad Nacional de La Plata, La Plata, Argentina

Consejo Nacional de Investigaciones Científicas y Técnicas, Rosario, Argentina
e-mail: agusgomez@fcaglp.unlp.edu.ar

C. N. Tocho
Facultad de Ciencias Astronómicas y Geofísicas, Universidad Nacional de La Plata, La Plata, Argentina

Comisión de Investigaciones Científicas de la Provincia de Buenos Aires, La Plata, Argentina

E. D. Antokoletz
Facultad de Ciencias Astronómicas y Geofísicas, Universidad Nacional de La Plata, La Plata, Argentina

H. J. Guagni · D. A. Piñón
Instituto Geográfico Nacional, Ministerio de Defensa, Buenos Aires, Argentina

1 Introduction

The International Height Reference System (IHRS; Ihde et al. 2017) was defined by the International Association of Geodesy (IAG) through its Resolution N° 1 from 2015 (Drewes et al. 2016). Its realisation through the establishment of the International Height Reference Frame (IHRF) is one of the main goals of modern geodesy. According to the definition of the IHRS, vertical coordinates are geopotential numbers

$$C(P) = W_0^{IHRF} - W(P), \quad (1)$$

where $W(P)$ is the Earth's gravity potential, $W_0^{IHRF} = 62636853.4$ m²s⁻² is the geopotential value of the reference surface (Sánchez et al. 2016), and P is a point located in space through three Cartesian coordinates that are referred to the International Terrestrial Reference Frame (ITRF; Petit et al. 2010). IHRS coordinates are expressed in SI units and given in the mean-tide concept (Mäkinen 2021).

There are various approaches to obtain IHRS coordinates of a given point P. One of them, known as vertical datum unification, focuses on integrating local height systems into the IHRF. A local vertical datum i is determined using sea level observations at a tide gauge located at a point P_{0i}. Vertical coordinates corresponding to the local height system are derived from local geopotential numbers

$$C_i(P) = W_{0i}^{LVD} - W(P), \quad (2)$$

where $W_{0i}^{LVD} = W(P_{0i})$ is the geopotential value of the local datum realized by the tide gauge located at point P_{0i} (Fig. 1). The integration of the local datum i into the IHRF requires the determination of the vertical datum parameter

$$\delta W_{0i} = W_0^{IHRF} - W_{0i}^{LVD}. \quad (3)$$

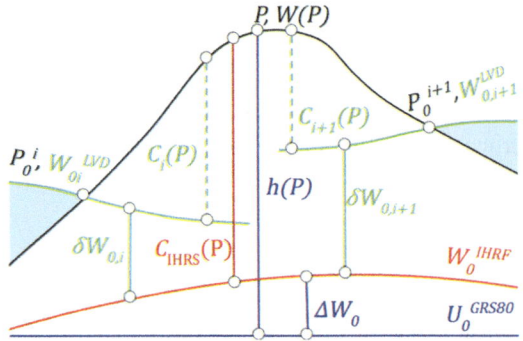

Fig. 1 Reference surfaces, vertical coordinates and vertical datum parameters. Black line: Topography. Green curves: Local vertical datums. Red curve: IHRF reference surface. Blue line: reference ellipsoid. Green solid vertical lines: vertical datum parameters of datums i and $i + 1$. Green dashed vertical lines: local geopotential numbers refered to datums i and $i + 1$

When δW_{0i} is known, local geopotential numbers can be shifted to the global datum via

$$C(P) = C_i(P) + \delta W_{0i}. \quad (4)$$

In Argentina, previous studies estimated either the vertical datum parameter or the zero-height geopotential value W_{0i}^{LVD}. Tocho and Vergos (2015) computed W_{0i}^{LVD} for the Argentinean height system official at that time, the National Vertical Reference System 1971 (SRVN71). The computation was performed using the Global Geopotential Model (GGM) EGM2008 (Pavlis et al. 2012). Since their estimation, new GGMs have been published, and a new Argentinean vertical reference network, called National Vertical Reference System 2016 (SRVN16; Instituto Geográfico Nacional 2016) has been adopted. A more recent estimation for 14 local vertical datums in the South American region is presented in Sánchez and Sideris (2017). Their study combined GNSS/Levelling data with geoid/quasigeoid models whenever possible or a GGM otherwise. In Argentina, EGM2008 was used.

In the present study, we determine the SRVN16 vertical datum parameter using a high-precision local gravimetric quasigeoid model and GNSS/Levelling data in a region of the Buenos Aires province, employing two different approaches.

2 Methodology

In this section, we describe two estimation methods for the vertical datum parameter, as well as the determination of a gravimetric quasigeoid model required for both approaches.

2.1 Method 1

This method uses the Geodetic Boundary Value Problem (GBVP) approach (Rummel and Teunissen 1988), which is described in Sánchez and Sideris (2017). It involves solving an extended version of the scalar-free GBVP following Molodensky's theory, which incorporates into the boundary conditions the vertical datum parameter as an additional unknown. The solution to this GBVP is (eq. (28) of Sánchez and Sideris 2017)

$$T_{i,GBVP}(P) = -\Delta W_0 + \delta W_{0i} + T_i(P) + \delta T_i(P), \quad (5)$$

where $T_{i,GBVP}$ is the disturbing potential, T_i can be recovered from a gravimetric quasigeoid model and its calculation is described in Sect. 2.3, and δT_i is known as the indirect bias term (Gerlach and Rummel 2013), which is given as

$$\delta T_i(P) = \frac{\delta W_{0i}}{2\pi} \iint_{\sigma_i} S(\psi) d\sigma, \quad (6)$$

where σ_i corresponds to the portion of the unit sphere that the datum i occupies. $S(\psi)$ is the Stokes function (eq. (2-305) of Hofmann-Wellenhof and Moritz 2006). This term is present because the gravity anomalies used in the GBVP solution refer to the local datum i instead of the IHRF geopotential surface. If T_i is computed with the remove-compute-restore (RCR) technique (Sansò and Sideris 2013), using a GGM of degree over 200, then the indirect bias term becomes negligible, as shown by Gerlach and Rummel (2013). Finally, $\Delta W_0 = W_0^{\text{IHRF}} - U_0^{\text{GRS80}} = -7.45 \text{ m}^2\text{s}^{-2}$ is the difference between the IHRS zero-height potential value and the normal potential value U_0^{GRS80} at the surface of the GRS80 reference ellipsoid (Moritz 2000).

The disturbing potential can also be computed with (eq. (30) of Sánchez and Sideris 2017)

$$T_{i,\text{GNSS}}(P) = \gamma h(P) - C_i(P), \tag{7}$$

where h is the ellipsoidal height of P, and γ is GRS80 normal gravity at the telluroid (eq. (2-215) of Hofmann-Wellenhof and Moritz 2006). Equalling Eqs. (5) and (7) leads to

$$\gamma h(P) - C_i(P) = T_i(P) + \delta T_i(P) - \Delta W_0 + \delta W_{0i}. \tag{8}$$

If, as previously stated, δT_i can be assumed to be negligible, then

$$\delta W_{0i} = \gamma h(P) - C_i(P) - T_i(P) + \Delta W_0 + \varepsilon(P), \tag{9}$$

where ε is the associated stochastic error.

At $k = 1, 2, \ldots, N$ GNSS/Levelling benchmarks located at points P_k, where their ellipsoidal heights $h(P_k) = h_k$ and local geopotential numbers $C_i(P_k) = C_{i,k}$ have been determined with GNSS and levelling techniques respectively, the vertical datum parameter can be estimated as a weighted average

$$\delta \hat{W}_{0i} = \sum_{k=1}^{N} w_k \delta W_{0i}^k, \quad w_k = \frac{1}{H_i^*(P_k)} \left(\sum_{j=1}^{N} \frac{1}{H_i^*(P_j)} \right)^{-1}, \tag{10}$$

where δW_{0i}^k is the vertical datum parameter computed at the k-th benchmark using Eq. (9). The weight factors w_k depend on the normal height H^* and assign less weight to data corresponding to higher altitudes, which are associated with greater observation errors that could propagate into the estimation (Grigoriadis et al. 2014). Since errors in ellipsoidal heights are usually much smaller than in physical heights, they were neglected in the definition of w_k. Errors in the gravimetric disturbing potential were also not considered.

2.2 Method 2

The method presented in this section relies on the computation of the geopotential value of the local datum W_{0i}^{LVD} by rearranging Eq. (2). This requires geopotential numbers and the gravity potential

$$W(P) = U(P) + T_i(P), \tag{11}$$

where $U(P)$ is the normal gravity potential generated by the GRS80 reference ellipsoid and can be computed using the ellipsoidal coordinates of P (eq. (2-126) of Hofmann-Wellenhof and Moritz 2006). T_i can be recovered from a gravimetric quasigeoid model, which is described in the next section.

Combining Eqs. (2), (3) and (11) the vertical datum parameter is given by

$$\delta W_{0i} = W_0^{\text{IHRF}} - C_i(P) - U(P) - T_i(P) + \varepsilon(P). \tag{12}$$

At $k = 1, 2, \ldots, N$ GNSS/Levelling benchmarks, the vertical datum parameter is now estimated by

$$\delta \hat{W}_{0i} = W_0^{\text{IHRF}} - \sum_{k=1}^{N} w_k [C_{i,k} + U(P_k) + T_i(P_k)], \tag{13}$$

where now δW_{0i}^k is computed at the k-th benchmark according to Eq. (12), and weights w_k are as defined in Eq. (13).

2.3 Quasigeoid Model Computation

The term T_i, required in both methods described above, corresponds to the solution of the scalar-free GBVP following Molodensky's approach using biased gravity anomalies, and is given by (eq. (2-357) of Hofmann-Wellenhof and Moritz 2006)

$$T_i(P) = \frac{G \delta M}{R} + \frac{R}{4\pi} \iint_{\sigma} (\Delta g_i + g_{1i}) S(\psi) d\sigma, \tag{14}$$

where Δg_i are gravity anomalies computed using heights referred to the local datum (i.e. biased gravity anomalies), g_{1i} is an analytical continuation term applied to the gravity anomalies (eq. (8-62) of Hofmann-Wellenhof and Moritz 2006), and R is a mean Earth radius. The term $G\delta M/R$, known as the zero-degree term of the anomalous potential (Section 2.17 of Hofmann-Wellenhof and Moritz 2006), is present due to the difference δM between the mass of the Earth (in practice obtained from a GGM) and the mass of the GRS80 reference ellipsoid.

$T_i(P)$ was recovered from a gravimetric quasigeoid model ζ_{mod} which is refered to an equipotential surface of the Earth's gravity field with value $W_0 = U_0^{\text{GRS80}}$ using the usual (non-generalized) Bruns formula (eq. (2-237) of Hofmann-Wellenhof and Moritz 2006)

$$T_i(P) = \gamma \zeta_{\text{mod}}(P). \qquad (15)$$

The quasigeoid model was computed with the remove-compute-restore (RCR) scheme. This scheme consists of removing the long and short wavelengths from the gravity anomalies Δg_i, to compute the residual gravity anomaly field

$$\Delta g_{\text{res}} = \Delta g_i - \Delta g_{\text{GGM}} - \Delta g_{\text{RTM}}, \qquad (16)$$

where Δg_{GGM} was obtained from a GGM, and Δg_{RTM} was computed through Residual Terrain Modelling (RTM) in the spatial domain (Forsberg 1984) using a high-precision Digital Terrain Model (DTM). The residual anomalies in Eq. (16) are then used as input for Molodensky's integral, computing the residual height anomalies ζ_{res}. Afterwards, the GGM and RTM effects are restored, and the complete quasigeoid model is given by

$$\zeta_{\text{mod}} = \zeta_{\text{res}} + \zeta_{\text{GGM}} + \zeta_{\text{RTM}}, \qquad (17)$$

where ζ_{RTM} is computed with the same DTM, and ζ_{GGM} is obtained with the same GGM used in the remove stage.

The computation of ζ_{res} was performed using the 1D FFT approach (Haagmans 1993) with a Wong-Gore modification of the Stokes kernel (Wong and Gore 1969). FFT techniques require the gravity data to be equally spaced on a grid. The gridding procedure described by Featherstone et al. (2000) was applied.

Finally, following the recommendations of Sánchez et al. (2021), all calculations regarding the quasigeoid determination were done in the zero-tide concept (Mäkinen 2021). All computations were performed using the Gravsoft software (Tscherning 1992).

3 Data Sets and Study Area

The approaches described in Sect. 2 require as data input geopotential numbers, ellipsoidal heights, and a gravimetric quasigeoid model, which in turn relies on gravity data, a GGM and a DTM. For this study, a region of Argentina with geographical limits $-36° \leq \phi \leq -32°$ in latitude and $-63° \leq \lambda \leq -61°$ in longitude was chosen (Fig. 2). It was selected due to its good gravity data coverage and regular distribution, and the fact that it is located nearby the tide gauge that realizes the local vertical datum, which implies that levelling data is less prone to systematic errors

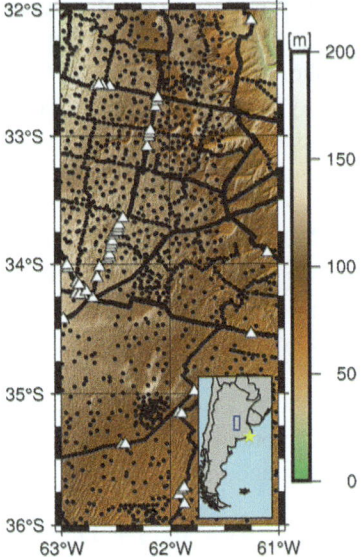

Fig. 2 Region of study and distribution of gravity data points. Yellow star: Zero-height reference point of SRVN16. Black points: measured gravity data. White triangles: GNSS/Levelling data. Background: Topography from MDE-Ar v2.1

that depend on the distance to the reference point. The gravity data set and the DTM were selected from a region one degree larger in every direction, in order to avoid edge effects in the quasigeoid computation.

The data set consists of 8018 points that belong to the Argentinean gravity reference networks (Antokoletz 2017), and 58 GNSS/Levelling-derived height anomalies. Geopotential numbers are referred to the Argentinean levelling network, SRVN16, have a precision in the order of 0.1 m^{2}s^{-2} and are given in the mean-tide concept (Instituto Geográfico Nacional 2016). Ellipsoidal heights are given in the official Argentinean geodetic reference frame Posiciones Geodésicas Argentinas 2007 (POSGAR07; Cimbaro et al. 2009), referred to IGS05 (Johnston et al. 2017), with a precision in the order of 1.5 cm (Instituto Geográfico Nacional, personal communication, 2023). Ellipsoidal coordinates were transformed to GRS80, in accordance with recommendations given by Sánchez et al. (2021). These heights were provided in the tide-free concept.

The official Argentinean DTM, denoted Modelo Digital de Elevaciones de la Argentina, (MDE-Ar v2.1; Instituto Geográfico Nacional 2021) was used to compute RTM effects in the RCR scheme. MDE-Ar has a 30 m resolution and is referred to the same vertical datum as SRVN16. The precision of the model in the Argentinean territory is estimated to be 2 m, while in the Buenos Aires province its precision is 1.6 m (Instituto Geográfico Nacional 2021).

The GGM chosen for this study was XGM2019e (Zingerle et al. 2020). It was selected because previous results showed that XGM2019e up to degree and order

2159 performed optimally in Argentina when compared to other high resolution GGMs (Tocho et al. 2022). The spherical harmonic coefficients were obtained through the International Centre for Global Earth Models (ICGEM; Ince et al. 2019).

The quasigeoid model was computed in the zero-tide concept. Gravity anomalies were computed using MDE-Ar heights which, since they are provided in the mean-tide (MT) concept, were previously converted to zero-tide (ZT) by adding to them the conversion term (eq. (21) of Mäkinen 2021)

$$\Delta H_T = -99.40 + 295.41 \sin^2 \phi + 0.42 \sin^4 \phi \text{ [mm]}, \quad (18)$$

where ϕ is the ellipsoidal latitude.

Vertical datum parameter estimations were performed in the MT concept (Sánchez and Sideris 2017). Height anomalies ζ_{ZT} derived from the quasigeoid model were converted to MT by subtracting from them the conversion term ΔH_T (Eq. 18), and ellipsoidal heights were converted from tide-free (TF) to MT through (eq. (23) of Mäkinen 2021)

$$h_{MT} = h_{TF} + 60.34 - 179.01 \sin^2 \phi - 1.82 \sin^4 \phi \text{ [mm]}. \quad (19)$$

4 Results and Discussion

4.1 Quasigeoid Model Determination

The quasigeoid model was determined according to Sect. 2.3. Free-air gravity anomalies Δg_i were computed at all gravity data points. GGM effects up to degrees 300 (Δg_{GGM}^{300}), 760 (Δg_{GGM}^{760}) and 2159 (Δg_{GGM}^{2159}) were computed using XGM2019e and then subtracted from Δg_i. Then, RTM effects Δg_{RTM} were subtracted. Results show that degree

Table 1 First row: original free-air anomalies at gravity stations. Second to seventh rows: effects computed with XGM2019e up to degree 300, 760 and 2159 and reduced effects. Eighth row: RTM effects. Ninth row: residual gravity anomalies computed with XGM2019e up to degree 760. Units: [mGal]

	Min	Max	Mean	Std. dev.
Δg_i	−17.57	48.38	7.62	10.46
Δg_{GGM}^{300}	−10.80	30.94	7.25	7.73
$\Delta g_i - \Delta g_{GGM}^{300}$	−29.68	27.34	0.37	7.70
Δg_{GGM}^{760}	−15.06	34.75	7.82	9.12
$\Delta g - \Delta g_{GGM}^{760}$	−24.54	22.97	−0.20	5.09
Δg_{GGM}^{2159}	−13.44	44.05	7.99	9.72
$\Delta g_i - \Delta g_{GGM}^{2159}$	−29.18	25.92	−0.37	5.54
Δg_{RTM}	−2.54	3.75	0.57	0.64
Δg_{res}	−26.60	20.33	−0.77	4.99

760 produced the best residual anomalies. The statistics of free air, reduced by GGM and residual anomalies, as well as GGM and RTM effects are presented in Table 1.

The quasigeoid model was computed on a grid with a 1' resolution. The absolute validation was done by comparing the GNSS/Levelling-derived height anomalies

$$\zeta_{h/H^*} = h_{GNSS} - H^* \quad (20)$$

with height anomalies derived from the quasigeoid model, ζ_{mod} (Fig. 3a). ζ_{h/H^*} was also compared with height anomalies derived from XGM2019e up to degree and order 760, denoted ζ_{GGM}. Table 2 shows the results of each comparison. A total of 3 outliers were detected according to a 3σ criterion and excluded from the validation.

The mean value of the differences in both cases can be explained by the fact that both the gravimetric quasigeoid model and the GGM refer to a geopotential surface whose value is $W_0 = U_0^{GRS80}$, as explained in Sect. 2.3, whereas the GNSS/Levelling-derived height anomalies refer to the Argentinean datum.

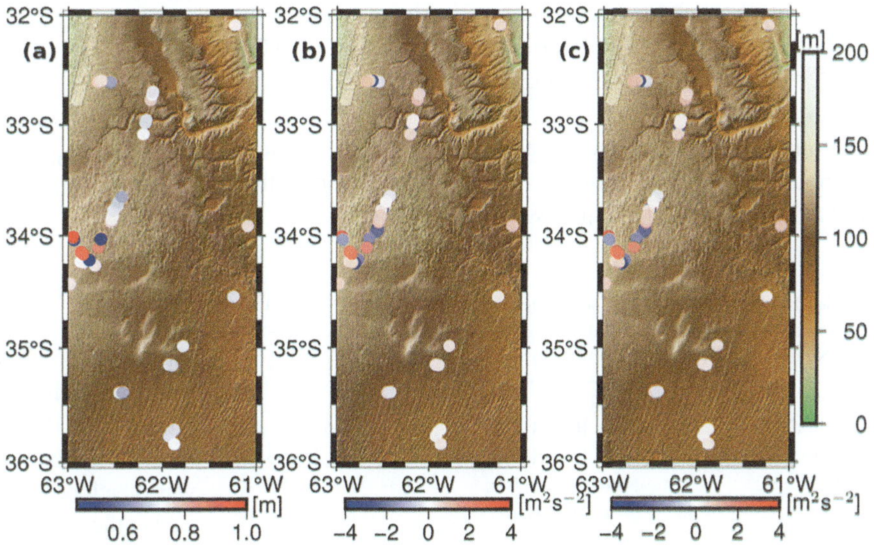

Fig. 3 (a) Difference between GNSS/Levelling-derived and quasigeoid model height anomalies. (b) Residuals from adjustment of method 1. (c) Residuals from adjustment of method 2. Circles: GNSS/Levelling points. Background: Topography from MDE-Ar v2.1

Table 2 Statistics of the difference between GNSS/Levelling-derived and modeled height anomalies (outliers are not considered). Units: [cm]

Difference	Min	Max	Mean	Std. dev.
$\zeta_{h/H^*} - \zeta_{mod}$	48.3	96.3	67.8	13.6
$\zeta_{h/H^*} - \zeta_{GGM}$	31.8	102.5	68.6	16.1

The standard deviation of the differences reflects the combination of the precision of the GNSS/Levelling-derived height anomalies and the quasigeoid model. The model's precision is limited by omission and commission errors (Sansò and Sideris 2013). The standard deviation of the differences decreases by more than 2 cm in the case of the quasigeoid model when compared with the GGM, showing the importance of accounting for local terrestrial gravity data in a quasigeoid model to reduce the omission error. Nevertheless, the accuracy of the quasigeoid can be further improved in various ways, for instance by including a larger and more dense gravity data set. As for the precision of GNSS/Levelling-derived height anomalies, it is mainly limited by the precision of normal heights, which is in the order of 10 cm.

4.2 Vertical Datum Parameter Estimation

The vertical datum parameter estimations are presented in Table 3, and the residuals of each method are depicted in Fig. 3b and c. A total of 55 observation equations were used in each method.

In the first method, the indirect bias term δT_{ind} was neglected from Eq. (9). This term depends on the area of study, the degree of the GGM applied in the quasigeoid modelling and δW_{0i} itself. It is not constant and, therefore, makes the observation equations more difficult to solve. To verify that it can be neglected without degrading the determination of δW_{0i}, δT_{ind} was estimated for the study region using the same Stokes kernel as in the quasigeoid modelling. The indirect bias term was found to be only 0.001 m²s⁻², meaning it can be disregarded from the observation equations.

Both methods agree very well with each other, while in method 2 the standard deviation is slightly lower than method 1. Method 1 not only requires a quasigeoid model for the computation of T_i but also the computation of height anomalies derived from GNSS/Levelling data (Eq. 7). Although the consistency of the permanent tide concept has been considered to agree with the IAG Resolution N° 1 from 2015, all other error sources constrain the quality of the vertical datum parameter (Sánchez 2012). In the second method, the precision of geopotential numbers (obtained with levelling and gravimetry) and the quasigeoid model computed in the region are the main limitations of this approach.

Prior to this study, two estimations of the Argentinean δW_{0i} were performed. The first one was carried out by Tocho and Vergos (2015) (δW_0^{TV} in Table 3) using 542 GNSS/Levelling benchmarks for the previous height system, SRVN71, instead of the current one SRVN16. However, it is worth noting that the datums of SRVN71 and SRVN16 are the same (Instituto Geográfico Nacional 2016) and, consequently, the vertical datum parameter agree very well with those determined in the present study. The estimated error of δW_0^{TV} is smaller than in our estimations, which can be mostly attributed to the number of GNSS/Levelling data points used in each case.

The second estimation was performed by Sánchez and Sideris (2017) (δW_0^{SS} in Table 3). Here, δW_0^{SS} was computed with method 1 for all the Argentinean territory, and it involved an adjustment of 663 observation equations corresponding to 14 vertical networks in South America. In Argentina, T_i was computed with EGM2008 instead of recovered from a regional gravimetric quasigeoid model. Moreover, an iterative approach was applied, where a preliminary δW_{0i} was obtained, and further used to recompute geopotential numbers and gravity anomalies. With the new information, method 1 was applied once again, repeating the computation until the difference between consecutive estimates was below the mm level.

The difference between δW_0^{SS} and our estimates is -7.03 m²s⁻², comparable to $\Delta W_0 = -7.45$ m²s⁻² (Eq. 5). The reasons behind this discrepancy are not clear, and further studies are required to explain it.

The difference between our error estimates and those associated to δW_0^{SS} can be explained by the iterative procedure applied by Sánchez and Sideris (2017), which was not applied in our study, and the greater number of GNSS/Levelling benchmarks used.

Table 3 Estimation results with method 1 and 2 (rows 1 and 2). Estimations done by Tocho and Vergos (2015) (δW_0^{TV}), and by Sánchez and Sideris (2017) (δW_0^{SS}). Units: [m²s⁻²]

Result	δW_0	W_0^{LVD}
Method 1	-0.46 ± 1.78	62636853.88
Method 2	-0.46 ± 1.37	62636853.87
δW_0^{TV}	-0.50 ± 0.14	62636853.90
δW_0^{SS}	6.51 ± 0.49	62636846.89

5 Conclusions

The Argentinean vertical datum parameter was successfully estimated with two methods, which require a quasigeoid model and GNSS/Levelling data. To minimise the omission error in the computation of δW_{0i}, a local quasigeoid model was developed. Both methods demonstrated consistency with each other, while method 2 has a lower standard deviation.

For future studies, it is recommended to apply method 2, as the observation equation derived for method 1 is only approximated theoretically, and it requires for the indirect bias term to be negligible. Results agree with the previous estimation of Tocho and Vergos (2015) but show differences with the one made by Sánchez and Sideris (2017). Reasons for this discrepancy are subject of further study. The accuracy of our estimation is directly related to the quality and quantity of the GNSS/Levelling data, meaning that the establishment of new GNSS/Levelling points is crucial for a more accurate estimation of the Argentinean vertical datum parameter.

A global height system can be realized in Argentina by relating SRVN16 heights to the IHRF using the estimated value of the Argentinean vertical datum parameter.

References

Antokoletz ED (2017) Red gravimétrica de primer orden de la República Argentina. Bachelor's thesis, Universidad Nacional de La Plata. http://sedici.unlp.edu.ar/handle/10915/60950

Cimbaro S, Lauría E, Piñón D (2009) Adopción del nuevo marco de referencia geodésico nacional. Instituto Geográfico Militar, Buenos Aires, Argentina

Drewes H, Kuglitsch FG, Adám J, et al (2016) The geodesist's handbook 2016. J Geodesy 90(10):907–1205. https://doi.org/10.1007/s00190-016-0948-z

Featherstone W, Kirby J (2000) The reduction of aliasing in gravity anomalies and geoid heights using digital terrain data. Geophys J Int 141(1):204–212. https://doi.org/10.1046/j.1365-246X.2000.00082.x

Forsberg R (1984) A study of terrain reductions, density anomalies and geophysical inversion methods in gravity field modelling, vol 5. Ohio State University, Department of Geodetic Science and Surveying

Gerlach C, Rummel R (2013) Global height system unification with GOCE: a simulation study on the indirect bias term in the GBVP approach. J Geodesy 87:57–67. https://doi.org/10.1007/s00190-012-0579-y

Grigoriadis V, Kotsakis C, Tziavos I, et al (2014) Estimation of the reference geopotential value for the local vertical datum of continental Greece using EGM08 and GPS/leveling data. In: Gravity, Geoid and Height Systems: Proceedings of the IAG Symposium GGHS2012, October 9–12, 2012, Venice, Italy. Springer, pp 249–255. https://doi.org/10.1007/978-3-319-10837-7_32

Haagmans R (1993) Fast evaluation of convolution integrals on the sphere using 1D FFT, and a comparison with existing methods for Stokes' integral. Man Geod 18:227–241

Hofmann-Wellenhof B, Moritz H (2006) Physical geodesy. Springer Science & Business Media, New York. https://doi.org/10.1007/978-3-211-33545-1

Ihde J, Sánchez L, Barzaghi R, et al (2017) Definition and proposed realization of the International Height Reference System (IHRS). Surv Geophys 38:549–570. https://doi.org/10.1007/s10712-017-9409-3

Ince ES, Barthelmes F, Reißland S, et al (2019) ICGEM–15 years of successful collection and distribution of global gravitational models, associated services, and future plans. Earth Syst Sci Data 11(2):647–674. https://doi.org/10.5194/essd-11-647-2019

Instituto Geográfico Nacional (2016) Nuevo Sistema Vertical de la República Argentina. Simposio SIRGAS 2016. Tech. rep., Argentina. https://ramsac.ign.gob.ar/posgar07pgweb/documentos/InformeRed_deNivelaciondelaRepublicaArgentina.pdf

Instituto Geográfico Nacional (2021) Modelo Digital de Elevaciones de la República Argentina versión 2.1. Tech. rep., Argentina. https://www.ign.gob.ar/archivos/InformeMDE-Arv2.130m.pdf

Johnston G, Riddell A, Hausler G (2017) The International GNSS Service. Springer International Publishing, Cham, pp 967–982. https://doi.org/10.1007/978-3-319-42928-1_33

Mäkinen J (2021) The permanent tide and the International Height Reference Frame IHRF. J Geodesy 95(9):106. https://doi.org/10.1007/s00190-021-01541-5

Moritz H (2000) Geodetic Reference System 1980. J Geodesy 74(1):128–133. https://doi.org/10.1007/s001900050278

Pavlis NK, Holmes SA, Kenyon SC, et al (2012) The development and evaluation of the Earth Gravitational Model 2008 (EGM2008). J Geophys Res Solid Earth 117(B4). https://doi.org/10.1029/2011JB008916

Petit G, Luzum B, et al (2010) IERS conventions. IERS Tech Note 36(1):2010. https://www.iers.org/IERS/EN/Publications/TechnicalNotes/tn36.html

Rummel R, Teunissen P (1988) Height datum definition, height datum connection and the role of the geodetic boundary value problem. Bull Géodésique 62:477–498. https://doi.org/10.1007/BF02520239

Sánchez L (2012) Towards a vertical datum standardisation under the umbrella of Global Geodetic Observing System. J Geodetic Sci 2(4):325–342. https://doi.org/10.2478/v10156-012-0002-x

Sánchez L, Sideris MG (2017) Vertical datum unification for the International Height Reference System (IHRS). Geophys J Int 209(2):570–586. https://doi.org/10.1093/gji/ggx025

Sánchez L, Čunderlík R, Dayoub N, et al (2016) A conventional value for the geoid reference potential W_0. J Geodesy 90:815–835. https://doi.org/10.1007/s00190-016-0913-x

Sánchez L, Ågren J, Huang J, et al (2021) Strategy for the realisation of the International Height Reference System (IHRS). J Geodesy 95(3):1–33. https://doi.org/10.1007/s00190-021-01481-0

Sansò F, Sideris MG (2013) Geoid determination: theory and methods. Springer Science & Business Media, New York. https://doi.org/10.1007/978-3-540-74700-0

Tocho C, Vergos G (2015) Estimation of the geopotential value W_0 for the local vertical datum of Argentina using EGM2008 and GPS/levelling data W_0^{LVD}. In: IAG 150 Years: Proceedings of the IAG Scientific Assembly in Postdam, Germany, 2013. Springer, New York, pp 271–279. https://doi.org/10.1007/1345_2015_32

Tocho CN, Antokoletz ED, Gómez AR, et al (2022) Analysis of high-resolution global gravity field models for the estimation of International Height Reference System (IHRS) coordinates in Argentina. J Geodetic Sci 12(1):131–140. https://doi.org/10.1515/jogs-2022-0139

Tscherning CC (1992) The GRAVSOFT package for geoid determination. In: Proc 1st IAG Continental Workshop of the Geoid in Europe, Prague, 1992

Wong L, Gore R (1969) Accuracy of geoid heights from modified Stokes kernels. Geophys J Int 18(1):81–91. https://doi.org/10.1111/j.1365-246X.1969.tb00264.x

Zingerle P, Pail R, Gruber T, et al (2020) The combined global gravity field model XGM2019e. J Geodesy 94:1–12. https://doi.org/10.1007/s00190-020-01398-0

Open Access This chapter is licensed under the terms of the Creative Commons Attribution 4.0 International License (http://creativecommons.org/licenses/by/4.0/), which permits use, sharing, adaptation, distribution and reproduction in any medium or format, as long as you give appropriate credit to the original author(s) and the source, provide a link to the Creative Commons license and indicate if changes were made.

The images or other third party material in this chapter are included in the chapter's Creative Commons license, unless indicated otherwise in a credit line to the material. If material is not included in the chapter's Creative Commons license and your intended use is not permitted by statutory regulation or exceeds the permitted use, you will need to obtain permission directly from the copyright holder.

Densification of the IHRF in Denmark, The Faroe Islands, and Greenland

Hergeir Teitsson, Laura Sánchez, and René Forsberg

Abstract

The International Association of Geodesy (IAG) introduced and defined, in 2015, the International Height Reference System (IHRS) as the conventional reference system for the global physical height determination. Following the conventions for the realisation of the IHRS, i.e. the determination of the International Height Reference Frame (IHRF), we utilize the existing GNSS reference stations in Denmark, The Faroe Islands, and Greenland to determine a local densification of the IHRF in these regions. The physical heights of these Danish, Faroese and Greenlandic GNSS reference stations have been transformed from the local Danish, Faroese, and Greenlandic height systems, DVR90, FVR09 and GVR16, respectively, to geopotential numbers and normal heights referring to the IHRF. The offset to the IHRF is found to be −44.0 ± 1.9 cm, −59.8 ± 5.1 cm and − 54.1 ± 11.3 cm for the DVR90, FVR09 and GVR16, respectively. This transformation relies on the existing precise local (quasi-)geoid models. This contribution describes the applied procedures in the IHRF densification and discusses the quality assessment of the results.

Keywords

Global unified vertical reference system · IHRF densification · International Height Reference Frame (IHRF) · World height system

1 Introduction

A globally unified height system is important for consistent, homogeneous, and long-term stable observations in the Earth system, e.g. monitoring sea level change and other changes to climate (Ihde et al. 2017). Most countries use individually constructed height systems, which are local or regional, based on local sea level, and using different types of heights, with discrepancies to adjacent height systems (Sánchez et al. 2021a). In 2015 the International Association of Geodesy (IAG) defined the IHRS as the global standard for physical height determination (IAG Resolution No. 1 (2015), Drewes et al. 2016).

The IHRS definition states that the vertical reference level is the equipotential surface with the conventional geopotential value $W_{0, IHRF} = 62636853.4$ m^2/s^2. The vertical coordinate is the geopotential number $C(P)$, i.e. the difference between the potential $W(P)$ of the Earth's gravity field at a point P and the conventional W_0. The spatial reference of P is given in the International Terrestrial Reference System (ITRS, Petit and Luzum 2010), and the coordinates should be related to the mean tide system.

The realisation of the IHRS is the IHRF and it will be implemented by a well-distributed global core network. National and regional densifications of the IHRF are necessary to provide the most optimal height system unification (Sánchez et al. 2021a) and to provide an easy access to the IHRF. The aim of the article is to show the whole process

H. Teitsson (✉) · R. Forsberg
DTU Space, Technical University of Denmark, Lyngby, Denmark
e-mail: herteit@space.dtu.dk

L. Sánchez
Deutsches Geodätisches Forschungsinstitut (DGFI-TUM), Technical University of Munich, Munich, Germany

of implementing the IHRF densification in the Kingdom of Denmark, i.e. Denmark, the Faroe Islands, and Greenland. From the choice of stations, acquiring coordinates in the International Terrestrial Reference Frame (ITRF, Altamimi et al. 2023) and gravimetric quasigeoids for the countries in the Kingdom of Denmark. Then, the determination of geopotential numbers in the IHRF and how it can differ, depending on the correction of the zero-degree term (ζ_0). An attempt at assessing the accuracy of the results is made by comparing to geopotential numbers from the global gravity models (GGM) EGM2008 (Pavlis et al. 2012, 2013) and XGM2019e (Zingerle et al. 2020), which is the alternative option to using the local/regional gravimetric quasigeoid.

Section 1 introduces the background and purpose of the article. The densification stations and gravimetric quasigeoid models are described in Sect. 2. The method of computing the IHRF geopotential numbers for the densifications is described in Sect. 3, both generally and the differences in computation for the densifications of the three countries in the Kingdom of Denmark. The results are shown and analysed in Sect. 4, followed by a discussion of the results in Sect. 5. The article is concluded in Sect. 6.

2 Data

2.1 Densification Stations

The Danish reference network consists of 14 Continuously Operating Reference Stations (CORS) distributed across Denmark (DK-net), providing Global Navigation Satellite System (GNSS) data. All of the CORS stations are co-located with absolute gravity stations. The Faroe Islands geometric reference network consists of 4 CORS-GNSS stations at locations distributed across The Faroe Islands (FO-net). The Greenlandic geometric reference network consists i.a. of 64 CORS-GNSS stations at 60 locations all around the coast of Greenland, which continuously monitor the land uplift on the island. These are called GNET, and 59 of the stations are included in the densification. The GNSS data from all stations is made available by the Danish Agency of Data Supply and Infrastructure (SDFI 2023) The tide system of the ITRF coordinates is assumed to be the tide-free system, as this is convention of the ITRF (Altamimi et al. 2023). In both Denmark and Greenland some CORS stations are not included, mainly because these were closely located to other CORS stations. In Denmark the CORS stations not co-located with absolute gravity were not included.

The convention of the IHRF refers to ITRF2014, epoch 2021.04. The ITRF coordinates of the stations are transformed to ITRF2014 (Altamimi et al. 2016) from other ITRF realisations, when necessary, with the standard seven-parameter transformation, which involves three shifts of the origin, three rotations and one scale

$$\begin{bmatrix} XS \\ YS \\ ZS \end{bmatrix} = \begin{bmatrix} X \\ Y \\ Z \end{bmatrix} + \begin{bmatrix} Tx \\ Ty \\ Tz \end{bmatrix} + \begin{bmatrix} D & -Rz & Ry \\ Rz & D & -Rx \\ -Ry & Rx & D \end{bmatrix} \begin{bmatrix} X \\ Y \\ Z \end{bmatrix}, \quad (1)$$

where X, Y, Z are the coordinates in e.g. ITRF2020, and XS, YS, ZS are the coordinates in ITRF2014. The transformation parameters Tx, Ty, Tz, Rx, Ry, Rz and D are derived from already published parameters in IERS Technical Notes and Annual Reports (ITRF-IGN 2023). Prior to analysis, positions and corrected using the simple linear model

$$P(t) = P(t_0) + \dot{P}(t - t_0) \quad (2)$$

converting the parameters P in eq. (X) from the reference epoch t_0, to the desired epoch t. The rate parameter $\dot{P}$ is also made available by the IERS.

2.2 Gravimetric Quasigeoid Models

The three countries in the Kingdom of Denmark have three different physical height systems and some details for each can be seen in Table 1.

For the Danish stations the *NKG2015 gravimetric quasigeoid* is used (Ågren et al. 2016), which is publicly available on the webpage of the International Service of the Geoid (Reguzzoni et al. 2021). It refers to the GRS80 ellipsoid (Moritz 2000) and to $W_{0,IHRF}$, but is shifted with a 1-parameter fit to GNSS-levelling height anomalies in the Nordic and Baltic countries ($\Delta h_{zero \to non-tidal}$ from Ekman 1989). Figure 1 shows the location of the DK-net stations on top of a plot of the NKG2015 gravimetric quasigeoid.

The physical height system in The Faroe Islands is based on the hybrid geoid FOGEOID2012, which in turn is based on the *FOGEOID2010 gravimetric quasigeoid* (Forsberg 2010, 2011) fitted to the local GNSS-levelling data existing over the region. The FOGEOID2010 was to computed as a quasigeoid. The quasigeoid model is not available online, but can be obtained from the Faroese Mapping Authority or DTU Space. Figure 2 shows the location of the FO-net stations on top of a plot of the FOGEOID2010 gravimetric quasigeoid.

The physical height system in Greenland is based on the geoid GGEOID16, defining a new geoid-based height system in Greenland, replacing some 78 local mean sea level (MSL) datums for individual towns and settlements. GGEOID16 is based on a gravimetric quasigeoid computation, followed by a geoid-quasigeoid separation correction (Forsberg 2016). The gravimetric quasigeoid can be made available from DTU Space or SDFI on request. The fit to Greenlandic GNSS-levelling in towns and settlements is rather poor, since local

Table 1 Details and specifications for the national height systems and gravimetric quasigeoid models used in the determination of the geopotential numbers in the IHRF

	Denmark	The Faroe Islands	Greenland
Physical height system	DVR90	FVR09	GVR16
Gravimetric quasigeoid model	NKG2015quasigeoid (epoch 2000)	FOGEOID2010	GGEOID16
Resolution of grav. Quasigeoid	1 km	0.5 km	2 km
No. stations	14	4	59
Reference ellipsoid	GRS80	GRS80	GRS80
GGM	GO_CONS_GFC_2_DIR_R5(Bruinsma et al. 2013)	EGM2008	EIGEN-6C4(Förste et al. 2014)
Uplift correction	NKG2016LU to epoch 2021.04(Vestøl et al. 2019)	Not applied	Not applied
1-p transf. Correction	$-(x_1 + \Delta h_{zero \to non-tidal}), x_1 = -0.4874\,m$	Not necessary	Not necessary
Permanent tide system	Zero-tide	Tide-free	Zero-tide
Bias to local GNSS-levelling ζ	−48.00 (after 1-p transf. and uplift correction)	−7.00 cm	−15.60 cm (shift to Nuuk MSL)
RMS with local GNSS-levelling ζ	1.70 cm	5.00 cm	10.00 cm

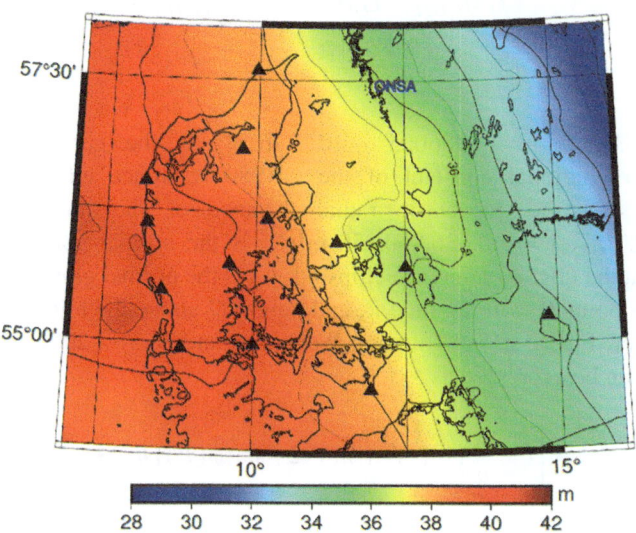

Fig. 1 The figure shows the local gravimetric quasigeoid NKG2015 over Denmark with 1 m contours. The locations of the 14 Danish IHRF densification network stations are shown as black triangles. The location of the IHRF global core station ONSA is indicated by the blue label

MSL systems are in use. Figure 3 shows the location of the GNET stations on a plot of the GGEOID16 gravimetric quasigeoid.

3 Method

3.1 ζ_0-Corrections

The method for computing the gravity potential values at the stations will be outlined below. A detailed description can be found in Sánchez et al. (2021a).

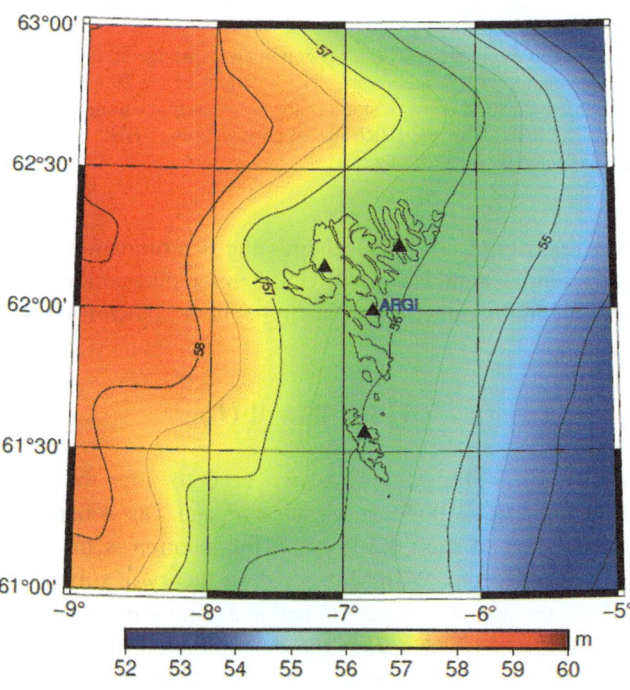

Fig. 2 The figure shows the local gravimetric quasigeoid FOGEOID2010 of The Faroe Islands, with 0.5 m contour lines. The locations of the 4 Faroese IHRF densification network stations are shown as black triangles on the plot. The location of the IHRF global core station ARGI is indicated by the blue label

The conventions of the IHRF prescribe the following conditions:

– The geopotential numbers must only come from the solution of the GBVP, i.e. no influence on the gravity potential values from GNSS-levelling data.
– The reference ellipsoid should be the GRS80.
– The geopotential numbers in IHRF are given in the mean-tide system.

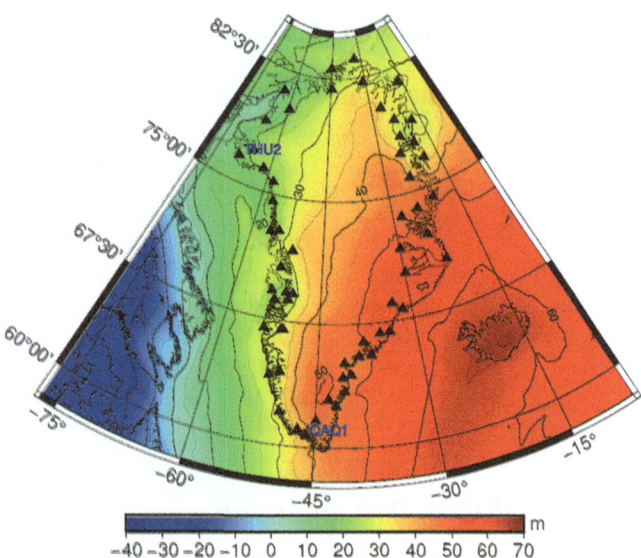

Fig. 3 The figure shows the local gravimetric quasigeoid GGEOID16 of Greenland, with 5 m contours. The locations of the 59 Greenlandic IHRF densification network stations are shown as black triangles on the plot. The location of the IHRF global core stations, THU2 and QAQ1 are indicated by the blue label

The realization of IHRS requires the determination of the geopotential numbers at the GNSS reference stations. The geopotential number C at a station P in the IHRF is defined as

$$C(P) = W_0 - W(P), \quad (3)$$

where $W_0 = W_{0,IHRF}$. Thus, we need to infer the IHRF potential values $W(P)$ at a station P and this depends on the solution of the GBVP, i.e. if the solution is the geoid or the quasigeoid. With Molodensky's theory of solving the GBVP the potential is determined at a point P on the surface of the Earth, which makes the determination of $W(P)$ and subsequently $C(P)$, rather straightforward. We work with the following solution of the GBVP (cf. Eq. 8–57, Hofmann-Wellenhof and Moritz 2006)

$$\zeta_{gravimetric} = \zeta_0 + \frac{R}{4\pi\gamma_0} \iint_\sigma \left[\Delta g - \frac{\partial \Delta g}{\partial h}(h - h_p) \right] S(\psi) \, d\sigma, \quad (4)$$

where ζ_0 is the zero-degree term, R is the radius of the earth, γ_0 is the normal gravity at the ellipsoid, Δg is the free-air gravity anomaly, h is the ellipsoidal height, h_p is the ellipsoidal height at a computation point P, and $S(\psi)$ is Stokes function. ζ_0 can be regarded as the distance between geoid with the geopotential W_0 and the reference ellipsoid with the normal potential U_0. It can be computed by the generalization of Bruns' formula (cf. eq. (2–182), Heiskanen and Moritz 1967)

$$\zeta_0 = \frac{GM_{GGM} - GM_0}{r\gamma} - \frac{W_0 - U_0}{\gamma}, \quad (5)$$

where GM_{GGM} is the geocentric gravitational constant of the GGM, GM_0 is the geocentric gravitational constant of the reference ellipsoid (normal gravity field), r is the radial distance, γ is normal gravity, W_0 is the gravity potential of the geoid, and U_0 is the normal potential at the surface of the reference ellipsoid. Notice that in cases where $W_0 = U_0$ the second term vanishes. Thus, Eq. (5) is used to correct the local gravimetric quasigeoid, based on the values of the properties of the GGM, to the conventional reference value $W_{0,IHRF}$ by (cf. Eq. (14) in Sánchez et al. 2021a)

$$\zeta_{0,IHRF} = \frac{GM_{GGM} - GM_{GRS80}}{r_P \gamma_Q} - \frac{W_{0,IHRF} - U_{0,GRS80}}{\gamma_Q}. \quad (6)$$

$GM_{GRS80} = 3.986005 \cdot 10^{14} \text{ m}^3\text{s}^{-2}$ is the geocentric gravitational constant of the GRS80 reference ellipsoid, r_P is the geocentric radial distance of the point P, $U_{0,GRS80} = 62{,}636{,}860.850 \text{ m}^2\text{s}^{-2}$ is the normal potential at the GRS80 reference ellipsoid. γ_Q is the normal gravity at a point P, with latitude φ, and height Q above the ellipsoid (cf. Eq. (2-215), Hofmann-Wellenhof and Moritz 2006). It can be expressed as

$$\gamma_Q = \gamma_0 \left[1 - \frac{2}{a}\left(1 + f + m - 2f\sin^2\varphi(P)\right) h_Q + \frac{3}{a^2} h_Q^2 \right], \quad (7)$$

where $h_Q = h(P) - \zeta(P)$, i.e. the ellipsoidal height h at point P minus the interpolated height anomaly ζ at point P, γ_0 is the normal gravity at the ellipsoid for the same latitude,

$$\gamma_0 = 9.780327 \\ \times \left(1 + 0.0053024\sin^2\varphi - 0.0000058\sin^4\varphi\right) \, [ms^{-2}]. \quad (8)$$

The GRS80 parameters (cf. Moritz 2000) are the semi-major axis, $a = 6378137$ m, the flattening, $f = 0.0035281068118$, and the unitless geodetic parameter $m = 0.00344978600308$.

The GRAVSOFT (Tscherning et al. 1992) program GEOCOL17, which is used in the computation of the FOGEOID10 and GGEOID16, uses $U_{0,GRS80}$ as reference ellipsoid and gravity potential on the geoid, i.e. $W_0 = U_0$. Thus a ζ_0 is already part of the mentioned gravimetric quasigeoids (and the same is the case for the GGMs which are used for comparison). This ζ_0 is then converted to the IHRF reference value by

$$\zeta_{0,IHRF} = -\frac{W_{0,IHRF} - U_{0,GRS80}}{\gamma_Q} \qquad (9)$$

Having applied this correction, we obtain correct $\zeta(P)$ and we can compute the provisional gravity potential value at P (cf. Eqs. (8) and (9) in Sánchez et al. 2021b):

$$W_{prov}(P) = W_0 - (h(P) - \zeta(P))\overline{\gamma}_{QQ_0}, \qquad (10)$$

Where

$$\overline{\gamma}_{QQ_0} = \gamma_0 \left(1 - \frac{1}{a}\left(1 + f + m - 2f\sin^2\varphi(P)\right)h_Q\right), \qquad (11)$$

is the mean normal gravity between the point Q on the telluroid and the reference ellipsoid.

$$\overline{\gamma}_{QQ_0} = \gamma_0 \left(1 - \frac{1}{a}\left(1 + f + m - 2f\sin^2\varphi(P)\right)h_Q\right), \qquad (12)$$

is the mean normal gravity between the point Q on the telluroid and the reference ellipsoid.

It is recommended to compute the geopotential values in the tide-free or zero-tide system and then afterwards transform into a mean-tide system, which is the convention of the IHRF (Sánchez et al. 2021a).

3.2 Permanent Tide Transformations

The IHRS conventions state that the all parameters in the IHRF should be in the mean-tide system. The reason is to support the geodetic monitoring of geophysical phenomena governed by fluids, e.g. oceanographic and hydrographic, which are best represented by a mean-tide system.

Both zero-tide and tide-free systems are being used for Earth's gravity field and crustal modelling, albeit that the IAG Resolution No. 16, 1983 (Tscherning 1984) states the zero-tide system should be used. There are two parts to consider in the permanent tide system transformation: The solution of the GBVP and the ITRF coordinates.

If the solution of the GBVP is in the zero-tide system a correction W_{T0} is necessary to transform $W_{prov}(P)$ to the mean-tide system. In that case we need the temporal average W_{T0} of the tide-generating potential. Written as function of geodetic latitude, and using that the geoid deviates from the GRS80 ellipsoid by a maximum of 100 m we have (Ihde et al. 2008):

$$W_{T0} = 0.9722 - 2.8841\sin^2\varphi - 0.0195\sin^4\varphi \ [m^2s^{-2}]. \qquad (13)$$

Geopotential numbers in the mean-tide system are defined as (cf. Eq. (19) Sánchez et al. 2021a)

$$C_{MT} = W_0 - (W_{ZT} + W_{T0}), \qquad (14)$$

where $W_{ZT} = W_{prov}$. If the solution of the GBVP is in the tide-free system, we need an additional correction, to first transform from the tide-free system to the zero-tide system, and then transform to the mean-tide system. This can be done by adding the correction (cf. Eq. (38) Mäkinen 2021):

$$\Delta \overline{W}^{GGM}(\varphi, h) = k_{20}\left(1 - \frac{3h}{a}\right)$$
$$\times \left(0.9722 - 2.8673\sin^2\varphi - 0.0690\sin^4\varphi\right) \ [m^2s^{-2}], \qquad (15)$$

where $k_{20} = 0.30190$ is the conventional Love number.

ITRF position coordinates are conventionally given in the tide-free system and it is necessary to restore the effects on the potential removed with the permanent tide. Another tide system related correction to W_{prov} must be added (cf. eq. (26) Mäkinen 2021):

$$\Delta W^{ITRF}$$
$$\approx -0.5901 + 1.7475\sin^2\varphi + 0.0273\sin^4\varphi \ [m^2s^{-2}]. \qquad (16)$$

In the case of both the solution of the GBVP and the ITRF coordinates being in the tide-free system we get

$$W_{ZT} = W_{prov} + \Delta\overline{W}^{GGM} + \Delta W^{ITRF} \qquad (17)$$

and then use Eq. (14) to compute the geopotential number $C(P)$. With the geopotential number we can determine the physical height, e.g. the normal height, H^*, of the station in the IHRF with the relation

$$H^* = \frac{C}{\overline{\gamma}_{QQ_0}} \qquad (18)$$

3.3 Determining Geopotential Numbers at the IHRF Densifications in Denmark, The Faroe Islands and Greenland

The computation of the geopotential numbers at the IHRF densifications for the three countries in the Kingdom of Denmark follows a similar process, but with some small differences between the three cases. The general scheme can be seen in Fig. 4, where the different steps are listed. We recommend this workflow in determining the geopotential numbers. Some of the first steps regarding the stations and the quasigeoid can be done simultaneously, but others

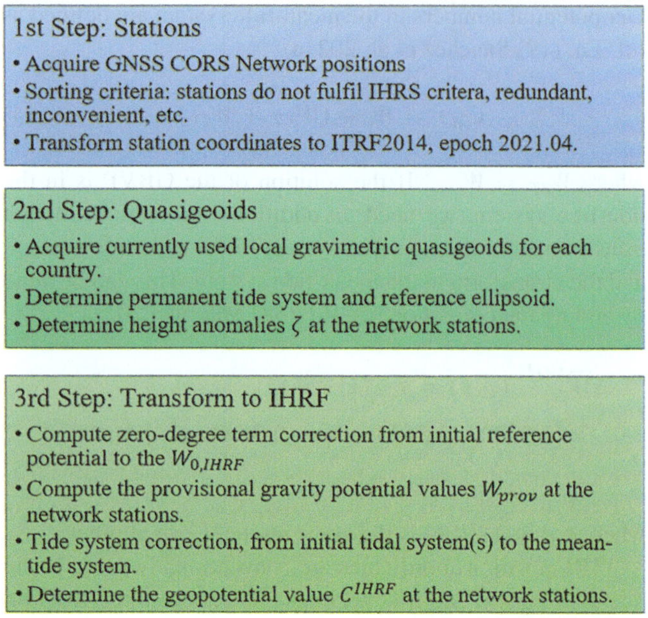

1st Step: Stations
- Acquire GNSS CORS Network positions
- Sorting criteria: stations do not fulfil IHRS critera, redundant, inconvenient, etc.
- Transform station coordinates to ITRF2014, epoch 2021.04.

2nd Step: Quasigeoids
- Acquire currently used local gravimetric quasigeoids for each country.
- Determine permanent tide system and reference ellipsoid.
- Determine height anomalies ζ at the network stations.

3rd Step: Transform to IHRF
- Compute zero-degree term correction from initial reference potential to the $W_{0,IHRF}$
- Compute the provisional gravity potential values W_{prov} at the network stations.
- Tide system correction, from initial tidal system(s) to the mean-tide system.
- Determine the geopotential value C^{IHRF} at the network stations.

Fig. 4 General working scheme to determine geopotential numbers, going from a national height system to the International Height Reference Frame. The steps in the scheme are a recommended workflow

require previous steps to be completed. To determine the geopotential numbers the position of the densifications in the ITRF2014 epoch 2021.04 was determined. Next, the height anomaly at each densification was determined by interpolation at the location. Table 3 shows the different corrections and transformations applied in the step "Transform to IHRF" in Fig. 4, for the three countries in the Kingdom of Denmark, along with an average of the total correction and a standard deviation. The actual correction at each station depends on the height and latitude of the station.

4 Results and Analysis

The full results of the densification of the IHRF for the 14 stations in Denmark, the 4 stations in The Faroe Islands and the 59 stations in Greenland are listed in the tables in the Appendix. It includes the station name, the ITRF2014, epoch 2021.04, position in ellipsoidal coordinates, the height anomalies interpolated from the local gravimetric quasigeoid, and the IHRF normal height, geopotential number and potential, inferred from the station coordinates and the height anomalies.

4.1 Offset from Local Height Systems to IHRF

The mean offset between the national height systems and the normal heights in the IHRF, is determined as the mean of the difference between the normal heights in the two systems. The normal heights H^* in the national systems are computed as the difference between the local height anomaly ζ from the gravimetric quasigeoid models and the ellipsoidal height h.

$$H^* = h - \zeta \qquad (18)$$

The normal heights in the IHRF are computed according to Eq. (18). Table 2 shows the results from computing the differences between the IHRF geopotential numbers and the local height systems in Denmark, The Faroe Islands and Greenland, respectively, in centimetres. The first row shows the difference to between the gravimetric quasigeoid used for computing the IHRF coordinates and the local height system, the second row shows the mean of the transformation correction from the local heights to IHRF heights, and the third row shows the final mean difference between the local height systems and the IHRF. These biases are the offsets, with corresponding uncertainties between the current local physical height systems in the countries and the IHRF, and produce the link, from said height systems, to all height systems connected to the IHRF. The results in the table indicate the differences in offsets between local height systems and the IHRF. This offset is very much dependent on the location and on the definition of the local height system (Table 3).

4.2 Comparison to EGM2008 and XGM2019

For the sake of comparison and validation of the results of the IHRF densification the height anomalies, ζ, at the DK-net, FO-net and GNET stations were computed with the global geopotential models of EGM2008 to d/o 2,160 and XGM2019e to d/o 719. The geopotential numbers in the IHRF were computed following the same scheme in the method section, using Eq. (9) to compute the ζ_0-correction, accounting for the different permanent tide-systems.

Finally, H^* at the densifications is computed from the geopotential number in the IHRF and subtracted from the corresponding H^* from the local gravimetric quasigeoids. Statistics of the comparison in the three countries can be seen in Table 4, which shows the mean, the standard deviation and the extrema. Overall the average differences are within ± 10 cm, but the standard deviations also show that there is a big variation in the differences, especially in Greenland. For the DK-net the GGMs agree well with the local gravimetric quasigeoid. The statistics of the differences, are quite close for the two cases. The standard deviation shows that there is a reasonably variation across the densifications. This small difference can be due to the topography in the region being quite flat, with high quality gravity data coverage.

Table 2 The biases, with corresponding uncertainties, between the local height systems in Denmark, The Faroe Islands, and Greenland, the gravimetric quasigeoids, and the IHRF conventional W_0. The mean of the full transformation to IHRF is also included. All units are in centimetres

Height system	DVR90	FVR09	GVR16
Bias to gravimetric quasigeoid	-48.00 ± 1.70 cm	-7.00 ± 5.00 cm	15.60 ± 10.00 cm
Mean of full IHRF transformation correction	-3.98 ± 0.15 cm	-66.77 ± 0.09 cm	-69.70 ± 0.71 cm
Bias from height system to IHRF	-44.20 ± 1.71 cm	-59.77 ± 5.00 cm	-54.10 ± 10.03 cm

Table 3 Correction to ζ_0, tide system transformation and average of the total correction, of the corresponding normal height, for the three different countries in the Kingdom of Denmark

	NKG2015quasigeoid	FOGEOID2010	GGEOID16
ζ_0 **correction**	Not necessary	$-\frac{W_{0,IHRF} - U_{0,GRS80}}{\gamma_Q}$	$-\frac{W_{0,IHRF} - U_{0,GRS80}}{\gamma_Q}$
Tide system transformation	$\Delta W^{ITRF} + W_{T0}$	$\Delta \overline{W}^{GGM} + \Delta W^{ITRF} + W_{T0}$	$\Delta W^{ITRF} + W_{T0}$
Average of total correction, in H^*	-3.98 ± 0.15 cm	66.80 ± 0.09 cm	69.65 ± 0.71 cm

Table 4 The statistics of the difference in normal height, inferred from C^{IHRF} from the Nordic-Baltic NKG2015 gravimetric quasigeoid, the Faroese FOGEOID2010, and the Greenlandic GGEOID16, to the corresponding normal heights from the global geopotential models EGM2008 and XGM2019, respectively

	Mean [cm]	St.dev. [cm]	Minimum [cm]	Maximum [cm]
NKG2015quasigeoid				
ΔEGM2008	-2.13	6.29	-11.80	8.90
ΔXGM2019	-1.82	6.28	-10.40	6.30
FOGEOID2010				
ΔEGM2008	-5.30	2.40	-7.60	-3.00
ΔXGM2019	-9.30	4.90	-14.20	-5.00
GGEOID16				
ΔEGM2008	7.40	28.40	-52.80	95.20
ΔXGM2019	3.10	25.10	-37.70	87.50

In the Faroe Islands the differences are ranging from -5.3 cm to -9.3 cm for FOGEOID2010, with standard deviations below 5 cm, though it should be noted that the statistical basis in the comparison is weak, with only four stations being compared. EGM2008 was used in the determination of the FOGEOID2010, and so it could be expected to give very similar results.

For the GNET stations the mean difference is 7.4 cm and 3.1 cm for the EGM2008 and XGM2019, respectively. The spread of the differences is quite a lot larger than for other countries, up to 28.4 cm. The differences are smaller for XGM2019 than for EGM2008, which could be due to the inclusion of more satellite data in the EIGEN-6C4, which was used in the determination for the gravimetric quasigeoid. The large differences could have been expected, because Greenland is the largest and most rugged country of the three.

5 Discussion

5.1 Estimated Accuracy of the IHRF Geopotential Numbers

The result of the comparison to the GGMs is an attempt at determining the accuracy of the densification networks. All mean differences are within ± 10 cm. The largest difference in Denmark and Greenland is to the EGM2008, while the difference to XGM2019 is largest in the Faroe Islands. The XGM2019 includes a substantial amount of satellite gravity data that was not available for the EGM2008. The GGMs used in NKG2015 and GGEOID16 also include much of the same satellite gravity data, e.g. data from the GOCE mission. Apart from that they should be based on largely the same terrestrial data for the three countries, as XGM2019 up to d/o 719 is based on 15′ gravity anomalies from NGA, which likely also have gone into the making of the EGM2008.

The question is then how accurate the inferred IHRF geopotential numbers are. The gravimetric quasigeoids are all modelled with the goal of optimising for respective regions that they cover. All three regions are to a large extent well covered by gravity data, and the comparison of the gravimetric quasigeoids to local GNSS-levelling data is mostly good. It would thus be expected that the geopotential numbers and physical heights inferred from the local gravimetric quasigeoids would have the advantage in accuracy over the corresponding results from the EGM2008 and XGM2019, and a careful estimate would be an accuracy on the same order as the local gravimetric quasigeoids used in the inference of the geopotential numbers. And this is also supported by the results of the comparison to EGM2008 and XGM2019.

Possibly the most delicate part of determining the geopotential numbers in IHRF is to figure out the correct way of determining the ζ_0 term or the correction to apply. Already existing national/regional geoids or quasigeoids are modelled and determined by different experts. Since there is not just a single solving the GBVP these models are constructed in different systems, with different references and conventions. Thus, some of the models might have a ζ_0 referring to one reference, while another model refers to different one. If one then strictly follows the method in Sánchez et al. (2021a) or (2021b) and determines the ζ_0 by Eq. (6) the inferred geopotential numbers will refer to a wrong W_0. In the current three cases we have exactly this problem that they cannot be treated in the same way. The ζ_0 of the NKG2015 gravimetric quasigeoid is already referring to $W_{0, IHRF}$. On the other hand, the FOGEOID10 and the GGEOID16 have a ζ_0 referring to the $U_{0, GRS80}$, and thus the ζ_0 has to be determined by Eq. (9). It is therefore essential, in the inference of the geopotential numbers in the IHRF, to be well informed on local geoid/quasigeoid models which are used, as it is otherwise easy to get erroneous results.

5.2 Outlook

The first IHRF solution is established, and these densifications contribute to the local establishments in Denmark, The Faroe Islands and Greenland. The densifications do not have the overall accuracy of the stationary coordinates (± 0.003 m) as targeted by the IHRS working group (Sánchez et al. 2021a), since the local gravimetric quasigeoid models don't have this accuracy. Rather the accuracy of the gravimetric quasigeoids is around an order of magnitude higher. However, this is a good target to work towards with future updates of the IHRF network and densifications.

With regards to the densifications in this study, there will certainly be possibilities of improvements in the near future. With the finalization of the FAMOS project and the release of the Baltic Sea Chart Datum 2000 (Schwabe et al. 2020, 2023; Liebsch et al. 2023), there will be a new and more accurate gravimetric quasigeoid for the Nordic-Baltic region, which in turn should result in more accurate stationary coordinates for the densifications in Denmark. Likewise, a new gravimetric quasigeoid of Greenland is under development at DTU Space. It will include new airborne gravity data, as well as updated topographic data, an updated ice thickness model, etc., which should result in an improved accuracy for the core IHRF network stations and the densifications in the region as well. As for the Faroe Islands, there is certainly possibilities of improvement. The FOGEOID2010 dates back to 2010 and more than a decade's worth of satellite gravity data is collected since then. Thus, a new and improved gravimetric quasigeoid for the Faroe Islands would likely result in an improved accuracy to the IHRF densifications in the region.

6 Conclusion

The IHRF densification is implemented for stations in Denmark, The Faroe Islands and Greenland, in accordance to the establishment of the first version of the IHRF. More specifically, IHRF coordinates are inferred for 14 CORS-GNSS stations in Denmark, 4 CORS-GNSS stations in The Faroe Islands and for 59 CORS-GNSS stations in Greenland, geographically distributed around the respective countries. Differences in the computation of the gravimetric quasigeoids, for the three countries, were regarded when inferring the geopotential numbers, specifically the correction of the ζ_0 term and the transformation of from initial tide system to the mean tide system.

Offsets from the current height systems (DVR90 in Denmark, FVR09 in The Faroe Islands, and GVR16 in Greenland) to the IHRF are on average -44.0 ± 1.9 cm, -59.8 ± 5.1 cm, and -54.1 ± 11.3 cm for the DVR90, FVR09 and GVR16, respectively.

Comparing to the global geopotential models EGM2008 and XGM2019 gives mean differences of up ± 10 cm. This indicates that the results agree with the global models, as well as there being a gain in accuracy from using the local gravimetric quasigeoid models.

Finally, the IHRF densification in Denmark, The Faroe Islands and Greenland provide the ability to link the local height systems to a global height system and to height systems in the neighbouring countries in Europe and North America.

Appendix: IHRF Coordinates of the Densifications

Tables 5, 6 and 7

Table 5 The table shows the result of the IHRF densification in Denmark, using the gravimetric quasigeoid NKG2015quasigeoid

Station	ITRF2014, epoch 2021.04			NKG2015quasigeoid, zero-tide	IHRF		
	Latitude, φ [°]	Longitude, λ [°]	Ellipsoidal height, h [m]	Height anomaly, ζ [m]	Normal height, H* [m]	Geopotential number, C [m²s⁻²]	Potential, W [m²s⁻²]
BUDP	55.73902	12.50003	94.430	36.531	57.939	569.258	62636284.142
ESBC	55.49357	8.45683	59.500	41.237	18.302	179.828	62636673.572
FER5	56.52302	8.11828	67.510	40.739	26.812	263.408	62636589.992
FYHA	54.99364	9.98627	46.300	40.551	5.787	56.866	62636796.534
FYNO	55.33527	10.74251	50.660	39.336	11.363	111.649	62636741.751
GESR	54.57443	11.92292	44.600	38.544	6.093	59.878	62636793.522
GREJ	55.75866	9.57119	136.220	40.373	95.887	942.103	62635911.297
HABY	55.97184	11.35531	62.180	37.623	24.597	241.664	62636611.736
HIRS	57.59110	9.96755	50.590	38.766	11.867	116.568	62636736.832
MOJN	54.94431	8.80538	56.990	40.888	16.140	158.596	62636694.804
RIKO	56.12398	8.17248	48.140	41.158	7.022	68.993	62636784.407
SKEJ	56.18759	10.17984	110.860	39.432	71.469	702.151	62636151.249
SUL5	56.84166	9.74213	120.050	39.004	81.088	796.587	62636056.813
TEJH	55.24842	14.83932	41.560	34.915	6.684	65.673	62636787.727

Table 6 The table shows the result of the IHRF densification The Faroe Islands, using the gravimetric quasigeoid FOGEOID2010

Station	ITRF2014, epoch 2021.04			FOGEOID10, tide-free	IHRF		
	Latitude, φ [°]	Longitude, λ [°]	Ellipsoidal height, h [m]	Height anomaly, ζ [m]	Normal height, H* [m]	Geopotential number, C [m²s⁻²]	Potential, W [m²s⁻²]
ARGI	61.99737	−6.78351	110.320	56.287	53.365	523.828	62636329.572
KLAV	62.22636	−6.58678	68.350	56.289	11.394	111.836	62636741.564
TVOR	61.56392	−6.84591	90.900	56.006	34.225	335.973	62636517.427
VEST	62.15359	−7.14957	67.940	56.806	10.467	102.736	62636750.664

Table 7 The table shows the result of the IHRF densification in Greenland, using the gravimetric quasigeoid GGEOID16

Station	ITRF2014, epoch 2021.04		Ellipsoidal height, h [m]	GGEOID16, zero-tide	IHRF		
	Latitude, φ [°]	Longitude, λ [°]		Height anomaly, ζ [m]	Normal height, H^* [m]	Geopotential number, C [m^{2}s^{-2}]	Potential, W [m^{2}s^{-2}]
AASI	68.71932	−52.79336	56.740	22.985	33.056	324.086	62636529.314
ASKY	75.72613	−58.25734	687.480	21.401	665.388	6515.585	62630337.815
BLAS	79.53861	−22.97474	484.260	29.755	453.817	4441.475	62632411.925
DANE	74.31195	−20.19983	177.250	48.006	128.552	1259.082	62635594.318
DGJG	71.78654	−29.85020	1494.300	51.275	1442.330	14132.817	62622720.583
DKSG	76.35162	−61.67768	609.880	20.789	588.400	5761.160	62631092.240
DMHN	76.77108	−18.65568	55.690	34.676	20.324	198.982	62636654.418
GMMA	77.80942	−19.65213	521.530	30.908	489.932	4796.026	62632057.374
GROK	78.44270	−22.90376	1046.410	29.944	1015.777	9942.747	62626910.653
HEL2	66.40116	−38.21571	425.460	48.687	376.071	3688.588	62633164.812
HJOR	63.41821	−41.14787	762.940	45.960	716.274	7029.164	62629824.236
HMBG	73.67598	−28.12908	1322.760	46.476	1275.591	12494.939	62624358.461
HRDG	81.87984	−44.51735	718.690	24.859	693.144	6781.987	62630071.413
ILUL	69.24042	−51.06075	55.340	24.119	30.523	299.220	62636554.180
ISOR	65.54665	−38.97492	84.090	45.542	37.845	371.252	62636482.148
JGBL	82.20875	−31.00422	753.580	30.491	722.402	7068.034	62629785.366
JWLF	83.11165	−45.11983	112.950	24.527	87.737	858.352	62635995.048
KAGA	69.22230	−49.81463	149.890	27.297	121.895	1194.951	62635658.449
KAGZ	79.13196	−65.85296	86.460	13.786	71.986	704.553	62636148.847
KAPI	64.43235	−50.27121	103.820	30.351	72.765	713.947	62636139.453
KBUG	65.14369	−41.15756	290.960	45.071	245.186	2405.386	62634448.014
KELY	66.98742	−50.94484	230.280	30.803	198.776	1949.428	62634903.972
KMJP	83.64324	−33.37708	85.020	26.960	57.374	561.280	62636292.120
KMOR	81.25271	−63.52739	203.880	11.166	192.027	1878.984	62634974.416
KSNB	66.86328	−35.57633	1721.200	54.971	1665.528	16334.482	62620518.918

KUAQ	68.58700	−33.05276	865.310	53.876	810.735	7948.666	62628904.734
KULL	74.58063	−57.22707	94.060	21.424	71.944	704.614	62636148.786
KULU	65.57934	−37.14937	67.890	49.958	17.229	169.013	62636684.387
LBIB	75.89381	−23.85293	1483.610	40.537	1442.382	14123.681	62622729.719
LEFN	80.45668	−26.29344	691.870	33.033	658.149	6440.564	62630412.836
LYNS	64.43048	−40.19805	173.870	44.409	128.757	1263.325	62635590.075
MARG	77.18704	−65.69463	670.500	19.797	650.013	6363.641	62630489.759
MIK2	68.14029	−31.45182	815.980	56.390	758.890	7440.979	62629412.421
MSVG	72.24082	−23.91285	88.490	51.442	36.353	356.185	62636497.215
NNVN	61.63188	−44.90107	2134.240	45.601	2087.930	20496.289	62616357.111
NORD	81.60015	−16.65545	69.350	28.669	39.994	391.328	62636462.072
NRSK	79.15503	−17.72543	348.010	29.366	317.955	3111.959	62633741.441
NUUK	64.18355	−51.73117	109.350	27.717	80.928	794.081	62636059.319
PAMI	62.01157	−49.67096	99.370	31.093	67.569	663.251	62636190.149
PLPK	66.89773	−34.03347	122.300	55.970	65.629	643.645	62636209.755
QAAR	70.74041	−52.68837	52.760	26.852	25.212	247.086	62636606.314
QAQ1	60.71527	−46.04776	110.970	36.845	73.415	720.790	62636132.610
QEQE	69.25263	−53.52233	49.180	23.194	25.288	247.901	62636605.499
RINK	71.84850	−50.99398	1337.580	32.236	1304.649	12783.596	62624069.804
SCBY	80.26013	−59.59364	543.900	18.172	525.040	5138.091	62631715.309
SCOR	70.48534	−21.95034	128.930	56.248	71.986	705.521	62636147.879
SENU	61.06958	−47.14133	666.890	35.487	630.693	6191.814	62630661.586
SISI	66.93431	−53.67287	113.200	26.820	85.679	840.276	62636013.124
SRMP	72.91068	−54.39372	370.960	24.583	345.683	3386.546	62633466.854
STNO	81.60015	−16.65545	69.350	28.669	39.994	391.328	62636462.072
THU2	76.53705	−68.82506	35.830	15.907	19.233	188.305	62636665.095
TIMM	62.53554	−42.28616	313.800	44.803	268.290	2633.271	62634220.129
TIN1	68.80124	−50.46319	537.830	26.895	510.236	5002.297	62631851.103
TREO	64.27707	−41.37509	122.050	44.971	76.374	749.386	62636104.014
UPVK	72.78828	−56.12800	164.720	23.125	140.901	1380.391	62635473.009
UTMG	62.92721	−43.30642	1471.100	48.425	1421.968	13955.716	62622897.684
VFDG	70.29993	−29.81765	1293.670	53.997	1238.976	12143.461	62624709.939
WTHG	73.95520	−24.30893	1109.590	47.965	1060.932	10391.782	62626461.618
YMER	77.43289	−24.32631	1069.980	34.390	1034.900	10131.341	62626722.059

References

Ågren J, Strykowski G, Bilker-Koivula M, Omang O, Märdla S, Forsberg R, Ellmann A, Oja T, Liepins I, Parseliunas E, Kaminskis J, Sjöberg L, Valsson G (2016) The NKG2015 gravimetric geoid model for the Nordic-Baltic region. https://doi.org/10.13140/RG.2.2.20765.20969

Altamimi Z, Rebischung P, Métivier L, Collilieux X (2016) ITRF2014: a new release of the international terrestrial reference frame modeling nonlinear station motions. J Geophys Res Solid Earth 121:6109–6131. https://doi.org/10.1002/2016JB013098

Altamimi Z, Rebischung P, Collilieux X, Métivier L, Chanard K (2023) ITRF2020: an augmented reference frame refining the modeling of nonlinear station motions. J Geod 97:47. https://doi.org/10.1007/s00190-023-01738-w

Bruinsma SL, Forste C, Abrikosov O, Marty JC, Rio MH, Mulet S, Bonvalot S (2013) The new ESA satellite-only gravity field model via the direct approach. Geophys Res Lett 40(14):3607–3612. https://doi.org/10.1002/grl.50716

Drewes H, Kuglitsch F, Adam J, Rózsa S (2016) The Geodesist's handbook 2016. J Geod 90(10):981–982. https://doi.org/10.1007/s00190-016-0948-z

Ekman M (1989) Impacts of geodynamic phenomena on systems for height and gravity. Bull Géod 63:281–296. https://doi.org/10.1007/BF02520477

Forsberg R (2010) Ny gravimetrisk geoide for Færøerne. Report of National Space Institute at the Technical University of Denmark, Copenhagen, Denmark (in Danish)

Forsberg R (2011) Endelig tilpasning af Færø geoiden til GPS og nivellement/vandstand: FOGEOID2011. Report of National Space Institute at the Technical University of Denmark, Copenhagen, Denmark (in Danish)

Forsberg R (2016) GGEOID16 - Opdateret geoide for Grønland - tilpasset havniveau i Nuuk. Report of National Space Institute at the Technical University of Denmark, Copenhagen, Denmark (in Danish)

Förste C, Bruinsma SL, Abrikosov O, Lemoine JM, Marty JC, Flechtner F, Balmino G, Barthelmes F, Biancale R (2014) EIGEN-6C4 The latest combined global gravity field model including GOCE data up to degree and order 2190 of GFZ Potsdam and GRGS Toulouse; GFZ Data Services, https://doi.org/10.5880/ICGEM.2015.1

Heiskanen WA, Moritz H (1967) Physical geodesy. W. H. Freeman Co., San Francisco

Hofmann-Wellenhof, B. and Moritz, H., 2006. Physical geodesy—Second Edition. Springer, Wien

Ihde J, Mäkinen J, Sacher M (2008) Conventions for the definition and realization of a European Vertical Reference System (EVRS)—EVRS Conventions 2007. IAG Sub-Commission 1.3a EUREF. https://evrs.bkg.bund.de/Subsites/EVRS/EN/References/Papers/papers.html. Last accessed 21 Mar 2024

Ihde J, Sánchez L, Barzaghi R, Drewes H, Förste C, Gruber T, Liebsch G, Marti U, Pail R, Sideris M (2017) Definition and proposed realization of the international height reference system (IHRS). Surv Geophys 38:549–570. https://doi.org/10.1007/s10712-017-9409-3

ITRF-IGN (2023) *Transformation parameters from ITRF2020 to past ITRFs*. Accessed on 16 Nov 2023. https://itrf.ign.fr/docs/solutions/itrf2020/Transfo-ITRF2020_TRFs.txt

Liebsch G, Schwabe J, Varbla S, Ågren J, Teitsson H, Ellmann A, Forsberg R, Strykowski G, Bilker-Koivula M, Liepiņš I, Parseliūnas E, Keller K, Vestøl O, Omang O, Kaminskis J, Wilde-Piórko M, Pyrchla K, Olsson P, Förste C, Ince ES, Somla J, Westfeld P, Hammarklint T (2023) Release note for the BSCD2000 height transformation grid. Int Hydrogr Rev 29(2):62–67. https://doi.org/10.58440/ihr-29-2-n11

Mäkinen J (2021) The permanent tide and the international height reference frame IHRF. J Geod 95:106. https://doi.org/10.1007/s00190-021-01541-5

Moritz H (2000) Geodetic reference system 1980. J Geod 74:128–133. https://doi.org/10.1007/s001900050278

Pavlis NK, Holmes SA, Kenyon SC, Factor JK (2012) The development and evaluation of the earth gravitational model 2008 (EGM2008). J Geophys Res 117:B04406. https://doi.org/10.1029/2011JB008916

Pavlis NK, Holmes SA, Kenyon SC, Factor JK (2013) Correction to "the development and evaluation of the earth gravitational model 2008 (EGM2008),". J Geophys Res Solid Earth 118:2633. https://doi.org/10.1002/jgrb.50167

Petit G, Luzum B (2010) IERS conventions 2010. IERS technical note 36. Verlag des Bundesamtes für Kartographie und Geodäsie, Frankfurt am Main

Reguzzoni M, Carrion D, De Gaetani CI, Albertella A, Rossi L, Sona G, Batsukh K, Toro Herrera JF, Elger K, Barzaghi R, Sansó F (2021) Open access to regional geoid models: the International Service for the Geoid. Earth System Sci Data 13:1653–1666. https://doi.org/10.5194/essd-13-1653-2021

Sánchez L, Ågren J, Huang J, Wang YM, Mäkinen J, Pail R, Barzaghi R, Vergos GS, Ahlgren K, Liu Q (2021a) Strategy for the realisation of the international height reference system (IHRS). J Geod 95(3). https://doi.org/10.1007/s00190-021-01481-0

Sánchez L, Huang J, Ågren J, Barzaghi R, Vergos GS (2021b) Recovering potential values from regional (quasi-)geoid models. Unpublished guidelines, IAG joint working group 0.1.3

Schwabe J, Liebsch G Ågren J, Mononen J, Andersen OB, Westfeld P, Hammarklint T (2020) THE BALTIC SEA CHART DATUM 2000 (BSCD2000) - Implementation of a common reference level in the Baltic Sea, International Hydrographic Review, Volume 23, https://ihr.iho.int/articles/the-baltic-sea-chart-datum-2000-bscd2000-implementation-of-a-common-reference-level-in-the-baltic-sea/

Schwabe J, Varbla S, Ågren J, Teitsson H, Ellmann A, Liebsch G, Forsberg R, Strykowski G, Bilker-Koivula M, Liepins I, Parseliunas E, Keller K, Omang O, Vestøl O, Kaminskis J, Wilde-Piorko M, Pyrchla K, Somla J, Westfeld P, Hammarklint T (2023) The Baltic Sea Chart Datum 2000 height transformation grid (Realization 2023). https://doi.org/10.58440/iho-bscd2000

SDFI, Styrelsen for Dataforsyning og Infrastruktur (2023) *Dataforsyningen*. Accessed on 16 Nov 2023. https://dataforsyningen.dk/data?filter=&view=gallery&search=GNSS

Tscherning CC (1984) The Geodesist's Handbook. Bull Géod 58:3

Tscherning CC, Forsberg R, Knudsen P (1992) The GRAVSOFT package for geoid determination. In: Holota P, Vermeer M (eds) Proc of 1st continental workshop on the geoid in Europe, Prague. Research Institute of Geodesy, Topography and Cartography, Prague, pp 327–335

Vestøl O, Ågren J, Steffen H et al (2019) NKG2016LU: a new land uplift model for Fennoscandia and the Baltic region. J Geod 93:1759–1779. https://doi.org/10.1007/s00190-019-01280-8

Zingerle P, Pail R, Gruber T, Oikonomidou X (2020) The combined global gravity field model XGM2019e. J Geod 94:66. https://doi.org/10.1007/s00190-020-01398-0

Open Access This chapter is licensed under the terms of the Creative Commons Attribution 4.0 International License (http://creativecommons.org/licenses/by/4.0/), which permits use, sharing, adaptation, distribution and reproduction in any medium or format, as long as you give appropriate credit to the original author(s) and the source, provide a link to the Creative Commons license and indicate if changes were made.

The images or other third party material in this chapter are included in the chapter's Creative Commons license, unless indicated otherwise in a credit line to the material. If material is not included in the chapter's Creative Commons license and your intended use is not permitted by statutory regulation or exceeds the permitted use, you will need to obtain permission directly from the copyright holder.

Part IV

Monitoring Sea Level Changes by Satellite and In-Situ Measurements

The Impact of Different Geophysical Corrections on Altimetry-Derived Sea Level Rise Estimates—Wet Troposphere

Denise Dettmering, Christian Schwatke, and Felix L. Müller

Abstract

Satellite radar altimetry has been providing sea surface heights on an almost global scale for the past 30 years. From this data, an average global mean sea level rise of 3-4 mm per year can be estimated. To determine these small changes with high accuracy, precise and stable measurements are required. Long-term data stability is particularly important for sea-level rise applications. This not only relates to the altimeter measurements themselves, but also to any geophysical correction applied to the data. Furthermore, consistency between different missions is essential to ensure a long time series that is useful for climate studies.

This contribution shows how global sea level rise estimates can be affected by geophysical corrections applied to satellite altimetry data and the importance of selecting the right datasets. The focus will be on atmospheric corrections, especially on different wet troposphere path delay corrections derived by models and observations. It will be shown that these corrections can introduce systematic errors in the order of 0.5 mm/year, which is the level of uncertainty currently assumed for the altimetry-derived global mean sea level trend.

Keywords

Geophysical corrections · Satellite altimetry · Sea level rise · Wet troposphere

1 Introduction

Sea level is among the most important Essential Climate Variables (ECV) as its monitoring provides evidence of climate change. An important variable to quantify sea level changes is the global mean sea level (GMSL) trend. It has been measured very precisely for more than 30 years by satellite altimetry. Currently, the trend for the altimeter era (1993 to 2021) is estimated to be 3.3 mm/year (Guérou et al. 2023). Trend uncertainty is also important in this context. Several studies deal with error analysis for global mean sea level trends, among them Ablain et al. (2019) and Guérou et al. (2023). The latter study states an uncertainty of 0.3 mm/year with 90% confidence, thus one order of magnitude better than the trend itself. One of the main contributors to this GMSL uncertainty on longer time scales (10 years and longer) is the wet troposphere correction (WTC) (Guérou et al. 2023).

In order to yield this high accuracy, the dataset used for sea level trend computation must be carefully selected. This holds for the measured altimeter range as well as all geophysical corrections required to calculate sea surface heights. Depending on the choice of corrections, the error in sea level trend can be much larger than 0.3 mm/year if systematic errors impact the estimations. This has been already shown years ago for the pole tide correction, where different models show large regional trend differences of up to 0.25 mm/year with a clear geographical pattern (Desai et al. 2015), even though the impact on global mean trend is negligible. Fernandes et al (2006) showed the impact

D. Dettmering (✉) · C. Schwatke · F. L. Müller
Deutsches Geodätisches Forschungsinstitut, Technische Universität München, Munich, Germany
e-mail: denise.dettmering@tum.de; christian.schwatke@tum.de; felix-lucian.mueller@tum.de

of several geophysical corrections as well as satellite orbit solutions on sea level studies. More recently, Dettmering and Schwatke (2022, 2023) have shown the impact of incorrectly scaled ionospheric correction models on sea level trend estimates that can be up to 1 mm/year on a global scale. This high value was shown for the Jason-1 period and becomes smaller for longer periods since it depends on the ionospheric activity and the 11-year solar cycle.

This study investigates global mean trend differences between different WTC products, which can be directly interpreted as trend errors for GMSL trend estimates. Using models and observations, it is shown that significant trend differences between WTC datasets exist that can introduce significant long-term errors in the estimated GMSL. Especially, when model data is used special care should be taken in choosing the best, i.e., most stable, dataset available for the period under investigation.

2 Data

This study is based on the so-called altimeter reference missions for sea level monitoring, namely TOPEX, Jason-1, Jason-2, and Jason-3, covering a period from 1992 to 2022. Each mission is investigated separately including their extended (interleaved) phases. Instead of adding the correction to the range measurement and computing trends in global sea level, the trend differences are analyzed separately in order to avoid impact from any other component (e.g., due to missing information necessary to compute sea surface heights). For that purpose, the along-track wet troposphere corrections are averaged over each 10-day repeat cycle of the missions. To avoid any influence from coastal or sea-ice regions, only open ocean data with a depth greater than 2 km and with a sea-ice concentration of 0% (taken from daily Sea Ice Concentrations from Nimbus-7 SMMR and DMSP SSM/I-SSMIS Passive Microwave Data, NSIDC-0051 (DiGirolamo et al. 2022)) is used.

A set of five different WTC are used, three of them from numerical weather models (NWM), and two based on observations. The formulae necessary to extract WTC from NWM can be found e.g. in Fernandes et al. (2014). The following WTC are used:

- measurements from the on-board microwave radiometer, MWR (from original Geophysical Data Records, GDR)
- model data taken from the original GDR (this is different for the missions, mainly the operational product from the European Centre for Medium-Range Weather Forecasts, ECMWF)
- model data from Vienna Mapping Function product, VMF3 (Landskron and Böhm 2018; Re3data.Org 2016)
- model data from latest reanalysis product of ECMWF, ERA5 (Hersbach et al. 2020)
- GNSS (Global Navigation Satellite System) derived Path Delay, GPD+ (Fernandes et al. 2015; Lázaro et al. 2020)

Moreover, the differences are compared to corrections based on stable water vapor climate data records (CDR) from two meteorological satellite missions HOAPS and REMSS (Barnoud et al. 2023), which are provided as cycle-mean corrections by AVISO. Namely, "correction_gmwtc_remss" and "correction_gmwtc_hoaps" from version 1.0 of the product have been used.

3 Results and Discussion

The radiometer-derived wet troposphere correction is the most accurate WTC over open ocean as it is measured directly together with the altimeter range (Fernandes et al. 2024). However, problems in the long-term stability of the radiometer corrections have been known for a long time, see for example Scharroo et al. (2004) or Brown et al. (2007). This can also be seen when comparing the radiometer correction of two overlapping missions.

As Fig. 1 shows, when comparing the corrections from TOPEX and Jason-1, one can clearly see a difference of about 1 mm/year between these two corrections. Of course, this number should be handled with caution since the period analysed here is only about two years long. Moreover, after the tandem flight of the first six months, TOPEX is moved into interleaved orbit and the observation points are no longer identical. However, this shows that the long-term behaviour of (at least one of the two) radiometers is not stable.

3.1 Differences Between Models

Figure 2 shows WTC model differences. Three different models have been compared. Based on the assumption that ERA-5 should be the best model currently available (Fernandes et al. 2024), differences are shown with respect to this model. One can see that the model corrections included in the GDR datasets show large differences to ERA-5, especially for TOPEX (light blue). Since 2004 this behaviour improved but still, systematic differences up to a few millimeters are visible. Moreover, a step in mid 2021 shows an inconsistency in the data. The usage of the external VMF3 model (orange line) improves the discrepancies in the first period and shows similar systematic differences to the GDR products from 2008. The VMF3 version used is based on ECMWF-interim until the end of 2007 and on ECMWF-operational from 2008 onward.

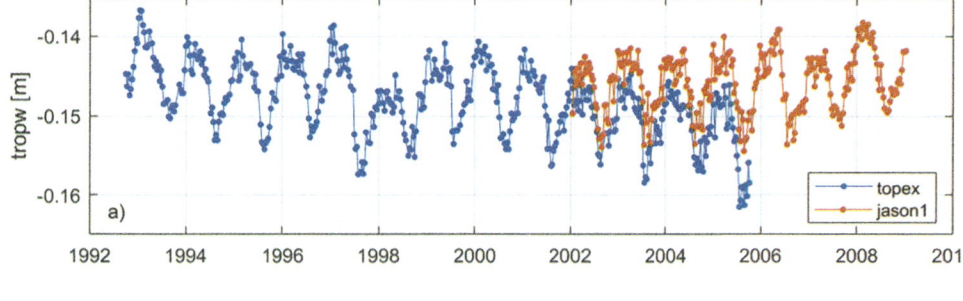

Fig. 1 Mean WTC from radiometers of TOPEX (blue) and Jason-1 (red) in subplot (**a**) and and their cycle-mean differences (black) with fitted linear trend (orange) in (**b**)

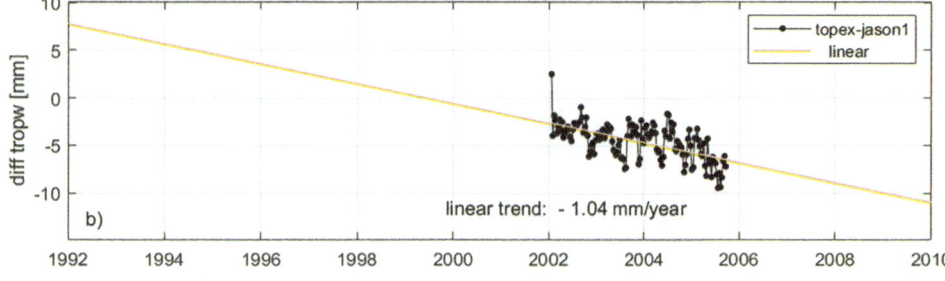

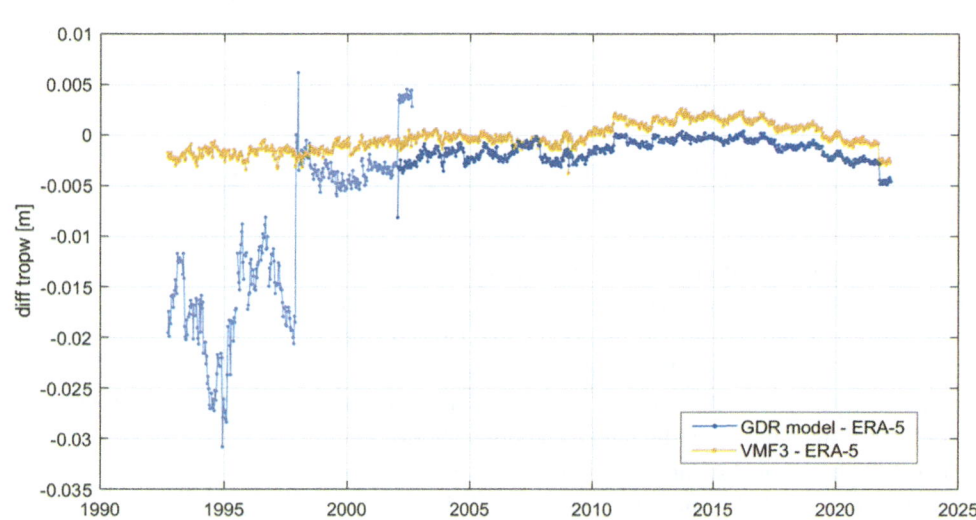

Fig. 2 WTC differences with respect to ERA-5: WTC model in GDR data (blue, TOPEX in light blue) and VMF3 (orange)

These results are in line with a recent study from University of Porto (Fernandes et al. 2024) analyzing different NWM for sea level trend differences.

3.2 Differences to GPD+

The GPD+ product is a combination of radiometer observations on board the altimeter satellites, GNSS-derived WTC, as well as measurements of MWR on other satellites (Lázaro et al. 2020). Model data are also used as first estimates in the data combination procedure, which are the final values if no observations are available. The correction is available for most of the current altimeter systems, for some also in the GDR products (e.g., CryoSat-2), and for others only as external products, e.g. from AVISO or directly from the University of Porto (where the correction has been developed).

Over open ocean, the product is mainly based on altimeter MWR observations since these provide the best short-scale information. Moreover, the product has been calibrated against data from several scanning imaging passive MWR (among them SSM/I and SSMIS sensors on various satellites), mainly to provide information over the ocean for altimetry missions not equipped with MWR but also to ensure long-term stability of the product.

Figure 3 shows the differences between the GPD+ correction and the radiometer correction of the reference missions.

One can see clear offsets between the different missions stemming from offsets in the MWR measurements. Moreover, systematic linear trends as well as annual variations are visible. Table 1 is summarizing the estimated trends and offsets for the different missions. The values for TOPEX, Jason-1, and Jason-2 fit reasonably well to the numbers provided in Lázaro et al. (2020).

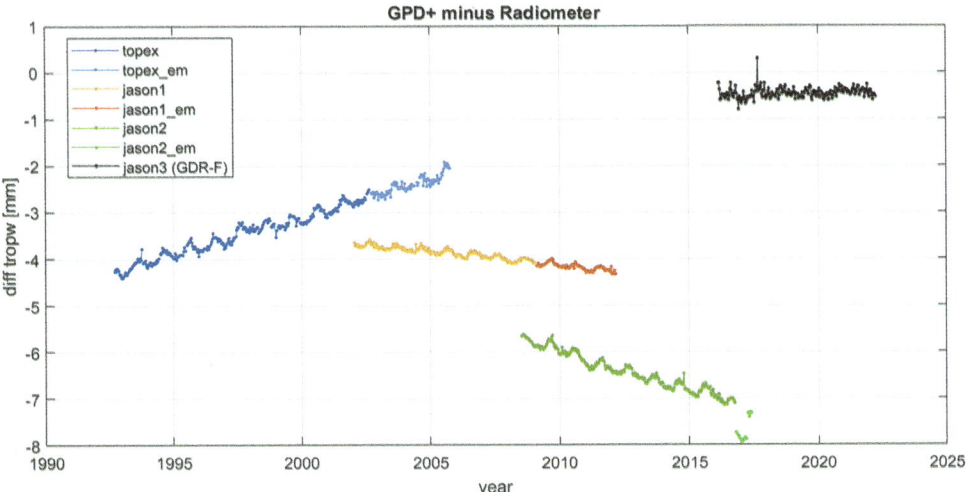

Fig. 3 Differences in WTC between GPD+ and radiometer solution for four different satellites (TOPEX in blue, Jason-1 in orange/red, Jason-2 in green, and Jason-3 in black)

Table 1 Estimated statistics from the difference between GPD+ and radiometer WTC; without extended orbit phases

Mission	Trend [mm/yr]	Mean offset [mm]	Length [years]
TOPEX	0.159±0.002	−3.392±0.477	9.87
Jason-1	−0.056±0.002	−3.850±0.130	7.00
Jason-2	−0.166±0.002	−6.451±0.406	8.20
Jason-3	0.010±0.004	−0.471±0.106	6.11

Only the Jason-3 differences seem to be quite stable and show only a small offset below one millimeter. One should note here, that this dataset is based on the latest GDR-F datasets, in which the MWR data have been long-term calibrated by NASA/JPL. The GPD+ data for this mission have been provided directly by the University of Porto (Fernandes 2023, personal communication). The GPD+ version available from AVISO is made for Jason-3 GDR-E.

The agreement between the Jason-3 GPD+ and radiometer WTC trends can be explained by the fact that, for this mission, no trend adjustment of GPD+ with respect to climate models has been performed (see Lázaro et al. 2020). However, it is known from previous studies (Barnoud et al. 2021) that the radiometer on board of Jason-3 is less stable than for other missions. Recently, a dedicated correction to account for this drift has been released that is not applied here (Brown et al. 2023; Brown 2013). Thus, both WTC (MWR and GPD+) show the same drift of about 2 mm over 8 years. A similar drift is visible between MWR and ERA-5 (see Fig. 4).

3.3 Comparison to Climate Data Records

In view of the documented trend differences and since it is known that the radiometer instruments on board the altimeter satellites might be prone to drifts, a comparison to highly

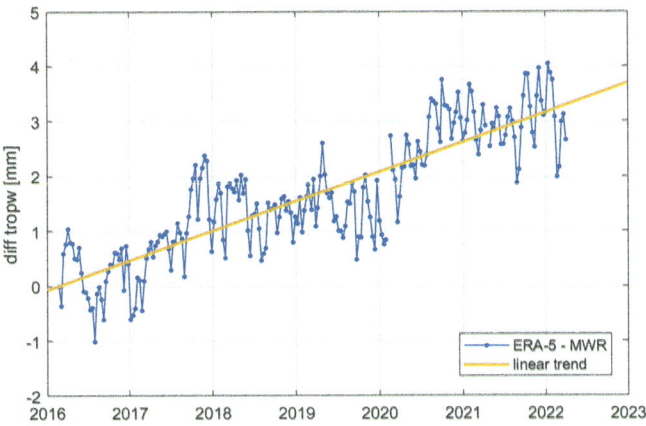

Fig. 4 Differences between ERA-5 and MWR WTC for Jason-3

stable climate data records of water vapor (converted to WTC) derived from independent MWR measurements on board meteorological satellites is strait forward. Instead of making its own comparisons to CDR, this study uses the pre-processed AVISO wet troposphere correction to the MWR WTC (Barnoud et al. 2023) for validation purposes.

Since in this study, no inter-mission offsets have been applied, only the trends can be compared while neglecting the absolute level of differences for each mission. Moreover, one should only compare the core phases of the missions (without interleaved phases) as this is also done in the AVISO products.

This comparison is mainly done visually based on Fig. 5, which shows the differences between GPD+ and MWR together with the CDR-based WTC correction from AVISO. One can see a nice similarity for Jason-1 and Jason-2 in terms of trend and offset. The trends of the CDR correction (for the mission periods) are −0.091 mm/yr for Jason-1 and −0.222 mm/yr for Jason-2 for REMSS and −0.018 and −0.386 mm/yr for HOAPS. These values are similar to the

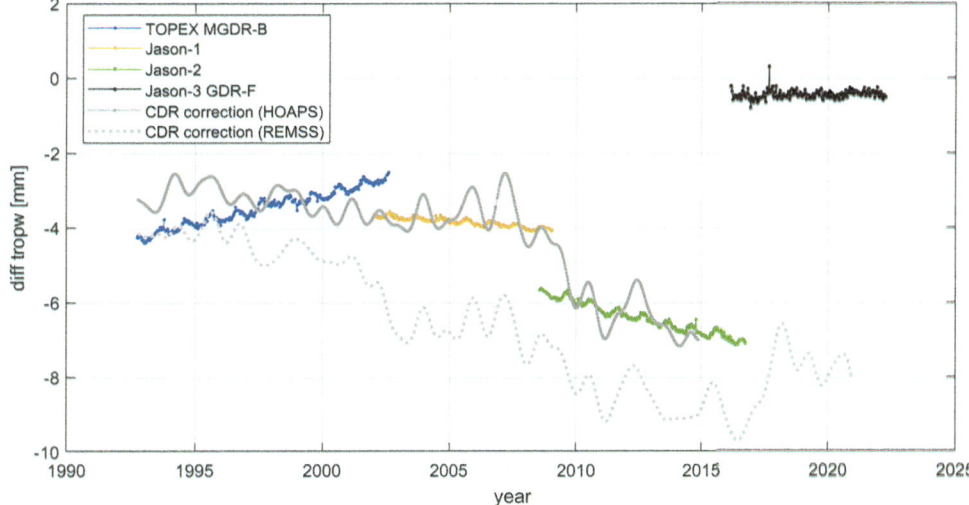

Fig. 5 CDR correction derived from long-term stable climate data records, namely MWR measurements from meteorological satellites HOAPS (grey) and REMSS (dashed light grey). The WTC differences from Fig. 3 (GPD+ minus MWR) are overlaid

trend differences between GPD+ and MWR as documented in Table 1. In terms of offsets, the GPD+ WTC agrees better to HOAPS than to REMSS.

For Jason-3 no trend is visible in the REMSS time series (as for the differences between GPD+ and MWR). However, the period for which data is available is too short to detect small trends. An offset of several millimeters is visible. That is due to a different version of Jason-3 GDR including an offset between the MWR corrections of about 7 mm (Fernandes 2023, personal communication).

According to Guérou et al. (2023), for TOPEX, the AVISO correction is not based on the MWR data but on an internal update. Thus, both time series can not be directly compared.

4 Conclusion

The long-term stability of the wet tropospheric correction for satellite altimetry is still an important challenge for sea level rise estimation and one of the most critical contributors to its uncertainty.

Model corrections given in the altimeter GDR datasets are often not long-term stable since most missions apply ECMWF operational products, which underlay constant changes and improvements. For sea level trend estimation, only reprocessed model data such as the ERA-5 dataset should be used if modeled WTC is required, e.g. in coastal areas or for missions without radiometer.

When compared with GPD+ corrections, the radiometer corrections for the reference missions show trend differences of up to ±0.2 mm/year. These differences are mainly due to drifts in the TOPEX, Jason-1, and Jason-2 on-board radiometers, as similar variability is observed when comparing MWR with CDR.

Since GPD+ is calibrated against passive imaging MWR it can be considered as long-term stable. Moreover, over open ocean, it contains the original altimetry MWR measurements and thus can be considered as a good product for global mean sea level studies (even though it has been developed with a focus on coastal areas). However, various versions of the product are available and they are not always consistent between different missions (as computed in different years with different standards). An improved documentation of the different versions (also from different data sources, e.g. AVISO or Uni Porto) would be very helpful here. When compared to the CDR-derived corrections provided by AVISO, good agreement of trends for the Jason satellites is found whereas larger discrepancies in the TOPEX period are visible. This is probably due to the fact that the GMSL CNES/AVISO+ record is based on a CLS update (GPD+) for TOPEX instead of the radiometer correction which is used for the Jason missions Guérou et al. (2023). An evaluation of the WTC in the newly processed TOPEX GDR-F data should be done in the future to update this study.

For Jason-3, MWR WTC includes a drift that should be accounted for by applying a dedicated correction (Brown et al. 2023). GPD+ is showing the same drift as no trend adjustment has yet been performed. These drifts show up in the comparison to ERA-5 with a trend differences of about 0.5 mm/year.

For climate applications, such as GMSL, an update to MWR WTC is recommended, either by using the AVISO CDR-correction or by applying GPD+. The latter has the advantage of being available for all missions whether or not an on-board radiometer is present.

Acknowledgements We thank the agencies in charge of maintaining the satellite altimetry missions and AVISO for distributing the GDR datasets. The GPD+ data used in this study were developed and validated by the University of Porto, and distributed by AVISO+, France.

The CDR-derived WTC product was produced by Magellium, Fluctus and LEGOS and distributed on the AVISO+ ODATIS portal (https://aviso.altimetry.fr) with support from CNES (https://doi.org/10.24400/527896/a01-2022.018 version 1.0). The VMF3 data are computed and distributed from TU Vienna based on Numerical Weather Models from the ECMWF.

References

Ablain M, Meyssignac B, Zawadzki L, et al (2019) Uncertainty in satellite estimates of global mean sea-level changes, trend and acceleration. Earth Syst Sci Data 11(3):1189–1202. https://doi.org/10.5194/essd-11-1189-2019

Barnoud A, Pfeffer J, Guérou A, et al (2021) Contributions of altimetry and argo to non-closure of the global mean sea level budget since 2016. Geophys Res Lett 48(14):e2021GL092824. https://doi.org/10.1029/2021GL092824

Barnoud A, Picard B, Meyssignac B, et al (2023) Reducing the uncertainty in the satellite altimetry estimates of global mean sea level trends using highly stable water vapor climate data records. J Geophys Res Oceans 128(3):e2022JC019378. https://doi.org/10.1029/2022JC019378

Brown S (2013) Maintaining the long-term calibration of the Jason-2/OSTM advanced microwave radiometer through intersatellite calibration. IEEE Trans Geosci Remote Sens 51(3):1531–1543. https://doi.org/10.1109/TGRS.2012.2213262

Brown S, Willis J, Fournier S (2023) Jason-3 wet path delay correction. Ver. F. PO.DAAC, CA, USA. https://doi.org/10.5067/J3L2G-PDCOR

Brown ST, Desai S, Lu W, et al (2007) On the long-term stability of microwave radiometers using noise diodes for calibration. IEEE Trans Geosci Remote Sens 45(7):1908–1920. https://doi.org/10.1109/TGRS.2006.888098

Desai S, Chander S, Ganguly D, et al (2015) Waveform classification and water-land transition over the Brahmaputra River using SARAL/AltiKa & Jason-2 altimeter. J Indian Soc Remote Sens 43(3):475–485. https://doi.org/10.1007/s12524-014-0428-y

Dettmering D, Schwatke C (2022) Ionospheric corrections for satellite altimetry - impact on global mean sea level trends. Earth Space Sci 9(4):e2021EA002098. https://doi.org/10.1029/2021EA002098

Dettmering D, Schwatke C (2023) Comparison of different methods to account for the plasmaspheric electron content in GNSS-derived ionospheric altimeter corrections and their impact on sea level trend estimation. Earth Planets Space 75(1). https://doi.org/10.1186/s40623-023-01764-0

DiGirolamo N, Parkinson C, Cavalieri D, et al (2022) Sea ice concentrations from nimbus-7 smmr and dmsp ssm/i-ssmis passive microwave data. https://doi.org/10.5067/MPYG15WAA4WX, https://nsidc.org/data/nsidc-0051/versions/2

Fernandes MJ, Barbosa S, Lázaro C (2006) Impact of altimeter data processing on sea level studies. Sensors 6(3):131–163. https://doi.org/10.3390/s6030131

Fernandes MJ, Lázaro C, Nunes A, et al (2014) Atmospheric corrections for altimetry studies over inland water. Remote Sens 6(6):4952–4997. https://doi.org/10.3390/rs6064952

Fernandes MJ, Lázaro C, Ablain M, et al (2015) Improved wet path delays for all esa and reference altimetric missions. Remote Sens Environ 169:50–74. https://doi.org/10.1016/j.rse.2015.07.023

Fernandes MJ, Vieira T, Aguiar P, et al (2024) How different ECMWF atmospheric models impact the estimation of sea-level trends. IEEE J Sel Top Appl Earth Obser Remote Sens 17:3069–3077. https://doi.org/10.1109/JSTARS.2023.3347089

Guérou A, Meyssignac B, Prandi P, et al (2023) Current observed global mean sea level rise and acceleration estimated from satellite altimetry and the associated measurement uncertainty. Ocean Sci 19(2):431–451. https://doi.org/10.5194/os-19-431-2023

Hersbach H, Bell B, Berrisford P, et al (2020) The era5 global reanalysis. Quarterly J Roy Meteorol Soc 146(730):1999–2049. https://doi.org/10.1002/qj.3803

Landskron D, Böhm J (2018) VMF3/GPT3: refined discrete and empirical troposphere mapping functions. J Geodesy 92(4):349–360. https://doi.org/10.1007/s00190-017-1066-2

Lázaro C, Fernandes MJ, Vieira T, et al (2020) A coastally improved global dataset of wet tropospheric corrections for satellite altimetry. Earth Syst Sci Data 12(4):3205–3228. https://doi.org/10.5194/essd-12-3205-2020

Re3data.Org (2016) VMF data server. https://doi.org/10.17616/R3RD2H, https://www.re3data.org/repository/r3d100012025

Scharroo R, Lillibridge JL, Smith WHF, et al (2004) Cross-calibration and long-term monitoring of the microwave radiometers of ers, topex, gfo, jason, and envisat. Marine Geodesy 27(1-2):279–297. https://doi.org/10.1080/01490410490465265

Open Access This chapter is licensed under the terms of the Creative Commons Attribution 4.0 International License (http://creativecommons.org/licenses/by/4.0/), which permits use, sharing, adaptation, distribution and reproduction in any medium or format, as long as you give appropriate credit to the original author(s) and the source, provide a link to the Creative Commons license and indicate if changes were made.

The images or other third party material in this chapter are included in the chapter's Creative Commons license, unless indicated otherwise in a credit line to the material. If material is not included in the chapter's Creative Commons license and your intended use is not permitted by statutory regulation or exceeds the permitted use, you will need to obtain permission directly from the copyright holder.

Bathymetry Estimation from ICESat-2 in a Region Swamped by Mud: A Case Story from Moreton Bay

Elisabet Anne Marie Hallström, Ole B. Andersen, Xiaoli Deng, and Richard Coleman

Abstract

The bathymetry of coastal bay environments, such as Moreton Bay near Brisbane in eastern Australia, is constantly reworked because of changes in energy dispersal and related sediment transport pathways. Updated and accurate bathymetric models are a crucial component for scientific, environmental, and ship safety studies.

NASA's Ice, Cloud, and Land Elevation Satellite-2 (ICESat-2) is equipped with a laser detecting system (green light) that penetrates the air-water interface. Under optimal conditions, it can provide shallow water bathymetry (depths <40 m). We attempted to use ICESat-2 measurements to study bathymetry and possible bathymetry changes from repeated tracks across Moreton Bay. We found that the water turbidity in Moreton Bay varies with time. More than half of the water area is affected by suspended sediment, which makes ICESat-2 difficult to obtain bathymetric measurements. In other areas, repeated ICESat-2 tracks performed consistently on the 1-meter level. This means that ICESat-2 can be used to update existing bathymetry in the region. We also devised a method to determine bathymetry in the shallower parts of the zone affected by mud.

Keywords

Altimetry · ATLAS · Bathymetry · Coastal bathymetry · Coastal region · ICESat-2 · Moreton Bay

1 Introduction

Accurate bathymetry is important in ship trafficking and safety, fish and marine industries, and coastal management. Near-shore bathymetry is important for engineering, coastal safety, and environmental monitoring.

In some regions, the bathymetry changes with time due to erosion and sediment transport. This is particularly important in shallow water bays containing large harbours. For such regions, the bathymetry must be mapped both frequently and accurately. Mapping the coastal bathymetry requires huge resources when using single or multibeam bathymetry mapping from ships. Consequently, optical methods are very attractive. Multi-spectral satellite imagery from satellites such as Landsat and Sentinel-2 is used in satellite-derived bathymetry (Nguyen et al. 2021; Zhang et al. 2022). Satellite-derived bathymetry is simpler, cheaper, and quicker but usually is less accurate than multibeam bathymetry (Ramnath et al. 2015). With the launch of ICESat and particularly ICESat-2, laser altimetry for bathymetry mapping has gained momentum (Le Quilleuc et al. 2021; Parrish et al. 2022).

Several studies have focused on determining bathymetry in shallow water from ICESat and ICESat-2, for which various methods have been used. Arsen et al. (2014) determined bathymetry in a lake from ICESAT. Parrish et al. (2019) showed the importance of applying a refraction correction and the capability of ICESat-2 to determine bathymetry at depths of 40 m. Xu et al. (2020) and Ma et al. (2020) used a machine-learning method based on a clustering algorithm to locate the water surface level and the bathymetry. Ranndal et al. (2021) used statistical methods to estimate bathymetry profiles. All these studies have demonstrated the great value of using ICESat-2 data for bathymetry mapping in regions with low water turbidity, and the optical properties are favorable.

In this study, we attempt to apply ICESat-2 data to a region with very turbid water, where suspended sediments render ICESat-2 problematic. In the turbid areas, the ICESat-2 photons are frequently scattered throughout the water column, making the bottom less visible.

Moreton Bay is located outside of Brisbane on the east coast of Australia (Fig. 1). The port of Brisbane is heavily trafficked, and all ships between the port and the Pacific Ocean must pass through Moreton Bay. The bay covers a water area of 1,500 km^2, and the water depth ranges up to 40 m. The ship routing depends on the dredged channel following the red line in Fig. 1 (left panel). Moreton Bay is a sandy bay that is constantly being reworked because of the changes in the patterns of energy dispersal due to storms and tides. As a result, dredging is performed annually (https://www.tmr.qld.gov.au/. Accessed 20 Oct 2023), and the port authority performs frequent bathymetric surveys. Detecting bathymetry changes with ICESat-2 would benefit the port authorities, dredging companies, etc. because it is much more cost-effective than surveys.

This paper investigates if NASA's Ice, Cloud, and Land Elevation Satellite-2 (ICESat-2) laser altimeter can assist in bathymetry estimation in the bay and its ability to detect possible bathymetry changes with time.

The following section introduces the Moreton Bay region and the ICESat-2 data. This is followed by a section on our method for the bathymetry estimation and a section presenting our results. In the end section, we sum up our conclusions.

2 Study Area and Data

2.1 Moreton Bay

The study case, Moreton Bay, has a very different coastal environment. Studies, including Eyre et al. (1998) and Lockington et al. (2017), show that Moreton Bay is seriously affected by suspended sediments in the water column, hampering optical visibility. The southern part of the bay has mainly a mud or sandy mud bottom, which reduces sunlight penetration, while sediments in the northern part of the bay are primarily clean sand, as described by Lockington et al. (2017). This study concluded that the presence of mud has increased in Moreton Bay and that the southern half of the bay (800 km^2) is affected (Fig. 1, right panel). This is a consequence of large floods over the past 45 years combined with land clearing. Sediment enters the bay via the Brisbane River estuary (Eyre et al. 1998). We used Fig. 1 (right panel) to define a "mud zone" with decreased visibility in the water column.

An existing bathymetry map, AusSeabed for the Great Barrier Reef region (Beaman 2017), with a resolution of 30 m, is used for comparison. We used an updated version from 2020. The AusSeabed bathymetry is shown in Fig. 2, with the various channels used for ship routing indicated (Fig. 1, left).

2.2 Icesat-2

ICESat-2 was launched on the 15th of September 2018. ICESat-2 carries the Advanced Topographic Laser Altimeter System (ATLAS), consisting of six laser beams: three strong and three weak beams. All the strong beams are paired with a weak beam with 90 m between them and 3.3 km between each pair. One beam emits about 20 trillion photons per pulse and 10,000 pulses per second, with a spatial resolution of 70 cm on the Earth's surface (Markus et al. 2017). This paper uses the full photon cloud from the ATL03 product, version 5. ICESat-2 has a repeat period of 91 days, resulting in 16 repeated cycles (until October 2022). The data is accessed through the National Snow and Ice Data Center (NSIDC)

Fig. 1 The left panel shows a map of Moreton Bay (Google Maps 2017). Most of the ship traffic between the ocean and the port of Brisbane follows the red line. The right panel shows the mud zone in Moreton Bay (https://www.uq.edu.au/news/article/2016/06/moreton-bay-being-swamped-mud. Accessed 20 Oct 2023)

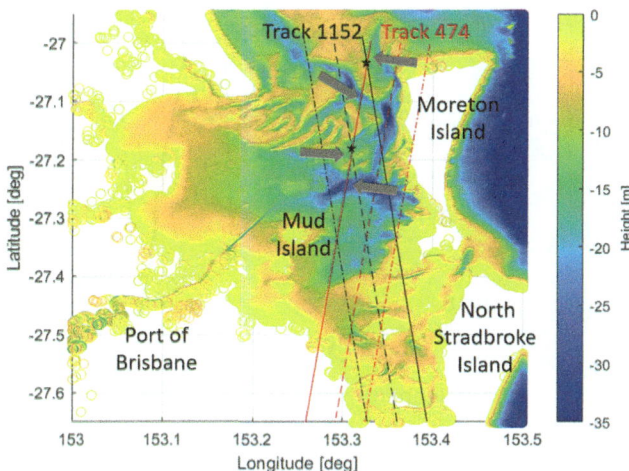

Fig. 2 AusSeabed bathymetry with the ICESat-2 Track 474 (red) and Track 1152 (black) with the beams gt1r (dash-dotted line), gt2r (dashed line), and gt3r (solid line). The four grey arrows illustrate from north to south the locations of the 'Northern Channel', the 'Central Channel', the 'Sub-Channel', and the 'Southern Channel' used in the following. The two stars indicate crossing locations where we compare bathymetry estimates in Sect. 4

(Neumann et al. 2021), including each photon's coordinate and height values.

There are two tracks from ICESat-2 crossing Moreton Bay. These are Track 474 and Track 1152, as shown in Fig. 2.

3 Bathymetry Determination

We determined bathymetry using a statistical approach based on the subsurface reflected photons.

To isolate subsurface photons, we initially applied the Ranndal et al. (2021) method to determine and remove the surface scattered photons. Here, we derive the sea surface height (SSH) by a moving median of 7,001 photons reflected between the heights -1.5 m and 1.0 m relative to the mean sea surface (Andersen et al. 2023).

Next, photons within the surface layer, defined as the upper 0.5 m of the water column, are removed following Ranndal et al. (2021) due to the noise of the reflected photons at the sea surface. In our analysis, we needed to increase the surface layer to 2 m for Moreton Bay due to the significant scattering from suspended sediments in the upper water column.

A simple refraction correction is applied (Ranndal et al. 2021) to account for the refraction that causes the actual bathymetry to be slightly shallower compared with the ICESat-2 observed bathymetry.

Ranndal et al. (2021) used a moving median filter with a window of 50 subsurface photons to estimate the bathymetry, as a moving median is more robust to outliers than other filters, e.g., a moving mean. Due to issues with scattering throughout the water column from suspended sediments (mud), we needed to increase the size of the window compared with Ranndal et al. (2021) from 50 to 181 subsurface photons to obtain stable results. We also found that updating the moving median to a moving quantile focusing on the lowest 22% gave the best bathymetry estimation in Moreton Bay compared with in-situ observations, and we consequently used this in the investigation.

Moreton Bay has significant tides and associated strong tidal currents. We used tidal observations from the Tangalooma tide gauge on Moreton Island inside Moreton Bay (Queensland Government 2023) to correct the tidal signal.

4 Results

ICESat-2 provides up to 16 possible cycles for bathymetry estimation for each of the two tracks covering the period from October 2018 to September 2022. We found that 1/3 of the cycles are completely unusable due to clouds, water vapor, etc. The remaining 'good' cycles are investigated for bathymetry signals. Due to the problem of atmospheric conditions and water turbidity, we defined a 'good' cycle as a cycle in which the central channel is visible and continuous in the ICESat-2 data. This limits the number of 'good' cycles to three for Track 474 (29th of October 2018, 26th of April 2020, and 24th of July 2021) and one for Track 1152 (9th of March 2021).

We estimated bathymetry for the western beam gt3r (strong) of track 474 in Fig. 3. This is shown in Fig. 4

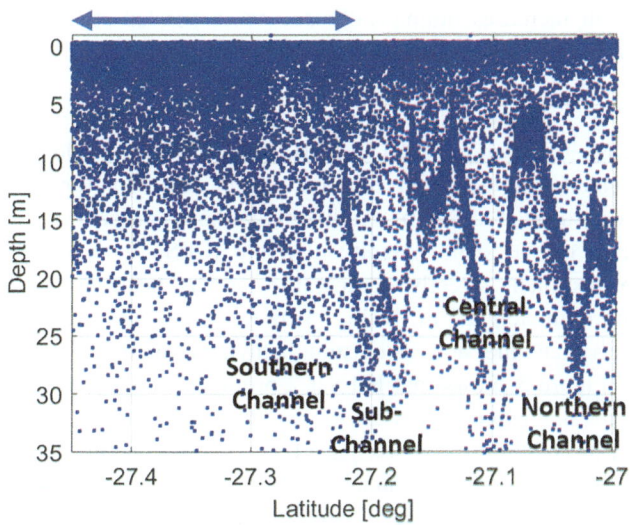

Fig. 3 Photon data (before any refraction correction) from track 474 on the 26th of April 2020 beam gt3r, the clearest bathymetry profile seen in Moreton Bay. The arrow indicates the mud zone (south of 27.23°S). The locations of the four bathymetry channels are also marked in Fig. 2

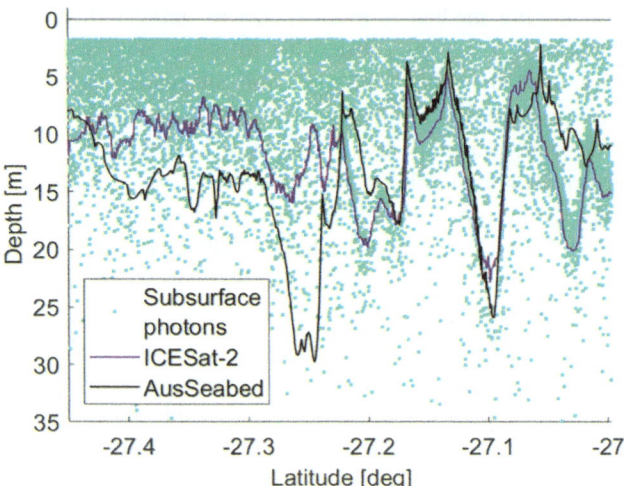
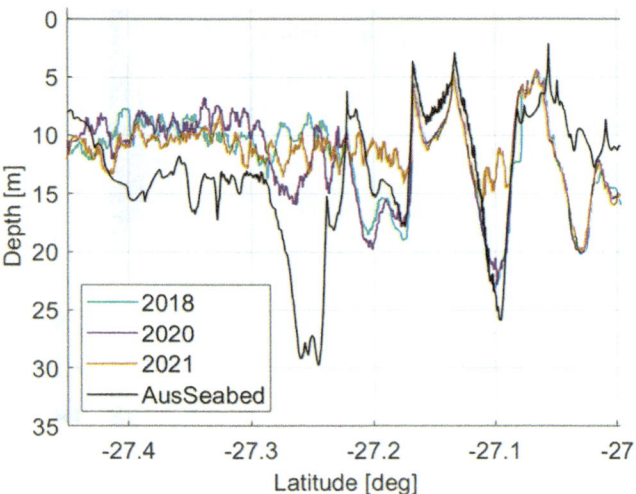

Fig. 4 The predicted bathymetry from track 474 on the 26th of April 2020 showing the photons after applying the refraction correction (light blue), the modeled bathymetry profile (purple), and the interpolated bathymetry profile from the AusSeabed (Beaman 2017) (black)

Fig. 5 The bathymetry models for the 'good' cycles from track 474 are shown in blue (2018), purple (2020), and brown (2021), and the interpolated bathymetry profile from AusSeabed (black)

for the track on the 26th of April 2020, which is the clearest cycle.

Figure 4 reveals that the bathymetry estimate is in close agreement with the AusSeabed bathymetry model outside the mud zone south of 27.23°S. Both the central and northern channels in Fig. 2 are seen from ICESat-2 data. The northern channel is particularly well resolved from ICESat-2. In the northern channel, there is a significant discrepancy between the location and depth of the channel from the interpolated AusSeabed. This is even though the ICESat-2 signal appears as strong for the northern channel in Fig. 4.

As soon as track 474 enters the mud zone south of 27.23°S, the number of scattered photons in the 0–10 m depth increases dramatically due to suspended sediments. This prevents bathymetry estimation for the southern channel region.

We subsequently estimated bathymetry for the repeated ICESat-2 cycles in 2018, 2020, and 2021. All cycles are located within 10 m horizontally in the Moreton Bay (Magruder et al. 2020). Additionally, there is a geolocation error of the same magnitude in the ICESat-2 data. This makes them directly comparable, as the beam width is approximately 20 m. In Fig. 5, these cycles are compared with each other and the existing bathymetry model.

Figure 5 shows that ICESat-2 profiles of 2018 and 2020 generally agree well with the existing bathymetry for the location of the central channel and mostly agree for the region between the channels. However, the deeper southern channel is not visible in any ICESat-2 cycles. ICESat-2 bathymetry from the three cycles agrees to be better than 1 m for the northern channel north of 27.05°S, but they differ by up to 10 m from AusSeabed. This indicates that some errors will likely be found in AusSeabed for at least the northern channel. South of the northern channel, the 2021 bathymetry estimate differs from the 2018 and 2020 estimates. The 2021 bathymetry does not resolve the central channel at 27.1°S, nor the sub-channel at 27.2°S. We found that increased mud coverage in 2021 in the area is a potential cause. This increased mud level can be seen by counting the number of subsurface reflected photons deeper and shallower than 10 m across the central and sub-channel. For the 2020 profile, 543 of 1,283 (42%) subsurface photons are reflected from water below 10 m. For 2021, the number of reflected subsurface photons from water below 10 m has decreased to 415 of 1,408 (29%). This indicates that the "mud zone" has increased from 2020 to 2021. The optical images from Sentinel-2 in Fig. 6 classify large parts of Moreton Bay as water in 2020 but as sand/rock in 2021 and confirm the increased extent of mud in Moreton Bay.

We compared the bathymetry estimates in Fig. 5 across the central channel between 27.13°S - 27.08°S. The ICESat-2 bathymetry estimates from 2020 agree with AusSeabed with a root mean square (RMS) of 2.1 m. Similarly, the bathymetry estimate from 2018 agrees with AusSeabed with an RMS of 2.2 m, while the RMS is 6.5 m for the 2021 model and the interpolated AusSeabed. The significant difference between our estimated seafloor and AusSeabed model may be caused by the interpolation error of the AusSeabed model. The 2018 and 2020 models agree with a RMS of 1.5 m within the channel. They agree with the 2021 model with RMS of 5.5 m and 5.0 m, respectively.

We investigated if it was possible to estimate bathymetry inside the mud zone by gradually increasing the depth of the surface layer within the mud zone south of 27.23°S to remove scattered photons. Increasing the upper surface layer

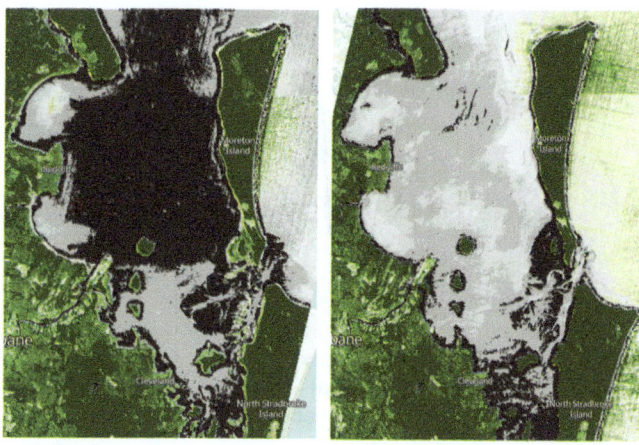

Fig. 6 Images from Sentinel-2 from 28th of April 2020 (left) and 24th of July 2021 (right) with an applied Normalized Difference Vegetation Index (NDVI). Black areas are classified as water, while brighter colors indicate, e.g., rock or sand. Images have been obtained from (EO Browser n.d.)

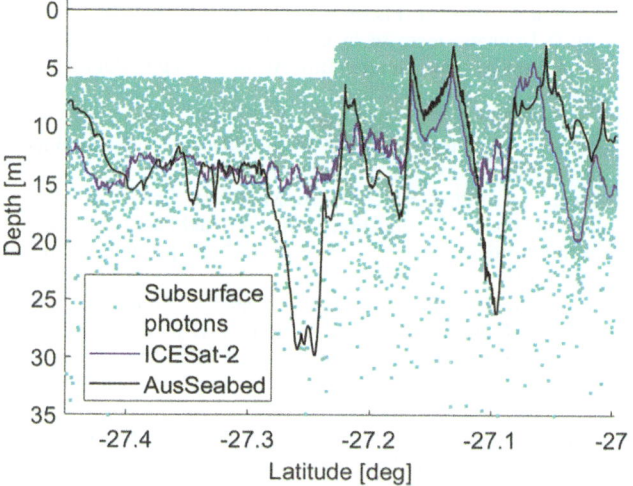

Fig. 7 The bathymetry estimate from the 24th of July 2021 from track 474, where the photons from the upper 5 m have been removed inside the "mud zone"

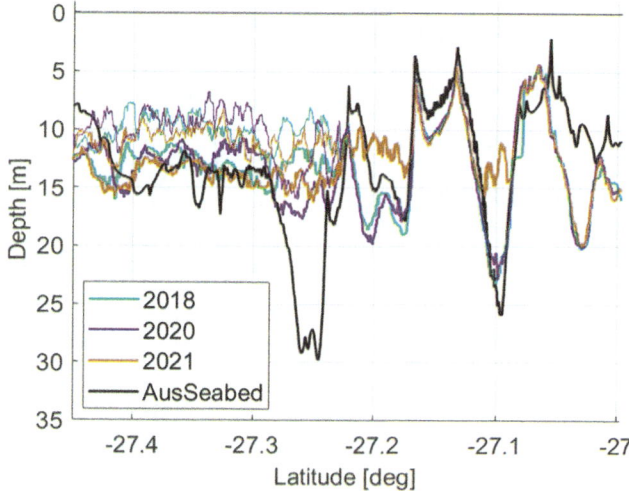

Fig. 8 ICESat-2 bathymetry profiles for 2018, 2020, and 2021, where the photons from the upper 5 m (thick solid lines) have been removed in the mud zone. The thin lines are the bathymetry models from Fig. 5 that are added for comparison

to 5 m reduced the RMS with AusSeabed from 4.3 m to 2.3 m in the shallow water region (depth of 5–15 m) south of 27.29°S. This is shown in the southern section in Figs. 7 and 8. Whatever method we applied, we could not estimate the bathymetry in the southern channel.

As a final validation of the ICESat-2 bathymetry estimation, we compared the 2020 model from track 474 with the crossing track 1152. In Fig. 2, the tiny stars illustrate two locations where the beams of the two tracks cross. Track 474 gt3r crosses track 1152 gt3r (at 27.05°S), and track 474 gt3r crosses track 1152 gt2r (at 27.18°S). Subsections of the tracks at the crossing locations are shown in Fig. 9.

Figure 9 illustrates that both estimates disagree with the existing bathymetry from AusSeabed at the crossing locations, but they agree with each other. At the northern crossing, the estimated depths are 19.1 m and 18.2 m from track 474 and 1152, respectively, versus 11.2 m from AusSeabed. At the southern crossing, the estimated bathymetry is 16.7 m and 17.6 m, respectively, compared with 15.2 m for AusSeabed. In both cases, the different bathymetry estimates at the crossing location from ICESat-2 agreed to 0.9 m. Still, they disagreed with existing bathymetry (by 2–7 m), illustrating the potential to update the bathymetry model using ICESat-2 data. The difference between the bathymetry estimates at the crossing locations will include waves, orbital, and ranging errors, as discussed by Wang and Sneeuw (2024).

5 Conclusions

Moreton Bay is a challenging environment in which to utilize ICESat-2 measurements due to varying ocean and river conditions and sediments in the water. We managed to estimate the bathymetry for only three out of 16 repeated cycles for track 474 due to the difficult conditions in the bay.

Due to high water turbidity, we modified the statistical method by Ranndal et al. (2021) and introduced a moving quantile of 22% for estimating the bathymetry. Results showed that the ICESat-2 cycles closely agree in large parts of Moreton Bay. The three available cycles agreed on the location of the central and northern channels. Interestingly, ICESat-2 disagrees with AusSeabed in the location and depth of the northern channel region, demonstrating the potential of using ICESat-2 to improve bathymetry estimates.

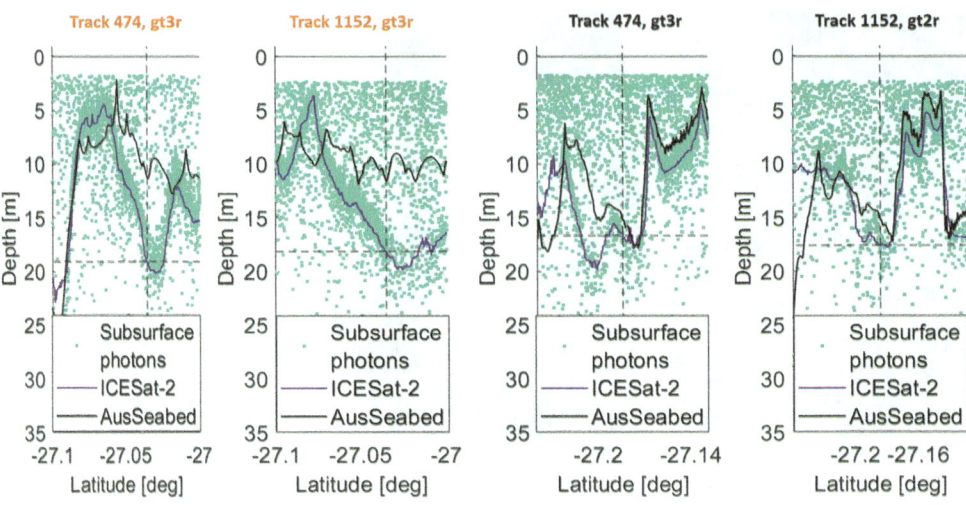

Fig. 9 Crossing ICESat-2 tracks in Moreton Bay. Track 474 gt3r (first panel) crosses Track 1152 gt3r (at 27.05°S), as illustrated by the two panels' dotted line. Similarly, track 474 gt3r (third panel) crosses Track 1152 gt2r (at 27.18°S). The dash lines highlight the exact crossover locations

The southern part of Moreton Bay is heavily affected by mud, and ICESat-2 could not detect the southern channel. In the mud zone south of the southern channel, we devised a method to remove surface photons affected by mud to enhance the estimation of bathymetry in the shallower parts of the mud zone, and the RMS was reduced by 47%.

Despite these efforts to enhance the bathymetry mapping in the shallower part of the bay, the optical conditions in Moreton Bay currently prevent us from obtaining consistent and reliable results on bathymetric changes. However, the results indicate that ICESat-2 can assist in mapping mud zones.

Acknowledgments The Australian Government supported this research through the Australian Research Council's Discovery Project funding scheme (project DP220102969). We thank NSDIC for providing ICESat-2 data and R. Beaman for assistance and advice on existing bathymetry in Moreton Bay.

References

Andersen OB, Rose SK, Abulaitijiang A, Zhang S, Fleury S (2023) The DTU21 global mean sea surface and first evaluation. Earth Syst Sci Data 15:4065–4075. https://doi.org/10.5194/essd-15-4065-2023

Arsen A, Crétaux JF, Berge-Nguyen M, Del Rio RA (2014) Remote sensing-derived bathymetry of Lake Poopó. Remote Sens 6:407–420

Beaman RJ (2017) High-resolution depth model for the great barrier reef - 30 m. Available as part of AusSeaBed from https://portal.ga.gov.au/persona/marine. https://doi.org/10.4225/25/5a207b36022d2

EO Browser (n.d.). https://apps.sentinel-hub.com/eo-browser/. Sinergise Ltd. Accessed 15 Jan 2024

Eyre B, Hossain S, McKee L (1998) A suspended sediment budget for the modified subtropical Brisbane River estuary, Australia. Estuar Coast Shelf Sci 47:513–522

Google Maps (2017) Moreton Bay. http://maps.google.com. Accessed 24 Oct 2023

Le Quilleuc A, Collin A, Jasinski M, Devillers R (2021) Very high-resolution satellite-derived bathymetry and habitat mapping using Pleiades-1 and ICESat-2. Remote Sens 14(1):133

Lockington J, Albert S, Fisher P, Gibbes B, Maxwell P, Grinham A (2017) Dramatic increase in mud distribution across a large subtropical embayment, Moreton Bay, Australia. Mar Pollut Bull 116(1-2):491–497

Ma Y, Xu N, Liu Z, Yang B, Yang F, Wang XH, Li S (2020) Satellite-derived bathymetry using the ICESat-2 lidar and Sentinel-2 imagery datasets. Remote Sens Environ 250:112047

Magruder LA, Brunt KM, Alonzo M (2020) Early ICESat-2 on-orbit geolocation validation using ground-based corner cube retro-reflectors. Remote Sens 12:3653. https://doi.org/10.3390/rs12213653

Markus T, Neumann T, Martino A, Abdalati W, Brunt K, Csatho B, Farrell S, Fricker H, Gardner A, Harding D et al (2017) The Ice, Cloud, and Land Elevation Satellite-2 (ICESat-2): science requirements, concept, and implementation. Remote Sens Environ 190:260–273

Neumann TA, Brenner A, Hancock D, Robbins J, Saba J, Harbeck K, Gibbons A, Lee J, Luthcke SB, Rebold T et al (2021) ATLAS/ICESat-2 L2A global geolocated photon data, version 5. ATL03, Boulder, Colorado. NASA National Snow and Ice Data Center Distributed Active Archive Center. https://doi.org/10.5067/ATLAS/ATL03.005. Accessed 15 Oct 2023

Nguyen VA, Ren H, Huang CY, Tseng KH (2021) Bathymetry derivation in shallow water of the South China Sea with ICESat-2 and Sentinel-2 data. J Appl Remote Sens 15(4)

Parrish CE, Magruder LA, Neuenschwander AL, Forfinski-Sarkozi N, Alonzo M, Jasinski M (2019) Validation of ICESat-2 ATLAS bathymetry and analysis of ATLAS's bathymetric mapping performance. Remote Sens 11:1634

Parrish CE, Magruder L, Herzfeld U, Thomas N, Markel J, Jasinski M, Imahori G, Hermann J, Trantow T, Borsa A, Stumpf R, Eder B, Caballero I (2022) ICESat-2 bathymetry: advances in methods and science, OCEANS 2022, Hampton Roads, Hampton Roads, VA, pp 1–6. https://doi.org/10.1109/OCEANS47191.2022.9977206

Queensland Government (2023). https://www.qld.gov.au/environment/coasts-waterways/beach/tide-sites. Accessed 20 Oct 2023

Ramnath V, Feygels V, Kalluri H, Smith B (2015) CZMIL (Coastal Zone Mapping and Imaging Lidar) bathymetric performance in diverse littoral zones, IEEE OCEANS 2015-MTS/IEEE, pp 1–10

Ranndal H, Sigaard Christiansen P, Kliving P, Baltazar Andersen O, Nielsen K (2021) Evaluation of a statistical approach for extracting shallow water bathymetry signals from ICESat-2 ATL03 photon data. Remote Sens 13(17):3548. https://doi.org/10.3390/rs13173548

Wang B, Sneeuw N (2024) Crossover adjustment of ICESat-2 satellite altimetry for the Arctic region. Adv Space Res 73(1):376–385. https://doi.org/10.1016/j.asr.2023.07.041

Xu N, Ma Y, Zhang W, Wang XH, Yang F, Su D (2020) Monitoring annual changes of lake water levels and volumes over 1984–2018 using landsat imagery and ICESat-2 data. Remote Sens 12(23):4004. https://doi.org/10.3390/rs12234004

Zhang X, Chen Y, Le Y, Zhang D, Yan Q, Dong Y et al (2022) Nearshore bathymetry based on ICESat-2 and multispectral images: comparison between Sentinel-2 Landsat-8 and testing Gaofen-2. IEEE J Sel Top Appl Earth Obs Remote Sens 15:2449–2462

Open Access This chapter is licensed under the terms of the Creative Commons Attribution 4.0 International License (http://creativecommons.org/licenses/by/4.0/), which permits use, sharing, adaptation, distribution and reproduction in any medium or format, as long as you give appropriate credit to the original author(s) and the source, provide a link to the Creative Commons license and indicate if changes were made.

The images or other third party material in this chapter are included in the chapter's Creative Commons license, unless indicated otherwise in a credit line to the material. If material is not included in the chapter's Creative Commons license and your intended use is not permitted by statutory regulation or exceeds the permitted use, you will need to obtain permission directly from the copyright holder.

Performance Analyses of Sentinel-3A and Sentinel-3B Over Lake Issyk Kul (Kyrgyzstan)

T. Schöne, J. Illigner, A. Zubovich, C. Zech, N. Stolarczuk, A. Sharshebaev, and M. Borisov

Abstract

As part of the European Copernicus program the radar altimetry satellites Sentinel-3A and Sentinel-3B were launched in 2016, and 2018 respectively. The satellites are one of the first operating in SAR mode allowing a much better height retrieval over the ocean and inland waters. The mission also benefits from the Open-Loop Tracking Command mode, where an a-priory elevation mask improves the performance over inland waters. This study analyses the performance and trends of the OCEAN and OCOG retracker functions in both, Ku and C band over Lake Issyk Kul. We make use of GNSS-derived lake profiles and information from shore-based tide gauges to analyze uninterrupted data series. We found biases of 2 ± 41 mm for Sentinel-3A and −45 ± 37 mm for Sentinel-3B for the OCEAN retracker and 307 ± 29 mm for Sentinel-3A and 345 ± 22 mm for Sentinel-3B using the OCOG retracker. Moreover, our results give evidence to small drifts for both satellites and also for both retracker.

Keywords

Radar altimetry · Validation · Retracker · Issyk Kul (Kyrgyzstan) · SAR · PLRM · Sentinel-3

1 Introduction

Over the past three decades radar altimetry (RA) provided a global, high-resolution, and regular monitoring of sea level and ocean circulation variations, information about coastal sea level, and also for inland water monitoring (Ablain et al. 2015; Crétaux et al. 2016; Song et al. 2015; Nerem et al. 2018). Starting with ERS-1 in 1991, an uninterrupted sequence of similar satellite RAs delivered heights of the ocean and inland water bodies on a 35-day repeat orbit (ESA 2011). Similarly, TOPEX/Poseidon (Fu et al. 1994) launched in 1992 was the first on a 10-day repeat orbit, continuing until today with Sentinel-6MF. Recently, the European Copernicus program (Copernicus 2015) launched a series of dedicated radar altimetry missions, with Sentinel-3A (S3A), launched in 2016, and Sentinel-3B (S3B), launched in 2018. Both satellites are on a 27-day repeat orbit.

Based on the radar principle, the satellites measure the travel time of a short pulse between the satellites center of gravity and the reflecting surface. Over the ocean from the leading edge of the returning waveform, the distance (range) can be estimated. It is usually assumed that the waveform over oceans follows the Brown model (Brown 1977), but this assumption fails near the coast or inland water bodies. To account for non-conforming surfaces (e.g., sea ice or land contamination) a so-called retracking is applied leading to improved height measurements (e.g., Martin et al. 1983; Rodriguez 1988; Laxon 1994; Dinardo et al. 2015). The pitfall are differences in the leading-edge detection between the different retracker models, causing inconsistencies in the

T. Schöne (✉) · J. Illigner · C. Zech · N. Stolarczuk
Deutsches GeoForschungsZentrum GFZ, Telegrafenberg, Potsdam, Germany
e-mail: tschoene@gfz-potsdam.de

A. Zubovich · A. Sharshebaev · M. Borisov
Central Asian Institute for Applied Geosciences CAIAG, Bishkek, Kyrgyzstan

derived water surface heights (e.g., Calmant et al. 2013), and, thus, to difficulties combining results from different missions and for using different retracker models in coastal areas.

Today, the majority of radar altimetry products provide results from different retracker models. Beside the OCEAN retracker (Brown 1977) also the offset center of gravity (OCOG, called ICE1 in prior missions) retracker (Wingham et al. 1986), where the range is defined through center of gravity of the returning waveform is a standard in altimetry. Dedicated other retracker ranges are also available for different RA missions (COPERNICUS n.d.), such as the SAMOSA retracker (Dinardo et al. 2015) for Sentinel-3.

An established method for the quality control and the estimation of biases between radar altimetry missions and different retracker is the ground truthing on dedicated monitoring sites. For ocean application, several calibration and validation sites already exist (Mertikas et al. 2020; Bonnefond et al. 2021; Esselborn et al. 2022; Zhou et al. 2023), but over inland waters sites are still rare. Therefore, Lake Issyk Kul (Kyrgyzstan) was established as a radar altimetry calibration site by Crétaux et al. (2009, 2011, 2018) and Quartly et al. (2020).

In the following, we analyze the retracker performance for the OCEAN and OCOG retracker of the Sentinel-3A and Sentinel-3B satellites, in particular the offsets (bias) and RMS and possible long-term trends in the stability of both altimeter satellites over Lake Issyk Kul with the recently established ground truth network as well as the newly acquired in-situ data.

2 The Issyk Kul Radar Altimetry Calibration Site

As part of a larger network of climate and water monitoring stations in Central Asia (Schöne et al. 2013) and in cooperation with the Central Asian Institute for Applied Geosciences (CAIAG), the Lake Issyk Kul observatory is developed since 2016 (Fig. 1) as part of GFZs Global Change Observatory Central Asia (GCOCA). The Lake Issyk Kul located in Kyrgyzstan is the second largest inland water body at an altitude of around 1600 m. The lake is surrounded by high mountains of the Tien Shan Mountain range. Several rivers supply water to the lake, mainly meltwater from glaciers and rain. The lake level is decreasing over time, likely due to global warming, but was much higher in the past (Gebhardt et al. 2017). Currently, the lake is endoreic, as the Chu River in the west is not reaching the lake anymore. The lake water is moderate saline and, therefore, neither used for irrigation nor freshwater supply. Also, the wind conditions in the lake area are moderate, exceeding 8 m/s only a few times per year.

In the southern and western part of the lake, two climate stations provide information about meteorological and hydrometeorological conditions such as wind velocity and

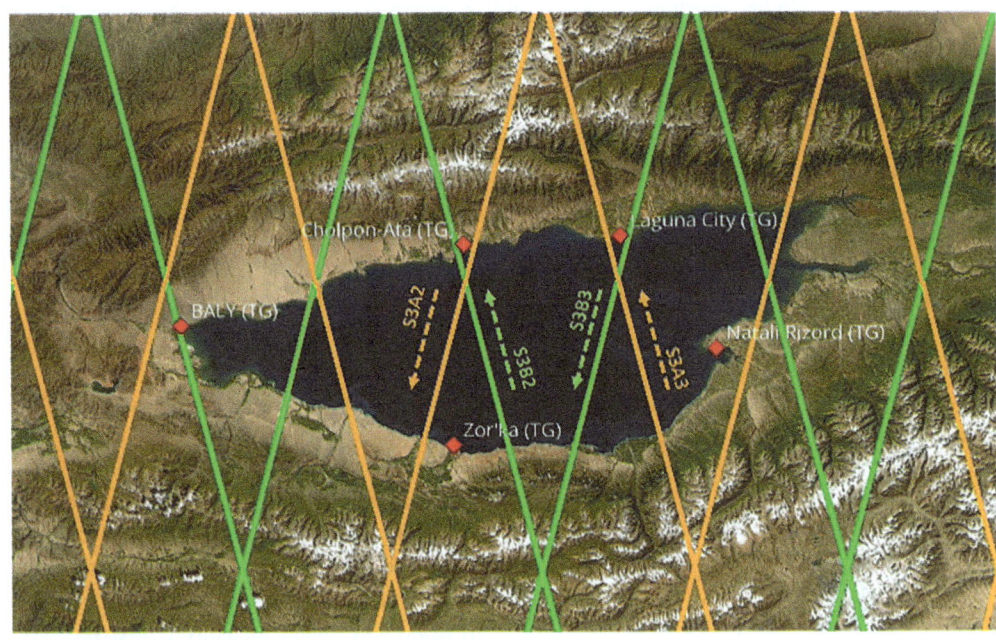

Fig. 1 Lake Issyk Kul (Kyrgyzstan) with the Sentinel-3A (orange) and -3B (green) tracks. In this study, only the central tracks are used (S3A2, S3A3, S3B2, S3B3) due to their sufficiently long water extent. BALY tide gauge is in the West, ZOKA in the South, CHOL and LAGU in the North, and NATI in the South-East. Source of background map: ESRI (World Imagery, 2023), Maxar, Earthstar Geographics, and the GIS User Community

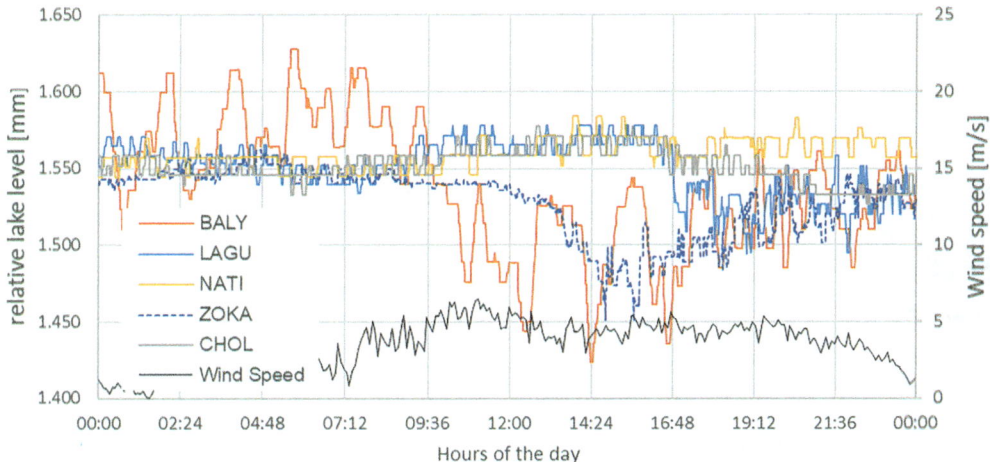

Fig. 2 Lake Issyk Kul (Kyrgyzstan) with a typical behavior of the tide gauge time series during strong winds. The wind directions are typical in W-E directions, changing with day-time

direction and air pressure (Zech et al. 2021). For in-situ water monitoring, four tide gauges have been installed distributed around the lake between 2016 and 2019. In 2022, a fifth tide gauge was added in the north-east to better characterize the internal lake state conditions, such as seiches and surges (Fig. 2), and in preparation of the Surface Water and Ocean Topography (SWOT) calibration/validation campaign. In addition, we obtained daily tide gauge data from the official gauge in Cholpon Ata on the northern shore, operated since 1927 by a branch of the national hydrometeorological service (KyrgyzHydromet).

Our tide gauge sites are equipped with a highly precise and stable radar gauge (VEGAPULS WL61), a CTD sensor (Conductivity, Temperature, Depth; D&K CTDSens300), and an air pressure probe (Setra CS100). All data is sampled minutely and transmitted at hourly basis. Internal level checks between the radar gauges demonstrate minor differential trends of about 1 mm/a. Therefore, we assume, all gauges deliver reliable long-term stable time series. In comparison the CTD sensors tend to drift and would require regular replacements and calibration, when including them for our validation. Therefore, we use the CTD data only for cross-checking. All tide gauge sites are regularly controlled by GNSS campaigns and spirit levelling. In the following, we use the tide gauge data from Cholpon Ata (CHOL) only, as we have the best control with two independent lake level time series.

The Lake Issyk Kul has favorite conditions for altimetry calibration, with mostly calm wave conditions. Only a few days per year show higher sea state conditions (during storms, 3...5 m). Due to the moderate salinity, the immense water body (lake depth 680 m), and cross-wind conditions, the lake is not freezing during wintertime and also mostly ice-float free. These conditions allow a year-around obser-

vation of RA. Due to the E-W stretching of the lake we occasionally observe seiches (bathtub effect) generated by rapid changes in wind force (from 2 m/s to 5 m/s) and direction (from W-E to E-W) measured at the monitoring station at Balykchy (western shore). The observed seiches show amplitudes at Balykchy (BALY) of up to ±10 cm and a wavelength of about 120 min. The seiches are not always visible in the tide gauge records of the other four gauges, but the surge effect can be detected at the southern tide gauges. The different amplitudes of the seiches could be addressed to shore shadowing effects (CHOL, NATI) or surge (ZOKA). The surge signal at LAGU may be caused by the bathymetry gradient (deep to shallow) near the gauge (ref. to Gebhardt et al. 2017, Fig. 1). The strong effect at BALY is most likely due to the narrowing and the shallow bathymetry of the lake at the eastern shore.

During satellite passages, this might have small effects on the instantaneous lake height departing from tide gauge readings. As another advantage, the Lake Issyk Kul is neither influenced by barotropic changes nor by tides, as typical for ocean calibration sites.

For our analyses we make use of GNSS lake profiles, which have been measured in 2017 and 2018 for the nominal S3A and S3B passes (Fig. 1) for both ascending and descending orbits. We equipped a local vessel (RV Multur) with a radar gauge (VEGA VEGAPuls 61) and a GNSS system (Septentrio AsteRX2c with a NAX3G+C antenna) mounted at the bow of the vessel (Fig. 3). The radar gauge measures the distance between the GNSS antenna reference point and the water surface. Both GNSS and radar gauge sample 1 Hz data. The GNSS data is processed in kinematic mode with the CSRS-PPP software (Banville et al. 2021). The ellipsoidal heights of the GNSS solution in combination with the radar distance measurements result in lake level heights (Fig. 4).

Fig. 3 Mounted GNSS antenna and radar gauge at the bow of RV Multur. The offset between the antenna reference point of the GNSS antenna and the zero point of the WL61 radar gauge is measured with millimeter accuracy

The derived height profiles are still noisy due to the constant motion (pitch and roll) of the vessel especially under higher sea state conditions. Therefore, we filter the results with a 120 second median filter, corresponding to roughly 350 m (at 10 km/h averaged speed of the vessel) on ground.

Since the Lake Issyk Kul level changes over time due to evaporation and rain in summer and glacier melting in spring, for operational reasons the profiles measured in 2018 are corrected to the medium lake level of our lake profiling in 2017 using the lake level difference measured by the Cholpon Ata tide gauge.

3 Conceptual Design for the Retracker Performance Analyses

For our analyses of the radar altimetry mission S3A and S3B over Lake Issyk Kul we follow a simple concept. The lake level profile along the satellite passage was either measured in or corrected to the 2017 measurement epoch. As the lake level changes with time, we use the changes in lake level heights from the reference tide gauge in Cholpon Ata to compute the lake level difference between the reference date in 2017 and the time of the satellite passage. Assuming that the geometrical form of the lake surface is primarily influenced by the geopotential and does not change with varying lake levels, we geometrical shift the reference track to the "instantaneous" lake level for each satellite passage by applying the tide gauge difference (Fig. 4). Here, we also assume that effects, such as surges, are small or neglectable (Fig. 2). The resulting reconstructed lake level profiles are still in the ITRF2014 reference frame (Altamimi et al. 2016) and can be used for a direct comparison with RA measurements. It can be noted from Fig. 4, that the mid-lake level is geometrically almost 2 m lower compared to the shore level. This can be explained by the gravity effects in this area, where the lake is surrounded by the high mountains of the Tien Shan with more than 5000 m altitudes and the lakes bathymetry, which has depth of up to -680 m.

Both Sentinel-3 satellites have the same orbital parameters, but with S3B shifted by half of a cycle. The orbit is near-polar and sun-synchronous at a nominal altitude of 814.5 km. The satellites operate at Ku and C band frequencies, the

Fig. 4 Example of Sentinel-3A descending track (westernmost) over Lake Issyk Kul for different dates (OCEAN retracker, 20 Hz data). The reference track is shifted in height by the lake level difference measured by the tide gauge in Cholpon Ata between the RA passage times and the reference time

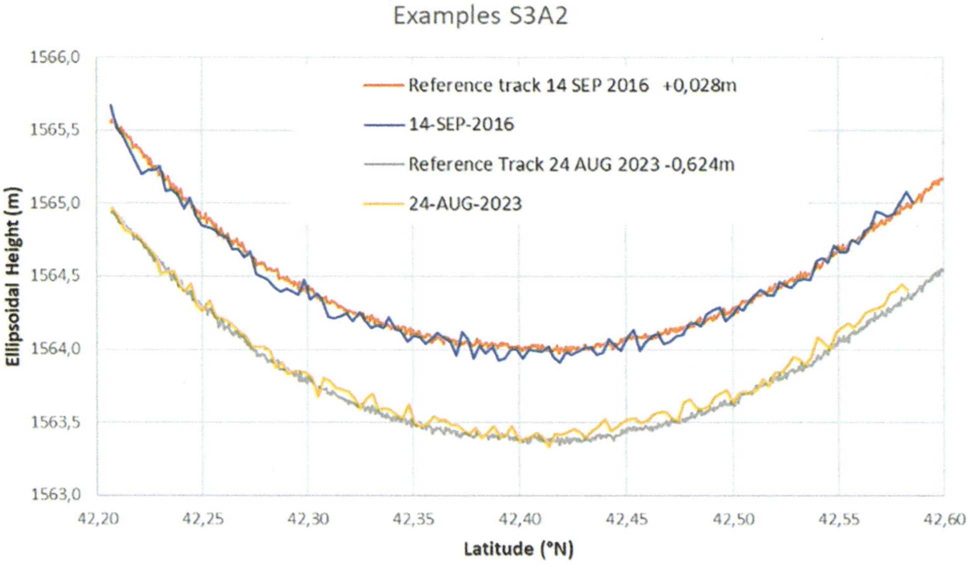

Table 1 Correction models applied for the instantaneous lake surface heights. No radiometer data is available; therefore, the ECMWF model output is used. Neither sea state bias nor inverse barometric corrections are applied over enclosed lakes. Tides are also not present at Lake Issyk Kul, the ocean loading is equal at the tide gauge and the lake surface, and thus not applied

Correction model	Source
Orbit	CNES, product internal
Ionosphere	GIM CODE (Dach et al. 2020)
Dry troposphere	ECMWF, product internal
Wet troposphere	ECMWF, product internal
Earth tides	IERS Conventions (Petit and Luzum 2010)
Pole tides	EOP 14C04 (Bizouard et al. 2019)

latter is important for computing the ionosphere attenuation correction for the Ku band altimeter measurements. The processing and distribution of Level 2 data is shared between EUMETSAT for data over ocean and ESA/Copernicus for data over land. Little effort has been made by ESA for a dedicated reprocessing, therefore, the data used is a mixture of two revisions (S3A_SR_2_LAN___ & S3B_SR_2_LAN___ in revision 3 and 4). Mid October 2023, the provision of this product was terminated and a new product was made available recently (S3A_SR_2_LAN_HY), but will not be discussed here. The land product is provided as netCDF files and contains all necessary parameters for computing the surface heights. In particular, we use orbital altitudes, geophysical and environmental correction models (Table 1), and range estimates for the different retracker. We performed the analysis using the OCEAN/SAMOSA (Dinardo et al. 2015; Copernicus 2022) and OCOG (Wingham et al. 1986) retracker ranges in Ku band and C band (COPERNICUS n.d.).

4 Retracker Performance Analyses in Ku Band

We use 20 Hz RA data in our analysis. In a first step, we apply the environmental corrections (Table 1) and compute the individual lake heights for each retracker in the frequency-band. S3B is additionally corrected for the USO drift (EUMETSAT 2021), which is not corrected in revision 003 and 004. The lake has no ocean tides or effects of barotropic changes, therefore, those corrections as well as corrections for sea state biases are not applied. Sea state biases may exist at Lake Issyk Kul but no model exist to compute this quantity over inland lakes. In a second step, we correct the reference track for the lake level changes and interpolate heights from our reference track for individual radar measurements. So far, we have to neglect any cross-track errors in our analyses, since we measured only single GNSS lake profiles. Aiming that the satellite heights have an offset in a first step, we estimated the mean offset of all cycles per track, which can be considered as a constant bias and is applied prior to the statistical analyses. Unfortunately, this value is not consistent for the ascending/descending reference tracks. The most likely reason is a small inconsistency in the reference heights. Although our reference profiles are in ITRF2014, we cannot ensure overall repeatability. Another reason could be found in the orbit determination, global analyses reveal also differences between ascending and descending tracks (e.g., Esselborn et al. 2018). For the OCEAN retracker (Ku Band) of S3A the mean offset difference between the ascending and descending track is -10 mm, for S3B 43 mm. For the OCOG retracker (Ku band), the offset differences are 20 mm and 34 mm respectively (Table 2). The mean offset values are applied

Table 2 Statistical parameters for S3A and S3B for the OCEAN and OCOG retracker in Ku band. Naming conventions for our ground tracks are in Fig. 1

Track	RMS mean deviation (mm)	Median RMS (mm)	Min RMS (mm)	Max RMS (mm)	Min number of 20 Hz	Median number of 20 Hz	Max number of 20 Hz	Trend (mm/a)	RMS of trends (mm/a)	OFFSET [mm] to lake level
OCEAN (Ku-Band)										
S3A2	35.8	61.0	38.0	140.0	13	132	133	4.24	1.92	7
S3A3	47.2	67.0	42.4	190.1	106	146	149	0.76	2.58	-3
S3B2	41.2	64.2	39.6	146.0	49	145	148	5.91	4.24	-66
S3B3	33.5	69.5	37.5	148.2	119	156	157	2.11	3.51	-23
OCOG (Ku-Band)										
S3A2	28.8	59.2	46.1	111.5	13	133	133	2.45	1.53	-317
S3A3	29.7	62.5	49.2	126.9	133	147	149	-0.28	1.67	-297
S3B2	24.5	67.9	51.4	115.4	78	147	148	1.64	2.27	-362
S3B3	20.4	62.1	50.8	117.1	128	156	157	-1.45	1.92	-328

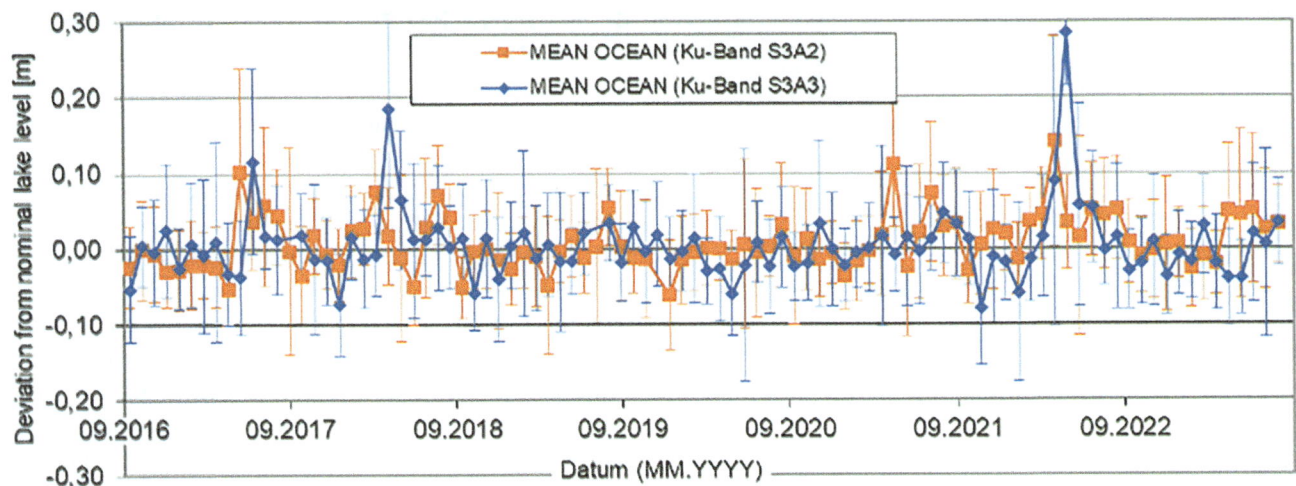

Fig. 5 Deviation of Sentinel-3A (S3A2 descending, S3A3 ascending) for Ku band from the reference track. The error bars are the RMS of all valid 20 Hz measurements per passage

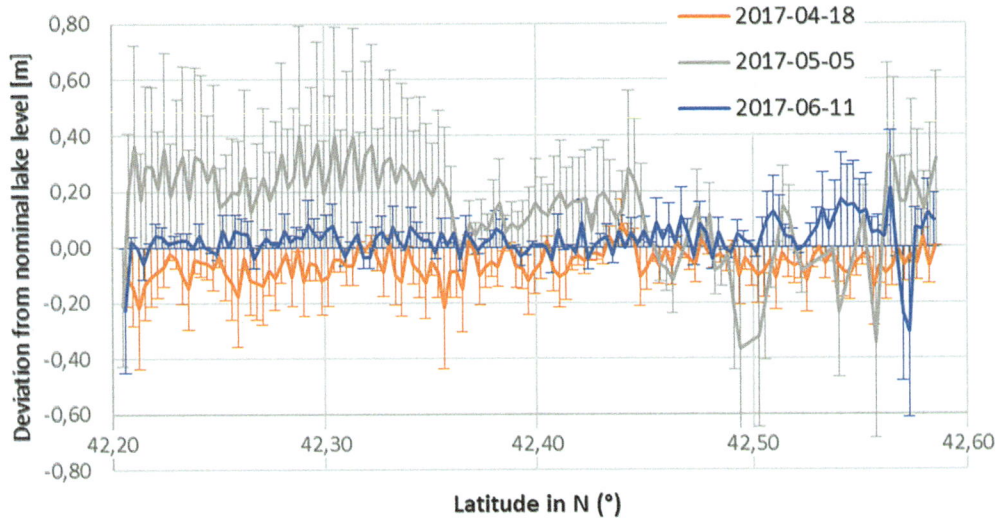

Fig. 6 Example of Sentinel-3A (descending). The heights are reduced by the reference track. The passage on 05-MAY-2017 is one of the few situations, where the retracking failed. Shown are also the passage before and after

individually for all tracks as a constant parameter. For outlier detection, radar altimetry heights more than 0.4 m apart from their reference track are eliminated (Schöne et al. 2018).

We define a number of quality criteria for the performance of a retracker/band combination. The primary criterion is the root mean square (RMS) of all 20 Hz measurements versus the reference track. For better characterization, Table 2 contains the minimum, maximum and median of the RMS for all cycles. The number of measurements is also shown in Table 2, again as minimum, maximum and median values. Finally, we estimate the trend of the time series and the RMS of the deviation from the mean value.

Sentinel-3A (OCEAN retracker, Fig. 5) shows stable performance over time. Except for very few passages where the retracker failed (Fig. 6), the number of valid 20 Hz measurements are close to the maximum return rate. The RMS for all passages is around 61 mm (S3A2) and 67 mm (S3A3), the mean offset from our nominal lake profile is 2 mm for S3A. While we explain the offset difference between the ascending and descending track with uncertainties in our nominal ground track, the trend difference between the descending and ascending track is notable. The descending track shows a trend of 4.2 mm/a (S3A2), the ascending track only 0.8 mm/a (S3A3). Reasons for this difference could either be orbital errors between ascending and descending tracks or the tropospheric correction. The passage times are around 11:23 AM (local time) for the descending and 22:18 PM (local time) for the ascending passage.

For S3B (Fig. 7) we see a similar performance. The RMS for all passages is around 64 mm (S3B2) and 69 mm (S3B3),

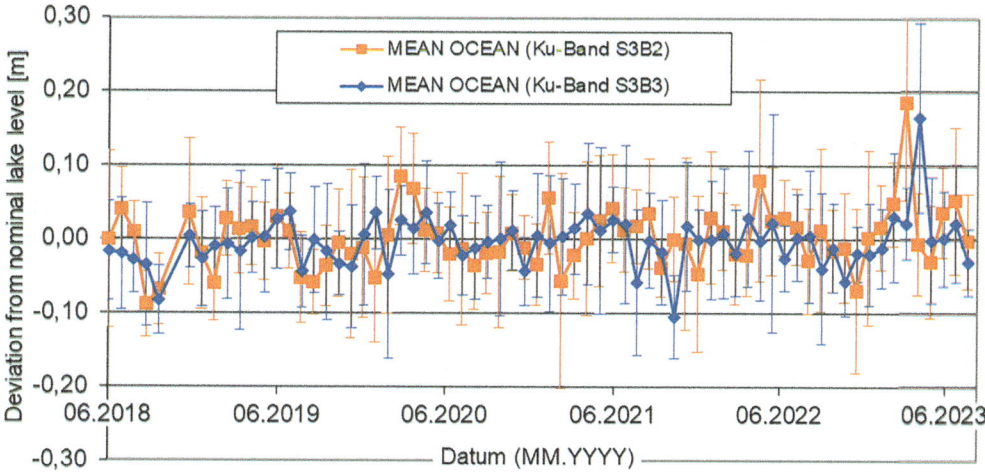

Fig. 7 Deviation of S3B (S3B2 ascending, S3B3 descending) for Ku band from the reference track. The error bars are the RMS of all valid 20 Hz measurements per passage

also the number of valid 20 Hz measurements is mostly close to the maximum. The offset is −45 mm but with a higher difference between the ascending and descending passages (Table 2). The trends compared to the reference tracks are slightly higher with 5.9 mm/a (S3B2) and 2.1 mm/a (S3B3), likely due to the shorter operation time compared to S3A. The local time of the S3B passages over Lake Issyk Kul is similar to S3A but for the opposed flight direction. The larger trends are for S3A2 (05:28LT) and S3B2 (16:19LT), thus, the troposphere correction is unlikely the reason for the trend difference.

It is also worthwhile to note, that the differences in the trends between the ascending and descending passages are similar for S3A and S3B (3.5 mm/a versus 3.8 mm/a). The reason remains unclear.

The OCOG retracker is widely used over inland waters (e.g., Calmant et al. 2013; Sulistioadi et al. 2014; Schöne et al. 2018; Gao et al. 2019). Over Lake Issyk Kul the OCOG retracker performs better than the OCEAN retracker, even the lake with its long passages is more ocean-like (Fig. 8). The RMS is smaller for both S3A and S3B, and the heights of the individual passages are more homogeneous over time (Table 2, "RMS of mean deviation"). The offsets to the reference profiles are higher, with −307 mm for S3A and −345 mm for S3B. The trends and the trend differences for the OCOG retracker to our reference tracks are lower, but not neglectable. The analysis also indicates no effect on the retracker performance in Ku Band for the OCEAN and OCOG retrackers for the release change between revision 3 and 4.

5 Retracker Performance Analyses C Band

Over oceans, the C band measurements are of particular importance since the dual frequency measurements are used estimating the ionospheric correction. The footprint is larger, but allow the estimation of the total electronic content and, thus, the ionospheric correction for the Ku band range measurements. For most of the inland water bodies the dual-frequency ionospheric correction is replaced by a Global Ionospheric Map value (Dach et al. 2020) value. The long tracks at Lake Issyk Kul allow also the C band validation (Fig. 9) and, therefore, are valuable to analyze using the same approach. Again, we analyze S3A and S3B separately for both retracker and for ascending and descending tracks. The individual RMS for the OCEAN(C) retracker are significantly higher than Ku band results, with up to 170 mm for both S3A and S3B. Again, the OCOG(C) retracker performs better, with RMS of around 140 mm (Table 3).

For the OCEAN(C) retracker the RMS is almost three times higher, for OCOG(C) retracker doubled in comparison to the Ku band values. The repeatability, trends and offsets are deteriorated since early 2022. The C band repeatability of both OCEAN(C) and OCOG(C) retracker for S3A and S3B is significantly more unstable after this date, but the RMS of the 20 Hz did not change. This performance degradation might be associated with the implementation of the processing baseline 3.04 but remains also for the latest processing baseline 3.05 (ESA 2023). For the release change between

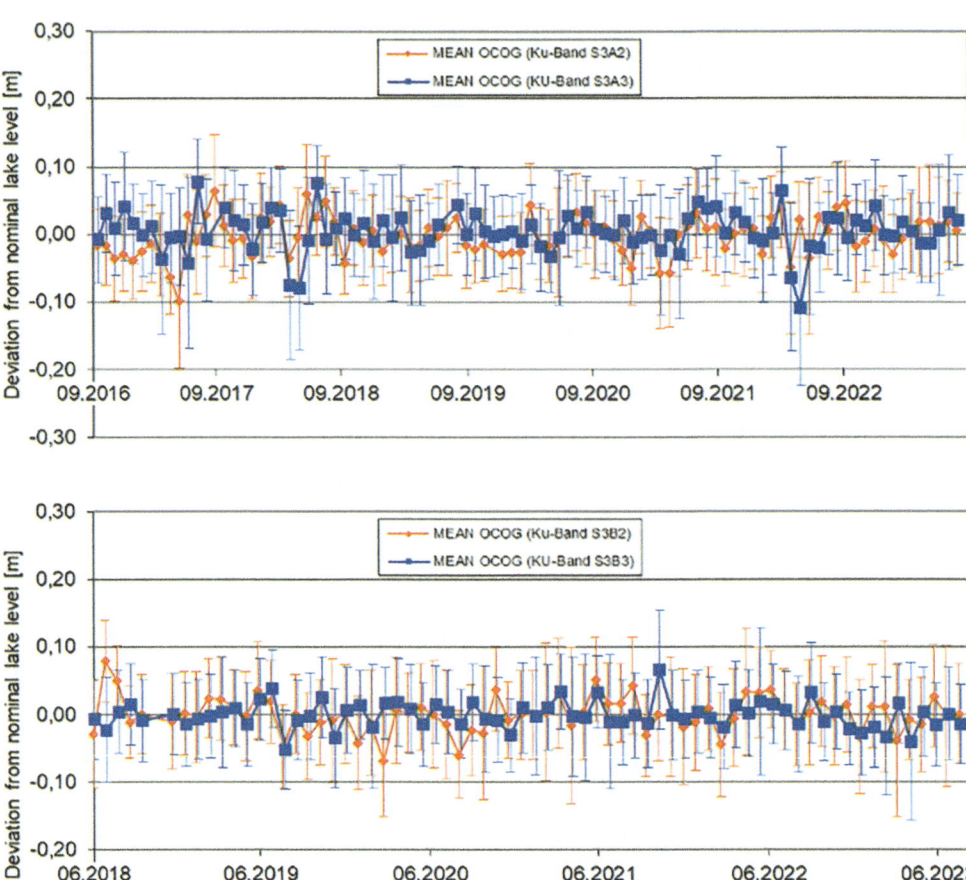

Fig. 8 Deviation from the reference tracks of S3A and S3B for the OCOG retracker in the Ku band. The error bars display the RMS of all valid 20 Hz measurements per passage

revision 3 and 4 the range estimates for the OCEAN and OCOG retrackers are stable, while the number of valid values for OCOG significantly increased for both satellites.

6 Retracker Performance Analyses in PLRM Mode (Ku Band)

One of the advantages of the Sentinel-3 satellites is the operation in SAR mode. This concept allows a much better resolution along-track and higher performance over inland water targets. Additionally, a pseudo low resolution mode (PLRM) is used for Sentinel-3 (Clerc et al. 2020) in Ku band which uses the individual echoes from the SAR mode data for the reconstruction of a classical LRM waveform. The Sentinel-3 land product provide two retracker based on this mode, also called OCEAN and ICE1 (similar to OCOG). In contrast to the OCEAN retracker in SAR mode, the Jason heritage ALT_RET_OCE_01 algorithm is used (CLS 2011). The RMS for each passage is significantly higher than their respective SAR mode results (Table 4). The repeatability (RMS of mean deviation) is better for OCEAN(PLRM) than for OCEAN in SAR mode, ICE1(PLRM) has a higher RMS. The trends for OCEAN(PLRM) are smaller than for OCEAN in SAR mode, while ICE1(PLRM) trends are higher with also a larger spread between ascending and descending tracks.

For the OCEAN(PLRM) results we find a similar performance compared to the OCEAN retracker in SAR mode, while ICE1(PLRM) performs less good compared to the OCOG retracker. Also, the analysis shows no effect for the revision change between revision 3 and 4.

7 Conclusions and Outlook

ESA/Copernicus Sentinel-3 is one of the first altimetry mission operating almost completely in a synthetic aperture radar (SAR) mode allowing a much better height retrieval over inland water bodies (Crétaux et al. 2023). Both satellites (S3A and S3B) operate in the open-loop tracking mode nearly globally, which leads to a better data return rate by using a predefinition of hydrological targets (Taburet et al. 2020). This allows a much better observation of inland water bodies but also enables us to study the quality of different retracker over inland water targets. We performed quality analyses of the OCEAN and OCOG retracker in SAR-mode for both frequency bands (Ku and C band) and in PLRM-

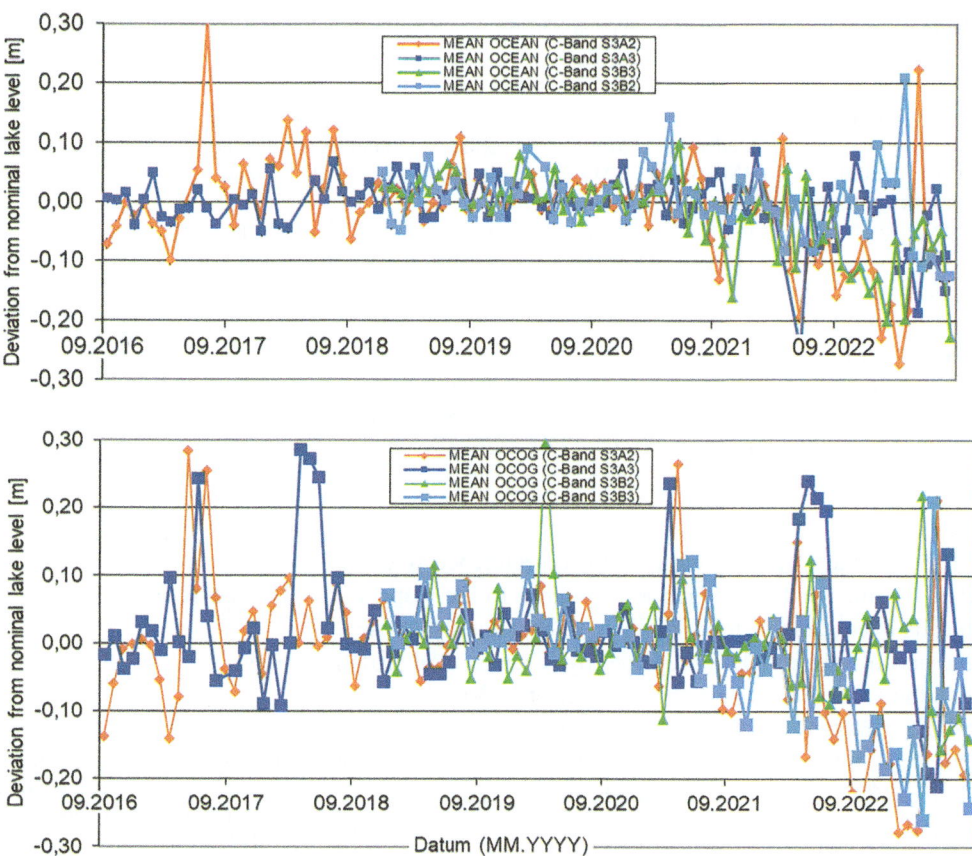

Fig. 9 Deviation of S3A and S3B for the OCEAN(C) and OCOG(C) retracker in the C band from the reference track. The error bars are not shown for clarity. Since early 2022 the performance is degraded

Table 3 Statistical parameters for S3A and S3B for the OCEAN(C) and OCOG(C) retracker in C band. Naming conventions for our ground tracks are in Fig. 1

Track	RMS of mean deviation (mm)	Median of RMS (mm)	Min of RMS (mm)	Max of RMS (mm)	Min number of 20 Hz	Median number of 20 Hz	Max number of 20 Hz	Trend (mm/a)	RMS of trends (mm/a)	OFFSET [mm] to lake level
OCEAN (C-Band)										
S3A2	87.1	169.6	88.5	264.5	2	100	120	−17.92	4.14	87
S3A3	53.1	176.6	86.5	251.0	2	101	132	−8.31	2.38	12
S3B2	59.1	171.6	70.5	242.4	2	103	128	−11.59	4.84	−116
S3B3	70.6	171.3	117.4	211.1	5	118	140	−34.52	4.71	−8
OCOG (C-Band)										
S3A2	106.6	124.8	61.1	177.9	12	112	128	−21.54	4.72	43
S3A3	89.3	140.0	58.6	227.7	32	112	142	−5.39	4.14	−15
S3B2	75.0	145.0	63.4	196.7	33	128	141	−14.27	6.04	−127
S3B3	91.3	134.8	92.0	161.8	70	144	151	−38.57	6.86	−21

mode in Ku band. Our results show small inconsistencies between the offsets for the ascending and descending tracks for both S3A and S3B, which are likely caused by our reference tracks rather than inherited by satellite performances. Further inconsistencies may arise from the deviation of the satellites from the nominal ground track, the large surface gradients at Lake Issyk Kul may result in a higher RMS of the repeatability when the satellite is not directly at our GNSS-derived reference track. With the recently launched SWOT altimetry missions, such cross-track gradients can be corrected.

Over Lake Issyk Kul the OCOG retracker performs better for both satellites than the OCEAN retracker. The RMS ("median of RMS" in Table 2) for both satellites and retracker is around 65 mm, but slightly smaller for the OCOG retracker. The repeatability ("RMS of mean deviation" in Table 2) is around 40 mm for the OCEAN retracker and around 30 mm for the OCOG retracker. The

Table 4 Statistical parameters for S3A and S3B for the OCEAN(PLRM_Ku) and ICE1(PLRM_Ku) retracker in Ku band. Naming conventions for our ground tracks are in Fig. 1

Track	RMS of mean deviation (mm)	Median of RMS (mm)	Min of RMS (mm)	Max of RMS (mm)	Min number of 20 Hz	Median number of 20 Hz	Max number of 20 Hz	Trend (mm/a)	RMS of trends (mm/a)	OFFSET [mm] to lake level
PLRM OCEAN (Ku-Band)										
S3A2	31.5	115.8	42.6	180.8	12	121	129	3.76	1.59	−1
S3A3	33.4	119.5	93.0	165.3	17	121	141	0.21	1.75	−14
S3B2	34.6	121.8	96.3	183.3	9	128	143	0.87	3.24	−90
S3B3	34.6	118.1	89.9	157.4	10	142	151	−0.15	3.23	−37
PLRM ICE1 (Ku-Band)										
S3A2	34.5	119.5	52.6	188.0	12	125	129	4.47	1.77	13
S3A3	45.3	123.7	52.6	207.8	21	132	142	2.39	2.09	0
S3B2	38.1	127.1	97.9	167.5	29	133	143	3.45	3.43	−65
S3B3	29.3	123.3	93.5	206.4	40	146	151	−4.17	3.51	−15

offset for the OCEAN retracker to the reference tracks is 2 ± 41 mm for S3A, which is very close to the responder results over ocean of Mertikas et al. (2019) but also close to the bias of −14 ± 20 mm reported by (Crétaux et al. 2018). For the OCOG retracker, the S3A offset is 307 ± 29 mm where Crétaux et al. (2018) reported 285 ± 20 mm. For S3B the offsets are −45 ± 37 mm (OCEAN) and 345 ± 22 mm (OCOG). Based on this assessment for inland water monitoring the OCOG retracker (SAR mode) is recommended.

The interpretation of the C band is more challenging, as the measurements are degraded in quality since 02/2022. We are addressing the degradation on performance for both satellites and both retracker to the change in the processing baseline. Therefore, we do not further discuss the values in Table 3. Prior to this change, the performance of C band shows higher RMS of the individual satellite passages but also higher RMS for the mean deviation. Also here, results from the OCOG retracker show smaller values than the OCEAN retracker.

For a next step, we plan to re-measure our ground truth reference tracks to get more valid information about the quality of our lake surface estimation. Also, replacing the tropospheric correction by results from shore-based continuous GNSS stations may improve the results.

Acknowledgements We are grateful to the captain and crew of RV Multur in Cholpon Ata and our colleagues at CAIAG in Bishkek who supported us prior, during and after our trips. We acknowledge the work of the Copernicus program providing the Level 2 altimetry land data. We thank the Hydrometeorological Service of Kyrgyzstan for providing the tide gauge data of Cholpon Ata. The owners of the piers kindly provided us the possibility installing the tide gauges at their premises. This work is part of GFZs Global Change Observatory Central Asia (GCOCA) and of the GreenCentralAsia program of the German Ministry of Foreign Affairs (AA7090002). Meteorological data can be accessed through https://sdss.caiag.kg. We also value the comments of the two anonymous reviewers, which helped to improve the manuscript.

References

Ablain M, Cazenave A, Larnicol G et al (2015) Improved sea level record over the satellite altimetry era (1993–2010) from the Climate Change Initiative project. Ocean Sci 11:67–82. https://doi.org/10.5194/os-11-67-2015

Altamimi Z, Rebischung P, Métivier L, Collilieux X (2016) ITRF2014: a new release of the International Terrestrial Reference Frame modeling nonlinear station motions. J Geophys Res Solid Earth 121:6109–6131. https://doi.org/10.1002/2016JB013098

Banville S, Hassen E, Lamothe P et al (2021) Enabling ambiguity resolution in CSRS-PPP. NAVIGATION 68(2):433–451. https://doi.org/10.1002/navi.423

Bizouard C, Lambert S, Gattano C et al (2019) The IERS EOP 14C04 solution for Earth orientation parameters consistent with ITRF 2014. J Geod 93(5):621–633. https://doi.org/10.1007/s00190-018-1186-3

Bonnefond P, Laurain O, Exertier P, Calzas M, Guinle T, Picot N, and the FOAM project Team (2021) Validating a new GNSS-based sea level instrument (CalNaGeo) at Senetosa Cape. Mar Geod. https://doi.org/10.1080/01490419.2021.2013355

Brown GS (1977) The average impulse response of a rough surface and its application. IEEE Trans Antennas Propag AP 25:67–73

Calmant S, da Silva JS, Moreira DM et al (2013) Detection of Envisat RA2/ICE-1 retracked radar altimetry bias over the Amazon basin rivers using GPS. Adv Space Res 51(8):1551–1564. https://doi.org/10.1016/j.asr.2012.07.033

Clerc S, Donlon C, Borde F et al (2020) Benefits and lessons learned from the Sentinel-3 tandem phase. Remote Sens 12(17):2668. https://doi.org/10.3390/rs12172668

CLS (2011) Surface Topography Mission (STM) SRAL/MWR L2 Algorithms Definition, Accuracy and Specification, CLS-DOS-NT-09-119, https://www-cdn.eumetsat.int/files/2020-04/pdf_s3_alt_level_2_adas.pdf. Accessed 24 Apr 2024

Copernicus (2015). https://www.esa.int/Applications/Observing_the_Earth/Copernicus/Sentinel-3A_on_its_way. Accessed 23 Oct 2023

COPERNICUS (2022) Sentinel-3 SRAL/MWR Land User Handbook, S3MPC-STM_RP_0038 Issue 1.1 – 05/12/2022, https://sentinel.esa.int/documents/247904/4871083/Sentinel-3+SRAL+Land+User+Handbook+V1.1.pdf. Accessed 18 Apr 2024

COPERNICUS (n.d.). https://sentinels.copernicus.eu/web/sentinel/technical-guides/sentinel-3-altimetry/level-2/re-tracking-estimates. Accessed 24 Oct 2023

Crétaux JF, Calmant S, Romanovski V et al (2009) An absolute calibration site for radar altimeters in the continental domain: Lake

Issykkul in Central Asia. J Geod 83:723–735. https://doi.org/10.1007/s00190-008-0289-7

Crétaux JF, Calmant S, Romanovski V et al (2011) Absolute calibration of Jason radar altimeters from GPS kinematic campaigns over Lake Issykkul. Mar Geod 34(3–4):291–318. https://doi.org/10.1080/01490419.2011.585110

Crétaux J-F, Abarca-del-Río R, Bergé-Nguyen M et al (2016) Lake volume monitoring from space. Surv Geophys 37(2):269–305. https://doi.org/10.1007/s10712-016-9362-6

Crétaux J-F, Bergé-Nguyen M, Calmant S et al (2018) Absolute calibration or validation of the altimeters on the Sentinel-3A and the Jason-3 over Lake Issykkul (Kyrgyzstan). Remote Sens 10(11):1679. https://doi.org/10.3390/rs10111679

Crétaux JF, Calmant S, Papa F et al (2023) Inland surface waters quantity monitored from remote sensing. Surv Geophys 44:1519–1552

Dach R, Schaer S, Arnold D, Kalarus MS, Prange L, Stebler P, Villiger A, Jäggi A (2020) CODE final product series for the IGS. IGS Datensatz Astron. Inst. Univ. Bern. https://www.aiub.unibe.ch/download/CODE. Accessed 24 Oct 2023

Dinardo S, Lucas B, Benveniste J (2015) Sentinel-3 STM SAR ocean retracking algorithm and SAMOSA model. In: IEEE International Geoscience and Remote Sensing Symposium (IGARSS), ISBN: 978-1-4799-7929-5, Milan, Italy, July 2015. https://doi.org/10.1109/IGARSS.2015.7327036

ESA (2011). https://www.esa.int/Applications/Observing_the_Earth/ERS_satellite_missions_complete_after_20_years. Accessed 24 Oct 2023

ESA (2023). https://sentinel.esa.int/documents/247904/2753172/Sentinel-3-Product-Notice-LAND-L1-L2-NRT-STC-NTC.pdf. Accessed 27 Oct 2023

Esselborn S, Rudenko S, Schöne T (2018) Orbit-related sea level errors for TOPEX altimetry at seasonal to decadal timescales. Ocean Sci 14:205–223. https://doi.org/10.5194/os-14-205-2018

Esselborn S, Schöne T, Illigner J et al (2022) Validation of recent altimeter missions at non-dedicated tide gauge stations in the southeastern North Sea. Remote Sens 14(1):236. https://doi.org/10.3390/rs14010236

EUMETSAT (2021). https://www-cdn.eumetsat.int/files/2021-12/S3B%20USO%20sign%20correction.pdf. Accessed 24 Oct 2023

Fu L-L et al (1994) TOPEX/POSEIDON mission overview. JGR 99(C12):24369–24381

Gao Q, Makhoul E, Escorihuela MJ et al (2019) Analysis of retrackers' performances and water level retrieval over the Ebro River Basin using Sentinel-3. Remote Sens 11(6):718. https://doi.org/10.3390/rs11060718

Gebhardt AC, Naudts L, De Mol L et al (2017) High-amplitude lake-level changes in tectonically active Lake Issyk-Kul (Kyrgyzstan) revealed by high-resolution seismic reflection data. Clim Past 13:73–92. https://doi.org/10.5194/cp-13-73-2017

Laxon S (1994) Sea ice altimeter processing scheme at the EODC. Int J Remote Sens 15(4)

Martin TV, Zwally HJ, Brenner AC, Bindschadler RA (1983) Analysis and retracking of continental ice sheet radar altimeter waveforms. J Geophys Res 88:1608–1616

Mertikas SP et al (2019) Absolute calibration of Sentinel-3A and Jason-3 altimeters with sea-surface and transponder techniques in West Crete, Greece. In: Mertikas S, Pail R (eds) Fiducial reference measurements for altimetry. International Association of Geodesy Symposia, vol 150. Springer, Cham. https://doi.org/10.1007/1345_2019_63

Mertikas S, Tripolitsiotis A, Donlon C et al (2020) The ESA permanent facility for altimetry calibration: monitoring performance of radar altimeters for Sentinel-3A, S3B and Jason-3 using transponder and sea-surface calibrations with FRM standards. Remote Sens 12:2642. https://doi.org/10.3390/rs12162642

Nerem RS, Beckley BD, Fasullo JT, Hamlington BD, Masters D, Mitchum GT (2018) Climate-change–driven accelerated sea-level rise detected in the altimeter era. Proc Natl Acad Sci U S A 115:2022–2025. https://doi.org/10.1073/pnas.1717312115

Petit G, Luzum B (eds) (IERS Technical Note No. 36) Technical Support: Stetzler, Beth; Bachmann, Sabine; Wolfgang, Dick R. Frankfurt am Main: Verlag des Bundesamts für Kartographie und Geodäsie, 2010. https://iers-conventions.obspm.fr/content/tn36.pdf. Accessed 24 Oct 2023

Quartly GD, Nencioli F, Raynal M, Bonnefond P, Nilo Garcia P, Garcia-Mondéjar A, Flores de la Cruz A, Crétaux J-F, Taburet N, Frery M-L, Cancet M, Muir A, Brockley D, McMillan M, Abdalla S, Fleury S, Cadier E, Gao Q, Escorihuela MJ, Roca M, Bergé-Nguyen M, Laurain O, Bruniquel J, Féménias P, Lucas B (2020) The roles of the S3MPC: monitoring, validation and evolution of Sentinel-3 altimetry observations. Remote Sens 12:1763. https://doi.org/10.3390/rs121111763

Rodriguez E (1988) Altimetry for non-Gaussian Oceans: height biases and estimation of parameters. J Geophys Res 93:14107–14120

Schöne T, Zech C, Unger-Shayesteh K et al (2013) A new permanent multi-parameter monitoring network in Central Asian high mountains - from measurements to data bases. Geosci Instrum Methods Data Syst 2(1):97–111. https://doi.org/10.5194/gi-2-97-2013

Schöne T, Dusik E, Illigner J, Klein I (2018) Water in Central Asia: reservoir monitoring with radar altimetry along the Naryn and Syr Darya Rivers. In: International Symposium on Earth and Environmental Sciences for Future Generations, (International Association of Geodesy Symposia; 147). Springer Verlag, Online 2016, Cham, pp 349–357. https://doi.org/10.1007/1345_2017_265

Song C, Huang B, Ke L (2015) Heterogeneous change patterns of water level for inland lakes in High Mountain Asia derived from multi-mission satellite altimetry. Hydrol Process 29:2769–2781. https://doi.org/10.1002/hyp.10399

Sulistioadi YB, Tseng K-H, Shum C et al (2014) Satellite radar altimetry for monitoring small river and lakes in Indonesia. Hydrol Earth Syst Sci Discuss 11:2825–2874. https://doi.org/10.5194/hessd-11-2825-2014

Taburet N, Zawadzki, Vayre M, Blumstein D, Le Gac S, Boy F, Raynal M, Labroue S, Cretaux J-F, Femenias P (2020) S3MPC: improvement on inland water tracking and water level monitoring from the OLTC onboard Sentinel-3 altimeters. Remote Sens 12:18, AN 3055. https://doi.org/10.3390/rs12183055

Wingham DJ, Rapley CG, Griffiths H (1986) New techniques in satellite altimeter tracking systems. Digest - International Geoscience and Remote Sensing Symposium (IGARSS) (1986), pp 1339–1344

Zech C, Schöne T, Illigner J et al (2021) Hydrometeorological data from a Remotely Operated Multi-Parameter Station network in Central Asia. Earth Syst Sci Data 13:1289–1306. https://doi.org/10.5194/essd-13-1289-2021

Zhou B, Watson C, Beardsley J, Legresy B, King MA (2023) Development of a GNSS/INS buoy array in preparation for SWOT validation in Bass Strait. Front Mar Sci 9:1093391. https://doi.org/10.3389/fmars.2022.1093391

Open Access This chapter is licensed under the terms of the Creative Commons Attribution 4.0 International License (http://creativecommons.org/licenses/by/4.0/), which permits use, sharing, adaptation, distribution and reproduction in any medium or format, as long as you give appropriate credit to the original author(s) and the source, provide a link to the Creative Commons license and indicate if changes were made.

The images or other third party material in this chapter are included in the chapter's Creative Commons license, unless indicated otherwise in a credit line to the material. If material is not included in the chapter's Creative Commons license and your intended use is not permitted by statutory regulation or exceeds the permitted use, you will need to obtain permission directly from the copyright holder.

Vision of a Clock-Based Network for Absolute Sea Level Monitoring

Asha Vincent and Jürgen Müller

Abstract

Global sea level shows an increasing trend for several decades driven mainly by climate change. Absolute Sea Level (ASL) changes can only be extracted from Relative Sea Level (RSL) measurements with proper reduction of vertical land movements of the bench marks. Atomic clocks at those tide gauges can potentially provide the absolute, near real-time physical height change. High-performance clocks with an uncertainty of 10^{-18} enable a height measurement with 1 cm accuracy. As RSL is related to regional tidal datums, one has to account for the local variations to obtain a globally consistent measurement of ASL. Hence, by incorporating land motion from clock observations, one can establish a consistent and uniform reference datum for assessing geoid-based absolute sea level changes worldwide.

Keywords

Absolute sea level · Atomic clocks · Physical height · Relative sea level · Relativistic geodesy

1 Introduction

The geoid, which represents the global mean sea level in rest, serves as a reference surface in geodesy. Precise measurements of the local relative sea level are typically carried out by comparing the level of the sea surface at a specific location to a fixed reference point on land (Wöppelmann and Marcos 2016). This reference point, often referred to as a benchmark or tidal datum, serves as a baseline for determining how the sea level at that location changes over time. At present GNSS (Global Navigation Satellite System) stations nearby tide gauge locations serve as benchmarks for estimating the relative land movements (Larson et al. 2013) in order to determine the Absolute Sea Level (ASL) (Peng et al. 2021).

The physical height at a point on the Earth surface depends upon the geoid at the time of the measurement (Ihde et al. 2017). As detailed in Dietrich (2014) ASL refers to the height of the sea surface above a fixed global reference without the effect of land motion. Unlike Relative Sea Level (RSL), which is determined by tide gauges with respect to a local or regional reference (tide gauge zero or a tidal datum), absolute sea level is referenced to a mean ellipsoid when using GNSS benchmarks as reference. Installation of high-performance atomic clocks at the tide gauge locations can provide the absolute physical height change which enables to directly obtain ASL changes with respect to the geoid. Philipp et al. (2020) give details on the relativistic definition of geoid and gravity potential.

2 Clocks to Replace GNSS Benchmarks

The physical height at a point on Earth's surface is continuously varying mainly due to external tidal effects and non-tidal mass distributions in the Earth system as explained in

A. Vincent (✉) · J. Müller
Institute for Geodesy, Leibniz University Hannover, Hannover, Lower Saxony, Germany
e-mail: vincent@ife.uni-hannover.de; mueller@ife.uni-hannover.de

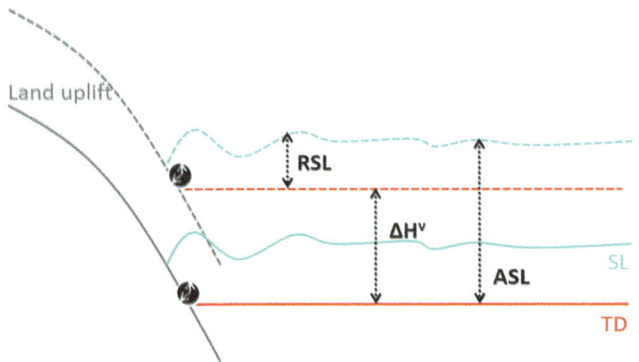

Fig. 1 Absolute sea level change (ASL) and relative sea level (RSL) with respect to a tidal datum (TD) in the case of land uplift ΔH^v and sea level (SL) rise

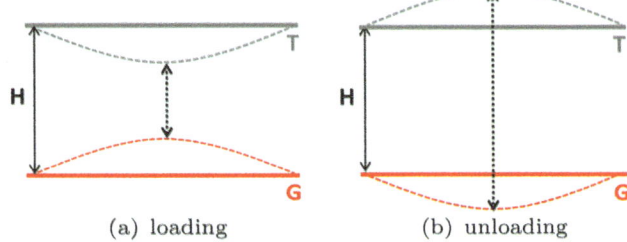

Fig. 2 Loading and unloading effects on physical height which depends upon geoid (G) height and vertical displacement of Earth surface (T)

Voigt et al. (2016). Tidal effects include solid-earth tides and ocean-load tides due to external bodies such as the Sun, Moon and other planets, centrifugal effects of polar motion result in pole tides and LOD (Length Of Day) tides arise from the variations in the Earth's angular velocity, atmospheric tides, etc. Non-tidal effects include mass redistributions in the geosphere, hydrosphere, atmosphere and biosphere. All these effects affect the geopotential at the point of interest as a result of mass change and the associated vertical displacements (Schröder et al. 2021). From terrestrial clock observations, the physical height changes can be inferred where GNSS receivers provide only ellipsoidal height changes. Thus, we can compute ASL changes with respect to the geoid by adding relative sea level changes with time-variable land motion (ΔH^v) from clock observations:

$$ASL = RSL + \Delta H^v. \qquad (1)$$

As RSL measurements are referenced to GNSS benchmarks near tide gauges, it doesn't necessarily guarantee consistency in the vertical datum across different locations. Vertical datums can vary regionally due to factors such as local geoid variations, tectonic movements, and other geophysical processes. Here, clocks offer an alternative by providing a direct reference to an equipotential surface such as the geoid. Using atomic clocks as benchmarks will be a more consistent way of deriving the ASL change globally by referencing both land and sea level measurements to a known equipotential surface, e.g., geoid. Then, we do not need to consider the regional tidal datum variations, as measurements will be taken with respect to the global geoid that preserves the uniformity of a reference surface without any local variations (Fig. 1).

If connected to a stable reference clock that is related to the geoid (W_0), terrestrial clocks provide physical heights H, which can also be represented in terms of the geopotential number, $C_p = W_0 - W_p$, where W_0 represents the potential on the geoid and W_p is the gravity potential at point p (Torge et al. 2023). Here, gravity potential changes can directly only be observed by clocks. And keep in mind, a potential change is independent of whether the mass change occurs above or below the Earth's surface, i.e. above or below the clock location. The gravitational potential change or the corresponding physical height difference between two clocks can be derived from the fractional frequency difference (Bjerhammar 1986; Müller et al. 2018; Wu and Müller 2019; Denker et al. 2018),

$$\frac{\Delta f}{f} \approx \frac{\Delta W}{c^2} \approx \frac{g \Delta H}{c^2} \qquad (2)$$

when using the GCRS (Geocentric Celestial Reference System) metric up to the first Newtonian order (orders of c^{-4} are omitted) where ΔW is the gravity potential difference between the two clock sites, c is the speed of light and, g is the mean gravity value. As illustrated in Fig. 2, tidal effects and non-tidal loading/unloading effects affect the physical heights represented as ΔH^v_{tidal} and $\Delta H^v_{non-tidal}$ respectively. According to Eq. (2), these time-variable changes are obtained by clocks (the static part cancels out for clocks at a fixed location),

$$\Delta H^v = \Delta H^v_{tidal} + \Delta H^v_{non-tidal}. \qquad (3)$$

In order to get the time-dependent variations in physical heights, there should be a stable reference clock somewhere on ground or space that is related to the geoid (Wu and Müller 2020; Philipp et al. 2023). As shown by Lisdat et al. (2016), the systematic uncertainty of the optical fibre link is of 10^{-19}. For our purpose, we propose only a single link for each tide gauge clock to the reference clock which is assumed to be in a geostationary satellite. We assume a space link accuracy in the 10^{-18} level. The major influence on the space link uncertainty is the clock error and velocity error as given in Shen et al. (2023). A link accuracy of 10^{-18} is sufficient to achieve a 1 cm accuracy in the ASL estimation.

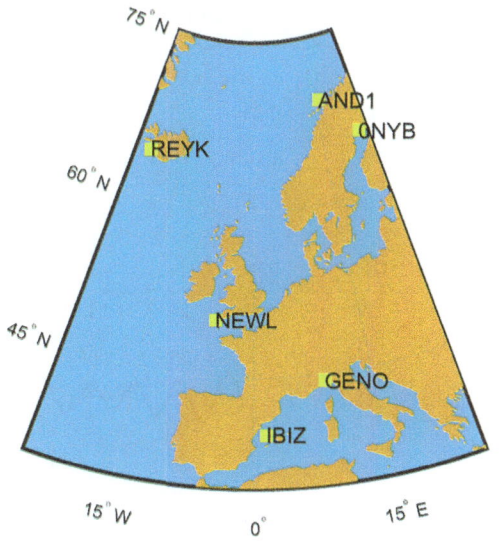

Fig. 3 Tide gauge locations considered in this study: NEWL (Newlyn), AND1 (Andenes), REYK (Reykjavik), IBIZ (Ibiza), GENO (Genova), and, 0NYB (Kalix)

3 Methodology

Tide gauge locations along the European coast are considered for this study (Fig. 3). A total of 6 sites – Newlyn (UK), Andenes (Norway), Reykjavik (Iceland), Ibiza (Spain), Genova (Italy), and Kalix (Sweden) are chosen based on the availability of nearby GNSS stations with longer time-series. The simulations are carried out for the time period 2006-2016 based on the availability of the needed datasets. The mean over the time period is reduced from monthly RSL RLR (Revised Local Reference) data obtained from PSMSL (Permanent Service for Mean Sea Level) (Holgate et al. 2013; Permanent service for mean sea level 2023) to derive the RSL change. Clock observations are simulated by combining the physical height change due to tidal and non-tidal effects (Eq. 3). Addition of RSL changes with the simulated effective vertical displacements or the clock observations generate the corresponding ASL changes at the selected tide gauge locations (Eq. 1).

4 Results and Discussion

4.1 Simulation of Tidal Signals as Clock Observations

Potential variations due to solid-earth tides, pole tides and LOD tides are computed on the deformable Earth surface using modified ETERNA34 (PREDICT program) Earth tide data processing package (Wenzel 2022) for the chosen time period. The HW95 (Hartmann and Wenzel 1995) tidal potential catalogue is used for deriving the hourly tidal potential values which are subsequently averaged to produce monthly values. Similarly, monthly averages of potential variations due to ocean tidal loading are determined with the SPOTL3.3.0.2 package (Agnew 2012) using the Empirical Ocean Tidal model (EOT11a) (Savcenko and Bosch 2012). Both ETERNA34 and SPOTL calculate the effective potential variations due to mass changes and land motion. Figure 4 shows the simulated major tidal values at a high-latitude site and a low-latitude site. The LOD tidal values are of the order of 0.001 m^2/s^2 which can be neglected. Similarly, atmospheric tidal values are also much less than the range of 10^{-18} (= 0.1 m^2/s^2) clock sensitivity which again can be neglected (Voigt et al. 2016).

The monthly averages of solid-earth tides range from approximately 10 cm at high-latitude sites, making it an important factor in total land motion. All the three major tidal contributions (Fig. 4) are combined to generate the total tidal effects (second subplots of Fig. 5). The observed negative long-term trend in the tidal values is caused by the 18.61 year lunar nodal cycle. According to Rochlin and Morris (2017), the declination of Moon's orbit is smallest from 2005 to 2015 which results in increasing tidal amplitudes that correspond to negative potential variations on the Earth surface.

4.2 Simulation of Non-Tidal Signals as Clock Observations

Non-tidal mass distributions include loading and unloading effects arising from various effects such as changes in Terrestrial Water Storage (TWS), atmospheric pressure, ocean bottom pressure, solid earth processes like GIA (Glacial Isostatic Adjustment), etc. (Schröder et al. 2021). The time series of geoid height variation due to the variation in the TWS can be determined from GRACE (Gravity Recovery And Climate Experiment) fully normalised monthly spherical harmonic coefficients (Kvas et al. 2019). Similarly, using the fully normalised monthly spherical harmonic coefficients (GAC) from Atmosphere and Ocean De-aliasing product (AOD1B RL06) (Dobslaw et al. 2017), the effect of atmosphere and ocean mass variability can be well estimated.

The surface deformations associated with these non-tidal mass distributions are given by NGL (Nevada Geodetic Laboratory) GNSS time-series solutions (Blewitt et al. 2018). A loading effect leads to a downward vertical displacement and vice versa. The effective gravitational potential variations can be computed by combining the potential variations due to mass changes and surface displacements (Vincent and Müller 2023). The separate effects of geoid height variations (GRACE+AOD) and the corresponding vertical displacements of the Earth surface (GNSS) are shown by the first subplots of Fig. 5.

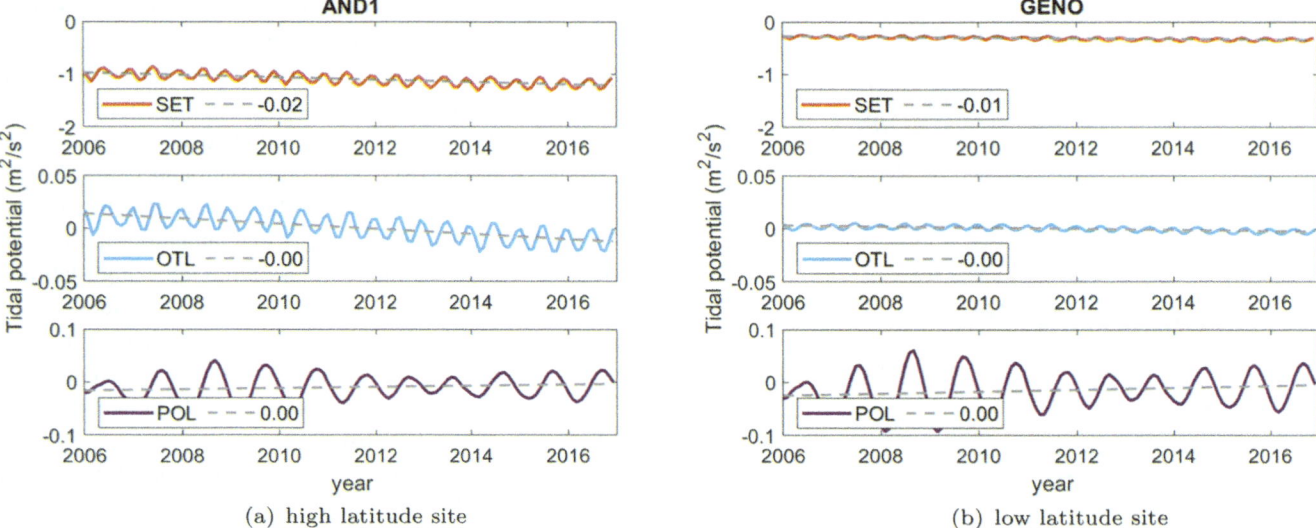

Fig. 4 Monthly averages of solid-earth tides (SET), ocean-load tides (OTL) and pole tides (POL) with their linear trends at Andenes (**a**) and Genova (**b**)

The simulated clock observations by combining the tidal and non-tidal potential variations (Eq. 3) at the chosen tide gauge locations are given by the third subplots of Fig. 5. 10^{-18} clocks can provide absolute land motion of 1 cm (i.e. corresponding gravity potential variations of 0.1 m^2/s^2) between the clock sites in near real-time. When one considers monthly observations over several years – as done in this study – the resulting trend accuracy is less than 1 mm/yr (Fig. 5).

4.3 Estimation of ASL Changes

As mentioned in Sect. 2, geoid-based ASL changes can be derived by combining clock observations with RSL data from tide gauges (Eq. 1). Clock observations of height variations provide a globally consistent and uniform reference for deriving ASL changes. High-performance clocks with fractional frequency uncertainty of 10^{-18} (McGrew et al. 2018; Takamoto et al. 2022) with same level of link accuracy can be utilised well in deriving ASL changes with high accuracy of 1 cm. The monthly variations of RSL are determined by reducing the mean over the chosen time-span. The clock observations which are simulated in terms of potential variations (m^2/s^2) are transferred into physical height variations (cm) by multiplying with g to combine with the RSL changes. Figure 6 shows the graphs of the estimated ASL changes with the long-term trend, clock observations in terms of physical height, and RSL changes. The measurement noise of RSL affects the estimation of ASL.

Present-day land uplift, e.g., at Kalix (0NYB) can fake a sea level fall. So, the land motion must be taken into account when deriving actual sea level changes. As seen in Figs. 5 and 6, the offsets from the monthly means of land motion affect the monthly averages of ASL change at the clock sites. On analyzing the linear trends at the tide gauge sites for the estimated time period, there is an overall increase in ASL with given accuracy.

5 Conclusions

Till now, land-based absolute sea level measurements are estimated with respect to GNSS benchmarks or local tidal datums that give ASL changes with respect to the reference ellipsoid. Further calculations are needed in order to derive a globally consistent measure of ASL. High-performance atomic clocks at tide gauge locations with better link uncertainty can provide more accurate physical height changes and thus, geoid-based ASL changes can be directly obtained. Hence, clock networks can be used for monitoring vertical land movements and providing uniform and consistent ASL changes world-wide.

We have estimated the ASL changes at some tide gauge locations along the European coast. The obtained long-term trend values indicate the geoid-based ASL changes per year for the given time-span. A maximum value of 0.71±0.49 cm/year was observed at Andenes (AND1) for the time period 2008–2016. All sites other than Ibiza (IBIZ) and Newlyn (NEWL) show an increasing trend. Significant land uplift at sites Kalix (0NYB), Andenes (AND1), and Reykjavik (REYK) should be effectively reduced for the accurate estimation of ASL changes. Also, for all the sites, tidal effects play an important role as even the monthly-averages can be as high as 10–20 cm. Thus, the accurate

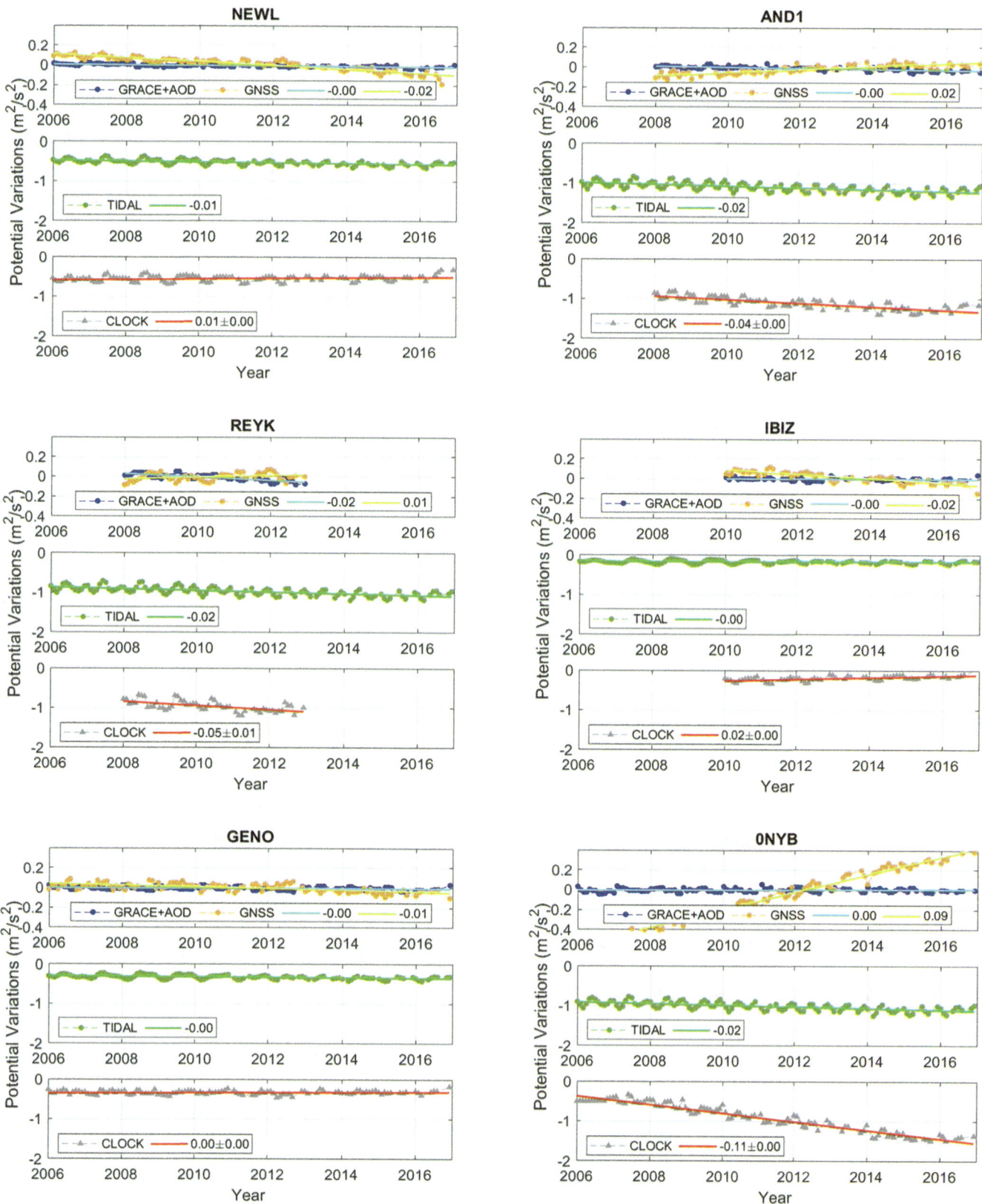

Fig. 5 Simulated clock observations (CLOCK) by combining the potential variations due to non-tidal effects (mass changes in TWS (GRACE), atmosphere and ocean (AOD) and associated vertical deformations (GNSS)) and tidal effects (TIDAL=SET+OTL+POL) with their linear trends (m^2/s^2yr) at the tide gauge sites

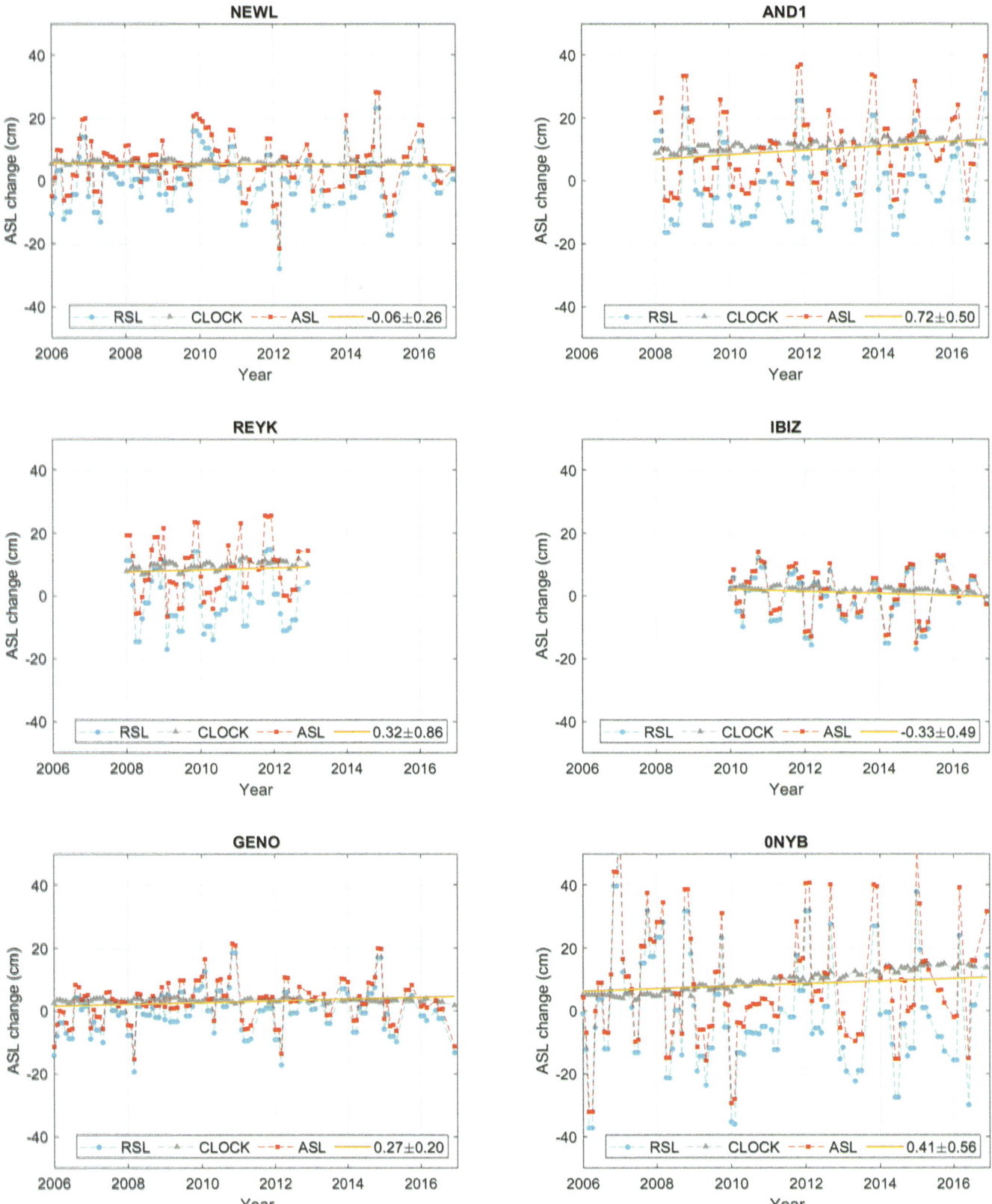

Fig. 6 Estimated monthly ASL changes (ASL) by reducing the clock observations (CLOCK) from RSL changes (RSL) at the tide gauge sites. The long-term trends of ASL changes (cm/yr) are also given

estimation of globally consistent geoid-based ASL changes favours the realization of clock-based networks in the near future.

Acknowledgements This study has been funded by the Deutsche Forschungsgemeinschaft (DFG, German Research Foundation) under Germany's Excellence Strategy EXC 2123 Quantum Frontiers – Project-ID 90837967 and the SFB 1464 TerraQ – Project-ID 434617780 within project C02.

References

Agnew DC (2012) Spotl: Some programs for ocean-tide loading. https://api.semanticscholar.org/CorpusID:127800542

Bjerhammar A (1986) Relativistic geodesy. U.S. Dept. of Commerce, National Oceanic and Atmospheric Administration, National Ocean Service, Charting and Geodetic Services, For sale by the National Geodetic Information Center, NOAA, Rockville, MD

Blewitt G, Hammond WC, Kreemer C (2018) Harnessing the gps data explosion for interdisciplinary science. Eos 99:485. https://doi.org/10.1029/2018EO104623

Denker H, Timmen L, Voigt C, Weyers S, Peik E, Margolis HS, Delva P, Wolf P, Petit G (2018). Geodetic methods to determine the relativistic redshift at the level of 10^{-18} in the context of international timescales: a review and practical results. J Geodesy 92(5):487–516. https://doi.org/10.1007/s00190-017-1075-1

Dietrich R (2014) Sea Level, pp 1Ű-9. Springer Netherlands, Dordrecht. ISBN 978-94-007-6644-0. https://doi.org/10.1007/978-94-007-6644-0_173-1

Dobslaw H, Bergmann-Wolf I, Dill R, Poropat L, Thomas M, Dahle C, Esselborn S, König R, Flechtner F (2017) A new high-resolution model of non-tidal atmosphere and ocean mass variability for de-aliasing of satellite gravity observations: AOD1B RL06. Geophys J Int 211(1):263–269. ISSN 0956-540X. https://doi.org/10.1093/gji/ggx302

Holgate SJ, Matthews A, Woodworth PL, Rickards L, Tamisiea ME, Bradshaw E, Foden PR, Gordon K, Jevrejeva S, Pugh J (2013). New data systems and products at the permanent service for mean sea level. https://api.semanticscholar.org/CorpusID:129718672

Ihde J, Sánchez L, Barzaghi R, Drewes H, Foerste C, Gruber T, Liebsch G, Marti U, Pail R, Sideris MG (2017) Definition and proposed realization of the international height reference system (ihrs). Surv Geophys 38:549–570. https://doi.org/10.1007/s10712-017-9409-3

Kvas A, Behzadpour S, Ellmer M, Klinger B, Strasser S, Zehentner N, Mayer-Gürr T (2019) Itsg-grace2018: Overview and evaluation of a new grace-only gravity field time series. J Geophys Res Solid Earth 124(8):9332–9344. https://doi.org/10.1029/2019JB017415

Larson KM, Ray RD, Nievinski FG, Freymueller JT (2013) The accidental tide gauge: A gps reflection case study from kachemak bay, alaska. IEEE Geosci Remote Sens Lett 10:1200–1204. https://api.semanticscholar.org/CorpusID:11995621

Lisdat C, Grosche G, Quintin N, Shi C, Raupach S, Grebing C, Nicolodi D, Stefani F, Al-Masoudi A, Dörscher S, et al. (2016) A clock network for geodesy and fundamental science. Nature Commun 7(1):1–7. https://doi.org/10.1038/ncomms12443

McGrew W, Zhang X, Fasano R, Schäffer S, Beloy K, Nicolodi D, Brown R, Hinkley N, Milani G, Schioppo M, et al. (2018) Atomic clock performance enabling geodesy below the centimetre level. Nature 564(7734):87–90. https://doi.org/10.1038/s41586-018-0738-2

Müller J, Dirkx D, Kopeikin SM, Lion G, Panet I, Petit G, Visser P (2018) High performance clocks and gravity field determination. Space Sci Rev 214(1):1–31. https://doi.org/10.1007/s11214-017-0431-z

Peng D, Feng L, Larson KM, Hill EM (2021) Measuring coastal absolute sea-level changes using gnss interferometric reflectometry. Remote Sens 13(21). ISSN 2072-4292. https://doi.org/10.3390/rs13214319

Permanent service for mean sea level (psmsl) (2023). tide gauge data, May 2023. http://www.psmsl.org/data/obtaining/

Philipp D, Hackmann E, Lämmerzahl C, Müller J (2020) Relativistic geoid: Gravity potential and relativistic effects. Phys Rev D 101:064032. https://doi.org/10.1103/PhysRevD.101.064032

Philipp D, Hackmann E, Hackstein JP, Lämmerzahl C (2023) General relativistic chronometry with clocks on ground and in space. https://doi.org/10.48550/arXiv.2310.11576

Rochlin I, Morris JT (2017) Regulation of salt marsh mosquito populations by the 18.6-yr lunar-nodal cycle. Ecology 98(8):2059–2068. https://doi.org/10.1002/ecy.1861

Savcenko R, Bosch W (2012) DGFI-Report No.89: Eot11a – empirical ocean tide model from multi-mission satellite altimetry. Technical report, DGFI-TUM

Schröder S, Stellmer S, Kusche J (2021) Potential and scientific requirements of optical clock networks for validating satellite-derived time-variable gravity data. Geophys J Int 226(2):764–779. https://doi.org/10.1093/gji/ggab132

Shen Z, Shen W, Xu X, Zhang S, Zhang T, He L, Cai Z, Xiong S, Wang L (2023) A method for measuring gravitational potential of satellite's orbit using frequency signal transfer technique between satellites. Remote Sens 15(14).. ISSN 2072-4292. https://doi.org/10.3390/rs15143514. https://www.mdpi.com/2072-4292/15/14/3514

Takamoto M, Tanaka Y, Katori H (2022) A perspective on the future of transportable optical lattice clocks. Appl Phys Lett 120(14):140502. https://doi.org/10.1063/5.0087894

Torge W, Müller J, Pail R (2023) Geodesy. De Gruyter Oldenbourg, Berlin, Boston. ISBN 9783110723304. https://doi.org/10.1515/9783110723304

Vincent A, Müller J (2023) Detection of time variable gravity signals using terrestrial clock networks. Adv Space Res. ISSN 0273-1177. https://doi.org/10.1016/j.asr.2023.07.058

Voigt C, Denker H, Timmen L (2016) Timevariable gravity potential components for optical clock comparisons and the definition of international time scales. Metrologia 53(6):1365. https://doi.org/10.1088/0026-1394/53/6/1365

Wenzel H-G (2022) Eterna – programs for tidal analysis and prediction. Karlsruhe Institute of Technology. https://doi.org/10.35097/746

Wu H, Müller J (202) Towards an international height reference frame using clock networks. In: Freymueller JT, Sánchez L (eds) Beyond 100: The Next Century in Geodesy, pp 3–10. Springer International Publishing, Cham. ISBN 978-3-031-09857-4. https://doi.org/10.1007/1345_2020_97

Wu H, Müller J, Lämmerzahl C (2019) Clock networks for height system unification: a simulation study. Geophys J Int 216(3):1594–1607. https://doi.org/10.1093/gji/ggy508

Wöppelmann G, Marcos M (2016) Vertical land motion as a key to understanding sea level change and variability. Rev Geophys 54(1):64–92. https://doi.org/10.1002/2015RG000502

Open Access This chapter is licensed under the terms of the Creative Commons Attribution 4.0 International License (http://creativecommons.org/licenses/by/4.0/), which permits use, sharing, adaptation, distribution and reproduction in any medium or format, as long as you give appropriate credit to the original author(s) and the source, provide a link to the Creative Commons license and indicate if changes were made.

The images or other third party material in this chapter are included in the chapter's Creative Commons license, unless indicated otherwise in a credit line to the material. If material is not included in the chapter's Creative Commons license and your intended use is not permitted by statutory regulation or exceeds the permitted use, you will need to obtain permission directly from the copyright holder.

Part V

Monitoring and Understanding the Dynamic Earth with Geodetic Observations

Towards Clock Ties for a Global Geodetic Observing System

Jan Kodet, Thomas Klügel, Christian Plötz, Willi Probst, Alexander Neidhardt, and Karl Ulrich Schreiber

Abstract

International reference frames play a pivotal role in metrological applications related to the Earth sciences, covering critical areas such as geodetic reference frames, global change research, deformation processes, and global mass transport. Despite substantial advancements in measurement precision over the past two decades, there are still discrepancies at the centimeter level presenting a persistent challenge. In this paper, we investigate a promising approach to address these subtle error sources, affecting critical parts of the measurement equipment despite the presence of calibration methods.

We have built a novel measurement constraint, based on a precise clock and active time delay compensation. By comparing the timing signals in a geodetic measurement system constantly against a precisely controlled optical ruler, we can identify variable system delays that were previously inaccessible. In this way, we introduce a novel tie to the geodetic measurement techniques that can even capture instrumental delay variations over several months.

Keywords

Common clock · GNSS · Multi-technique combination · SLR · Time in geodesy · VLBI

1 Introduction

The combination of the space geodetic techniques is essential in defining both the International Terrestrial and Celestial Reference Frames. Fundamental stations, integrating multiple space geodetic techniques, play a pivotal role in this matter as they provide geometric reference points, which allow the connection of these techniques (Seitz et al. 2022). A particular challenge arises when systematic biases of unknown origin occur between the large measurement systems, especially when the geodetic reference point doesn't align with the signal reference point.

To overcome this challenge and enhance measurement consistency, our main objective is to tie these points together. We have developed an innovative Optical Time and Frequency Distribution System at the Geodetic Observatory Wettzell (GOW). This system provides actively delay-compensated timing signals, which are unaffected by variable propagation delays and environmental changes. It thus provides a common clock to all measurement systems, which is stable to ± 1 ps. The system generates ultra-stable and low-noise electric timing signals at the end of each link terminal. The mode-locked laser is the central component housed in the TWIN radio-telescope building and serves as a stable time and frequency reference for all the instrumentation of space geodesy. The timing system provides phase coherence over long distances, which enables tight control over systematic measurement errors in instruments as large as a telescope in closure over time.

J. Kodet (✉) · A. Neidhardt · K. U. Schreiber
Forschungseinrichtung Satellitengeodäsie, Geodätisches Observatorium Wettzell, Technische Universität München, Munich, Germany
e-mail: jan.kodet@tum.de

T. Klügel · C. Plötz · W. Probst
Geodätisches Observatorium Wettzell, Bundesamt für Kartographie und Geodäsie, Frankfurt am Main, Germany

The overarching goal is the realization of a long-term stable system reference, which recovers variable biases.

2 Optical Timing System

The optical time and frequency distribution system is the critical component for the geodetic measurement systems, providing a solution for the reduction of systematic biases. It provides a reference scale against which variations show up.

The design features a mode-locked fs-pulse laser referenced to a hydrogen maser. The ultrashort optical pulse train is distributed to the measurement system via an active delay-compensated optical fiber (Schreiber and Kodet 2018; Kodet et al. 2018), providing a fixed relationship for the time markers, with a stability of 1 ps over many days across the entire observatory. In 2022, the installation was completed, solving the most challenging problem of including the moving VLBI antennas, where the cable twister introduces considerable delay variations. The core component of the active delay compensation is an optical-microwave phase detector (BOMPD) transferring the radio frequency (RF) based reference frequency from a hydrogen maser to the optical domain. This allows the locking of a femtosecond laser source without compromising the RF quality of the source (Peng et al. 2014).

The BOMPD's fundamental function lies in precisely measuring the phase delay between optical and microwave signals. This is achieved through the conversion of optical into electrical signals, ensuring continuous alignment of the phases in both domains. The system achieves an impressive low drift of 10 fs over several days. This tight synchronization is vital for continuous high coherence at each endpoint of the time and frequency distribution system.

Such a laser pulse train is further distributed around the observatory using delay-compensated and drift-free optical links. The active stabilization of link delays is achieved through interferometry. The pulse train is split into two parts, with one part serving as a short reference arm installed in a temperature-stabilized front end. The signal arm includes the optical fiber and delivers the optical pulses to the measurement systems. A mirror at the end of the line reflects most of the light back to the interferometer. The pulse traverses first a dispersion-compensated fiber, ensuring that the pulse width at the optical cross-correlator remains unaffected by optical fiber dispersion. The optical cross-correlator performs an interferometric comparison, as depicted in Fig. 1, gauging the delay between a pulse from the reference arm and the link pulse. A feedback loop drives the fiber stretcher to ensure a precise overlap of 150 fs pulses in width. The stabilization has a bandwidth from DC to approximately 1 kHz, allowing it to cancel out any fiber delay fluctuations within this range.

The result is an exact copy of the input signal at the end of the fiber with a constant phase offset. However, the signals of interest are electrical signals, namely, 1 pulse per second (PPS) and typical reference frequencies, i.e., 10 and 100 MHz. To obtain high delay stability, it is crucial to generate signals with extremely low-temperature coefficients. This is because temperature variations have a significant impact on timing stability. While it is often claimed that sufficient temperature stabilization of electronics is possible, this is not fully accurate. In fact, a large temperature coefficient of the electronic components can result in unrealistic high requirements for temperature stabilization. For instance, the temperature coefficients of filters used in frequency multipliers can easily reach 50 ps/K. In order to achieve 1 ps stability

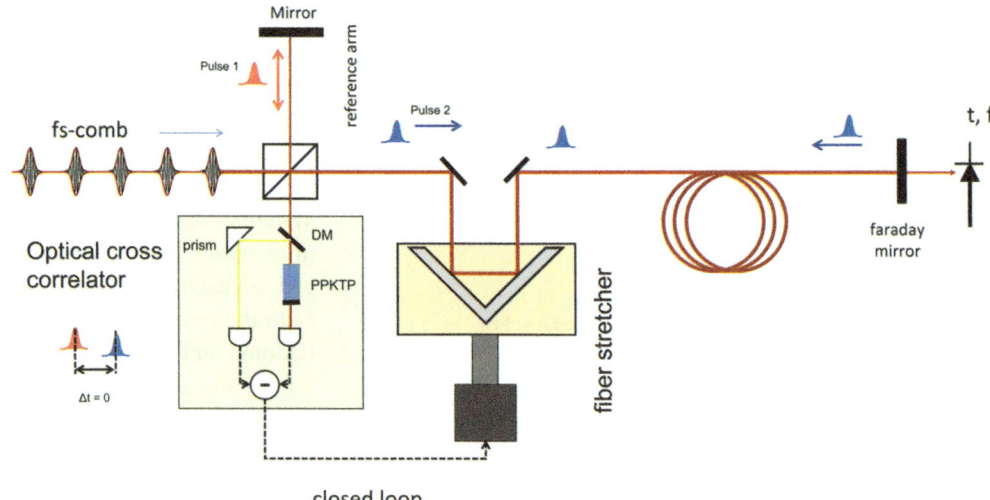

Fig. 1 Design of fiber-link stabilization unit utilizing optical cross-correlator and a feedback loop operating a fiber stretcher

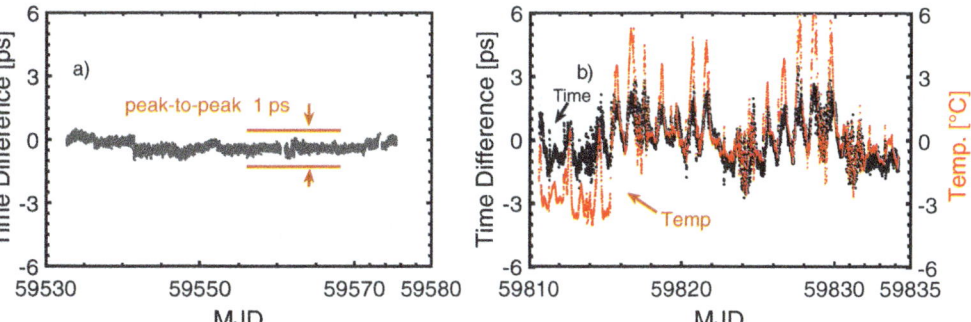

Fig. 2 Timing difference of electrical signals generated at the end of two stationary links (**a**), and time difference measured between two radiotelescopes, Radiotelescope Wettzell (Wz) and TWIN Wettzell-North (Wn) (**b**). 59,530 corresponds to 11.12.2021, and 59,810 corresponds to 19.08.2022

for long-term operations, a temperature stability of 20 mK would be required. Therefore, we have invested a significant effort into designing electronics with very low-temperature coefficients to achieve the desired high performance. A typical temperature coefficient of electrical signals provided by the electronic at each link end (back-end) is 1 ps/K.

To validate the quality of the time distribution units, two event timers implementing a two-way time transfer concept (TWOTT) for widely separated locations were employed (Kodet et al. 2016a, b). This technique achieves sub-picosecond timing jitter and timing stability, obtained from the time deviation evaluation (TDEV), reaching 50 fs.

We conducted an extensive investigation involving two types of links in operation at the GOW. The first type involves the assessment of link stability for a stationary link installed in a rack mount within a laboratory. Figure 2a illustrates the delay stability observed over a period exceeding 1 month. Notably, the stability of electrical signals between the front-end and back-end exhibited a negligible increase, staying within the 1 ps range.

The evaluation of the signal delay for terminals installed inside the elevation cabinets of the radiotelescopes is more difficult due to temperature variations. The advantage of low-temperature coefficients turned out to be very important since the temperature difference between the two radio telescopes reached up to 8 K.

In Fig. 2b, we observe residual timing variations that are highly correlated with temperature. Based on the underlying data set, the temperature coefficient was determined to be 0.42 ps/K, and we must consider three sources for these variations: the back-end of the timing system, the TWOTT event timing modules, and the interconnecting RF cables. The back-end electronics are additionally temperature stabilized within a fractional degree of Celsius, so any temperature variations in this component are expected to be attenuated. The TWOTT timers have a temperature coefficient of 1.4 ps/K, and the interconnecting RF cables type LMR-240 have a temperature coefficient of 0.5 ps/K per meter. A notable aspect is the system's capability to achieve precise

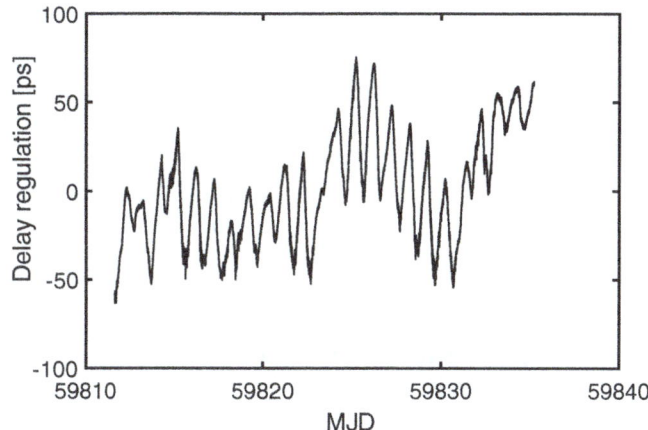

Fig. 3 The optical link delay is actively controlled; the figure illustrates the amount of adjustment required to achieve 1 ps stability

timing even in dynamic conditions, as demonstrated by links in radiotelescopes experiencing telescope movements and temperature variations. Figure 3 illustrates the effectiveness of the delay compensation by showing the error signal for a measurement over several days.

The precision of timing signals is presented through TDEV values, revealing that, in most links, the precision is close to 50 fs, likely limited by TWOTT measurements. The comparison is depicted in Fig. 4.

Radiotelescope links, which are subject to dynamic regulation, have achieved a time deviation (TDEV) as low as 200 fs. Among all the links, the one located in the Cesium clock room is the worst, with the highest TDEV. This is due to poor thermal stability. To address this issue, plans are in place to relocate the Cesium clock facility to a more stable room, considering the impact of environmental factors on timing signals.

In summary, the optical timing system delivers unprecedented timing stability, especially in challenging conditions. The TDEV analysis and comparisons highlights the system's capability to provide high time coherence right at the locations where critical system calibrations are performed.

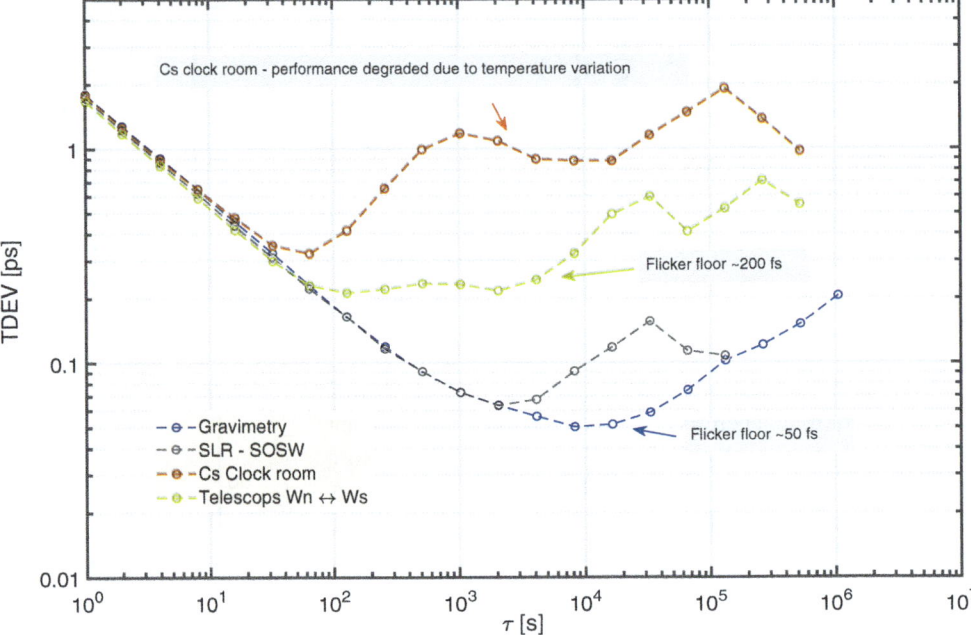

Fig. 4 A TDEV analysis demonstrates link stability of 50 fs for a laboratory measurement and 200 fs between the radiotelescopes

3 Application of the Optical Timing System in VLBI

This section explores the practical application of the optical timing system in VLBI and shows the case where more stable clock distribution improves VLBI products. In the initial stages of developing the optical timing system in 2015, we could synchronize two local radiotelescopes in the following way. The 20 m Radiotelescope Wettzell (Wz) was locally referenced to hydrogen masers and equipped with a phase calibration unit and cable calibration unit. In contrast, the Wn (TTW1) telescope, while referenced to another maser, lacked these calibration units. Synchronization of both masers was achieved using the TWOTT technique, connecting close to a maser as a synchronization point. Therefore, not all delays were properly captured during the observation. The baseline between these radio-telescopes spanned 123 m and over 100 days, TWOTT measurement data covered 18 R1 and R4 VLBI sessions in the period of (Sep 2015–Feb 2016).

The Levika software (Schüler et al. 2015) was employed for the VLBI analysis, solving for the local VLBI baseline. The local tie measurements provided the coordinates for the positions of both telescopes. Additional constraints were put on the ionosphere and troposphere, considering they are the same for both telescopes. This approach facilitated the estimation of VLBI clock differences from each quasar observation, providing a unique opportunity to analyze VLBI clocks with high resolution during 24-h sessions.

The distinctions between a VLBI clock estimation and the TWOTT measurements are described in (Kodet et al. 2016a, b). Here, we can take a different perspective on these measurements. According to (Whitney 1974) precision, determined by observing the group delay between two telescopes can be expressed as:

$$\sigma_{\tau_g} = \frac{1.24 * 10^3}{B_s \mathrm{Jd}_1 d_2} \sqrt{\frac{T_{s_1} T_{s_2}}{\varepsilon_1 \varepsilon_2 \mathrm{WT}}}, \qquad (1)$$

where ε_i is antenna efficiency, d_i is antenna diameter, B_s is spanned bandwidth, W is channel bandwidth, T_{si} is total system noise temperature, and T is integration time. Typical values for VGOS or S/X observations yield a 10–30 ps precision. The above formula assumes only white-phase noise of the observable group delay between the telescopes, which is given (Nothnagel et al. 2018) as

$$\tau_{obs} = \tau_{geom} + \tau_{iono} + \tau_{atm} + \tau_{clk} + \tau_{inst} + \tau_{def} + \epsilon. \qquad (2)$$

Here, we focus on τ_{clk} and τ_{inst} only, since the rest of the variables are constrained for this specific data set. τ_{clk} should contain the phase difference of the reference frequency generated by involved stations clocks and τ_{inst} contains instrumental delays, which should be calibrated. While the antenna moves from one source to another, the antenna delay may change and must be well-calibrated. In classical VLBI analyses the clock delay is parameterized and estimated using piece-wise linear functions with a resolution of 20 min to 1 h. Challenges arise when other noise sources interfere with white-phase noise averaging. This could be well observed in our data set, where each quasar observation was used for the τ_{clk} and τ_{inst} estimation.

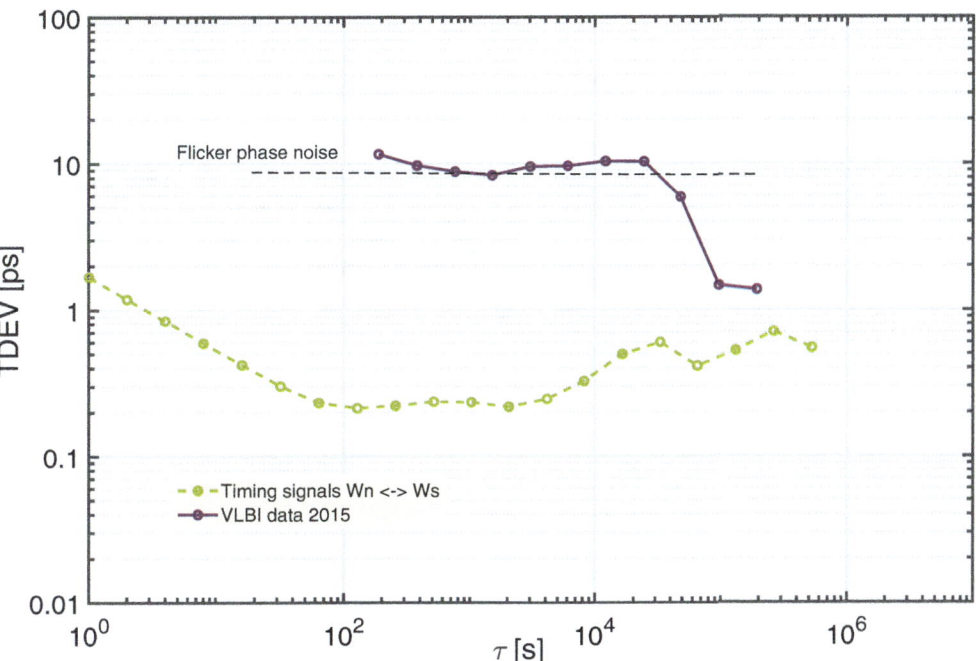

Fig. 5 Clock offsets expressed as TDEV between Wz and Wn antennas were obtained from a VLBI analysis in 2015, including TWOTT measurements. The flicker floor limits the VLBI instrumental resolution much earlier than the clocks. The TDEV of timing signals difference from optical time and frequency distribution as measured in 2022 is plotted for comparison

We used this dataset from 2015, which allowed us to look at these results from the perspective of TDEV analyses and determine which type of noise dominates VLBI observation. Figure 5 illustrates the results, offering insight into the limitations of white-phase noise averaging. The VLBI data starts averaging at around 13 ps, aligned with the formula (1), but the group delay does not improve with increasing averaging time. The precision reaches the flicker floor and levels off at around 10 ps, suggesting that increasing the number of observations or integration time does not provide higher resolution for the baselines. Points beyond a 24-h integration time exhibit a downward step, which appears to be an artifact from the TDEV analysis due to an insufficient number of observations. Therefore, any enhancement achieved through the use of averaging cannot be implemented. It appears that the VLBI flicker floor limitation is mainly due to variable instrumental delays, which are not adequately addressed in the calibration during the experiment. This finding motivated the development of the coherent optical time and frequency distribution system. In 2022, we completed the installation of the delay-compensated terminals in all three radio telescopes at GOW.

4 First VLBI Local Experiments with the Optical Timing System

In a test run in 2022, we conducted two local experiments utilizing Wz and Wn radiotelescopes in this new setup. In these experiments, each digitizer was locked to its own optical time distribution terminal, while the Local Oscillators (LO) and phase calibration units (pCal) in the antenna were connected to another delay-compensated back-end. The experimental setup is sketched up in Fig. 6.

For verification purposes, we evaluated TWOTT during a 1-h experiment. It is necessary to point out that these experiments were primarily focused on compatibility checks of VLBI instruments with all relevant signals generated from the optical time and frequency distribution system. While VLBI provides the delay between radiotelescopes, Fig. 7 illustrates the signal differences. The amount of delay removed by active stabilization is together with the stability of the difference between the 100 MHz local oscillator generation and the phase calibration injection point, as measured by TWOTT.

We observed 31 radio sources in a 1-h experiment. That does not provide sufficient data for a full statistical analysis of the group delay estimation. Nevertheless, it provides important evidence, if all instrumental delays are sufficiently

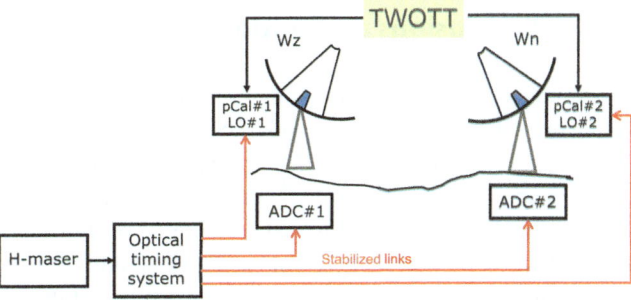

Fig. 6 Local experimental setup for the Wz and Wn radiotelescopes, utilizing 4 stable delay-compensated back-ends

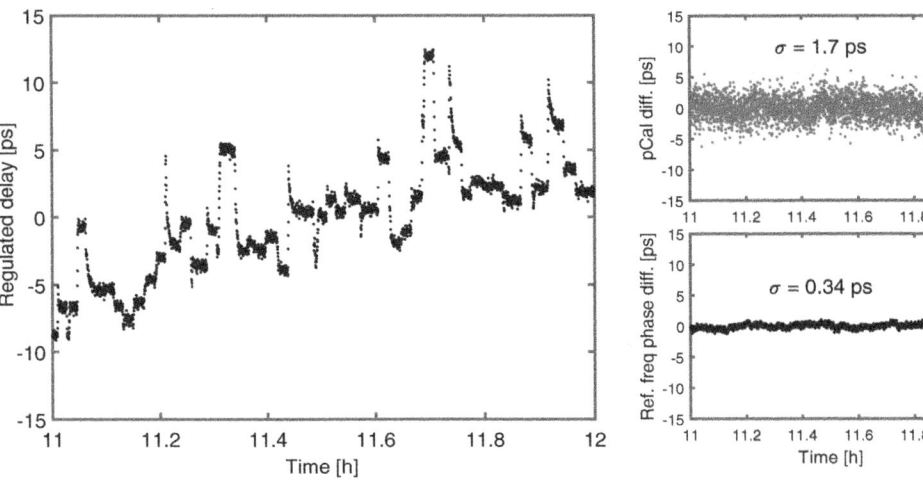

Fig. 7 The active stabilization aims to bring a stable phase of the reference frequency as the timing signal for pCal generation into the radiotelescopes elevation cabins. As an example, the figure illustrates for the duration of 1 h the amount of delay difference (Wn-Wz), which was actively regulated (left side), as well as the difference of pCal and reference frequency phase difference between the antennas (right side)

well calibrated and whether the VLBI delay estimation is contaminated by an additional noise source. A noticeable difference in the TDEV is observed when a phase calibration signal is applied to the measurement. Figure 4 supports the argument that the timing signals to which the telescopes are referenced do not increase the VLBI flicker floor. However, TDEV computed from VLBI clocks offsets does not exhibit the $1/\sqrt{\tau}$ slope, which is indicative of the presence of white phase noise. This could originate from the application of pCal phase calibration in the fringe fit after correlation, as detailed in (Nothnagel 2023). The phase calibration correction is applied as a single value for the duration of one observation, meaning that each quasar observation has its unique instrumental phase delay correction. The pCal phase delay estimation is used to equalize different observation channels. However, it is not used to relate individual scans to the same phase reference. Therefore, when the antenna shifts position to point to another source, the changes in instrumental delay, and consequently, the phase delay is estimated with respect to another phase reference. This aspect is currently not contained in the fringe-fitting process. It is currently assumed that the delay change behaves as white-phase noise which will be absorbed during parameter estimation. This appears to be not the case.

5 Conclusions and Outlook

In accordance with the overarching objectives of the Global Geodetic Observing System (GGOS) to provide high-accuracy reference frames. Our research effort focuses on reducing systematic errors, which influence geodetic measurements from fundamental stations. We introduce active delay compensation to increase the stability of the measurement systems. At the same time, we operate the instrumentation with a common clock and a very tight phase relationship. From this actively controlled delay, the goal is to establish an additional tie between the different instruments, which is based on time coherence. Measurements on the VLBI systems in Wettzell allow us to observe biases caused inside the instrumentation, which limit the performance of the technique and, hence, its long-term stability.

Early experiences with the newly implemented time and frequency distribution system are promising. The delay compensation system provides exceptional stability, achieving levels well below 1 mm when interpreting time delay as a distance. The successful performance of VLBI experiments further underscores the potential of deriving common clocks through delay drift-free links in VLBI applications. While the initial VLBI experiments were conducted over a relatively short duration only, they have provided valuable insights. Currently, we still have to resolve a technical issue, which causes the presence of low-level spurious frequencies in the calibration signal.

Acknowledgments The authors acknowledge support from the Wettzell VLBI team. This work was supported by the Deutsche Forschungsgemeinschaft contract SCHR645/8-1; project number 423159159.

References

Kodet J, Pánek P, Procházka I (2016a) Two-way time transfer via optical fiber providing subpicosecond precision and high temperature stability. Metrologia 53(1):18

Kodet J, Schreiber U, Panek P, Prochazka I, Männel B, Schüler T (2016b) Optical two-way timing system for space geodesy applications. In: 2016 European Frequency and Time Forum (EFTF), pp 1–6

Kodet J, Schreiber KU, Eckl J, Plötz C, Mähler S, Schüler T, Klügel T, Riepl S (2018) Co-location of space geodetic techniques carried out at the Geodetic Observatory Wettzell using a closure in time and a multi-technique reference target. J Geod. https://doi.org/10.1007/s00190-017-1105-z

Nothnagel A (2023) Elements of geodetic and astrometric very long baseline interferometry. https://www.vlbi.at/data/publications/Nothnagel_Elements_of_VLBI.pdf

Nothnagel A, Nilsson T, Schuh H (2018) Very long baseline interferometry: dependencies on frequency stability. Space Sci Rev 214(3):66. https://doi.org/10.1007/s11214-018-0498-1

Peng MY, Kalaydzhyan A, Kärtner FX (2014) Balanced optical-microwave phase detector for sub-femtosecond optical-RF synchronization. Opt Express 22(22):27102–27111. https://doi.org/10.1364/OE.22.027102

Schreiber KU, Kodet J (2018) The application of coherent local time for optical time transfer and the quantification of systematic errors in satellite laser ranging. Space Sci Rev 214(1):22. https://doi.org/10.1007/s11214-017-0457-2

Schüler T, Kronschnabl G, Plötz C, Neidhardt A, Bertarini A, Bernhart S, la Porta L, Halsig S, Nothnagel A (2015) Initial results obtained with the first TWIN VLBI radio telescope at the Geodetic Observatory Wettzell. Sensors 15(8):18767. https://doi.org/10.3390/s150818767

Seitz M, Bloßfeld M, Angermann D, Seitz F (2022) DTRF2014: DGFI-TUM's ITRS realization 2014. Adv Space Res 69(6):2391–2420. https://doi.org/10.1016/j.asr.2021.12.037

Whitney AR (1974) Precision geodesy and astrometry via very-long-baseline interferometry. PhD thesis, MIT

Open Access This chapter is licensed under the terms of the Creative Commons Attribution 4.0 International License (http://creativecommons.org/licenses/by/4.0/), which permits use, sharing, adaptation, distribution and reproduction in any medium or format, as long as you give appropriate credit to the original author(s) and the source, provide a link to the Creative Commons license and indicate if changes were made.

The images or other third party material in this chapter are included in the chapter's Creative Commons license, unless indicated otherwise in a credit line to the material. If material is not included in the chapter's Creative Commons license and your intended use is not permitted by statutory regulation or exceeds the permitted use, you will need to obtain permission directly from the copyright holder.

Assessment of the Tropospheric Delay Coefficients at Co-located Sites with VGOS and GNSS

Anastasiia Walenta, Claudia Flohrer, Daniela Thaller, Rolf Dach, Stefan Schaer, Gerald Engelhardt, and Dieter Ullrich

Abstract

An assessment of the tropospheric parameters independently derived from the analysis of Very Long Baseline Interferometry (VLBI) and Global Navigation Satellite Systems (GNSS) data serves as a cross-validation of the two space geodetic techniques on the parameter level. Time series of the tropospheric parameters are studied for the most frequently observed VLBI stations at 7 co-located sites covering a time span between 2019 and middle of 2023. These sites are equipped in total with 10 small and fast-slewing antennas that have been specifically built to satisfy the concept of the VLBI Global Observing System (VGOS). Next to the VGOS antennas, 5 legacy VLBI antennas are located providing an additional source of the VLBI observations for comparison. VLBI conducts observations on a session-wise basis of 24 hours at least twice a week, whereas GNSS observes continuously. As a consequence, the paired tropospheric parameters are restricted to epochs where VLBI data are available. The closest GNSS receivers are chosen next the VGOS or VLBI antennas to ensure the same path propagation delays. For the sake of a meaningful comparison, the parameterization of the troposphere is homogenized between the two techniques in favor of the VLBI analysis: the VLBI observations are scheduled to provide even sky-coverage within every hour. After omitting modelled offsets between the reference points of the VLBI and GNSS antennas expected due to the height differences, the obtained tropospheric estimates of two independent techniques show a good agreement level, which lays within their scatter. The remaining tropospheric variations are averaged at the level of 4-6 mm in terms of root mean square differences. A larger scatter of these tropospheric variations is obtained for a few stations at the level between 8–10 mm. These extreme cases can be explained by specific issues at each individual station, i.e. the short paired time series.

All the authors have contributed equally to this work.

A. Walenta (✉)
Federal Agency for Cartography and Geodesy (BKG), Frankfurt am Main, Germany

Federal Office of Metrology and Surveying (BEV), Vienna, Austria
e-mail: anastasiia.walenta@bev.gv.at

C. Flohrer · D. Thaller · G. Engelhardt · D. Ullrich
Federal Agency for Cartography and Geodesy (BKG), Frankfurt am Main, Germany
e-mail: claudia.flohrer@bkg.bund.de; daniela.thaller@bkg.bund.de

R. Dach
Astronomical Institute of the University of Bern (AIUB), Bern, Switzerland
e-mail: rolf.dach@unibe.ch

S. Schaer
Astronomical Institute of the University of Bern (AIUB), Bern, Switzerland

swisstopo, Federal Office of Topography, Wabern, Switzerland
e-mail: stefan.schaer@swisstopo.ch

Keywords

Co-location sites · Combination at the parameter level · GNSS · Troposphere · VGOS · VLBI

1 Introduction

The highly variable tropospheric zenith path delays are one of the key parameters evaluated in the analysis of the microwave-based space-geodetic data. The Global Navigation Satellite Systems (GNSS) and Very Long Baseline Interferometry (VLBI) are microwave techniques dealing with signals at frequencies, where similar propagation characteristics can be expected. If VLBI and GNSS signals pass through the same troposphere the related path propagation delays are expected to be the same. Some unmodelled effects in the observational model, e.g., zenith-angle dependent weaknesses of the GNSS antenna calibration, might be absorbed by the estimation of the tropospheric path delay. Nevertheless, a good agreement of the obtained parameters derived independently confirms the physical meaning of the tropospheric products and serves also as a prerequisite for a more rigorous combination of the two techniques.

In the past years, there have been several studies comparing troposphere parameters estimated independently by different space geodetic techniques, e.g., Steigenberger et al. (2007). All studies revealed that (He et al. 2023; Ning et al 2012), on the one hand, there are co-location sites where the troposphere estimates fit very well, but on the other hand, there are also co-location sites where significant biases between the techniques become visible. First attempts to combine troposphere parameters from VLBI and GNSS on the level of normal equations have been undertaken, e.g., Krügel et al (2007). They could demonstrate that the combination of the troposphere is a valuable additional link between microwave-based space geodetic techniques, apart from local ties (combination of station coordinates) and global ties (combination of Earth rotation parameters).

The signal propagation delay through the troposphere is taken into account by introducing the modelled variations of the very well-predictable hydrostatic component as provided by Vienna Mapping Function VMF3 (Landskron and Böhm 2018; re3data.org 2020) and estimating the extremely variable wet delay part together with the horizontal gradient components. The obtained results of the designated operational analysis are shared in the geodetic community in the form of dedicated geodetic products, where the tropospheric product is represented by the time series of the zenith total delay as well as horizontal gradients in the North and East directions. The sum of a hydrostatic part and wet part constitutes the zenith total delay (ZTD), on which this paper is focused in order to establish the working concept of the consistent troposphere evaluation derived from VLBI and GNSS observations.

In this paper, we review the tropospheric products derived independently during VLBI analysis and from the analysis of the GNSS data with the solid purpose to endorse our operational analysis. The tropospheric products as a result of the operational VLBI analysis at BKG follow the guidelines given by the International VLBI Service for Astronomy and Geodesy (IVS, Nothnagel et al. 2017). Similarly, the International GNSS Service (IGS, Johnston et al. 2017) provides the tropospheric product. Due to the large GNSS network, the corresponding solution provides the estimates of the GNSS stations selected by the related Analysis Center (AC), while a VLBI solution delivers the estimates for all stations of the respective session. For this reason, the Center for Orbit Determination in Europe (CODE, Dach et al 2023) – acting as AC of IGS – has prepared additionally Precise Point Positioning (PPP) solutions for the subset of IGS stations co-located to the VLBI stations. This paper is aimed to summarize the results of the work of the CODE consortium, to which the institutions of the authors of this paper contribute. Thus, the operational IVS solution labelled bkg2023a (Engelhardt et al 2023) and available at the IVS Data Centers, i.e. BKG Data Center (Walenta 2023), is validated against the CODE GNSS solution prepared for this paper.

The considered time series of the tropospheric parameters are restricted to the most frequently observing VLBI stations, since VLBI does not provide continuous observations on a regular basis like GNSS. This selection implies the consideration of the VLBI sessions (Nothnagel et al. 2017) observing 24 hours at least twice a week. The other source of promptly growing VLBI time series is the broadband VGOS observations (Petrachenko et al. 2012). While at this stage VGOS observations have a lack of a cadence similar to the legacy VLBI observations, the considerably higher density of VGOS observations is expected to allow (Petrachenko et al. 2012; Schuh and Böhm 2013) for better tropospheric estimation within a 24-hour slot than the legacy sessions. The dense sky coverage is enabled already at the step of observation scheduling (Schartner and Böhm 2019) to deal with the common dependency (Schuh and Böhm 2013) introduced by the simultaneous estimation of the troposphere and station position at the same sites. This dependency is

Table 1 The pairs of VLBI and GNSS stations are ordered by their height differences Δh. The spatial distance Δl between corresponding VLBI and GNSS stations as well as the number of their common epochs N, at which the tropospheric estimates time series derived from each technique individually are available, serve as further parameters to characterise VLBI-GNSS pairs. The $Bias_{theor.}$ defines the offset, which is introduced by the height differences and calculated as a rule of thumb (Böhm 2004). The bias $\Delta ZHD(h)$ of the time series is different from the rough estimation $Bias_{theor.}$, because the hydrostatic part is derived from a more comprehensive approach: the modelled variations are used as provided by VMF3, where a sophisticated layer modeling is employed. The $Bias_{est.}$ represents the impact of the height differences on the estimated troposphere differences before taking into account the offset $\Delta ZTD(h)$ and afterwards in terms of $Bias_{res.}$ and $\overline{RMS}$

VLBI-GNSS	N	Δl, m	Δh, m	$Bias_{theor.}$ ΔZHD mm	Bias $\Delta ZHD(h)$ mm	$Bias_{est.}$ ΔZTD mm	$Bias_{res.}$ ΔZTD mm	$\overline{RMS}$ ΔZTD mm
GGAO12M – GODN	380	95.27	−0.7	−0.2	−5.7	−1.7	4.5	5.8
RAEGSMAR – RAEG	215	60.30	1.2	0.4	−5.0	−2.9	3.2	10.5
KOKEE12M – KOKB	2622	19.65	−1.2	−0.4	2.5	−3.1	−4.9	10.6
WETTZELL – WTZR	4761	139.43	−3.1	−0.9	−4.7	−6.9	−1.2	4.9
NYALES20 – NYA1	3729	106.35	−3.1	−0.9	−2.9	−4.2	−0.5	6.7
RAEGYEB – YEBE	1887	158.79	−4.2	−1.3	−6.5	−4.9	2.6	5.0
WETTZ13N – WTZR	1972	93.97	−6.5	−2.0	−4.2	−7.4	−1.8	4.5
WETTZ13S – WTZR	2346	98.60	−6.5	−2.0	−4.3	−7.5	−2.2	4.3
ONSA13NE – ONSA	2613	394.89	−7.6	−2.3	−7.2	−6.8	1.3	7.8
ONSA13SW – ONSA	2075	465.25	−7.6	−2.3	−7.2	−6.3	1.7	5.1
KOKEE – KOKB	4202	46.48	−9.2	−2.8	−0.2	−9.4	−7.4	8.6
ONSALA60 – ONSA	2838	79.57	−13.7	−4.1	−7.2	−8.2	−0.0	5.4
YEBES40M – YEBE	663	152.75	−16.1	−4.8	−6.4	−7.1	−0.5	7.9
NYALE13N – NYAL	160	1470.11	24.9	7.5	7.8	6.2	−1.3	4.7
NYALE13S – NYAL	3652	1470.11	24.9	7.5	7.7	5.2	−2.0	11.3

resolved in GNSS observations by tracking multiple satellites (Hadas et al 2020).

The subset of GNSS stations is chosen to be located next to the VLBI antennas as close as possible in order to ensure, that the processed signals propagate through the troposphere of the most similar characteristics: pressure, temperature and humidity. The height differences between GNSS and VLBI reference points are varied from a few meter level up to almost 25 meter differences as shown in Table 1. The horizontal differences deviate to a higher extent as listed in Table 1, where the shortest distance between KOKEE12M and KOKB is approximately 20 meters and the longest between NYALE13N or NAYLE13S and NYAL is almost 1.5 kilometre. The observations of the closest GNSS station NABG to the twin antennas provide no common pairs of estimates with VLBI analysis results, thus other GNSS station NYAL appears to be close to the twin antennas than NYA1. For the consistent troposphere estimation, the common interval length of 1 hour is applied in the special CODE solution, whereas it is the standard resolution for the VLBI solution. We review the results derived from the GNSS and VLBI analysis starting with 2019, when VGOS sessions were commenced, and finishing with the most recent processed data of the middle of the year 2023.

In this paper, Sect. 2 presents the methods of the present comparison by describing each solution, i.e., GNSS and VLBI, and the offset of the tropospheric estimates due to the height difference of reference points of GNSS and VLBI antennas. Section 3 summarizes the results obtained from the comparisons between VLBI and GNSS. The paper finishes in Sect. 4 with a summary and an outlook.

2 Methods

The tropospheric products derived independently from the different space geodetic techniques, i.e., VLBI and GNSS, are inter-compared in a straightforward manner: identical epochs in the troposphere time series for the closest co-location sites between VLBI and GNSS are used to compute the difference between the techniques. The offsets are still visible in these differences and originate from the height differences between the neighboring GNSS and VLBI stations. Apart from the offsets, the corresponding RMS values are compared to evaluate the similarities in both time series.

2.1 GNSS Solution

The selection of the GNSS core station network is one of the crucial steps of the GNSS analysis tackling a global network. The co-location to VLBI stations as well as to the other space-geodetic techniques, thus, is only one of the criteria. The others are, e.g., support of various GNSS, availability of antenna calibrations, and homogeneous station distribution. This implies that not all co-located GNSS-VLBI stations are included in the CODE GNSS network solution. Thanks to the PPP approach a consistent densification of the GNSS network solution is allowed for any GNSS station that is not included in the global network solution. As follows, the PPP solution requires the best possible consistency with respect to the global network solution. For this reason, it is beneficial, if the PPP solution is provided by the same AC to ensure a compatible processing setup using the same software package, i.e. (Bernese, Dach et al 2015). The solution provided by the CODE IGS AC is used in this paper with the setup from the latest reprocessing effort repro3 (Dach et al 2021), which also defines the IGS contribution to the ITRF2020, until the operational solution of the IGS was switched to the IGS20 reference frame on November 28, 2022 (Rebischung 2023). For the reprocessing period, the coordinates are computed in a specific reference frame (named IGS14R3), which is comparable to the scale as defined by the Galileo satellite antenna calibrations. After November 2022, the scale is adjusted to the ITRF2020. The difference corresponds to about 7 mm on the Earth's surface (Altamimi et al 2023). The VMF3 (Landskron and Böhm 2018; re3data.org 2020) is used to model tropospheric delay for each observation. The tropospheric parameters are evaluated with hourly resolution in time in case of PPP solution to match VLBI solution parameterization, whereas the original GNSS network solution uses a two-hourly resolution in order to limit the number of estimated parameters.

2.2 VLBI Solution

The tropospheric product of the IVS AC at BKG is derived from a multi-session stacked solution. The stacked solution represents a combination of the normal equations of all pre-processed single sessions. The station positions, source coordinates and Earth Orientation Parameters (EOP) are estimated globally over the course of all stacked sessions. Contrary, the clocks and troposphere parameters are estimated on a session-wise basis. The clocks and ZTD are obtained with 1-hour resolution and horizontal gradients are estimated once per session. The piecewise linear functions are applied to session-wise parameters, where the station-wise clock parameters are evaluated as polynomials up to degree 2 in addition. The relative constraints are applied in the VLBI analysis for both: troposphere (1.5 cm/h) and horizontal gradients (0.5 mm for offset and 2.0 mm/day for the rate), while the hourly troposphere estimates are generated without relative constraints in GNSS solution. The results[1] are distributed within IVS by the IVS Data Centers on a session-wise basis in the form of tropospheric SINEX files (IVS Pilot Project 2001). The VLBI solution utilizes the latest realizations of the reference systems: ITRF2020 for the station coordinates and ICRF3 for the source positions. Also, the most recent version of a priori time series IERS 20 C04 is applied for the EOP. The troposphere is modelled as provided by the appropriate analysis software routine of the used in the analysis software package (Calc/Solve, Gipson 2018) based on VMF3 Data (Landskron and Böhm 2018) provided at (VMF Data Server, re3data.org 2020), which is not limited only to the mapping function but also include zenith hydrostatic and wet delays – the same as the GNSS solution in order to achieve the highest consistency. The only difference in VMF3 Data application is that VLBI analysis uses the station-wise corrections and GNSS – grid data. Additionally, the VLBI solution includes the atmospheric tidal loading corrections of the harmonics S_1 and S_2 as well as the atmospheric non-tidal loading corrections as the recommended in the last version of the IERS Conventions 2010 (Gérard and Brian 2010). The corresponding geophysical corrections are adopted as provided at the ESMGFZ server (Dill and Dobslaw 2018) and consistent with VMF3, since both are driven by the same numerical weather model ESMWF.[2]

2.3 Height Reduction

In order to generate the tropospheric products, the total zenith delay (ZTD) is treated as a sum of hydrostatic (ZHD) and wet (ZWD) parts, where ZHD can be modelled with the high confidence level as defined by the Saastamoinen model using in-situ atmospheric pressure measurements and station heights or numerical weather models, such as in case with VMF3 Data (re3data.org 2020), it is ERA-Interim Numerical Weather Models as well as atmospheric pressure loading corrections [2]. Contrary, the ZWD part is poorly predictable. Hence, the geodetic analysis can only include ZHD in the set of applied apriori reductions. The estimated tropospheric parameters describe, thus, mostly the observed ZWD variations. The tropospheric products deliver the total value ZTD as a result. A comparison of these products derived from VLBI and GNSS solutions requires that one needs to take into account the contribution due to the height differences of the reference points of GNSS and VLBI antennas, where ZHD is the major part contributing to the corresponding offset. The difference between ZHD at the VLBI and GNSS stations $\Delta ZHD(h)$ is straight forward when assuming GNSS and VLBI receiver registering signals propagating thorough

---

[1] The corresponding bkg2023a time series are available at the official IVS Data Centers (@IVSDC, https://ivscc.gsfc.nasa.gov/about/org/components/dc-list.html) with path https://ivscc.gsfc.nasa.gov/products-data/products.html#trop.

[2] European Centre for Medium-Range Weather Forecasts (ECMWF) https://www.ecmwf.int/.

Table 2 Total zenith delay differences: without accounting for the height differences as well as with accounting for the height difference by considering the contribution of the hydrostatic and wet parts separately

VLBI-GNSS	Δh, mm	ΔZTD [VLBI − GNSS]			ΔZTD [VLBI − GNSS] −ΔZHD(h)			ΔZTD [VLBI − GNSS] −ΔZHD(h) − ΔZWD(h)		
		Bias, mm	$\overline{\text{RMS}}$, mm	RMS, mm	Bias, mm	$\overline{\text{RMS}}$, mm	RMS, mm	Bias, mm	$\overline{\text{RMS}}$, mm	RMS, mm
GGAO12M − GODN	0.7	1.7	21.3	21.7	4.5	5.8	7.4	4.5	5.8	7.4
RAEGSMAR − RAEG	1.2	2.9	27.5	28.3	3.2	10.5	11.2	3.2	10.5	11.2
KOKEE12M − KOKB	1.2	3.1	37.0	37.3	5.0	10.6	11.5	5.0	10.6	11.5
WETTZELL − WTZR	3.1	6.9	80.7	81.4	1.3	4.9	5.1	1.3	4.9	5.1
NYALES20 − NYA1	3.1	4.2	49.1	49.8	0.5	6.7	6.7	0.5	6.7	6.7
RAEGYEB − YEBE	4.2	4.9	135.1	135.5	2.6	5.0	5.7	2.6	5.0	5.7
WETTZ13N − WTZR	6.5	7.4	33.6	35.9	1.9	4.5	5.0	1.9	4.5	5.0
WETTZ13S − WTZR	6.5	7.5	87.8	88.6	2.3	4.3	4.9	2.3	4.3	4.9
ONSA13NE − ONSA	7.6	6.8	53.5	54.6	1.2	7.8	7.8	1.2	7.8	7.8
ONSA13SW − ONSA	7.6	6.3	29.4	31.3	1.6	5.1	5.3	1.6	5.1	5.3
KOKEE − KOKB	9.2	9.4	82.3	83.6	7.5	8.6	11.4	7.5	8.6	11.4
ONSALA60 − ONSA	13.7	8.2	26.8	29.9	0.2	5.4	5.4	0.2	5.4	5.4
YEBES40M − YEBE	16.1	7.1	40.5	41.7	0.6	7.9	7.9	0.5	7.9	7.9
NYALE13N − NYAL	24.9	6.2	12.2	12.8	0.9	4.7	4.9	0.9	4.7	4.9
NYALE13S − NYAL	24.9	5.2	20.2	20.8	1.7	11.3	11.4	1.7	11.3	11.4

the atmosphere of the similar characteristics:[3]

$$\Delta\text{ZHD}(h) = \text{ZHD}_{\text{VLBI}} - \text{ZHD}_{\text{GNSS}}, \quad (1)$$

where the corresponded parts ZHD_{VLBI} and ZHD_{GNSS} are available from bkg2023a VLBI solution and CODE GNSS solution internally for this paper. VMF3 Data (re3data.org 2020; Landskron and Böhm 2018) provides the appropriate basis for both solutions, i.e. ZHD_{VLBI} and ZHD_{GNSS} for VLBI and GNSS correspondingly. Even though the ZWD cannot be modelled at the same level as ZHD, the empirical equation is often used (i.e. by Thaller (2008)) to describe the corresponding variations $\Delta\text{ZWD}(h)$ for the station height differences. The remaining differences ΔZTD are calculated from the differences of the estimated zenith total delay ΔZTD [VLBI − GNSS] between VLBI and GNSS and zenith total delay corrections $\Delta\text{ZTD}(h)$ due to the height differences as given by equation 1 and empirical approach $\Delta\text{ZWD}(h)$:

$$\Delta\text{ZTD} = \Delta\text{ZTD}\,[\text{VLBI} - \text{GNSS}] \\ - (\Delta\text{ZHD}(h) + \Delta\text{ZWD}(h)). \quad (2)$$

3 Results

The time series of the obtained differences in tropospheric estimates ΔZTD are compared in terms of root mean square

---
[3] It is true for the chosen co-location stations and as long as the distance between receivers is not longer than cluster size of the chosen numerical weather model.

(RMS) and bias. The bias of the time series is defined as the median value of the corresponding differences ΔZTD. Its reduction (see Tables 1 and 2) confirms the fact, that the offset due to the height differences is taken into account. The impact of the height differences for the set of chosen stations is calculated approximately using a rule of thumb (Böhm 2004, (Bias$_{\text{theor.}}$ ΔZHD in Table 1)) as well as deliberately using modelled corrections of the hydrostatic part of the zenith delay (Bias $\Delta\text{ZHD}(h)$ in Table 1) based on VMF3. The corrections $\Delta\text{ZHD}(h)$ in sum with empirically derived $\Delta\text{ZWD}(h)$ are employed to calculate the ultimate Bias$_{\text{res.}}$. The expected bias reduction is demonstrated for the chosen pairs of stations with exceptions: GGA012M-GODN, RAEGSMAR-RAEG and KOKEE12M-KOKB, where the small amount of observations available for the pairs is considered to be insufficient to deliver reliable results. The results of shortest time series NYALES13N-NYAL with newcomer NYALES13N into the routine VLBI observations is based on only 5 VLBI sessions, those it requires including most recent observations. A larger unbiased $\overline{\text{RMS}}$ ΔZTD is calculated for these co-location stations and also for the other pairs: NYALES20-NYA1, ONSA13NE-ONSA, KOKEE-KOKB, YEBES40M-YEBE and NYALES13S-NYAL. The level of the corresponding scatter ($\overline{\text{RMS}}$) varies between 5 and 10 mm, thus a recognition of smaller signals in the obtained unbiased residuals (Bias$_{\text{res.}}$) is limited by this value for each considered pair in Table 2. An attempt of inclusion of such small effects of the next order contribution as a modeled wet part of zenith delay $\Delta\text{ZWD}(h)$ illustrates in Table 2, where two last sets consisted of 3 columns each are almost identical, that corresponding impact can not be

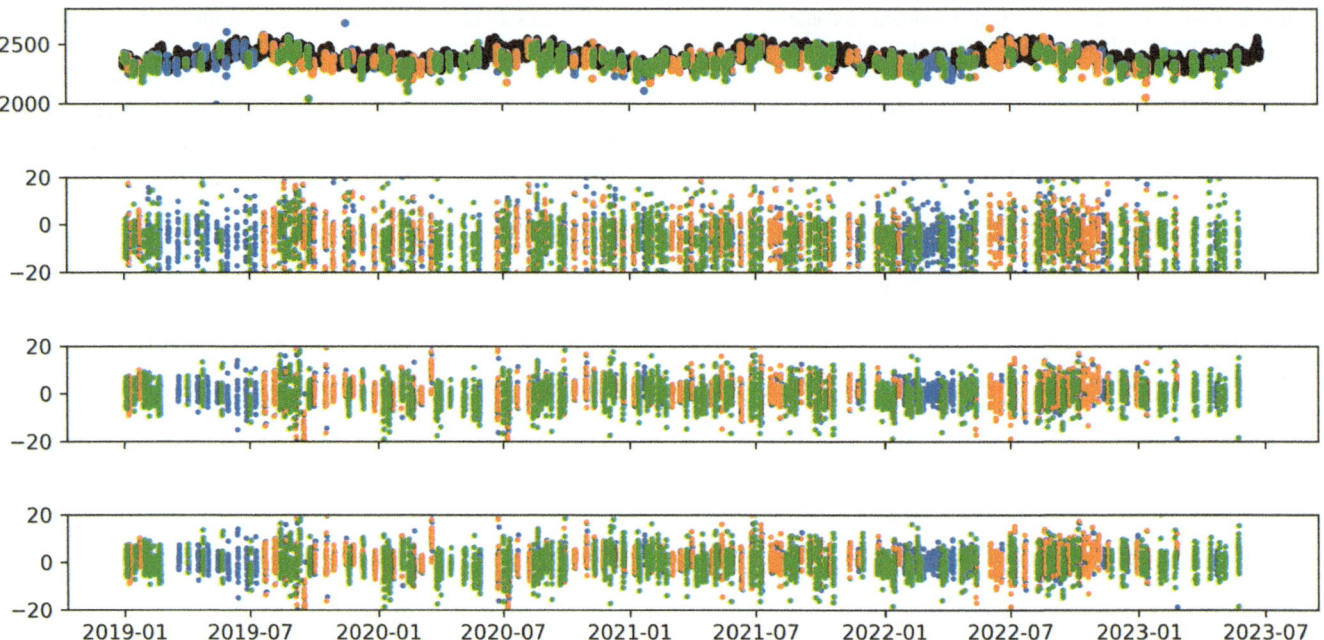

Fig. 1 The panels present troposphere estimates and their comparison for the co-location site Onsala: GNSS antenna (black) and the three VLBI antennas, i.e., legacy ONSALA (green), ONSA13NE (blue), ONSA13SW (orange). The total zenith delay estimates (mm) are shown in the upper panel. The three lower panels represent the differences (mm) between the GNSS antenna ONSA and VLBI antennas: before and after taking into account the height differences by subtracting the hydrostatic part only. The lowest panel illustrates the remaining variations (mm) after including a smaller contribution of $\Delta ZWD(h)$ in addition to the hydrostatic differences as defined by the equation (2)

detected at the obtained level. Dealing with the scatter of the unbiased residuals is to be the major part of further studies.

The obtained tropospheric variations, i.e., the total values ZTD, demonstrate a good agreement between the independently derived results for the GNSS and VLBI antennas. Figure 1 demonstrates this agreement for the Onsala site as an example. In order to quantify the similarities in these ZTD time series the corresponding differences are shown in the lower panels as calculated between GNSS antenna ONSA and VLBI antennas, i.e., ONSALA60 and twin antennas ONSA13NE and ONSA13SW. As follows, the corrections due to the height differences are introduced on a step-wise basis as described in Sect. 2.3. Since the paired time series are compared at the same epochs, the direct subtraction of ZTD values between VLBI and GNSS (step 1) illustrates the effect of the unaccounted height differences in the second panel of Fig. 1. The third panel shows smaller differences, less scattered and closer to zero, because the corrections derived from the antenna-dependent hydrostatic part differences $\Delta ZHD(h)$ are applied (step 2). The wet corrections $\Delta ZWD(h)$ are subtracted in addition to $\Delta ZHD(h)$. The remaining differences (step 3) are shown in the bottom panel. These final differences satisfy the full budget of the modelled height differences defined by equation (2). The bias and RMS values and unbiased RMS values of the troposphere differences are collected in Table 2 for the above-mentioned steps. The simple difference of the total values of ZTD (step 1) is given in the first set of three columns in Table 2). The second set of three columns shows the corresponding values for step 2, and the third set of three columns for step 3. A reduction of the bias can be observed from step 1 to 2, which is also illustrated by the last two panels in Fig. 1. The step 3 corrections impact the RMS and bias values insignificantly as expected.

The VLBI analysis is aimed to deliver the best possible formal errors obtained by stacking sessions during 40 years of observations, i.e., approx. 6500 VLBI 24-hour sessions. In contrast, the single session analysis of the VLBI observations usually leads to global parameter estimates, which are only worse by a factor of 2-3 (depending on the station) compared to the 40-year long-term solution. No correlations are considered between single observations, since the scheduling of observations is assumed to be a sufficient tool to avoid this kind of dependency. VLBI analysis refers to the global parameters as to the parameters estimated once over the course of considered observations, namely station coordinates, EOP, source positions and etc. The tropospheric parameters are estimated always as local or session-wise parameters in a global or session-wise solution, nevertheless they benefit from the solving a global solution as well. By this reason, the global solution was chosen to deliver the IVS tropospheric products. In general, the obtained formal errors of the corresponding tropospheric estimates for the VGOS stations are remarkably good for VLBI analysis: at the same locations, where the long time series are available, the formal errors obtained for the legacy stations are worse by a factor of two: RAEGYEB (1.0 mm) and YEBES40M (2.1 mm), ONSALA60 (1.7 mm) and twin antennas (ONSA13SW and ONSA13NE, 0.9 mm), KOKEE (2.7 mm) and KOKEE12M (2.0 mm). The sky coverage obstruction at the Kokee site might be the reason for higher formal errors among the other VGOS antennas (i.e., 2 mm compared to 1 mm). Also one of the twin antennas which has observed in VGOS mode, i.e., WETTZ13S next to the legacy antenna WETTZELL, shows twice better standard deviations. The other VGOS antenna WETTZ13N, which has observed in S/X mode only until the beginning of 2023, shows standard deviations at the level of the legacy antenna. Similar to Wettzell, in Ny-Alesund one of the twin VGOS antennas (i.e., NYALE13S) has observed in S/X mode. The corresponding formal errors are relatively high (5.9 mm) notwithstanding the comparable length of time series with the legacy antenna results. The second twin VGOS antenna (i.e., NYALE13N) has observed only five sessions in VGOS mode for the time span considered in our study. Nevertheless, the results (2.0 mm) overcome already the level of the legacy antenna performance (2.6 mm), although the length of the NYALES13N time series is too short to draw a reliable conclusion. The remaining VGOS antennas which have observed in VGOS mode seem to provide a consistent level of formal errors of approximately 1 mm.

4 Conclusion and Outlook

The results of the analysis of two independent geodetic techniques are discussed in this paper for their cross-validation at the parameter level. The obtained tropospheric estimates demonstrate a good agreement after omitting the offsets between the reference points of the VLBI and GNSS antennas. Similar results have been shown by Haas et al (2023) for a subset of stations located in Onsala and Kokee observatories. The detailed comparisons of the tropospheric estimates carried out in the studies preseted here are the basis for a combination of VLBI and GNSS tropospheric parameters, either on parameter level or on normal equation level. The latter procedure is foreseen as the next step for our investigations as we have both space-geodetic techniques, i.e., GNSS and VLBI available in-house data analysis. The pairs of VLBI and GNSS stations have been selected to be as close to each other as possible, while otherwise the large discrepancies can be seen for the stations at Westford (Wang et al. 2023). The differences in the tropospheric estimates between VLBI and GNSS can be fitted better if more common epochs of the time series are available. This can be seen when comparing, e.g., twin antennas in Wettzell (with many common epochs) and Ny-Alesund (with only very few common epochs). The other reasons for the larger differences might be seen in the lack of sky coverage, i.e. stations at Kokee. The formal errors of the tropospheric estimates for the VGOS stations are obtained at the level of up to a factor of 2 smaller than for legacy stations. Considering the VGOS antenna design, the analysis of the tropospheric estimates seems to serve as a good measure to qualify the VGOS antenna performance. The obtained results are promising and encourage further investigations with higher temporal resolution of the tropospheric estimates derived from VLBI and GNSS solutions. Especially the high-quality results of those sites operating a VGOS antenna have a huge potential for VLBI-based troposphere estimates becoming an important data source for geoscience applications (Haas et al 2022). In this context, the homogenization of the VLBI and GNSS tropospheric estimation – as it was applied for this paper – is crucial for providing precise geodetic product.

References

Altamimi Z, Rebischung P, Métivier L, Chanard K (2023) ITRF2020: an augmented reference frame refining the modeling of nonlinear station motions. J Geodesy 97(47): article number 47. https://doi.org/10.1007/s00190-023-01738-w

Böhm J (2004) Troposphärische Laufzeitverzögerungen in der VLBI. Geowiss. Mitt. 68, Inst. für Geod. und Geophys, Vienna, PhD Thesis

Dach R, Lutz S, Walser P, Fridez P (2015) Bernese GNSS software version 5.2. User manual. Astronomical Institute, University of Bern, Bern Open Publishing, Bern. https://doi.org/10.7892/boris.7229

Dach R, Selmke I, Villiger A, Arnold D, Prange L, Schaer S, Sidorov D, Stebler P, Jäggi A, Hugentobler U (2021) Review of recent GNSS modelling improvements based on CODEs Repro3 contribution. Adv Space Res 68(3):1263–1280. https://doi.org/10.1016/j.asr.2021.04.046

Dach R, Schaer S, Arnold D, Kalarus M, Prange L, Stebler P, Villiger A, Jäggi A, Brockmann E, Ineichen D, Lutz S, Willi D, Nicodet M, Thaller D, Klemm L, Rülke A, Söhne W, Bouman J, Hugentobler U (2023) CODE analysis center: IGS technical report 2022. In: Dach R, Brockmann E (eds) International GNSS service: technical report 2022, pp. 45–64. IGS Central Bureau. online. https://doi.org/10.48350/179297

Dill R, Dobslaw H (2018) ESMGFZ: operational model products for geodetic applications. In: Geophysical research abstracts; Vol. 20, EGU2018-2772. General Assembly European Geosciences Union (Vienna 2018). https://meetingorganizer.copernicus.org/EGU2018/EGU2018-2772.pdf?EGUsphere

Engelhardt G, Walenta A, Ullrich D, Flohrer C (2023) BKG VLBI Analysis Center: IVS biannual report 2022. In: KL Armstrong DB, Baver KD (eds) International VLBI service for geodesy and astrometry 2021+2022 biennial report. NASA/TP-20230014975, online

Gérard P, Brian L (ed) (2010) IERS conventions (2010). Verlag des Bundesamts für Kartographie und Geodäsie, Frankfurt am Main

Gipson J (2018) Calc/Solve: https://space-geodesy.nasa.gov/techniques/tools/calc_solve/calc_solve.html

Hadas T, Hobiger T, Hordyniec P (2020) Considering different recent advancements in GNSS on real-time zenith troposphere estimates. GPS Solut 24:99. https://doi.org/10.1007/s10291-020-01014-w

Haas R, Varenius E, Schartner M, Matsumoto S (2022) Combining VGOS, legacy S/X and GNSS for the determination of UT1. J Geodesy 96:55–58. https://doi.org/10.1007/s00190-022-01648-3

Haas R, Elgered G, Johansson J, Nilsson T, Ning T (2023) Atmospheric parameters derived from VGOS sessions observed with the Onsala twin telescopes. In: 26th European VLBI Group for Geodesy and Astronomy Working Meeting, Bad Kötzting Germany, 11–15 June 2023

He C, Pollet A, Coulot D, Schott-Guilmault V, Perosanz F (2023) Towards the tropospheric ties in the GPS, DORIS, and VLBI combination analysis during CONT14. J Geodesy 97:111. https://doi.org/10.1007/s00190-023-01803-4

IVS Pilot Project (2001) File format is SINEX V2.0: https://ivscc.gsfc.nasa.gov/products-data/sinex_v2_trop.pdf. https://ivscc.gsfc.nasa.gov/publications/gm2002/boehm2.pdf. Online notes

Johnston G, Riddell A, Hausler G (2017) The international GNSS service. In: Teunissen PJG, Montenbruck O (eds) Springer handbook of global navigation satellite systems, pp 967–982. Springer, Cham, Switzerland. https://doi.org/10.1007/978-3-319-42928-1

Krügel M, Thaller D, Tesmer V, Rothacher M, Angermann D, Schmid R (2007) Tropospheric parameters: Combination studies based on homogeneous VLBI and GPS data. J Geodesy 81(6-8):515–527. https://doi.org/10.1007/s00190-006-0127-8

Landskron D, Böhm J (2018) VMF3/GPT3: refined discrete and empirical troposphere mapping functions. J Geodesy 92:349–360. https://doi.org/10.1007/s00190-017-1066-2

Ning T, Haas R, Elgered G, Willén U (2012) Multi-technique comparisons of 10-years of wet delay estimates on the west coast of Sweden. J Geodesy 86:565. https://doi.org/10.1007/s00190-011-0527-2

Nothnagel A, Artz T, Behrend D, Malkin Z (2017) International VLBI Service for Geodesy and Astrometry – Delivering high-quality products and embarking on observations of the next generation. J Geodesy 91:711–721. https://doi.org/10.1007/s00190-016-0950-5

Petrachenko WT, Niell AE, Corey BE, Behrend D, Schuh H, Wresnik J (2012) VLBI2010: Next generation VLBI system for geodesy and astrometry. In: Kenyon S, Pacino MC, Marti U (eds) Geodesy for planet earth, pp 999–1005. Springer, Berlin, Heidelberg

re3data.org (2020) VMF Data Server; editing status 2020-12-14; re3data.org - Registry of Research Data Repositories. http://doi.org/10.17616/R3RD2H. https://vmf.geo.tuwien.ac.at/

Rebischung P (2023) Reference frame working group: IGS technical report 2022. In: Dach R, Brockmann E (eds) International GNSS service: technical report 2022, pp 217–226. IGS Central Bureau, online. https://doi.org/10.48350/179297

Schartner M, Böhm J (2019) Viesched++: A new vlbi scheduling software for geodesy and astrometry. Publ Astronom Soc Pacific 131(1002):084501. https://doi.org/10.1088/1538-3873/ab1820

Schuh H, Böhm J (2013). In: Xu Guochang (ed) Very long baseline interferometry for geodesy and astrometry, pp. 339–376. Springer, Berlin, Heidelberg. https://doi.org/10.1007/978-3-642-28000-9_7

Steigenberger P, Tesmer V, Krügel M, Thaller D, Schmid R, Vey S, Rothacher M (2007) Comparisons of homogeneously reprocessed GPS and VLBI long time-series of troposphere zenith delays and gradients. J Geodesy 81(6-8):503–514. https://doi.org/10.1007/s00190-006-0124-y

Thaller D (2008) Inter-technique combination based on homogeneous normal equation systems including station coordinates, Earth orientation and troposphere parameters. https://doi.org/10.2312/GFZ.b103-08153. PhD Thesis

Walenta M, Goltz A (2023) BKG data center: IVS biannual report 2022. In: KL Armstrong DB, Baver KD (eds) International VLBI service for geodesy and astrometry 2021+2022 biannual report. NASA/TP-20230014975, online

Wang J, Glaser S, Balidakis K, Ge M, Heinkelmann R, Schuh H (2023) Applying tropospheric ties in GNSS and VLBI integrated solution: Impact of deterministic and stochastic uncertainty. XXVIII General Assembly of the International Union of Geodesy and Geophysics (IUGG) (Berlin 2023)

Open Access This chapter is licensed under the terms of the Creative Commons Attribution 4.0 International License (http://creativecommons.org/licenses/by/4.0/), which permits use, sharing, adaptation, distribution and reproduction in any medium or format, as long as you give appropriate credit to the original author(s) and the source, provide a link to the Creative Commons license and indicate if changes were made.

The images or other third party material in this chapter are included in the chapter's Creative Commons license, unless indicated otherwise in a credit line to the material. If material is not included in the chapter's Creative Commons license and your intended use is not permitted by statutory regulation or exceeds the permitted use, you will need to obtain permission directly from the copyright holder.

Real-Time GNSS Integrated Water Vapor Sensing Based on Time Series Correction Deep Learning Models

Duo Wang, Peng Yuan, and Hansjörg Kutterer

Abstract

In the past three decades, GNSS-based Integrated Water Vapor (IWV) retrieval has been intensively investigated, and its products have been widely used in meteorology like severe weather event monitoring. The physical model for the inversion of IWV from the tropospheric Zenith Total Delay (ZTD) requires meteorological data at the location of the GNSS station, such as the surface pressure and the atmospheric weighted mean temperature. However, real-time acquisition of the meteorological data is a very challenging task for most GNSS stations. While proposed empirical models such as Global Pressure and Temperature 3 (GPT3) can provide the meteorological data based on their historical information, larger estimation distortions are found in specific mid- and high-latitude regions. Moreover, we analyzed the seasonal variations in GPT3 prediction errors. In view of the above-mentioned problems, this study implements an IWV conversion model based on a feedforward Deep artificial Neural Network (DNN) and Long Short-Term Memory Network (LSTM) network, which learns historical data from GNSS stations and allows real-time ZTD to IWV conversion without the need of actual meteorological observation but of values only GPT3. Results at four selected mid- and high-latitude GNSS stations show that the Root Mean Square Error (RMSE) of the proposed deep learning method decreases from an average of 3.97 mm to 2.84 mm compared to GNSS IWV retrieved from GPT3. The proposed model provides a broad applicability in real-time GNSS IWV prediction without the availability of real-time measured meteorological data.

Keywords

Deep learning · GNSS · Integrated water vapor (IWV) · Long short-term memory network (LSTM) · Zenith total delay (ZTD)

D. Wang · H. Kutterer
Geodetic Institute, Karlsruhe Institute of Technology, Karlsruhe, Germany

P. Yuan (✉)
Geodetic Institute, Karlsruhe Institute of Technology, Karlsruhe, Germany

GFZ German Research Centre for Geosciences, Potsdam, Germany
e-mail: pyuan@gfz-potsdam.de

1 Introduction

Water vapor is a significant component of the Earth's atmosphere in terms of energy transport by latent heat and radiative forcing. It has a vital role in the water and energy cycles (Worden et al. 2007), climate change (e.g., Karl and Trenberth 2003; Yuan et al. 2021) and the understanding of many extreme weather phenomena (e.g., Zhu and Newell 1994; Jiang et al. 2017). Due to the rapid spatiotemporal variations of water vapor, its real-time acquisition with high spatiotemporal resolution remains a challenge.

Typically, water vapor can be quantified using integrated water vapor (IWV) in units of kg/m^2, representing the mass of water vapor within a 1 m^2 atmospheric column. Since the 1990s, with the development of satellite and remote sensing technologies, researchers have successfully retrieved IWV by employing high-spectral and multispectral remote sensing techniques, such as the Moderate Resolution Imaging Spectroradiometer (MODIS, King et al. 1992), and atmospheric reanalyses (Schröder et al. 2018; Yuan et al. 2023a). However, these techniques still face limitations in water vapor estimation when it comes to conducting real-time, all-weather, and high accuracy. For example, multispectral remote sensing technologies like MODIS are susceptible to weather conditions, particularly cloud cover and diurnal changes (Vaquero-Martínez et al. 2017). Reanalysis data, such as the fifth generation ECMWF atmospheric reanalysis (ERA5) (Hersbach et al. 2023), can provide medium-resolution gridded data with a latency of about 5 days or more, this limiting their availability for real-time or near real-time access. In addition, the traditional radiosondes, which utilize meteorological balloons, can offer high-precision vertical distribution of water vapor, but it is constrained by the limited number of measurements (Durre and Yin 2008).

The Global Navigation Satellite Systems (GNSS) has emerged as a unique technique for IWV retrieval. It boasts outstanding characteristics, including high accuracy, high temporal resolution, and all-weather capability (Bevis et al. 1992). The GNSS signals, transmitted from satellites to ground receivers, experience delay as they traverse through the Earth's atmosphere. This delay can be measured by GNSS stations and is referred to as Zenith Tropospheric Delay (ZTD). ZTD consists of two components (Böhm and Schuh 2013), the Zenith Hydrostatic Delay (ZHD) and Zenith Wet Delay (ZWD). ZHD can be modelled with pressure data. The remaining component, ZWD, can be used to estimate IWV using temperature and water vapor pressure data (Bevis 1994). Due to significant improvements in recent years, some GNSS stations can even provide real-time or near-real-time ZTD products. However, there are still significant limitations in the acquisition of meteorological data in terms of both space and time.

GNSS is very promising technique in water vapor sensing, but it requires meteorological data which is usually unavailable in real-time. To obtain real-time weather data predictions, Landskron and Böhm (2018) put forward the Global Pressure and Temperature 3 (GPT3) model by analysing 10 years mean monthly pressure level data from ERA-Interim. GPT3 is composed of a series of spherical harmonic functions. By inputting the time and location information, it predicts meteorological data such as temperature, pressure, and water vapor pressure. The GPT3 model is widely acknowledged for its effectiveness; however, it exhibits limitations associated with geographical considerations, particularly in regions with high latitudes or experiencing significant climatic variations (Yang et al. 2021). Furthermore, due to its inherent design, the model fails to adequately account for diurnal fluctuations, consequently impeding its ability to accurately forecast the rapid and high-frequency changes inherent in diurnal meteorological data. To overcome this limitation, this work proposes a conversion model by using deep learning, a machine learning technique that leverages deep artificial neural networks (DNNs). It has achieved success in various fields such as computer vision, natural language processing, and often represents the state-of-the-art in these domains (Lecun et al. 2015; Goodfellow et al. 2016). Thanks to the powerful feature extraction capabilities of deep neural networks, deep learning can often achieve end-to-end implicit feature learning without the need to design features manually. This feature has made deep learning widely embraced by researchers in the remote sensing field (Zhu et al. 2017a, b; Wang et al. 2020, 2022). In GNSS meteorology research, scholars have also conducted relevant research on ZTD or IWV using machine learning technology. For example, Zheng et al. (2022) used a stacked ensemble algorithm to map ZTD to IWV. This model regresses IWV using multiple machine learning models and uses a two-layer neural network to ensemble the regression results. This method maps ZTD to IWV directly but lacks time series modeling, so the contextual information of the IWV time series may be ignored. Shangguan et al. (2023) proposed the WLA model to predict IWV time series. This model takes into account the contextual connection of IWV but does not introduce ZTD-measured information, therefore it may not reflect the real-time IWV changes. Long Short-Term Memory Networks (LSTM) are a type of DNN with a recurrent structure (Hochreiter and Schmidhuber 1997), capable of effectively learning patterns in sequences. In this paper, we propose combining the use of feedforward DNN and LSTM to implicitly learn spatiotemporal patterns in meteorological data, enabling us to perform ZTD to IWV conversion without the need of actual meteorological observation but of values only GPT3. It is a combined use, not a real combination, to cover different characteristics in different ranges of IWV values. The details are given in Sect. 2. Due to LSTM-based learning that can capture the features from the history sequence of IWV, the method presented in this paper is expected to mitigate high-frequency retrieved errors in ZTD-IWV inversion without current meteorological data.

In the following sections, we will introduce the empirical analytical model for GNSS IWV retrieval, and then describe the deep learning techniques we used. Subsequently, we employ case studies in Europe to compare their performances. Based on the analysis above, we draw the conclusions in Sect. 4.

2 Data and Methodology

2.1 IWV Retrieved from GNSS ZTD

Detailed information on obtaining water vapor products from GNSS can be found in (Bevis 1994). In brief, GNSS positioning involves measuring the time delay of microwave signals transmitted from satellites to ground stations. However, the signals suffer a series of delays during propagation. For the ionospheric component, more than 99.9% can be modelled and removed using dual-frequency measurements (Hernández-Pajares et al. 2007). In addition to the ionospheric delay, the GNSS signal is also delayed by the troposphere, which can be quantified ZTD, indicating the troposphere delay at zenith over the station. Approximately 90% of the ZTD comes from its hydrostatic component (ZHD) (Leick et al. 2015). The ZHD is relatively stable, and it can be accurately estimated with station pressure as follows (Saastamoinen 1972):

$$\text{ZHD} = 2.2768 * \frac{p_s}{1 - 2.66 * 10^{-3} * \cos(2 * LAT) - 2.8 * 10^{-7} H_s}, \quad (1)$$

where p_s is the station air pressure (in hPa), LAT is the latitude of the station and H_s is the orthometric height (in m) of the station. The wet delay component can be calculated by:

$$\text{ZWD} = \text{ZTD} - \text{ZHD}, \quad (2)$$

The diffusion of water vapor primarily occurs in the troposphere. It is the primary cause of wet delays in microwave transmission through the troposphere. According to Bevis (1994), IWV can be obtained by multiplying ZWD by a conversion coefficient Π as follows:

$$\text{IWV} = \Pi * \text{ZWD}, \quad (3)$$

$$\Pi = \frac{10^6}{R_v * \left[k_2' + \frac{k_3}{T_m} \right]}, \quad (4)$$

$$T_m = \frac{\int_{H_s}^{H_{top}} \frac{e(h)}{T(h)} dh}{\int_{H_s}^{H_{top}} \frac{e(h)}{T(h)^2} dh}, \quad (5)$$

Where $R_v = 461.522 \, J/(kg*K)$ (Kestin et al. 1984) is the specific gas constant for water vapor. $k_2' = 22.1 \, K/hpa$ and $k_3 = 373\,900 \, K^2/hpa$ are atmospheric refractivity constants (Bevis 1994), T_m is the weighted mean temperature calculated by the integration ratio of water vapor pressure e and temperature T from the height of the station to the top of atmosphere. In this paper, the GNSS ZTD were provided by Nevada Geodetic Laboratory (NGL, Blewitt et al. 2018, http://geodesy.unr.edu/index.php, latest access date 2024 May). The meteorological variables were calculated using the ERA5 pressure level product by vertical correction and horizontal interpolation to the location of each GNSS station. For the details on the calculations, readers are referred to Yuan et al. (2023b).

2.2 Deep Learning Approach

Due to the challenges in obtaining real-time and accurate meteorological data for GNSS stations, we propose using a deep neural network to implicitly learn the meteorological information and directly map the ZTD to IWV. According to Sect. 2.1, IWV can be considered as a function of ZTD, latitude LAT, station height H_s, temperature T, water vapor pressure e and pressure p_s. As T, e, p_s exhibit temporal and spatial variabilities, these quantities can be modeled based on location and time. GPT3 is one of the newest empirical model that maps location and time information to various meteorological data such as T, e, p_s. Although the GNSS IWV can be retrieved by using the p_s and T_m from GPT3, however, its errors could be significant compared to the GNSS IWV retrieved with ERA5, especially in mid-to-high latitude regions. In this work, we utilize the GPT3 results as part of the network input for training. This will help learn from prior knowledge and allow faster convergence.

Our work is inspired by the limitations of the GPT3 in retrieving with P_s and T_m. Since the GPT3 cannot respond effectively to the rapidly changing diurnal meteorological conditions, it is necessary to analyze the correlation between the IWV retrieval error caused by using P_s and T_m from GPT3 and the water vapor activity. We found that the IWV retrieved error is related to its retrieved IWV value, that is, the water vapor activity level. Besides, when the IWV value is high (usually in summer, depending on the location), the correlation between retrieved IWV and the retrieved error of this IWV is low; on the contrary, when the IWV value is low (usually in winter and spring, depending on the location), the retrieved error is correlated with the IWV value. An example from KIRU station (see Fig. 3 and Table 1 for more details) shows this property in Fig. 1. When the GNSS IWV is low, the IWV value retrieved using GPT3 has a strong linear correlation with its error (from January 1 to May 25, 2022, $R^2 = 0.77$). However, when the IWV value is high, this linear correlation declines rapidly (from January 1 to May 26 to August 4, 2022, $R^2 = 0.13$). For the specific value of IWV values, please see Fig. 5.

With this property, we propose to build different models for the active (High IWV value) and inactive periods (Low IWV value) of water vapor and use them in combination to improve performance. The classification of high/low

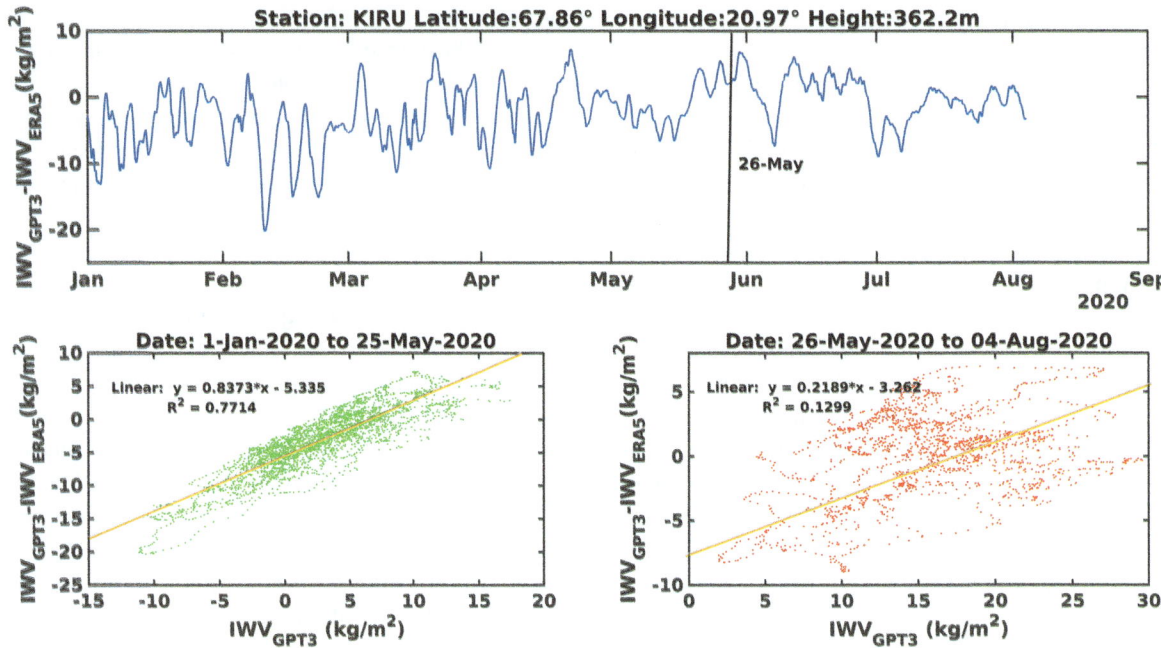

Fig. 1 The correlation between the IWV retrievals and its associated error. Before 26-May-2020, the R^2 between IWV and retrieved error is 0.77, while after is 0.13. Data from September to December 2020 at KIRU is not available because of missing observation files

Table 1 Coordinates of the GNSS stations selected for model validation

GNSS station	Longitude (°)	Latitude (°)	Height (m)
KIRU	20.9684	67.8574	362.2
DZYL	6.9404	53.3200	15.7
DZY1	6.9360	53.3228	8.2
FFMJ	8.6650	50.0906	130.1

boundaries can be obtained through change point detection from Bayesian Estimator of Abrupt Change, Seasonality, and Trend (BEAST, Zhao et al. 2019), or simply using empirical value. It is not necessary to find a very precise boundary because neural networks have the ability to generalize. For our case study, 20 kg/m^2 is a recommended threshold. We discussed the boundaries effect in the next section.

For high IWV data, we construct a training set using all features that can be obtained in real-time to train a k-layer feedforward DNN f_{DNN} with weight parameter θ, as described as follows:

$$X_{highIWV} = \{ZTD, LAT, LON, H_s, DOY, HOD, T_{GPT3}, p_{sGPT3}, e_{GPT3}\} \quad (6)$$

$$\hat{y}_{highIWV} = f_{DNN}(X_{highIWV}, \theta) = ReLU(\theta_k \cdot h_{k-1} + b_k) \quad (7)$$

$$h_l = ReLU(\theta_l \cdot h_{l-1} + b_l), l = 2, 3, \ldots, k-1 \quad (8)$$

$$h_1 = ReLU(\theta_1 \cdot X_{highIWV} + b_1) \quad (9)$$

$$ReLU(x) = \begin{cases} x \text{ if } x > 0, \\ 0 \text{ otherwise}. \end{cases} \quad (10)$$

where ZTD, LAT, LON, H_s, DOY, HOD, T_{GPT3}, p_{sGPT3}, e_{GPT3} represent the GNSS-measured ZTD, latitude, longitude, and altitude of GNSS station, day-of-year, hour-of-day, and the estimated temperature, pressure, and water vapor pressure from GPT3 used for training, respectively. $X_{highIWV}$ is the input vector constructed by features above from the period of high IWV, $\hat{y}_{highIWV}$ is the output of the network, representing the predicted IWV corresponding to its period, θ_l and b_l are the weight and bias vector of l-th layer neurons, h_l is the output of l-th layer neurons. We use rectified linear unit ($ReLU$) as the nonlinear activation function of the network.

The network's training is performed using the backpropagation algorithm (Rumelhart and Hintont 1986) with the Adam optimizer (Kingma and Ba 2015) to optimize the following loss function:

$$argmin_\theta \frac{1}{N} \sum \left(f_{DNN}(X_{highIWV}, \theta) - y_{highIWV}\right)^2 \quad (11)$$

where $y_{highIWV}$ is the ground-truth GNSS IWV retrieved with P_s and T_m from ERA5 in high IWV period. It should be noted that the ERA5 data is only used in the training process but not in its prediction process.

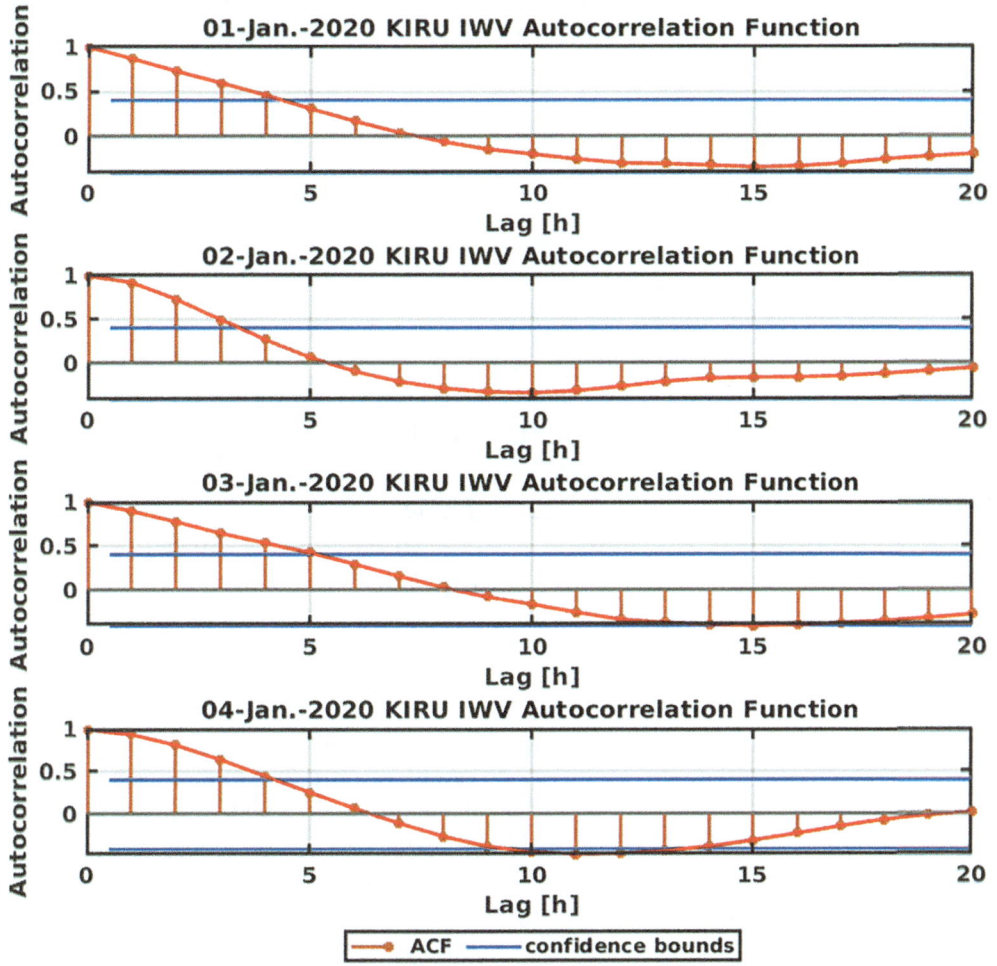

Fig. 2 ACF of diurnal GNSS IWV of KIRU, 01-Jan.-2020 to 04-Jan.-2020

For low IWV data, we have found that there is short-term autocorrelation in water vapor variations, an example is showed in Fig. 2, which allowing the water vapor retrieval errors to be further corrected by time series models.

LSTM is a deep neural network capable of effectively learning time series patterns. Since the output of an LSTM network is not only related to the current input features but also to its historical status, we employ an LSTM network for low water vapor periods IWV prediction. Unlike feedforward DNNs, LSTM learns from time series data. This means that training LSTM requires constructing a training set composed of a series of consecutive acquisitions. The length of the series is recommended to be determined based on the maximum lag in which the autocorrelation function (ACF) (Geurts et al. 1977) is larger than the approximate upper confidence bound of the diurnal historical IWV. For our case study, there is typically a significant lag of 3–5 h in the IWV diurnal sequences during low water vapor periods. Therefore, we divide the acquired data into sequences of 5 consecutive hours to build the training set. Time series data are described using a sequence partial order relationship to represent temporal information. Hence, only GNSS ZTD and the GNSS IWV (denote as $IWV_{GPT3,t}$) retrieved with GPT3 were taken as input features (Eq. 12). Experimental results show that for LSTM, training with only ZTD and IWV achieves better performance than using all features obtained. By giving the time series data from low IWV periods, the prediction IWV series by using LSTM are described as follows:

$$X_{lowIWV,t} = \{ZTD_t, IWV_{GPT3,t}\}, t = 1, 2 \ldots 5 \quad (12)$$

$$f_t = \sigma_{sig}\left(W_f \cdot X_{lowIWV,t} + U_f \cdot \hat{y}_{lowIWV,t-1} + b_f\right) \quad (13)$$

$$i_t = \sigma_{sig}\left(W_i \cdot X_{lowIWV,t} + U_i \cdot \hat{y}_{lowIWV,t-1} + b_i\right) \quad (14)$$

$$o_t = \sigma_{sig}\left(W_o \cdot X_{lowIWV,t} + U_o \cdot \hat{y}_{lowIWV,t-1} + b_o\right) \quad (15)$$

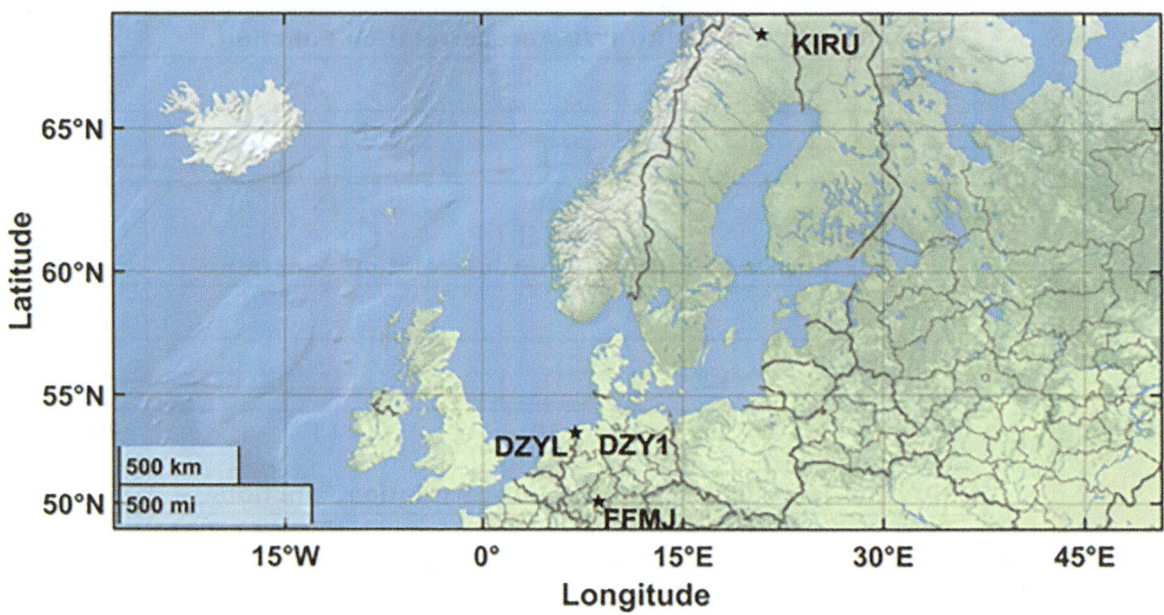

Fig. 3 The locations of GNSS stations KIRU, DZYL, DZY1 and FFMJ. Note that both DZYL and DZY1 are located in Delfzijl, Netherlands, and they are too close to be distinguished from the map

$$\tilde{c}_t = \sigma_{tanh}\left(W_c \cdot X_{lowIWV,t} + U_c \cdot \hat{y}_{lowIWV,t-1} + b_c\right) \quad (16)$$

$$c_t = f_t \odot c_{t-1} + i_t \odot \tilde{c}_t \quad (17)$$

$$\hat{y}_{lowIWV,t} = o_t \odot \sigma_h(c_t) \quad (18)$$

where $f_t, i_t, o_t, \tilde{c}_t, c_t$ are the forget gate's activation vector, input gate's activation vector, output gate's activation vector, cell input activation vector, and cell state vector, respectively. $W, U,$ and b represent the weights and biases of the neurons which need to be learned associated with the respective gates (Hochreiter and Schmidhuber 1997). In LSTM, there are two nonlinear activation functions σ_{sig} and σ_{tanh}, which are Sigmoid function and hyperbolic tangent function as described below:

$$\sigma_{sig}(x) = \frac{1}{1+e^{-x}} \quad (19)$$

$$\sigma_{tanh}(x) = \frac{e^{2x}-1}{e^{2x}+1}. \quad (20)$$

The $\hat{y}_{lowIWV,0}$ and c_0 are initialized as 0. $\odot$ denotes the Hadamard product.

3 Results and Discussion

For our case study analysis to validate the effectiveness of the proposed method, we selected the following GNSS stations (see Fig. 3 and Table 1): KIRU in Kiruna, Sweden, DZYL and DZY1 in Delfzijl, Netherlands, and FFMJ in Frankfurt, Germany, to represent different climate types in the mid-to-high latitude regions of Europe. Data from 2016 to 2019 were used for training, whereas the data in 2020 were used as independent test set. Various metrics are used in the evaluation, such as Root Mean Square Error (RMSE), Mean Bias (MB), Mean Absolute Percentage Error (MAPE) and coefficient of determination (R^2):

$$\text{RMSE} = \sqrt{\tfrac{1}{N}\sum\left(IWV_{pre} - IWV_{ERA5}\right)^2} \quad (21)$$

$$MB = \tfrac{1}{N}\sum\left(IWV_{pre} - IWV_{ERA5}\right) \quad (22)$$

$$MAPE = \tfrac{1}{N}\sum\left|\tfrac{IWV_{pre}-IWV_{ERA5}}{IWV_{ERA5}}\right| * 100\% \quad (23)$$

$$R^2 = \frac{\left[\sum(IWV_{pre}-\overline{IWV_{pre}})*(IWV_{ERA5}-\overline{IWV_{ERA5}})\right]^2}{\sum\left(IWV_{pre}-\overline{IWV_{pre}}\right)^2 * \left(IWV_{ERA5}-\overline{IWV_{ERA5}}\right)^2} \quad (24)$$

where IWV_{pre} is the predicted IWV, and IWV_{ERA5} is the ground-truth GNSS IWV retrieved with ERA5 as reference.

The proposed model is trained station by station. For each station, we randomly select 90% of its data from 2016 to 2019 for training, and the remaining part from 2016 to 2019 is used as a validation set for parameter adjustment to select the model with the best generalization ability. To evaluate the performance of different models in different water vapor period, we compared GPT3, the DNN only, and the combined model with DNN and LSTM. In the combined model, a DNN model was developed for high IWV values

Table 2 RMSE, MB, MAPE and R^2 at different GNSS stations. Note that CMB denotes the combined model proposed in this paper

Metrics	KIRU	DZYL	DZY1	FFMJ
RMSE (GPT3) (kg/m^2)	4.93	3.84	3.86	3.25
RMSE (CMB) (kg/m^2)	2.24	3.20	3.15	2.78
MB (GPT3) (kg/m^2)	−2.00	−0.21	−0.20	−0.05
MB (CMB) (kg/m^2)	0.27	−0.07	−0.11	−0.16
MAPE (GPT3)	62.29%	25.20%	25.63%	21.06%
MAPE (CMB)	32.05%	20.57%	20.54%	17.93%
R^2 (GPT3)	0.69	0.76	0.78	0.84
R^2 (CMB)	0.85	0.84	0.84	0.88

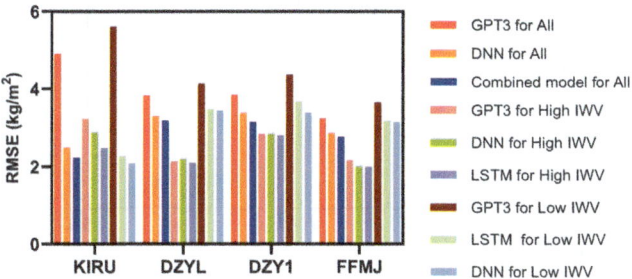

Fig. 4 RMSE comparison at different GNSS stations

whereas a LSTM model was used for low IWV values. In DNN only model and the one in the combined model, their network structures are the same, with 4 layers with 200 neurons in each layer. Training started with an initial learning rate of 0.01 and decayed by a factor of 0.7 every 500 epochs until it reached the max of 3,000 epochs. For the LSTM part, it had been constructed with 256 neurons for each gate, and the learning strategy was: Initial learning rate 0.001, decay by a factor 0.8 every 500 epochs until it reached the max epoch 3,000. To evaluate our approach, we compared the matrices RMSE, MB, MAPE and R^2 between our approach and GPT3 showed in Table 2. Besides, we also evaluated the RMSE for the entire year, high IWV period and low IWV period as shown in Fig. 4.

From the evaluation results, it can be observed that the combined model proposed in this paper consistently achieves smaller RMSE compared to GPT3 and the DNN at all times. For the four GNSS stations KIRU, DZYL, DZY1, and FFMJ, the RMSE of the combined model is 2.24, 3.20, 3.15, and 2.78 kg/m^2, respectively, representing improvements of 55%, 17%, 18%, and 15% compared to the GNSS IWV retrieved from GPT3. Since the proposed model is trained station by station, similar results were obtained at the DZYL and DZY1 stations located on both sides of the EMS canal to prove the stability of the proposed model. The improvement in retrieval performance of those four stations can mainly be attributed to the use of LSTM for IWV sequence correction during periods of low IWV. Compared to the GNSS IWV retrieved from GPT3, the combined model proposed in this paper achieves retrieval improvements of 63%, 17%, 22%, and 14% during low IWV periods, significantly enhancing IWV retrieval in the Northern European region. In fact, Ding and Chen (2020) has indicated that the stability of GPT3 model predictions for temperature and pressure decreases with increasing latitude. During periods with low IWV in high-latitude regions, where water vapor variations are not drastic, the substantial distortions in GPT3 temperature and pressure predictions can result in unstable IWV retrievals based on the physical model described in Sect. 2.1, leading to significant errors. By using the LSTM time series model, we can effectively correct the stability of the sequence to address this issue, as shown in Figs. 5, 6, 7, and 8.

From Fig. 5 to Fig. 8, it can be observed that in Low IWV periods, the GPT3 model exhibits significant random errors during this period (red dots), while the model proposed in this paper significantly corrects this issue (blue dots). When the IWV values are higher, all three models make similar predictions. It is noteworthy that the difference of DNN and LSTM between classification boundaries for high and low IWV is not overly significant, as illustrated in Figs. 5, 6, 7, and 8, where the blue and red points near the black line are close. While discontinuities at the edges of high/low IWV cases exist, their impact on inversion results remains within acceptable bounds. This can be attributed to the first step of LSTM (right edge) the $\hat{y}_{lowIWV,0}$ and c_0 are initialized as 0. Following the completion of the first iteration, the model proceeds normally. Therefore, this discontinuity will only occur 5 h after the first bound, and will not exert a significant influence.

4 Conclusions and Outlook

In this study, we proposed a deep learning method jointly using LSTM and feedforward DNN to perform real-time ZTD-IWV conversion without the need of actual meteorological observation but of values only GPT3. We evaluated the proposed method at four different GNSS stations in the mid-to-high-latitude regions of Europe, obtaining RMSE values of 2.24, 3.20, 3.15, and 2.78 kg/m^2, respectively. Compared to the GNSS IWV retrieved from GPT3 with RMSE values of nearly 4 kg/m^2, our method improves the real-time retrieval results at these stations.

Since climate patterns change slightly every year, we recommend using at least 3 years of data from the year to be investigated for training the model to maintain stability. Increasing the training data period may be beneficial, depending on the similarity of the year to be investigated to its history. Considering that the stratospheric wind changes approximately every 2.5 years (Quasi-biennial oscillation) and the El Niño phenomenon occurs approximately every 3–

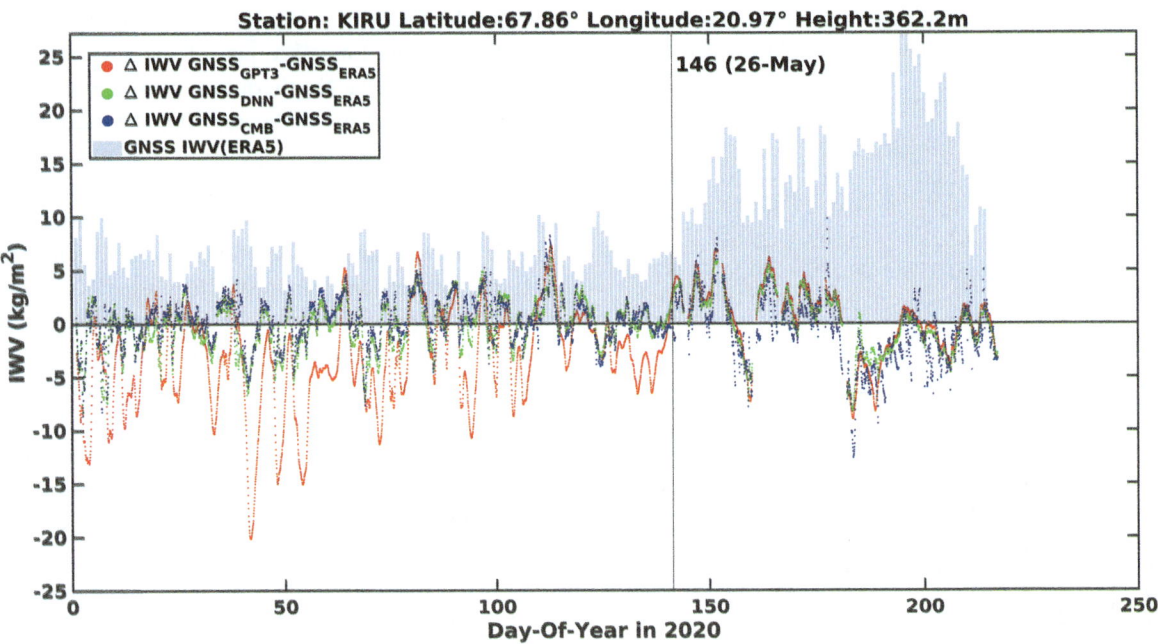

Fig. 5 IWV prediction difference comparison of KIRU, 2020. Before 26-May is the Low IWV period and after is the High IWV period. Note that CMB denotes the combined model proposed in this paper

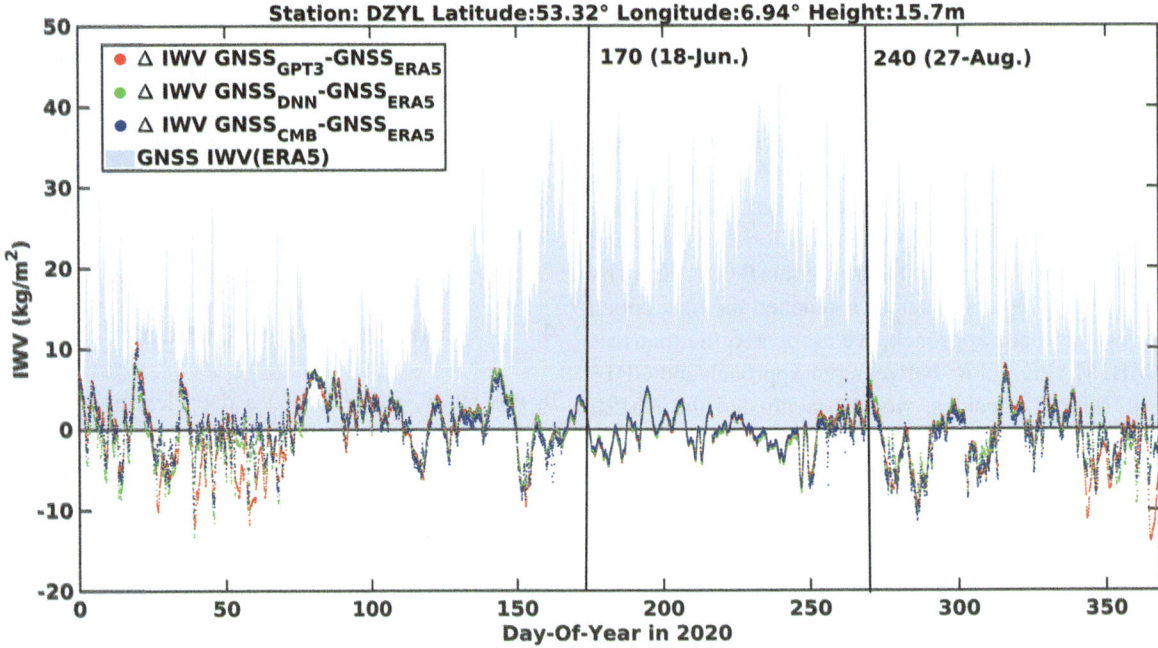

Fig. 6 IWV prediction difference comparison of DZYL, 2020. From 18-Jun. to 27-Aug. is the High IWV period and other days belong to the Low IWV period. Note that CMB denotes the combined model proposed in this paper

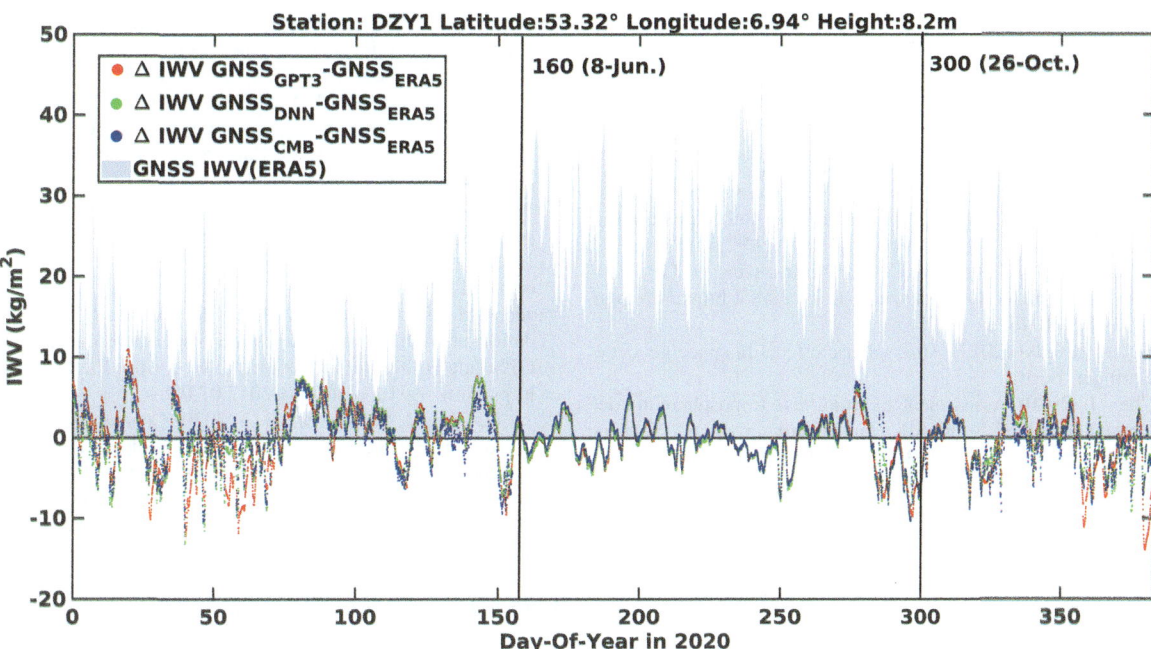

Fig. 7 IWV prediction difference comparison of DZY1, 2020. From 8-Jun. to 26-Oct. is the High IWV period and other days belong to the Low IWV period. Note that CMB denotes the combined model proposed in this paper

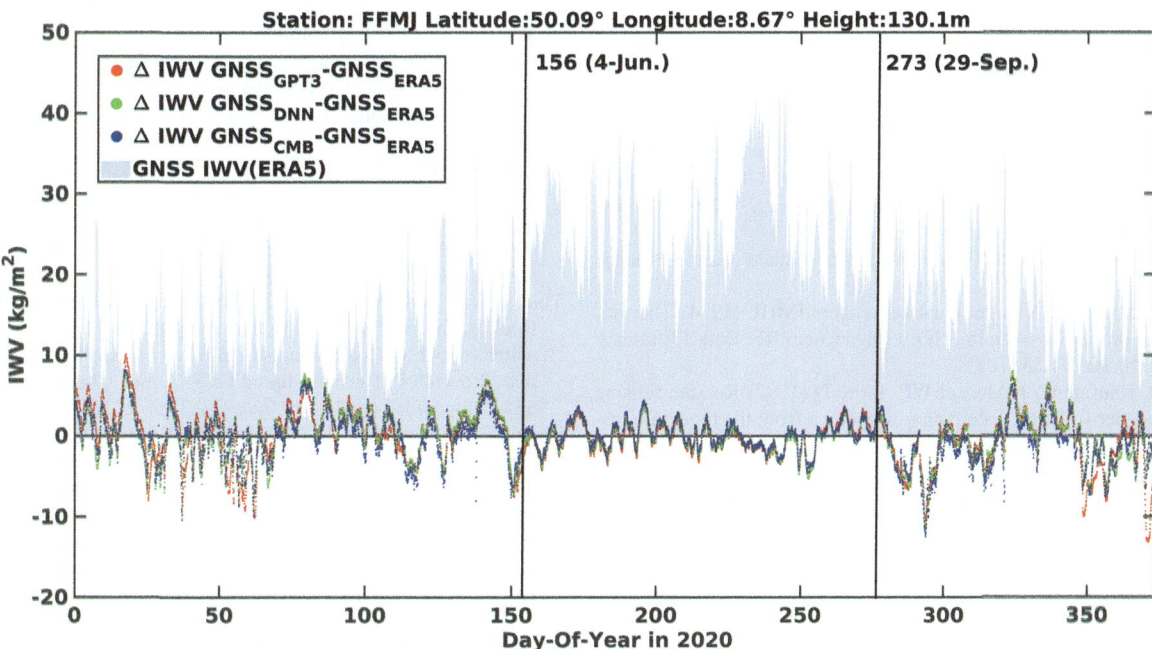

Fig. 8 IWV prediction difference comparison of FFMJ, 2020. From 4-Jun. to 9-Sep. is the High IWV period and other days belong to the Low IWV period. Note that CMB denotes the combined model proposed in this paper

5 years, we do not recommend using data that is too long or too old for training.

Currently, this method relies on the water vapor temporal patterns specific to the stations being retrieved. To broaden its generalizability, we aim to integrate regional or global water vapor cycle information in the future. In addition, ERA5 forecasts data could also be an alternative for real-time GNSS IWV monitoring, but the evaluation on its quality is beyond the scope of this work.

Acknowledgements We thank NGL and ECMWF for providing the GNSS ZTD and ERA5 data. D.W. is funded by the China Scholarship Council (CSC). P.Y. is funded by the German Federal Ministry for the Environment, Nature Conservation, Nuclear Safety and Consumer Protection (BMUV) project (EKAPEx, project number: 67KI32002C).

References

Bevis M (1994) GPS meteorology: mapping zenith wet delays onto precipitable water. J Appl Meteorol 33. https://doi.org/10.1175/1520-0450(1994)033{\mathsurround=\opskip$<$}0379:GMMZWD{\mathsurround=\opskip$>$}2.0.CO;2

Bevis M, Businger S, Herring TA et al (1992) GPS meteorology: remote sensing of atmospheric water vapor using the global positioning system. J Geophys Res 97. https://doi.org/10.1029/92jd01517

Blewitt G, Hammond WC, Kreemer C (2018) Harnessing the GPS data explosion for interdisciplinary science. Eos 99:485. https://doi.org/10.1029/2018EO104623

Böhm J, Schuh H (eds) (2013) Atmospheric effects in space geodesy, vol 5. Springer, Berlin

Ding J, Chen J (2020) Assessment of empirical troposphere model GPT3 based on NGL's global troposphere products. Sensors (Switzerland) 20. https://doi.org/10.3390/s20133631

Durre I, Yin X (2008) Enhanced radiosonde data for studies of vertical structure. Bull Am Meteorol Soc 89(9):1257–1262

Geurts M, Box GEP, Jenkins GM (1977) Time series analysis: forecasting and control. J Mark Res 14. https://doi.org/10.2307/3150485

Goodfellow I, Bengio Y, Courville A (2016) Deep learning an MIT press book

Hernández-Pajares M, Juan JM, Sanz J, Orús R (2007) Second-order ionospheric term in GPS: implementation and impact on geodetic estimates. J Geophys Res Solid Earth 112(8). https://doi.org/10.1029/2006JB004707

Hersbach H, Bell B, Berrisford P et al (2023) ERA5 hourly data on single levels from 1940 to present. Copernicus climate change service (C3S) climate data store (CDS). https://doi.org/10.24381/cds.adbb2d47

Hochreiter S, Schmidhuber J (1997) Long short-term memory. Neural Comput 9. https://doi.org/10.1162/neco.1997.9.8.1735

Jiang W, Yuan P, Chen H, Cai J, Li Z, Chao N, Sneeuw N (2017) Annual variations of monsoon and drought detected by GPS: a case study in Yunnan, China. Sci Rep 7:5874. https://doi.org/10.1038/s41598-017-06095-1

Karl TR, Trenberth KE (2003) Modern global climate change. Science 1979:302

Kestin J, Sengers JV, Kamgar Parsi B, Sengers JMHL (1984) Thermophysical properties of fluid H2O. J Phys Chem Ref Data 13. https://doi.org/10.1063/1.555707

King MD, Kaufman YJ, Menzel WP, Tanré D (1992) Remote sensing of cloud, aerosol, and water vapor properties from the moderate resolution imaging spectrometer (MODIS). IEEE Trans Geosci Remote Sens 30

Kingma DP, Ba JL (2015) Adam: a method for stochastic optimization. 3rd international conference on learning representations, ICLR 2015 - conference track proceedings, pp 1–15

Landskron D, Böhm J (2018) VMF3/GPT3: refined discrete and empirical troposphere mapping functions. J Geod 92. https://doi.org/10.1007/s00190-017-1066-2

Lecun Y, Bengio Y, Hinton G (2015) Deep learning. Nature 521:436–444. https://doi.org/10.1038/nature14539

Leick A, Rapoport L, Tatarnikov D (2015) GPS satellite surveying, 4th edn

Rumelhart DE, Hintont GE (1986) Learning representations by back-propagating errors. Nature 323:533–536. https://doi.org/10.7551/mitpress/1888.003.0013

Saastamoinen J (1972) Atmospheric correction for the troposphere and stratosphere in radio ranging satellites. In: The use of artificial satellites for geodesy, vol 15, pp 247–251

Schröder M, Lockhoff M, Fell F et al (2018) The GEWEX water vapor assessment archive of water vapour products from satellite observations and reanalyses. Earth Syst Sci Data 10. https://doi.org/10.5194/essd-10-1093-2018

Shangguan M, Dang M, Yue Y, Zou R (2023) A combined model to predict GNSS precipitable water vapor based on deep learning. IEEE J Sel Top Appl Earth Obs Remote Sens 16:4713–4723. https://doi.org/10.1109/JSTARS.2023.3278381

Vaquero-Martínez J, Antón M, Ortiz de Galisteo JP et al (2017) Validation of MODIS integrated water vapor product against reference GPS data at the Iberian Peninsula. Int J Appl Earth Obs Geoinf 63. https://doi.org/10.1016/j.jag.2017.07.008

Wang D, Wang J, Scaioni M, Si Q (2020) Coarse-to-fine classification of road infrastructure elements from mobile point clouds using symmetric ensemble point network and Euclidean cluster extraction. Sensors (Switzerland) 20. https://doi.org/10.3390/s20010225

Wang D, Even M, Kutterer H (2022) Deep learning based distributed scatterers acceleration approach: distributed scatterers prediction net. Int J Appl Earth Obs Geoinf 115

Worden J, Noone D, Bowman K et al (2007) Importance of rain evaporation and continental convection in the tropical water cycle. Nature 445. https://doi.org/10.1038/nature05508

Yang F, Guo J, Meng X, Li J, Zou J, Xu Y (2021) Establishment and assessment of a zenith wet delay (ZWD) augmentation model. GPS Solutions 25:1–11

Yuan P, Hunegnaw A, Alshawaf F, Awange J, Klos A, Teferle FN, Kutterer H (2021) Feasibility of ERA5 integrated water vapor trends for climate change analysis in continental Europe: an evaluation with GPS (1994–2019) by considering statistical significance. Remote Sens Environ 260:112416. https://doi.org/10.1016/j.rse.2021.112416

Yuan P, Blewitt G, Kreemer C, Hammond WC, Argus D, Yin X, Van Malderen R, Mayer M, Jiang W, Awange J, Kutterer H (2023a) An enhanced integrated water vapour dataset from more than 10 000 global ground-based GPS stations in 2020. Earth Syst Sci Data 15:723–743. https://doi.org/10.5194/essd-15-723-2023

Yuan P, Van Malderen R, Yin X, Vogelmann H, Jiang W, Awange J, Heck B, Kutterer H (2023b) Characterisations of Europe's integrated water vapour and assessments of atmospheric reanalyses using more than 2 decades of ground-based GPS. Atmos Chem Phys 23:3517–3541. https://doi.org/10.5194/acp-23-3517-2023

Zhao K, Wulder MA, Hu T, Bright R, Wu Q, Qin H et al (2019) Detecting change-point, trend, and seasonality in satellite time series data to track abrupt changes and nonlinear dynamics: a Bayesian ensemble algorithm. Remote Sens Environ 232:111181

Zheng Y, Lu C, Wu Z, Liao J, Zhang Y, Wang Q (2022) Machine learning-based model for real-time GNSS precipitable water vapor sensing. Geophys Res Lett 49(3):e2021GL096408. https://doi.org/10.1029/2021GL096408

Zhu Y, Newell RE (1994) Atmospheric rivers and bombs. Geophys Res Lett 21. https://doi.org/10.1029/94GL01710

Zhu XX, Tuia D, Mou L et al (2017a) Deep learning in remote sensing: a review. https://doi.org/10.1109/MGRS.2017.2762307

Zhu XX, Tuia D, Mou L et al (2017b) Deep learning in remote sensing: a comprehensive review and list of resources. IEEE Geosci Remote Sens Mag 5

Open Access This chapter is licensed under the terms of the Creative Commons Attribution 4.0 International License (http://creativecommons.org/licenses/by/4.0/), which permits use, sharing, adaptation, distribution and reproduction in any medium or format, as long as you give appropriate credit to the original author(s) and the source, provide a link to the Creative Commons license and indicate if changes were made.

The images or other third party material in this chapter are included in the chapter's Creative Commons license, unless indicated otherwise in a credit line to the material. If material is not included in the chapter's Creative Commons license and your intended use is not permitted by statutory regulation or exceeds the permitted use, you will need to obtain permission directly from the copyright holder.

Analyzing the 3D Deformation Induced by Non-tidal Loading in GNSS Time Series in Finland

Yohannes Getachew Ejigu, Jean-Paul Boy, Arttu Raja-Halli, Fatemeh Khorrami, Jyri Naranen, and Maaria Nordman

Abstract

Improving our understanding of non-tidal loading (NTL) in geodetic time series, especially at regional and local scales, holds paramount importance. This deeper comprehension enables accurate modeling and effective removal of NTL effects from the time series, consequently enhancing the overall stability and reliability of geodetic observations. In this study, we compared the performance of different loading products and investigated their impact on the 20-year time series of four permanent GNSS stations within the Finnish permanent GNSS network (FinnRef). We employed original GNSS time series data products generated by four different analysing centers. We qualitatively compared NTL corrections involving ten different combinations of different hydrological, non-tidal atmospheric, and non-tidal oceanic loading models to see how various loading configurations operate and how they affect the noise characteristics of GNSS 3D time series, and ultimately to figure out which models are the most realistic in Finland. We observed weighted RMS reduction rates of up to 20% for the vertical coordinate and up to 10% for the horizontal coordinate. Additionally, we identified a maximum annual amplitude reduction rate of 87.2%. The results demonstrate a substantial improvement through the integration of hydrological loading products derived from GRACE satellites in our study conducted over Finland.

Keywords

Annual amplitude and phase · GNSS time series · Non-tidal loading · RMS reduction rate · Trend and trend uncertainty

1 Introduction

At different scales in space and time, the Earth's surface is subject to various perturbations, including tectonic movements, volcanic eruptions, and landslides. Mass changes in oceans, atmosphere, groundwater, and other natural systems induce detectable changes in the Earth's shape and gravity field (Blewitt et al. 2001). Responding to mass redistribution occurring on sub-daily to inter-annual timescales, which are not due to gravitational forces from Sun and Moon, is referred to as non-tidal loading (NTL). In theory, over seasonal timescales, non-tidal atmospheric, and hydrological loading have the potential to induce deformations of up to

30 mm in the vertical coordinate (van Dam et al. 2001; Schuh et al. 2003).

Global Navigation Satellite Systems (GNSS) have been used to explore geophysical phenomena, including the NTL surface deformation (e.g., Nordman et al. 2009, 2015, van Dam et al. 2012, Mémin et al. 2020. GNSS position time series often exhibit unforeseen broadband fluctuations (such as a nonlinear motion) depending on unknown deterministic parameters (Bevis and Brown 2014), and unknown stochastic parameters (Bos et al. 2020), characterized by power-law and white noise models (Mao et al. 1999; Williams et al. 2004; Santamaria-Gomez et al. 2011). Interpreting interannual variations in GNSS position time series and their common modes across a region remains a challenging task due to the correlation between NTL and noise variations (Williams et al. 2004). Moreover, GNSS serves as a pivotal geodetic technique, providing essential input to the International Terrestrial Reference Frame (ITRF) for the precise computation of coordinates (Altamimi et al. 2016). Correcting the NTL could contribute to the stability of the ITRF by improving the GNSS positioning quality, and statistical modelling of geodetic time series.

The NTL typically encompasses non-tidal atmospheric loading (NTAL), non-tidal ocean loading (NTOL), and hydrological loading (HYDL). In their recent study, Nicolas et al. (2021) showed the impact of river storage-induced HYDL on the substantial enhancement of both vertical and horizontal GNSS displacements in regions across South America. Klos et al. (2021) found that HYDL contributes to GNSS displacements within the seasonal band, while NTAL exhibits a positive correlation with GPS displacements across a range of temporal resolutions. Li et al. (2020) conducted a comparative analysis of surface mass loading corrections in GNSS time series, with a reduction in the RMS value reaching up to 20% in the vertical component. Nordman et al. (2009) employed tide gauge data to assess the impact of NTOL on the vertical positioning of GPS stations in the Baltic Sea. Overall, numerous researchers have delved into the deformation resulting from NTL. However, the majority of these studies focus on the separation of NTL into distinct components such as NTAL (e.g., Tregoning and van Dam 2005; Tregoning and Watson 2011), NTOL (e.g, Nordman et al. 2009; van Dam et al. 2012; Geng et al. 2012; Williams and Penna 2011, Gobron et al. 2021), and HYDL (van Dam et al. 2001, 2007; Davis et al. 2004; Tregoning et al. 2009). This study, however, is focused on examining the combined contributions of NTAL, NTOL, and HYDL in long-period GNSS time series data.

The NTL displacement products are available from different Earth System Modeling groups, including the School and Observatory of Earth Sciences (EOST)[1] loading service from the University of Strasbourg, and the German Research Center for Geosciences in Potsdam (ESMGFZ).[2] We conducted a comprehensive performance comparison of the EOST and ESMGFZ loading model products while also examining their influence on the scatter of GNSS time series data across four permanent GNSS stations of the FinnRef network. Furthermore, we delved into different loading configurations and their respective impacts on the noise characteristics of GNSS vertical displacement trend time series. Ultimately, our objective is to identify the most optimal loading models within the context of the Finnish region.

In this chapter, we first present a concise overview of the GNSS processing strategy and introduce the NTL computation, including the use of different models (Sect. 2). Section 3 presents results obtained before and after implementing NTL corrections. The chapter concludes in Sect. 4 with a summary of findings and an outlook for future research.

2 Data and Methods

2.1 GNSS Data and Processing

In this study, we focused on four GNSS stations within the FinnRef network in Northern Europe: JOEN in East Finland, METS in South Finland, SODA in North Finland, and VAAS in West Finland (for locations, see Fig. 3). Our analysis encompassed a 20-year (2002–2022) dataset obtained from three distinct GNSS analysis centers including the Nevada Geodetic Laboratory (NGL, Blewitt et al. 2018), the Jet Propulsion Laboratory (JPL, specifically 'JPL-2018a', Bertiger et al. 2020), and the IGS CNES-CLS analysis center (hereafter referred to as CNE, Michel et al. 2021). Furthermore, we created our own solution, processing GNSS data using PRIDE-AR ver 2.2 GNSS software (Geng et al. 2019a). Both NGL and JPL position timeseries products were estimated using the JPL GipsyX software, while CNES utilized the GINS software.

In both NGL and PRIDE (PRI), the apriori tropospheric delays are sourced from the Vienna Mapping Function (VMF1) grids (Böhm et al. 2006b). Also, the hydrostatic and wet zenith delays are mapped to observation elevations using VMF1. The JPL and CNE solutions employ the global mapping function (GMF) (Böhm et al. 2006a) for tropospheric modeling, along with the global pressure and temperature empirical function GPT2 (Lagler et al. 2013) to estimate tropospheric delays.

All four GNSS (actually GPS-only) datasets share a common processing strategy based on precise point positioning (PPP) (Zumberge et al. 1997) with carrier phase ambiguity resolution (JPL and NGL (Bertige et al. 2010), CNE (Loyer et al. 2012) and PRI (Geng et al. 2019b)). The final

---

[1] http://loading.u-strasbg.fr/.

[2] http://rz-vm115.gfz-potsdam.de:8080/repository.

daily coordinates position time series are expressed in the IGS14 reference frame (Altamimi et al. 2016). The impact of the first-order ionospheric effect was removed using the ionospheric-free linear combination. A second-order calibration of the remaining ionosphere effects was implemented using IGS's global ionospheric maps in conjunction with the International Geomagnetic Reference Field (IGRF-12)'s magnetic field model; the solid Earth tide and pole tide were corrected according to the IERS 2010 Conventions, and the ocean tide loading effect were corrected using the FES2004 model (Lyard et al. 2006). The NTL corrections were not applied to these timeseries products. The resulting time series are in CF frame at non-secular time scale (Dong et al. 2003).

2.2 Non-tidal Loading Time Series Data

We used NTL-induced surface deformation timeseries from two Earth System Modeling groups: EOST and ESMGFZ Loading Service. Both modeling groups employ Green's functions to calculate NTL displacements; however, they employ slightly different procedures for computing the integrals. The EOST group utilizes the rheological parameters from the PREM (Preliminary Reference Earth Model, Dziewonski and Anderson 1981) Earth model to compute displacement, whereas the GFZ employs the Elastic Earth model "ak135" (Kennett et al. 1995) for their calculations. In both EOST and GFZ products, we averaged time series into daily time series to harmonize their temporal resolutions. In this study, the NTL products utilized are in CF, and the summary including their providers is shown in Table 1.

Table 1 Loading products provided by EOST and GFZ. The different abbreviations in the loading source and models columns refer to different products and data centers providing the data

Dataset/Load	Models	Spatial/Temporal Res.
EOST/NTAL	ERA5/IB	0.25° × 0.25°/1 h
	ERA5/TUGO-m[a]	0.25° × 0.25°/1 h
	ECMWF OP/IB	0.5° × 0.5°/3 h
	MERRA2/IB	0.5° × 0.5°/1 h
GFZ/NTAL	ECMWF OP	0.5° × 0.5°/3 h
EOST/NTOL	ECCO2	0.25° × 0.25°/24 h
GFZ/NTOL	MPIOM	1° × 1°/3 h
EOST/HYDL	ERA5	0.25° × 0.25°/1 h
	GLDAS2/Noah	0.5° × 0.5°/3 h
	MERRA2	0.50 × 0.625°/1 h
	GRACE[a, b]	1° × 1°/monthly
GFZ/HYDL	LSDM	0.5° × 0.5°/24 h
NASA/HYDL[c]	GRACE/GSFC/Mascons[b]	1° × 1°/monthly

For more detailed explanations see websites of EOST[1] and ESMGFZ[2]
[a]TUGO-m model is the sum of NTAL and NTOL circulation model (such as ECCO2) [1]. EOST GRACE is the sum of HYDL and NTOL[1]
[b]We used GRACE-derived data provided by EOST and NASA (Argus et al. 2022). For more details, see websites[1,3]
[c]https://grace.jpl.nasa.gov/data/get-data/jpl_global_mascons/

3 Results and Discussion

3.1 Optimal Fusion of NTL

We undertake a comprehensive quantitative evaluation to assess the cumulative impact resulting from three distinct categories of NTL-induced displacements, namely NTAL, NTOL, and HYDL. Consequently, we establish a total of eight combinations within the EOST category and the two combinations within GFZ. The summary of these combinations can be found in Table 2.

Moreover, we applied these NTL correction combinations to our four original (reference) time series for the four stations under study without interpolating any gaps.

In general, we observed a reduction in the scattering of the original displacement time series (not shown here) after applying the correction. The RMS scatter exhibits improvement, with RMS difference between the original and corrected RMS scatter ranging from −0.93 mm to 4.5 mm, positive numbers meaning reduction in the scatter and negative numbers meaning increase in the scatter. The extent of this improvement varies based on the specific station and the type of NTL correction combination implemented. For the up component, the RMS differences between the original and corrected RMS scatter range from −0.93 mm to 3.51 mm for JPL time series, −0.71 mm to 2.73 mm for CNE time series, −0.73 mm to 2.85 mm for NGL time series, and −0.72 mm to 4.52 mm for PRI time series, as shown in Fig. 1 (third column). The RMS difference in the horizontal components is minimal, with a positive deviation of up to +0.5 mm in the east and a negative deviation of up to −0.5 mm in the north (Fig. 1, first and second columns).

A comprehensive overview and assessment of the effectiveness of various environmental loading products were obtained through a computation of the weighted RMS (WRMS) percentage reduction rates between the original GNSS position time series and the combined NTL corrected time series. The WRMS was computed as,

Table 2 Combinations of NTL correction

	Combination
EOST1	ERA5IB + ECCO2 + ERA5hyd
EOST2	ERA5TUGO + ERA5hyd
EOST3	ECMWF-OP/IB + ECCO2 + ERA5hyd
EOST4	MERRA2 + ECCO2 + GLDAS
EOST5	MERRA2 + ECCO2 + MERRA2h
EOST6	ERA5IB + GRACE
EOST7	MERRA2IB + GRACE
EOST8	MERRA2IB + ECCO2 + GRACE NASA mascon
GFZ1	GFZ NTAL + GFZ NTOL + GFZ HYDL
GFZ2	GFZ NTAL + GFZ NTOL + GRACE NASA mascon

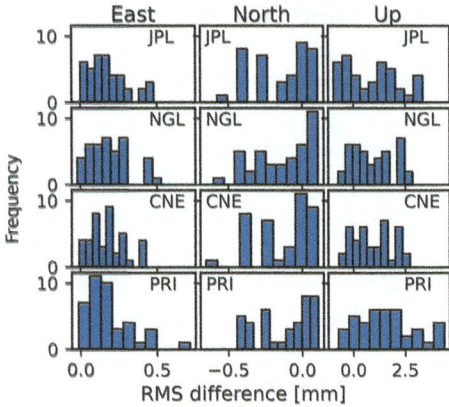

Fig. 1 RMS difference between the original and ten NTL corrected time series for four stations. The y-axis frequency refers to the data from these four stations multiplied by the ten different NTL corrections

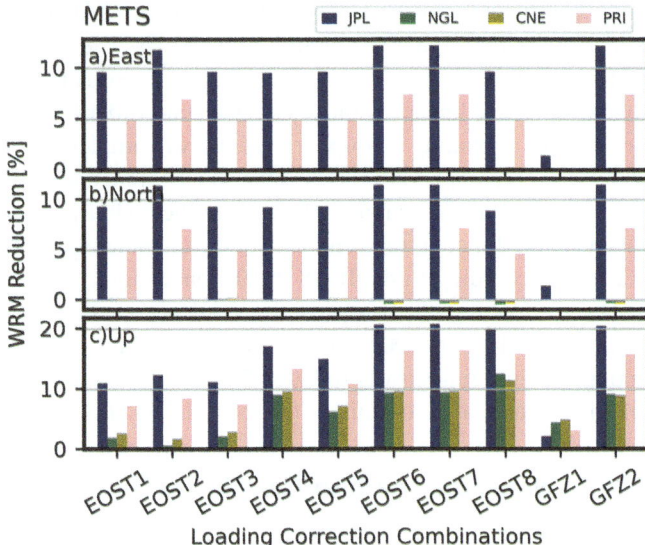

Fig. 2 Percentage reduction of WRMS for 3D deformation (East-North-Up from top to bottom) with NTL corrections applied at station METS, for the period from 2002 to 2022

$WRMS = \sqrt{\frac{1}{N-1} \sum_{i=1}^{N} \left(\frac{(Ts_i - \overline{Ts_i})^2}{\sigma_i^2} \right) / \sum_{i=1}^{N} \sigma_i^2}$, where Ts_i is the daily coordinate displacement solution, N is the number of data points of Ts_i, and σ_i is the standard deviation (formal error) of Ts_i. However, we merely considered the formal error information from the GNSS time series solution and did not include the NTL formal error information. The weighted average value, $\overline{Ts_i}$, was calculated as, $\overline{Ts_i} = \sum_{i=1}^{N} \left(\frac{Ts_i}{\sigma_i^2} \right) / \sum_{i}^{N} \frac{1}{\sigma_i^2}$. The WRMS percentage reduction rate is given by, $Red_{WRMS} = (1 - WRMS_c/WRMS_o) \times 100\%$, where $WRMS_o$ and $WRMS_c$ refer to the original and NTL corrected WRMS, respectively.

As an example, Fig. 2 shows WRMS percentage reduction rates for both the horizontal (East and North) and vertical (Up) coordinate time series at station METS. The RMS reduction rates in all three coordinate time series exhibited positive values for all solutions, with considerable variation depending on the GNSS solution products and stations. Maximum reductions reached approximately 10% in the horizontal direction and about 20% in the vertical direction. In all NTL corrections, both the JPL solution and our PRI solution consistently demonstrate substantial reductions. In respect to combinations of NTL correction, we observed high reductions between the original data and the corrections applied in EOST7 and EOST8 across all solution products and stations compared to other corrections. It's important to note that our EOST7 and EOST8 corrections incorporate the HYDL obtained from GRACE. This aligns with the findings of Springer et al. (2019), who showed that HYDL-induced deformation obtained from GRACE can explain up to 25% of the deformation in daily GNSS height time series in Southeastern Europe. On the other hand, the corrections from EOST showed consistently higher RMS reduction rates in all solution products and stations when compared to those obtained from GFZ. Specifically, the correction derived from the internal combination at GFZ, referred to as GFZ1, exhibited the lowest reduction in RMS error. This again demonstrates that the GRACE hydrological loading significantly contributes to reducing the RMS error in NTL corrections. In prior research, Nordman et al. (2015) noted a maximum 16% reduction in variance in the vertical coordinate for inter-station vectors among GNSS stations in the Baltic Sea, specifically for NTOL corrections during the period from 2008 to 2012.

3.2 Annual Signal and Frequency Content

To understand the time series behaviour better, we analyzed the spectral characteristics, including annual amplitude, phase, and trends, of GNSS coordinate time series both in their original form and after applying corrections, using Hector software version 1.9 (Bos et al. 2021). During the estimation of these parameters in Hector, the noise combination of power-law and white noise (PL + WN) model was utilized in the analysis of stochastics characteristic of GNSS coordinate time series. This noise model was applied to the residual time series and allowed for the simultaneous estimation of the non-integer spectral index of the residual series.

Figure 3 illustrates a comparison between the annual amplitude and phase changes for the vertical component, highlighting notable discrepancies in the annual signal amplitudes across various stations. We found that the peak annual amplitude and phase of the considered stations generally occur between July and October, except for the West Finland station (VAAS), where the maximum annual amplitude and phase extend from January to April. Upon applying corrections, the annual amplitude decreased in

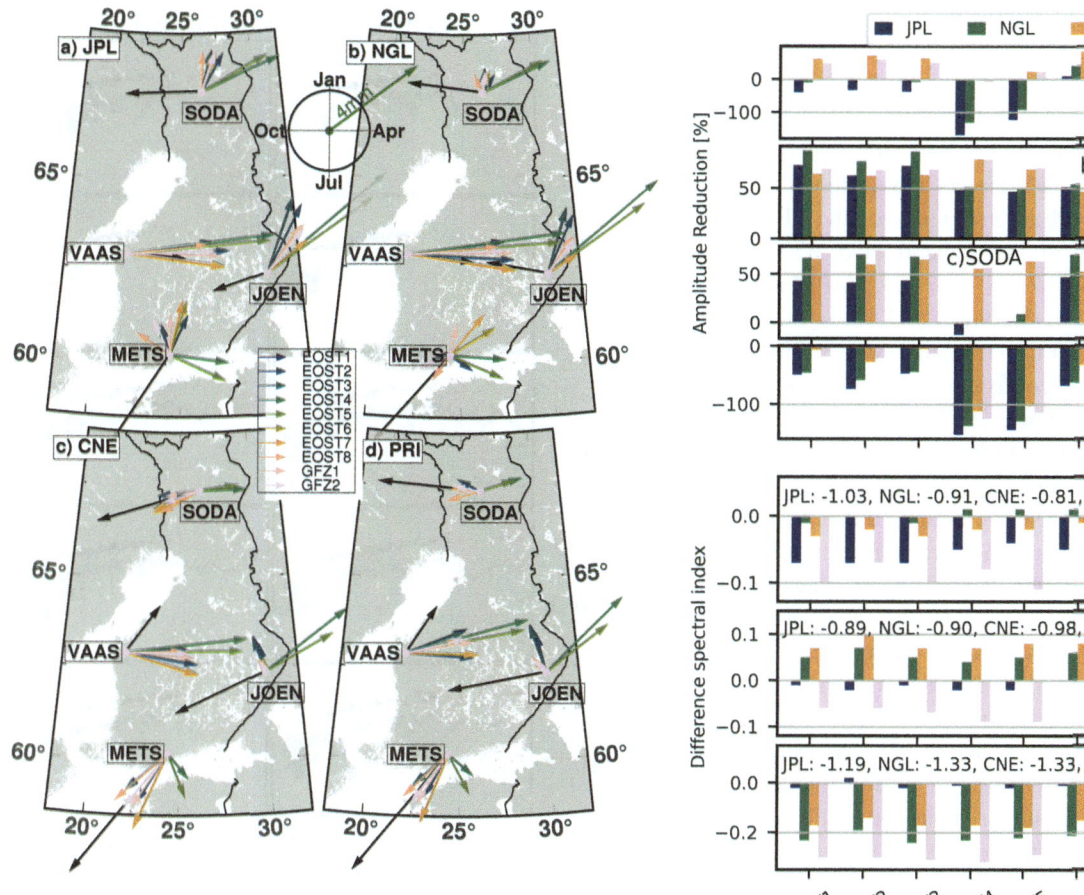

Fig. 3 Comparison of amplitude and phase maps for the vertical component annual signals between the Original GNSS time series (black) and results with NTL corrections (other colours) at four selected stations. The annual amplitude is represented by the length of the arrows. The direction of each arrow indicates the peak-value month, measured in a clockwise direction from the North, starting with January

Fig. 4 (**a–d**) Percentage reduction in spectral annual amplitude for Up component deformation with NTL corrections applied at stations JOEN, METS, SODA and VAAS, and (**e–g**) the spectral index difference (original minus corrected) for 3D deformation (East-North-Up from top to bottom). The number in (**e–f**) panel shows the spectral index estimated from the original time-series

all combinations for stations METS and SODA, and in combinations EOST 4-8, as well as GFZ1-2 for station JOEN. Both the JPL and NGL series display a out-of-phase variation in the annual phases for three of the considered stations (JOEN, METS, and SODA). In contrast, the CNE and PRI series consistently maintain the annual phase pattern across the four GNSS series. Notably, at station METS, the annual signal phase of the JPL and NGL series diverges by 90 to 180° (typically equivalent to 091 to 182 days of year) from the original signal's annual phase, pointing in different directions.

To statistically evaluate the impact of NTL corrections on the annual signal, we also employ the amplitude reduction rate, $Red_{Amp} = (1 - Amp_c/Amp_o) \times 100\%$, where Amp_o and Amp_c refers to the orginal and NTL corrected annual amplitudes, respectively. As depicted in Fig. 4 (a–d), the amplitude generally decreases after applying the correction for most NTL corrections. However, there are exceptions, such as in the case of station JOEN, where some corrections,

and station VAAS where almost all corrections result in a negative reduction (meaning the amplitude is increased). The maximum reduction rate observed is 87.2% for the PRI solution in relation to the EOST6 and EOST7 corrections. For METS, the minimum annual amplitude reduction rate is 46.6% observed for the EOST5 correction in the JPL solution, while the maximum reduction rate reaches 86.7% for the EOST1 correction in the NGL solution. Regarding SODA, overall observe positive reduction rates were observed, except for the EOST4 correction in the JPL solution. The highest reduction rate, reaching 74.4%, is seen in the PRI solution. However, for VAAS, we found negative reduction rates, with the maximum reduction rate being −1.4% for the EOST7 and EOST8 corrections in the CNE and PRI solutions.

The presence of noise in the GNSS coordinate time series can have a substantial impact on the accuracy and induced error in the geodetic time series of the annual signal (Blewitt et al. 2001; Collilieux et al. 2007). To examine the effect of NTL deformation in the GNSS time series on the annual signal, we computed the average Lomb-Scargle periodogram of the coordinate residuals before and after applying NTL corrections. Figure 4 (e–g) depicts the difference in the stacked median estimated spectral index of the PL + WN noise model between its original and corrected. We observed a significant reduction in the spectral index for PRI solutions in the East (−1.05 to −0.93), North (−1.06 to −0.97) and Up (decreasing from −1.4 to −1.08) after correction. Particularly noteworthy was the consistent reduction in the spectral index for the North component exclusively in PRI solutions across all EOST corrections. Additionally, notable reductions were observed for NGL and CNE solutions in the East and Up components. In contrast, corrections did not affect the spectral index in the JPL solution, resulting in contrasting outcomes in WRMS comparisons. This indicates that the WRMS reduction is due to the high portion of white noise in the time series. The exemplary stacked median spectral density (PSD) estimated for vertical coordinate residuals time series for the CNE solution is shown in Fig. 5. In the PSD, it can be observed that the annual peak of the time series is weakened to some extent after the correction for NTL. However, it is important to note that the NTL corrections do not completely eliminate the annual peaks. Nevertheless, there is a large reduction in frequency bands in the annual and between roughly 3–6 months, particularly noticeable in EOST7, EOST8, and in both GFZ1 and GFZ2 corrections. This would agree with Klos et al. (2021), who showed that the frequency bands of annual and between 4–9 months exhibit notable significant RMS reductions (>40%) after correcting for HYDL. This suggests a robust sensitivity to hydrological variations, elucidating a substantial portion of the seasonal signals in GNSS displacements.

To analyze the effect of NTL correction on the trend (velocity) and trend uncertainty in our regional stations, we conducted further assessments. The up-coordinate trends observed for the stations under consideration, JOEN, METS, SODA, and VAAS, fall within the ranges of 3.44–3.93 mm/year, 4.06–4.63 mm/year, 7.13–7.69 mm/year, and 9.18–9.89 mm/year, respectively. To illustrate this, we present the statistical results after correction for the up-coordinate trend and trend uncertainty for station JOEN in Fig. 6. It is evident that after applying the loading correction, the overall trend shows a decrease, with the exception of the EOST1, EOST2, EOST3 and GFZ1 corrections. For the EOST6, EOST7, EOST8, and GFZ2 corrections, the effect on trend values can decrease by up to 0.6 mm/year for the NGL solution and 0.4 mm/year for the other three solutions compared to the original solution trend (see Fig. 6). Similar results were observed for other stations, although they are not

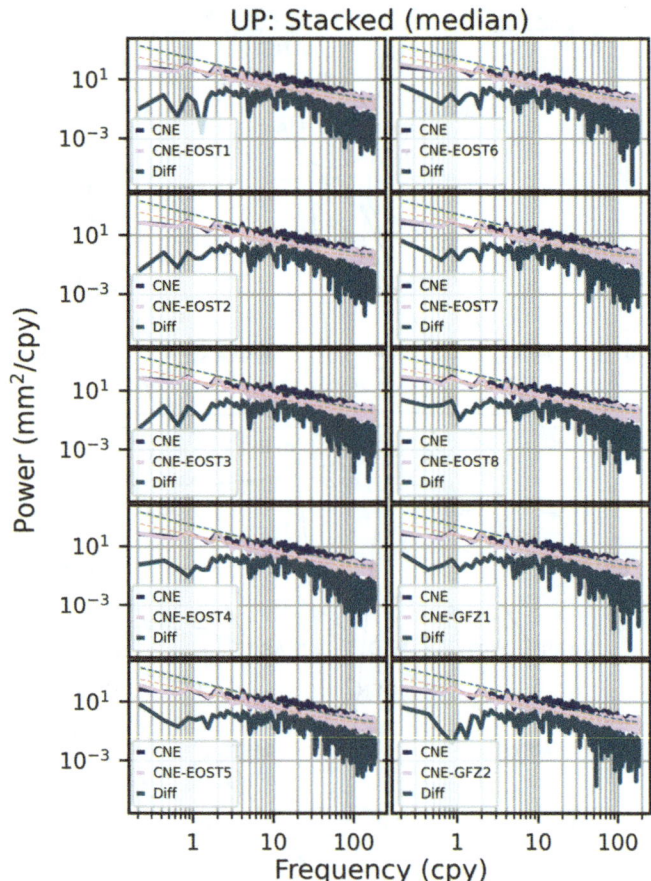

Fig. 5 Stacked power spectral densities (PSDs) plotted for vertical component of CNE solution. The original GNSS displacements are represented in blue, the NTL-corrected displacements are illustrated in pink, and the differences are highlighted in green. The dashed green and red lines indicate the estimated PL + WN noise model for the original and corrected displacements, respectively

shown here. This result is in line with the study of Gobron et al. (2021), who demonstrated that trend uncertainties were reduced by 70% at high latitudes after applying combinations of NTAL and NTOL correction in up-displacements.

In general, when comparing different NTL correction combinations, EOST1-3 seem to produce smallest improvements. The NTAL corrections for these three combinations are from ERA5, whereas other models are used for other combinations. Integrating HYDL data of GLDAS and MERRA2h in EOST4 and EOST5 improves WRMS reduction rates, while impacts on annual amplitude vary by station. Notable differences between EOST6 and EOST7 are in their NTAL components (ERA5IB and MERRA2IB), with EOST7 showing significant enhancements. Difference in HYDL components between GFZ1 and GFZ2 are the ESMGFZ hydrology model and GRACE product, respectively. Overall, GFZ2, EOST7 and EOST8 results that have GRACE-derived HYDL, outperform others.

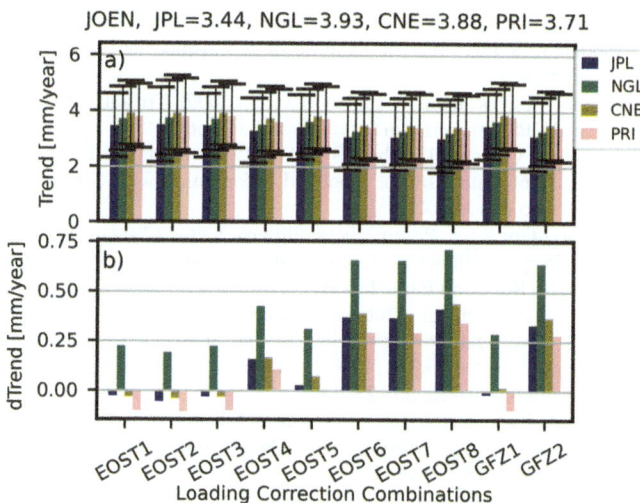

Fig. 6 Trend and uncertainties after NTL Correction for JOEN. (**a**) the trend of the over 20-year time series. (**b**) the difference in trend before and after NTL correction. The error bars in the top panel represent the uncertainty in the trend estimation. The numbers on top of the top panel header denote the trends in the original time series in mm/year

4 Conclusions and Outlook

We compared NTL corrections for four GNSS stations in Finland, using data from our PPP PRIDE GNSS processing and three other centers. While processing strategies were similar, a key difference was found in the tropospheric delay model and mapping functions.

The WRMS reduction in GNSS time series varies with different combinations of NTL products. Variations in reduction rates underscore the importance of selecting appropriate GNSS solution products and considering station characteristics. Stations like SODA, JOEN, and VAAS demonstrate diverse behaviors. Both JPL and PRI solutions consistently achieve substantial WRMS reductions across various NTL correction combinations, indicating their reliability. Differences between EOST and GFZ corrections highlight the importance of careful model selection for GNSS processing. The qualitative comparison of various combinations of NTL corrections, notably, the combinations involving EOST7 and EOST8 NTL corrections exhibited significant reductions in WRMS compared to other combinations.

Our analysis revealed a general decrease in amplitudes, spectral index, trend (velocity) and trend uncertainty after NTL corrections. Some corrections even led to negative reduction rates for the annual amplitude (i.e., increase). In most of the considered stations, annual amplitude reduction rates of the solution product were positive, indicating that the corrections tended to reduce the amplitude of the data, potentially improving data accuracy. After applying corrections, particularly in the case of JPL and NGL data, noticeable out-of-phase variations in annual phases were observed at some stations, while the spectral index generally exhibited a decreasing in the annual peak. In contrast, CNE and PRI data consistently preserved the annual phase pattern across all stations.

Integrating GLDAS and MERRA2h data from HYDL improves WRMS reduction rates, while the extent of improvement in the annual amplitude reduction rate varies from station to station. This suggests that the inclusion of these datasets exhibits different characteristics depending on the local stations. The GRACE-dervied HYDL notably reduces RMS error, annual amplitude, trend and trend uncertainties in EOST7, EOST8, and GFZ2, emphasizing the importance of dedicated gravity missions like GRACE and GRACE-FO in improving GNSS corrections. Further analysis is needed to refine correction methods and enhance data accuracy for regional and local geodetic stations.

Acknowledgements The authors would like to express their gratitude to JPL, NGL, and CNES for GNSS coordinate data, as well as to GFZ, EOST, and NASA for loading model data publicly available. We are also grateful to the two anonymous reviewers for their constructive suggestions during the review of this manuscript.

References

Altamimi Z, Rebischung P, Metivier L, Collilieux X (2016) ITRF2014: a new release of the international terrestrial reference frame modeling nonlinear station motions. J Geophys Res Solid Earth 121(8):6109–6131

Argus DF, Martens HR, Borsa AA, Knappe E, Wiese DN, Alam S, Gardiner WP (2022) Subsurface water flux in California's Central Valley and its source watershed from space geodesy. GRL 49(22)

Bertige W, Desai SD, Haines B, Harvey N, Moore AW, Owen S, Weiss JP (2010) Single receiver phase ambiguity resolution with GPS data. J Geod 84:327–337

Bertiger W, Bar-Sever Y, Dorsey A et al (2020) Gipsyx/rtgx, a new tool set for space geodetic operations and research. Adv Space Res 66(3):469–489

Bevis M, Brown A (2014) Trajectory models and reference frames for crustal motion geodesy. J Geod 88:283–311

Blewitt G, Lavalle D, Clarke P et al (2001) A new global mode of earth deformation: seasonal cycle detected. Science 294(5550):2342–2345

Blewitt G, Hammond W et al (2018) Harnessing the GPS data explosion for interdisciplinary science. Eos 99

Böhm J, Niell A, Tregoning P et al (2006a) Global mapping function (GMF): a new empirical mapping function based on numerical weather model data. Geophys Res Lett 33(7)

Böhm J, Werl B, Schuh H (2006b) Troposphere mapping functions for GPS and very long baseline interferometry from european Centre for medium-range weather forecasts operational analysis data. J Geophs Res: solid earth 111(B2)

Bos MS, Montillet JP, Williams SD et al (2020) Introduction to geodetic time series analysis. Geodetic Time Series Analysis Earth Sci:29–52

Bos M, Fernandes R, Bastos L (2021) Hector user manual version 1.9. Tech. Rep., tech. Rep. SEGAL Universidade da Beira Interior, Covilha

Collilieux X, Altamimi Z, Coulot D, Ray J, Sillard P (2007) Comparison of very long baseline interferometry, GPS, and satellite laser ranging height residuals from ITRF2005 using spectral and correlation methods. J Geophys Res Solid Earth 112(B12)

Davis JL, Elosegui P, Mitrovica JX, Tamisiea ME (2004) Climate driven deformation of the solid earth from GRACE and GPS. Geophys Res Lett

Dong D, Yunck T, Heflin M (2003) Origin of the international terrestrial reference frame. J Geophys Res Solid Earth 108(B4)

Dziewonski AM, Anderson DL (1981) Preliminary reference Earth model. Physics of the earth and planetary interiors. 25(4):297–356.

Geng J, Williams SDP, Teferle FN, Dodson AH (2012) Detecting storm surge loading deformations around the southern North Sea using subdaily GPS. Geophys J Int 191(2):569–578

Geng J, Chen X, Pan Y et al (2019a) PRIDE PPP-AR: an open-source software for GNSS PPP ambiguity resolution. GPS Solutions 23:1–10

Geng J, Chen X, Pan Y, Zhao Q (2019b) A modified phase clock/bias model to improve PPP ambiguity resolution at Wuhan University. J Geod 93(10):2053–2067

Gobron K, Rebischung P, Van Camp M, Demoulin A, de Viron O (2021) Influence of aperiodic non-tidal atmospheric and oceanic loading deformations on the stochastic properties of global GNSS vertical land motion time series. J Geophys Res Solid Earth 126:e2021JB022370

Kennett BL, Engdahl ER, Buland R (1995) Constraints on seismic velocities in the Earth from traveltimes. Geophys J Int 122(1):108–124

Klos A, Dobslaw H, Dill R et al (2021) Identifying the sensitivity of GPS to non-tidal loadings at various time resolutions: examining vertical displacements from continental eurasia. GPS Solutions 25(3):89

Lagler K, Schindelegger M, Bohm J et al (2013) GPT2: Empirical slant delay model for radio space geodetic techniques. Geophys Res Lett 40(6):1069–1073

Li C, Huang S, Chen Q et al (2020) Quantitative evaluation of environmental loading induced displacement products for correcting GNSS time series in cmonoc. Remote Sens 12(4):594

Loyer S, Perosanz F, Mercier F, Capdeville H, Marty JC (2012) Zero-difference GPS ambiguity resolution at CNES–CLS IGS analysis center. J Geod 86:991–1003

Lyard F, Lefevre F, Letellier T. et al (2006) Modelling the global ocean tides: modern insights from FES2004. Ocean Dynamics 56:394–415. https://doi.org/10.1007/s10236-006-0086-x

Mao A, Harrison CG, Dixon TH (1999) Noise in GPS coordinate time series. J Geophy Res Solid Earth 104(B2):2797–2816

Mémin A, Boy JP, Santamaria-Gomez A (2020) Correcting GPS measurements for non-tidal loading. GPS Solutions 24:1–13

Michel A, Santamaria-Gomez A, Boy JP et al (2021) Analysis of gnss displacements in europe and their comparison with hydrological loading models. Remote Sens 13(22):4523

Nicolas J, Verdun J, Boy JP et al (2021) Improved hydrological loading models in South America: analysis of gps displacements using m-ssa. Remote Sens 13(9):1605

Nordman M, Makinen J, Virtanen H et al (2009) Crustal loading in vertical GPS time series in fennoscandia. J Geodyn 48(3–5):144–150

Nordman M, Virtanen H, Nyberg S et al (2015) Non-tidal loading by the Baltic Sea: comparison of modelled deformation with GNSS time series. Geo Res J 7:14–21

Santamaria-Gomez A, Bouin MN, Collilieux X et al (2011) Correlated errors in gps position time series: implications for velocity estimates. J Geophy Res Solid Earth 116(B1)

Schuh H, Estermann G, Cretaux JF, Berge-Nguyen M, van Dam T (2003) Investigation of hydrological and atmospheric loading by space geodetic techniques. In: C Hwang, CK Shum, JC Li (eds) IAG Proceedings, vol 126, pp 123–132

Springer A, Karegar MA, Kusche J, Keune J, Kurtz W, Kollet S (2019) Evidence of daily hydrological loading in GPS time series over Europe. J Geod 93:2145–2153

Tregoning P, van Dam TM (2005) Atmospheric pressure loading corrections applied to GPS data at the observation level. Geophys Res Lett

Tregoning P, Watson C (2011) Correction to "atmospheric effects and spurious signals in GPS analyses". J Geophys Res

Tregoning P, Watson C, Ramillien G, McQueen H, Zhang J (2009) Detecting hydrologic deformation using GRACE and GPS. Geophys Res Lett

van Dam T, Wahr J, Milly PCD, Shmakin AB, Lavalle BD, Larson KM (2001) Crustal displacements due to continental water loading. Geophys Res Lett 28(4):651–654

van Dam T, Wahr J, Lavallée D (2007) A comparison of annual vertical crustal displacements from GPS and Gravity Recovery and Climate Experiment (GRACE) over Europe. J Geophys Res 112:B03404. doi:10.1029/2006JB004335.

van Dam TM, Collilieux X, Altamimi Z, Ray J (2012) Non-Tidal Ocean loading: amplitudes and potential effects in GPS height time series. J Geod 86(11):1043–1057

Williams SDP, Penna NT (2011) Non-Tidal Ocean loading effects on geodetic GPS heights. Geophys Res Lett

Williams SD, Bock Y, Fang P, Jamason P, Nikolaidis RM, Prawirodirdjo L, Miller M, Johnson DJ (2004) Error analysis of continuous GPS position time series. J Geophys Res Solid Earth 109(B3)

Zumberge J, Heflin M, Jefferson D et al (1997) Precise point positioning for the efficient and robust analysis of GPS data from large networks. J Geophys Res Solid Earth 102(B3):5005–5017

Open Access This chapter is licensed under the terms of the Creative Commons Attribution 4.0 International License (http://creativecommons.org/licenses/by/4.0/), which permits use, sharing, adaptation, distribution and reproduction in any medium or format, as long as you give appropriate credit to the original author(s) and the source, provide a link to the Creative Commons license and indicate if changes were made.

The images or other third party material in this chapter are included in the chapter's Creative Commons license, unless indicated otherwise in a credit line to the material. If material is not included in the chapter's Creative Commons license and your intended use is not permitted by statutory regulation or exceeds the permitted use, you will need to obtain permission directly from the copyright holder.

A Geodetic Analysis of the Volume Transport in the ACC Region Based on Satellite Data

Juan A. Vargas-Alemañy, M. Isabel Vigo, David García-García, and Ferdous Zid

Abstract

Geostrophic currents, driven by the Coriolis and pressure gradient forces, are crucial for understanding ocean circulation. The Antarctic Circumpolar Current (ACC) in the Southern Ocean, which surrounds Antarctica, has a significant global impact, and its volume transport (VT) remains a challenge to measure. We use satellite data, combining altimetry and gravity satellite missions, to estimate VT within the ACC region. Our study provides a comprehensive spatial and temporal analysis, including both barotropic and baroclinic VT components. The spatial analysis reveals a mean VT of 210.44 ± 3.4 Sv for the entire study area, with maxima near critical choke points. Focusing on the time-varying component, we identify a mean VT of 15.86 ± 0.05 Sv per 1° grid cell, a linear trend of −0.007 ± 0.002 Sv per month, and significant seasonal and biannual signals. The baroclinic component drives low-frequency variability, while the barotropic component controls high-frequency variability. We propose a specific ACC zonal VT of 201.63 ± 0.71 Sv. We validate our results with in situ measurements from the Drake Passage. In conclusion, our satellite-based approach provides valuable insights into the ACC VT. This methodological extension improves our understanding of the ocean circulation dynamics of the ACC and demonstrates the utility and robustness of satellite data in oceanographic research.

Keywords

Antarctic circumpolar current · Satellite altimetry · Satellite gravity · Volume transport

1 Introduction

Geostrophic currents (GC) are the movements of ocean water that are primarily influenced by the balance between two important forces: the Coriolis force and the pressure gradient force. The Coriolis force, caused by the Earth's rotation, deflects moving objects, such as water, to the right in the Northern Hemisphere and to the left in the Southern Hemisphere. The pressure gradient force, on the other hand, is the result of variations in pressure over an area. The balance between these two forces gives rise to GC, which play a crucial role in understanding ocean circulation patterns.

The Antarctic Circumpolar Current (ACC) is located in the Southern Ocean (SO) region. The ACC is the largest and most powerful current system in the world, encircling the entire continent of Antarctica (Carter et al. 2008). It is characterized by strong and continuous eastward flow, driven by the combined effects of wind, density differences, and the Earth's rotation. The ACC area is of great scientific interest because of its influence on global climate, ocean circulation, and marine ecosystems.

The ACC covers a large latitudinal range, extending from approximately 40°S to 60°S. Within this region, the ACC is known for its distinct fronts, including the Subantarctic Front, the Polar Front, and the Southern ACC Front (Orsi et al. 1995; Sokolov and Rintoul 2009; Tarakanov 2021). These

J. A. Vargas-Alemañy (✉) · M. I. Vigo · D. García-García · F. Zid
Department of Applied Mathematics, University of Alicante, Alicante, Spain
e-mail: juan.vargas@ua.es

fronts mark the boundaries between different water masses and play a crucial role in the exchange of heat, carbon, and nutrients between the ocean and the atmosphere.

The remote and challenging nature of the ACC presents a considerable challenge when it comes to obtaining direct measurements of its VT. However, the Drake Passage (DP) provides a critical corridor. It is situated between the southern tip of South America (Cape Horn) and the northernmost point of the Antarctic Peninsula, where the ACC flows from the Pacific Ocean into the Atlantic Ocean. This location serves as the most suitable gateway for the study of the ACC, resulting in numerous monitoring programs in the region aimed at estimating its VT (Whitworth III 1983; Whitworth III and Peterson 1985; Cunningham et al. 2003; Firing et al. 2011; Chidichimo et al. 2014; Koenig et al. 2014; Donohue et al. 2016). These studies have reported ACC transport across the DP in the range of 136–173 Sv.

Estimating oceanic VT using satellite data has several advantages. Satellite observations provide a valuable tool for studying VT and their variability on large time and spatial scales. These satellite-based techniques allow continuous monitoring of VT and ocean circulation dynamics. By integrating altimetry and gravity satellite data, we can estimate the GC and their associated VT (Wunsch and Gaposchkin 1980). Previous studies (Kosempa and Chambers 2014; Vigo et al. 2018; Vargas-Alemañy et al. 2023), have successfully used this methodology to analyze GC and VT in the SO region.

In this study, we extend the application of this methodology to perform a thorough analysis of VT within the ACC region. This analysis includes both spatial and temporal aspects, as well as an investigation of the barotropic and baroclinic components of VT. To ensure the validity of our estimates, we compare them with in situ measurements in the DP region as provided by Cunningham et al. (2003).

2 Methodology

Utilizing the same methodology as Vigo et al. (2018), we estimate the geostrophic ocean flow by synergizing space data, incorporating measurements from satellites for altimetry and gravity, along with in situ data (Wunsch and Gaposchkin 1980). First, we define the Absolute Dynamic Topography (ADT) as follows:

$$ADT(x, y, t) = SSH(x, y, t) - N(x, y), \quad (1)$$

where N is a time-averaged geoid, and x and y are longitude and latitude.

Here, ADT represents the instantaneous sea height above the geoid and can be computed, as shown in Eq. (1), by taking the difference between Sea Surface Height (SSH) and the geoid. SSH, in turn, can be derived from satellite altimetry data by combining Sea Level Anomalies (SLA) with the corresponding Mean Sea Surface (MSS):

$$SSH(x, y, t) = SLA(x, y, t) + MSS(x, y), \quad (2)$$

where t is time.

Furthermore, by calculating the disparity between the geoid and the MSS, the Mean Dynamic Topography (MDT) can be determined:

$$MDT(x, y) = MSS(x, y) - N(x, y). \quad (3)$$

If both MDT and SSH are referenced to the same MSS, we can express ADT, according to Eqs. (1), (2), and (3), as:

$$ADT(x, y, t) = MDT(x, y) + SLA(x, y, t). \quad (4)$$

We will use Eq. (4) to compute the ADT by integrating the satellite gravity estimates of the MDT with the satellite altimetry estimates of the SLA. Both MDT and SLA products are referenced to the same MSS. A description of the datasets used is given in Sect. 3.

The Surface Geostrophic Current (SGC) can be computed by determining the directional derivatives of the ADT by applying of the geostrophic equation. This equation describes the balance between the pressure gradient force and the Coriolis force at the ocean surface:

$$\begin{aligned} u_s(x,y,t) &= -\tfrac{g(y)}{f}\tfrac{\delta ADT}{\delta y}(x,y,t), \\ v_s(x,y,t) &= \tfrac{g(y)}{f}\tfrac{\delta ADT}{\delta x}(x,y,t), \end{aligned} \quad (5)$$

where u_s is the zonal surface velocity (positive eastward) and v_s is the meridional surface velocity (positive northward). The parameter $f = 2\omega \sin(y)$ represents the Coriolis parameter, where w is the mean angular rate of the Earth's rotation and g is the gravitational acceleration (latitude dependent).

In order to derive GC at different depths, it is first necessary to establish the concept of Relative Dynamic Topography (RDT):

$$RDT(x, y, z, t) = \frac{1}{g(y)} \int_{P(z)}^{0} \frac{dP}{\rho(x, y, z, t)}, \quad (6)$$

where $P(z)$ is the pressure at depth z (in Pascal units), and ρ is the water density.

Using the SGC as the reference level and considering the directional derivatives of the RDT, the value for the GC and any depth z_i can be obtained as follows:

$$\begin{aligned} u_s(x,y,t) &= -\tfrac{g(y)}{f}\tfrac{\delta RDT}{\delta y}(x,y,z_i,t) + u(x,y,z_i,t), \\ v_s(x,y,t) &= \tfrac{g(y)}{f}\tfrac{\delta RDT}{\delta x}(x,y,z_i,t) + v(x,y,z_i,t). \end{aligned} \quad (7)$$

By vertically integrating the three-dimensional geostrophy from a depth D to the surface within a cell of a regular grid, we can determine the volume of water transported by the geostrophic flow from the surface to that specific depth. For accurate volume calculations, it is essential to consider the width of the grid cell perpendicular to the flow direction.

$$VT_u(x,y,t) = w_{NS}\int_{-D}^{0} u(x,y,z,t)\,dz,$$
$$VT_v(x,y,t) = w_{EW}(y)\int_{-D}^{0} v(x,y,z,t)\,dz, \quad (8)$$

where VT_u is the zonal VT, which is positive eastward, while VT_v is the meridional VT, which is positive northward. w_{NS} and w_{EW} denote the North-South and East-West widths of the grid cell, respectively. It is important to note that w_{EW} depends on the latitude y, while w_{NS} remains constant. The VT is measured in Sverdrups (Sv), where 1 Sv is equivalent to $10^6 m^3/s$.

To distinguish between the barotropic and baroclinic components of the VT, we follow the definition provided by Fofonoff (1962). According to this definition, the barotropic transport refers to the part of the VT attributed to a water column moving uniformly and at the same speed as the bottom current. Conversely, the baroclinic transport represents the remaining component of VT, which accounts for the non-uniform movement of the water column.

To estimate the barotropic component for the zonal VT, we identify $z_{max}(x,y)$ as the depth z of the deepest current at each point (x,y). We then define u_{BT}, the geostrophic component of the bottom current relative to the barotropic component:

$$u_{BT}(x,y,t) = u(x,y,z_{max}(x,y),t). \quad (9)$$

For each depth z_i, the geostrophic current relative to the baroclinic components is obtained as:

$$u_{BC}(x,y,z_i,t) = u(x,y,z_i,t) - u_{BT}(x,y,t). \quad (10)$$

The barotropic component of the VT at point (x,y) from surface to a depth $z_{max}(x,y)$ is obtained as:

$$VT_{u_{BT}}(x,y,t) = w_{NS}\int_{-z_{max}(x,y)}^{0} u_{BT}(x,y,t)\,dz$$
$$= w_{NS}\cdot u_{BT}(x,y,t)\cdot z_{max}(x,y), \quad (11)$$

while the baroclinic component is given by:

$$VT_{u_{BC}}(x,y,t) = w_{NS}\int_{-z_{max}(x,y)}^{0} u_{BC}(x,y,z,t)\,dz. \quad (12)$$

Note that with this definition, the following equality follows:

$$VT_u(x,y,t) = VT_{u_{BC}}(x,y,t) + VT_{u_{BT}}(x,y,t). \quad (13)$$

The barotropic and baroclinic components of the meridional VT are obtained analogously.

Following Vargas-Alemañy et al. (2023), we define the ACC region as the grid points within the SO region where the mean zonal full-depth VT exceeds 12 Sv, with certain outliers removed. The boundaries of this region, as defined by these criteria, are outlined in Fig. 1.

3 Data

In this study, we calculated the ADT according to Eq. (4), using both the MDT and the SLA with reference to the high-resolution MSS model DTU18MSS – developed by the Danish National Space Center. This model is based on a 25-year dataset collected from various multi-mission satellite altimeters, including a 3-year record from Sentinel-3A and an enhanced 7-year record from Cryosat-2 LM. Further details can be found in the work of Andersen et al. (2018).

For the MDT, we used on the DTUUH19MDT geodetic model, also developed by the Danish National Space Center. This model is based on the OGMOG geoid model, expanded with EIGEN-6C4 coefficients up to degree and order 2,160, along with the aforementioned DTU18MSS mean sea surface model. This integration incorporates drifter data to enhance the MDT resolution; see Knudsen et al. (2019) for comprehensive information.

To derive the SLA, we used the CCI-Sea Level Project (http://www.esa-sealevel-cci.org) product of sea-level maps available as a monthly merged solution from different altimetry satellites (Jason 1 and 2, TOPEX/Poseidon, Envisat, ERS-1 and -2, and GEOSAT-FO). These maps, with a spatial resolution of 0.25°, cover the period from 1 January 1993 to 31 December 2015 (version v2.0, downloaded in December 2019), and are presented as anomalies with respect to the same DTU18MSS model used for the MDT.

For the calculation of the RDT, we used the EN4.1.1 objective analyses dataset, of subsurface ocean temperature (T) and salinity (S) data sourced from the Met Office Hadley Centre. These profiles include T and S measurements incorporating ARGO data and extend to depths of 5,500 m, allowing us to compute near full-depth VT. Within this dataset, T and S measurements have been optimally interpolated onto a regular 1° × 1° grid across 42 depth layers (see Good et al. 2013 for more information). The EN.4.1.1 data were obtained from https://www.metoffice.gov.uk/hadobs/en4/ (© British Crown Copyright, Met Office, [2021]) and are available under a Non-Commercial Government License (http://www.nationalarchives.gov.uk/doc/non-commercial-government-licence/version/2/). We use the seawater state equation provided by the Gibbs Seawater Oceanography Toolbox (Mcdougall and Barker

Fig. 1 Bathymetry (in meters) of the SO region from https://download.gebco.net/ (GEBCO Compilation Group 2020). The area delineated as the ACC region is highlighted in black

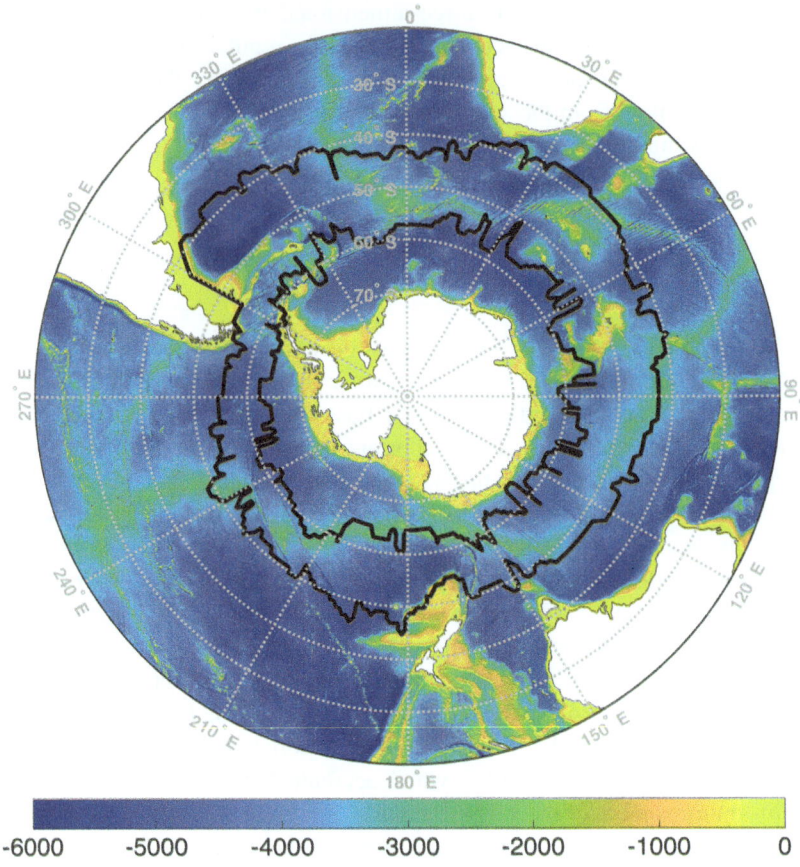

2011) to derive density based on ocean temperature, salinity, and pressure.

4 Results

To deepen our comprehension of the studied area, Fig. 1 illustrates the bathymetry of the SO. We have demarcated the identified ACC region with black dots for clarity.

In Fig. 2a, we present the mean GC speeds for the entire study period. Notably, maximum values of up to 51.99 cm/s are observed in close proximity to the Agulhas Current ([40°S] × [30°E, 60°E]) and the Brazil-Malvinas Current ([40°S, 50°S] × [300°E, 330°E]).

Figure 2b illustrates the depth dependent decrease in mean speed from 5 m to 2,000 m. This was achieved by applying a linear fit to each data point based on its depth level. The slope of the fitted model is shown for each data point

The blue values represent a decrease with depth, which is to be expected. However, there are some positive (red) values, indicating an increase in speed with depth. These positive values represent only 1.96% of the total (the value 0 is at the 98th percentile). These positive values indicate regions characterized by the presence of meandering strong currents.

The mean slope for the entire region is $-0.0034 \pm 3.8 \times 10^{-5}$ (cm/s)/m, with an average GC speed of 11.01 ± 0.11 cm/s at 5 m depth and 5.55 ± 0.14 cm/s at 2,000 m depth. These averages are expressed in cm/s per 1° cell and are latitude weighted.

In Fig. 3 we present two different plots of the time-averaged VT. In Fig. 3a, the colored background indicates the vector norm, while the black arrows representing direction indicate the mean VT vector, calculated by taking the temporal average of each component. In addition, at each grid point we compute a monthly time series of VT vector norms, and Fig. 3b displays the temporal average of this series, which we refer to as the monthly VT norm.

It's important to note the differences between these two visualizations. The mean VT vector (**a**) becomes null, indicating no VT at all, when there are two opposite vectors of equal magnitude at the same grid point for two consecutive months. Conversely, the monthly VT norms (**b**) will show the norm of the two vectors, representing the mean VT through the grid point in both directions. The monthly VT norms will never be less than the norm of the mean VT vector.

The mean VT for the entire region is 23.7 ± 0.3 Sv per 1° cell, and the mean of the monthly VT norms is 51.68 ± 0.5 Sv per 1° cell. These means are latitude-weighted.

Fig. 2 (**a**) Mean (GC) for the whole study period (2004–2015) at a depth of 5 m (cm/s). Maximum values reach 51 cm/s. Color saturated at 30 cm/s to better visualize the results. (**b**) Depth dependent mean speed decrease from 5 m to 2,000 m

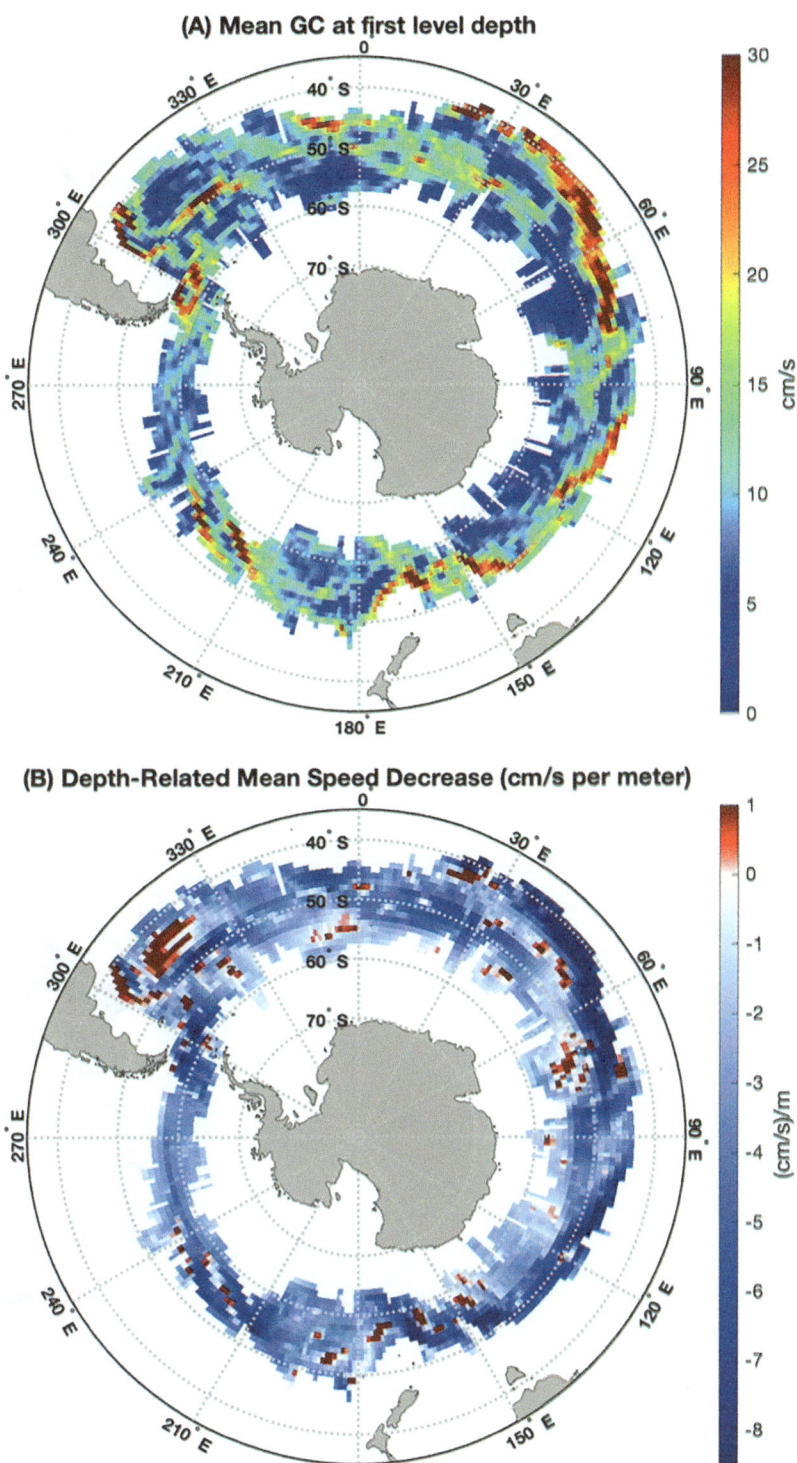

Fig. 3 Mean geostrophic volume transport (2004–2015): (**a**) Arrows depict the mean vectors, with color indicating their magnitude. Arrows are only shown for mean vectors with a magnitude greater than 15 Sv for enhanced clarity. Units are expressed in Sv, with the color scale capped at 100 Sv, while the highest values can extend to 193 Sv. (**b**) Mean of monthly vector magnitudes. Each grid point has a monthly time series of VT norms, and this plot represents the mean of these monthly time series. Units are in Sv, and the color scale is capped at 170 Sv, with maximum values reaching up to 338 Sv

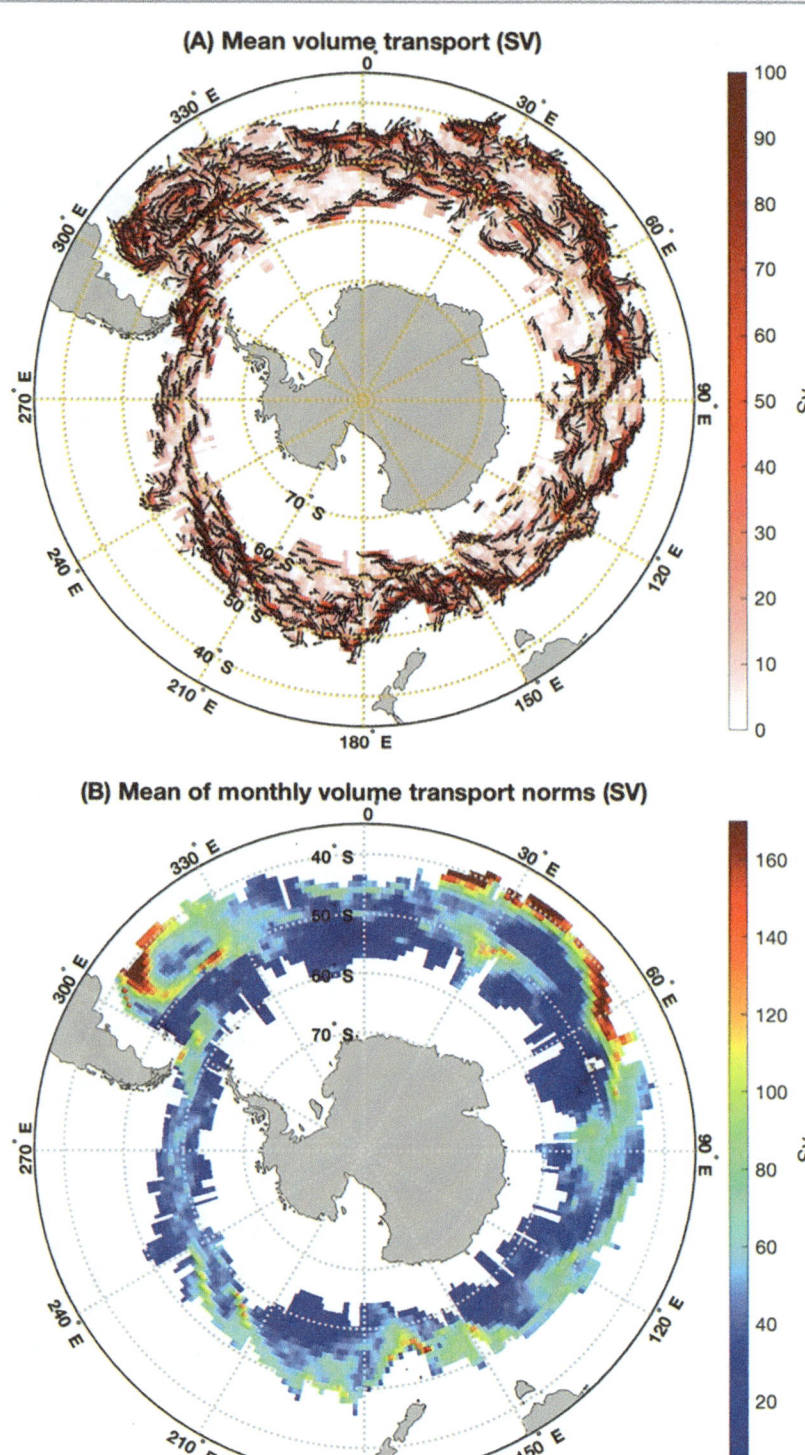

Fig. 4 Longitudinal series of total (indicated by the black line with asterisks), zonal (represented by the red curve with triangles), and meridional (depicted as the blue curve with dots) VT. These longitudinal series are generated by averaging the data over time and integrating them latitudinally at each longitude across the ACC region

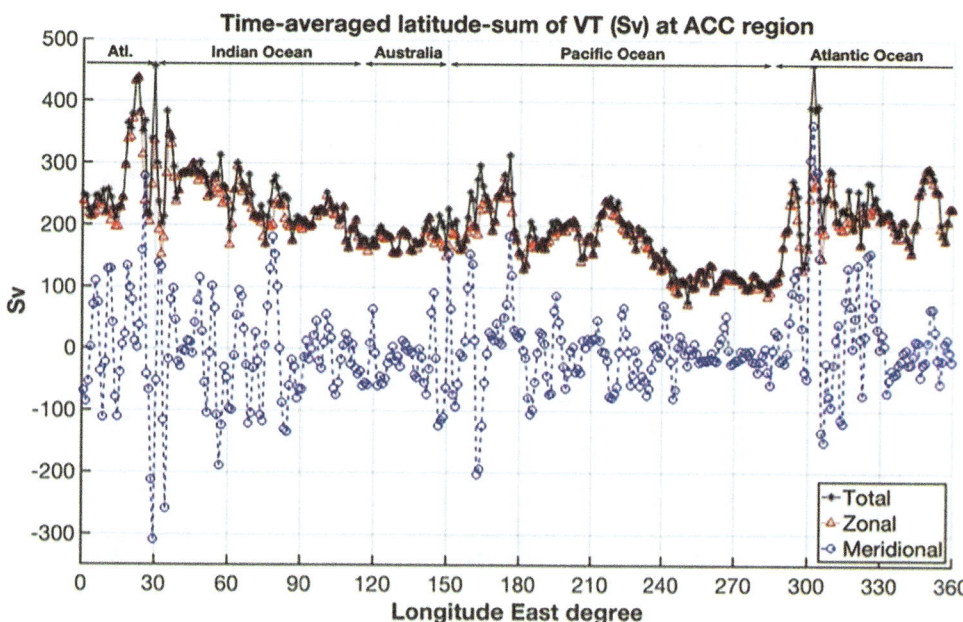

For the monthly VT norms, maximum values can reach up to 248.93 Sv and are mainly concentrated near the Brazil-Malvinas and the Agulhas currents. For the mean VT, maximum values also reach up to 193.04 Sv and are also concentrated near the Brazil-Malvinas and the Agulhas currents. However, there are also some points in the [50°S, 60°S] × [150°S, 240°S] region that can reach these high values.

For a more comprehensive analysis of the spatial variability of VT, the longitudinal series are shown in Fig. 4. This series is derived by first averaging the data over time and then, for each longitude, integrating all latitudes within the ACC region. This process is applied to the total, zonal, and meridional VT.

The total VT exhibits an average of 210.44 ± 3.4 Sv. In particular, three prominent maxima are observed near 30°E, 170°E, and 300°E, corresponding to the choke points near South Africa, South Australia, and the DP. These choke points essentially mark the boundaries between the three major ocean basins. There is also a gradual decrease from west to east.

The zonal VT, with a mean of 198.47 ± 2.95 Sv, accounts for 94.31% of the total VT and mainly influences the long-wave variations. On the other hand, the meridional VT, which primarily contributes to the high-frequency spatial variations, has a mean of -2.03 ± 4.06 Sv, indicating that it is not significantly different from zero. This phenomenon can be attributed to the north-south shifts that occur in the ACC due to the meandering of its branches.

To analyze the barotropic and baroclinic components of VT, Fig. 5 shows the contributions of the barotropic (green dashed line) and baroclinic (magenta dashed line) components for both the zonal and meridional variables presented in Fig. 4.

Regarding the zonal component (Fig. 5a), the baroclinic component accounts for 69.53% of the zonal VT and has a mean signal of 137.99 ± 2.29 Sv. Meanwhile, the barotropic component has a mean signal of 60.48 ± 2.38 Sv.

Both components show a strong correlation with the zonal VT, with a correlation coefficient of 0.65 for the baroclinic component and 0.61 for the barotropic component, both statistically significant (p-value $\ll 0.05$). However, their contributions to the total zonal VT are different; the baroclinic component mainly influences long-wave spatial variability, while the barotropic component plays a key role in driving high-frequency spatial variability.

The barotropic and baroclinic components of the meridional VT have mean values of -0.23 ± 4.02 and -1.79 ± 1.71, respectively. These values are not significantly different from zero, mirroring the behavior of the meridional component itself.

For a more detailed analysis of the temporal variation in the ACC, Fig. 6 shows the time series of the mean VT per 1° grid cell for the total (black asterisks), zonal (red triangles), and meridional (blue circles) components.

The total VT shows a mean signal of 15.86 ± 0.05 Sv, with a linear trend of -0.007 ± 0.002 Sv per month. It also shows an annual signal with an amplitude of 0.42 ± 0.11 Sv peaking in late May, and a biannual signal with an amplitude of 0.12 ± 0.11 Sv peaking in the 5th month of the 24-month period.

The zonal component, with a mean signal of 15.85 ± 0.05 Sv, accounts for 99.98% of the total VT, as expected from to the zonal nature of the ACC driving circulation.

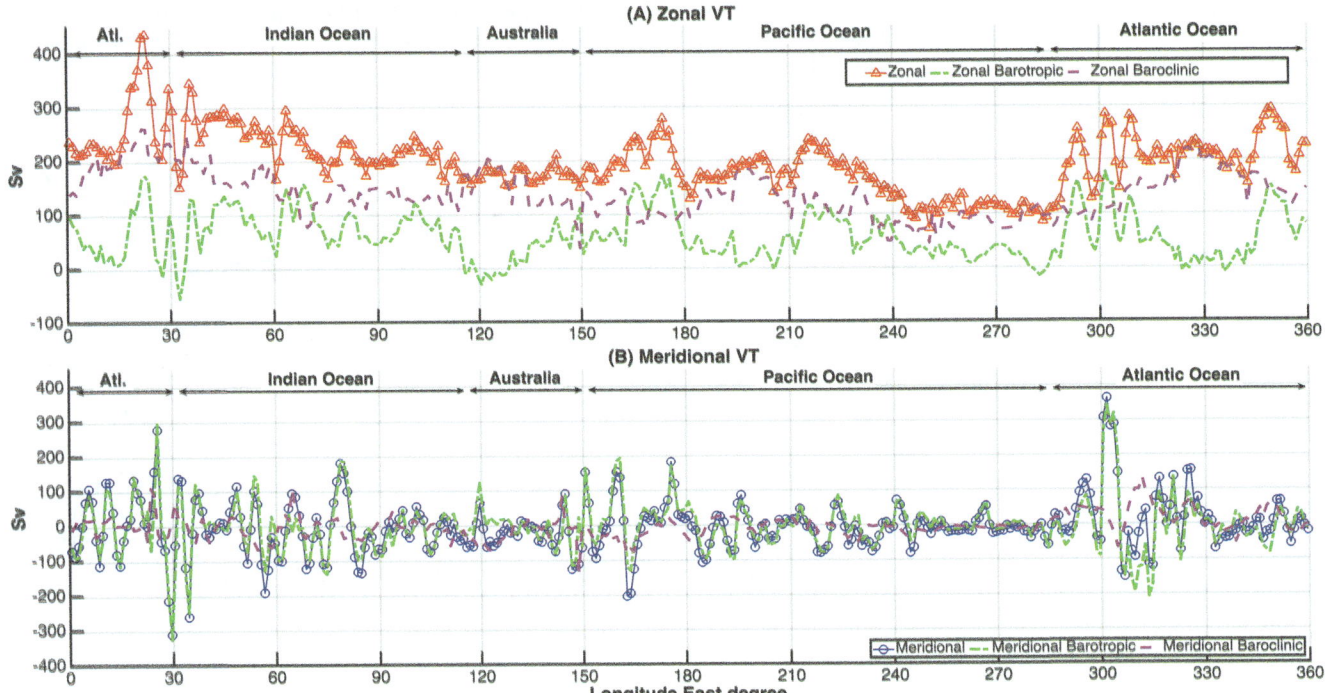

Fig. 5 Longitudinal series data obtained using the same methodology as in Fig. 4. (**a**) zonal VT (represented by the red curve with triangles), and (**b**) meridional VT (depicted as the blue curve with dots). Each of these series includes both its barotropic component (green dashed line) and its baroclinic component (magenta dashed line)

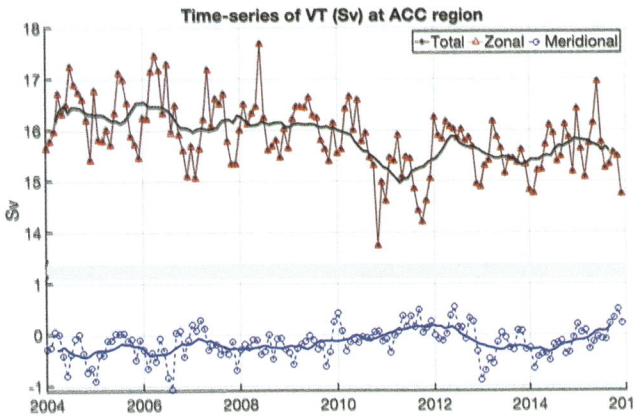

Fig. 6 VT within the ACC region per 1° grid, including total (black asterisks), zonal (red triangles), and meridional (blue circles) VT. Thick lines represent a 12-month running mean, with units in Sv

Conversely, the meridional VT maintains a mean value of -0.15 ± 0.02, which is not significantly different from zero. This behavior, as seen in Fig. 5, is due to north-south shifts within the ACC.

From this time series, we can derive an estimate of the zonal VT at any given meridional section within the ACC. This is done by multiplying the zonal value at a given time by the number of grid points corresponding to that section. This method allows us to calculate the mean zonal VT for any section at a given longitude. Importantly, for the entire ACC region, where the average number of grid points for a given longitude is 12.7, our estimate indicates that the mean zonal VT across the entire ACC zone is 201.63 ± 0.71 Sv.

To examine the barotropic and baroclinic components, Fig. 7 shows these components desegregated for both the zonal (Fig. 7a) and meridional (Fig. 7b) components.

The baroclinic component of the zonal VT maintains a mean value of 11.02 ± 0.01 Sv, which represents 69.49% of the total zonal VT. Conversely, the barotropic component, with a mean value of 4.84 ± 0.06 Sv, plays a pivotal role in driving the variability and shows a strong correlation with the zonal component (correlation coefficient of 0.99, p-value $\ll 0.05$).

Regarding the meridional component, the baroclinic component has a mean value of -0.15 ± 0.01 Sv, while the barotropic component has a mean value of 0.003 ± 0.031 Sv. Neither value is significantly different from zero. Similar to the zonal component, the variability of the meridional component is primarily influenced by the barotropic transport, with a correlation coefficient of 0.83 (p-value $\ll 0.05$).

To validate our results, we performed a comparative analysis with the results of Cunningham et al. (2003). They provided an estimate for the ACC based on in situ data focusing on measurements from the DP. The DP, located between South America and Antarctica, acts as a natural passage, which makes the ACC narrower. This has led to considerable

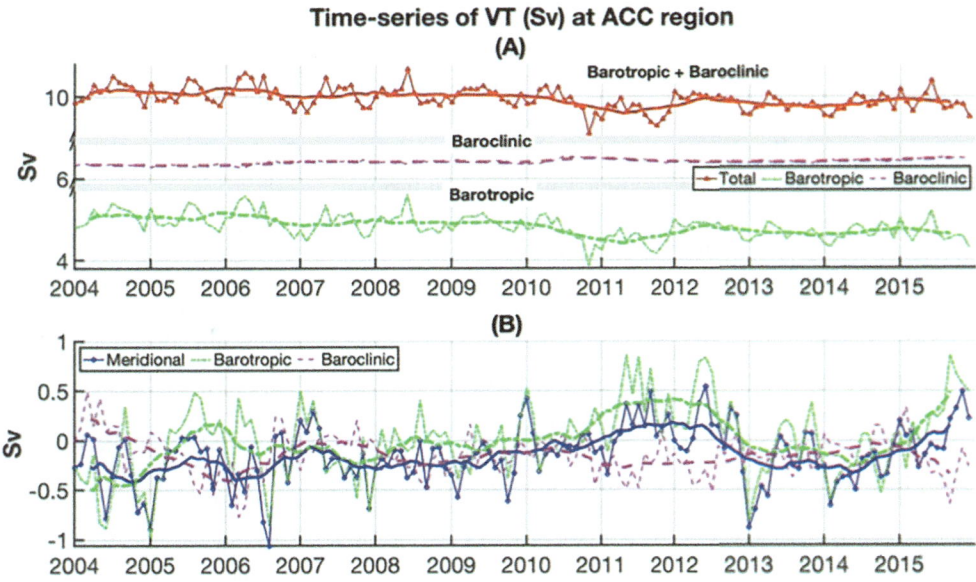

Fig. 7 VT within the ACC region per 1° grid as in Fig. 6. (**a**) Zonal VT (represented by the red curve with triangles), and (**b**) Meridional VT (represented by the blue curve with dots). Each of these series includes both its barotropic component (green dashed line) and its baroclinic component (magenta dashed line). Thick lines represent 12-month running means

interest in studying ocean circulation in this area. Transport through this critical point has been extensively investigated by several monitoring programs. Cunningham et al. (2003) reported a canonical estimate of 136.7 ± 7.8 Sv for the ACC VT.

In the same region, our method gave an estimate of 142.7 ± 41 Sv. Although our estimate aligns with the in situ results, it shows greater variability. The comparison between our estimate and Cunningham et al.'s results highlights the consistency of our approach with the in situ data and reinforces the validity of our satellite-based results, especially within the specific region of the DP.

5   Conclusions

In this study, we apply a geodetic methodology to investigate the full-depth VT within the ACC region. This approach relies on a fusion of data derived from altimetry and gravity satellite missions, allowing the determination of the SGC. In addition, we use T and S profiles to assess flows at different depths, providing a comprehensive analysis of the dynamics of the ACC. By using these different datasets and methods, our study provides a thorough investigation of the ocean circulation in the ACC.

In our analysis, we found that the spatial variability of the total VT remains relatively steady at 210.44 ± 3.4 Sv, characterized by three prominent maxima near the choke points around South Africa, South Australia, and the DP. Examining the temporal variability, we found a mean VT of 15.86 ± 0.05 Sv per 1° grid cell, along with a linear trend of −0.007 ± 0.002 Sv per month. Furthermore, we observed a periodic signal with an amplitude of 0.42 ± 0.11 Sv, peaking in late May, and a biannual signal with an amplitude of 0.12 ± 0.11 Sv, peaking in the 5th month of the 24-month period.

Both analyses confirm that the total VT is mainly influenced by the zonal VT, while the meridional VT does not significantly deviate from zero. When examining the barotropic and baroclinic components, it becomes evident that the baroclinic component is responsible for low-frequency variations, whereas the barotropic component drives high-frequency variations.

To validate our results, we compare them with the established canonical value of 136.7 ± 7.8 Sv provided by Cunningham et al. (2003), which is derived from in situ measurements in the DP region. Using our results, we obtain a similar estimate of 142.7 ± 41 Sv for the same area, which is in good agreement with the in situ results.

A major advantage of our satellite-based geodetic approach is its capability to estimate the full-depth VT over the entire ACC region, in contrast to previous studies that have often focused only on the DP. This extended spatial coverage is crucial for a more comprehensive understanding of the ACC dynamics, as it enables the characterization of VT variability along its entire path, rather than in isolated regions. By leveraging this extensive spatial analysis, we propose a mean value for the ACC region zonal VT of 201.63 ± 0.71 Sv. This estimate is calculated by multiplying the mean zonal VT per 1° grid cell by the average number of grid points covering the entire ACC region.

In addition, it is important to acknowledge the potential sources of uncertainty inherent in our methodology. The paucity of observations, particularly at depths below 2,000 m, introduces uncertainties in the T and S data in the ACC region (Kosempa and Chambers 2016). We are aware of the limitations imposed by the restricted availability of observational data in this region. Despite these challenges, our study provides valuable insights into the dynamics of the VT within the ACC region. Addressing these uncertainties remains a key area for further refinement in future research.

In summary, our study has employed a geodetic methodology using satellite-based gravity and altimetry data to estimate VT within the ACC region. This approach has provided a comprehensive spatial and temporal analysis of the dynamics of the ACC. Furthermore, our results are in close agreement with in situ estimates, reinforcing the robustness of our methodology. This study contributes valuable insights into the understanding of ocean circulation patterns within the ACC, and highlights the power of satellite data in advancing our knowledge of this critical region.

Ackowledgments The authors acknowledge the supported of all data providers: ESA in the frame of the CCI Sea Level Project for Altimetry data; DTU SPACE from the Danish National Space Center for MDT and MSS products; and the Met Office Hadley Center for EN4.1.1 data product.

Funding This research is supported by grant PID2021-122142OB-I00 funded by MCIN/AEI/10.13039/501100011033, grant PROMETEO/2021/030 funded by Generalitat Valenciana, and grant GVA-THINKINAZUL/2021/035 funded by Generalitat Valenciana and "European Union NextGenerationEU/PRTR".

References

Andersen O, Knudsen P, Stenseng L (2018) A new DTU18MSS Mean Sea Surface—improvement from SAR altimetry. In: Abstract from 25 years of progress in radar altimetry symposium, Portugal, September 2018

Carter L, McCave IN, Williams MJM (2008) Chapter 4. Circulation and water masses of the Southern Ocean: a review. In: Antarctic climate evolution. Elsevier, pp 85–114. https://doi.org/10.1016/s1571-9197(08)00004-9

Chidichimo MP, Donohue KA, Watts DR, Tracey KL (2014) Baroclinic transport time series of the Antarctic circumpolar current measured in Drake passage. J Phys Oceanogr 44(7):1829–1853. https://doi.org/10.1175/jpo-d-13-071.1

Cunningham SA, Alderson SG, King BA, Brandon MA (2003) Transport and variability of the Antarctic circumpolar current in Drake passage. J Geophys Res 108:8084. C5 ISSN 0148-0227. https://doi.org/10.1029/2001jc001147

Donohue KA, Tracey KL, Watts DR, Chidichimo MP, Chereskin TK (2016) Mean Antarctic circumpolar current transport measured in Drake passage. Geophys Res Lett 43(22):0094–8276. https://doi.org/10.1002/2016gl070319

Firing YL, Chereskin TK, Mazloff MR (2011) Vertical structure and transport of the Antarctic circumpolar current in Drake passage from direct velocity observations. J Geophys Res 116:C08015. https://doi.org/10.1029/2011JC006999

Fofonoff NP (1962) Dynamics of ocean currents. In: Physical oceanography, The sea, vol 1. Wiley-Interscience, Hoboken, NJ, pp 323–395

GEBCO Compilation Group (2020) GEBCO Gridded Bathymetry Data. https://doi.org/10.5285/e0f0bb80-ab44-2739-e053-6c86abc0289c

Good SA, Martin MJ, Rayner NA (2013) EN4: quality controlled ocean temperature and salinity profiles and monthly objective analyses with uncertainty estimates. J Geophys Res Oceans 118(12):6704–6716. https://doi.org/10.1002/2013JC009067

Knudsen P, Andersen O, Maximenko N, Hafner J (2019) A new combined mean dynamic topography model - DTUUH19MDT. In Anonymous living planet symposium, Milan, May 2019

Koenig ZE, Provost C, Ferrari R, Sennéchael N, Rio M (2014) Volume transport of the Antarctic circumpolar current: production and validation of a 20 year long time series obtained from in situ and satellite observations. J Geophys Res Oceans 119(8):5407–5433. https://doi.org/10.1002/2014jc009966

Kosempa M, Chambers DP (2014) Southern Ocean velocity and geostrophic transport fields estimated by combining Jason altimetry and Argo data. J Geophys Res Oceans 119. https://doi.org/10.1002/2014JC00985

Kosempa M, Chambers DP (2016) Mapping error in Southern Ocean transport computed from satellite altimetry and Argo. J Geophys Res Oceans 121. https://doi.org/10.1002/2016JC011956

Mcdougall TJ, Barker PM (2011) Getting started with TEOS-10 and the Gibbs seawater (GSW) oceanographic 423 Toolbox. Scor/Iapso WG 532, 1–28. Version 3.0 (R2010a). ISBN 978-0-646-55621-5

Orsi AH, Whitworth T, Nowlin WD (1995) On the meridional extent and fronts of the Antarctic circumpolar current. Deep Sea Res Part I Oceanogr Res Pap 42(5):641–673. https://doi.org/10.1016/0967-0637(95)00021-w

Sokolov S, Rintoul SR (2009) Circumpolar structure and distribution of the Antarctic circumpolar current fronts: 1. Mean circumpolar paths. J Geophys Res 114:C11018. https://doi.org/10.1029/2008jc005108

Tarakanov RY (2021) Multi-jet structure of the Antarctic circumpolar current. In: Morozov EG, Flint MV, Spiridonov VA (eds) Antarctic peninsula region of the Southern Ocean, Advances in polar ecology, vol 6. Springer, Cham. https://doi.org/10.1007/978-3-030-78927-5_2

Vargas-Alemañy JA, Vigo MI, García-García D, Zid F (2023) Updated geostrophic circulation and volume transport from satellite data in the Southern Ocean. Front Earth Sci 11:1110138. https://doi.org/10.3389/feart.2023.1110138

Vigo MI, García-García D, Sempere M, Chao BF (2018) 3D geostrophy and volume transport in the Southern Ocean. Remote Sens 10:715. https://doi.org/10.3390/rs10050715

Whitworth T III (1983) Monitoring the transport of the Antarctic circumpolar current at Drake passage. J Phys Oceanogr. https://doi.org/10.1175/1520-0485

Whitworth T III, Peterson RG (1985) Volume transport of the Antarctic circumpolar current from bottom pressure measurements. J Phys Oceanogr 15:810–816. https://doi.org/10.1175/1520-0485(1985)015{\mathsurround=\opskip$<$}0810:vtotac{\mathsurround=\opskip$>$}2.0.co;2

Wunsch C, Gaposchkin EM (1980) On using satellite altimetry to determine the general circulation of the oceans with application to geoid improvement. Rev Geophys 18(4):725–745. https://doi.org/10.1029/rg018i004p00725

Open Access This chapter is licensed under the terms of the Creative Commons Attribution 4.0 International License (http://creativecommons.org/licenses/by/4.0/), which permits use, sharing, adaptation, distribution and reproduction in any medium or format, as long as you give appropriate credit to the original author(s) and the source, provide a link to the Creative Commons license and indicate if changes were made.

The images or other third party material in this chapter are included in the chapter's Creative Commons license, unless indicated otherwise in a credit line to the material. If material is not included in the chapter's Creative Commons license and your intended use is not permitted by statutory regulation or exceeds the permitted use, you will need to obtain permission directly from the copyright holder.

A Pipeline to Explore Transient Signals in GNSS Data: A Preliminary Approach Applied to the Cascadia Subduction Margin

Cristian Garcia, Benjamin Männel, Susanne Glaser, and Jonathan Bedford

Abstract

Throughout the earthquake cycling at fault zones, Earth's crust undergoes deformations. GNSS coordinate time series record linear tectonic motion, seismic displacements, post-seismic decays, and periodic signatures such as non-tidal loading. Any additional motions can be classed as transient tectonic signals, i.e., unexpected accelerations with respect to the standard trajectory model. As the number of permanent stations increases and as time series grow, we are increasingly able to recognise transient tectonic signals. Since some of these suspected tectonic transients have subtle magnitudes or sometimes unusual spatiotemporal features, we need to develop methods for determining which transients are artifacts and which are of tectonic origin.

Here, we investigate the impact of certain GNSS processing choices and how they affect the appearance of transients in the GNSS displacement time series solutions. In this study, we choose data from Cascadia, a region for which the occurrence of transient signals in the GNSS time series is well known. We processed data based from 189 selected stations in network mode for the time span 2015 to 2019. After producing coordinate time series, we then built a pipeline to isolate processing artifacts and tectonic transients, using the regression model-based algorithm known as GrAtSiD (Greedy Automatic Signal Decomposition).

The residuals of GNSS observations show that most sites have a precision of 5 mm to 10 mm. Using the GrAtSiD algorithm, we detected transient signals with velocities exceeding 0.3 mm/day near the ALBH station.

Keywords

GNSS · Processing artifacts · Tectonic geodesy · Transient motion

C. Garcia (✉) · B. Männel
GFZ German Research Centre for Geosciences, Potsdam, Germany
e-mail: cristian.garcia@gfz-potsdam.de

S. Glaser
GFZ German Research Centre for Geosciences, Potsdam, Germany

University of Bonn, Faculty of Agricultural, Nutritional and Engineering Sciences, Institute of Geodesy and Geoinformation, Bonn, Germany

J. Bedford
Institute of Geology, Mineralogy and Geophysics, Ruhr University Bochum, Bochum, Germany

1 Introduction

Tectonic geodesy utilises measurements of surface deformation to track active plate motions and the underlying processes driving earthquakes. Initially reliant on classical methods, advancements in space geodesy, particularly in the Global Navigation Satellite System (GNSS), over the past decades have propelled this field to the forefront of active tectonics, providing insights into various surface displacement phenomena beyond earthquakes (Bürgmann et al. 2013). Earthquake research has categorised various types of earthquakes that produce crustal deformation: tectonic earthquakes, volcanic earthquakes, induced earthquakes, tectonic tremor or nonvolcanic tremor (NVT) according to Obara (2002), and slow earthquakes (Stein and Wysession 2009). Furthermore, slow earthquakes are distinguished by their prolonged duration in comparison to regular seismic events. It is precisely this extended time frame that characterises them as slow earthquakes (Beroza and Ide 2011). There exists a classification system, which differentiates various types driven by time scale (Obara and Kato 2016). Low-frequency tremors (LFEs) and very-low-frequency (VLF) earthquakes produce subtle vibrations and are detected through seismographic recordings. In contrast, slow slip events (SSEs) can be of either long duration, spanning from six months to several years, or short-term, lasting only a few days. Over the past few decades as predominantly the campaign-mode networks have given way to an increasing density of continuous networks at (Stenmark 2014), our capacity to detect SSE and the coupled episodic tremor and slip (ETS) phenomena at plate boundaries has accelerated. These events exhibit a diverse range of durations, magnitudes, and recurrence patterns, and have been observed in various subduction zones worldwide. For instance, studies have provided crucial insights into the nature of SSEs, which can cause the fault to move by tens or hundreds of centimetres over the course of days or years, in regions such as southwest Japan (Hirose et al. 1999), the Cascadia Subduction Zone (Dragert et al. 2001), Alaska (Freymueller et al. 2008), New Zealand (Wallace and Beavan 2010), Costa Rica (Jiang et al. 2012), Chile (Klein et al. 2018), and Greece (Saltogianni et al. 2021). Additionally, the occurrence of ETS has been documented in regions such as the Cascadia subduction zone (Rogers and Dragert 2003) and Japan (Obara et al. 2004), often accompanied by NVT. These observations, made possible by GNSS technology, have significantly advanced our understanding of these complex and intriguing seismic phenomena.

While the term 'transient motion' lacks a universally accepted definition, Riel (2017) suggests that transient deformation signals encompass non-periodic, non-secular accumulation of strain in the crust. Moreover, it is important to note that the origin of transient motion, whether tectonic or anthropogenic, is still a subject of ongoing debate within the scientific community. Therefore, this critical definition should be established based on observed deviations in relation to the trajectory model proposed by Bevis and Brown (2014), who delve into the application and evolution of trajectory models in crustal motion. As discussed in their research, recent breakthrough have led to the development of more versatile trajectory models adept at accommodating accelerating patterns of displacement, including phenomena like postseismic transient deformation. In this context, it is noteworthy that our present study employs a modified extended trajectory model ETM (Eq. 1) as defined by Bedford and Bevis (2018) that allows for the modelling of unexpected transient motions alongside the ETM.

The displacement x over time t is modeled using a combination of linear, seasonal, step, and decay basis functions. The linear part represents the secular velocity m and the constant component d. Seasonal oscillations are captured by coefficients (S_k and C_k), typically with two frequencies (n_k), w_1 and w_2, corresponding to annual and semi-annual periods. Step functions, denoted by H, account for predetermined jump times (t_j) with associated coefficients (b_j). Additionally, GrAtSiD can also identify step functions even if they are not given a-priori. Some jumps may also include decays with coefficients (a_i) and decay constants (T_i) at specific onset locations (n_i), where T_i is the onset time of the decay. The noise (high frequency scatter) in the data, $\xi(t)$, is assumed to follow a normal distribution. In this reformulation, it is adopted exponential transients instead of logarithmic ones and eliminate the explicit reliance on a reference time.

$$x(t) = mt + d + \sum_{k=1}^{n_k}[S_k sin(w_k t) + C_k cos(w_k t)]$$
$$+ \sum_{j=1}^{n_j} b_j H(t-t_j) + \xi(t) + \sum_{r=1}^{n_r}\sum_{i=1}^{n_i} a_i (1 - e^{(-(t-t_r)/T_i)})$$
(1)

In this study, we delve into the influence of specific GNSS data processing strategies on the manifestation of transients in GNSS displacement time series solutions. Cascadia serves as our focal region, renowned for its well-documented occurrences of transients in GNSS records. Our dataset encompasses chosen GNSS station data from a network of 189 stations, both reference and monitoring, spanning the period from 2015 to 2019. This selection should afford us the ability to discern between tectonic transients and potential artifacts introduced through GNSS data processing.

Following this initial overview, Sect. 2 outlines the station selection process and details the chosen processing approach. In Sect. 3, we delve into the derived time series and the estimated horizontal transient velocities. Additionally, Sect. 3 delves into the transient signals detection and their potential classification either tectonic or non tectonic. Finally, in Sects. 4 and 5, we wrap up with key findings, suggestions for further research, and conclusions.

2 Data and Method

2.1 GNSS Data Analysis

The Pacific Northwest Geodetic Array (PANGA) comprises approximately 220 continuous GNSS (cGNSS) sites operated by CWU (Central Washington University) (Array 1996) and the Geodetic Facility for the Advancement of Geoscience (GAGE), a project supported by the National Science Foundation and operated by Earthscope. Both networks were engineered with the aim of capturing the three-dimensional, spatiotemporal variations in crustal strain along the boundary of the North American and Pacific plates. Thus, 123 out of 189 stations are used as monitoring station (Fig. 1) for this study can be accessed from publicly available repositories.[1,2] Moreover, for a well realised geodetic datum we incorporated 66 worldwide selected stations provided by the International GNSS Service (IGS) (Johnston et al. 2017).

For the processing, a network approach was chosen with ambiguity fixing according to Ge et al. (2005) but without orbit determination. Therefore, to reduce the computational and personnel workload we introduced the orbit and clock products provided in the GFZ repro3 solution (Männel et al. 2020; Männel et al. 2021). To ensure consistency, the processing strategy followed the IGS repro3 settings[3] and is described in detail in Table 1. In addition, during the data processing a datum realisation is necessary to overcome the rank defect in the normal equations (NEQs). In this study, the geodetic datum we realised by applying only the No-Net-Rotation (NNR) condition in 3D with respect to the IGS14b positions and velocities of the selected core stations. This realisation of the geodetic datum corresponds to a minimum constraint solutions (e.g. Glaser et al. 2015) and ensures that the relative geometry by the geodetic observations remains.

To assess the effectiveness of our processing strategy, we consider two key factors: the precision of each monitoring station and the reliability of our datum definition. In Fig. 1, we also present the precision of the daily-coordinate estimation for each monitoring station, determined by the mean observation residual over the specified period for each site, i.e., the difference between the observed phase signal and the combination of a priori information and estimated parameters. While the majority of sites exhibit an precision within the range of 5 mm to 10 mm, three of them (BRN3, PUPU, and NINT) have an precision exceeding 1 cm. This lower quality can be attributed to non-systematic errors, specifically multipath effects arising or the presence of tree canopies in forested areas. Nevertheless, it is important to highlight that, for every processed monitoring station, outliers accounted for less than 2.5% of the daily observations over the period observed.

The reliability of our datum definition is crucial for ensuring the accuracy and consistency of our positioning system. This aspect pertains to the stability and precision of the reference frame used to define the positions of monitoring stations. In Fig. 2, we illustrate the variations in datum stability by the translation parameters (TX, TY and TZ) estimated from a 7-parameter Helmert transformations between a priori coordinates and the estimated coordinates at the core stations when the NNR condition is applied to them. Thus, the translation parameters TX, TY, and TZ are shifted on average by −1.9 mm, −2.1 mm, and −2.5 mm respectively from the reference frame origin. As a result, our translation parameter estimates between the reference frame IGS14b used in our products (Table 1) and the combined solution ITRF2020 (Altamimi et al. 2023) are biased by 0.5 mm, 1.2 mm, and −0.7 mm, respectively, with respect to the Helmert parameters estimated between ITRF solutions.[4]

The observed apparent seasonality is expected due to the geocenter motion, which is caused by fluctuations in the mass distribution. However, why this possible seasonal effect can-

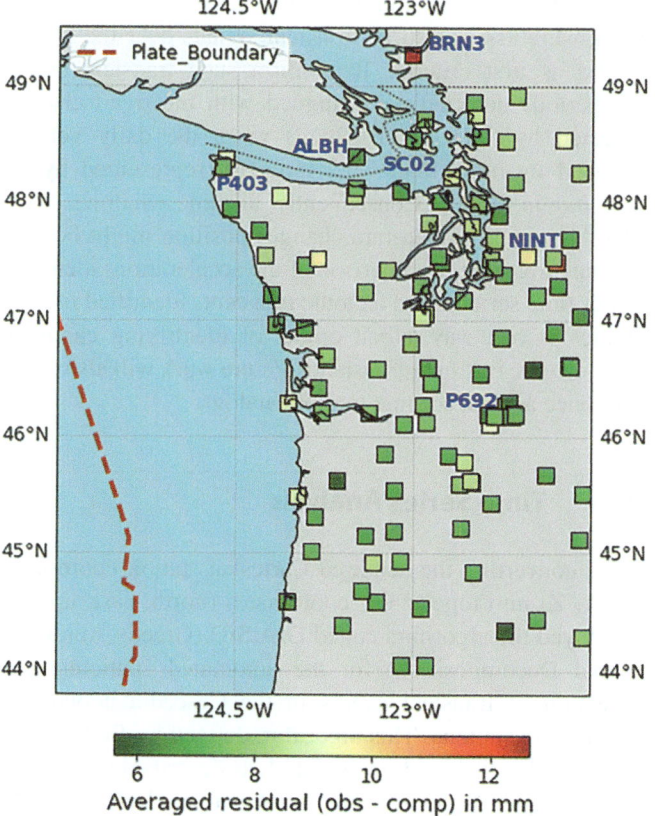

Fig. 1 Station selection: Regional distribution of the selected monitoring station (*squares*) in Cascadia area for testing proposal

---

[1] http://data.unavco.org/archive/gnss/rinex/obsEarthscope.
[2] http://www.geodesy.cwu.edu/pub/data/PANGA.
[3] https://www.igs.org/acc/reprocessing/, last accessed January 2023.
[4] https://itrf.ign.fr/en/solutions/transformations.

Table 1 Summary of estimation and processing strategy by using GFZ software EPOS.P8; time span 2015–2019

Modeling and a-priori information	
Observations	Ionosphere-linear combination formed by undifferenced GPS observations
A priori products	Orbits, clock corrections, Earth rotation parameters from GFZ repro3 solution (Männel et al. 2020; Männel et al. 2021)
Tropospheric correction	Troposphere delays computed with Saastamoinen, mapped with VMF (Böhm et al. 2006)
Ionospheric correction	1st order effect considered with ionosphere-free linear combination, 2nd order correction applied
GNSS phase center	Corrections from dedicated repro3 ANTEX applied (igsR3_2135.atx, http://ftp.aiub.unibe.ch/users/villiger/igsR3_2135.atx)
Gravity potential	GOCO6s up to degree and order 12 (Kvas et al. 2019)
Solid Earth tides	According to IERS 2010 Conventions (Petit and Luzum 2010)
Permanent tide	Conventional tide free
Ocean tide model	FES2014b (Lyard et al. 2006)
Ocean loading	Tidal: FES2014b (Lyard et al. 2006)
Atmospheric loading	Tidal: S_1 and S_2 corrections (Ray and Ponte 2003)
High-frequent EOP model	Desai-Sibois model (Desai and Sibois 2016)
Mean pole tide	Linear mean pole as adopted by the IERS in 2018
Parametrization	
Station coordinates	NNR constraints applied to 66 datum stations (reference frame IGS14b as described in IGS-mail 8026 (https://lists.igs.org/pipermail/igsmail/2021/008022.html))
Troposphere	25 zenith delays; VMF; two gradient pairs per station and day
Receiver clock	Pre-eliminated every epoch
GNSS ambiguities	Ambiguity fixing according to Ge et al. (2005)
Products	
Station coordinates	Provided in daily SINEX files
Troposphere	ZTD and gradients are provided in daily TROP SINEX files (version 2.00)

not be seen any more from 2018 in TX and Ty component is unclear. Furthermore, applying a No-Net-Translation (NNT) condition to the network alignment is not needed, otherwise it may risk overconstraining the solution.

2.2 Forming Time Series

As discussed in Sect. 1, Cascadia has already been extensively investigated in previous studies for the detection of transient motion. To prepare our time series for further analysis, we compared our results using the IGS14b reference frame, with those obtained from the Nevada Geodetic Laboratory (NGL) (Blewitt et al. 2018). We conducted two steps for the purpose of direct comparison. Firstly, we applied a standard Hampel filter (Pearson et al. 2016) with a 15-day window to our entire GNSS displacement time series at each monitoring station to remove outliers. As a result, the daily observation numbers in our dataset for the monitoring stations was reduced by an average between 3% and 5%.

In Fig. 3, we illustrate the difference between the GFZ solution and the NGL solution for the ALBH station located at ($-123.29°$E, $48.23°$N). When comparing the average root mean square (rms) of both solution, the discrepancies amount to 0.02 mm, and 0.43 mm in the east, and north components, respectively after removing the common mode error (CME). These variances suggest that our solutions are equivalent in terms of repeatability to the NGL. Changes in position with respect to the linear motion larger than 5 mm within one week were asserted to be transient signals over the studied period as first criteria. To confirm such transient signal, the periods detected were aligned with the trend changes detected by Saux et al. (2022) when the daily velocity dropped by more than -0.2 mm/day (represented by the *blue bar* in Fig. 3). Consequently, all time windows out of blue bars with sudden rate changes position might be non-tectonic transient. Comparison of the accelerations identified in our time series to the tectonic transients identified in other studies is one way, albeit crude, of identifying candidate non-tectonic (artifact) transients. Future work will also cross-reference against seismic tremor catalogs.

3 Time Series Analysis

After converting the acquired Cartesian station coordinates (X, Y, Z) into topocentric coordinates (north, east, up), we employed the algorithm called GrAtSiD (Greedy Automatic Signal Decomposition) for the automated spatiotemporal detection of transient events in the Cascadia network of monitoring stations. This algorithm defined by Bedford and Bevis (2018), is based on the ETM, as described in Sect. 1.

Consequently, for our time series analysis, by using GrAtSiD,[5] according to Bedford and Bevis (2018) transient signals are identified by using a regression-based approach to decompose time series data into three parts: background seasonal motion, linear and transient motion, and the residual

---
[5] https://github.com/TectonicGeodesy-RUB/Gratsid.

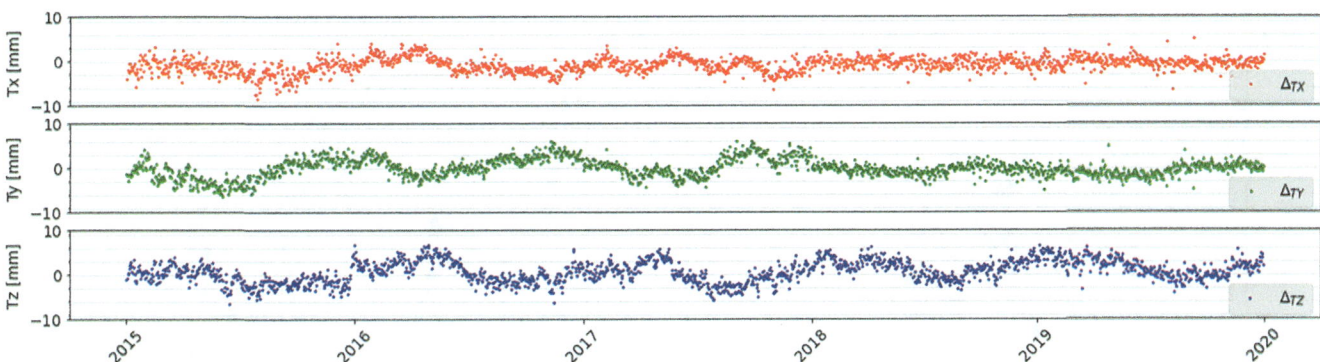

Fig. 2 Time series of the translation parameters TX (red), TY (green), and TZ (blue) computed from a 7-parameter Helmert transformation between a priori coordinates and the derived coordinates for core stations based on daily solutions between 2015 and 2019

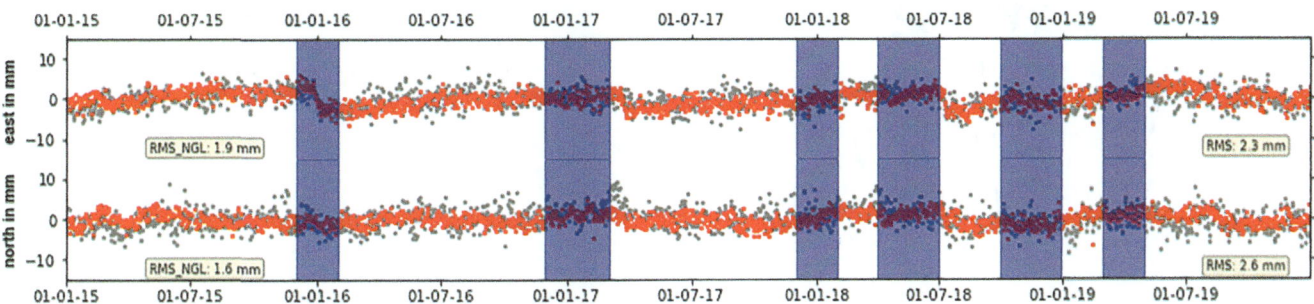

Fig. 3 Comparison of detrended GNSS time series displacement between the NNR solution (grey) after Hampel filter, and the NGL solution (red) for ALBH station, Canada. Besides, (*blue bar*) highlight automatic detection of position reversals inferred to reflect transient signals

noise. The algorithm utilises a greedy approach to fitting the time series using a minimum number of multitransients (sparse functions) in addition to the permanent time functions in a linear regression. Moreover, it adopts the multitransient as a versatile function for modeling transient motion over a range of time scales, allowing the extraction of transient signals with varying characteristic time scales and onset times. This method enables the algorithm to effectively detect and extract the transient signals present in the data.

For tectonic discontinuities, we utilized information from the USGS earthquake catalogue, considering earthquakes with a magnitude larger than 5 Mw within the area of investigation. For non-tectonic discontinuities, i.e., changes related to GNSS antenna or receiver, we obtained the necessary information directly from the log files retrieved from the NGL web site.[6]

Following the process outlined in the pipeline in Fig. 4, we estimated the velocities with respect to tectonic motion, which is the combination of the polynomial and transient signals, and subsequently differentiated. In our case, to estimate the velocity at each station, we computed the median of those 10 solutions obtained for each time series. Thus, the deviation for each day from the average velocity over a certain time period corresponds to the transient velocity.

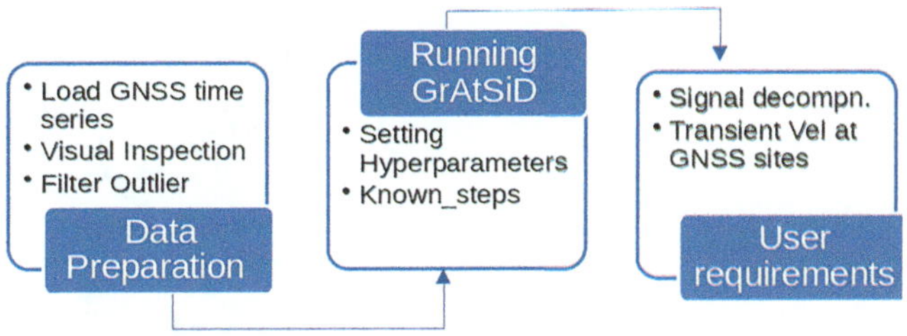

Fig. 4 Pipeline for automatic transient detection by GrAtSid

On 7 January 2016, a date contained in the Saux catalogue, most of the monitoring stations within buffer of no more than 30 km from the ALBH station (−123.487°W, 48.390°N) exhibit a displacement variation in the position reversals to the interseismic motion (Fig. 5a). Hence, the occurrence of a transient motion in this area is confirmed. However, in the stations below a latitude of 47°N, we observe a position reversals toward the south. It may be plausible to consider the possibility that these velocities are linked to transient movements along the fault due to underlying geodynamic processes. On the other hand, for the second time window, in January 2017, (Fig. 5b), transient velocities are observed at some stations (e.g. P692) close to the Mount St. Helens National Volcanic Area, which has a

---

[6]http://geodesy.unr.edu/NGLStationPages/steps.txt.

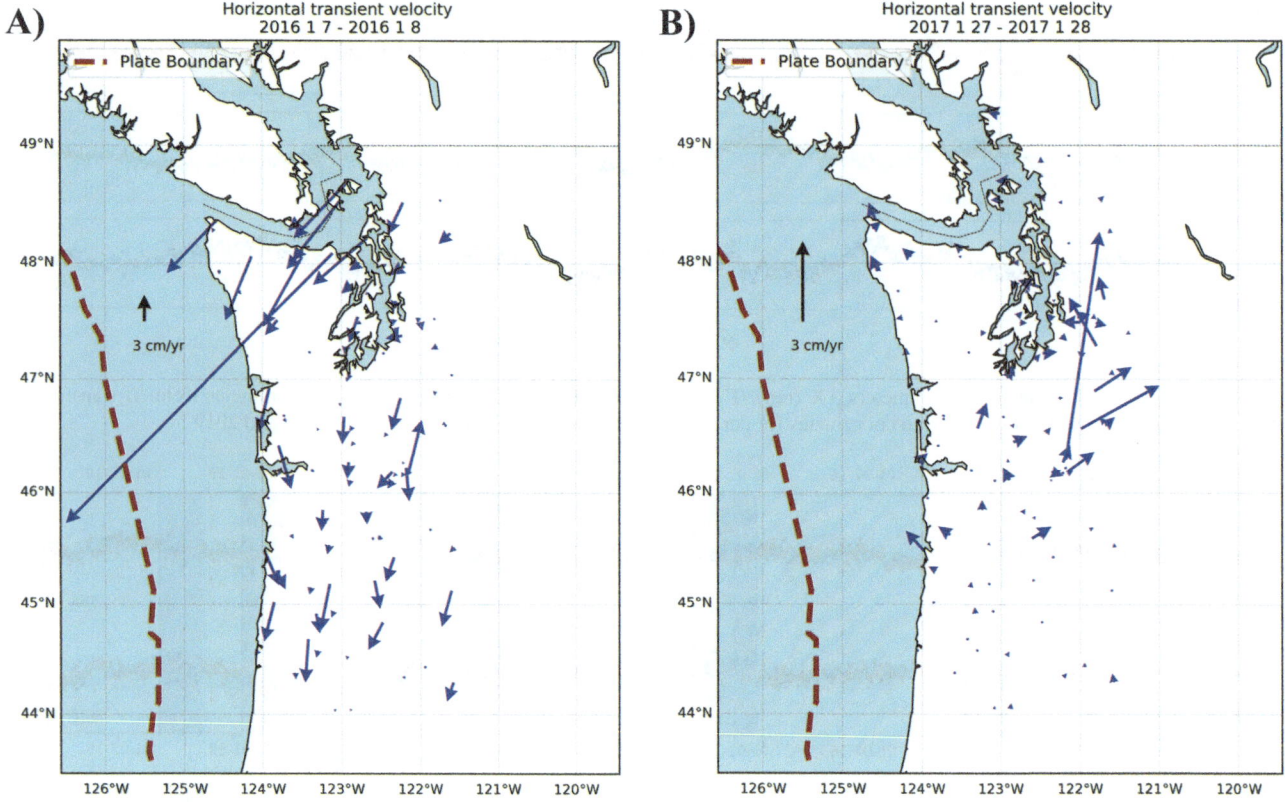

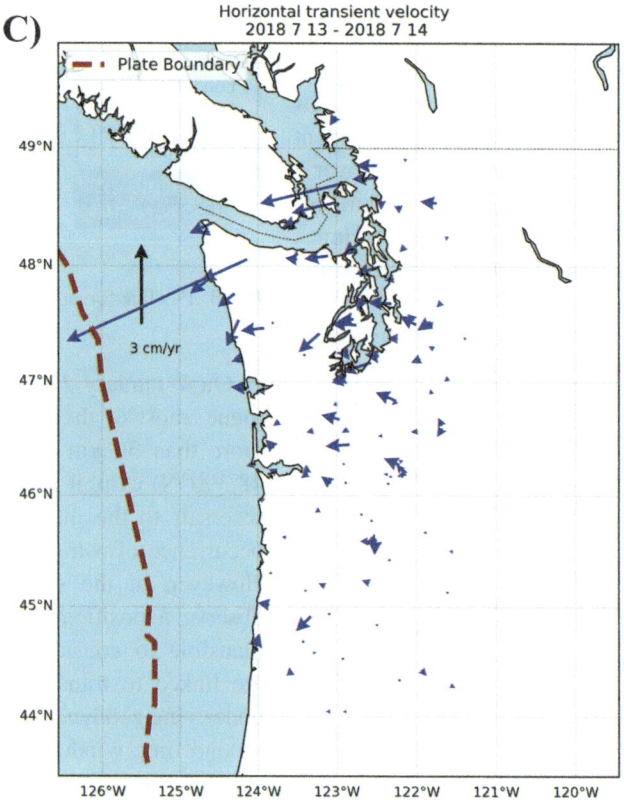

Fig. 5 Tectonic transient signal automatically detected by GrAtSiD

tectonically driven displacement rate of 6 mm/yr,[7] so they could be explained by some volcanic deformation. Finally, in November 2018 (Fig. 5c), besides ALBH, the stations P403, and SC02 show both a velocity of 0.3 mm/day and epoch which is equivalent to that found by Saux et al. (2022) for ALBH; we confirm that such sites show a tectonic transient.

In contrast to the tectonic-driven transient motion discussed earlier, Fig. 6a offers a distinct perspective on non-tectonic transient movements within the observed region. The plot clearly shows that all stations have a positional reversal towards the south from beginning of August in 2015. This displacement may be attributed to a non-tectonic transient motion. Furthermore, upon watching the available animation,[8] we can observe that the effect is less pronounced in November 2015. However, it becomes visible again in December 2015 and persists until the middle of February 2016, when most monitoring stations experience a positional reversal towards the west, along with an increase in transient velocity, signifying a tectonic transient motion as was identified in that period (Fig. 4).

In the case of Fig. 6b, in September 2017, only a group of stations located along to the Cascade Mountains show an increase in transient velocity directed towards the south. Meanwhile, in Fig. 6c, at the beginning of May 2019 the transient velocities larger than 0.2 mm/day are observed in both the mountainous area and stations close to the coastline south of 46°N. Such an epoch was not included in the catalogue defined by Saux. Thus, it could be assumed to be as non-tectonic transient due to some degradation in the quality of the GNSS data, which will be part of the next stage of our investigation.

4 Discussion

Our study encompasses from the processing of GNSS raw data to the tectonic analysis of the trends apparent in the processed time series. The analysis of GPS observation residuals, averaging at approximately 7 mm and devoid of discernible clusters, underscores the critical importance of factoring in the local environment in our assessments.

It is imperative to recognise the central role of the alignment of the regional network to the global reference by the conditions established in the geodetic datum definition. Grounded in the concept of plate tectonics, these models provide a crucial framework for interpreting changes in station positions and enabling precise alignment with the global reference frame.

---

[7]https://www.usgs.gov/volcanoes/mount-st.-helens/science/deformation-monitoring-mount-st-helens, accessed May 2024.

[8]Refer to the supplementary material for the video titled 'cascadiadailyvelo2023259.webm 5.

While our study achieves the detection of transient signals in GNSS data, it does not yet accomplish their discrimination. Distinguishing between tectonic signals, indicative of geodynamically-driven changes in the Earth's crust, and non-tectonic signals, originating from processing artifacts, represents a central activity. This filtering process is essential to avoid misinterpretations and provide an accurate insight into the Earth's crustal evolution.

Our findings are in line with previous research emphasising the importance of environmental factors in geodetic analyses, as well have GNSS-related indicators (data quality, observation residuals, troposphere delay) to identify potential GNSS artifacts appearing as transient signals. This consistency strengthens the validity of our results and contributes to the growing body of knowledge in this field.

While our study has provided valuable insights, it is not without limitations. Future research requires investigating specific aspects, such as enhancing processing strategies or exploring additional environmental variables that may influence GNSS data to define a robust criterion for identifying non-tectonic transient. Additionally, the integration of complementary geophysical methods could offer a more comprehensive understanding of crustal dynamics.

5 Conclusion

In the course of our investigation, we undertook outlined a processing pipeline, spanning from the initial acquisition of raw GNSS data to the derivation of crucial geophysical parameters.

Furthermore, drawing an analogy to a supply chain, it's evident that geodesists play a pivotal role in the upstream phase (Fig. 7). This involves tasks ranging from station setup in the field to the meticulous collection and subsequent processing of RINEX data, culminating in the generation of displacement time series. Throughout this stage, a range of criteria, processing strategies, and diverse models are implemented to transform observed GNSS data into a time series reflecting changes in station positions. This process is crucial for the accurate interpretation of tectonic signals, particularly transient signals, which can be simplistically understood as unexpected fluctuations in the Earth's crust over a short duration.

Looking ahead, future research could further improve our understanding of transient plate boundary processes by conducting a thorough analysis of the identified transients GNSS displacement signals and, cross-referencing them with relevant GNSS-related metrics, including but not limited to data quality assessments, residual observations, and tropospheric delays. Through this process, we aim to pinpoint any potential instances of GNSS-related anomalies that could be mistaken for genuine transient signals.

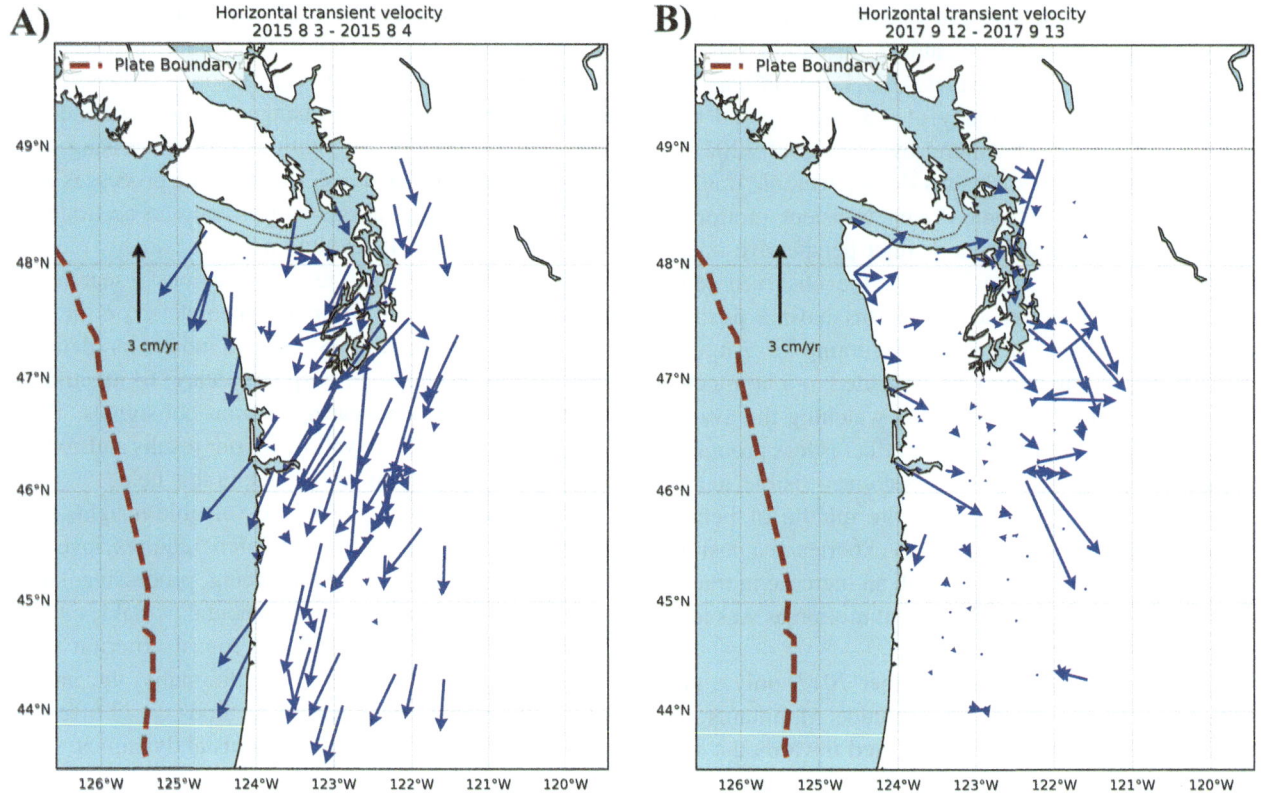
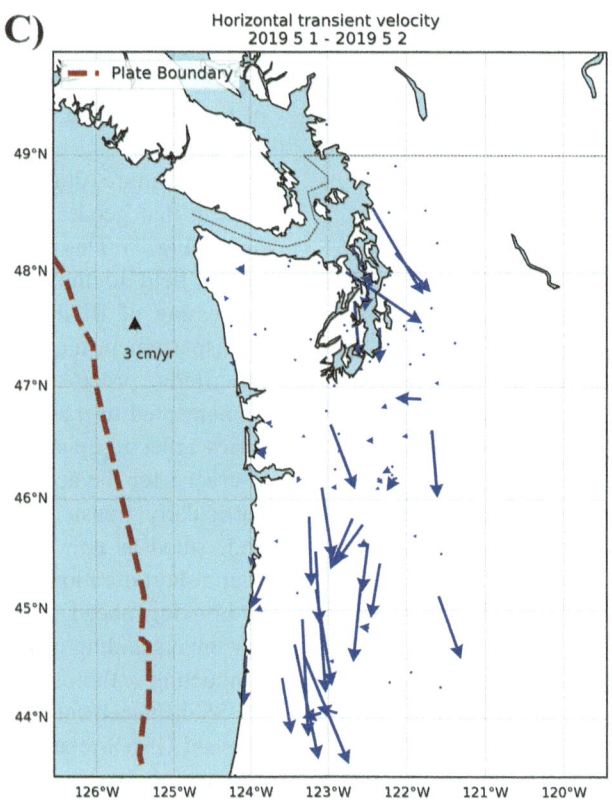

Fig. 6 Non tectonic transient signal automatically detected by GrAtSiD

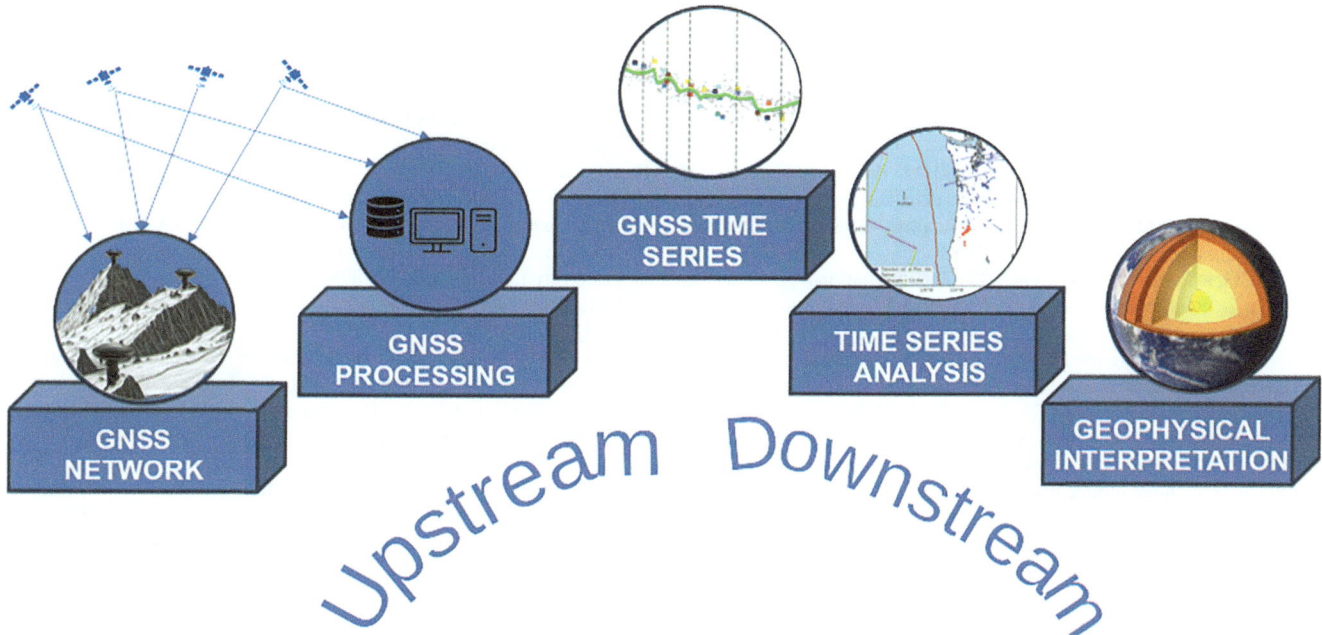

Fig. 7 Upstream and downstream operations for detection of transient signal in Tectonic-Geodesy. Include element from "Earth cutaway.png" by Charles, licensed under [CC BY-SA 3.0] (https://creativecommons.org/licenses/by-sa/3.0/)

Acknowledgements We would like to thank the IGS, the Geodesy laboratory of the CWU and EarthScope for providing data and products. This investigation is part of TectoVision project funded by the European Union (ERC, TectoVision, 101042674). Views and opinions expressed are however those of the author(s) only and do not necessarily reflect those of the European Union or the European Research Council Executive Agency. Neither the European Union nor the granting authority can be held responsible for them.

Author Contribution CG, BM, JB and SG designed the study. BM and CG developed the processing framework. CG implemented the displacement time series analysis and prepared the discussion. CG wrote the paper with contributions from BM, JB and SG. All the authors joined the research discussion and gave their feedback on the manuscript.

Data Availability Statement All GNSS data are available at GFZ, PANGA, EarthScope, and NGL data centers, respectively.

Supplementary Material The supplementary material for this study can be found online at: https:nextcloud.gfz-potsdam.de/s/r5FKikfEai3yi5f.

References

Altamimi Z, Métivier L, Rebischung P, Collilieux X, Chanard K, Barnéoud J (2023) Itrf2020 plate motion model. Geophys Res Lett 50(24):e2023GL106,373

Array PNG (1996) Gps/gnss network and geodesy laboratory: Central washington university, other/seismic network. Int Fed of Digital Seismograph Networks. https://doi.org/10.7914/SN/PW

Bedford J, Bevis M (2018) Greedy automatic signal decomposition and its application to daily gps time series. J Geophys Res Solid Earth 123(8):6992–7003

Beroza GC, Ide S (2011) Slow earthquakes and nonvolcanic tremor. Annu Rev Earth Planet Sci 39:271–296

Bevis M, Brown A (2014) Trajectory models and reference frames for crustal motion geodesy. J Geodesy 88:283–311

Blewitt G, Hammond WC, Kreemer C (2018) Harnessing the GPS data explosion for interdisciplinary science. Eos 99. https://doi.org/10.1029/2018EO104623

Böhm J, Werl B, Schuh H (2006) Troposphere mapping functions for GPS and VLBI from European Centre for medium-range weather forecasts operational analysis data. J Geophy Res 111(B2):B02,406. https://doi.org/10.1029/2005JB003629

Bürgmann R, Thatcher W, Bickford M (2013) Space geodesy: A revolution in crustal deformation measurements of tectonic processes. Geologic Soc Am Special Paper 500:397–430

Desai SD, Sibois AE (2016) Evaluating predicted diurnal and semidiurnal tidal variations in polar motion with gps-based observations. J Geophys Res Solid Earth 121(7):5237–5256. https://doi.org/10.1002/2016JB013125

Dragert H, Wang K, James TS (2001) A silent slip event on the deeper cascadia subduction interface. Science 292(5521):1525–1528

Freymueller JT, Woodard H, Cohen SC, Cross R, Elliott J, Larsen CF, Hreinsdóttir S, Zweck C, Haeussler P, Wesson R, et al (2008) Active deformation processes in alaska, based on 15 years of gps measurements. Active Tecton Seismic Potential Alaska 179:1–42

Ge M, Gendt G, Dick G, Zhang FP, Reigber C (2005) Impact of GPS satellite antenna offsets on scale changes in global network solutions. Geophys Res Lett 32(6):L06,310. https://doi.org/10.1029/2004GL022224

Glaser S, Fritsche M, Sośnica K, Rodríguez-Solano C, Wang K, Dach R, Hugentobler U, Rothacher M, Dietrich R (2015) A consistent combination of GNSS and SLR with minimum constraints. J Geod, 1–16. https://doi.org/10.1007/s00190-015-0842-0

Hirose H, Hirahara K, Kimata F, Fujii N, Miyazaki S (1999) A slow thrust slip event following the two 1996 hyuganada earthquakes beneath the bungo channel, southwest Japan. Geophys Res Lett 26(21):3237–3240

Jiang Y, Wdowinski S, Dixon TH, Hackl M, Protti M, Gonzalez V (2012) Slow slip events in Costa Rica detected by continuous GPS observations, 2002-2011. Geochem Geophys Geosyst **13**, Q04006. https://doi.org/10.1029/2012GC004058

Johnston G, Riddell A, Hausler G (2017) The international gnss service. Springer handbook of global navigation satellite systems, pp 967–982

Klein E, Duputel Z, Zigone D, Vigny C, Boy JP, Doubre C, Meneses G (2018) Deep transient slow slip detected by survey gps in the region of atacama, chile. Geophys Res Lett 45(22):12–263

Kvas A, Mayer-Gürr T, Krauss S, Brockmann JM, Schubert T, Schuh WD, Pail R, Gruber T, Jäggi A, Meyer U (2019) The satellite-only gravity field model goco06s. https://doi.org/10.5880/ICGEM.2019.002. http://dataservices.gfz-potsdam.de/icgem/showshort.php?id=escidoc:4081892

Lyard F, Lefevre F, Letellier T, Francis O (2006) Modelling the global ocean tides: modern insights from FES2004. Ocean Dynam 56(5–6):394–415. https://doi.org/10.1007/s10236-006-0086-x

Männel B, Brandt A, Bradke M, Sakic P, Brack A, Nischan T (2020) Status of IGS Reprocessing Activities at GFZ, Springer, Berlin, Heidelberg, pp 1–7. https://doi.org/10.1007/1345_2020_98

Männel B, Brandt A, Bradke M, Sakic P, Brack A, Nischan T (2021) GFZ repro3 product series for the International GNSS Service (IGS). https://doi.org/10.5880/GFZ.1.1.2021.001. GFZ Data Services

Obara K (2002) Nonvolcanic deep tremor associated with subduction in southwest Japan. Science 296(5573):1679–1681

Obara K, Kato A (2016) Connecting slow earthquakes to huge earthquakes. Science 353(6296):253–257

Obara K, Hirose H, Yamamizu F, Kasahara K (2004) Episodic slow slip events accompanied by non-volcanic tremors in southwest Japan subduction zone. Geophys Res Lett **31**, L23602. https://doi.org/10.1029/2004GL020848

Pearson RK, Neuvo Y, Astola J, Gabbouj M (2016) Generalized hampel filters. EURASIP J Adv Signal Process 2016:1–18

Petit G, Luzum B (2010) IERS Conventions (2010). IERS Technical Note 36. Verlag des Bundesamts für Kartographie und Geodäsie, Frankfurt am Main. iSBN 3-89888-989-6

Ray R, Ponte R (2003) Barometric tides from ECMWF operational analyses. Ann Geophys 21(8):1897–1910

Riel B (2017) Automatic decomposition of geodetic time series for studies of ground deformation. California Institute of Technology

Rogers G, Dragert H (2003) Episodic tremor and slip on the cascadia subduction zone: The chatter of silent slip. Science 300(5627):1942–1943

Saltogianni V, Mouslopoulou V, Dielforder A, Bocchini GM, Bedford J, Oncken O (2021) Slow slip triggers the 2018 mw 6.9 zakynthos earthquake within the weakly locked hellenic subduction system, greece. Geochem Geophys Geosyst 22(11):e2021GC010,090

Saux JP, Molitors Bergman EG, Evans EL, Loveless JP (2022) The role of slow slip events in the cascadia subduction zone earthquake cycle. J Geophys Res Solid Earth 127(2):e2021JB022,425

Stein S, Wysession M (2009) An introduction to seismology, earthquakes, and earth structure. Wiley

Stenmark J (2014) Precise to a fault: How gps revolutionized seismic research. Earth 59(5):32–39

Wallace LM, Beavan J (2010) Diverse slow slip behavior at the Hikurangi subduction margin. New Zealand J Geophys Res **115**, B12402. https://doi.org/10.1029/2010JB007717

Open Access This chapter is licensed under the terms of the Creative Commons Attribution 4.0 International License (http://creativecommons.org/licenses/by/4.0/), which permits use, sharing, adaptation, distribution and reproduction in any medium or format, as long as you give appropriate credit to the original author(s) and the source, provide a link to the Creative Commons license and indicate if changes were made.

The images or other third party material in this chapter are included in the chapter's Creative Commons license, unless indicated otherwise in a credit line to the material. If material is not included in the chapter's Creative Commons license and your intended use is not permitted by statutory regulation or exceeds the permitted use, you will need to obtain permission directly from the copyright holder.

Emphasizing the Value of Geodesy to Science and Society Through IAG-GGOS

Martin Sehnal, Laura Sánchez, Detlef Angermann, Allison Craddock, Basara Miyahara, and Lena Steiner

Abstract

Without geodesy – the science that determines the shape of the Earth, its gravity field, and its rotation as functions of space and time – accurate positioning would not be possible. Geodetic observation techniques, analysis infrastructure, and products provided by the International Association of Geodesy (IAG) are fundamental to Earth system research and are the backbone for location-based applications. IAG's Global Geodetic Observing System (GGOS) is a collaborative effort of the global geodesy community aimed at providing consistent and openly accessible geodetic Earth observations. In addition, GGOS supports activities and projects that promote the importance of geodesy to science and society. This paper summarizes recent GGOS initiatives and achievements in strengthening the awareness of the value of geodesy.

Keywords

Film · Geodesy · Geodetic · GGOS · IAG · Observations · Outreach · Portal · Website

1 Value of Geodesy to Science and Society

Geodesy, the scientific discipline dedicated to determining Earth's shape, gravity field, and rotation across space and time, utilizes advanced instruments and analysis techniques to provide essential information about the Earth. Tools generated by international collaborations in geodesy enable the detection of temporal variations, ranging from large-scale phenomena to subtle deformations, with increasing spatial and temporal precision, heightened accuracy, and reduced latency. When used to improve climate change monitoring is evident, geodesy enables precise tracking of sea level changes, ice cap melting, groundwater variations, weather forecasting, and monitoring shifts in the Earth's surface, such as tectonic motions, landslides, mudslides, or permafrost retreat (see Fig. 1).

To ensure the interoperability, comparability, stability, and consistency of geodetic measurements and products, the establishment of geodetic reference frames is imperative. These reference frames serve as the cornerstone for

M. Sehnal (✉)
BEV, Austrian Federal Office of Metrology and Surveying, Vienna, Austria
e-mail: martin.sehnal@bev.gv.at

L. Sánchez · D. Angermann
DGFI-TUM, Deutsches Geodätisches Forschungsinstitut, Technical University of Munich, Munich, Germany
e-mail: lm.sanchez@tum.de; detlef.angermann@tum.de

A. Craddock
NASA Jet Propulsion Laboratory, California Institute of Technology, Pasadena, CA, USA
e-mail: allison.b.craddock@jpl.nasa.gov

B. Miyahara
GSI Japan, Geospatial Information Authority, Osaka, Japan
e-mail: miyahara-b96ip@mlit.go.jp

L. Steiner
TU Wien, Technical University of Vienna, Vienna, Austria
e-mail: e12009660@student.tuwien.ac.at

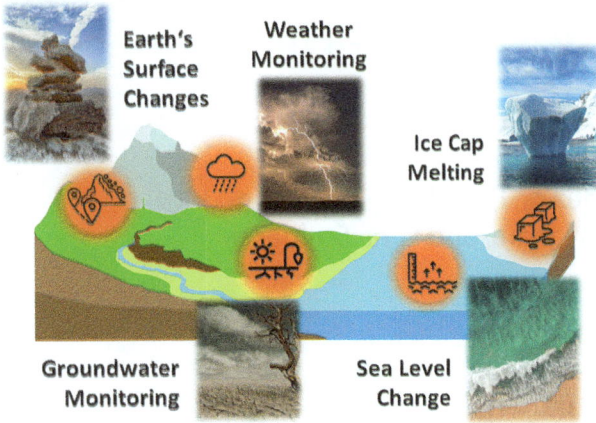

Fig. 1 Contribution of geodesy to climate change monitoring

determining time-dependent coordinates of Earth's points or objects and describing the Earth's movement in relation to space. The International Association of Geodesy (IAG) plays a pivotal role by providing a range of geodetic products to the public free of charge. This foundational support underscores the collaborative efforts within the geodetic community to contribute valuable data and insights crucial for scientific understanding and societal benefit.

2 Importance of IAG's Global Geodetic Observing System (GGOS)

GGOS stands as a collaborative initiative within the global geodesy community, dedicated to the observation and monitoring of the Earth System. Its pivotal work facilitates the generation and sharing of essential Earth observations crucial for monitoring, mapping, and comprehending changes in Earth's shape, rotation, and mass distribution. Emphasizing the global geodetic reference frame as the foundational bedrock for both monitoring and accurately interpreting global change processes, GGOS actively advocates for the requisite geospatial infrastructure to ensure global coherence and sustainability.

In the pursuit of sustaining the enduring significance of these fundamental geodetic contributions, GGOS engages in close collaboration with various components of its parent organization, the IAG. The technical services of the IAG play a crucial role in operating the monitoring infrastructure and providing solutions that form the backbone of all GGOS contributions. Simultaneously, the IAG Commissions, Inter-Commission Committees, and dedicated projects contribute expertise and theoretical support, addressing critical scientific topics (Fig. 2).

Moreover, GGOS extends its support to the IAG by fortifying external and interdisciplinary relationships, actively participating in broader geospatial information communities. This involvement encompasses key entities such as the United Nations (UN) Committee of Experts on Global Geospatial Information Management (UN-GGIM), its Sub-committee on Geodesy, and the UN Global Geodetic Center of Excellence (UN-GGCE). GGOS takes on a primary role in endorsing initiatives and projects that effectively communicate the significance of geodesy to society. Additionally, it strives to contribute to identifying, understanding, and resolving challenges faced by the global geospatial community.

3 Outreach in the Focus: The GGOS Coordinating Office

The GGOS Coordinating Office (CO), has been hosted by the Austrian Federal Office of Metrology and Surveying (BEV, *Bundesamt für Eich- und Vermessungswesen*) in Vienna since 2016. The CO is responsible for all administrative aspects of GGOS, ensuring the flow of information, maintaining documentation of GGOS activities, and managing specific support functions that improve harmonization across all areas of GGOS (Fig. 2). This includes coordinating interactions with other IAG components as well as the organization and support of multi-stakeholder scientific and business meetings to ensure that GGOS components provide consistent and ongoing input to IAG and its stakeholder community.

The GGOS CO also actively sustains external relations and engagement with various stakeholders, either representing GGOS or by appointment of the IAG. These stakeholders include significant entities like the Group on Earth Observations (GEO), the Committee on Earth Observation Satellites (CEOS), the International Organization for Standardization (ISO), and the International Science Council (ISC) World Data System (WDS). In this capacity, the GGOS CO seeks avenues to establish connections between geodesy and key United Nations frameworks and other policy engagement mechanisms. This includes initiatives such as the Sendai Framework for Disaster Risk Reduction, the UN Integrated Geospatial Information Framework, and UN Sustainable Development Goals.

The GGOS CO is also responsible for developing, maintaining, and managing the GGOS website and social media channels, as well as other public outreach activities. Significant progress has been made in recent years, culminating in a new website (www.ggos.org) designed to optimize usability and create a comprehensive geodetic information platform (Fig. 3) to serve a diverse and growing user community.

The implementation of an improved communications strategy has significantly elevated GGOS's visibility. This enhanced approach involves utilizing various media outlets,

Emphasizing the Value of Geodesy to Science and Society Through IAG-GGOS

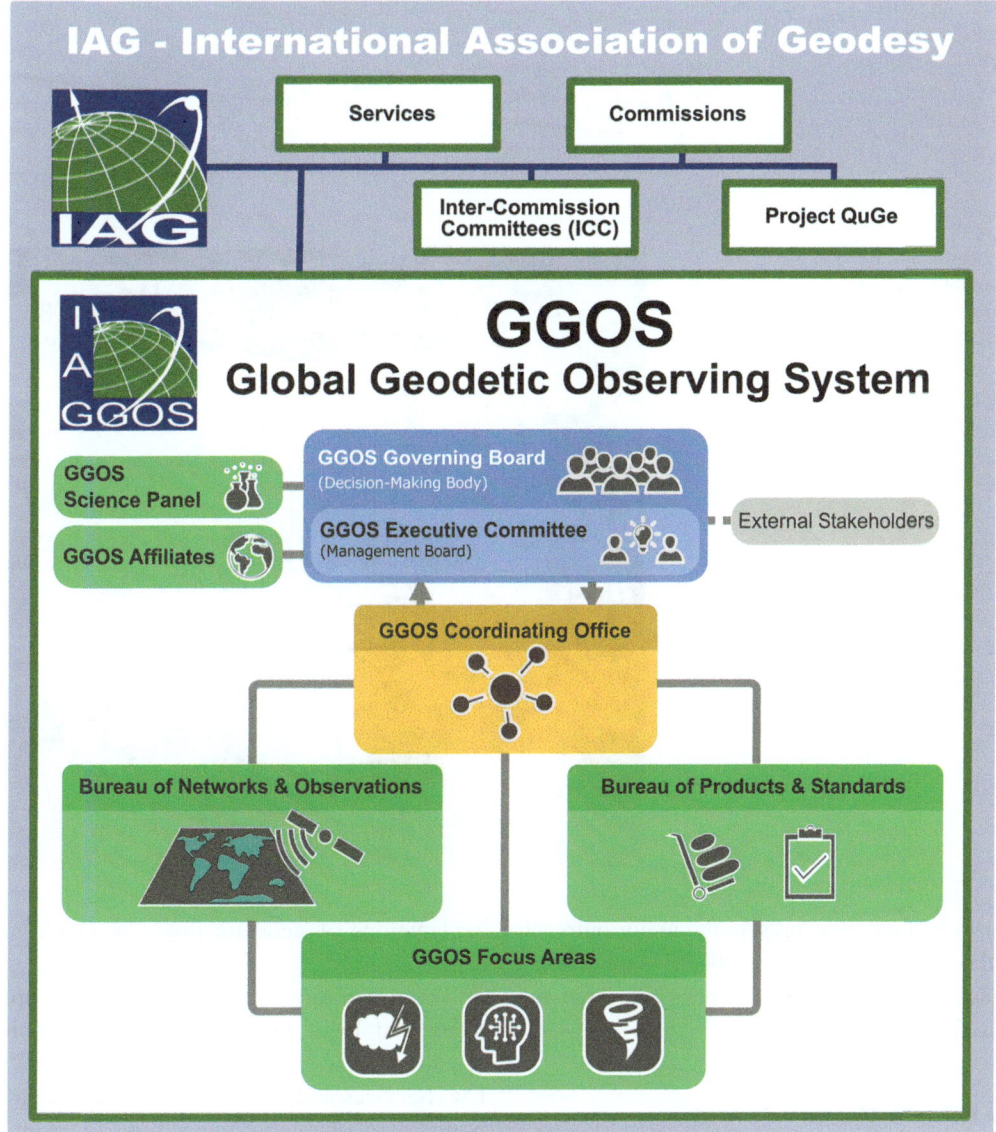

Fig. 2 Organizational structure of IAG and GGOS

Fig. 3 Geodetic information platform www.ggos.org

including the GGOS blog, GGOS newsletter, and platforms such as X/Twitter (@IAG_GGOS), LinkedIn (iag-ggos), YouTube (@iag-ggos), and Facebook (iagGGOS). This has led to a substantial surge in awareness of GGOS. Over the past years, both website traffic and the number of registered users across social media channels have witnessed a sevenfold increase, as depicted in Fig. 4. This underscores the effectiveness of the revised strategy in engaging a wider audience and amplifying GGOS's reach.

In addition to these online outreach successes, GGOS has raised the awareness of geodesy within other geosciences at international conferences. For instance, GGOS collaborated with the IAG Communication and Outreach Branch (COB) to host a booth at the International Union of Geodesy and Geophysics (IUGG) 2023 General Assembly exhibition in Berlin (Fig. 5). As a result, scientific stakeholders from outside of the geodesy community were introduced to the field and the benefits of geodesy through the guidance of IAG and GGOS.

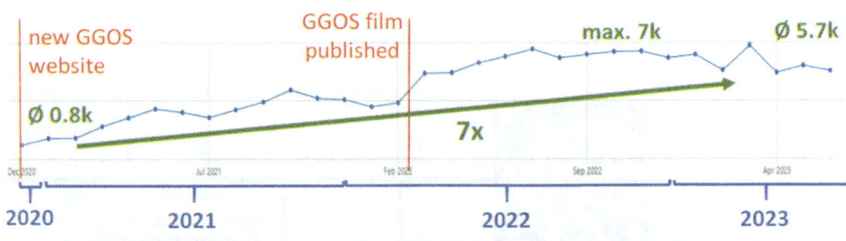

Fig. 4 Visitors by month on the GGOS website www.ggos.org over the last 2.5 years

Fig. 5 IAG-GGOS booth at the IUGG General Assembly 2023 in Berlin, Germany (from left to right: Martin Sehnal, Yannic Öhlknecht)

Sections 4 through 6 will now present the major outreach projects led by the GGOS CO that have been – or will be – critical to increasing the visibility of geodesy through the work of IAG and GGOS.

4 GGOS Information Platform

In late 2020, the GGOS website (www.ggos.org) underwent a redesign aimed at enhancing the discoverability and usability of geodetic data and products. The primary goal was to support activities and initiatives that effectively communicate the value of geodesy to both science and society.

The website serves as a comprehensive information platform, continually refined to provide deep insights into the IAG's geodetic observation system and its structure. This iterative process ensures that the website remains a dynamic resource, centered on key aspects such as the IAG's observation techniques (ggos.org/obs), geodetic products (ggos.org/products), and IAG Services (ggos.org/services). To enhance user experience, visually appealing images (Fig. 6) are strategically placed to guide users to illustrative explanations of geodetic products or observation methods (Fig. 7).

The descriptions of observations and products are complemented by a large selection of web links leading to scientific descriptions and data repositories provided by the IAG Services and other additional data sources. This is supported by clickable reference links to scientific explanations and data repositories provided, by IAG Services and other data sources to further enhance the descriptions of geodetic observations and products.

As the custodian of the GGOS website, the GGOS CO has collaborated extensively with members of the GGOS Bureau of Products and Standards (BPS) (Angermann et al. 2022), the GGOS Bureau of Networks and Observations (BNO), the GGOS Science Panel, and other key members of the geodetic science community to establish and launch this information platform. The contributions of IAG Services and other geodetic product vendors are gratefully acknowledged. In this way, the GGOS website helps to raise awareness of geodesy and to promote the IAG and GGOS globally and across disciplines (Sehnal et al. 2023).

Fig. 6 Interactive overview graphic of Earth components to which geodesy contribute (ggos.org/products)

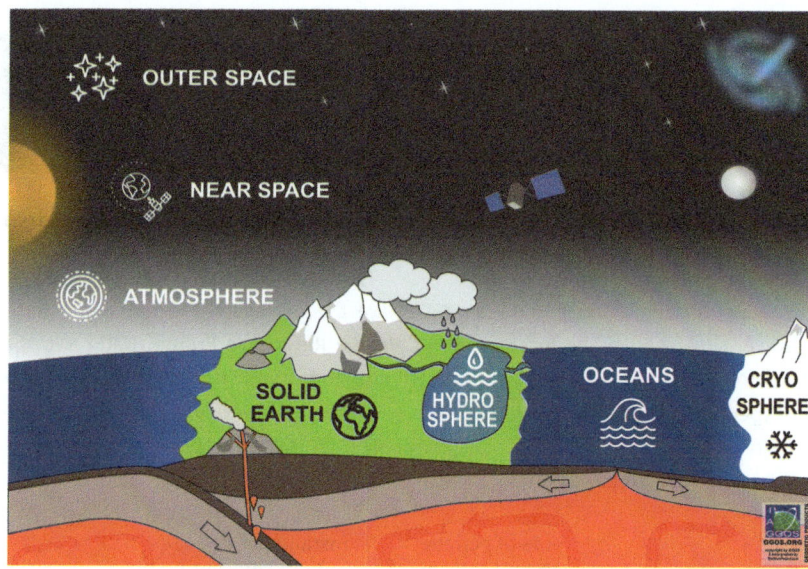

5 GGOS Films

Effective public relations hinge on reaching the target audience through diverse channels and media. In 2021, GGOS embarked on broadening its outreach efforts by utilizing its YouTube channel (youtube.com/@iag-ggos) to share both pre-existing and newly created films centered around geodesy.

A significant milestone in 2022 was the collaborative effort between the GGOS CO and key GGOS contributors to produce the short film "Discover GGOS and Geodesy" (Fig. 8). This video aimed to articulate the primary objectives of GGOS and elucidate the importance of geodesy to non-geodesists (Sehnal et al. 2022). With the dedicated support of numerous volunteers, the film was successfully translated into 14 languages, including English, Chinese, Spanish, French, Arabic, Portuguese, German, Farsi, Japanese, Italian, Dutch, Bulgarian, Turkish, and Greek. This substantial response, with all versions collectively amassing over 19,000 views and garnering numerous positive comments, underscores the widespread interest in such educational films.

Buoyed by this success and propelled by a commitment to expanding outreach initiatives, GGOS has made the strategic decision to develop additional films illustrating geodetic products and observation methods in the future. Taking the initial step, a film elucidating one of the fundamental products of geodesy – the terrestrial reference frame – was recently produced. Titled "Terrestrial Reference Frames - Connecting the World through Geodesy" (see Fig. 9), this film was launched on the GGOS YouTube Channel in September 2023 and is accessible in multiple languages, marking another milestone in GGOS's extensive international efforts to provide high-quality educational content about geodesy.

6 GGOS Portal: One-Stop Shop for Geodetic Data and Products

The IAG Technical Services provide critical and valuable geodetic data, information, and data products for Earth System research, observations, and decision-making. This spans from monitoring worldwide changes to applications such as satellite navigation, surveying, mapping, engineering, and geospatial information systems. Considering that geodesy is often "hidden in plain sight", many users in the past found it difficult to get an overview of all accessible geodetic data and products made available by the IAG and its components. GGOS intends to address this need by creating the GGOS Portal ggos.org/portal.

The planned GGOS Portal will complement the previously discussed communications and outreach on the GGOS website and social media platforms, serving as a one-stop shop for geodetic data and products. Through this new resource, data and products provided by the IAG and its components will be described and catalogued by extensive metadata, and providing direct links to the data in the respective data centers of each IAG service provider. Further design considerations for the GGOS Portal will include a

Earth Orientation Parameters

Why are days getting longer and Earth is wobbling?

The Earth Orientation Parameters (EOP) are the parameters representing the rotational part of the **transformation between** the current releases of the **International Celestial Reference Frame (ICRF)** and the **International Terrestrial Reference Frame (ITRF)**. Accordingly, the EOP describe the **change of the orientation of the Earth's surface**, on which the observatories are located, **with respect to a space fixed reference frame**. Earth rotation refers to the solid Earth component of the dynamic Earth system excluding deformations, such as tides and continental drift. Through interactions of individual components, angular momentum can be transferred between the atmosphere, ocean, and solid Earth, or between the Earth's core and mantle due to the coupling mechanisms.

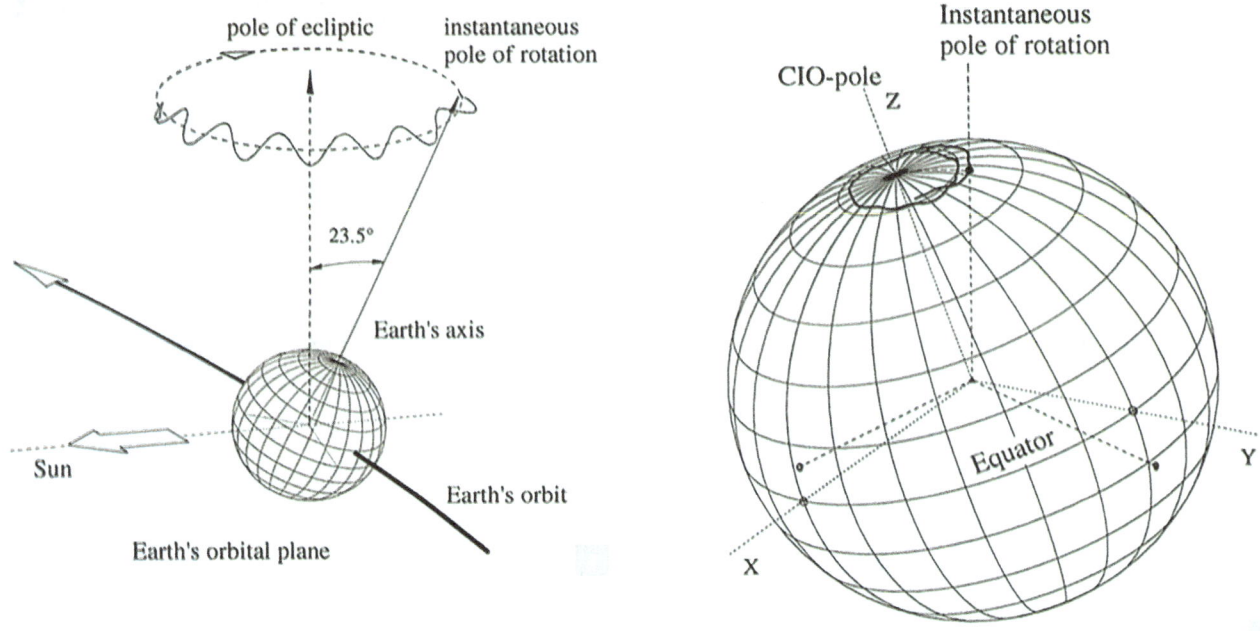

Earth orientation parameters – **Precession** and **nutation** (left) and **polar motion** (right)

Fig. 7 Example of a product description on the GGOS website. (Screenshot: ggos.org/item/earth-orientation-parameter/, Graphics credits: Andreas Verdun, Astronomical Institute, University of Bern)

collection of tools for structured knowledge search, including visualizations to aid in the discovery and selection of relevant information, data, and products.

While various geodetic data portals do already exist, the GGOS Portal will be much more than just a data portal for IAG geodetic data. Current plans include both easy-to-understand product and observation descriptions coupled with a complementary and comprehensive set of detailed geodetic metadata (Sehnal et al. 2023).

In March and April 2023, the GGOS-CO conducted a survey across the geodetic and geoscience communities to gain an understanding of the present availability of data products and associated metadata. This questionnaire collected feedback from geodetic data users on topics such as data availability and visibility, as well as essential user interface features, to inform design of a comprehensive and easy-to-use GGOS Portal. In total, 195 people from 57 different countries participated in this survey (Fig. 10), with 49% of

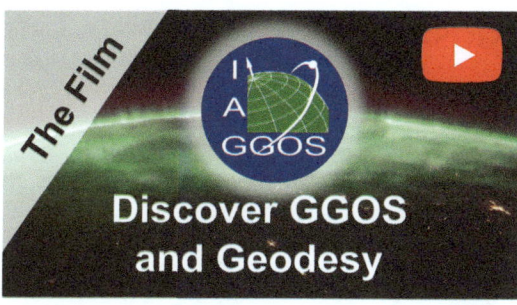

Fig. 8 GGOS film "Discover GGOS and Geodesy" (youtu.be/Jwqz097N2IY) (Background Image from Earth Science and Remote Sensing Unit, NASA Johnson Space Center – https://eol.jsc.nasa.gov/)

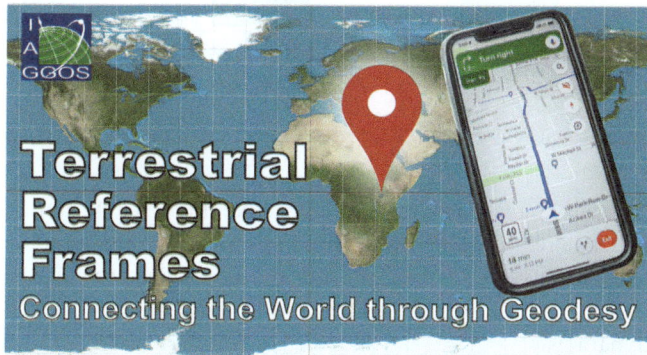

Fig. 9 Film about Terrestrial Reference Frames (youtu.be/vvNXv05646M) (Background Image: credit by Strebe – no changes made – https://shorturl.at/jkop5)

the respondents identifying as participating in one or more IAG component.

Analysis of the aggregated answers indicated that the majority of data providers currently participating in GGOS were in favor of publishing their data via the GGOS Portal in the future (43%) or were willing to consider this as a form of data dissemination (49%). In addition, survey respondents indicated an interest in the availability of not only IAG geodetic data via this future platform, but also geospatial data from stakeholder or related organizations, if possible (Fig. 11). The outcomes of this survey (Sehnal and Steiner 2023) will inform the architecture of this metadata platform with key requirements of the geoscience community.

Currently, the GGOS CO is collaborating with the Vienna University of Technology (TU Wien) on the development of the GGOS Portal. As part of this partnership, a feasibility study is currently underway to explore possible implementation options. Potential software packages (GeoNetwork and CKAN) are being evaluated for their appropriateness to realize this metadata platform. In addition, efforts are being made to apply the existing metadata of geodetic products within the IAG and the built-in metadata harvesting functions are being tested to synchronize the metadata automatically. A further goal of this feasibility study is to develop suggestions for data providers and provide recommendations for improved data and metadata dissemination.

With this future metadata portal, GGOS aspires to further increase the visibility and availability of geodetic data to the scientific community and raise awareness of geodesy and its useful products in other disciplines and in society.

7 Conclusions and Outlook

The comprehensive overview of the International Association of Geodesy's Global Geodetic Observing System (GGOS) presented in this paper underscores the pivotal role GGOS plays in both understanding and communicating

Fig. 10 Statistics of the participants in the GGOS Portal Survey (from Sehnal and Steiner 2023, licensed under CC-BY 4.0)

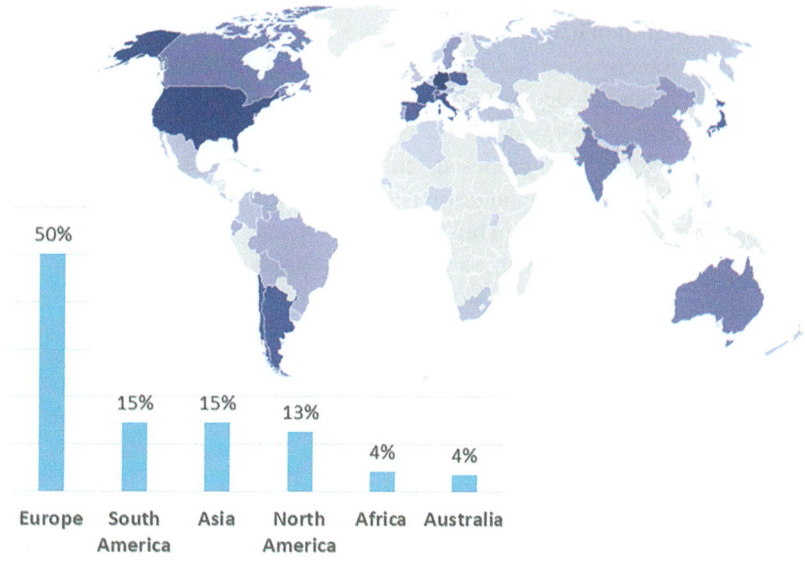

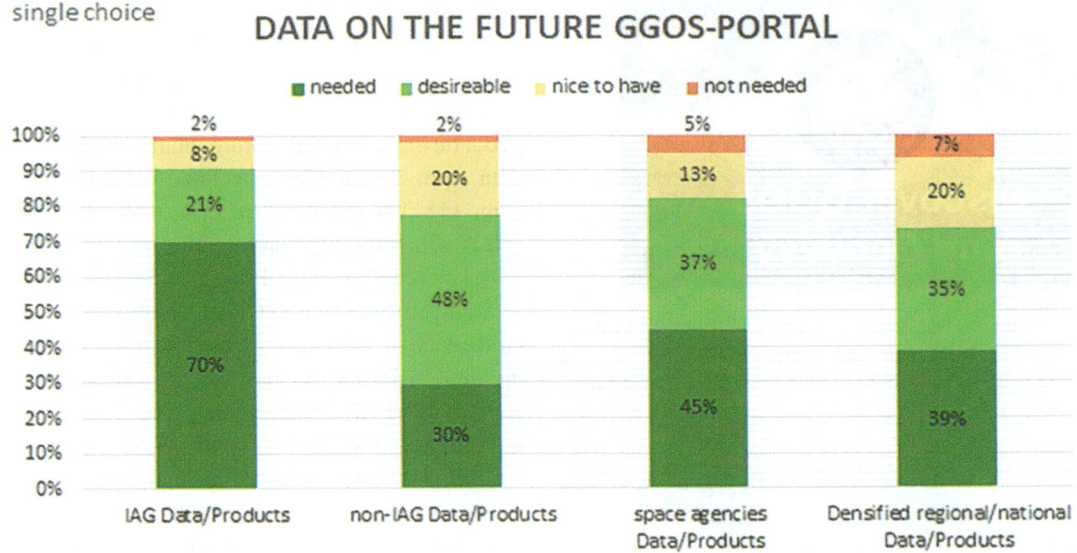

Fig. 11 Survey result: required data provided at the future GGOS Portal (from Sehnal and Steiner 2023, licensed under CC-BY 4.0)

the importance of Earth's shape, gravity field, and rotation. The work supported by GGOS not only enables precise monitoring of various phenomena, from sea level changes to tectonic motions, but also contributes significantly to climate change monitoring and our understanding of global geospatial dynamics.

Over the past decade, the GGOS Coordinating Office (CO) has played a central role in managing administrative aspects, fostering external relations, and spearheading impactful public outreach. Engaging with international stakeholders, participating in key United Nations frameworks, and enhancing GGOS's online presence have notably increased awareness, supported by statistical evidence of heightened website traffic and social media engagement. The CO's strategic initiatives, such as the GGOS website and multilingual informative films, exemplify a proactive approach to make geodetic data accessible. The ongoing development of the GGOS Portal signifies a commitment to creating a comprehensive one-stop shop, addressing the need for structured knowledge search and enhancing visibility within the scientific community.

Looking ahead, the proposed collaboration between GGOS and the IAG Communication and Outreach Branch is poised to strengthen public relations further. This commitment, coupled with multilingual outreach and innovative platform development like the GGOS Portal, underscores GGOS's dedication to global collaboration, accessibility, and increasing awareness of the crucial role geodesy plays in our understanding of Earth's complex dynamics.

References

Angermann D, Gruber T, Gerstl M, Heinkelmann R, Hugentobler U, Sánchez L, Steigenberger P, Gross R, Heki K, Marti U, Schuh H, Sehnal M, Thomas M (2022) GGOS bureau of products and standards: description and promotion of geodetic products. In: Freymueller JT, Sánchez L (eds) Geodesy for a sustainable earth. International Association of Geodesy Symposia, vol 154. Springer, Cham. https://doi.org/10.1007/1345_2022_144

Sehnal M, Steiner L (2023) GGOS-Portal – Survey Results. https://doi.org/10.5281/zenodo.8113241

Sehnal M, Craddock A, Angermann D, Sánchez L, Miyahara B, Pearlman M, Barzaghi R, Jäggi A, Altamimi Z, Gross R, Heki K, Otsubo T, Soudarin L, Sehnal K (2022) Discover GGOS and Geodesy. Film available at YouTube: youtube.com/@iag-ggos. https://doi.org/10.5446/56288

Sehnal M, Angermann D, Sánchez L, Miyahara B (2023) GGOS contribution to promote geodesy and increase its visibility in science and society. Journal VGI - Österreichische Zeitschrift für Vermessung & Geoinformation 111:100–104. OVG Österreichische Gesellschaft für Vermessung und Geoinformation

Open Access This chapter is licensed under the terms of the Creative Commons Attribution 4.0 International License (http://creativecommons.org/licenses/by/4.0/), which permits use, sharing, adaptation, distribution and reproduction in any medium or format, as long as you give appropriate credit to the original author(s) and the source, provide a link to the Creative Commons license and indicate if changes were made.

The images or other third party material in this chapter are included in the chapter's Creative Commons license, unless indicated otherwise in a credit line to the material. If material is not included in the chapter's Creative Commons license and your intended use is not permitted by statutory regulation or exceeds the permitted use, you will need to obtain permission directly from the copyright holder.

EPOS-GNSS Data Quality Monitoring Web Portal

Fikri Bamahry, Juliette Legrand, Carine Bruyninx, and Andras Fabian

Abstract

The European Plate Observing System (EPOS) is a large and complex European e-infrastructure that facilitates the integrated use of multi-disciplinary datasets and services for Solid Earth research. EPOS' GNSS (Global Navigation Satellite Systems) component provides access to GNSS data and products. This paper introduces a new EPOS web portal (https://gnssquality-epos.oma.be) that has been developed with the aim to provide the necessary information to monitor the availability and quality of daily GNSS data that are discoverable through EPOS. Currently, the web portal includes the tracking performances of more than 1600 GNSS stations. Several GNSS data quality indicators (DQIs), such as the number of observed versus expected observations, the number of missing epochs, the number of observed satellites, the maximum number of observations, the number of cycle slips, the Standard Point Positioning results, and the multipath values on code observation are monitored and their plots are available online. These DQIs provide helpful information that can be used to detect a potential degradation of the quality of the GNSS observations. Here, we will present the status of the web portal, the considered data quality indicators, and their benefits for GNSS data users.

Keywords

Data quality monitoring · EPOS · GNSS

1 Introduction

The European Plate Observing System (EPOS) is a European research infrastructure that brings European nations together with the goal to enhance the integration, accessibility, and utilization of solid Earth science data, products, services, and facilities. EPOS maintains a sustainable and cross-disciplinary central data portal (https://www.ics-c.epos-eu.org/) that offers open access to standardized metadata and quality-controlled data from various Earth science disciplines, accompanied by essential tools for analysis and modeling. (EPOS-ERIC 2017)

EPOS is organised in Thematic Core Services (TCS), such as TCS seismology, TCS geomagnetism, or TCS GNSS data and products (also called EPOS-GNSS). Each TCS organises the contribution of their specific community to EPOS. EPOS-GNSS is dedicated to delivering GNSS data, metadata, products, and software to EPOS (Fernandes et al. 2022). Within the scope of its participation in EPOS-GNSS, the Royal Observatory of Belgium is developing the EPOS-GNSS Data Quality Monitoring Service (DQMS) with as frontend the EPOS data quality monitoring web portal (https://gnssquality-epos.oma.be/). The primary goal of the DQMS is to oversee and evaluate the availability and quality of daily GNSS data that are discoverable through the EPOS-GNSS architecture. Similar GNSS data quality assessment procedures are also provided by EUREF Permanent Network (Bruyninx et al. 2019, 2022), Nevada Geodetic

F. Bamahry (✉) · J. Legrand · C. Bruyninx · A. Fabian
Royal Observatory of Belgium, Brussels, Belgium
e-mail: fikri.bamahry@oma.be

Laboratory (Blewitt 2021), and International GNSS Service (Johnston et al. 2017).

This paper describes the current status of the EPOS-GNSS data quality monitoring web portal and illustrates with some practical examples how the information provided by the portal can help interpreting GNSS data products targeting geodetic or geophysical applications.

2 EPOS-GNSS Data Dissemination Concepts

GNSS data become discoverable and accessible within EPOS after going through several procedures which aim at applying the same standard to GNSS (station and file) metadata and also ensuring the quality of GNSS data.

Figure 1 illustrates the EPOS-GNSS data dissemination workflow to deliver GNSS station and file metadata to the EPOS data portal. Currently, there are ten EPOS-GNSS data nodes that operate on top of GNSS data repositories. These data nodes use the open-source Geodetic Linkage Advance Software System (GLASS) (Fernandes et al. 2022), a set of softwares developed within EPOS-GNSS, to establish a virtualization layer on top of the data repository. GLASS performs several tasks, including indexing GNSS data files (in RINEX format), data validation with respect to station metadata from M^3G (https://gnss-metadata.eu/) (Fabian et al. 2021), data quality checks using G-Nut/Anubis software (Vaclavovic and Dousa 2015), and it stores all relevant information in the local node database. Furthermore, GLASS ensures the synchronization of file metadata to the EPOS-GNSS Data Gateway (DGW), where all EPOS-GNSS data are accessible through an Application Program Interface (API) and a web portal.

3 EPOS-GNSS Data Quality Monitoring Portal

3.1 Introduction

The backend of the EPOS-GNSS DQMS retrieves from each of the EPOS-GNSS data nodes every day the list and metadata of available daily GNSS data as well as their data quality metrics. This information is then stored in the DQMS database and used to compute the GNSS data quality indicators (DQIs) from which a selection is presented and made available on the data quality monitoring portal (Bamahry et al. 2022).

This portal is structured into two main sections. The first section focuses on RINEX data availability. It allows to check the distribution and the availability of the GNSS data that are discoverable through EPOS and filter them by EPOS-GNSS data node, GNSS network, or M^3G metadata maintainer. For each station, the resulting number of available daily GNSS data files are then graphically represented on a map (see Fig. 2).

The second section focuses on RINEX data quality and provides plots of several GNSS DQIs, such as:

(a) The percentage of observed vs. expected observations computed as the ratio of the number of actual observations with respect to the number of expected observations for each constellation. This DQI plot provides the ratio on at least one frequency (1freq) or at least two frequencies (2+freq). 2+freq is more relevant for geodetic positioning applications as it is needed to reduce the effect of the ionospheric delay. Additionally, the lowest elevation cut-off observed is also provided on this plot to help identifying data quality degradations at low elevations.

Fig. 1 The EPOS-GNSS data dissemination workflow

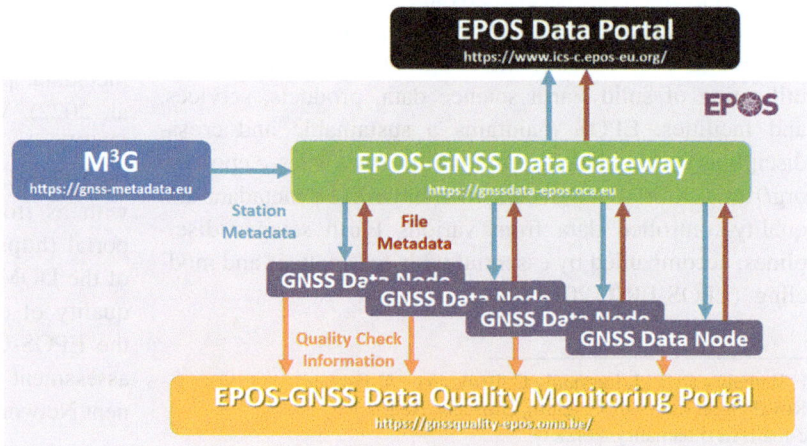

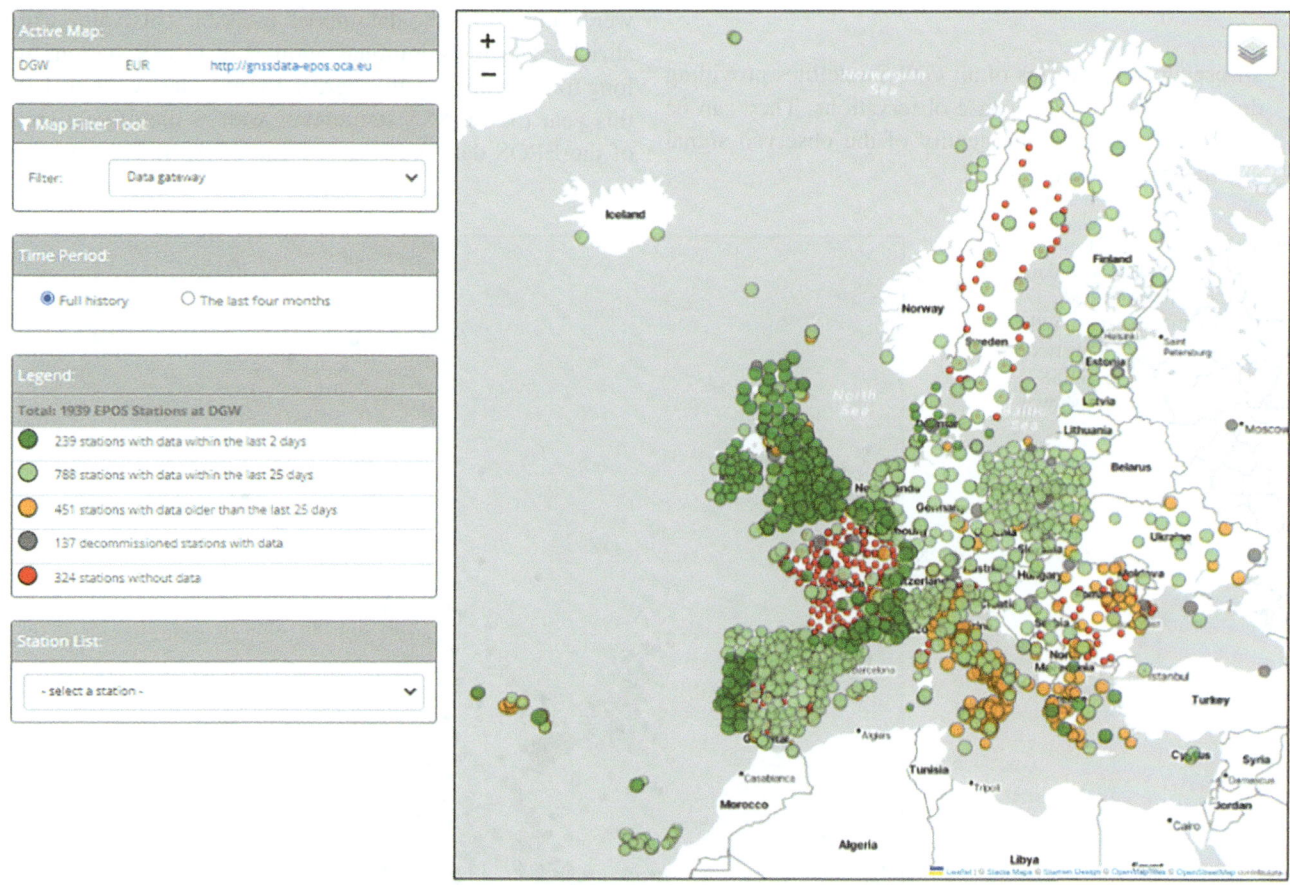

Fig. 2 Web page of the GNSS data quality monitoring web portal showing the map of the RINEX data availability

(b) The number of epochs without observations.
(c) The number of observed satellites for each constellation tracked by the station.
(d) The maximum number of observations counted for each frequency and each constellation tracked by the station.
(e) The number of cycle slips represents the ratio of the number of identified phase cycle slips × 1000 with respect to the number of phase observations for each constellation tracked by the station.

(f) The Standard Point Positioning (SPP) coordinates, representing the daily mean XYZ coordinates estimated from the code observations obtained from a specific satellite constellation. The plot shows the deviation of the estimated SPP coordinates with respect to their median value over the station history.
(g) The multipath values on code observations represent the daily mean of code multipath per frequency band and satellite constellation. These values are calculated for

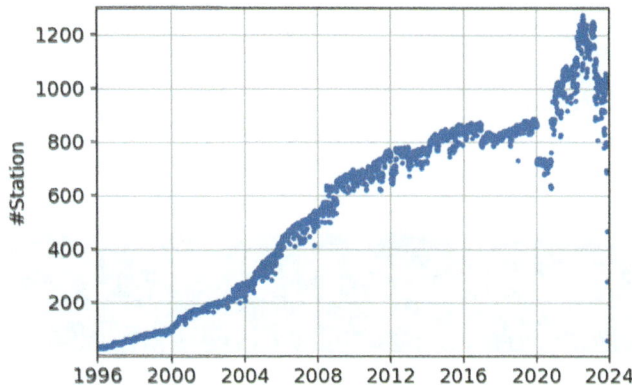

Fig. 3 Number of daily RINEX files discoverable from the EPOS-GNSS Data Gateway as function of time

3.2 The Availability of EPOS-GNSS Data

Based on the information gathered within the DQMS database, the number of daily GNSS data discoverable through the EPOS-GNSS Data Gateway increases clearly over the time (see Fig. 3).

Currently there are 1631 stations with data distributed throughout ten active EPOS-GNSS data nodes (see Fig. 4). Almost 40% of GNSS data that are discoverable from the EPOS-GNSS Data Gateway are coming from the ROB-EUREF data node (see Fig. 5), which makes the daily RINEX data of the EUREF stations that agreed to share their data within EPOS available.

A significant number (35%) of the daily GNSS data files were indexed by the data nodes in 2023. This value can be attributed to the additional number of GNSS stations with long tracking history that agreed to share data with the EPOS this year (327 GNSS stations) as well as the official launch of the EPOS data portal at the end of April 2023. Figure 4

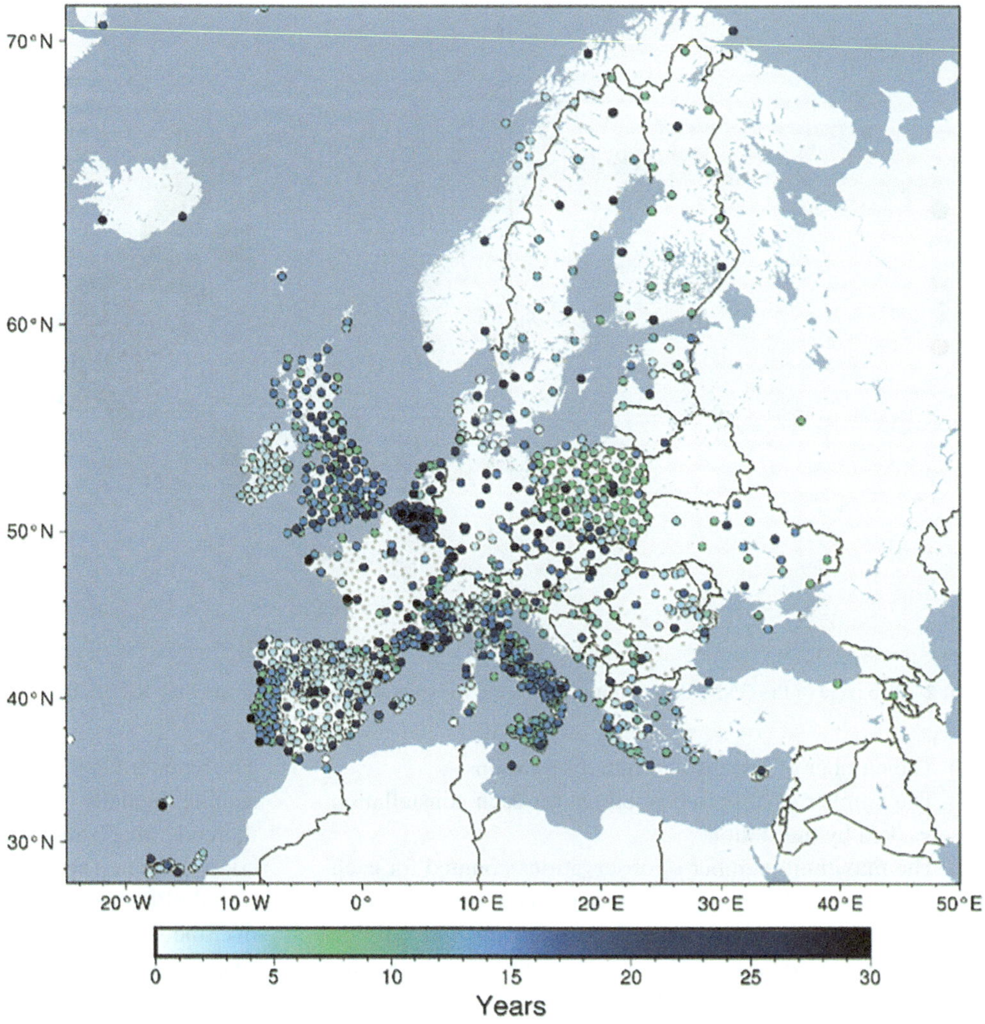

all pseudo-range codes of all GNSS satellites providing dual-frequency carrier-phase observations. They can be used to characterize the quality of the observed signal and the site environment.

Fig. 4 Map of the 1631 EPOS-GNSS stations with daily RINEX data; the color code shows the number of days with data quality metrics (converted in years). The 315 EPOS-GNSS stations without data are shown in grey

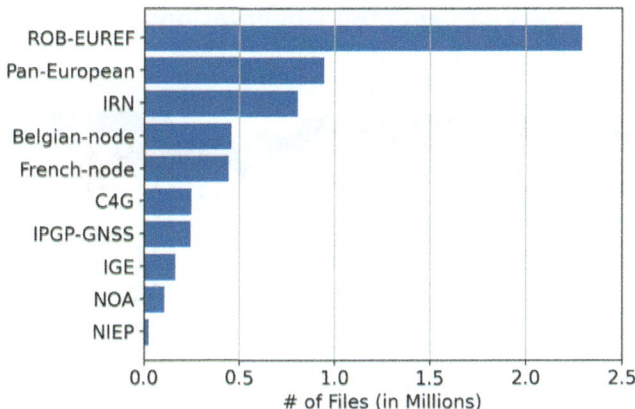

Fig. 5 Number of daily RINEX observation files discoverable from ten active data nodes

shows that there are 315 EPOS-GNSS stations without data, primarily because some EPOS-GNSS data nodes are still in the process of populating their databases.

3.3 The Quality of EPOS-GNSS Data

The evolution of GNSS data quality over time was calculated in this study to evaluate the data quality performance of all EPOS-GNSS stations. We determine the daily median value for constellation tracking performance: 2+freq above 0°, 1freq above 0°, and 1freq above 10° over all EPOS-GNSS stations. These values were smoothed over three-month periods to visualise long-term trends in tracking performance for each constellation.

In Fig. 6, we observe that the values for GPS 2+freq above 0° over all EPOS-GNSS stations have been increasing 20% over the past 28 years. Conversely, GLONASS 2+freq above 0° exhibited an upward trend only until 2016. From 2016 on, the performance of GLONASS 2+freq above 0° slowly degraded until mid of 2021. This phenomenon was explained by Bruyninx et al. (2019), attributing the degradation of GLONASS 2+freq above 0° observed in EPN stations to several GLONASS satellites with impaired L2 observations. In contrast, both Galileo and BeiDou demonstrated an improvement in their performance at 2+freq above 0° after mid 2019. In addition, we also can see that the number of stations that track GLONASS observations has dramatically decreased in 2010. Based on our investigation, this feature is due to a bug in the software used by EPOS-GNSS and this period is currently being reprocessed by the nodes in order to solve the issue.

The degraded quality of daily GNSS data can impact the accuracy of GNSS products. We used the detrended GNSS position time series from Nevada Geodetic Laboratory (Blewitt et al. 2018) of the three stations (ACOR00ESP, BUCK00GBR, TRO100NOR) to illustrate how GNSS DQIs plots can help to correctly interpret GNSS position time series. Figure 7 depicts the degraded tracking observed at low elevation at ACOR00ESP station in 2003, resulting in spurious subsidence in its position time series. Similarly, undocumented elevation cut off changes in 2014 at the BUCK00GBR station have a modest effect on its position time series (see Fig. 8). As response to these issues, we are presently developing an algorithm for automatically detecting changes in elevation cut-off angle using Online

Fig. 6 Historical trends of observed vs. expected observations for each constellation. Upper panel: 3 months moving averages of the daily median overall station of 2+freq above 0°, 1freq above 0°, 1freq above 10° for GPS (blue), GLONASS (GLO, red), Galileo (GAL, green), and BeiDou (BDS, yellow). Lower panel: number of EPOS stations used to calculate these trends

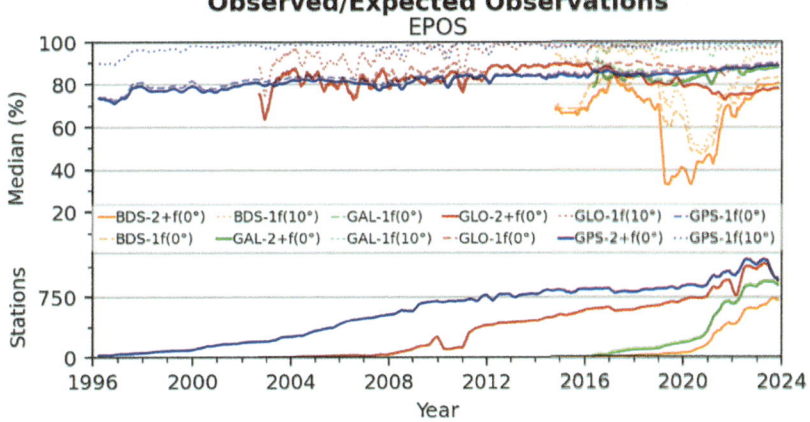

Fig. 7 Degraded tracking at low elevation in 2003 caused spurious subsidence at the station ACOR00ESP. Upper panel: the percentage of the observed with respect to expected observations, minimum elevation is shown in purple. Lower panel: detrended position time series

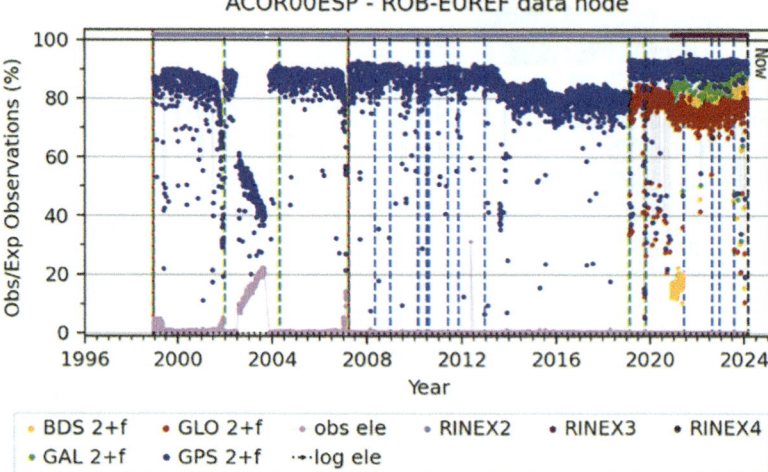

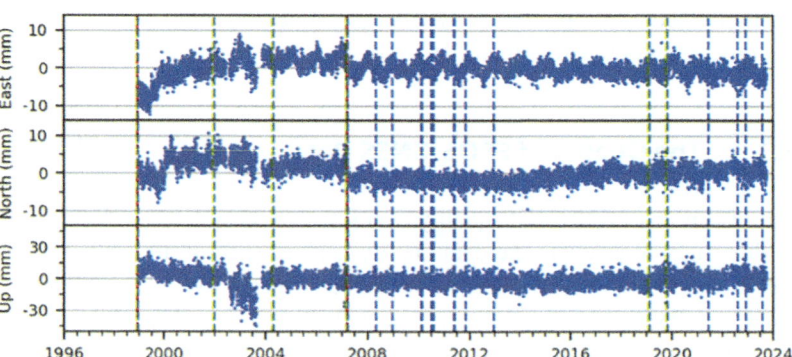

Bayesian method (Adams and Mackay, 2007) and provide the information to station owners with the final goal to correct station metadata.

In Fig. 9, a strong correlation is evident between anomalies in the cycle slips, multipath values, and the position time series at TRO100NOR station in 2000–2004. There is no specific information available concerning the cause of this quality degradation. Nevertheless, given the seasonal signals in the DQI, the data quality degradation could be attributed to environmental conditions in the vicinity of the antenna.

Fig. 8 Example of an undocumented change affecting the cut off angle in 2014 for the station BUCK00GBR. Upper panel: the percentage of observed with respect to expected observations, minimum elevation is shown in purple. Lower panel: detrended position time series, a purple vertical line in 2014-05-16 shows an un-documented change of the cut off angle observed

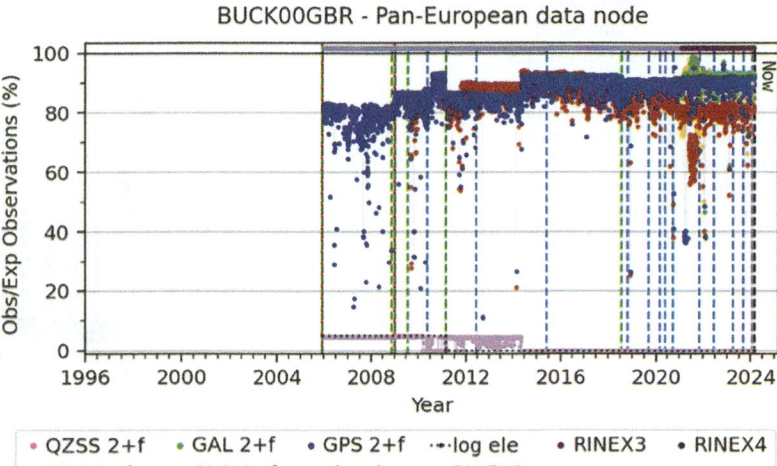

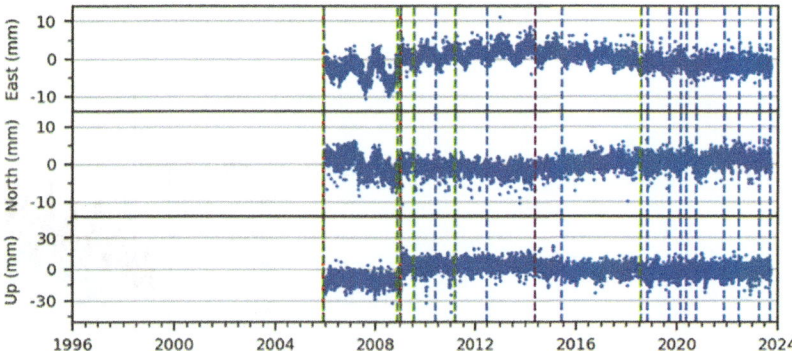

4   Summary and Future Works

The new DQMS web portal monitors the availability and quality of the daily GNSS data that are discoverable through the EPOS-GNSS Data Gateway and EPOS Data Portal. In this paper, we outlined several use cases and analyzed GNSS data quality indicator plots available from this web portal. Drawing from our experience at ROB, these data quality indicators are employed to gain insights into outliers or irregularities within position time series. However, it is important to note that the correlation between these indicators and position time series is very complex. Consequently, we are also exploring the utilization of artificial intelligence to better comprehend the complex correlation between DQIs and position time series, enabling us to automatically identify degraded GNSS data quality that might be contributing to outliers or anomalies in position time series (Bamahry et al. 2023).

Fig. 9 Degraded tracking performance between 2000 and 2004 affecting the number of cycle slips, the multipath values and the position time series for the station TRO100NOR. Upper panel: number of cycle slips per number of observations. Middle panel: daily mean of code multipath per frequency band and satellite constellation. Lower panel: detrended position time series

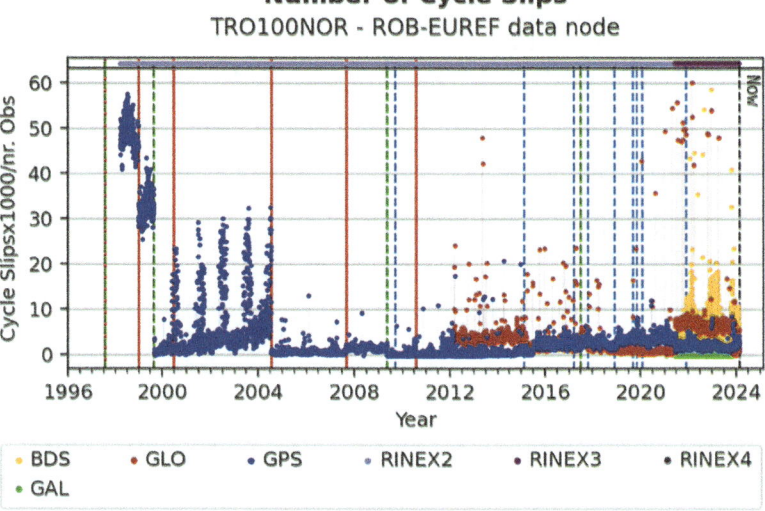

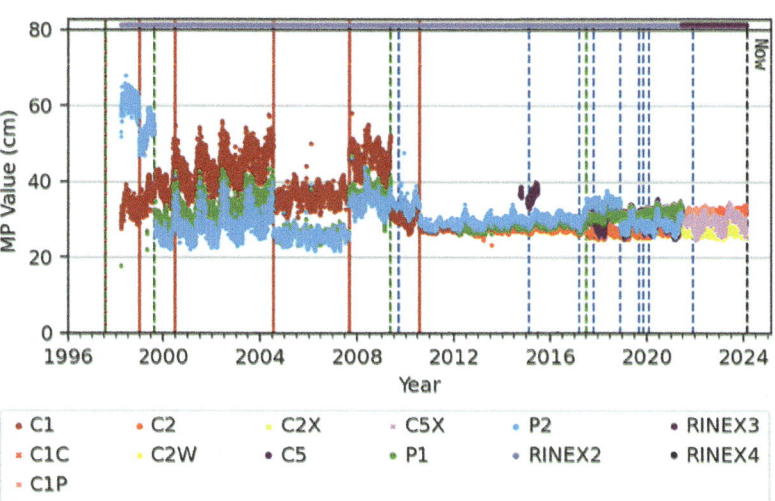

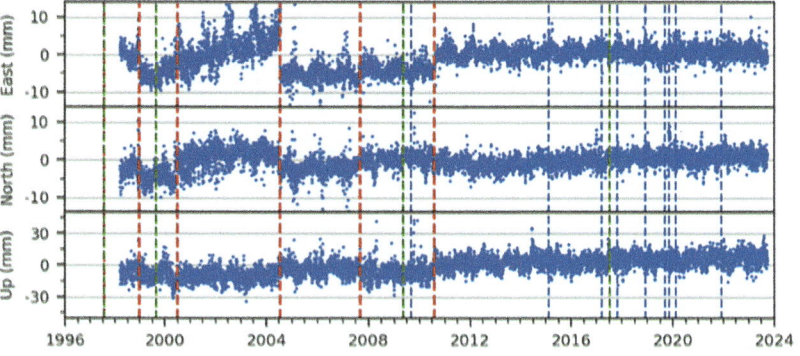

Acknowledgements This work is supported by Belgian Science Policy Office (BELSPO) – Grant nr. EF/211/SERVE: Strengthening the Provision of Core Services and Data to the EPOS. We acknowledge EPOS-GNSS Data Gateway, all EPOS-GNSS data nodes (Belgian-node, C4G, French-node, IGE, IPGP-GNSS, IRN, NIEP, NOA, Pan-European, and ROB-EUREF), M^{3}G and EPOS-GNSS software group. Special thanks to the G-Nut software s.r.o. that developed G-Nut/Anubis software that is used by all EPOS-GNSS data nodes.

References

Adams RP, MacKay DJC (2007) Bayesian Online Changepoint Detection. arXiv 0710.3742. https://doi.org/10.48550/arXiv.0710.3742

Bamahry F, Legrand J, Bruyninx C, Mesmaker D, Fabian A (2022) EPOS GNSS Data Monitoring, Available from Royal Observatory of Belgium. https://doi.org/10.24414/ROB-EPOS-EGDM

Bamahry F, Legrand J, Pottiaux E, Bruyninx C, Fabian A (2023) Using supervised machine learning based on data quality indicators for automated data cleaning of GNSS position time series. In: XXVIII General Assembly of the International Union of Geodesy and Geophysics (IUGG). https://doi.org/10.57757/IUGG23-3291

Blewitt G (2021) A Guide to QA Files for Practical Quality Assessment of GPS Data. [Internet-cited 2023Oct.26]. Available from: http://geodesy.unr.edu/gps_timeseries/QA.pdf

Blewitt G, Hammond WC, Kreemer C (2018) Harnessing the GPS data explosion for interdisciplinary science. Eos 99. https://doi.org/10.1029/2018EO104623

Bruyninx C, Legrand J, Fabian A, Pottiaux E (2019) GNSS metadata and data validation in the EUREF permanent network. GPS Solut 23:106. https://doi.org/10.1007/s10291-019-0880-9

Bruyninx, C., Legrand, J., Mesmaker, D., Moyaert, A., Fabian, A. (2022): EUREF Permanent Network Central Bureau (EPN CB) Information System. https://doi.org/10.24414/ROB-EUREF-EPNCB

EPOS-ERIC (2017) Scientific and technical description of the European plate observing system. (EPOS) Research Infrastructure

Fabian A, Bruyninx C, Miglio A, Legrand J (2021) M^{3}G - Metadata management and distribution system for multiple GNSS networks. https://doi.org/10.24414/ROB-GNSS-M3G

Fernandes R, Bruyninx C, Crocker P, Menut J-L, Socquet A, Vergnolle M, Avallone A, Bos M, Bruni S, Cardoso R, Carvalho L, Cotte N, D'Agostino N, Deprez A, Andras F, Geraldes F, Janex G, Kenyeres A, Legrand J, Ngo K-M, Lidberg M, Liwosz T, Manteigueiro J, Miglio A, Soehne W, Steffen H, Toth S, Dousa J, Ganas A, Kapetanidis V, Batti G (2022) A new European service to share GNSS data and products. Ann Geophys 65(3):DM317. [Internet-cited 2023Oct.23]. Available from: https://www.annalsofgeophysics.eu/index.php/annals/article/view/8776

Johnston G, Riddell A, Hausler G (2017) The international GNSS service. In: Teunissen PJG, Montenbruck O (eds) Springer handbook of global navigation satellite systems, 1st edn. Springer International Publishing, Cham, Switzerland, pp 967–982. https://doi.org/10.1007/978-3-319-42928-1

Vaclavovic P, Dousa J (2015) G-Nut/Anubis: open-source tool for multi-GNSS data monitoring with a multipath detection for new signals, frequencies and constellations. In: Rizos C, Willis P (eds) IAG 150 Years. International Association of Geodesy Symposia, vol 143. Springer, Cham. https://doi.org/10.1007/1345_2015_97

Open Access This chapter is licensed under the terms of the Creative Commons Attribution 4.0 International License (http://creativecommons.org/licenses/by/4.0/), which permits use, sharing, adaptation, distribution and reproduction in any medium or format, as long as you give appropriate credit to the original author(s) and the source, provide a link to the Creative Commons license and indicate if changes were made.

The images or other third party material in this chapter are included in the chapter's Creative Commons license, unless indicated otherwise in a credit line to the material. If material is not included in the chapter's Creative Commons license and your intended use is not permitted by statutory regulation or exceeds the permitted use, you will need to obtain permission directly from the copyright holder.

The GGXF Standard File Format for Gridded Geodetic Data

Chris Crook, Kevin M. Kelly, Roger Lott, and Richard Stanaway

Abstract

Monitoring the Earth is undertaken in numerous geometric and physical reference frames and coordinate reference systems. Analysis often requires transformation of coordinates between these frames. As the accuracy of positioning and geospatial data improves, the use of gridded data to describe the quantities used in these coordinate transformations is increasing. Rectangular grids also provide an efficient means of disseminating other geodetic data. IAG Commission 1 Working Group 1.3.1 in association with the Open Geospatial Consortium (OGC) have developed a Geodetic data Grid eXchange Format (GGXF) for quantifying and disseminating gridded geodetic data. GGXF was developed in conjunction with a functional model for crustal deformation (FMCD) including support for time-dependent changes, but has been designed to support any type of regularly-gridded geodetic data including but not limited to geoid models, offsets between reference frames (of one, two or three dimensions), velocity grids, tidal surfaces, etc. The purpose of GGXF is to provide a single comprehensive, efficient distribution format through which producers can disseminate gridded geodetic data and users can exchange and apply this information. This paper presents an overview of the GGXF format.

Keywords

Coordinate reference system · Coordinate transformation · Deformation · Geodesy · Geoid · Grid · Reference frame

1 Introduction

Grid files have been used in geodetic applications for several decades. Grids of two- and three-dimensional positional differences for use in coordinate transformations provide superior modelling of network distortion compared to linear transforms. Advances in some of the methodologies used in geodesy, including global and regional velocity models and space-borne gravity sensing (Johannessen et al. 2003), have significantly increased the volume of gridded data being made available for various geodetic content. Geoid and tidal surface modelling through grids of heights above an ellipsoid is now a standard tool. In deforming regions increased data collection has led to modelling of the motion of the Earth's crust due to glacial isostatic adjustment, tectonic stress and resultant earthquakes, and human activity. More and more geodetic data sets are being produced and described through gridded data. This trend is expected to continue.

C. Crook
Toitū Te Whenua Land Information New Zealand, Wellington, New Zealand

K. M. Kelly
Environmental Systems Research Institute, Redlands, CA, USA

R. Lott (✉)
International Association of Oil and Gas Producers, London, UK

R. Stanaway
Quickclose Pty. Ltd., Melbourne, VIC, Australia

Department of Surveying and Land Studies, Papua New Guinea University of Technology, Lae, Papua New Guinea

We find that different agencies use different formats for the same gridded geodetic data product. Furthermore, different gridded products use different formats. In the context of geoid models the situation is summarized by (Neacsu 2010) as:

> Geoid files are presumed to be simple stuff as they contain just one value (undulation – the separation between ellipsoid and geoid) in a regular, rectangular grid pattern. You would have expected that the few hundred people who are involved in generating those models would be able to agree on a format and follow that standard. Alas, the reality is far different. For us the story began at some point in the '90s when we implemented the GEOID96 model for continental US. Figuring that there wasn't much merit in inventing a new file format, we implemented the NGS format at the time: a binary format with the extension. GEO. Not long after, NGS decided to change the format and came out with another one called BIN. It didn't take long for other organizations to come out with new formats, some ASCII, some binary, some going from North to South, others going from South to North, from East to West and from West to East.

A similar situation to geoid models exists for other gridded geodetic data products such as (but not limited to) deformation models, velocity grids, and grids of two- and three-dimensional positional differences for use in coordinate transformations. For these gridded geodetic products intra-agency and inter-agency inconsistency prevails.

This non-standardization of formats has led to the following situation:

a) Grid producers have often had to supply software to allow external users to read and interpolate their gridded data.
b) Developers of geospatial software have had to repeatedly write programs to import the different grid formats. This has often resulted in each developer separately reformatting the grid file content to their proprietary internal format.
c) Large corporations, such as international energy producing companies, usually have strict procedures for the installation of software on company computers, causing difficulty in utilising grid producer's software. If users in such corporations are to use the gridded data, it needs to be through one of their internally authorised geospatial software applications. These users are unable to use gridded data until it has been incorporated into the software application. At best this causes a time delay, sometimes years. At worst there may be no economic benefit to the application developer to incorporate specialized grids into their product and the end user community is unable to access the gridded data.

From a computing perspective, the mechanisms for reading and processing data from a gridded data set are independent of the intrinsic meaning of that data. Therefore it makes sense to seek a solution that would be general to many different geodetic data types. A generic geodetic data grid exchange format could then meet a wide variety of potential applications.

The remainder of this paper discusses requirements for and challenges of defining the Geodetic data Grid eXchange Format, GGXF. The work was conducted jointly under the auspices of IAG Commission 1 Working Group 1.3.1 (Time-dependent transformations in deforming regions) in association with the Open Geospatial Consortium (OGC) Coordinate Reference System Working Group. The combined project team brought together representatives of data producers, geomatic application developers and end user communities. The work was integrated with a parallel project developing a functional model for crustal deformation (Stanaway et al. 2024; OGC 2024a).

2 Existing Formats

The U.S. National Geodetic Survey's NADCON grids of geodetic latitude and longitude coordinate differences were introduced in 1989 (NOAA 1990). A later version of NADCON, NADCON5, introduced a different file format catering to ellipsoidal coordinate differences in two or three dimensions (NOAA 2017). NADCON and NADCON5 use multiple files for the shifts in each coordinate dimension, rather than a single file holding all shifts. A single file yields more efficient and faster access to the data.

Other countries have taken alternative approaches utilizing grid files, with different content and different format. Examples include the Ordnance Survey National Grid Transformation (OSTN) grid in Great Britain (OS 2004) and the GR3DF97A format in France (IGN 1997).

Natural Resources Canada's National Transformation (NT) format followed shortly after NADCON, with a second version, NTv2, not long afterwards (Junkins and Farley 1995). Led initially by Australia and followed by several others, there has been further adoption of the NTv2 format which has become a *de facto* standard outside of the US for two-dimensional transformations using grids of geodetic latitude and longitude coordinate differences.

There has been no similar adoption of one format for other gridded geodetic data products. As noted by Neascu (2010) for geoid and height correction models, almost every producer of these models has defined their own file format and some producers have used multiple formats. Different formats are used for the U.S. National Geospatial Intelligence Agency's Earth Gravitational Model geoid grid files of various epochs (EGM84, EGM96 and EGM2008) (NGA 2013, 2014), and for the French encoding to geoid files for both metropolitan France and French overseas departments and territories (IGN 2012). The US NOAA/NGS Vertical Datum Transformation (VDatum) with its GTX file format (NOAA 2012) and UK Vertical Offshore Reference Frame (VORF) (UCL 2012) applications which integrate hydrographic sounding datums, gravity-related (orthometric/normal) heights and ellipsoid heights take different grid format approaches. The International Service for the Geoid, a repository of geoid models made available through the Polytechnic University of Milan, has developed a standard format for the distribution of its geoid model data (Reguzonni et al. 2021). For its generic coordinate transformation software the Open Source Geospatial Foundation (OSGeo) has adopted the NTv2 format for two-dimensional grid transformations of ellipsoidal coordinates and the NGS GTX format for vertical data (PROJ 2023a). The GeoTIFF format does not directly support the complexity of a deformation model format, although OSGeo has also developed a deformation model format using a JSON master file and multiple GeoTIFF files (PROJ 2023b). Although arbitrary XML metadata can be encoded into a GeoTIFF file, the GeoTIFF format does not require or specify a format for this metadata.

None of these existing formats have sufficient descriptive information within the file to identify the data, its intended use and source, or even its file format description. This means that if a user obtains a copy of only the file, it cannot be used without additional external information.

In contrast to the geodetic community, the scientific geophysical community uses altogether different grid formats such as Network Common Data Form (netCDF) and Hierarchical Data Format (HDF). netCDF is a set of software libraries and machine-independent data formats that support the creation, access, and sharing of array-oriented scientific data. It is also a community standard for sharing scientific data. netCDF has grown into a *de facto* standard within, for example, the atmospheric, meteorological and oceanographic communities. HDF is a data model, library, and file format for storing and managing data. It supports an unlimited variety of data types, and is designed for flexible and efficient input and output (I/O) and for high volume and complex data. HDF is portable and extensible, allowing applications to evolve in their use of HDF. The HDF Technology suite includes tools and applications for managing, manipulating, viewing, and analysing data in the HDF format. HDF allows hierarchical data objects to be expressed in a very natural manner, in contrast to the tables of relational database. Whereas relational databases support tables, HDF supports n-dimensional datasets and each element in the dataset may itself be a complex object. Relational databases offer excellent support for queries based on field matching, but are not well suited for sequentially processing all records in the database or for subsetting the data based on coordinate-style lookup. netCDF and HDF comprise application interfaces (API) to read and write scientific data files.

3 GGXF Features

The GGXF project team established a set of criteria for features required in a standard file format for gridded geodetic data. A wide range of existing formats including those noted in the previous section were evaluated against these criteria. No existing format had all of the features identified by the GGXF project team as being needed in a standard file format for gridded geodetic data. Modifying or extending one of the existing formats was considered but rejected for being incomplete or inefficient. GGXF was developed to fulfil all of the following criteria.

Complete and Informative Header in a Single File GGXF allows and encourages a full description of the data contained in the file, including but not limited to its provenance. Most existing grid format headers provide little or no information about what the gridded data represents. A user must rely on external documentation for this description; if it is missing or unavailable the grid data is rendered meaningless.

User Definable Content GGXF allows producers to define the type of data represented by the grid. A predefined list of well known geodetic data types is part of the format specification. Users are able to propose additions to this list which can readily be added to the specification.

Flexible GGXF permits grids to carry any number of parameters at grid nodes. Most existing geodetic grid formats permit only pre-defined parameters at each grid node. The GGXF format allows producers to define a set or tuple of numeric data at each grid node, $(d_1, d_2, ..., d_i)$ where d_i is the *ith* data value in the tuple. For example, three-dimensional velocity data could be defined at each grid node as the sequential set (V_X, V_Y, V_Z) in an Earth-Centered, Earth-Fixed Cartesian reference frame such as ITRF2014 or equivalently in ellipsoidal geodetic coordinates as $(V_\varphi, V_\lambda, V_h)$ where φ, λ, h are geodetic latitude, geodetic longitude and ellipsoidal height. In both cases accuracy data may also be included in a number of parameterisations, for example as standard error values ($\sigma_X, \sigma_Y, \sigma_Z$, or $\sigma_\varphi, \sigma_\lambda, \sigma_h$, respectively).

Multi-dimensional A survey of current implementations showed that the grid carrying the parameters (the interpolation coordinate reference system) is spatially two-dimensional. GGXF is capable of handling grids in two or more dimensions. In addition GGXF provides the ability to describe changes with time.

Hierarchical GGXF permits overlapping grids and when grids overlap it allows the definition of priority of use.

Multi-resolution GGXF permits multiple levels of data resolution using sub-grids in an analogous way to the Canadian NTv2 format.

Interpolation GGXF allows for the specification of the required or recommended interpolation method, e.g. bilinear, bicubic, etc. This facilitates application software applying the interpolation algorithm intended by the grid producer.

Efficient GGXF supports a platform independent binary file providing a compact encoding that can be used efficient direct computation of the encoded geodetic parameters. It permits fast access to any part of very large files.

Easy to Use The GGXF project team strived to ensure that the format is straightforward for grid producers to create and easy and efficient for application developers to implement. A well documented description, complete with examples and an open-source reference implementation for developers accompanies the format. The file format and metadata syntax is published and versioned.

Recognizable The GGXF file format is identified by file name extension as well as by identifiers within the file.

4 GGXF File Structure

A GGXF file consists of (i) a header block containing metadata about the content, provenance and applicability of the data in the whole file; (ii) one or more groups of grids each with (a) a header block containing metadata applicable to the group and (b) one or more grids. Each grid has (a) a header block containing metadata applicable to the grid and (b) an array of nodes at which one or more parameter values are given.

The full GGXF structure supports complex data sets such as crustal deformation caused by a number of earthquakes. Each event or component of an event is described by a separate group. Most gridded geodetic data sets do not require the full complexity of this file structure. They may be represented with a single group, and often with a single grid.

Different groups in a GGXF file represent complementary facets of the content. For example, in a deformation model one group could represent horizontal deformation and another vertical deformation. Within each group, multiple grids may be used to densify data locally, or to cover irregularly shaped areas. The GGXF file unambiguously identifies which grid of the group applies at any location using a combination of a numerical priority and parent-child relationships among the grids. Together the grids implement a spatial function that defines each of the parameters of the group over the extents of the grids.

Missing data at grid nodes can be identified by a special parameter value, although this is discouraged because it may result in differing approaches to grid interpolation and different interpolated values. The statistical uncertainty of parameter values may be defined in the group header or more specifically at each grid node. A producer may include structured check data in the GGXF file header with which a user can verify that software is correctly calculating and applying the parameter values.

5 GGXF Conventions

The GGXF headers (file header, group header and grid header) fully define the file content. The headers use key-value pairs to unambiguously identify each attribute. In these pairs the keyword is defined through the GGXF "Conventions", so-called following the precedent set in the climate and forecasting community's use of netCDF. Approximately 250 attributes are defined in the GGXF Conventions.

6 GGXF Encoding

The GGXF format specifies requirements and recommendations for two encodings: binary and text. The binary encoding includes all header records and all grids in a single file and is the authoritative GGXF file format used for data exchange. It utilises the capabilities of netCDF and HDF5. There are several versions of netCDF: GGXF uses netCDF-4. The Unidata Program Center supports and maintains netCDF programming interfaces for C, C++, Java, and Fortran. Programming interfaces are also available for Python, IDL, MATLAB, R, Ruby, and Perl.

The GGXF text file is envisaged primarily as an intermediate stage optionally used by some file producers in the assembling of grid data. The text file is not intended for public distribution. The text format is both human and machine readable, though human reading will be practical only for reviewing the file metadata and for very small grids. The GGXF text file uses syntax in compliance with YAML version 1.2 (Ben-Kiki et al. 2009). YAML is Unicode based and human-readable.

The GGXF text format has an option for defining the grid data in separate text files rather than including the grid data directly within the GGXF YAML file. The header information in the GGXF YAML file is then consolidated and easier to read and edit. The GGXF text grid data format is based on the commonly used comma separated value (CSV) file format, but it allows the file to use either comma, space or tab as the separator. Many data analysis software packages can handle grid data in this format. Also CSV is easy to use for data visualisation in geographic information systems. The GGXF grid data CSV format is defined through the GGXF Conventions.

Scripts for converting between the text and binary GGXF formats are available on the OGC GGXF Github web site (OGC 2023).

7 GGXF Documentation

GGXF documentation (OGC 2024b) gives a detailed description of the format including its metadata provisions and conformance requirements. The documentation includes the GGXF Conventions as well as examples of several content types such as coordinate offsets, geoid models, velocity grids, deformation models and deviation of the vertical data. Further examples are available on the OGC GGXF Github web site (OGC 2023).

8 Challenges

The format will be judged successful if it becomes extensively used, thereby removing the problems of the multitude of existing formats. By far the greatest and most immediate challenge is to encourage grid producers to distribute their grids in the GGXF format. Few existing format headers are as information-rich as the GGXF header. Any conversion of data from existing formats into GGXF will require some human intervention to complete missing header information.

Developers of new applications using gridded geodetic data should be able to adopt GGXF for their internal data handling without difficulty, eliminating the need for proprietary formats for one or more different data types. The intent and certainly the long term hope of the GGXF project team is that software applications will use the format directly without reformatting GGXF into some other external or proprietary format, with the consequent risks of misrepresentation and multiple non-authoritative version of data sets.

For end users, a binary format is inaccessible. Utilities to convert the header into readable text are required. netCDF to YAML routines for this already exist. End users may also require small subsets of a grid to be converted into readable text, and for the grid data to be available in graphic form. These supporting capabilities need to be made available.

9 Conclusions and Outlook

No global standard file format has existed for gridded geodetic data, yet these data are widely used within the geomatics and GIS communities. Usage is likely to increase with more data and modelling of intra-frame crustal deformations becoming available and with the increasing adoption of vertical datums defined by geoid models. GGXF offers a path to improved efficiency in the use of gridded geodetic data, for data used in coordinate transformations and exchanged for other purposes. Initially the GGXF standard will be maintained by OGC. There is an established pathway for OGC standards being moved into ISO, and this could be pursued. Future extensions of GGXF functionality beyond regularly-gridded data might be developed to include handling of sparse data, as well as other spatial representations of parametric functions such as triangulated networks.

References

Ben-Kiki O, Evans C, Ingerson B (2009) YAML Ain't Markup Language (YAML™) Version 1.2, Revision 1.2.2 (2021-10-01), https://yaml.org/spec/1.2.2/, Accessed 26 Jan 2024

IGN (1997) Grille de paramèters de transformation de coordonnées GR3DF97A, Notice d'utilisation, Institut Géografique National, April 1997

IGN (2012) Grilles de conversion altimetric. Institut Géografique National, July 2012

Johannessen J et al (2003) The European gravity field and steady-State Ocean circulation explorer satellite Mission its impact on geophysics. Surv Geophys 24(4):339–386. https://doi.org/10.1023/B:GEOP.0000004264.04667.5e

Junkins D, Farley S (1995) NTv2 National Transformation Version 2: developer's guide, geodetic survey division, geomatics Canada, Natural Resources Canada

Neacsu M (2010) Information about Geoid File Formats, Hypack Inc. Newsletter, May 2010

NGA (2013) Earth Gravitational Model 2008 (EGM2008). National Geospatial-Intelligence Agency, May 06, 2013

NGA (2014) NGA/NASA EGM96 Earth Gravitational Model. National Geospatial-Intelligence Agency, November 18, 2014

NOAA (1990) NADCON – the application of minimum-curvature-derived surfaces in the transformation of positional data from the North American Datum of 1927 to the North American Datum of 1983. NOAA NOS NGS Technical Manual 50

NOAA (2012) Manual for development and support of NOAA's vertical datum transformation tool, National Geodetic Survey Center for Operational Oceanographic Products and Services, June 2012

NOAA (2017) NADCON 5.0: geometric transformation tool for points in the National Spatial Reference System. NOAA Technical Report NOS NGS 63

OGC (2023) https://github.com/opengeospatial/CRS-Gridded-Geodetic-data-eXchange-Format/tree/master/examples, Accessed 13 Oct 2023

OGC (2024a) Abstract specification topic 24 - deformation model functional model. https://doi.org/10.62973/22-010r4. https://docs.ogc.org/as/22-010r4/22-010r4.html. Accessed 5 Jun 2024

OGC (2024b) The GGXF geodetic data grid exchange format v1.0. Open Geospatial Consortium, 29th April 2024. https://doi.org/10.62973/22-051r7. https://docs.ogc.org/is/22-051r7/22-051r7.pdf. Accessed 5 Jun 2024

OS (2004) OSTN02: a new definitive transformation from GPS-derived coordinates to National Grid coordinates in Great Britain. Surv Rev 37(293):502–519

PROJ (2023a) PROJ coordinate transformation software library. Open Source Geospatial Foundation. URL https://proj.org/. https://doi.org/10.5281/zenodo.5884394

PROJ (2023b) Geodetic TIFF Grids (GTG), PROJ specifications, Open Source Geospatial Foundation. https://proj.org/en/9.3/specifications/geodetictiffgrids.html, Accessed 13 Oct 2023

Reguzzoni M et al (2021) Open access to regional geoid models: the International Service for the Geoid. Earth System Sci Data 13:1653–1666. https://essd.copernicus.org/articles/13/1653/2021/

Stanaway R, Crook C, Kelly KM, Lott R (2024) A functional model for quantifying deformation in reference frame transformations, International Association of Geodesy Symposia, Berlin 2023. https://doi.org/10.1007/1345_2024_247

UCL (2012) Vertical Offshore Reference Frames, University College London, May 26, 2012

Open Access This chapter is licensed under the terms of the Creative Commons Attribution 4.0 International License (http://creativecommons.org/licenses/by/4.0/), which permits use, sharing, adaptation, distribution and reproduction in any medium or format, as long as you give appropriate credit to the original author(s) and the source, provide a link to the Creative Commons license and indicate if changes were made.

The images or other third party material in this chapter are included in the chapter's Creative Commons license, unless indicated otherwise in a credit line to the material. If material is not included in the chapter's Creative Commons license and your intended use is not permitted by statutory regulation or exceeds the permitted use, you will need to obtain permission directly from the copyright holder.

Signal Decomposition with InSAR Displacement Time Series Above a Storage Cavern Field: Example Epe (NRW, Germany)

Alison Seidel, Malte Westerhaus, Markus Even, and Hansjörg Kutterer

Abstract

Time series of interferometric SAR (InSAR) images offer the potential to detect and monitor surface deformation with high spatial resolution, even for slow deformation processes. However, many different sources contribute to phase changes which are used in InSAR to estimate displacements. Complex displacement mechanisms or strong atmospheric contributions can complicate the separation of these contributions and even cause problems when unwrapping the phase. A preliminary model of expected displacements can support this process but requires information about all involved deformation processes. However, as these processes are often the main subject of the investigation, they are not sufficiently understood in advance.

In this contribution, we approach this issue by analyzing InSAR time series results of regions with complex deformation behavior with the established statistical methods of principal and independent component analysis to identify dominant displacement patterns. We study Sentinel-1 InSAR data from 2015 to 2022 above the storage cavern field Epe in North Rhine Westphalia, Germany. Epe displays a spatially and temporally complex surface deformation field, which was described in previous studies as consisting of a linear signal relating to the cavern convergence as well as of seasonal and cavern pressure-dependent contributions. Our resulting displacement components can be clearly separated and appointed to different sources. This is supported by ground truth data and supplemental measurements of cavern pressure levels and groundwater levels. We also find that the previously described linear parametrization of displacements related to cavern convergence is no longer sufficient for longer time series.

Our results show that we can obtain source-dependent displacement models from long and complex InSAR time series when using ICA. These can then either be used to refine time series processing or to describe the physical processes causing to the surface displacements with a geophysical source model. Both will be the subject of future investigations.

Keywords

Component analysis · Distributed scatterers · Gas storage caverns · InSAR · Persistent scatterers · Radar interferometry · Satellite geodesy · Signal decomposition · Surface displacements

A. Seidel (✉) · M. Westerhaus · M. Even · H. Kutterer
Karlsruhe Institute of Technology, Karlsruhe, Germany
e-mail: alison.seidel@kit.edu

1 Introduction

Time series of the satellite-based imaging method of interferometric SAR (InSAR) can be used to detect and monitor surface displacements efficiently. InSAR utilizes the reflection of radar waves on the ground and measures the change of phase of the wave between two acquisitions to determine surface displacements along the line of sight (LOS) of the satellite in magnitudes of millimeters. Thus, InSAR complements the better-established methods of GNSS and leveling measurements. While GNSS provides 3D displacements with high temporal resolution, at usually very sparsely distributed stations, and leveling can measure vertical displacements on a denser spatial resolution than GNSS, but with measurements often once per year or even rarer, InSAR observes displacements both frequently and spatially densely.

However, other spatially correlated signals, especially atmospheric contributions can often not be removed completely from InSAR results. Moreover, rapid changes in surface displacements can also lead to parts of the displacements being falsely identified as spatially correlated noise (scn). Consequently, they are removed from the final displacement result as shown by Even et al. (2020). Even if the atmospheric contributions can be sufficiently modeled with high spatial and temporal coverage, multiple displacement phenomena will remain which need to be separated. Although the respective causes might neither be connected nor be of particular interest, they will have impact on the analysis results as they superpose each other. This can complicate the interpretation of the displacement map.

If knowledge about the respective local displacement processes is available, individual time series can be analyzed manually as a first option. Then, only such scatterers are selected that exclusively contain the desired signal. However, this can lead to a loss of information in scatterers that contain multiple signals contributions.

If a deformation process is sufficiently understood, it is – as a second option – also possible to make predictions for future displacements that follow this process and to detect unexpected changes in surface behavior quickly. Subsequently, one can then deduce if the effect could relate to the source of interest or if it is caused by other processes such as surface response to precipitation. Thus, it is important to be able to differentiate displacements from different superposing source mechanisms.

Here, we analyze InSAR time series with statistical component analysis methods to separate the different signals in a complex displacement field statistically. By this, we explore a method to obtain displacements originating from individual sources from complex conditions without prior knowledge.

Our study area is the gas storage cavern field Epe where Even et al. (2020, 2022) described a complex surface displacement field with a 3-year InSAR time series study. We utilize their method for joint processing of Persistent and Distributed Scatterers (PS and DS) for a longer, 7-year time series and examine these results with the component analyses. In doing so, we gain a deeper understanding of the source processes of the surface displacements, and we can show that the components can provide a better displacement model than the parametrization used by Even et al. (2020, 2022).

2 Area of Study: Epe Gas Cavern Field

2.1 Overview

The gas storage cavern field Epe in NRW, Germany currently has 114 caverns located in the Zechstein salt layer in depths between 1,000 and 1,200 m which are used for brine production, oil and gas storage. More than 50 caverns, primarily located in the central and eastern part of the field (see Fig. 1), are filled with natural gas and are operated by six different gas provider companies. The companies operate the storage independently from each other, but do follow a similar cycle of filling the caverns in summer and emptying them in winter.

The cavern field is regularly monitored with annual leveling campaigns that cover over 500 measurement points across the entire field and three GNSS stations that were installed in 2018 (see Fig. 1). There are also several groundwater measurement points (GWMPs) available where water levels are measured regularly. Cavern filling levels are published online at the Aggregated Gas Storage Inventory (AGSI) as daily mean values for each provider since 2016 (longer for some providers). Due to these ground truth measurements and supplemental data, this region is ideal to explore and validate data-driven ways to differentiate complex displacements from different sources and obtain models, as results and predictions can be verified.

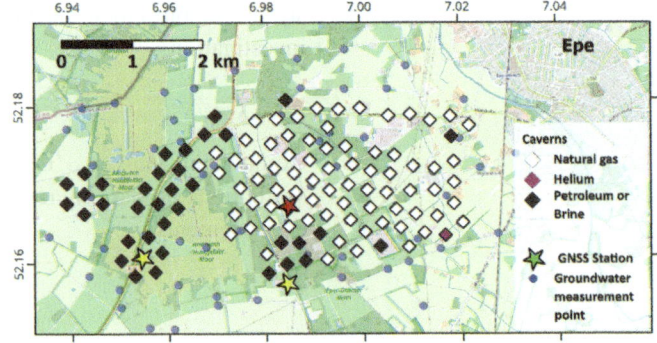

Fig. 1 Overview of the gas storage cavern field Epe. White diamonds indicate the location of natural gas caverns; black diamonds indicate liquid filled caverns

2.2 Surface Displacements and Known Sources

Convergence of gas caverns in salt bodies is a well-known process that occurs due to the lower pressure of the gas inside the caverns compared to the surrounding rock. This process usually causes a subsidence bowl in the area above the caverns which has been described at a number of places (e.g., Tajduś et al. 2021; Xie et al. 2018; Wang et al. 2014). For Epe and InSAR, Even et al. (2020, 2022) were able to develop a simple geophysical source model from 3 years InSAR time series for describing the mean subsidence rate through cavern convergence. They also observed and modeled a cyclic signal corresponding to the withdrawal and filling of the caverns. Finally, they found a strong seasonal signal superposing the other two effects. This occurs mainly in the western part of the cavern field above a fen, related to ground water levels. With the aid of groundwater measurements and the cavern filling levels, Even et al. (2020) could separate and model three main signals in the InSAR time series and described them as follows (Figs. 2 and 3):

1. An approximately linear subsidence above all caverns. The extent depends mainly on the cavern filling type.
2. Cyclic displacements which are a delayed response (approx. 90 days) to the withdrawal and filling of the gas caverns only.
3. A seasonal, periodic displacement over the fen area as a surface response to ground water changes.

They also parameterized the composition of the displacement field by the following equation, defining three spatially varying parameters p for the three different source mechanisms:

$$d_{\text{Epe}} = a + p_{\text{lin}}\, t + p_{\text{pres}}\, m_{\text{pres}}(t) + p_{\text{fen}}\, m_{\text{fen}}(t) \quad (1)$$

with d_{Epe} being the observed displacements at an arbitrary position in the cavern field, $a + p_{\text{lin}}\, t$ describing the linear trend of displacements, $m_{\text{pres}}(t)$ the time-dependent model curve describing the surface response due to pressure changes and $m_{\text{fen}}(t)$ the time-dependent model curve for groundwater caused displacements in the fen.

2.3 Data

For this study, we processed long time series from April 2015 to December 2022 of Sentinel-1 SAR acquisitions of the two tracks ascending 15 and descending 139. Both tracks cover the complete area of interest so that we obtained time series of the respective displacements in line of sight (LOS) of both satellite view geometries. Orbit combination is only performed in one specific case, and all other analysis is done on the LOS-data. We performed our analysis on the processed tracks of both orbits separately. They generally show the same described phenomena and are in good agreement with supplemental measurements.

3 Methodology

3.1 Procedure Overview

To obtain the InSAR time series, we use an augmented version of the Stanford Method for Persistent Scatterers (StaMPS). This augmentation was developed by Even (2019) to include the selection and joint processing of Distributed Scatterers, similar to the SqueeSAR algorithm by Ferretti et al. (2011). We follow the processing settings specified in Even et al. (2020), except that we only use a simple linear model for aiding the estimation of spatially correlated noise and unwrapping. This is to simulate processing time series without having prior knowledge of the involved displacement processes. Even in these complex displacement conditions, this proved to be sufficient for the resulting displacement time series to not contain major unwrapping errors. We intentionally keep the estimated scn-phase contributions in our

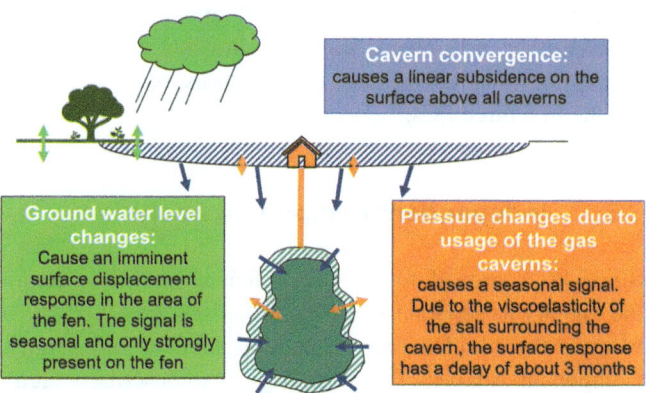

Fig. 2 Schematic depiction of the three main processes that cause surface displacements in Epe at the example of a single gas cavern

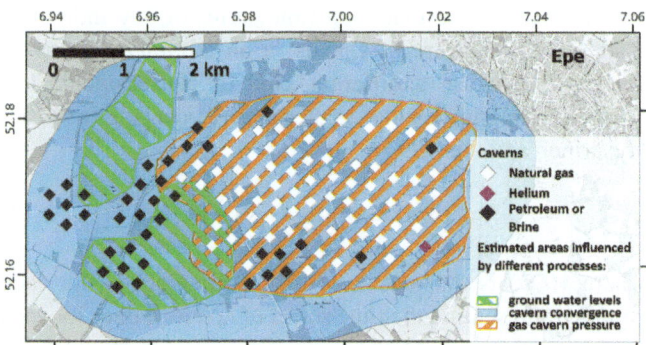

Fig. 3 Approximate areas that are affected by the different signals, derived from the results of Even et al. (2020)

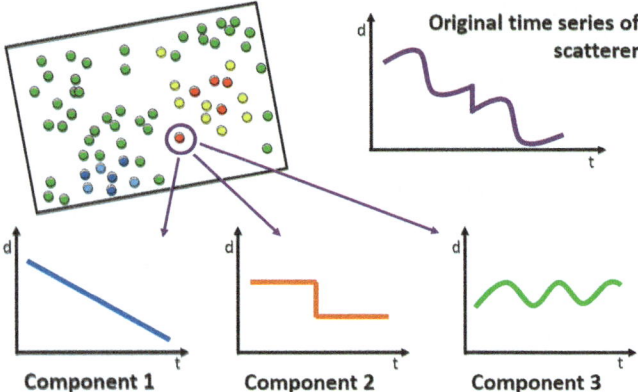

Fig. 4 Schematic depiction of three component time series that were obtained from the inverted components of a single scatterer. As comparison, the original time series is also depicted

final StaMPS result to preserve potentially leaked displacement contributions and only remove the estimated orbital error and spatially correlated look angle error contributions.

In the following, Principal Component Analysis (PCA) is applied first to our displacement time series result and then Independent Component Analysis (ICA). Other studies that utilize PCA or ICA in the InSAR context (see, e.g., Ebmeier 2016; Gaddes et al. 2018; Lin et al. 2010; Vajedian 2020) perform this kind of analysis on whole unwrapped interferograms, often in an SBAS network to keep temporal decorrelation low. Although this has the benefit of retaining the full interferograms for the analysis, this is not applicable in Epe where the displacements are small and noisy. For this reason, decorrelation in interferograms with long temporal baselines is high due to large areas of vegetation and unwrapping the interferograms individually was not successful.

As the component analysis methods reduce the dimensions of our data, we invert the obtained independent components individually back to the original spatiotemporal dimension space to obtain time series of the components (in the following called "component time series") for each scatterer (see Fig. 4). This process will be further explained in Sect. 3.3.

We can now analyze the temporal curves of the components as well as the spatial distribution and magnitude of contribution of every component in each scatterer. After we have successfully identified components that correspond to our expected source mechanisms, we use them as displacement model curves in a second InSAR time series processing to improve our result.

3.2 Principal Component Analysis

Principal component analysis (PCA) is commonly used in various scientific fields to reduce multidimensional datasets to the most relevant dimensions. It is a linear transformation which transforms the data into a new coordinate system, so that the dimension of the greatest variation in the data is used as the first principal component (PC). Accordingly, the second PC describes the dimension with second greatest variance, etc. All PCs are mutually orthogonal. PCA has been used in InSAR data analysis before: to filter out uncorrelated signals, and to differentiate signals from various sources (see, e.g., Lin et al. 2010). We perform PCA on our first result of InSAR displacement time series, where we use all time series of all scatterers of one track as input, with each SAR image acquisition time acting as individual measurement for each scatterer. Thus, we obtain PCs that describe a specific displacement behavior over the observation time, with a parameter value that describes the magnitude of this behavior in each scatterer (see Sect. 4.2, Figs. 8 and 9 for visualization). Ideally, such an analysis would result in individual components for each displacement source, as well as one for the scn term, consisting mainly of the atmospheric contributions.

We found for Epe that the first three PCs explain about 80% of the total variance in the data in both tracks. Most of the residual signal (not described by the first three PCs) is located in the area of the fen where the displacement is dominated by the surface response to ground water level changes. Here, we can expect a more complex deformation pattern that might be related to different soils and fields. Since further components show very low but overall similar magnitudes of total variance contributions, we disregard all but the first three PCs.

3.3 Independent Component Analysis

Since PCA proved to be not sufficient in differentiating the signals to separate source mechanisms (see Sect. 3.2), we use the blind source separation method Independent Component Analysis (ICA). ICA aims to describe a signal as linear mixture of various statistically independent sources or components that each has a non-Gaussian probability distribution (Comon 1994; Hyvärinen and Oja 1997). For InSAR, we assume that each time series of any individual scatterer is a mixture of signals of an arbitrary number of sources that cause a different kind of phase signal. The source and mixture relation is then described as follows (after Hyvärinen and Oja 1997):

$$x(t,n) = A(m,t)\ s(n,m) \qquad (2)$$

where $x(t,n)$ is the matrix of observed signals, with each column representing the time series of t phase values for each of the n scatterers. Matrix $s(n,m)$ consists of m unknown independent source components (ICs) at each of the n scatterers. $A(m,t)$ then, is the unknown mixing matrix that contains coefficients that describe the temporal behavior for each of the m components.

$$s(n, m) = W(m, t)\, x(t, n) \qquad (3)$$

As $A(m, t)$ is unknown, we use its inverse $W(m, t)$, the so-called unmixing-matrix, as in Eq. (3), to transform the observation vectors to their independent sources, by maximizing the non-Gaussianity of the presumed sources. We use an implementation of the FastICA algorithm (Hyvärinen and Oja 2000) which starts with centering and whitening the data, while removing each observed signals mean value and ensuring that the signal consists of uncorrelated variables. Due to this, the order and sign of the resulting components are no longer significant. As Ebmeier (2016) describes in detail, ICA can be applied to multitemporal InSAR in two ways: by maximizing either spatial or temporal independence of the sources. We use the former approach, by treating each phase value at each SAR image acquisition time as individual observation for each scatterer, as detailed in Eq. (2). The algorithm then solves for A and s, so that the rows of A contain the "model time series" of the independent sources and s contains parameters (ICs), that describe the magnitude of how much each of these "model time series" describes the signal in each scatterer.

We can invert the ICs back individually by multiplication of $W(m, t)$ with $s(n, m)$, and setting all rows of $s(n, m)$ to zero, except the one corresponding to the desired component, and then reverse whitening and centering. Thus, we obtain component time series in the original space and time dimensions.

In general, the optimum number of ICs is unknown. Ebmeier (2016) suggests as many components as there are interferograms. However, we perform ICA on selected scatterers of a time series analysis result, which were already cleaned from most non-displacement phase contributions, rather than on whole interferograms. The exception to this might be atmospheric contributions that were not already removed through PCA. Still, we do expect that our data is mainly a mixture of different displacement sources and thus we can reduce the number of components significantly. Also, since PCA showed that only three principal components describe most of our data, we found that only two independent components already gave good results.

3.4 Parameter Fit and Reprocessing

For obtaining even more information about the spatial distribution of the signal of our displacement processes and improving our time series results, we perform a second StaMPS processing, with the same set up and data as before, but we now use our component time series as models for the displacements of the different parameters instead of a linear model. The application of this model during processing is given in Even et al. (2020). We follow the parametrization for Epe's displacement composition as described in Eq. (1) and use the normalized component time series of our two independent components as model curves for $m_{\text{pres}}(t)$ and $m_{\text{fen}}(t)$, respectively. Since both ICs still contain a negative trend, we fit a linear trend to them and remove it from the curves. Thus, we can fit all three parameters (p_{lin}, p_{pres}, and p_{fen}) from Eq. (1) individually to the InSAR time series through a parameter space search. We deliberately choose a lower temporal phase coherence threshold for the DS (see Even et al. 2020) to obtain a higher number of scatterers. So, the signal parts described with our models are estimated even for less-trusted scatterers.

4 Results

4.1 InSAR Displacement Field

The results of the original processing of the ascending and descending tracks, as can be seen in Fig. 5, show strong negative displacements (i.e., subsidence) in the center of the cavern field with displacements of up to 20 cm over the 7½ years of the time series. The maximum of the displacements seems to be located more to the west in the ascending than in the descending track, which indicates the presence of significant horizontal displacements. In the area of the fen, the displacements look more similar between ascending and descending.

Time series of scatterers in the fen show a negative trend that is superposed by a distinct seasonal sinusoidal signal (see Fig. 6, top). Scatterers located above the gas caverns display an even stronger negative trend, but also show cyclic behavior. In all scatterers above the gas caverns, there is a prominent negative displacement in a short period of time in 2018, while a similarly (but later reverting) strong negative displacement appears in the scatterer in the fen.

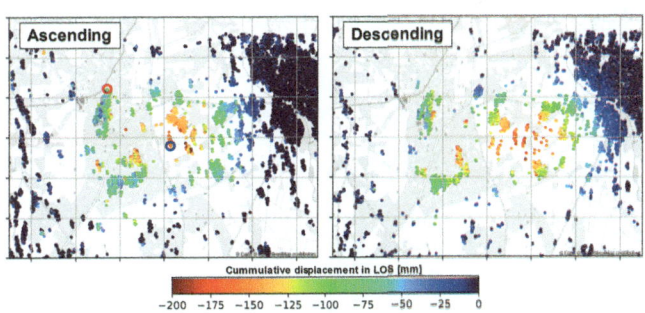

Fig. 5 Displacement field in line of sight, cumulative displacement of ascending (left) and descending (right) tracks. Two scatterer positions are marked in the left plot, of which we display time series in Fig. 6

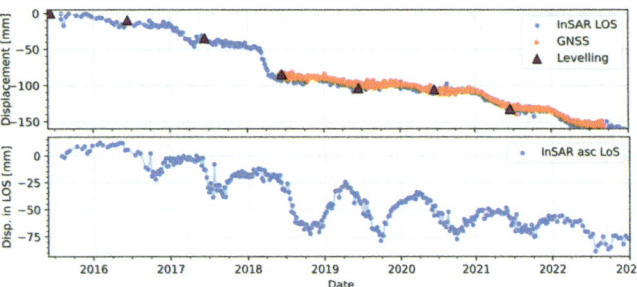
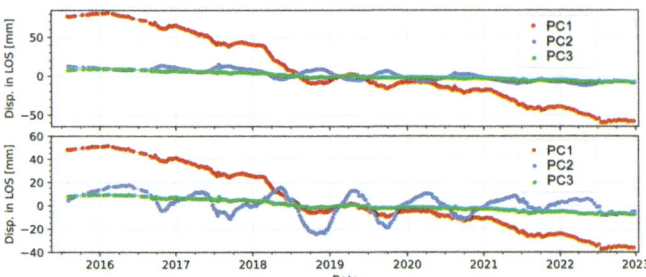

Fig. 6 InSAR time series of two scatterers of the ascending track. Both scatterers were chosen so that they are most likely only affected by the convergence caused subsidence and either the ground water-dependent displacement in the cavern pressure-dependent displacement (top) or in the fen (bottom). Both are marked in Fig. 5. Time series of GNSS station data and yearly leveling measurements (projected to LOS), located close to the first scatterer, agree with the InSAR result

Fig. 8 Component time series of the two scatterers of the ascending track from the first three PCs. Top: in the center area of the gas caverns, bottom: in the upper fen area

However, this rapid displacement occurs in March 2018 above the gas caverns, and 3–6 months later in the fen, implying that even though appearing similar, the causes for these displacements are not likely to be the same. We validate our displacement time series, by transforming GNSS and leveling measurements at similar positions to scatterers to LOS and find a good agreement between the results (see Fig. 6, top).

4.2 Components from PCA and ICA

4.2.1 PCA

As mentioned before, the PCA shows that most of the displacement time series data can be described by three principal components. When displaying their magnitudes corresponding to each scatterer spatially (see Fig. 7), one can see that the first PC contains a signal which is present in the entire cavern field, including the fen, having its maximum in the center of the field. Outside of the cavern field, it shows slightly negative values. The second PC has a maximum in the fen and a minimum of a similar magnitude but with a reversed sign above the gas caverns. The third PC shows a patchy behavior and is mostly present in the lower fen area.

After inverting each component individually back to the original dimension space, we can plot the component time series at the individual scatterers, as seen in Fig. 8. We show two component time series of the same scatterers as in Fig. 6, as examples, where we expect the pressure-related surface response, and respectively the ground water-dependent surface response to dominate. We see that the second PC, which is relevant in both chosen scatterers, as shown in Fig. 7, displays a periodic, seemingly seasonal behavior. Now, the first component shows a rather linear trend, but also has a cyclic component, which is inverse to PC2 at the gas caverns.

4.2.2 ICA

When trying to determine the optimum number of ICs, we find that we obtain the best results with two ICs. As the resulting components can differ when performing ICA on the same dataset multiple times, we choose those two components that statistically occurred most often. In both InSAR tracks, one of the two components only affects scatterers in the fen, while the other component has its maximum above the gas caverns (Fig. 9).

In addition, we inverted the components individually and looked at the component time series of those scatterers that display the strongest magnitude of the components (see Fig. 10). Here, we found that both shows a behavior very similar to the InSAR time series in Fig. 6. which were manually picked to contain only one signal contribution, while the respective other component shows almost no contribution in these scatterers.

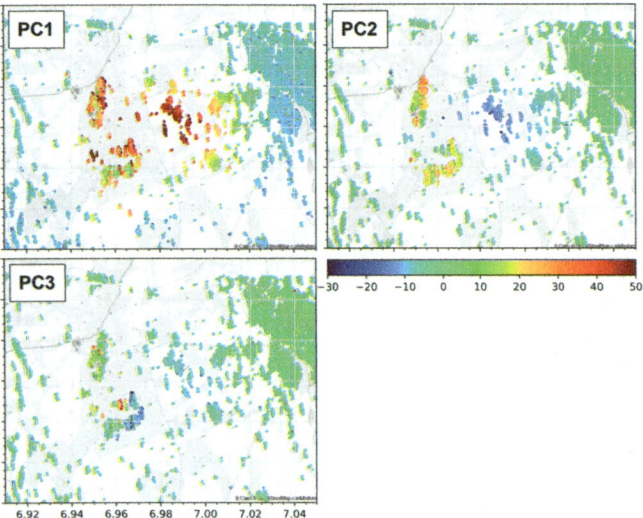

Fig. 7 Results of the PCA of the ascending track with three PCs

Signal Decomposition with InSAR Displacement Time Series Above a Storage Cavern Field: Example Epe (NRW, Germany)

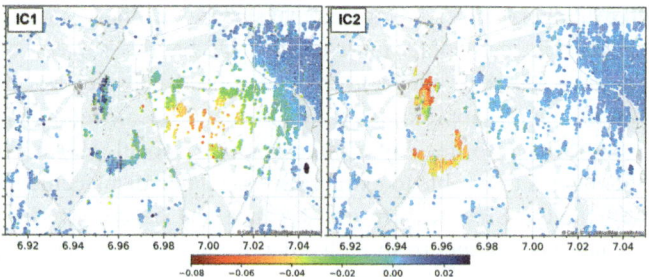

Fig. 9 Results of two ICs of the descending track. The first IC displays its absolute maximum in the center of the cavern field, while the second component only shows up in scatterers in the fen area

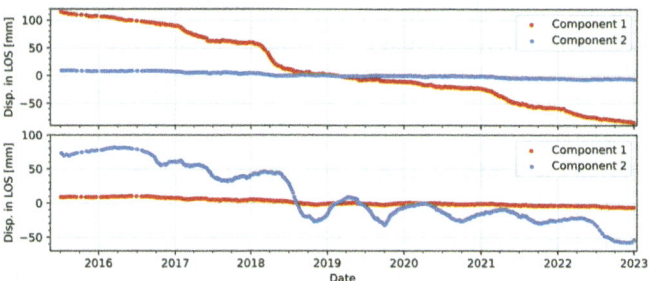

Fig. 10 Component time series from two independent components from the descending track. (**a**) At the maximum of the second component, in the center area of the gas caverns, (**b**) at the maximum of the first component, in the northern fen area

4.3 Parameter Space Search and Fit

The distribution of p_{pres} and p_{fen} in the now denser number of scatterers shows high values above the gas caverns with slowly declining values outward, and high values above the fen, with defined edges between high and low parameter values at the edges of the fen, respectively (see Fig. 11). Their maximum total displacement is 40 mm and 60 mm, respectively. However, p_{lin} shows less clearly distributed values and even displays high values outside of the cavern field. Displacement rates reach up to 50 mm/year in the center of the cavern field, but also for individual scatterers outside.

5 Discussion

5.1 IC Relation to Gas Cavern Filling Levels

Our component time series show very strong similarity with the original time series of the scatterers which we identified as being mostly caused by either the fen or the pressure-related displacements. The spatial distribution and the temporal behavior of the independent components support this observation. Therefore, we can confidently assign IC1 to describing gas cavern behavior and IC2 to the ground water-related displacement. We prove this claim by comparing the time series of IC1 with the filling levels of the gas caverns of Epe in the observation time (see Fig. 12).

For this purpose, we determined the mean filling level of Epe by compiling filling data of all providers (AGSI 2023) and accounted for how many gas caverns are used by each provider. We observe the cyclic signal throughout the component time series having stronger magnitudes where the filling levels were lower (better visible in Fig. 6). This is especially distinct in early 2018, where due to a cold winter, the filling levels dropped down to only 20% resulting in very low pressure inside the caverns and a strong negative displacement of about 5 cm in the component time series. The visible time delay between the emptying of the caverns and the surface response is caused by the viscoelasticity of the salt body in which the caverns lie, as modeled by Even et al. (2020, 2022). As the component time series is dominated by a strong negative trend, a removal of the fitted trend would presumably result in a similar model curve as

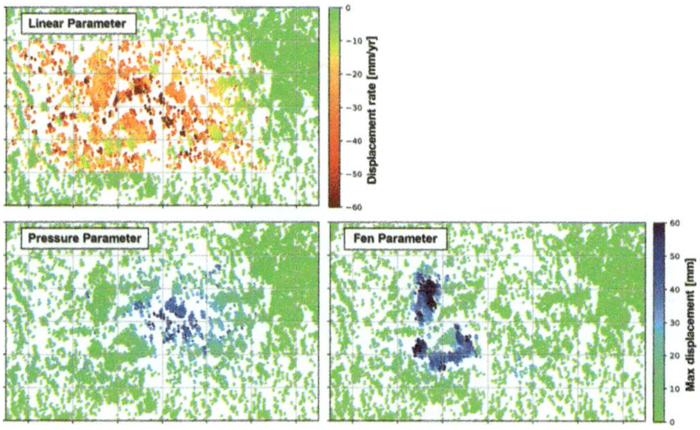

Fig. 11 Fitted parameters in the second InSAR time series result for the track of the ascending orbit. The linear parameter is displayed in mm/year, the pressure and fen parameter as maximum total displacement in mm

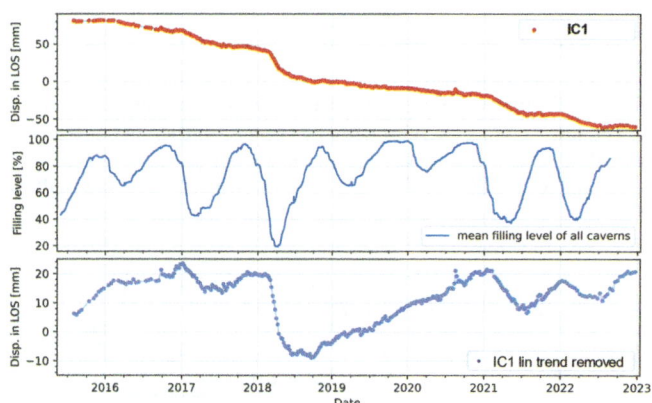

Fig. 12 Displacement time series of IC1 in LOS of the ascending track in the center of the gas caverns (top), component time series of mean filling levels of the gas caverns (source: AGSI 2023) (middle), and trend removed displacement time series of IC1 (bottom)

the filling levels. We find however, that this assumption is only approximately true for the timespans between 2015–2018 and 2021–2023 but does not hold for the years from 2018 to 2021 as displayed in Fig. 12 bottom).

We conclude that the cavern convergence and therefore the resulting subsidence vary with different years, presumably depending on the level of depletion of the caverns in the year. To check this hypothesis, we perform a simple orbit combination (as described in Wright et al. 2004) for one scatterer of both tracks. With this we obtain the vertical displacements in one scatterer of the component time series. We disregard the north component due to lack of sufficient view geometries, being aware that this leads to errors in the estimated vertical displacements. Subsequently, we determined the dates in every year, where the gas caverns reached their highest filling level and added 90 days to account for the delayed surface response described in Sect. 2.2. We then calculate annual linear trends between the displacements at those points in time (see Fig. 13). Subtraction of these trends from the IC1 component time series yields a much better fit with the filling levels (see Fig. 14). The delayed surface response can also be seen clearly now.

Residual discrepancies between the curves can have many causes, such as the influence of the disregarded north displacement component during orbit combination, or the differing depletion levels between different providers and the position of the scatterer regarding to caverns of these providers. Additionally, as described in Liu et al. (2019), the relation between cavern depletion and convergence is even more complex and depends on additional parameters like temperature. And again, the surface response to this process is also more intricate. It will be the subject of future research if the convergence caused subsidence can be sufficiently approximated by an annual linear trend or if we will have to develop a more complex model. However, ICA seems to be able to extract cavern-related displacements sufficiently. Since cavern convergence and cavern pressure are not independent from each other, it is reasonable that ICA cannot separate their effects in surface displacement. Also, in our case, ICA cannot differentiate between gas cavern signals of different providers, which is albeit probably rather related to the close proximity of these caverns to each other and only small differences in filling levels between the providers. Such differences in surface response are too small to be detected with InSAR.

5.2 IC Relation to Groundwater Levels

A similar comparison as in Sect. 4.1 can be made with IC2 and its assumed ground water level relation. We compare component time series of IC2 at the point of its maximum with groundwater measurements (GWM) located near these scatterers. The irregular, yet seasonal shape of the water levels is also visible in the component time series of IC2 (see Fig. 15). After removing a linear trend from the curve, this relation becomes even more clear, proving that IC2 indeed describes the surface displacement related to groundwater changes, which primarily occurs in the fen.

We see that, again between 2018 and 2021, the curve of the trend-removed graph seems to match the GWM a bit less, similar to what we observed for IC1. This indicates that this area may also be affected by yearly varying trends, even though not as prominently as the area above gas caverns. However, it is less likely that these trends are actually caused

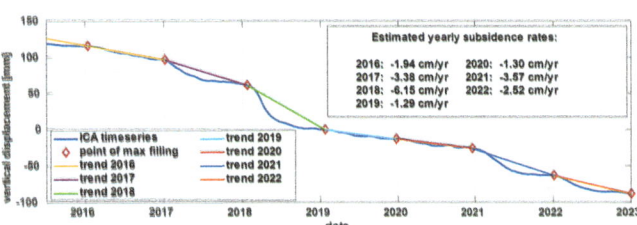

Fig. 13 Component time series of vertical displacements, derived from IC1 in the center of the subsidence bowl and estimated trends between the points of maximum cavern filling of each year shifted by 90 days (as described in Even et al. 2020) to consider for the viscoelasticity of the salt

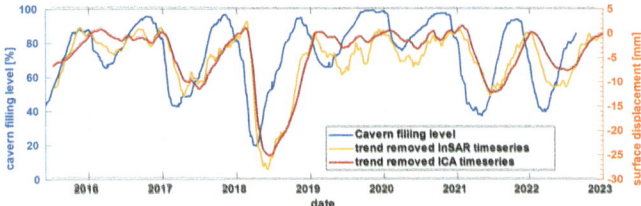

Fig. 14 Vertical displacement curve of original InSAR and IC1 component time series when a annually varying linear trend is removed together with a time series of the mean of the cavern filling levels of all companies

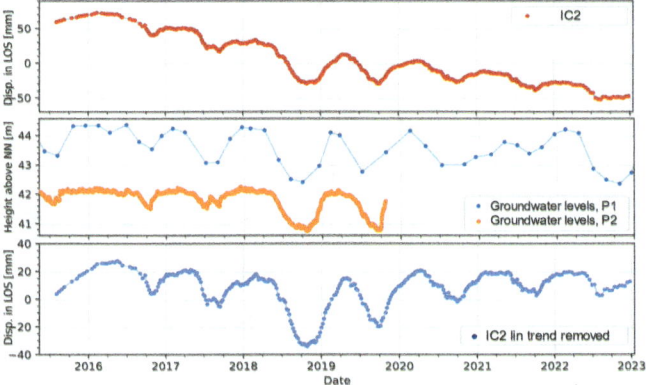

Fig. 15 Component time series of IC2 of the ascending track in the upper fen (top), time series of groundwater levels at two points in the fen (middle) and trend removed component time series of IC2 (bottom). One of the GWMPs only has bi-monthly (or less) measurements, while the other one has daily measurements, but only continues to late 2019, which is why both were included

by the gas caverns, as they are mostly too far away to affect all scatterers in the fen. Rather we presume that these subsidence trends are a combination of the (assumed) linear trends caused by convergence of liquid filled caverns and subsidence that occurs due to the ongoing drying of the fen.

5.3 Parameter Discussion

By utilizing the normalized component time series of the components as displacement models for further time series processing and allowing for DS with lower temporal coherence, we are able to retrieve spatially dense information about the distribution of the signal for the cavern pressure and precipitation-related displacements. The spatial pattern and estimated parameters match with our expectations in terms of magnitude and location of fen boundaries and gas cavern positions and agree with the findings of Even et al. (2020, 2022). However, in case of the linear displacement parameter, the results do not show the expected distribution described by Even et al. (2020, 2022). Even though this parameter seems to be generally higher in the center of the cavern field, there are many scatterers with high values outside of the cavern field. Here we would not expect any relation to cavern convergence. This is presumably related to the inclusion of scatterers with lower temporal coherence. It could also be due to a simple linear trend not being distinct enough to only occur in the subsidence bowl of the cavern field as there is a variety of possible causes for linear trends in scatterers. However, since a linear trend cannot describe displacements caused by cavern convergence for the gas caverns anyway (as shown and discussed in Sect. 4.1), we expect this issue to be resolved when we develop a more complex model for this part of the signal and include it in the parameter estimation instead.

6 Conclusions and Outlook

We used component analysis methods on InSAR time series data of a complex displacement field and showed at the example of the gas cavern field Epe that the retrieved components through ICA can be associated to different source mechanisms. Thus, we show that signal separation in such a complex deformation field is possible, even without using a geophysical source model or a-priori information about the distribution of the different types of displacement.

For other complex displacement fields, where less knowledge about the nature and distribution of the displacement processes is available, this statistical approach of source separation can support extracting and differentiating dominant displacement patterns. Studying this statement in settings with stronger, more complex atmospheric contributions than those occurring in Epe need to be addressed in future studies.

In the case of Epe, the two components identified as the dominant displacements can be assigned to cavern-related and groundwater-related displacements. We were not able to separate displacements originating from cavern convergence and cyclic pressure change with ICA. Trying to fit and remove a linear trend manually proved to not sufficiently describe both displacements. We found that we can better approximate the cavern convergence trends with annual linear trends, depending on the cavern usage in that year. In doing so, we improve upon the parameterized model of the cavern field displacements described in Even et al. (2020, 2022). Our work provides a necessary basis for developing an accurate geophysical source model of the cavern field. This will then help gain a deeper understanding of the processes involved that relate cavern depletion and subsequent convergence to subsequent surface displacement.

Acknowledgments We thank Stefan Meyer from SGW (Salzgewinnungsgesellschaft Westfalen) for providing the GNSS and leveling data, as well as the groundwater measurement data for Epe. We also thank the German Federal Ministry of Economics and Climate Protection (BMWK) for funding this work as part of the SAMUH2 project.

References

Aggregated Gas Storage Inventory (AGSI). https://agsi.gie.eu/#/. Accessed 10 Sept 2023

Comon P (1994) Independent component analysis, a new concept? Signal Process 36:287–314. https://doi.org/10.1016/0165-1684(94)90029-9

Ebmeier SK (2016) Application of independent component analysis to multitemporal InSAR data with volcanic case studies. JGR Solid Earth 121:8970–8986. https://doi.org/10.1109/TGRS.2023.3322595

Even M (2019) Adapting stamps for jointly processing distributed scatterers and persistent scatterers. In: Proceedings of the IGARSS 2019–2019 IEEE international geoscience and remote sensing symposium, Yokohama, Japan, 28 Jul–2 Aug 2019, pp 2046–2049. https://doi.org/10.1109/igarss.2019.8897808

Even M, Westerhaus M, Simon V (2020) Complex surface displacements above the storage cavern field at Epe, NW-Germany, observed by multi-temporal SAR-interferometry. Remote Sens 12:3348. https://doi.org/10.3390/rs12203348

Even M, Westerhaus M, Seidel A (2022) Konvergenz- und druckabhängige Oberflächenverschiebungen über einem Kavernenspeicherfeld in NW-Deutschland, beobachtet mit Methoden der Radarinterferometrie. In: DGMK/ÖGEW Frühjahrstagung 2022 – Geo-energy-systems and subsurface technologies – key elements towards a low carbon world. Deutsche Wissenschaftliche Gesellschaft für Erdöl, Erdgas und Kohle e.V. (DGMK), pp 33–41

Ferretti A, Fumagalli A, Novali F, Prati C, Rocca F, Rucci A (2011) A new algorithm for processing interferometric data-stacks: SqueeSAR. IEEE Trans Geosci Remote Sens 49:3460–3470. https://doi.org/10.1109/TGRS.2011.2124465

Gaddes ME, Hooper A, Bagnardi M et al (2018) Blind signal separation methods for InSAR: the potential to automatically detect and monitor signals of volcanic deformation. JGR Solid Earth 123. https://doi.org/10.1029/2018JB016210

Hyvärinen A, Oja E (1997) A fast fixed-point algorithm for independent component analysis. Neural Comput 9:1483–1492. https://doi.org/10.1162/neco.1997.9.7.1483

Hyvärinen A, Oja E (2000) Independent component analysis: algorithms and applications. Neural Netw 13:411–430. https://doi.org/10.1016/S0893-6080(00)00026-5

Lin YN, Kositsky AP, Avouac J-P (2010) PCAIM joint inversion of InSAR and ground-based geodetic time series: application to monitoring magmatic inflation beneath the Long Valley Caldera. Geophys Res Lett 37. https://doi.org/10.1029/2010GL045769

Liu H, Zhang M, Liu M, Cao L (2019) Influence of natural gas thermodynamic characteristics on stability of salt cavern gas storage. IOP Conf Ser Earth Environ Sci 227:042021. https://doi.org/10.1088/1755-1315/227/4/042021

Tajduś K, Sroka A, Misa R et al (2021) Surface deformations caused by the convergence of large underground gas storage facilities. Energies 14:402. https://doi.org/10.3390/en14020402

Vajedian S (2020) Incorporating independent component analysis and multi-temporal SAR techniques to retrieve rapid postseismic deformation. ISPRS Ann Photogramm Remote Sens Spatial Inf Sci 3:173–178. https://doi.org/10.5194/isprs-annals-V-3-2020-173-2020

Wang T, Yang C, Yan X et al (2014) Dynamic response of underground gas storage salt cavern under seismic loads. Tunn Undergr Space Technol 43:241–252. https://doi.org/10.1016/j.tust.2014.05.020

Wright TJ, Parsons BE, Lu Z (2004) Toward mapping surface deformation in three dimensions using InSAR. Geophys Res Lett 31:1–5. https://doi.org/10.1029/2003GL018827

Xie P, Wen H, Wang G (2018) An analytical solution of stress distribution around underground gas storage cavern in bedded salt rock. J Renew Sustain Energy 10. https://doi.org/10.1063/1.5004238

Open Access This chapter is licensed under the terms of the Creative Commons Attribution 4.0 International License (http://creativecommons.org/licenses/by/4.0/), which permits use, sharing, adaptation, distribution and reproduction in any medium or format, as long as you give appropriate credit to the original author(s) and the source, provide a link to the Creative Commons license and indicate if changes were made.

The images or other third party material in this chapter are included in the chapter's Creative Commons license, unless indicated otherwise in a credit line to the material. If material is not included in the chapter's Creative Commons license and your intended use is not permitted by statutory regulation or exceeds the permitted use, you will need to obtain permission directly from the copyright holder.

Reviewers

Agnieszka Jenerowicz
Alejandro Blazquez
Alexa Putnam
Alexander Kehm
Alexandre Couhert
Ambrus Kenyeres
Amir Khodabandeh
Andreas Bauch
Anna Klos
Anno Löcher
Axel Nothnagel
Bart Root
Benjamin Männel
Bert Wouters
Blažej Bucha
Bo Wang
Charon Birkett
Christian Bizouard
Christoph Foerste
Christopher Kotsakis
Corné Kreemer
Don Chambers
E. Sinem Ince
Ernst Schrama
Felix Landerer
Florian Kunzi
Frank Lemoine
Georgios S. Vergos
Gino Tuccari
Gregor Möller
Hana Krasna
Hansjörg Kutterer
Hans-Jürgen Götze
Harald Schuh
Heiner Denker
Hugo Lecomte
Hussein A. Abd-Elmotaal
Jan Martin Brockmann
Jean-Francois Cretaux
Jean-Paul Boy
Jean-Phillipe Montillet
Jeff Freymueller

Jianliang Huang
Joana Fernandes
Johannes Böhm
Jürgen Kusche
Justyna Sliwinska-Bronowicz
Kirsten Elger
Krzysztof Nowel
Krzysztof Sosnica
Lennard Huisman
Lisa Kern
Lucia McCallum
Manon Dalaison
Manuela Seitz
Martin Lidberg
Mathis Bloßfeld
Matt A. King
Matthias Glomsda
Maxime Mouyen
Michael Craymer
Mike Chin
Miltiadis Chatzinkos
Oleg Titov
Oliver Montenbruck
Pascale Defraigne
Pavel Novák
Riccardo Barzaghi
Róbert Cunderlík
Roland Pail
Rowena Lohman
Salim Masoumi
Santiago Belda
Takuya Nishimura
Thomas Gruber
Thomas Herring
Tilo Schöne
Tomasz Liwosz
Urs Marti
Vincenza Tornatore
Xavier Collilieux
Xiaopeng Gong
Zohreh Adavi

© The Author(s) 2025
J. T. Freymueller, L. Sánchez (eds.), *Together Again for Geodesy*,
International Association of Geodesy Symposia 157, https://doi.org/10.1007/978-3-031-91167-5

Author Index

A
Altamimi, Z., 103
Andersen, O.B., 303
Angermann, D. 391
Antokoletz, E.D., 227, 273
Arnold, D., 179, 199, 207

B
Bachmann, S., 165
Balasubramanian, N., 77, 131
Balidakis, K., 227
Bamahry, F., 401
Barzaghi, R., 253, 263
Bedford, J., 381
Belda, S., 165
Bizouard, C., 157
Böhm, J., 3, 25, 77
Böhm, S., 131
Borisov, M., 311
Boy, J.-P., 361
Bruyninx, C., 401

C
Calvés, G.M., 17
Chen, J., 103
Cheng, Y., 157
Choi, E.-K., 245
Choi, S., 245
Ciuban, S., 111
Coleman, R., 303
Collilieux, X., 103
Courde, C., 103
Craddock, A., 391
Crook, C., 63, 411

D
Dach, R., 341
Dahle, C., 119
Deng, X., 303
Dettmering, D., 297
Dikshit, O., 77, 131
Dobslaw, H., 227
Drewes, H., 53
Du, Z., 103

E
Ejigu, Y.G., 87, 361
Engelhardt, G., 341
Even, M., 417

F
Fabian, A., 401
Ferguson, S., 219
Flechtner, F., 119
Fletling, N., 187
Flohrer, C., 139, 165, 341
Foroughi, I., 219
Forsberg, R., 281
Frederikse, T., 69
Furhmann, T., 103

G
Garcia, C., 381
García-García, D., 369
Gavriilidou, G., 213
Geisser, L., 207
Gisinger, C., 103
Glaser, S., 33, 119, 381
Goli, M., 219
Gómez, A.R., 273
Gou, J., 147
Gruber, T., 103
Guagni, H.J., 273

H
Halilovic, D., 165
Hallström, E.A.M., 303
Hellmers, H., 139, 165
Herrera Pinzón, I.D., 235
Hippenstiel, R., 103
HosseiniArani, A., 187

I
Illigner, J., 311

J
Jäggi, A., 179, 199, 207

K
Kelly, K.M., 63, 411
Kern, L., 25
Khorrami, F., 87, 361
Kiani Shahvandi, M., 147
Kim, S.-W., 245
Klemm, L., 139, 165
Klügel, T., 227, 333
Knabe, A., 187
Koch, J.A., 235

Kodet, J., 333
König, D., 165
König, R., 33, 119
Krásná, H., 25, 77, 131
Kupriyanov, A., 187
Kutterer, H., 349, 417

L
Laha, A., 131
Lambert, S., 157
Lasser, M., 179, 199
Legrand, J., 69, 401
Lott, R., 63, 411

M
Madzak, M., 25
Mammadaliyev, N., 119
Männel, B., 381
Marti, U., 235
McCallum, L., 17
Meyer, U., 179, 199, 207
Miyahara, B., 391
Modiri, S., 145, 165
Müller, F.L., 297
Müller, J., 187, 323
Müller, L., 9
Müller, V., 187

N
Naranen, J., 361
Näränen, J., 87
Neidhardt, A., 333
Neumayer, K.H., 33, 45, 119
Nordman, M., 87, 361
Nothnagel, A., 25

P
Pagiatakis, S., 219
Piñón, D.A., 273
Plötz, C., 333
Poreh, D., 103
Probst, W., 333

R
Raja-Halli, A., 87, 361
Raut, S., 33
Rebischung, P., 69, 103
Reinhold, A., 45
Reis, A., 187
Richard, J.-Y., 157
Rietbroek, R., 69
Romeshkani, M., 187
Rothacher, M., 9, 235

S
Sánchez, L., 53, 253, 263, 281, 391
Santamaría-Gómez, A., 69

Sato, Y., 103
Schaer, S., 341
Schartner, M., 147
Schöne, T., 311
Schreiber, K.U., 333
Schreiner, P., 33, 45, 119
Schuh, H., 33
Schunck, D., 17
Schwatke, C., 297
Sehnal, M., 391
Seidel, A., 417
Seitz, M., 53
Sharshebaev, A., 311
Shin, Y., 245
Singh, S., 77
Soja, B., 9, 147, 235
Stanaway, R., 63, 411
Steiner, L., 391
Stolarczuk, N., 311

T
Teitsson, H., 281
Teunissen, P.J.G., 111
Thaller, D., 139, 165, 341
Tiberius, C.C.J.M., 111
Tocho, C.N., 273
Tsoulis, D., 213

U
Ullrich, D., 139, 341

V
van Noort, B.G., 111
Vargas-Alemañy, J.A., 369
Vergos, G., 263
Vigo, M.I., 369
Vincent, A., 323

W
Walenta, A., 139, 165, 341
Wang, D., 349
Westerhaus, M., 417
Willi, D., 235
Wolf, H., 3
Wziontek, H., 227

Y
Yin, C., 111
Yuan, P., 349

Z
Zaminpardaz, S., 111
Zbinden, J., 199
Zech, C., 311
Zid, F., 369
Zubovich, A., 311

Subject Index

A
Absolute sea level (ASL), 323–329
Accelerometer, 33–41, 122, 123, 126, 180, 181, 188–192, 195, 200
Advanced Topographic Laser Altimeter System (ATLAS), 237, 304
Airborne gravimetry, 220, 222–224, 257, 288
Altimetry, 45–50, 121, 124, 195, 239, 258, 297–301, 303, 313, 319, 370, 377, 378
Analytical gravity signal, 213–216
Annual amplitude and phase, 364, 365, 367
Antarctic circumpolar current (ACC), 369–378
Argentinean vertical reference system, 273–279
Atmospheric loading, 78, 84, 87–100, 230, 384
Atomic clocks, 323, 324, 326

B
Bathymetry, 303–308, 313, 314, 372

C
Center for Space Research (CSR), 182, 183, 203, 207, 208
CERN, 235–242
Clock, 5, 6, 9, 13, 26, 28, 34, 37–40, 121, 125, 141, 238, 256, 264, 323–329, 333–338, 344, 383, 384
Coastal bathymetry, 303
Coastal region, 220, 304
Co-location, 6, 17, 36, 122, 256, 343, 345
Co-location sites, 3, 17, 103–108, 342, 343, 346
Combination, 5, 11, 13, 17, 20, 25, 26, 28, 36, 37, 53, 70–74, 88, 112, 118, 122, 123, 125, 140, 143–145, 166, 167, 172, 192, 194, 204, 208–211, 215, 220, 239, 254, 257–260, 264–266, 268, 270, 278, 299, 313, 316, 333, 342, 344, 347, 350, 363–367, 383–385, 414, 419, 424, 425
Combination at the parameter level, 347
Common clock, 333, 338
Component analysis, 420–421, 425
Control of quality, 25, 312, 401
Coordinate reference system, 413
Coordinate transformation, 66, 67, 411, 413, 415
COST-G FSM, 209–211
Crustal motion, 382

D
Data quality monitoring, 401–408
Deep learning, 349–357
Deflections of the vertical (DoV), 238–239, 242, 267
Deformation, 46, 54, 63, 71, 78, 87, 104, 112, 121, 228, 256, 325, 361, 382, 393, 412, 418
Deformation model, 53–61, 63–67, 413, 415
Density modelling, 242, 249, 250
Detection, Identification and Adaptation (DIA), 117, 118

DIA-estimator, 111–118
Distributed scatterers (DS), 418, 419, 421, 425
Doppler Orbitography and Radiopositioning Integrated by Satellite (DORIS), 3, 4, 6, 17, 36, 45–50, 53, 55, 59, 104–107, 120–122, 125, 126, 140, 165, 168

E
Earth orientation parameters (EOPs), 6, 26, 28, 36, 46, 54, 60, 78, 120, 132, 133, 135, 136, 139–145, 148, 151, 154, 155, 165–174, 315, 344, 346, 384
Earth Parameter and Orbit System (EPOS), 71, 88, 120, 384
Earth rotation parameters (ERPs), 34, 37, 39–41, 48–50, 58, 125, 126, 131–136, 157–163, 166–167, 208–211, 342
Earth surface deformation, 325
Effective angular momentum functions, 147–155
EOST, 70–72, 78–80, 82–85, 88, 90, 95, 99, 362–367
EPOS-OC software, 36, 37, 45, 46, 119–126
ESMGFZ, 79–85, 344, 362, 363, 366
European Plate Observing System (EPOS), 401–408

F
Forecasting, 147–155, 391, 414
Free core nutation (FCN), 131–136, 166, 167, 172
Future Circular Collider (FCC), 235–242
Future gravimetry missions, 187–195

G
GAMIT/GLOBK, 87–100
Gas storage caverns, 417–425
Genesis, 3, 4, 6–7, 17, 18, 20–23, 247
GENESIS mission, 3, 6, 17–23, 125
Geocenter, 34, 37, 39–41, 121, 125, 126, 207, 208, 383
Geodetic datum, 26, 27, 30, 65, 383, 387
Geodetic grid exchange format (GGXF), 63, 67, 411–415
Geodynamics, 64, 66, 385
Geoid, 182, 190, 219, 235, 254, 264, 274, 282, 323, 370, 411
Geoid determination, 219–224, 276
Geophysical corrections, 77–85, 297–301, 315
Geophysical models, 54
Global geodetic observing system (GGOS), 34, 41, 78, 90, 120, 141, 253–260, 264, 269, 270, 333–338, 392–398
Global geodetic reference frame (GGRF), 63, 140, 392
Global Navigation Satellite Systems (GNSS), 3, 9, 17, 33, 46, 53, 63, 69, 78, 87, 104, 112, 120, 140, 157, 165, 208, 236, 254, 267, 274, 282, 298, 313, 323, 341, 349, 361, 381, 401, 418
Global unified height system, 255, 264
Global unified vertical reference system, 254
GNSS time series, 69, 87–100, 105, 325, 361–367, 385
GNSS velocities, 69–74

GRACE, 9, 34, 120, 179, 188, 199, 208, 268, 325, 363
GRACE follow-On, 180, 183, 184, 199–205, 209
Gradiometer, 126, 189–192, 195
Gravimetric quasigeoid model, 241, 274–278, 282–284, 286–290
Gravitational potential uncertainties, 214–217
Gravity, 6, 10, 36, 46, 78, 120, 179, 188, 199, 207, 213, 219, 227, 237, 245, 254, 264, 274, 281, 311, 323, 361, 370, 384, 391, 411
Gravity field, 6, 36, 46, 120, 179, 188, 200, 207, 213, 219, 240, 245, 255, 264, 276, 281, 391
Gravity field interpretation, 245–250
Gravity field models, 122, 124, 125, 180–184, 189, 190, 192, 195, 200, 207–211, 255–259, 265–268, 270
Gravity field recovery, 179–184, 190–195, 200, 203, 204

H
High time-resolution ERP, 157

I
NASA's Ice, Cloud, and Land Elevation Satellite-2 (ICESat-2), 303–308
IHRF Coordination Centre, 260, 269–271
IHRF densification, 281–291
Integrated water vapor (IWV), 349–357
Interferometric SAR (InSAR), 104, 105, 417–425
International Association of Geodesy (IAG), 63, 64, 70, 73, 104, 120, 125, 166, 254–256, 258–260, 264, 265, 267–270, 274, 278, 281, 285, 392–398, 412
International Gravity Field Service (IGFS), 255, 256, 259, 260, 264, 268–270
International Height Reference Frame (IHRF), 255–260, 263–271, 273–279, 281–291
International Height Reference System (IHRS), 255–260, 263–271, 274, 275, 281, 284, 285, 288
International Laser Ranging Service (ILRS), 46, 78, 120, 121, 125, 168, 172, 208–211
International Mass Loading Service (IMLS), 78–80, 82–85
International Terrestrial Reference Frame 2020 (ITRF2020), 3, 28, 45–50, 54–56, 59, 70–73, 78, 105–107, 133, 142, 144, 282, 344, 383
Inverse problems, 113, 222
Isolation forest, 201, 203–204

L
LAGEOS-1/2, 208–211
LARES, 125, 208–211
Lithosphere plate model, 55, 59
Local tie, 3, 6, 7, 17, 104, 105, 108, 268, 336, 342
Long short-term memory network (LSTM), 350, 353–355
Low Earth orbit (LEO), 9–15, 20, 120, 124, 188

M
Machine learning (ML), 148, 151, 152, 154, 172, 199–205, 304, 350
Meteorite impact, 245–250
Multi-modal probability distribution, 118
Multi-technique combination, 140
Mutual information (MI), 200–202, 204

N
NEQ level, 145
No-net rotation (NNR), 6, 27–31, 37, 48, 53, 54, 56–60, 64, 65, 67, 133, 158, 208, 383–385
Non-tidal atmospheric loading (NTAL), 78–83, 87–100, 362, 363, 366
Non-tidal loading (NTL), 78, 84, 87, 88, 90, 324, 344, 361–367

Non-tidal ocean loading (NTOL), 78–81, 83, 87, 228, 230–232, 362–364, 366
Non-tidal surface loading (NTSL), 78–85

O
Observations, 3–7, 9–15, 17–23, 25–27, 31, 34, 36, 37, 40, 46, 47, 53–55, 59, 69, 70, 74, 77, 80, 88, 114, 116, 120, 122, 124, 131–133, 141, 145, 157–163, 172, 180, 181, 190, 195, 200, 203, 204, 208, 210, 219–224, 227–232, 238, 242, 248, 255, 258, 274, 281, 298, 299, 305, 324–328, 336, 342, 345, 370, 377, 382–384, 387, 392, 395, 402–408
Outlier detection, 112, 199–205
Outreach, 392–395, 398

P
Parameter estimation, 11, 55, 112, 114, 207, 208, 277, 278
Persistent scatterers, 104, 419
Physical height, 254–256, 258, 264, 266, 267, 282, 283, 285–287, 323–326
Plate rotation, 56
Polyhedral modelling, 213
Portal, 105, 395–398, 401–408
Positional integrity, 111–118
Precise orbit determination (POD), 11, 34, 37, 46–48, 50, 120, 122–125
Predictions, 34, 57, 140, 145, 148–151, 154–155, 165–174, 211, 352, 353, 355–357, 418
Processing artifacts, 387
Pseudo low resolution mode (PLRM), 318, 320

Q
Quality of control, 25, 312

R
Radar altimetry, 311–314, 316
Radar interferometry, 417
Real time, 141, 148, 165, 166, 326, 349–357
Reference frames, 6, 7, 26, 28, 33–41, 45–50, 53–61, 63–67, 69, 71, 88–90, 104, 107, 120, 123–125, 131, 140, 147, 157, 209, 210, 237, 254–257, 265, 269, 276, 314, 338, 344, 363, 383, 384, 387, 391, 392, 395, 413
Reference level W_0, 255, 256, 258
Reference systems, 53, 64, 104, 165, 239, 254–256, 258, 260, 264, 265, 344, 412, 413
Regional geoid, 236, 238–240, 242, 256, 259, 268–270, 288
Relative sea level, 323, 324
Relativistic geodesy, 323
Retracker, 311, 312, 314–320
RMS reduction rate, 364

S
Satellite altimetry, 258, 297, 301, 370
Satellite constellation, 9, 125, 189, 403, 408
Satellite formation flights GRACE, 188–192, 194, 195
Satellite geodesy, 34, 35, 119–126
Satellite gravity, 219, 287, 288, 370
Satellite laser ranging (SLR), 3, 4, 6, 17, 36, 46–47, 53, 104, 105, 107, 120, 122, 125, 126, 140, 145, 165, 168, 172, 207–211, 268, 336
Scale, 5, 17, 26, 34, 37, 39, 49, 50, 53, 65, 70–72, 87, 91, 95, 106, 120, 147, 148, 157, 179, 180, 188, 194, 199, 200, 207, 217, 242, 248, 254, 259, 282, 297–299, 334, 344, 361, 363, 370, 374, 382, 385, 391

Sea level rise, 297–301
Sentinel-3, 312, 314, 318
Signal decomposition, 384, 417–425
Space geodetic techniques, 4, 6, 7, 17, 25, 33, 46, 104, 120, 121, 125, 140, 165–174, 256, 333, 342, 343, 347
Spherical harmonic (SH) coefficients, 122, 124, 183, 184, 190, 194, 208, 209, 213–217, 277, 325
Static gravity field, 122, 179–184, 187, 189, 190, 195, 207, 208, 210
Surface displacements, 60, 79, 325, 418, 419, 424, 425
Survey flight design, 219
Synthetic aperture radar (SAR), 104, 105, 107, 108, 318, 418–421

T
Tectonic geodesy, 381
Terrestrial gravity time series, 232
Terrestrial reference frame (TRF), 6, 7, 25, 28, 29, 34, 39–40, 55, 59, 104, 132, 139, 140, 142, 157, 395, 397
Tidal effects, 157–163, 323, 324, 326, 327
Time-dependent transformation, 63–65, 67, 412
Time in geodesy, 333
Time-variable gravity field, 124, 179–184, 191, 199, 200, 208, 210, 211
Transient motion, 382, 384, 385, 387
Trend and trend uncertainty, 366
Troposphere, 36, 37, 46, 147, 317, 336, 342–344, 346, 347, 351, 384, 387

U
UT1-UTC, 4, 5, 132–136, 157, 165, 210, 211

V
Validation, 45–47, 49, 124, 125, 144, 209–211, 238–239, 241, 242, 258, 267, 268, 270, 277, 286, 300, 307, 312, 313, 317, 352, 354, 402
Vertical datum parameter, 258, 269, 270, 273–279
Vertical datum unification, 256, 258, 274
Very Long Baseline Interferometry (VLBI), 3–7, 17–23, 25–31, 46, 53, 55, 59, 77–85, 104, 105, 107, 120, 121, 125, 126, 131–136, 139–145, 157, 165, 167, 168, 172, 334, 336–338, 342–347
VieAPL, 79, 80, 83
VieVS, 5, 6, 18, 28, 80, 83, 84, 125, 133
Visibility, 7, 18, 20–23, 304, 392, 394, 396–398
VLBI Global Observing System (VGOS), 5–7, 18, 20–23, 132, 133, 135, 136, 140–145, 336, 341–347
VLBI transmitter, 3–7, 18, 20, 23
Volume transport (VT), 369–378

W
Wet troposphere, 297–301, 315
World height system, 255, 264

Y
Yangju-circular structure (YCS), 245–250

Z
Zenith total delay (ZTD), 342–345

The manufacturer's authorised representative in the EU is Springer Nature Customer Service Centre GmbH, Europaplatz 3, 69115 Heidelberg, Germany. If you have any concerns regarding our products, please contact ProductSafety@springernature.com

Printed and bound by CPI Group (UK) Ltd, Croydon, CR0 4YY

26/03/2026

02078998-0002